20th Argentine Bioengineering Congress (SABI 2015) and the XI Conference of Clinical Engineering

Journal of Physics: Conference Series Volume 705

San Nicolas de los Arroyos, Argentina
28 - 30 October 2015

ISBN: 978-1-5108-2416-4
ISSN: 1742-6588

Printed from e-media with permission by:

Curran Associates, Inc.
57 Morehouse Lane
Red Hook, NY 12571

CURRAN ASSOCIATES INC. proceedings.com

Some format issues inherent in the e-media version may also appear in this print version.

Printed by Curran Associates, Inc. (2016)

For permission requests, please contact the Institute of Physics
at the address below.

Institute of Physics
Dirac House, Temple Back
Bristol BS1 6BE UK

Phone: 44 1 17 929 7481
Fax: 44 1 17 920 0979

techtracking@iop.org

Additional copies of this publication are available from:

Curran Associates, Inc.
57 Morehouse Lane
Red Hook, NY 12571 USA
Phone: 845-758-0400
Fax: 845-758-2633
Email: curran@proceedings.com
Web: www.proceedings.com

Table of contents

Volume 705

April 2016

20th Argentinean Bioengineering Society Congress, SABI 2015 (XX Congreso Argentino de Bioingeniería y IX Jornadas de Ingeniería Clínica)28–30 October 2015, San Nicolás de los Arroyos, Argentina

Accepted papers received: 24 March 2016 Published online:
Preface

Papers

Bioinformatics and Molecular Biology

Bioinstrumentation , Biosensors , bioMEMS and Nanotechnology

Biomechanics , Robotics and Motion Analysis

Biomedical Image Processing

Biomedical Modelling and Simulation Systems

Biomedical Signal Processing

Clinical Engineering and Management of Medical Technologies

Educational and professional experiences

Medical Informatics and Telemedicine

Neuro Engineering

Products and Medical Devices

Rehabilitation Engineering and Prosthetics

20th Argentinean Bioengineering Society Congress - SABI 2015

Abstract

From 28 to 30 October 2015, the twentieth edition of the Argentine Bioengineering Congress and the IX Conference of Clinical Engineering, SABI 2015, took place in the Colonial Hotel in the city of San Nicolas, Buenos Aires, Argentina.
The present Congress was organized by the Biomedical Research Group (GADIB) of San Nicolas Universidad Tecnológica Nacional, Facultad Regional San Nicolás and Argentine Bioengineering Society (SABI).

The objectives of this edition of the Congress were:

• To provide an academic and scientific dissemination and discussion of work and experience in the field of bioengineering.
• To disseminate in scientific, healthcare community and society technological solutions that impact on the health and on the quality of life.
• To establish relationships of exchange and cooperation between students and bioengineering professionals in Argentina and Latin America, coordinating activities that support research, technological development, transfer and innovation.
• To promote the exchange of teaching experiences and knowledge in the area of Bioengineering and Clinical Engineering.
• To promote discussion of norms, legal references, and procedures of the biomedical technologies applications.
A total of 152 scientific papers were received, of which 135 were accepted on the following 14 thematic areas:

- Bioinformatics and Molecular Biology
- Bioinstrumentation, Biosensors, BioMems and Nanotechnology
- Biomaterials, Tissue Engineering and Artificial Organs
- Biomechanics, Robotics and Motion Analysis
- Biomedical Image Processing
- Biomedical Modelling and Simulation Systems
- Biomedical Signal Processing

- Clinical Engineering and Management of Medical Technologies
- Educational and professional experiences
- Medical Informatics and Telemedicine
- Neural Engineering
- Products and Medical Devices
- Radiobiology, Radiotherapy and Medical Physics
- Rehabilitation Engineering and Prosthetics

Prestigious speakers of national and international relevance, who spoke about Signal Processing, Clinical Engineering, Neuroscience and Bioengineering, among others, attend this edition of the Congress.

Rodrigo Quian Quiroga holds a Research Chair at the University of Leicester, UK. He is the director of the Centre for Systems Neuroscience and the Head of Bioengineering at the University of Leicester. He gave us two conferences named: "The use of Wavelets in Neuroscience" and "Concept cells", highlighting the rise of bioengineering in the discovery of the human brain understanding.

Furthermore, we have received the prestigious visit of Dr. Ravi Sankar Professor of Electrical Engineering at the University of South Florida, Tampa and Distinguished Lecturer for the IEEE Engineering in Medicine and Biology Society (EMBS). His conference "Advanced Wearable Sensors and Technologies for Advancing Healthcare: From Myth to Reality" opened a discussion area in one of the main research themes worldwide.

The third conference showed the main topic of the congress, "Medical Advancement on patient safety", presented by Ing. Edwit Tironi, Director of Clinical Engineering at Children's Health Systems of Miami, responsible of the management of medical equipment of the entire system, including telemedicine.

National speakers:

Dr. Arnaldo Medina

Medical degree from the Universidad Nacional de La Plata. Master in Health Economy of the Universidad ISALUD. Executive Director of "Mi Pueblo" Hospital, of Florencio Varela and Undersecretary of Health Planning in the Ministry of Health of Buenos Aires Province. At present, he is the Executive Director of El Cruce, "Nestor Carlos Kirchner" High Complexity Hospital, Director of the Instituto de Ciencias de la Salud and Vice rector of the Arturo Jauretche National University.

Dr. Walter Legnani

Doctor in Physics of the UBA (University of Buenos Aires). He did the Post-Doctoral of Adaptive Meshing Models, in the Department of Applied Mathematics and Theoretical Physics, Cambridge University. Science, Technology and Post graduated Secretary of Universidad Tecnológica Nacional. Titular Profesor and Director of Chair of Fundamentos para el Análisis de Señales of the Electrical Engineering department of the UTN – FRBA.

Lic. Fernando Peirano

Degree in Economics by the University of Buenos Aires. In 2001 he did a posgraduate in economy, innovation and technological politics in the Complutense University of Madrid. Actually, he is Undersecretary in the National Ministry of science, technology and productive innovation.

Dr. Miguel Ángel Laborde

Vice-president of technological topics of CONICET . PhD. in Chemical Science. Member of the academic council of Natural Sciences Faculty, UBA. Evaluator of engineering, master to doctorate careers in CONEAU.

In addition to receiving enthusiastic students in the fifth edition of SABI Student, a new event was included in the Congress, called InnovaSABI, where prototypes and designs related to bioengineering devices were presented, made both by students of the career or related to it, as well as researchers and general public. This activity favored the inclusion of prototyping and design experience from any educational level and its presentation in the field of the congress. Also we added, simultaneously, an exhibition of companies in the sector and a Bioinformatics Track.
The event was held at the Colonial Hotel, which is located on the outskirts of the city, surrounded by a huge park, swimming pool, spa, and galleries to promote friendship and relationships among the participants.
San Nicolas de los Arroyos gave us the magic of its colonial architecture, the warmth of its people, and the charm of the Parana River.

Organizing Committee:

Ing. Sergio Ponce – Chair

Ing. Marcelo Lencina - Co Chair

Ing. Débora Rubio

Dra. Débora Arce

Ing. Mario Cordisco

Ing. Martín Roberti

Ing. Francisco Madrid

Ing. Gustavo Schuemer

Ing. Gustavo Veleche

Dra. Natalia López

Ing. Bernardo Lina Angeli

Ing. Mauricio Caggioli

Ing. Germán Plattoni

Sr. Bruno Contartese

Sr. Marco Bertolaccini

Sr. Alexis Sparapani

Sr. Camilo Garcia

Sr. David Piccinini

Sr. Wilson Wayton

Scientific Committee:

Chair: Dr. Carlos D'Attellis

Co Chair: Dr. Walter Legnani

Msc. Bioing. Rubén Carlos Acevedo

Dr. Ricardo Armentano

Dr. Ariel Andrés Braidot

Ing. Enrique Mario Ávila Perona

Dra. Ing. Virginia Laura Ballarín

Ing. Diego Antonio Beltramone

Sr. José Luis Ciani

Mg. Ing. Ana María Echenique

Msc. Bioing. Eduardo Filomena

Ing. José María Flores

Bioing. Melisa Antonella Frisoli

Dr. Ing. Juan Pablo Graffigna

Bioing. Alejandro Hadad

Ing. Guillermo La Mura

Ing. Marcelo Horacio Lencina

Dra. Bioing. Natalia López

Ing. Marco Antonio López Ibarra

Dr. Bioing.César Ernesto Martínez

Dr. Ing. Jorge Mazzeo

Dr. Gustavo Javier Meschino

Dra. Mónica Teresita Miralles

Dr. Renato García Ojeda

Mg. Juan Manuel Olivera

Esp. Bioing. Emilce Preisz

Mg. Viviana Inés Rotger

Dr. Franco Simini

Dr. Enrique Spinelli

Ing. Ricardo A. M. Taborda

Ing.Oscar Rodolfo Vanella

Bioing. Rosa María Weisz

Ing. Carlos Marcelo Gomez

Dra. Adriana Gabriela Scandurra

Dra. Teresita Raquel Cuadrado

Ing. Fernando Ballina

Bioing. Pedro Tomiozzo

Dr. Anselmo Frizera Neto

Dra. Silvia Kochen

Ing. Graciela M Salum

Ing. María Eugenia Gómez Marrello

Ing. Eduardo Fernández Sardá

Ing. Florencia Montini Ballarin

Ing. Juan Carlos Perfetto

Dr. Marcelo Risk

Dr. Gustavo Abraham

Dr. Pedro Arini

Dr. Maximo Eugenio Valentinuzzi

Dra. Rossana Madrid

Dr. Juan Ignacio Pastore

Dr. Cesar Martínez

Dr. Leonardo Rufiner

Dr. Pedro Pablo Escobar

Msc. Andres Valdez

Ing. Pedro Escobar

Msc. Andrés Valdez

Dra. Myriam Herrera

Ing. Martin Roberti

Ing. Inti Anabella Pagnuco

Ing. Diego Sebastian Comas

Ing. Agustina Bouchet

Ing. Gabriel Gentiletti

Dr. Alejandro Uriz

Mg. Carolina Tabernig

Dr. Ruben Spies

Dr. Diego Evin

Mg. Julio Politti

Bioing. Paola Catalfamo

Bioing. Emiliano Ravera

Dra. Silvia E. Rodrigo

Ing. Diana M. Casas

Mst. Bioing. Adrián Salvatelli

Bioing. Gustavo Bizai

Lic. Luis Rodriguez Perasso

Ing. Sergio Ponce

Bioing. Débora Rubio

Ing. Pablo Solarz

SABI 2015

Extracellular recordings

INNOVASABI Winners

Peer review statement

All papers published in this volume of *Journal of Physics: Conference Series* have been peer reviewed through processes administered by the proceedings Editors. Reviews were conducted by expert referees to the professional and scientific standards expected of a proceedings journal published by IOP Publishing.

Journal of Physics: Conference Series **705** (2016) 012001 doi:10.1088/1742-6596/705/1/012001

Formal modeling of Gene Ontology annotation predictions based on factor graphs

Flavio Spetale[1,2], Javier Murillo[1,2], Elizabeth Tapia[1,2], Débora Arce[3], Sergio Ponce[3], Pilar Bulacio[1,2]

[1]CIFASIS-Conicet, 27 de Febrero 210 bis, S2000EZP Rosario, Argentina.

[2]Facultad de Ciencias Exactas, Ingeniería y Agrimensura, Universidad Nacional de Rosario, S2000EZP Riobamba 245 bis, Rosario, Argentina.

[3] Facultad Regional San Nicolás, Universidad Tecnológica Nacional, Colón 332, San Nicolás, Argentina.

E-mail: spetale@cifasis-conicet.gov.ar

Abstract.
Gene Ontology (GO) is a hierarchical vocabulary for gene product annotation. Its synergy with machine learning classification methods has been widely used for the prediction of protein functions. Current classification methods rely on heuristic solutions to check the consistency with some aspects of the underlying GO structure. In this work we formalize the GO *is-a* relationship through predicate logic. Moreover, an ontology model based on Forney Factor Graph (FFG) is shown on a general fragment of Cellular Component GO.

1. Introduction

The high-throughput sequencing technologies entail new challenges in data processing. As a result, the use of machine learning algorithms has become relevant in many bioinformatics applications [1, 2]. In particular, for Automated Functional protein Prediction (AFP) based on Gene Ontology (GO), ensemble methods consider the ontological structure (DAG, directed acyclic graph) through a hierarchical classification [3, 4]. The design of ensemble methods is made in two steps: *i*) In the first one, a set of binary classifiers is built to predict GO-terms (classes), *ii*) In the second one, consistency of GO-DAG relationship is done. Focusing on this last step, different solutions based on heuristics [5, 6] have been proposed such as the True Path Rule (TPR) algorithm [5] where the *is-a* DAG relationship is fulfilled implementing the rule: *"If the child term describes the gene product, then all its parent terms must also apply to that gene product"*.

Heuristic solutions may be a good accuracy-effort trade-off, but a step beyond to attempt logical formalization for checking the DAG restrictions. In this paper we propose that formalization through Forney Factor Graph (FFG) model [7], since it allows a TPR restriction representation by the logical factorization of functions of several variables associated with GO-terms. The achieved model, called FFG-GO, is able to infer functional predictions of genes by using the sum-product algorithm [8].

In the next Section a brief background of FFG is presented. Then, in Section 3 TPR restrictions are formalized thought predicate logic, in Section 4 FFG-GO model is described. In the last Section, a subgraph of Cellular Component GO is modeled by FFG-GO.

2. Background

An FFG is a model that represents the factorization of a function for several variables. Briefly, the FFG diagram (Fig. 1) has nodes, ordinary edges, and leaf-edges interpreted as follows: each factor f_i (also called function) is represented by a *node*; each state variable S_j is between two factors and is represented by an *ordinary-edge*; and each input variable A_k must be involved in just one factor and represented by a *leaf-edge*. The global function (f) is factorized into factors as a product of local functions, where each factor depends on a subset of variables of f. For instance, the function $f(A_1, A_2, A_3, A_4, S_1, S_2, S_3)$ can be factorized as:

$$f(A_1, A_2, A_3, A_4, S_1, S_2, S_3) = f_1(A_1, S_1)f_2(A_2, S_1, S_2, S_3)f_3(A_3, S_2)f_4(A_4, S_3) \quad (1)$$

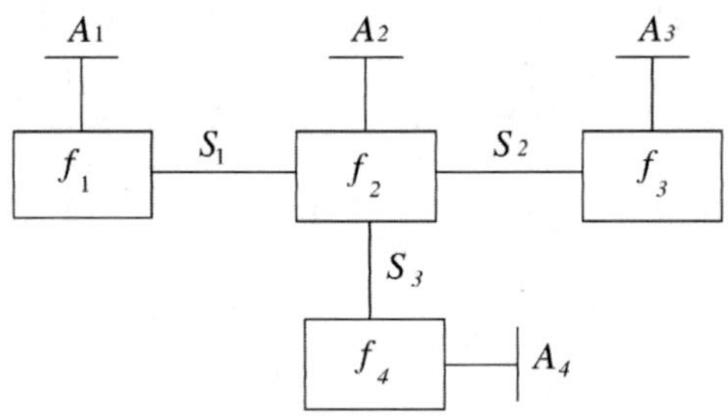

Figure 1: FFG diagram. Boxes are factors, f_i theirs functions; ordinary-edges, S_j theirs state variables; and leaf-edges, A_k theirs input variables.

3. Formalizing GO and TPR restrictions

Aiming to include the GO constraint into FFG model, the TPR restriction is formulated by predicate logic. Next, GO-DAG is denoted by $G = (V, \leq)$, where V is a set of GO terms and $\leq$ is a binary relation on V. GO terms are represented by GO nodes $(GO_1, \cdots, GO_m)$ where m is the cardinality of V. $GO_l \leq GO_z$ denotes that GO_z is a parent of GO_l, written in predicate logic as $is\_a(GO_l, GO_z)$.

The Gene Ontology Consortium has defined the TPR restriction to guarantee the *is_a* GO-DAG consistency as follows: "An annotation for a class in the hierarchy is automatically transferred to its ancestors, while unannotated genes for a class cannot be annotated for its descendants". Note that classes are GO terms and annotation is the process of assigning GO terms to gene products.

Based on [9], we rewrite the TPR by two rules of predicate logic: one related to the parents (Eq.2) and the other related to offspring (Eq.3).

$$r_1 : \forall GO_k \ \Big(is\_a(GO_l, GO_k) \wedge pos(GO_l) \rightarrow pos(GO_k)\Big) \quad (2)$$

$$r_2 : \forall GO_k \ \Big(is\_a(GO_k, GO_z) \wedge \neg pos(GO_z) \rightarrow \neg pos(GO_k)\Big) \quad (3)$$

where $pos(GO_k)$ means a gene annotation with the GO_k term.

4. FFG-GO model

Regarding the GO behavior modeling under FFG, some GO issues are briefly recalled. The gene ontology is composed of nodes representing gene functions, i.e., the k-th node and its function are jointly called *GO:k*. The GO nodes are connected in a DAG structure through edges that characterize node relationships. Without losing generality, in this work we focus on *is-a* relation. In Fig. 2(a) a comprehensive GO example that considers multiple offspring-parents is illustrated, while Fig. 2(b) shows its FFG-GO counterpart. GO nodes are matched to FFG model by input variables for root and leaves nodes, and by input state variables for inner nodes. For instance, the leaf *GO:5* matches to A_5, and the inner node *GO:2* matches to A_2, S_1, and S_2. The DAG relationships together with GO restrictions are represented by FFG functions that depend on state and input variables. For instance, r_{12} and r_{13} relation match to the function f_1.

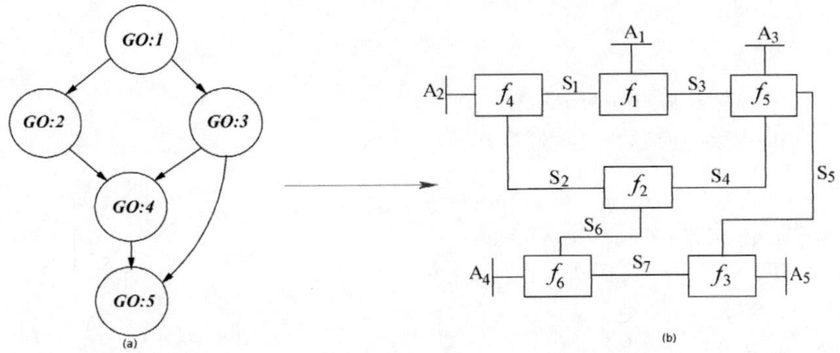

Figure 2: (a). The GO-DAG example. (b). The FFG-GO counterpart. *GO:k* node is translated to A_k/S_j variables, GO relationship r are $f_i(S_j, A_k)$.

The strength of FFG model is the design of functions f_i through logical expressions [10] to describe both the native FFG constraints and the GO-DAG restrictions. In the FFG-GO approach we identify three kinds of factors: *equality*, *multiple offspring*, and *multiple inheritance*. Note that the equality is a native FFG factor, but the inclusion of the multiple offspring together with multiple inheritances allows the formal implementation of TPR.

Equality constraint: Also called identity function and symbolized with $f_=$ forces all its variables to be equals. It is required in the FFG building [7, 11]. Fig. 3 shows the identify function block, Table 1 describes its truth table, where as expected $f_= = 1$ when all its variables $(A_2, S_1, S2)$ are equals.

Table 1: Truth table for equality constraint.
Inputs: A_2, S_1 and S_2; Output: $f_=$

.

A_2	S_1	S_2	$f_=(A_2, S_1, S_2)$
0	0	0	1
0	0	1	0
0	1	0	0
0	1	1	0
1	0	0	0
1	0	1	0
1	1	0	0
1	1	1	1

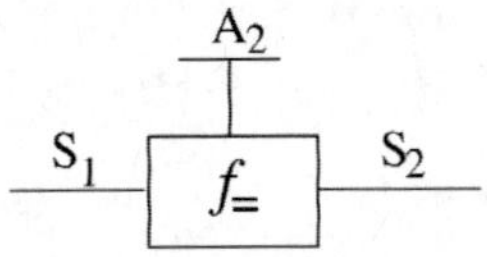

Figure 3: Identity function in graphic FFG-GO model

The $f_=$ Boolean expression is $f_=(A_2, S_1, S_2) = A_2 \cdot S_1 \cdot S_2 + \overline{A_2} \cdot \overline{S_1} \cdot \overline{S_2}$

The general form for equality constraint is:

$$f_{=}(A_k, S_1, \ldots, S_n) = \begin{cases} 1 & pos(A_k) \wedge pos(S_1) \wedge \ldots \wedge pos(S_j) \vee \neg pos(A_k) \wedge \neg pos(S_1) \wedge \ldots \wedge \neg pos(S_j) \;\; j = 1, \ldots, n \\ 0 & \text{otherwise} \end{cases} \tag{4}$$

Multiple offspring: Symbolized with $f_\wedge$, it describes the allowed node states depending on its multiple children states. See Fig. 4 where *GO:1* matches to A_1, *GO:2* to A_2 and S_1, and *GO:3* to A_3 and S_3; Table 2 describes its truth table according to the predicate logic of Eq. 3.

Table 2: Truth table for multiple offspring. Inputs: A_2, S_1 and S_3; Output: $f_\wedge$

.

A_1	S_1	S_3	$f_\wedge(A_1, S_1, S_3)$
0	0	0	1
0	0	1	0
0	1	0	0
0	1	1	0
1	0	0	1
1	0	1	1
1	1	0	1
1	1	1	1

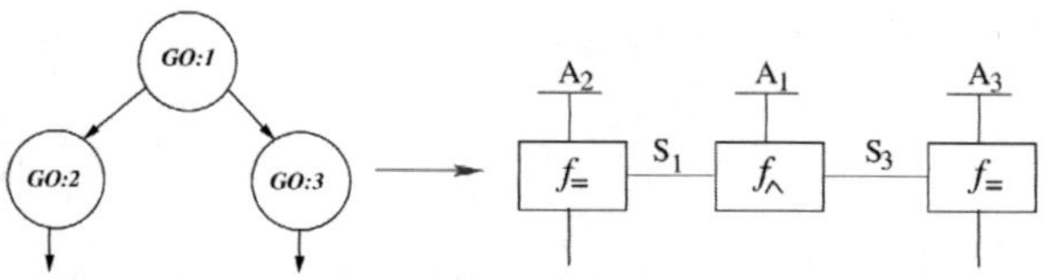

Figure 4: Multiple offspring in FFG-GO graphic model.

The $f_\wedge$ Boolean expression is $f_\wedge(A_1, S_1, S_3) = A_1 + \overline{A_1} \cdot \overline{S_1} \cdot \overline{S_3}$

The general form for multiple offspring constraint is:

$$f_\wedge(A_k, S_1, \ldots, S_n) = \begin{cases} 1 & pos(A_k) \vee \neg pos(A_k) \wedge \neg pos(S_1) \wedge \ldots \wedge \neg pos(S_j) \quad j = 1, \ldots, n \\ 0 & \text{otherwise} \end{cases} \tag{5}$$

The A_k is the FFG input variable that represents the parent GO node, and S_1 to S_n are variables that represent its children.

Multiple inheritance: Symbolized with $f_\vee$, it describes the allowed node states depending on its multiple parent states. See Fig. 5 where *GO:4* matches to A_4 and S_6, *GO:2* to A_2 and S_2, and *GO:3* to A_3 and S_4; Table 3 describes its truth table according to the predicate logic of Eq. 2.

Table 3: Truth table for multiple inheritance. Inputs: S_6, S_2 and S_4; Output: $f_\vee$

.

S_6	S_2	S_4	$f_\vee(S_6, S_2, S_4)$
0	0	0	1
0	0	1	1
0	1	0	1
0	1	1	1
1	0	0	0
1	0	1	0
1	1	0	0
1	1	1	1

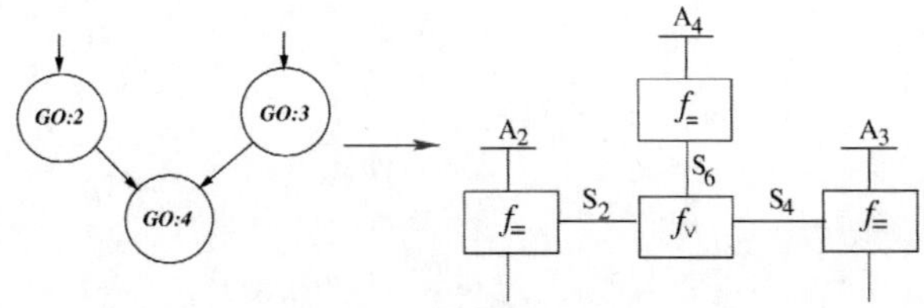

Figure 5: Multiple inheritance in FFG-GO graphic model.

The $f_\vee$ Boolean expression is $f_\vee(S_6, S_2, S_4) = \overline{S_6} + S_6 \cdot S_2 \cdot S_4$

The general form for multiple inheritance constraint is as follow:

$$f_\vee(A_k, S_1, \ldots, S_n) = \begin{cases} 1 & \neg pos(A_k) \vee pos(A_k) \wedge pos(S_1) \wedge \ldots \wedge pos(S_j) \quad j = 1, \ldots, n \\ 0 & \text{otherwise} \end{cases} \tag{6}$$

The S_1 to S_n are the variables that represents the parent GO nodes, and A_k is the input variable that represent its son.

Summarizing the GO to FFG-GO matching of the multiple offspring-inheritance example, Fig. 2, is shown in Fig. 6.

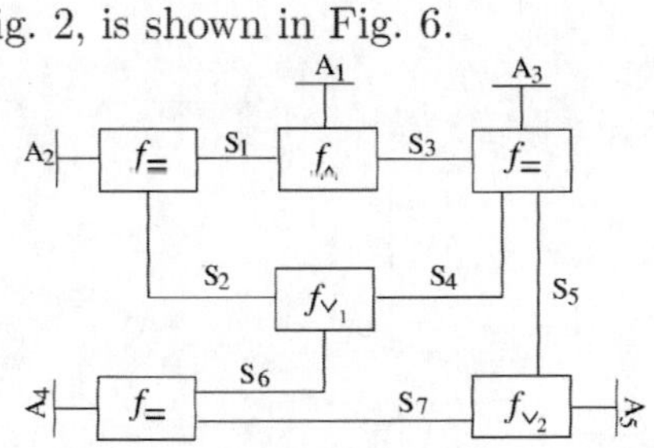

Figure 6: FFG-GO graphic model of multiple-offspring/inheritance, Fig. 2.

$$f_\wedge = A_1 + \overline{A_1} \cdot \overline{S_1} \cdot \overline{S_2}$$

$$f_{\vee_1} = \overline{S_6} + S_2 \cdot S_4 \cdot S_6$$

$$f_{\vee_2} = \overline{A_5} + A_5 \cdot S_5 \cdot S_7$$

5. Results and discussion

In order to show the strength of the proposed FFG-GO, a model on a rich fragment of GO Cellular Component (CC) is done. This CC subgraph has all the TPR restrictions discussed above, see Fig. 7. For simplicity we have renamed GO nodes, *GO:1* is the CC root *GO:0005575*, and *GO:2,. . .*, *GO:13* are *GO:0032991*, *GO:0043226*, *GO:0043228*, *GO:0005623*, *GO:0032993*, *GO:0043229*, *GO:0044464*, *GO:0043232*, *GO:0032993*, *GO:0044426*, *GO:0044422*, *GO:0044427*, *GO:0030894* respectively.

Fig. 8 shows FFG-GO model associated with CC fragment in Fig. 7. The f shows that the GO CC and FFG-GO matching, i.e., GO-terms match directly to A_k input variable. In one hand, the CC root *GO:1* matches to A_1 input variable, and has always associated with the multiple offspring function $f_\wedge$. On the other hand, leaves (*GO:9*,*GO:13*) match to A_9, A_{13} input variables, and have always associated with the multiple inheritance function $f_\vee$. We should note

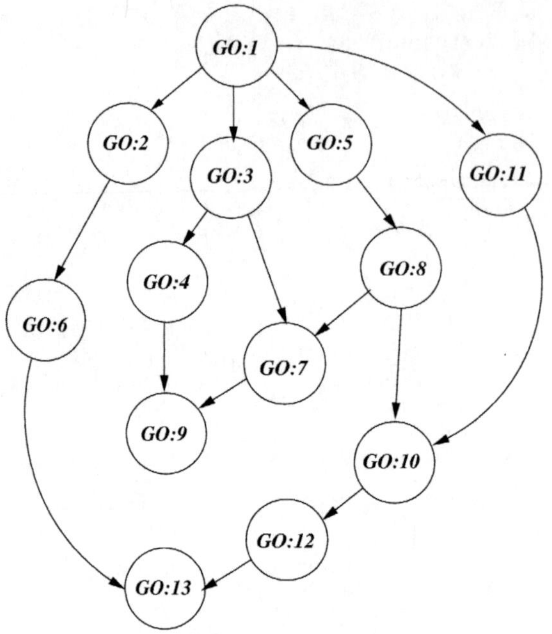

Figure 7: Cellular Component subgraph

that TPR restrictions are associated just with FFG-GO functions, hence, a restriction change requires a function modification but not a FFG-GO model rebuilding as happens with other approaches.

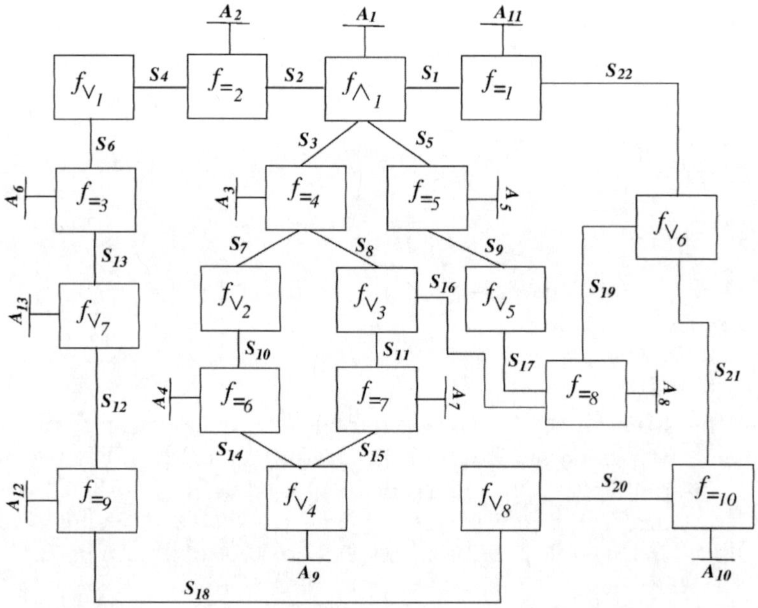

Figure 8: Cellular Component ontology in FFG-GO

The proposed FFG-GO model allows the formalization of *is-a* DAG relationship of GO ontology. Without loss of generality, the inclusion of new DAG relationships or restrictions requires just a redesign of functions f without changing the core of FFG-GO model. Likewise, the insertion of new GO-terms is simple and easy to interpret graphically. In order to complete the GO-term annotation prediction for each sample (protein), a dynamic inference process must takes place on the FFG-GO model. This inference process is carried out by the sum-product algorithm, but its description is out of the scope of this work, see [8] for details.

6. Conclusion

In ensemble classification methods, GO-term predictions computed by base binary classifiers are leveraged by checking the consistency of predefined GO relationships. Both formal leveraging strategies, with main focus on annotation precision, and heuristic alternatives, with main focus on scalability issues, have been described in literature.
Focusing on formal strategies, the formalization of TPR restrictions by predicate logic. The further work is to embody this formalization in a hierarchical classification method based on graphical models.
Along this paper we have focused our attention on is_a relationship of GO. However, this approach may be extended to another types of transitive relationships, such as *part of*, *has part*.

Acknowledgments

The authors were supported by projects PID INI3600; PICT 0253; PICT 2513, Argentina.

References

[1] Altschul S F, Gish W, Miller W, Myers E W and Lipman D J 1990 *Journal of Molecular Biology* **215** 403–410 ISSN 0022-2836
[2] Edgar R C 2004 *Nucleic Acids Research* **32** 1792–1797
[3] Cheng L, Lin H, Hu Y, Wang J and Yang Z 2014 *PLoS ONE* **9** e107187
[4] Schietgat L, Vens C, Struyf J, Blockeel H, Kocev D and Dzeroski S 2010 *BMC Bioinformatics* **11** 2
[5] Valentini G 2011 *Computational Biology and Bioinformatics, IEEE/ACM Transactions on* **8**
[6] Stojanova D, Ceci M, Malerba D and Dzeroski S 2013 *BMC Bioinformatics* **14** 285 ISSN 1471-2105
[7] Forney GD J 2001 *Information Theory, IEEE Transactions on* **47** 520–548 ISSN 0018-9448
[8] Kschischang F R, Frey B J and Loeliger H A 2001 *IEEE Trans. Inf. Theor.* **47** 498–519
[9] Burger A, Davidson D and Baldock R A 2004 *Bioinformatics* **20** 259–267
[10] Morcos F, Sikora M, Alber M, Kaiser D and Izaguirre J 2010 *Information Theory, IEEE Transactions on* **56** 742–755
[11] Loeliger H 2004 *IEEE Signal Processing Magazine* **21** 28–41 ISSN 1053-5888

Analysis of genetic association in Listeria and Diabetes using Hierarchical Clustering and Silhouette Index

Inti A. Pagnuco [1,2,3], Juan I. Pastore [1,2,3], Guillermo Abras [1], Marcel Brun [1,2], Virginia L. Ballarin [1]

[1]Digital Image Processing Group, School of Engineering, UNMdP
[2] Department of Mathematics, School of Engineering, UNMdP
[3] CONICET

E-mail: intipagnuco@fi.mdp.edu.ar

Abstract. It is usually assumed that co-expressed genes suggest co-regulation in the underlying regulatory network. Determining sets of co-expressed genes is an important task, where significative groups of genes are defined based on some criteria. This task is usually performed by clustering algorithms, where the whole family of genes, or a subset of them, are clustered into meaningful groups based on their expression values in a set of experiment.

In this work we used a methodology based on the Silhouette index as a measure of cluster quality for individual gene groups, and a combination of several variants of hierarchical clustering to generate the candidate groups, to obtain sets of co-expressed genes for two real data examples. We analyzed the quality of the best ranked groups, obtained by the algorithm, using an online bioinformatics tool that provides network information for the selected genes.

Moreover, to verify the performance of the algorithm, considering the fact that it doesn't find all possible subsets, we compared its results against a full search, to determine the amount of good co-regulated sets not detected.

1. Introduction

New technologies for genomic analysis measure, simultaneously, the expression of thousand of genes. An important goal in some studies is to find sets of co-expressed genes, to study co-regulation and biological functions. This task is usually performed by clustering algorithms, where the whole family of genes, or a subset of them, are clustered into meaningful groups based on their expression values in a set of experiment. These techniques provide insight on the possible co-regulation between genes, under the hypothesis that co-expression may indicate evidence for co-regulation, and these hypotheses of co-regulation must eventually be corroborated or be rejected by further experiments. Existing clustering algorithms are based, usually, on a global approach [1, 2, 3, 4, 5, 6], or on just pairwise analysis [7]. While some of them provide methods for the selection of the best number of clusters [2], they result in very large number of clusters, depending on the number of groups required or determined by the algorithm. The large size of these sets makes improbable their use as hypothesis generator.

A more comprehensive approach consist on the analysis of co-expression of all possible sub-sets of genes. While simple, this approach fails because of the need to analyze an exponential number of candidate sets. This search of meaningful set would be impractical if the number of available genes for the analysis is large, which it is usual in this field.

In this work we present a new algorithm, that combine the particularities of clustering hierarchical algorithm with individual cluster validation based on the Silhoutte index, to generate a ranked list of gene group, avoiding the exhaustive search, but providing high quality results. As in [2] we use the Silhouette index as a quality/homogeneity measure, but instead of using it to generate the partitions, we use it to select the best sets, from several partitions, obtained by different variants of hierarchical clustering.

This approach permits the use of many different clustering algorithms, combining the best results of each one of them, avoiding the use of a full search approach, while still providing, as shown in the results, a good approach to the optimal results. In the next section we present a) an introduction to pattern recognition tools, b) the proposed algorithm, c) the result in genomic data. The final conclusions show that this algorithms could result in a useful tool to the researchers as preliminar technique to data analysis.

2. Approach

2.1. Pattern Recognition

Pattern Recognition techniques have been widely used to identify object according to different type of features [8]. There are two basic types of pattern classification, supervised and unsupervised classification. Supervised classification uses samples previously classified to design the classifier, to be applied to new patterns. Unsupervised classification groups unclassified patterns based on similarity. Clustering algorithms is a collection of techniques for unsupervised classification [9]. The clusters are formed according a similarity measure, which is usually defined as the proximity of the points according to a distance function. One of the distance function often used is the Euclidean distance:

$$d(x,z) = \| x - z \| = \sqrt{(x-z)^t(x-z)} \tag{1}$$

where x and z are two vectors. In this case, when minor is the distance between two elements, major is their *similarity*.

In this work, the objects to be classified are genes. The data consists of a set of m samples $S_1, S_2, .., S_m$ and n genes $g_1, g_2, ..g_n$ that are normally represent by a matrix of two dimension M where $M(i,j)$ represent the expression of gen g_i for the sample S_j. The expression of each gen g_i, across all samples, correspond to a row of matrix M, and it is represented by a feature vector $X_i(x_{i_1}, x_{i_2}, .., x_{i_m})$, where each value x_{i_j} represent the expression of geng$_i$ for the sample S_j. The feature vector is then composed by m samples, so that the regions of the partition H are in space R_m.

Each gen can be assigned to one of k possible groups, and the result of a clustering algorithm is a partition of the set of genes, as a set $W = W_1, W_2, .., W_k$, where each W_i is a group of genes with certain degree of similarity.

2.2. Hierarchical Clustering

The Hierarchical Clustering is one of most used clustering algorithms in bioinformatics [10]. In this algorithm, it is not necessary to know previously the number of classes, to divide the total set of elements according their characteristics, since it generates a partitioning tree that can be later used to generate any amount of clusters. At each iteration, the two closest clusters are joined together (based on some measure of distance) and form a new one. If the number of classes is known beforehand, the algorithm ends when you reach the K number of classes searched. If the number of classes is not previously known, the algorithm can be continue its process until there is only one large cluster with all the elements on it. The staggered manner in which the clusters are gathering can be displayed on a tree diagram called dendrogram, where successive junctions between groups are showed with increasing distance from the root. Given a set of N items to be clustered, and an $N \times N$ distance (or similarity) matrix, the basic process of hierarchical clustering [11] is described here:

(i) Start by assigning each item to a different cluster, so that for N items, there are N clusters, each containing just one item. Let the distances (similarities) between the clusters be the same as the distances (similarities) between the items they contain.

(ii) Find the closest (most similar) pair of clusters and merge them into a single cluster, so that now there is one cluster less.

(iii) Compute distances (similarities) between the new cluster and each of the old clusters.

(iv) Repeat steps 2 and 3 until all items are clustered into a single cluster of size N.

Step 3 can be done in different ways, which is what distinguishes single-linkage from complete-linkage and average-linkage clustering. In single-linkage clustering (also called the connectedness or minimum method), the *distance* between two clusters is the shortest distance from any member of one cluster to any member of the other cluster. If the data consist of similarities, the *similarity* between two cluster is the largest similarity from any member of one cluster to any member of the other cluster. In complete-linkage clustering (also called the diameter or maximum method), the distance between two clusters is the largest distance from any member of one cluster to any member of the other cluster. In average-linkage clustering, the distance between two clusters is the average distance from any member of one cluster to any member of the other cluster.

2.3. Quality measures

A common problem of clustering algorithms is the validation of results. There are basically two types of validation. Internal validation, which is based on calculations performed on the resulting clusters such as the separation between them, its roundness or closely packed. This type of validation does not require additional information to that reviewed. The other is the comparison of partitions, which can be obtained compared with other partitions generated by the same algorithm with different parameters (relative validation), or with the actual partition of the original data (external validation) [12]. One of the methods of internal validation used is Silhouette index, which measures the quality of clustering as the average quality of its elements [13]. The Silhouette value for an element ranges between -1 and 1. A high Silhouette value indicates that the element is closer to its own cluster elements than to the ones that do not belong to its own cluster.

To define the Silhouette index for the points, two measures need to be defined first. Lets N be the number of clusters. Let $X \in C_k$ a point (element) that belongs to the cluster C_k, and let n_k the number of elements of C_k, then the first measure $a(X)$, the average distance of X to the points of C_k is defined by:

$$a(X) = \frac{1}{n_k - 1} \sum_{Y \in C_k, Y \neq X} d(X, Y)$$

The second measure, $b(X)$, the average distance to the nearest cluster, for a point $X \in C_k$, is defined by:

$$b(X) = \min_{h=1..N, h \neq k} \frac{1}{n_h} \sum_{Y \in C_h} d(X, Y)$$

Finally, the Silhouette index $S(X)$ is defined by:

$$S(X) = \frac{b(X) - a(X)}{max(a(X), b(X))}$$

Then, the Silhouette index for a cluster is defined as the average Silhouette of its points:

$$S(C_h) = \frac{1}{n_k} \sum_{X \in C_k} S(X)$$

It can be seen that if $b(X) \gg a(X)$ (the point is closer to points in the same cluster than points in other clusters) then $S(X)$ is close to 1. If this happens for most of the points in a cluster C, then the Silhouette $S(C)$ is also close to one, indicating that it is isolated from other clusters.

In the other hand, if $b(X) \approx a(X)$, (there are points outside the cluster closer to X than points in the same cluster), then $S(X)$ is close to 0. If this happens for most of the points in a cluster C, then the Silhouette $S(C)$ is also close to zero, indicating that it is mixed with some other clusters.

For our analysis, we use a slightly modified measure of cluster silhouette. In this case, for each set C of points, to compute its Silhouette we assume that all the points that do not belong to C belong to a second cluster C^c, and the *modified* silhouette index is computed based on this 2-clusters situation:

$$b(X) = \frac{1}{n^c} \sum_{Y \in C^c} d(X, Y)$$

3. Methods

3.1. Proposed algorithm

The objective of this study is to select subsets of genes highly correlated. In the expression profile space of genes (expression through all samples), provide by a distance function, where each genetic profile is represented with a point of R_n, this corresponds to find compact sets that are separated to other points sets. A naive way to find these compact and separate groups would consist in *measuring* the compactness of all possible subset of the N profiles. A suitable measure to measure compactness may is the Silhouette Index [13], successfully used by the authors for other purposes [14, 15].

One problem of this approach is that the amount of subsets grows up in a combinatorial form with the number of genes. For example for a set of 2000 genes, the number of subsets to evaluate is 2^{2000}. For this reason it is required some technique to find the best subsets (or many of them) avoiding the exhaustive search. To solve this problem, in this paper we propose to limit the family of subsets where the search, using hierarchical clustering to generate a family of candidate subsets, and then evaluate the *modified* Silhouette index only on those subsets. An application of the hierarchical clustering algorithm generates a total of $2 * N$ subsets to process. Because there are so many variants of hierarchical clustering, an important problem is to determine the best variants of hierarchical clustering for the task of detecting co-expressed genes on genomic data.

In a previous work [16] we studied the ability of variants of hierarchical clustering (complete-linkage, average-linkage, single-linkage) to detect the best subsets using Silhouette index as quality measure. To this purposes we used four different sets of simulated data with known result. Keeping the size of the artificial sets small, we were able to analyze all possible groups, sorting them by Silhouette Index, and use that information as gold standard to evaluate the performance of variants of hierarchical clustering to detect the best subset.

As an additional approach, we combine all the variants, forming a large sets of candidate groups, still smaller than the maximum 2^N sets, and select the best groups from this set. This approach extends the search space, relative to the use of only one algorithm, maintaining the computational requirements still low. Another advantage of this approach is that new clustering algorithms may be added to the pool, generating more candidate groups, therefore improving the overall quality of the result.

The main steps of the proposed algorithm are:

- Apply to the data some variants of hierarchical clustering.
- For each algorithm, register all the possible branches in the dendrogram three (one-gene groups are not processed).
- Compute the Silhouette index for each group (against the rest of the genes)
- Select those groups with Silhouette index above a threshold

The important step in this method, that sets it apart from previous methods, is that it does not use just one clustering (partition) of the genes, based on an algorithm, but a larger set of groups, defined by the dendrogram tree from the hierarchical clustering algorithms. Allowing the evaluation (via the modified Silhouette index) of intermediate groups (ones not showing on a *optimal* partition) avoids issues were a large group is not part of that partition, but would be a good candidate because of its compactness.

In the previous work, the analysis was aimed at finding which variant of hierarchical method detected a major proportion of the groups with highest Silhouette (determined by a full search in simulated contexts). In these analysis, and additional ones realized for this work, it was visible that no variant has better performance, except when using a combination of all the variants.

In the next sections we apply this approach, based on the combination of variants to generate the candidate clusters, and the modified Silhouette index to rank, to search for correlated genes in two real data situations.

4. Application Examples

In this section we describe two application examples, where the algorithm is applied to two different sets of data: microarray based expression, for diabetes, and QTLs, for listeria. In both cases we applied the algorithm to detect significant sets of genes/QTLs, and analyzed their significance, based on existing knowledge about them.

4.1. Listeria

In the first application example, we used data of mouse susceptibility to monocytogenes listeria [17, 18]. This study consists in the analysis of the relation between QTLs (*Quantitative trait Loci*, stretches of DNA containing or linked to the genes that underlie a quantitative trait) and the *survival* time of 120 mouses that have been infected with listeria, where *survival* is defined by a survival time of more than 240 hours. The dataset consists in 35 surviving mice and 85 un-surviving, analysing 133 QTLs for each mouse. After filtering those QTLs with missing data, only 28 QTLs were retained.

It is important to note that here we are using QTLs instead of genes as target for our search of correlation. Because of the nature of the model used here is not restricted to expression information, the same analysis can be applied to discrete QTL studies.

To verify the performance of the algorithm in this context, considering the small amount of markers under study, we were able to perform a full search of QTLs groups (up to 7 elements), computing the Silhouette index of all of these subsets. The performance of the search algorithm (based on clustering) is measured based on the number of best subsets detected.

Table 1 shows the comparison of results between full search and combination of variants of hierarchical clustering. We can see in the table that the 8 most significant groups were properly detected by the algorithm. Still more, the top groups of 2, 3, 4, 5 and 6 elements, were successfully detected by the algorithm.

Table 1. Comparison of results between full search and combination of variants of hierarchical clustering. The best 8 groups were successfully detected by combination algorithm.

Id_All	Sil	Id_combination	Num_elem	Groups Member
1	0.90661	1	2	8,9
2	0.86923	2	2	3,4
3	0.86788	3	2	1,2
4	0.814008	4	2	25,26
5	0.736231	5	3	8, 9, 10
6	0.701954	6	3	24, 25, 26
7	0.672453	7	2	11,12
8	0.66783	8	4	1,2,3,4
18	0.592628	9	2	14,15
21	0.563286	10	2	17,18
24	0.554229	11	5	8,9,10,11,12
37	0.49119	12	3	16,17,18
44	0.452103	13	6	8,9,10,11,12,13
69	0.393421	14	5	14,15,16,17,18
71	0.387993	15	4	23,24,25,26
327	0.293433	17	6	14,15,16,17,18,20
346	0.290286	18	6	6,14,15,16,17,18

4.2. Diabetes

In this second example of biological application we run the algorithm on microarray data. This data was obtained from a previous study that analyzed the expression profiles of obese and thin subject [19]. This study presents the expression profile to 18 subjects, 13 with obesity and 5 without, using a U133A chip of Affymetrix. From the original dataset, with 22283 genes, only the most variable 1000 genes were preselected for the analysis. In this case, it is not possible to compare the resulting groups with a full search, since there are 2^{1000} possible subsets. Therefore, the resulting groups/sets were studied based on actual biological knowledge about them.

We restricted the analysis to the top eight ranked sets, described in Table 2. We verified if the genes found on these groups, have assigned similar biological functions.

It should be noted that in group 1, even if there are two different Affymetrix probes 204550*_x_at* and 215333*_x_at*, they make reference to the same gene. The same situation is repeated for group 6.

Table 2. Sets detected with higher Silhouette using a combination of variants of hierarchical algorithms.

Group	Silhouette	Size	Probes (Affy Ids)	Analysis
1	0.9548	2	204550_x_at , 215333_x_at	Unique gene
2	0.9429	3	204418_x_at , 204550_x_at , 215333_x_at	2 Genes from the same family.
3	0.82689	2	207831_x_at , 207907_at	No relationship found.
4	0.80904	2	201639_s_at , 201904_s_at	No relationship found.
5	0.79401	2	205175_s_at , 213670_x_at	Indirect metabolic Relationship
6	0.79394	2	200966_x_at , 214687_x_at	Unique gene
7	0.79345	2	200991_s_at , 202676_x_at	Common function:Protein binding
8	0.79343	3	201379_s_at , 207831_x_at , 207907_at	Common function:Protein binding

The most interesting case identified is Group 2. In this group there are 3 probes, 2 of them reference to the same gene called GSTM1 (Glutathione S-transferase mu 1). The relation between these 2 probes and the third one, which references to gene GSTM2 (Glutathione S-transferase mu 2) is that all of them are members of the same family, and they are involved in metabolic process.

We analized these genes with the GeneMania bioinformatic tool, using the Co-expression option. GeneMania searches large biological datasets to find related genes, including protein-protein, protein-DNA and genetic interactions, pathways, reactions, gene and protein expression data, protein domains and phenotypic screening profiles [20] . Figure 1 show the result of the analysis of co-expression for groups 2, 5 and 66. The figure 1 shows a large amount of linkage between the genes of the selected groups. Most of the associations are obtained from the Gene Expression Omnibus database (GEO), but GeneMania tool only collect data associated with publications. To group 2 some functions associated are glutathione transferase activity, glutathione derivative metabolic process, peptide metabolic process, peptide binding, modified amino acid binding, etc.

To third and forth group have not documented relationships. From this analysis, they could be good candidates for further analysis of co-regulation, or other biological relationship.

The fifth group shows interesting relationships. The first Affymetrix Identifier (205175_s_at)reference to a locus NSUN5P1. This locus represents a transcribed pseudo gene of a nearby locus on chromosome 7, which encodes a putative methyltransferase. Diseases associated with NSUN5P1 include Williams-Beurem Syndrome. The other group member references to the KHK gene (213670_x_at). The KHK gene encodes ketohexokinase that catalyzes conversion of fructose to fructose-1-phosphate. The product of this gene is the first enzyme with a specialized pathway that catabolizes dietary fructose (GeneCard information [21]). Due to the fact that NSUN5P1 is a pseudo gene, we can not do an analysis of relation with the KHK gene (probe 213670_x_at). For this reason we search genes associated to Williams-beuren disease with pathways related to the KHK gene.

In this search we found the MLXIPL gene that encodes a basic helix-loop-helix leucine zipper transcription factor. This protein forms a herodimeric complex and binds and activates, in a glucose-dependent manner, carbohydrate response element (ChoRE) motifs in the promoters of triglyceride synthesis genes. This gene is deleted in Williams-Beurem Syndrome. Both genes are involved in metabolism process. Figure 1 show the result of a co-expression analysis between MLXIPL and KHK gene, using the GeneMania bioinformatic tool.

For groups 7 and 8 we used GeneCards for analysis. GeneCards is a searchable, integrated database of human genes that provides information on all known and predicted human genes [21]. For both groups their members have a common function, that is protein binding, according to GeneCards.

Other case analyzed was a big group with eleven elements, this group had been ranked in the position 66, and its Silhouette index is 0.7044, large value based on our past experience with the Silhouette index. We searched the elements and ran a co-expression analysis with GeneMania. In figure 1 show the relation between some elements. To verify the quality of the result we select randomly 11 genes and ran the same analysis, the result obtained was a graph more dispersed (Graphic not included here).

5. Discussion

In this work we apply a simple but powerful method to identify groups of genes/markers that are compact and well separated from other genes/markers, using the Silhouette index to rank the sets, and a combination of several variants of hierarchical clustering to search the best sets. In one case, due to the small size, the resulting groups were compared to the results from full search. In the second case, the resulting groups were analyzed using standard bioinformatics tools, verifying strong relationship between the genes in the top groups.

The algorithm provides a balance between search time and detection rate. It avoids the full

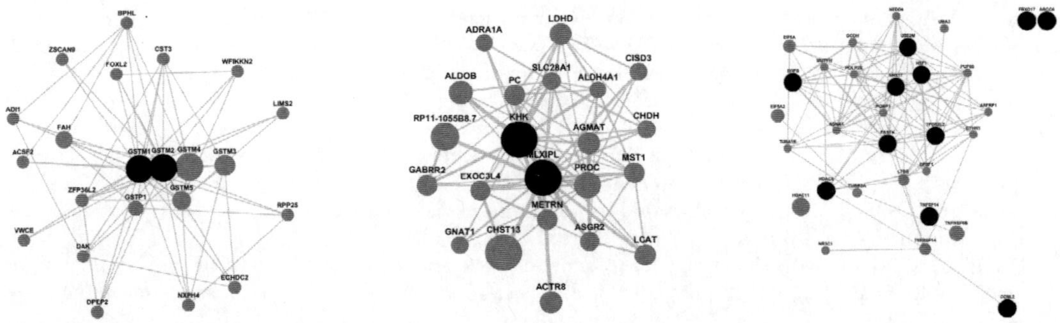

(a) (b) (c)

Figure 1. Results from GeneMania tool: a) Graph for group 2, where GSTM1 and GSTM2 are closely related; b) Graph for group 5, where KHK and MLXIPL gene are closely related; c) Graph for group 66, where the 11 elements are related

search, which can be impractical for large number of genes, with the cost of missing some good sets, but it is able to detect most of the top ranked sets, which is not usually possible by using only one clustering algorithm.

The effectiveness of this algorithm is related to the ability of the Silhouette index to score properly compact groups, which are at the same time separated from other groups of genes.

6. Conclusion

With this analysis we verified the results of the proposed tool, and we considered that is useful for biologists or researchers in computational biology interested in generating new hypotheses about the co-expression of genes, or genomic markers like QTLs, which are not provided in most standard analysis tools. This algorithm will generate quickly a set of good groups on base of Silhouette index, and combines the advantages of each variant of hierarchical clustering algorithm. Future work includes the analysis of other indices of group quality, besides Silhouette, and the application to new datasets, including SNPs.

Acknowledgement

This work was partially supported by CONICET and FONCYT.

7. References

[1] L. Heyer, S. Kruglyak, and Shibu Yooseph. Exploring expression data: Identification and analysis of coexpressed genes. *Genome Research*, 1999.

[2] K. S. Pollard and M. J. van der Laan. New methods for identifying significant clusters in gene expression data. *Joint Statistical Meetings - Biometrics Section-to include ENAR & WNAR*, 2002.

[3] H. Peng, F. Long, M. B. Eisen, and E. W. Myers. Clustering gene expression patterns of fly embryos. *IEEE, ISBI*, 2006.

[4] T. Nguyen and R. Nowakowski I. Androulakis. Unsupervised selection of highly coexpressed and noncoexpressed genes using a consensus clustering approach. *OMICS*, 2009.

[5] T. Nguyen, J. Mattick, Q. Yang, M. Orman, M. Ierapetritou, F. Berthiaume, and I. Androulakis. Bioinformatics analysis of transcriptional regulation of circadian genes in rat liver. *BMC Bioinformatics*, 2014.

[6] W. De Mulder, M. Kuiper, and R. Boel. Clustering of gene expression profiles: creating initialization-independent clusterings by eliminating unstable genes. *Journal of Integrative Bioinformatics*, 2010.

[7] A. Feltus, S. Ficklin, S. Gibson, and M. Smith. Maximizing capture of gene co-expression relationships through pre-clustering of input expression samples: an arabidopsis case study. *BMC System Biology*, 2013.

[8] *Pattern Clasification and Scene Analysis.* Wiley-Interscience; 2 edition (November 9, 2000), 2000.
[9] *Clustering: revealing intrinsic dependencies in microarray data.* 2005 Hindawi Publishing Corporation, 2005.
[10] D. Chiang, P. Brown, and M. Eisen. Visualizing associations between genome sequences and gene expression data using genome-mean expression profile. *Bioinformatic*, 2001.
[11] Johnson S. C. Hierarchical clustering schemes. *Psychometrika*, 1967.
[12] Dalton L., Ballarin V., and Brun M. Clustering algorithms: On learning, validation,performance, and applications to genomics. *Current Genomics*, 2009.
[13] P. Rousseeuw. Silhouettes: A graphical aid to the interpretation and validation of cluster analysis. *Journal of Computational and Applied Mathematics*, 1987.
[14] J. Pearson and et al. Identification of the genetic basis for complex disorders by use of pooling-based genomewide single nucleotide polymorphism association studies. *The American Journal of Human Genetics*, 2007.
[15] J. Hua, D. Craig, M. Brun, J. Webster, W. Tembe, K. Joshipura, M. Huentelman, E. Dougherty, and D. Stephan. Sniper-hd: improved genotyped calling accuracy by an expectation-maximization algorithm for high-density snp arrays. *Bioinformatic*, 2007.
[16] G. Abras, J. Pastore, M. Brun, and V. Ballarin. Deteccin de conjuntos significativos de genes via silohuette. *CAIS 2010, 1er Congreso Argentino de Informtica y Salud, 38 JAIIO*, 2010.
[17] V. L. Boyartchuk, K. W. Broman, R. Mosher, S. D'Orazio, M. Starnbach, and W. F. Dietrich. Multigenetic control of listeria monocytogenes susceptibility in mice. *Nature Genetics*, 2001.
[18] K. W. Broman, V. L. Boyartchuk, and W. F. Dietrich. Mapping time-to-death quantitative trait loci in a mouse cross with high survival rates. *Technical Report MS00-04, Department of Biostatistics, Johns Hopkins University*, 2000.
[19] J. Pihlajamki, T. Boes, and Eun-Young Kim et al. Thyroid hormone-related regulation of gene expression in human fatty liver. *J Clin Endocrinol Metab*, 2009.
[20] S. Mostafavi, D. Ray, D. Warde-Farley, C. Grouios, and Q. Morris. Genemania: a real-time multiple association network integration algorithm for predicting gene function. *Genome Biology*, 2008.
[21] M. Rebhan, V. Chalifa-Caspi, J. Prilusky, and D. Lancet. Genecards: A novel functional genomics compendium with automated data mining and query reformulation support. *Bioinformatics*, 1998.

The impact of quality control in RNA-seq experiments

Gabriela A Merino[1] , Cristóbal Fresno[1], Frederico Netto [2]. Emmanuel Dias Netto[2], Laura Pratto[3] and Elmer A Fernández[1]

[1] CONICET, Universidad Católica de Córdoba, Córdoba, Argentina
[2] Medical Genomics Group, A.C. Camargo Hospital, São Paulo, Brasil
[3] Universidad Nacional de Villa María, Villa María, Córdoba, Argentina

E-mail: gmerino@bdmg.com.ar

Abstract. High throughput mRNA sample sequencing, known as *RNA-seq*, is as a powerful approach to detect differentially expressed genes starting from millions of short sequence reads. Although several workflows have been proposed to analyze RNA-seq data, the experiment quality control as a whole is not usually considered, thus potentially biasing the results and/or causing information lost. Experiment quality control refers to the analysis of the experiment as a whole, prior to any analysis. It not only inspects the presence of technical effects, but also if general biological assumptions are fulfilled. In this sense, multivariate approaches are crucial for this task.

Here, a multivariate approach for quality control in RNA-seq experiments is proposed. This approach uses simple and yet effective well-known statistical methodologies. In particular, Principal Component Analysis was successfully applied over real data to detect and remove outlier samples. In addition, traditional multivariate exploration tools were applied in order to asses several controls that can help to ensure the results quality. Based on differential expression and functional enrichment analysis, here is demonstrated that the information retrieval is significantly enhanced through experiment quality control. Results show that the proposed multivariate approach increases the information obtained from RNA-seq data after outlier samples removal.

1. Introduction

Since their appearance, high throughput sequencing (HTS) technologies have had a great impact on genomic and transcriptomic researches. One application of HTS is the sample mRNA sequencing (*RNA-seq*). This application overcomes many of the limitations of previous technologies, such as the dependence on prior knowledge of the organism, as required for microarrays. In addition, RNA-seq promises to unravel previously inaccessible transcriptome complexities, such as novel promoters and isoforms.

In general, the RNA-seq data analysis implies the identification of expressed genes through of some experimental conditions, using millions of reads produced by a HTS machine. The huge and complex output is processed in several steps and requires careful consideration of many aspects. In this kind of experiments many variation sources exists, including technical and random effects, which should be properly accounted in order to reduce results bias [1]. At present, several methodologies and tools are available for a quality control step [2, 3]. But, in general, they are only focused on the reads assessment or mapping quality instead on the

experiment as a whole.
For instance, in gene expression analysis, only a small fraction of genes are expected to show Differential Expression (DE) between experimental conditions. However, a global overview of all the analyzed genes should provide, if it exists, some evidence of the evaluated experimental conditions. For example, sample replicates (technical or biological) should behave similarly. On the contrary, an opposite behaviour is expected from samples of different conditions. These effects cannot be seen in a sample by sample exploration, thus requiring a whole sample analysis by means of multivariate approaches.
Here, a multivariate approach for RNA-seq experiment quality control is proposed. The procedure is based on simple and yet effective well-known methodologies, such as Principal Component Analysis (PCA), density, scatter and boxplots. The combination of these methods allows performing appropiate experiment quality control. In particular, its utility is demonstrated here in an RNA-seq control-treatment experiment.

2. Materials and Methods

2.1. RNA-seq database

mRNA samples of human Oral Squamous Cell Carcinoma (OSCC) were studied (unpublished data). Five of these were from patients with lymph node metastasis and five from nonmetastatic subjects. These experimental conditions, hereafter, are called *treatment* and *control* respectively. Patients, all men, were collected from Hospital A.C. Camargo, São Paulo, Brazil. mRNA samples were sequenced using 75-35 pair-end RNA-seq protocol using ABI5500 SOLID® platform. Two sequencing runs were done, involving six lanes in each other. In order to perform the sequencing, samples were divided in two groups, five per run, each containing both experimental conditions. In each sequencing group, samples were split into six aliquots, one for each lane and sequenced. Pair sequenced read files of each lane were separately mapped to the Ensembl reference Human Genome (GRCh37/hg19, [4]) using LifeScope® software. Then, all alignments of each sample were paired and combined, resulting in one alignment file per sample. These ten files represent the starting input for the proposed approach.

2.2. Quality Control Workflow

The *Quality Control Workflow* involves several steps that must be carried out in order to perform the global control.

2.2.1. Data Filtering: This step is focused on gene filtering. Low mapping quality, low counts and not interesting genes will be excluded after this step.

(i) Reads filtering

Low quality mapping value (MAPQ, [5]) and non-uniquely aligned reads can result in false expression value estimation. For this reason, reads with MAPQ bigger than 20 and read pairs uniquely aligned to a genome feature were counted using HTSeq [6] software with the *intersection strict* mode. Here, a gene was considered as a genome feature, where each gene is assumed as the union of all its exons. Then, the R platform [7] was used to join count tables in order to build a GxP count matrix, *CM*, (G: # genes, P: # samples) that will be used in the downstream analysis.

(ii) Gene filtering

Usually noisy data removal is essential in order to reduce possible results bias. In the context of RNA-seq data analysis, only reliable measured genes are required. In addition, if functional analysis will be performed, un-annotated and non-coding genes should be removed. In this sense, genes with expression counts lower than 10 in at least one condition (non-reliable genes), genes with less than 200 base pairs (small RNAs) and genes without annotation in the functional analysis DAVID platform ([8, 9]) were removed. The resulting expression count matrix is named CM_1.

2.2.2. Global Quality Control: Before gene expression analysis performing, several aspects like samples comparability and compliance of global experimental design assumptions must be checked. Keeping in mind that the work involves a set of samples a multivariate approach is necessary in order to explore them together.

(i) Examining sample distributions

In a general biological context, all genes cannot be affected by the experiment. Therefore, only a small fraction of them are expected to be differentially expressed. Thus, gene counts boxplots and density functions plots should be similar in shape and position. If this does not occur, a between sample normalization process is necessary. In the RNA-seq context, it is well-known that library size seems to be one of the main factors to account for distribution variability [10]. The normalization method selection process is beyond the scope of this work. The edgeR-RLE (Relative Log Expression) implementation is suggested to calculate size factors of count matrix [11]. Finally normalized samples density and boxplots should be obtained and re-assessed.

(ii) Exploring sample separability and filtering

This is an important and crucial step. The samples gene expression global separability into two experimental groups must be assured. In this context, a multivariate approach could be used over the count matrix. Principal Component Analysis (PCA) is a simple and effective descriptive technique that allows checking this statement [12]. PCA results often reveal data relationships that were not previously suspected and thereby allow interpretations that would not ordinarily result. The analysis starts from the covariance ($\boldsymbol{\Sigma}$) or correlation ($\boldsymbol{\rho}$) matrix of p random variables $\boldsymbol{X_1, X_2, ..., X_p}$. The covariance matrix's eigenvalues ($\boldsymbol{\lambda_i}$) and corresponding eigenvectors ($\boldsymbol{e_i}$) are computed obtaining the uncorrelated linear combinations given in equation 1 which define the *principal components*.

$$\begin{aligned} \boldsymbol{Y_1} &= \boldsymbol{e_1'X} = & e_{11}\boldsymbol{X_1} + e_{12}\boldsymbol{X_2} + ... + e_{1p}\boldsymbol{X_p} \\ \boldsymbol{Y_2} &= \boldsymbol{e_2'X} = & e_{21}\boldsymbol{X_1} + e_{22}\boldsymbol{X_2} + ... + e_{2p}\boldsymbol{X_p} \\ &\vdots & \vdots \\ \boldsymbol{Y_p} &= \boldsymbol{e_p'X} = & e_{p1}\boldsymbol{X_1} + e_{p2}\boldsymbol{X_2} + ... + e_{pp}\boldsymbol{X_p} \end{aligned} \tag{1}$$

The principal components variables are defined as those linear combinations which have maximum variance and are ordered in a descending way. Then, the first two principal components could be explored in order to explore sample variances using a scatter plot. In RNA-seq case-control experiments, two groups of sample are involved, in which genes are the variables of interest. Given the fact that an experiment only affects a small group of genes, sample variances should be related to them. Thus, it is expected that the replicate from the same experimental condition appear grouped together in the principal components space and different condition samples appear as far apart as possible.
Since metastatic vs. non-metastatic global sample separability must holds, any sample that invalidate this assumption is considered as an outlier sample. The presence of outlier samples could bias the results, therefore they should be identified and removed from the analysis. After that, *gene control* and *filtering* must be carried out again, in order to obtain a new count

matrix, CM_2. If any outlier sample is not removed, neither sample separability nor differential expression analysis would be possible.

2.2.3. Impact on downstream analysis: Metastatic and non-metastatic OSCC have very different outcome, thus quite different gene expression profiles are expected. In order to demonstrate the importance of experiment quality control and outlier sample removal, downstream analysis results obtained using both count matrices (CM_1 and CM_2) were contrasted. In this sense, gene expression analysis was conducted using DESseq2 R package [13]. This package provides methods to detect and correct those genes having extreme values affecting gene model fit, called count outlier genes. Count values of them are imputed and recovered for DE analysis. DE results were compared in terms of amount of Differentially Expressed Genes (DEG) and in terms of the amount of count outlier genes . After DEG determination, functional enrichment analysis using DAVID platform was done ([8,9]). In particular, Gene Ontology (GO) terms related to Biological Process (BP) category were explored [14]. Functional enrichment analysis results were validated by means of literature evidence.

3. Results

3.1. Data Filtering

The overall raw count expression matrix consists in *G=54.664* genes and *P=10* samples. From them, the *gene filtering* process removed:

(i) 34.657 noncoding and unannotated genes
(ii) 828 small RNAs
(iii) 5.988 low counts genes

This yields to the expression count matrix CM_1 consisted in 13.191 rows and 10 columns.

3.2. Global Quality Control

Distributional characteristics of the samples for raw and normalized expression counts are shown in Figure 1. For instance, median expression bias is observed over different samples (panel **a)**) and also observed in density distribution (panel **b)**). Data normalization using edgeR-RLE method removed this bias, providing homogeneity for both density and boxplot distribution of expression values over samples (panel **c)** and **d)**). No evidence of outlier samples can be observed beyond the fact that the normalization process improves distributional assumptions. Then, PCA was performed over raw and normalized CM_1 matrix. The two first principal components are shown in the scatter plots of panel **a)** and panel **b)** in Figure 2. It is observed that samples are not completely separated between metastatic (triangles) and non-metastatic (circles) as expected. One non-metastatic (metastatic) sample, *R2C1* (*R2C2*), tends to be closer to the opposite experimental condition indicated by arrows. This suggests that they are outlier samples rather than a technical effect and they could be removed from analysis.

After outlier samples removal, *Data filtering* step was carried out again, obtaining the second count matrix, CM_2 (13.482x8). It worth to note that this matrix has a greater number of genes than CM_1. Indeed, when only considering 8 samples the low count filtering criterion is modified. Then, edgeR-RLE normalization and PCA were performed. In Figure 3, PCA scatter plots of the two first principal components, for both raw (panel **a)**) and normalized (panel **b)**) CM_2 are shown. Now, separation between groups was achieved. It can be observed that outlier samples removal improves the experimental conditions separation in the principal component

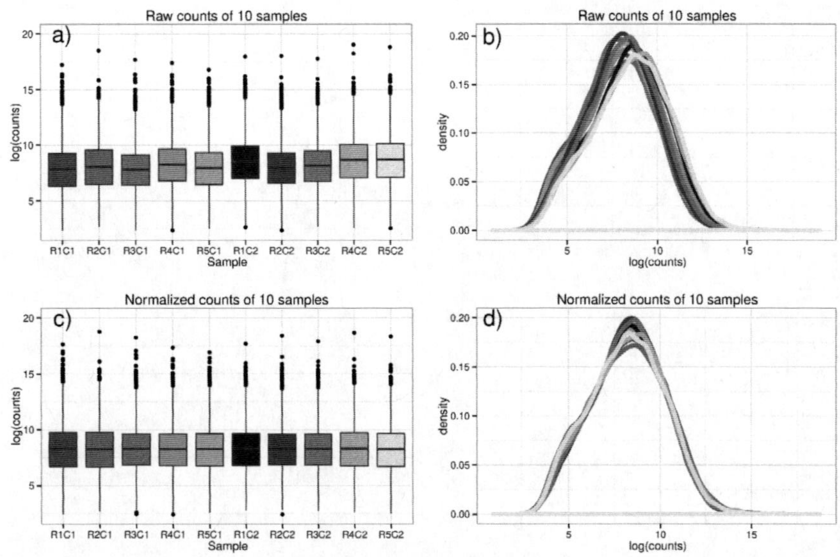

Figure 1. Sample counts boxplot (left) and density plot (right). Raw counts (**a)** and **b)**).After normalization (**c)** and **d)**). RXCY stands for: RX (replica number), CY (condition number) where C1/2 is nonmetastatic/metastatic sample respectively.

space. In addition, it is remarkable that normalization process does not affect or correct sample separability suggests that is not a technical effect as stated.

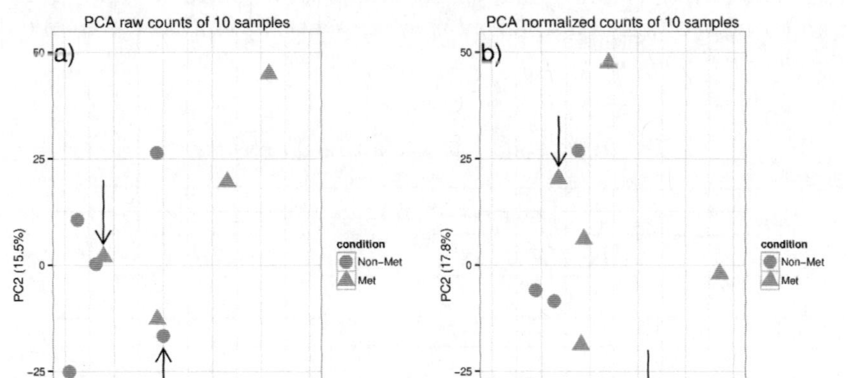

Figure 2. Scatterplots of the two first principal components over counts matrices: **a)** raw CM_1 contains 10 samples; *b)* raw CM_2 contains 8 samples.

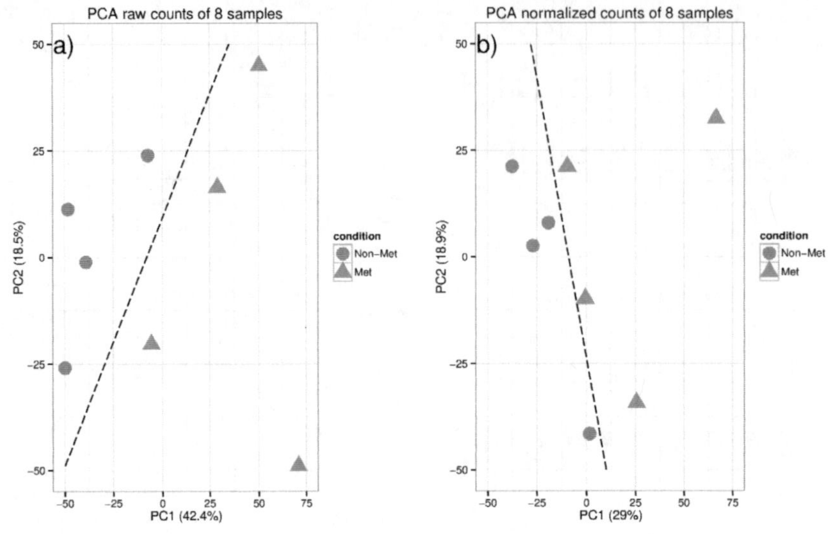

Figure 3. Scatterplots of the two first principal components over counts matrices: **a)** raw CM_1 contains 10 samples; *b)* raw CM_2 contains 8 samples.

3.3. Impact on downstream analysis

Table 1 summarizes the obtained DE results for both CM_1 and CM_2 matrices. The *Outlier genes* columns indicate the DESeq2's count outlier results [13]. It can be observed that the number of count outliers in CM_1 is greater than CM_2. In the first case, the 6.35% of the analyzed genes presents extreme count values, recovering 88% of them after DESeq2's imputation strategy. When CM_2 was analyzed, the 5.5% of analyzed genes presents extreme value. From them, 95% were recovered after imputation. This percentage represents 7% more genes than in the previous case, improving information gain.

Table 1. Differential expression results.

	Genes with counts outlier		
Count matrix	Detected	Final	DE genes
CM_1	837	93	91
CM_2	737	38	424

The most important impact of the proposed approach is shown in the *DE genes* column in Table 1. This contains the number of DEG found with both matrices. It is observed that the nonseparability of CM_1 samples directly impacts on the number of DEG detected allows finding only 91 genes. However, when the two outlier samples were excluded, the number of DEG detected increases more than 465% times, achieving a total of 424 genes.

It is not enough to only compare the number of total DEG found in both cases, but also to establish how many of these genes are shared between CM_1 and CM_2 results. A simple Euler diagram was built to determine this issue. Figure 4 shows the overlap of detected DEG using both CM_1 and CM_2. In addition, the circles size illustrates the greater amount of DEG found when using CM_2 (DEG_2). This result suggests that the inclusion of outlier samples could causes information lost (78.5 % of DEG not detected) which is recovered when they are removed.

Logarthmic fold changes and adjusted p-values, of the overlapped genes (81), obtained using

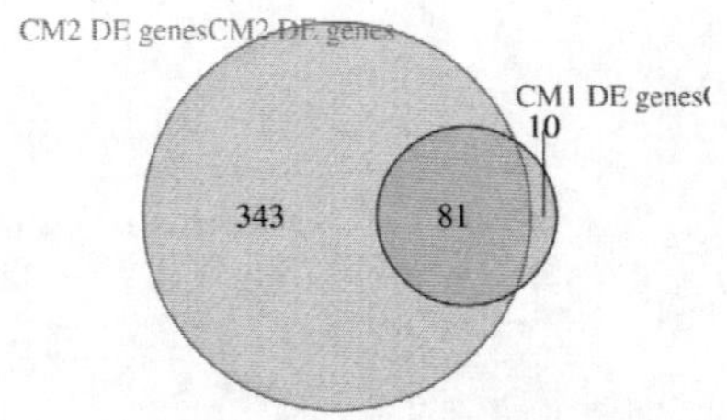

Figure 4. Euler diagram for differentially expressed genes found in CM_1 and CM_2 sets.

CM_1 (x axis) and CM_2 (y axis) were explored. Their corresponding scatter plots are showed in panel **a)** and **b)** of the Figure 5. Black line indicates the identity line and allows detecting fold changes differences. Thus, when outlier samples are removed, absolute fold changes values were higher in both up (pink dots, above identity line) and down (cyan dots, under identity line) regulated gene groups. Concordantly, the corresponding adjusted p-values were mostly lower after outlier samples removal. Absolute fold change differences were significant (p-value $4.957e^{-7}$) using the Wilcoxon non-parametric test.

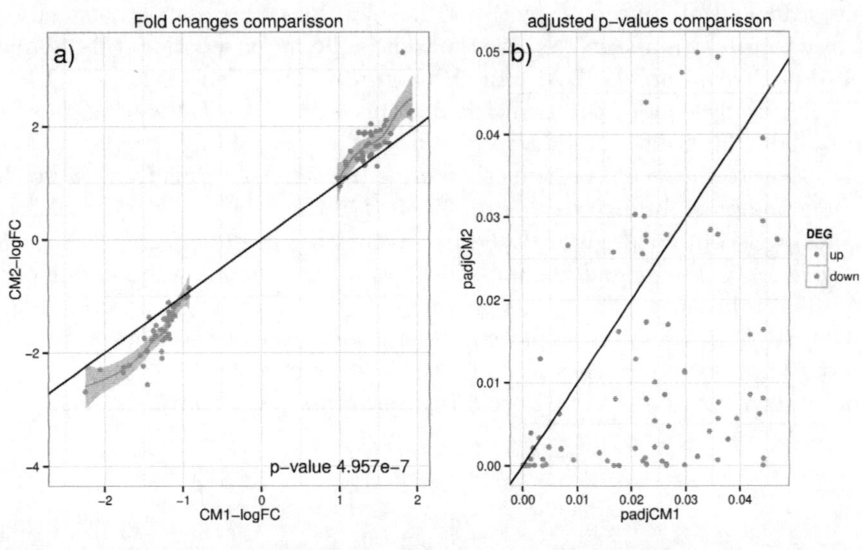

Figure 5. Scatter plots for 81 DEG found using both CM_1 and CM_2 expression matrices. **a)** Logarithmic fold change values and **b)** Adjusted p-values. After outlier samples removal,the absolute fold change values increased and the adjusted p-values decreased in both up and down regulated gene groups.

Journal of Physics: Conference Series **705** (2016) 012003 doi:10.1088/1742-6596/705/1/012003

Outlier samples removal also impacts on functional analysis. The Figure 6 shows the comparison between GO's BP terms found using both count matrices. Enrichment analysis of DEG_1 yields 122 enriched BP terms, where 82 of them (67%) were also found on the 287 enriched BP terms using DEG_2. This increment represents 69.29% of more enrichment when CM_2 was analyzed.

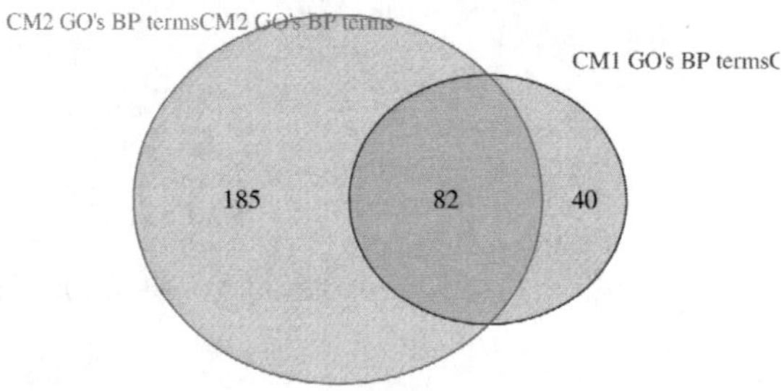

Figure 6. Euler diagram for enriched GO's BP terms found in CM_1 and CM_2 sets.

Figure 7 illustrates the enriched GO's BP terms tree Direct Acyclic Graph (DAG). Each DAG node represents a biological concept (term) and has several genes associated to it. The nodes are organized in a hierarchical manner where the root is the most generic term and more specific ones are found downstream on the DAG. In the Figure 7, green nodes are the BP terms only enriched by DEG_1, orange nodes those enriched terms from CM_2 (DEG_2). Those nodes that were enriched by both DEG lists are showed in red. It is possible to observe that most of the green nodes are closer to the tree root, thus they are more general terms than those closer to the leave nodes. Only one green branch can be observed. On the other hand, several branches (red and orange nodes) were obtained with DEG_2 list. Terms like *positive regulation of angiogenesis* and *positive regulation vascular endothelial growth factor production* were enriched using both CM_1 and CM_2. This is because DEG related to them were in majority found using both matrices. But, when GO's BP terms graph is explored, several new orange branches appear when CM_2 DEG were used. These new branches are related to several biological processes known to be related to cancer, such *immune response*, *cell differentiation*, *cell migration* and *angiogenesis*.

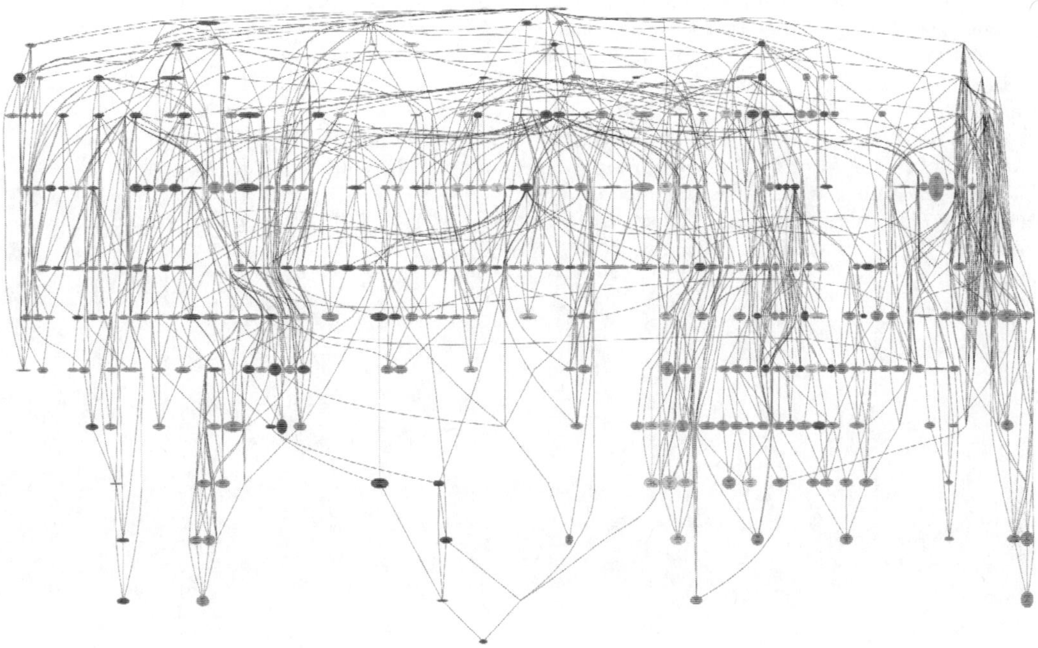

Figure 7. GO graph for enriched terms. In green node terms enriched only using CM_1 DEG and in orange only using CM_2 DEG. In red, enriched terms using CM_1 or CM_2 DEG genes sets.

4. Conclusion

Experiment quality control is a crucial step that should be performed before any expression analysis and biological interpretation.Here it is shown that simple multivariate principal component analysis allows identifying outlier samples that hide biological information. Its utility in a real RNAseq context was demonstrated, where under a traditional approach only 91 genes were detected as differentially expressed, whereas the proposed approach increased this number up to 424 genes due to outlier samples removal. Functional analysis results were also improved when the proposed approach was applied. In particular, 185 new GO's BP terms were enriched when outlier samples were removed. Here is demonstrated that outlier samples presence could lead to wrong biological conclusions. In this context, experiment quality control directly impacts in downstream analysis results. For this reason, it must not be considered neither lightly nor a trivial step in the analysis.

References

[1] Oshlack A, Robinson M D, Young M D *et al.* 2010 *Genome biol* **11** 220
[2] Andrews S *et al.* 2010 *Reference Source*
[3] Tarazona S, Garcia-Alcalde F, Dopazo J, Ferrer A and Conesa A 2011 *Genome research* **21** 4436
[4] Cunningham F, Amode M R, Barrell D, Beal K, Billis K, Brent S, Carvalho-Silva D, Clapham P, Coates G, Fitzgerald S *et al.* 2015 *Nucleic acids research* **43** D662–D669
[5] Li H, Handsaker B, Wysoker A, Fennell T, Ruan J, Homer N, Marth G, Abecasis G, Durbin R *et al.* 2009 *Bioinformatics* **25** 2078–2079

[6] Anders S, Pyl P T and Huber W 2014 *Bioinformatics* btu638
[7] R Core Team 2015 *R: A Language and Environment for Statistical Computing* R Foundation for Statistical Computing Vienna, Austria URL `http://www.R-project.org/`
[8] Huang D W, Sherman B T and Lempicki R A 2008 *Nature protocols* **4** 44–57
[9] Huang D W, Sherman B T and Lempicki R A 2009 *Nucleic acids research* **37** 1–13
[10] Maza E, Frasse P, Senin P, Bouzayen M and Zouine M 2013 *Communicative & Integrative Biology* **6** e25849
[11] Robinson M D, McCarthy D J and Smyth G K 2010 *Bioinformatics* **26** 139–140
[12] Jackson J E 2005 *A user's guide to principal components* vol 587 (John Wiley & Sons)
[13] Anders S and Huber W 2010 *Genome biol* **11** R106
[14] Ashburner M, Ball C A, Blake J A, Botstein D, Butler H, Cherry J M, Davis A P, Dolinski K, Dwight S S, Eppig J T *et al.* 2000 *Nature genetics* **25** 25–29

Heterogeneous expression pattern of tandem duplicated sHsps genes during fruit ripening in two tomato species

Arce DP[1, 2*] Krsticevic FJ[3*] Ezpeleta J[3] Ponce SD[2] Pratta GR[1] Tapia E[3]

[1]Cátedra de Genética. Departamento de Biología. Facultad de Ciencias Agrarias. Universidad Nacional de Rosario.
[2]Grupo de Análisis, Desarrollos e Investigaciones Biomédicas. Facultad Regional San Nicolás. Universidad Tecnológica Nacional.
[3]Centro Internacional Franco Argentino de Ciencias de la Información y de Sistemas. CONICET- UNR- AMU.
*These authors contributed equally to this work
Corresponding author's: debora.arce@gmail.com

Abstract. The small heat shock proteins (sHSPs) have been found to play a critical role in physiological stress conditions in protecting proteins from irreversible aggregation. To characterize the gene expression profile of four sHsps with a tandem gene structure arrangement in the domesticated *Solanum lycopersicum* (Heinz 1706) genome and its wild close relative *Solanum pimpinellifolium* (LA1589), differential gene expression analysis using RNA-Seq was conducted in three ripening stages in both cultivars fruits. Gene promoter analysis was performed to explain the heterogeneous pattern of gene expression found for these tandem duplicated sHsps. *In silico* analysis results contribute to refocus wet experiment analysis in tomato sHsp family proteins.

1. Introduction

Small heat shock proteins (sHsps) are low molecular weight (12-40 kDa) proteins that play a critical role in physiological stress conditions protecting proteins from irreversible aggregation and stress tolerance. Furthermore, some sHsps are expressed during certain stages of development, such as embryogenesis, germination, pollen development, and fruit maturation [1]. The special importance sHsp family in plants is suggested by their unusual abundance and diversity. Small Hsps are found in distinct cellular compartments like cytosol, chloroplast, endoplasmic reticulum, mitochondria and membranes [2]. The sHsp family possess a conserved domain of approximately 80 to 100 amino acids called alpha-crystalline domain (ACD), flanked by N- and C-terminal regions. The number of genes in the sHsp family is overestimated because the same conserved domain IPR008978 (ACD or HSP20 domain) has become incorporated into a large number of unrelated proteins, known as ACD-like or HSP20-like [3].

Tomato has become the model system to study organ expansion, maturity, ripening, shelf-life and fruit nutritional quality [4]. Tomato, the centerpiece of the Solanaceae family, has emerged as a model of fleshy fruit development, primarily because this is the species for which the genetic and molecular toolkits are most advanced. The inbred tomato *S. lycopersicum* cultivar 'Heinz 1706' and its wild relative *S. pimpinellifolium* LA1589 were sequenced and estimated divergence between the wild and domesticated genomes is 0.6% (5.4 million single nucleotide polymorphisms distributed along the chromosomes). Consequently, due ancestrally similar gene structure and expression profiles are expected for sHsps family in both cultivars. Transcriptome analysis by next generation sequencing technology (NGS), such as high-throughput RNA sequencing (RNA-Seq) have led to the characterization of previously unrecognized mechanisms of gene regulation, in-depth sequence analysis, transcriptional structure and expression level for each gene in these two tomato cultivars [5]–[7].

Analysis of the tomato *S. lycopersicum* cv Heinz 1706 transcriptome indicated a total of 2459 loci that could be annotated as transcription or transcriptional regulators of which 223 show expression patterns that change substantially between mature unripe and ripening fruit and thus potentially influencing ripening processes [7]. From a preliminary analysis combining phylogeny and RNA-Seq expression patterns in the sHsp family, we found a subset of four sequences. These related sequences were originated in two independent events at chromosome 1 (Solyc01g098790 and Solyc01g098810) and 4 (Solyc04g082720 and Solyc04g082740) by tandem duplication event, showed an heterogeneous pattern of expression (unpublished Krsticevic et al, personal communication) and maybe the promoter regulation could be explain this fact.

The aim of this work was to analyze gene promoter region to further characterize sHsps transcriptional regulation during maturation in the genome of domesticated *S. lycopersicum* cv. Heinz 1706 and its wild close relative *S. pimpinellifolium*. Furthemore, we compare gene expression profile of four sHsps in *S.pimpinellifolium* with the already data obtained for the commercial variety. Differential sHsps gene expression analysis using RNA-Seq was conducted in three ripening stages (G, BR and RR) in *S.lycopersicum* cv Heinz 1706 and *S. pimpinellifolium* LA1589 fruits. Finally, 1000 bp upstream of gene promoter region for each gene was performed and putative transcriptional regulation during maturation was explained.

2. Materials and methods

S. lycopersicum and S. pimpinellifolium transcriptome Illumina datasets

The Illumina RNA-Seq reads files from two biological replicates of *S. lycopersicum* (cv Heinz 1706) mature green or G (SRR404324, SRR404325), breaker or BR (SRR404326, SRR404327), red ripening fruits or RR (SRR404328, SRR404329) and *S. pimpinellifolium* (LA 1589) G (SRR404331, SRR404333), BR (SRR404334, SRR404336) and RR (SRR404338, SRR404339) were downloaded from the Sequence Read Archive (SRA from DDBJ, http://trace.ddbj.nig.ac.jp/dra/index_e.html). These data were used by The Tomato Genome Consortium (Supplementary material, 2.3.5 Transcriptome sequencing pp 46). The SRA files were converted to FASTQ files using the fastq-dump utility of the SRA toolkit (http://eutils.ncbi.nih.gov/Traces/sra/?view=software). The nomenclature used to describe the different stages of tomato fruit development is G: mature green, BR: mature orange-breaker, RR: mature red ripening stages [9].

Bowtie alignment and counting

The Bowtie short read aligner (v. 1.0.1) was used to map the reads against a set of reference genomic sequences [10]. To assess the expression levels of each gene in the different samples (G, BR, RR), the obtained raw counts (Table of counts) after Bowtie software alignment were used for differential gene expression analysis.

Alignment : The 4 sHsp tandem duplicated sequences plus the HSP70 as positive control of maturation RR stage and other two sHsps sequences which have previous experimental evidences for up or down-regulation at RR stage were used to build the reference library. Whenever possible, we tried to include ~1 kb up- and down-stream of the CDC region (~3 kb), taking care to prevent overlapping with other CDC gene regions. This procedure helped us ensure that all the sHsps reads (even those aligned to the 5′ and 3′ ends) were included in our analysis. The genomic regions contained sHsps CDC of Solyc01g098790, Solyc01g098810, Solyc04g082720, Solyc04g082740, Solyc12g042830, Solyc08g062450, Solyc09g011710, HSP70 and their ortologs in *S. pimpinellifolium* genomic sequences. HSP70 gene region was used as a positive control for the *in silico* experiment, because is known to be up-regulated during the fruit maturation process [11]. The other sHsps used as controls with experimental evidences included were mitochondrial Solyc12g042830 (up-regulated in *S.*

lycopersicum RR stage), class II Solyc08g062450 (up-regulated in *S. lycopersicum* RR) and class I Solyc09g011710 (down-regulated in *S. lycopersicum* RR). The nomenclature used to describe the sHsps in *S. lycopersicum* is S01g08790_SL, S01g098810_SL, S04g082720_SL and S04g082740; in *S. pimpinellifolium* sHsps the letter "L" change for a "P". The number of reads that aligned to each gene provided a quantification of how many RNA transcripts of that gene were present in the sample. Mapped reads were quantified from SAM output files using GNU commands to obtain a Table of counts to further statistical analysis.

Transcript abundance
The Fragments Per Kilobase of transcript per Million (FPKM) mapped fragments were calculated manually after Bowtie alignment to determine transcript abundance (mRNA absence or presence) for each gene analyzed. Mapped reads were obtained as described above and average FPKM were calculated for the two biological replicates of each sample and for each gene. Genes were considered expressed when the average FPKM of both biological replicates was greater than or equal to 2, as described by Steijger et al [12] calculated that fragments per kilobase of gene model per million mapped reads (FPKM) values of 1 to 2 correspond to ~1 mRNA molecule per cell [13].

Statistics for differential gene expression analysis
Descriptive and RNA-Seq statistics were calculated using the edgeR Bioconductor package in R, with two modifications [14]. The standard edgeR pipeline calculates library sizes as the sum of read counts for each gene, assuming that the library is mapped to the entire genome. In our work, reads were instead mapped to a reference constructed with a set of genomic regions which were extracted using the Solgenomics tomato genome browser (Jbrowse, genome version *S. lycopersicum* ITAG2.40 release and *S. pimpinellifolium* draft genome WGS Contigs cshl_1.00). Therefore, to prevent edgeR from incorrectly assuming that the library sizes were equal to the sum of read counts for each gene, we manually input the correct library sizes, which were first calculated by aligning the reads to the complete tomato reference genome. In addition, the calcNormFactors function was omitted because it is also intended for genome-wide expression studies, as it performs a normalization which assumes that most genes are *not* differentially expressed. Instead, the present study focuses on a subset of genes, a significant number of which *are* expected to be differentially expressed, which makes the above correction not only unnecessary but also counterproductive. The G stage was taken as baseline and edgeR's exact test was applied to calculate the $\log_2$ fold changes (log FC) of the BR and RR stages relative to this baseline. Positive $\log_2$ FC values indicate up-regulation, negative values indicate down-regulation and zero indicates constant gene expression relative to the G stage. Transcripts whose absolute $\log_2$fold change ($\log_2$FC) were greater than 1 and whose p-value was lower than 0.01 for either BR or RR were included in the plots.

Gene promoter analysis
Gene promoter sequences were extracted 1000 pb upstream of each gene transcription start site (TSS) Sequences were obtained from Solgenomics (http://solgenomics.net/). For gene promoter extraction in *S. pimpinellifolium* genome bed tools were also used (http://bedtools.readthedocs.org/en/latest/). Plant CARE database was used to find specific TFBS for tomato and Arabidopsis plants [15].

3. Results

Transcript abundance: FPKM quantification
Transcript abundance from four *S. lycopersicum* (SL) and *S. pimpinellifolium* (SP) sHps (S04g082720, S04g082740, S01g098790, and S01g098810) was estimated by counting Illumina reads from mature green (G), breaker (BR) and mature red ripening (RR) fruit stages aligned to their genomic regions using short read aligner Bowtie1. Results are shown as FPKM quantification in Table 1. Variability in

Journal of Physics: Conference Series **705** (2016) 012004 doi:10.1088/1742-6596/705/1/012004

transcript levels was observed from G to RR maturing stages. Small Hsps with previous experimental evidences with different subcellular localization were also used as controls (up-regulated: mitochondrial S12g042830, class II S08g062450; or down-regulated: class I S09g011710). The highest FPKM values were observed in S04g082720 in *S. lycopersicum* and S08g062450 (control) in both cultivars. The S04g082740 sequence showed an opposite transcript profile with a decreased in FPKM levels as development advanced in *S. lycopersicum* and *pimpinellifolium*. The columns RR-G in Table 1 represent differences in transcript abundances between RR and G ripening stages, showing positive values as up-regulated and negative values as down-regulated genes.

Table 1. Small Hsps FPKM quantification from RNA-Seq in ripening tomato fruits. Expression transcript levels were quantified as FPKM for three ripening fruit stages: mature green (G), breaker (BR) and red ripe (RR) in *S. lycopersicum* and *pimpinellifolium* tomato fruits.

	S. lycopersicum				*S. pimpinellifolium*			
Gene ID	G	BR	RR	RR-G	G	BR	RR	RR-G
S01g098790	0.26709592	0.37646615	0.33152983	0.06443391	1.10390168	0.61207048	0.34932573	-0.75457595
S01g098810	9.28909591	8.97380435	5.98045634	-3.30863957	4.37726509	2.98985227	2.11367002	-2.26359508
S04g082720	44.0794412	52.4421379	103.57674	59.4972988	29.2965039	48.006047	78.6167481	49.3202442
S04g082740	5.50757926	2.29629047	0.55631675	-4.95126251	4.75406219	2.86131773	1.16123017	-3.59283201
S08g062450	64.853396	421.062959	819.664401	754.811005	72.2412884	113.097962	441.333315	369.092026
S09g011710	186.614016	115.135764	58.9089129	-127.705103	270.269463	161.215473	99.9142164	-170.355247
S12g042830	8.47584812	44.4285468	87.5254453	79.0495972	58.2405858	293.787118	202.543895	144.303309

Differential gene expression analysis

To evaluate if these four sHsps might be involved specifically in any of the three ripening stages, we compared sHsps expression levels in RR and BR stages using G ripening stage as baseline levels. Figure 1 show sHsps transcript levels in BR and RR referred to G transcript levels using edgeR package. Genes with $\log_2$FC values lower than 1 mean that transcript levels were not differentially expressed (NDE) in BR or RR stages compared to G stage. Tandem chromosome 1 sHsps transcripts (S01g098790 and S01g098810) were NDE in the BR and RR stages relative to the G fruit stage in *S. lycopersicum* and *pimpinellifolium* except for S01g098810 which was down-regulated at both stages BR and RR in *S. pimpinellifolium* (Fig. 1). Chromosome 4 sHsp (S04g082740) was also NDE in *S. pimpinellifolium* but it was down-regulated at BR and RR stages in *S. lycopersicum*. Only one sHsp (S04g082720) from these four tandem genes analyzed showed up-regulation in BR and RR ripening stages compared to G fruits in both cultivars. Because tandem genes are expected to show the same expression pattern, gene promoter regions might explain this distinct expression profile (see below in Discussion). Small Hsps used as controls with previous experimental evidences showed the expected differential gene expression profiles: S08g062450 and S12g042830 were up-regulated in both cultivars. HSP70 was up-regulated except at BR stage in *S. pimpinellifolium*.

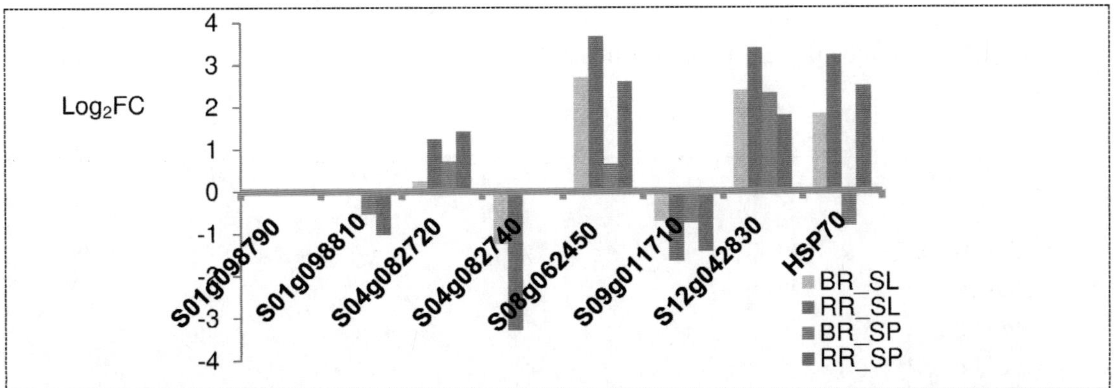

Figure 1. sHsps differential gene expression during tomato fruit ripening. Expression transcript levels were quantified as $\log_2$FC (base 2) for three ripening stages in *S. lycopersicum* (SL) and *S. pimpinellifolium* (SP) fruits: mature green (green \`x\` axis), breaker (BR) and red ripe (RR). Significative statistical differences calculated with edgeR are shown in $\log_2$FC (absolute values greater than 1) for BR (light and dark orange columns) and RR (light and dark red columns) stages compared to the G stage with a p-value lower than 0.01.

Small Hsps gene promoter analysis

Results from gene promoter analysis are shown in Table 2. Note that column with differential gene expression results comparing RR versus G stage (DGE RR) from Fig. 1 were included showing up-regulation (Up), down-regulation (Do), and not differentially expressed (NDE) genes in Table 2. Two gene relative to chromosome 1 (S01g098810_SL, S01g098810_SP), and two from chromosome 4(S04g082720_SL, and S04g082720_SP) showed the presence of heat stress motif (heat shock element, HSE). Interestingly, tandem chromosome 1 sHsps showed NDE despite the presence or absence of HSE in *S. lycopersicum*. Contrarily, the promoter of the down-regulated S01g098810 showed the presence of HSE in *S. pimpinellifolium*. Mitochondrial S12g042830 in *S.pimpinellifolium* does not show HSE. All gene promoters of sHsps used as controls have HSE motifs. We also looked for motifs which belong to different stress and hormonal regulation pathways like abscisic acid or ABA (ABRE), ethylene (ethylene response, ERE), gibberellins (GA) and transcription factors (MYB, WRKY) in sHsps gene promoter sequences. These motifs were searched because they are over-represented in gene promoter regulatory sequences from genes differentially expressed during stress, maturation and development processes [16], [17]. Variability in these last motifs was observed in the sequences analyzed.

Pairwise sequence analysis between promoters of orthologues sHps in chromosome 1 and 4 were also analyzed with Needle (http://www.ebi.ac.uk/Tools/psa/emboss_needle/nucleotide.html) and identity and similarity (as percentage) are shown in column "ID/Sim (%)". The highest values were observed for pairs S01g098790_SL/S01g098790_SP, S04g082740_SL/S04g082740_SP and S04g082720_SL/S04g082720_SP. In opposite, pair S01g098810_SL/S01g098810_SP showed the lowest ID/Sim (%) with a value of 43.8%.

Table 2.sHsps gene promoter analysis in *S.lycopersicum* (SL) and *pimpinellifolium* (SP)

Promoter	ERE	MYB	ABRE	AUXIN	GA	WRKY	HSE	DGE RR	ID/Sim (%)
S01g098790_SL	-	+	+	+	+	+	-	NDE	99.1/ 99.1
S01g098790_SP	-	+	+	-	+	-	-	NDE	
S01g098810_SL	+	+	-	-	-	-	+	NDE	43.8/ 43.8
S01g098810_SP	-	-	+	-	+	-	+	Do	
S04g082720_SL	+	+	-	-	-	-	+	Up	81.2/ 81.2
S04g082720_SP	-	-	-	-	-	-	+	Up	
S04g082740_SL	-	+	-	-	+	+	-	Do	99.6/ 99.6
S04g082740_SP	-	-	-	-	+	+	-	NDE	
S08g062450_SL	-	-	-	-	-	-	+	Up	ND
S08g062450_SP	-	-	-	-	-	-	+	Up	
S12g042830_SL	+	-	+	-	-	-	+	Up	ND
S12g042830_SP	+	-	-	+	+	-	-	Up	
S09g011710_SL	-	+	-	-	+	-	+	Do	ND
S09g011710_SP	-	+	-	-	+	-	+	Do	

4. Discussion

In the present work differential gene expression and promoter analysis was conducted for four sHsps with tandem gene structure. This gene group has heterogeneous pattern of expression. The chromosome 1 gene member S01g098790 is NDE in both cultivars and has not HSE in the promoter regions. These members might represent pseudogenes because its transcript abundance levels are very low in all tissues assayed as few reads were observed in the Bowtie alignment. Anyway, they show a similar expression profile (NDE) as is expected for copies with high percentage of identity. The typical ACD domain is absent in their protein structure (data not shown) which does not mean that a pseudogene cannot be expressed. Contrarily, some pseudogenes can also be expressed and their expression levels tend to be lower compared with annotated genes in Arabidopsis and rice [18].Furthermore, tandem chromosome 1 sHsps in *S. pimpinellifolium* showed a different pattern of expression where S01g098810 gene promoter showed a combination of HSE, GA and ABRE motifs that might be enough to trigger down-regulation of gene expression.

Contrarily, S01g98810_SL and S01g98810_SP have heterogeneous expression pattern since one member is NDE and the other is down-regulated. Differences in gene expression could be explained by pairwise gene promoter alignment showing an ID/Sim of 43.8%. This value is low and might be indicating gene sub functionalization due to differences at their promoter regions.

Interestingly, S04g082720_SL and S04g082720_SP show similar up-regulation expression profiles, as was expected by their high percentage of identity in their genomic sequences. These two chromosome 4 gene copies have HSE in their promoter regions and present similar expression profiles (Up-regulation) like it observed by the positive controls used in this work (S08g062450_SL and S12g042830 in both cultivars). All these findings support the hypothesis that despite S04g082720_SL and S04g082720_SP don not have the typical ACD domain, they act as if they were functional sHsps in the RR stage, indicating that S04g082720 could be specifically involved in *S. lycopersicum* and *pimpinellifolium* fruit ripening process. These results are according to previous findings by Matas et al.

who showed that S04g082720 referred as TU143174 is differentially expressed specifically in the collenchyma and inner epidermis of the *S. lycopersicum* fruit (cv Ailsa Craig) [19].

On the other hand, S04g082740_SL and S04g082740_SP show down-regulation and NDE in RR stage compared to G stage expression profiles. Although gene promoter analysis showed a high percentage of identity (99.6%) they do not have HSE. MYB motif TFBS might explain this results since its presence in S04g082740_SL is the only difference between these two promoter sequences.

HSE were identified and these TFBS are always present in the promoter region of heat shock genes. The HSE is the binding target site for recognition of the heat shock factors (Hsfs) [21]. The essential role of HSE for heat-dependent transcriptional activation in plants has been established by promoter deletion assays in tomato [20], [21]. Moreover, HSE present in ascorbate peroxidase 1 (apx1) gene promoter is involved in antioxidative response in Arabidopsis indicating that heat and oxidative pathways could be related and share transcriptional regulation mechanisms [22]. Note in our results that none specific combination of ERE, ABRE, GA, MYB, WRKY and HSE motifs is necessary to trigger gene expression but suggest that heat response and hormone mediated stress signaling may share common pathways where oxidative stress is implicated. Moreover, recent study with a similar strategy (promoter analysis and regulatory network prediction of non-redundant DREB genes stored in RefSeq database in Arabidopsis) was accomplished by use of Plant CARE program and GeneMANIA web tool, respectively. The results indicated that DREB genes have diverse motifs involved in abiotic stress responses and cooperation of phytohormones like abscisic acid, ethylene, salicylic acid and methyl jasmonate signaling is crucial for the regulation of the expression of abiotic stress (drought and cold responses) through *DREB* transcription factors [23].

Further promoter analysis, protein structure analysis, phylogenetic analysis and experimental evidences should be conducted to elucidate the functionality of these tandem genes in chromosome 1 and 4 in tomato genomes. There is a need to exploit the transcription factors associated with sHsps and its regulation and differential expression in order to use it for developing more resistant cultivars to stress and better fruit quality.

5. Conclusions

RNA-Seq transcriptome analysis combined with structural protein and gene promoter analysis carried out in this study contributed to interpret the transcriptional regulation cascade of 4 sHsps in tandem arrangement during ripening in *S. lycopersicum* (Heinz 1706) and *S. pimpinellifolium* (LA1589). Chromosome 4 S04g082720 act as a functional sHsps in the RR stage in both cultivars, indicating that it could be specifically involved in *S. lycopersicum* and *S. pimpinellifolium* fruit ripening process.

Promoter analysis conducted in this work might explain some gene regulation aspects of sHsps but wet experiments like Chip-Seq should be considered to clarify the functionality of sHsps and its gene regulation mechanism. Gene network during fruit ripening is a complex process mediated by different counterparts and probably the sHsp family might have some role in this networking process.

References

[1] C. Prasinos, K. Krampis, D. Samakovli, and P. Hatzopoulos, "Tight regulation of expression of two Arabidopsis cytosolic Hsp90 genes during embryo development.," *J. Exp. Bot.*, vol. 56, pp. 633–644, 2005.

[2] W. X. Wang, B. Vinocur, O. Shoseyov, and A. Altman, "Role of plant heat-shock proteins and molecular chaperones in the abiotic stress response," *Trends Plant Sci.*, vol. 9, pp. 244–252, 2004.

[3] R. L. van Montfort, E. Basha, K. L. Friedrich, C. Slingsby, and E. Vierling, "Crystal structure and assembly of a eukaryotic small heat shock protein.," *Nat. Struct. Biol.*, vol. 8, pp. 1025–1030, 2001.

[4] H. J. Klee and J. J. Giovannoni, "Genetics and control of tomato fruit ripening and quality attributes.," *Annu. Rev. Genet.*, vol. 45, pp. 41–59, Jan. 2011.

[5] K. Aoki, K. Yano, A. Suzuki, S. Kawamura, N. Sakurai, K. Suda, A. Kurabayashi, T. Suzuki, T. Tsugane, M. Watanabe, K. Ooga, M. Torii, T. Narita, T. Shin-I, Y. Kohara, N. Yamamoto, H. Takahashi, Y. Watanabe, M. Egusa, M. Kodama, Y. Ichinose, M. Kikuchi, S. Fukushima, A. Okabe, T. Arie, Y. Sato, K. Yazawa, S. Satoh, T. Omura, H. Ezura, and D. Shibata, "Large-scale analysis of full-length cDNAs from the tomato (*Solanum lycopersicum*) cultivar Micro-Tom, a reference system for the Solanaceae genomics.," *BMC Genomics*, vol. 11, p. 210, Jan. 2010.

[6] K. Aoki, Y. Ogata, K. Igarashi, K. Yano, H. Nagasaki, E. Kaminuma, and A. Toyoda, "Functional genomics of tomato in a post-genome-sequencing phase.," *Breed. Sci.*, vol. 63, no. 1, pp. 14–20, Mar. 2013.

[7] T. Tomato and G. Consortium, "The tomato genome sequence provides insights into fleshy fruit evolution.," *Nature*, vol. 485, pp. 635–41, 2012.

[8] F. Zambelli, G. Pesole, and G. Pavesi, "Motif discovery and transcription factor binding sites before and after the next-generation sequencing era.," *Brief. Bioinform.*, vol. 14, no. 2, pp. 225–37, Mar. 2013.

[9] P. Pesaresi, C. Mizzotti, M. Colombo, and S. Masiero, "Genetic regulation and structural changes during tomato fruit development and ripening.," *Front. Plant Sci.*, vol. 5, no. April, p. 124, Jan. 2014.

[10] B. Langmead, C. Trapnell, M. Pop, and S. L. Salzberg, "Ultrafast and memory-efficient alignment of short DNA sequences to the human genome.," *Genome Biol.*, vol. 10, no. 3, p. R25, Jan. 2009.

[11] S. Fragkostefanakis, S. Simm, P. Paul, D. Bublak, K.-D.Scharf, and E. Schleiff, "Chaperone network composition in S olanumlycopersicum explored by transcriptome profiling and microarray meta-analysis," *Plant. Cell Environ.*, vol. 38, no. 4, pp. 693–709, Apr. 2015.

[12] T. Steijger, J. F. Abril, P. G. Engström, F. Kokocinski, M. Akerman, T. Alioto, G. Ambrosini, S. E. Antonarakis, J. Behr, P. Bertone, R. Bohnert, P. Bucher, N. Cloonan, T. Derrien, S. Djebali, J. Du, S. Dudoit, M. Gerstein, T. R. Gingeras, D. Gonzalez, S. M. Grimmond, R. Guigó, L. Habegger, J. Harrow, T. J. Hubbard, C. Iseli, G. Jean, A. Kahles, J. Lagarde, J. Leng, G. Lefebvre, S. Lewis, A. Mortazavi, P. Niermann, G. Rätsch, A. Reymond, P. Ribeca, H. Richard, J. Rougemont, J. Rozowsky, M. Sammeth, A. Sboner, M. H. Schulz, S. M. J. Searle, N. D. Solorzano, V. Solovyev, M. Stanke, B. J. Stevenson, H. Stockinger, A. Valsesia, D. Weese, S. White, B. J. Wold, J. Wu, T. D. Wu, G. Zeller, D. Zerbino, and M. Q. Zhang, "Assessment of transcript reconstruction methods for RNA-seq.," *Nat. Methods*, vol. 10, pp. 1177–84, 2013.

[13] T. Hart, H. K. Komori, S. Lamere, K. Podshivalova, and D. R. Salomon, "Finding the active genes in deep RNA-seq gene expression studies," 2013.

[14] M. D. Robinson, D. J. McCarthy, and G. K. Smyth, "edgeR: a Bioconductor package for differential expression analysis of digital gene expression data.," *Bioinformatics*, vol. 26, no. 1, pp. 139–40, Jan. 2010.

[15] B. Mohanty, S. P. T. Krishnan, S. Swarup, and V. B. Bajic, "Detection and preliminary analysis of motifs in promoters of anaerobically induced genes of different plant species.," *Ann. Bot.*, vol. 96, no. 4, pp. 669–81, Sep. 2005.

[16] V. Petrov, V. Vermeirssen, I. De Clercq, F. Van Breusegem, I. Minkov, K. Vandepoele, and T. S.Gechev, "Identification of cis-regulatory elements specific for different types of reactive oxygen species in Arabidopsis thaliana.," *Gene*, vol. 499, no. 1, pp. 52–60, May 2012.

[17] E. Picot, P. Krusche, A. Tiskin, I. Carré, and S. Ott, "Evolutionary analysis of regulatory sequences (EARS) in plants.," *Plant J.*, vol. 64, no. 1, pp. 165–76, Oct. 2010.

[18] C. Zou, M. D. Lehti-Shiu, F. Thibaud-Nissen, T. Prakash, C. R. Buell, and S.-H.Shiu, "Evolutionary and expression signatures of pseudogenes in Arabidopsis and rice.," *Plant Physiol.*, vol. 151, no. 1, pp. 3–15, Sep. 2009.

[19] A. J. Matas, T. H. Yeats, G. J. Buda, Y. Zheng, S. Chatterjee, T. Tohge, L. Ponnala, A. Adato, A. Aharoni, R. Stark, A. R. Fernie, Z. Fei, J. J. Giovannoni, and J. K. C. Rose, "Tissue- and cell-type specific transcriptome profiling of expanding tomato fruit provides insights into metabolic and regulatory specialization and cuticle formation.," *Plant Cell*, vol. 23, no. 11, pp. 3893–910, Nov. 2011.

[20] L. Nover, K. D. Scharf, D. Gagliardi, P. Vergne, E. Czarnecka-Verner, and W. B. Gurley, "The Hsf world: classification and properties of plant heat stress transcription factors.," *Cell Stress Chaperones*, vol. 1, pp. 215–223, 1996.

[21] S. K. Baniwal, K. Bharti, K. Y. Chan, M. Fauth, A. Ganguli, S. Kotak, S. K. Mishra, L. Nover, M. Port, K.-D. Scharf, J. Tripp, C. Weber, D. Zielinski, and P. von Koskull-Döring, "Heat stress response in plants: a complex game with chaperones and more than twenty heat stress transcription factors.," *J. Biosci.*, vol. 29, pp. 471–487, 2004.

[22] S. Storozhenko, P. De Pauw, M. Van Montagu, D.Inzé,S.Kushnir,"The heat-shock element is a functional component of the Arabidopsis APX1 gene promoter" *Plant Physiol.*, vol. 118, no. 3, pp. 1005–1014, 1998.

[23] S. Sazegari, A. Niazi, and F. S. Ahmadi, "A study on the regulatory network with promoter analysis for Arabidopsis DREB -genes," vol. 11, no. 2, pp. 2–7, 2015.

Journal of Physics: Conference Series **705** (2016) 012005 doi:10.1088/1742-6596/705/1/012005

Noise generator for tinnitus treatment based on look-up tables

Alejandro J. Uriz[1], Pablo Agüero[2], Juan C. Tulli[2], Jorge Castiñeira Moreira[1], Esteban González[2], Roberto Hidalgo[2], Manuel Casadei[2]

[1] CONICET / Laboratorio de Comunicaciones - Facultad de Ingeniería - Universidad Nacional de Mar del Plata

[2] Laboratorio de Comunicaciones - Facultad de Ingeniería - Universidad Nacional de Mar del Plata

E-mail: {ajuriz,casti}@fi.mdp.edu.ar

Abstract. Treatment of tinnitus by means of masking sounds allows to obtain a significant improve of the quality of life of the individual that suffer that condition. In view of that, it is possible to develop noise synthesizers based on random number generators in digital signal processors (DSP), which are used in almost any digital hearing aid devices. DSP architecture have limitations to implement a pseudo random number generator, due to it, the noise statistics can be not as good as expectations. In this paper, a technique to generate additive white gaussian noise (AWGN) or other types of filtered noise using coefficients stored in program memory of the DSP is proposed. Also, an implementation of the technique is carried out on a dsPIC from Microchip®. Objective experiments and experimental measurements are performed to analyze the proposed technique.

1. Introduction

The noise is usually considered an undesirable signal which should be removed from the desired signal. However, noise is also an useful tool for the characterization of systems [1, 2]. It can be used for instance for the analysis of communication systems [3, 4] or for characterising filters [1].

In recent years, the noise has been used for medical purposes such as treatment of tinnitus [5, 6, 7]. In this way, a noise synthesizer is used to mask the tinnitus and, consequently, improve the quality of life of the individual who suffers from this condition [6, 7]. Thus, it is convenient to implement a noise generator (based on a pseudoaleatory numbers generator) in assistive listening devices, which actually are based on a digital signal processor (DSP). In this sense, it is possible find digital hearing aid devices[8, 9] which implement noise generators based on fractals or more complex noise generators.

One of the limitations found in performing this task is the complexity of the algorithms necessary to synthesize white noise with the appropriate statistical properties. In some cases, white noise must be filtered through a digital filter to obtain the appropriate response to mask tinnitus, which generates an additional computational load to the system, which may not be acceptable because this functionality is usually not the principal task in assistive listening devices. There are papers in the literature [10, 11] where noise generators are described, but those require a DSP exclusively dedicated to their implementation, and they were not developed for tinnitus treatment and the considerations required [6, 7] for this application.

In this work, a technique for generating sequences of pseudo-random numbers in a microcontroller using tables stored in the program memory is presented. The proposal technique improves system performance mainly because filters are not necessary into the DSP to obtain noise with specific features, and furthermore the castings between numbers formats are reduced.

Objective and subjective experiments and, finally, experimental measurements are carried out to analyse the performance of the technique. Moreover, an implementation performed in a Microchip® dsPIC [12] is described.

This paper is organised as follows. In Section 2, the most important aspects of the tinnitus masking techniques are presented. In Section 3, a new method is proposed. In Section 4, the results of the objective and subjective tests are presented and discussed. Finally, the main conclusions are summarised in Section 5.

2. Tinnitus masking using sounds

Tinnitus is a hearing condition that can be very rarely cured, even using medical treatments and surgical methods. In recent years methods to improve the quality of life of individual with tinnitus were developed. One of the most used approaches are methods based on sound therapies, which reduce the perception of tinnitus in the individual. There are various types of therapies based on sounds, in the following subsection one of the most commonly used methods are presented.

2.1. Tinnitus treatment using masking sounds

This kind of therapy can be divided into two subgroups: techniques based on zen music, and based on colored noise [5], [6], [7]. In the first subgroup of therapies techniques based on natural sounds such as the sound of rain or waterfalls, among others, are included. These sounds can be recorded or, as in the case of some high-end assistive devices[8], they may be generated using mathematical algorithms.

The second subgroup includes techniques based on colored noise. These use the spectral masking phenomenon between components[13],[14]. Masking is defined as the process taking place in the auditory system of a person, where a sound of a certain energy can reduce the perception of another sound which is spectrally adjacent to it.

There are three possible situations defined between the tinnitus and the sound generated for masking it. Firstly, if the sound does not change the perception of tinnitus, there is not masking between components. Moreover, if the sound level generated a partial alteration of tinnitus is said that the tinnitus is partially masked by the external stimulus. Finally, in cases where tinnitus is not perceived due to the external stimulus, it is said that first component has been completely masked by the masker. Note that when applying these therapies, care should be taken that the external stimulus not causes damage to the auditory system of the patient.

Therapies based on colored noise attempts to partially or fully mask the tinnitus using synthesised noise. Actually, this technique is included in high-end assistive listening devices[9], where the colored noise is obtained by filtering AWGN. In these devices, the filtering is done by means of an equalizer with up to 16 subbands, which are also divided into four bands located between *100Hz-500Hz, 500Hz-1400Hz, 1400Hz-3500Hz* and *3500Hz-8000Hz.* In recent years, our researching Group has developed a digital hearing aid device, which has been patented [18]. Simultaneously, a tinnitus treatment method has been presented [17]. Thus, in the next section, a method to implement a pseudo-random number generator in the architecture of the DSP used in the developed hearing aid (or another microcontroller) is presented.

3. Proposed Technique

In cases where according the needs of the individual, filtered noise is necessary, this signal is obtained by filtering a white noise source. In these cases, the hearing aid device must be capable to compute the filter operation, which in several cases requires a considerable amount of resources.

Thus, the main goal of this work is to implement a noise synthesiser in a hearing assistance device, with a substantial reduction of DSP operations. Also, the system must perform simultaneously other functions such as audibility extender, noise reduction, etc. Furthermore, it should be noted that a microcontroller has more program memory (EPROM) than data memory (RAM), which usually is a limitation in this kind of design. Therefore, there is a tendency to store constant data (such as coefficients for windowing, or constants for computing a Fast Fourier Transform (FFT), etc.) in EPROM. In this paper, a pseudo-random number generator which uses coefficients stored in the program memory of the microcontroller is proposed.

The first step of the technique is to generate segments containing pseudo-random values on a PC using a pseudo-random number generator (for example the `rand()` or `randn()` functions of MATLAB®), which is choosen depending on the nature of the noise to be synthesised, additionally the signal could be filtered using MATLAB® according to the needs of the individual. This filtering can be done using the `filter()` function of MATLAB® or a more complex subroutine.

Then, the generated segments of noise are stored in the program memory of the DSP. The coefficients are stored in program memory in a matrix of 256 columns and M rows, which in the implementation in a microcontroller, will be replaced by a table with $256xM$ elements. Then, a random pointer, pointed to a specific program memory location (which is generated using the architecture of the microcontroller), can choose which of the rows of the table will be used to synthesise each segment. Consequently, the order of appearance of each row in the synthesised signal will depend on the jump generator.

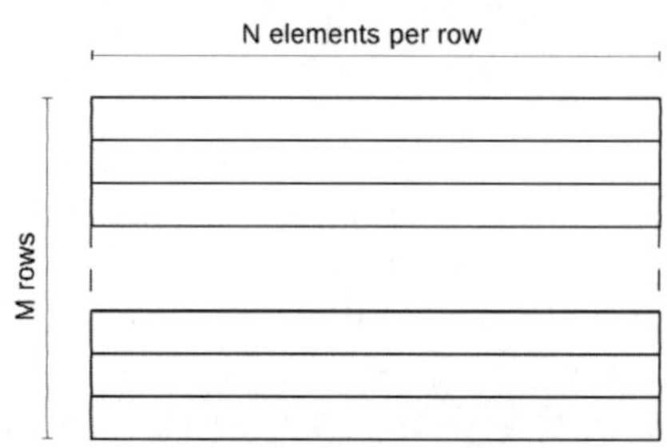

Figure 1. Architecture of the proposed table of data to generate noise.

One of the advantages of the proposed method is that it can be implemented in a microcontroller **without an architecture dedicated to processing mathematical operations**. This is because, if it is desired to synthesise filtered white noise through a specific filter, segments of the desired signal will be stored in the program memory of the microcontroller. Firstly, it avoids noise filtering on chip, according to the needs of the user and, secondly, the casting between numbers formats is also avoided. Thereby, once the microcontroller was programmed, the operation of the proposed noise generator is limited to generate a random integer which defines which row vector of the matrix of coefficients will be selected to synthesise each signal segment.

Then, in this paper it is assumed that vectors of length $N = 256$ are generated to analyse the generator. Then coefficients generated using MATLAB® are stored in a table that contains 256 columns. The number of rows M of the table is adjusted depending on the amount of memory available in the microcontroller. Figure 1 shows a diagram of the proposed implementation. It can be seen that the matrix of coefficients generated contains $N = 256$ columns and M rows. The proposed technique works according to the scheme shown in Figure 2.

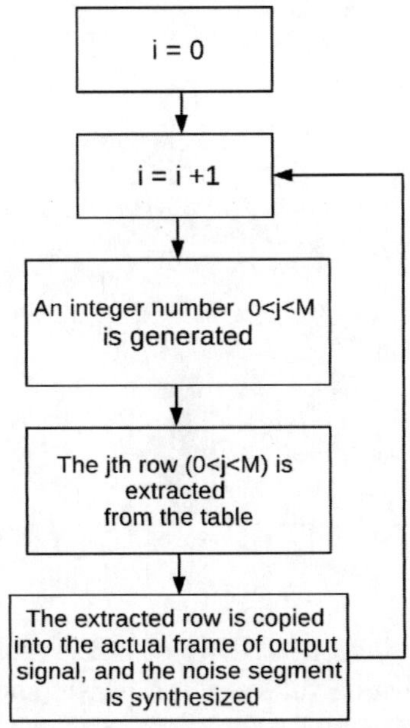

Figure 2. Block diagram of the proposed algorithm.

For each segment of the signal to synthesise, an integer $0 < j < M$ is randomly generated. Then, this value represents the row of the matrix used for synthesising the frame. This is performed in the microcontroller by means of a pointer to the starting position according to the row $(POS + j * 256)$, where the constant POS represents the initial position of the table in the program memory, then the following 256 elements are used to synthesise the signal.

Note that by increasing the value of M a noise generator with better performance is obtained from the statistical point of view, since when having more values per column, the repetition period of each value is also increased. Another interesting aspect of the proposed technique is that the type of noise of the synthesiser can be modified by simply modifying the values stored in the table. Thus, if for example, `randn()` is used instead of using MATLAB® function `rand()`, the pseudo-random number generator output signal is obtained with Gaussian distribution. Finally, as was described previously, another option to change the features of the output signal (which is appropriated to set the noise features to the specific needs of an individual who suffer tinnitus) is to change the coefficients of the filter used to filter the noise. In the next section the experiments and measurements carried out to analyse the performance of the system are presented.

4. Experiments

With the aim of testing the performance of the developed method, a set of objective and subjective experiments, and finally experimental measurements, were performed. As was previously described, the main goal of this work is to implement a filtered noise generator in a hearing aid device based on a dsPIC33EP256MU806 of Microchip® [12], which has 256 kB of program memory. Previous works [17, 19] establish that only 10% of program memory is required to implement the most important functions of a hearing aid in the chosen dsPIC. Thus, as a design rule, it is established that the proposed pseudo-random generator is implemented using up to 80% of EEPROM, in this manner the last 10% of EEPROM will be reserved to future updates of the software. In these conditions, there are 209715 bytes assigned to implement the

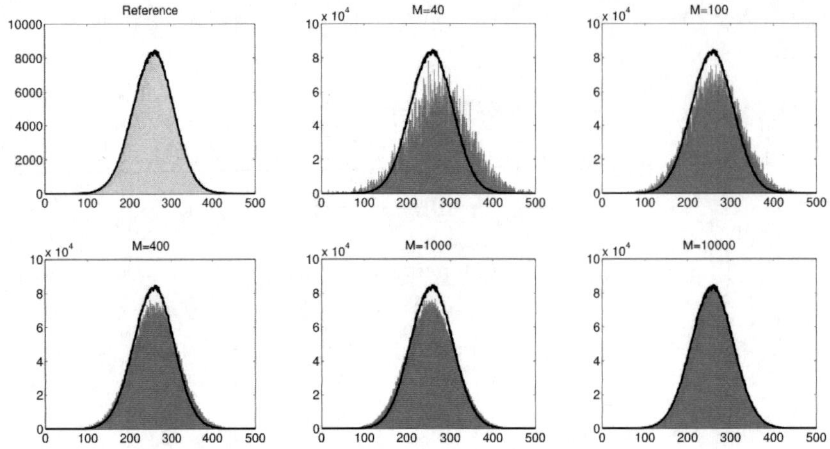

Figure 3. Probability Density Functions (PDF) of noise with gaussian distribution generated using the proposed method for several values of M. Also, the reference PDF is shown using a black line.

generator, then, by setting $N = 256$, and 16 bits data, the maximum value of M is 409. A second hearing aid device implemented on a dsPIC33EP512MU806 of Microchip® [12], which has 512 kB of program memory was also included in the experiment, taking into the account the same considerations than for the other device, the maximum value of M is 972.

Based on this limitation, configurations of the proposed technique with $M = 40$, $M = 100$, $M = 400$, $M = 950$ and $M = 10000$ will be used in the experiments.

4.1. Analysis of the obtained Probability Density Functions

To appreciate one of the effects of choosing the value of M, Figures 3 and 4 show the Probability Density Function (PDF) of two implementations of the pseudo-random number generator. Figure 3 shows in red colour the PDFs of the system configured as an AWGN generator for the six values of M previously adopted, and in green colour for the function `randn()` of MATLAB®. As a reference, using a black line, the ideal PDF function of a AWGN was also drawn in each figure. It is possible to see that, according to expectations, for bigger values of M, the PDF obtained tends to the reference PDF.

In Figure 4 the experiment is repeated, but in this case, the generator was configured for a uniform distribution, and the function `rand()` of MATLAB® was adopted as reference. The observed tendency is similar to the results obtained for the Figure 3.

4.2. Analysis using the Marsaglia toolbox

In this Subsection a set of objective experiments are performed in order to analyze the statical features of the pseudo-random number generator. For this proposal the Marsaglia toolbox [15], a set of statistical tools, was used. This toolbox is widely used to test pseudo-random number generators and it is disposal in the website of the Florida State University [21]. This tool allows to perform 15 tests, which give as result p-values which are in the range $0 < p < 1$[15].

In this case, the pseudo-random number generator was configured as an uniform distribution noise generator without filtering, and several values of M were used to study the influence of this parameter in the noise signal obtained. The results are obtained for the reference method and for the proposed method with $M =$ *100, 400, 1000, 10000, 100000, 400000* and *1000000* were

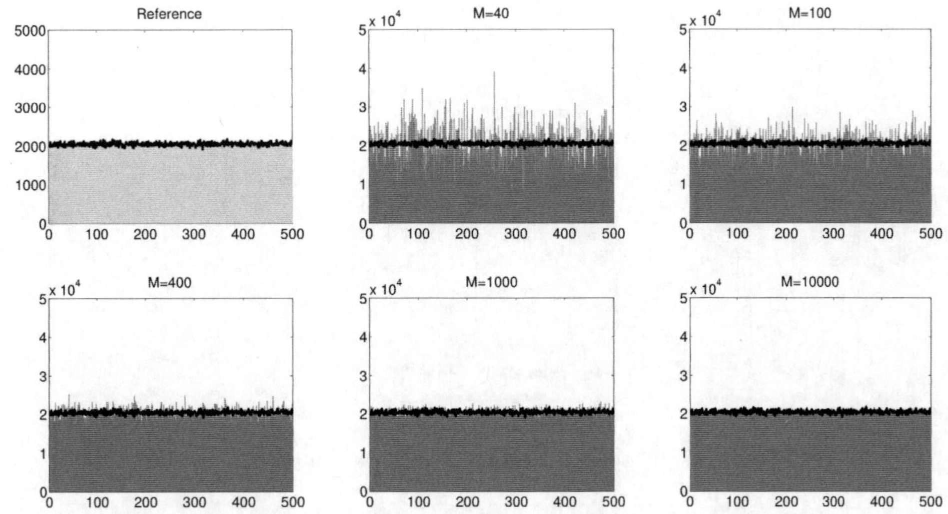

Figure 4. Probability Density Functions (PDF) of noise with uniform distribution generated using the proposed method for several values of M. Also, the reference PDF is shown using a black line.

obtained. In this case, it was verified that with larger values of M, the median value tends to 0.5 and the first and third quartile tends to 0.25 and 0.75, respectively. Then, a large value of M is necessary to ensure a proper performance of the method, which exceed the limit established previously ($M = 972$). But, in order to analyze the feasibility of using the proposed method for tinnitus treatment, in the next subsection a set of experimental measurements are presented.

4.3. Experimental measurements: Implementation on a dsPIC

With the aim of testing the performance of the system, it was implemented in a dsPIC33EP256MU806 [12] of Microchip®, which was adopted in the developed hearing aid[17, 19]. The language tool-suite choosen was C30 [22], which has libraries that contain functions appropriated to digital signal processing, then this toolbox can be used to implement the most important functions of a hearing aid device in a very efficient way.

The proposed technique was successfully implemented, in this case segments of noise of $N = 256$ were generated and a sampling frequency of $F_s = 16384Hz$ was used. Then, several experimental measurements were done. First, an acquisition time $15.63ms$ was measured for each frame. Consequently, in order to operate the system in real-time, the processing time must be lower than this limit. The measured processing time was $3.32ms$, which is according to expectations. Also, the program and data memory usages, were 79% and 40%, respectively. Finally, direct memory access (DMA) was used to acquire and synthesise data in a more efficient way. In Figure 6 the block diagram of the developed system is shown. It can be seen that the system is composed by two main blocks: a digital signal processor (in this case: dsPIC33EP512MU806) and an audio codec PCM3052 [16], which acquires the analog data, digitalises it and sends it to the DSP. Then, once the information is processed, it is returned to the codec, which by means of a delta-sigma DAC synthesises it. In this experiment the input signal is zero, thus the output signal only contains the noise synthesised.

In order to test the performance of the system, the testing bench shown in Figure 5 was mounted. In this case, the system's output was measured with a PDF meter [23]. The outputs

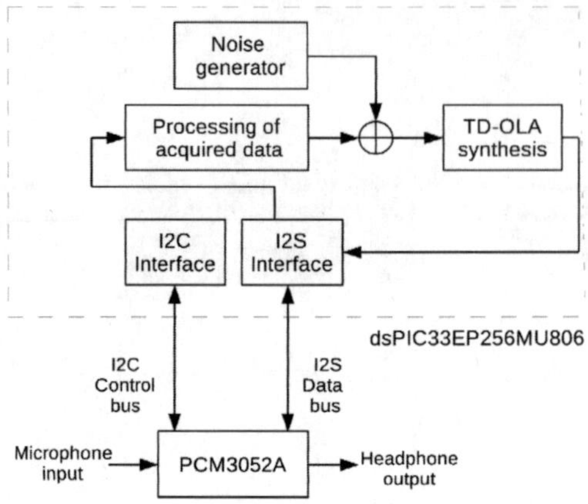

Figure 5. Blocks diagram of the implemented system.

of the PDF meter must be analysed using a oscilloscope, which must be configured in XY mode.

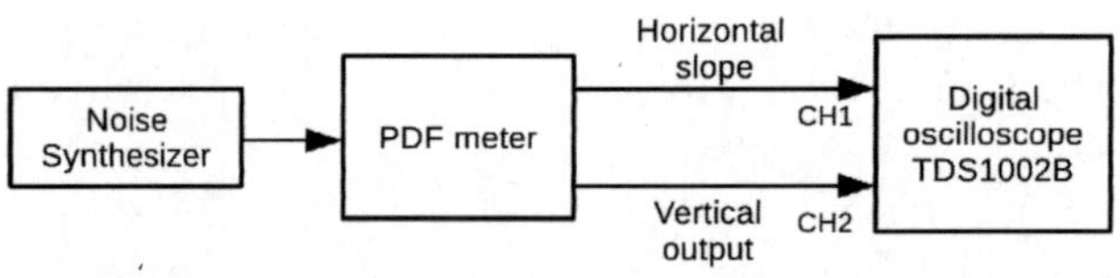

Figure 6. Block diagram of the testing bench.

For the experiment, a $2Vpp$ output noise wave was synthesised using two devices: the system under study (with $M = 400$ and $M = 950$) and a computer (using WAVE files). The sampling frequency and the data path of each device were the same ($16kHz$, 16 bits). The output of each system was compared using the PDF meter, whose output is shown in Figure 7. In the figure is seen that both probability density functions are similar.

Finally, the output of the system and the computer were compared using a spectrum analyser. Results obtained using four filtering configurations are shown in Figs. 8, 9 and 10 where it is

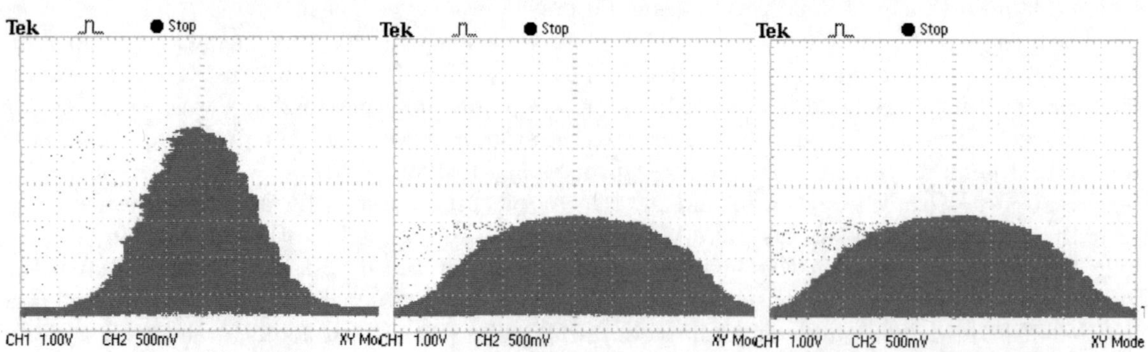

Figure 7. Screen captures of an oscilloscope for the probability density functions measured. Left figure shows the PDF for a WAVE file which contains a signal generated using the `randn()` MATLAB®. Middle and right figures show the measured PDF obtained the proposed technique for AWGN without filtering with $M = 400$ and $M = 950$, respectively.

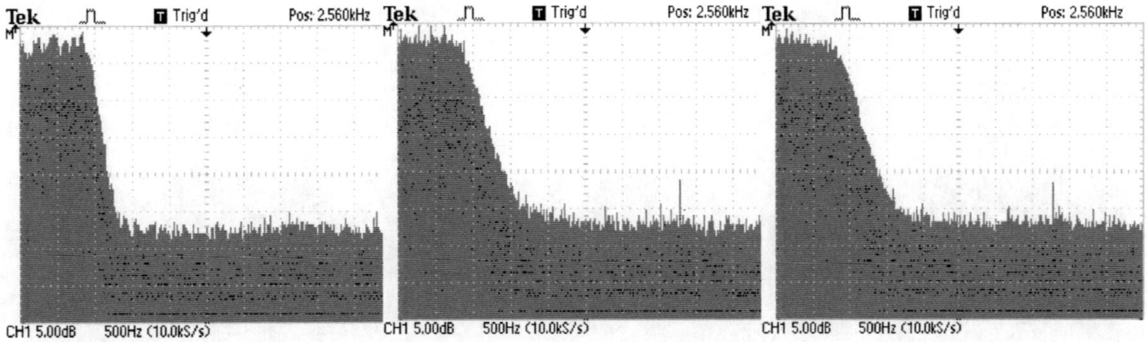

Figure 8. Spectrum of output signals using a LPF with $f_C = 1000Hz$. Left figure shows the reference (using the `randn()` MATLAB® function. Middle and right figures show the cases where $M = 400$ and $M = 1000$.

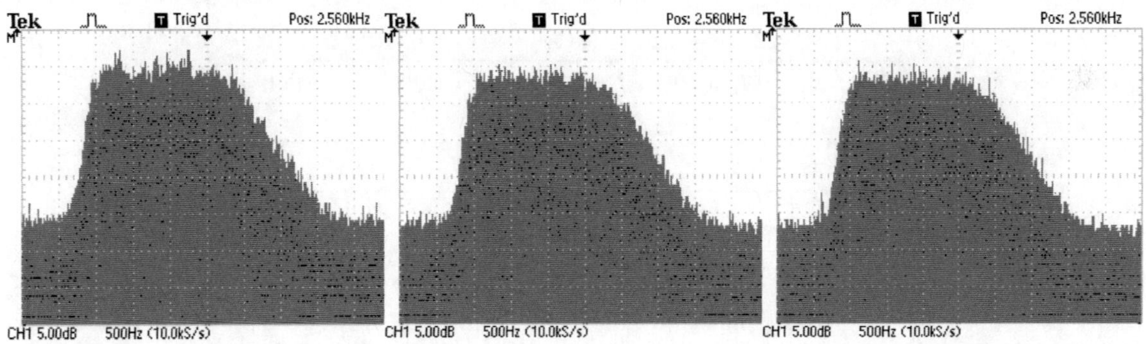

Figure 9. Spectrum of output signals using a BPF with $f_{C1} = 1000Hz$ and $f_{C2} = 3000Hz$. Left figure shows the reference (using `randn()`). While, middle and right figures sketch the spectrum obtained with $M = 400$ and $M = 1000$.

possible to seen that the spectra of the sound generated by means of the computer is similar to the generated using the hearing aid device. In Figure 8 are shown the spectrum of the output signal using a low pass filter (LPF) with $f_C = 1000Hz$. In Figure 9 the spectrum of the output signal using a band pass filter (BPF) with $f_{C1} = 1000Hz$ and $f_{C2} = 3000Hz$. In the last place, in Figure 10 the spectrum of the output signal using a BPF with $f_{C1} = 1000Hz$ and $f_{C2} = 2000Hz$.

4.4. Subjective experiments

Subjective experiments were done in order to test the performance of the proposed algorithm with values of M in the implementable range ($M < 950$). 15 volunteers (which accepted an informed consent) were asked to listen to the synthesised algorithms in random order. Listeners were asked to judge the similarity between two noises using a 5-point scale, from 1 (noises are distinguishable) to 5 (sounds are identical). The aim of this test is to analyse if a human can perceive a difference between a sound synthesised using the MATLAB `randn()` and `filter()` functions and the proposed technique, in this last case for several values of M. Although, 15 volunteers is not the amount of population needed to determine a conclusion, it is possible to obtain a trend. As a futher work, more complex subjective experiments with a bigger population will be carried out.

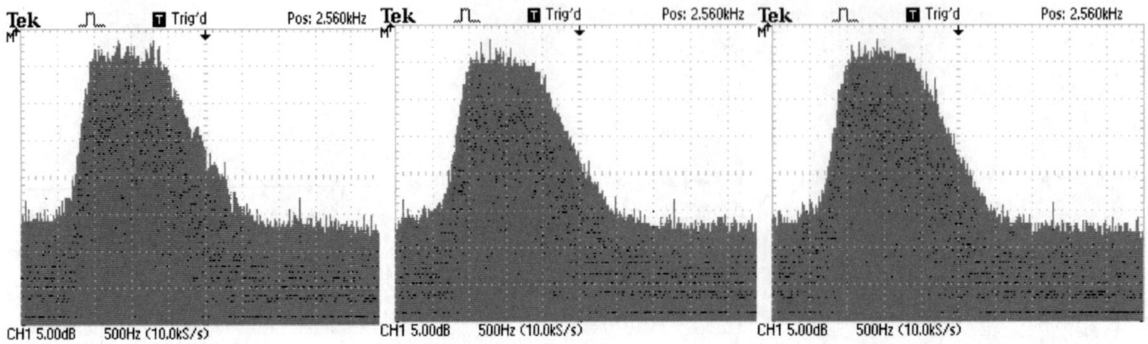

Figure 10. Spectrum of output signals using a BPF with $f_{C1} = 1000Hz$ and $f_{C2} = 2000Hz$. Left figure shows the reference (using `randn()`). Middle and right figures presents $M = 400$ and $M = 1000$.

Table 1. Results of MOS-S for each method.

Method	$M = 40$	$M = 100$	$M = 400$	$M = 950$
MOS-S	4.10	4.35	4.80	4.95

In Table1 it is possible to see that the value of the similarity mean opinion scores (MOS-S) is increased with the value of M, which is according to expectations. Also, all the MOS-S values obtained in the experiment are bigger than 4. Thus, even in the lower adopted values of M, the listeners were not able to distinguish between signals. Thus, although the method only obtains good scores in the Marsaglia test for values of $M > 400000$, human listeners are not able to distinguish the difference between noise synthesized using the `randn()` function, and the proposed technique (even for values of M in the required interval).

5. Conclusions

In this work, a technique to generate filtered noise by means of a pseudo-random number generator is a microcontroller is presented. The main aim of this method is to simplify the design of noise generators in a DSP and, consequently, it allows help people who suffer tinnitus.

It was demonstrated that the technique allows to successfully generate filtered noise reducing the amount of floating point operations (the most computational expensive task). Also, the method was implemented in a hearing aid device developed by the group. In order to validate the successful operation of the system, a set of subjective and objective test were carried out, and also experimental measurements were performed.

Firstly, objective experiments were developed to analyse and validate the performance of the technique. Then, once theoretical aspects of the technique were validated, the effectiveness of masking tinnitus was analysed to mask tinnitus. In this case, a set of volunteers heard a set of audios with a simulated tinnitus, and also different types of colored noise were added. At a first glance, the proposed technique obtains scores similar to other baseline methods. However, a set of more complex subjective experiments will be necessary to determine if the method can accomplish the expectations.

Finally, a set of experimental measurements were performed in order to analyse the operation of the technique in the hearing aid. In this experiment it was demonstrated that the system operates in real time, which is essential for this application. As a further work, it is necessary to carried out more extensive subjective experiments, in order to validate the performance of the proposed method for tinnitus treatment. Also, it is desired to study and implement in a DSP a set of most complex noise generators oriented to tinnitus treatment.

Acknowledgement

The authors would like to thank to CONICET and and National University of Mar del Plata for the funding received through projects *Secuencias caóticas digitales en procesamiento y encriptado de señales* and *Análisis de señales destinados a mejorar la calidad de vida*, respectively. The authors would like to thank the cooperation of PhD. Luciana de Micco for her assistance in the work.

References

[1] B.M. Oliver and J.M. Cage, *Electronic Measurement and Instrumentation.* McGraw Hill, 1971.

[2] A. Papoulis, *Probability, Random Variables and Stochastic Processes.* Mc Graw Hill. 2002.

[3] J.L. Danger, A. Ghazel. E. Boutillon and H. Laamari, *Efficient FPGA Implementation of Gaussian Noise Generator for Communication Channel Emulation.* IEEE ICECS. Vol. 1. pp. 366–369. 2000.

[4] A. Ghazel, E. Boutillan, I.L. Danger, G. Gulak and H. Laamari. *Design and Performane Analysis of a High Speed AWGN Communication Channel Emulator.* IEEE PACRIM Conference. pp. 374–377. 2001.

[5] J. A. Henry, T. L. Zaugg, P. J. Myers and M. A. Schechter, *Using therapeutic sound with progressive audiologic tinnitus management*, Trends on amplification, Vol. 12, No. 3, pp.188-209, 2008.

[6] A. J. Schleuning and R. M. Johnson, *Use of masking for tinnitus*, The International Tinnitus Journal, Vol.3, No.1., pp. 25-29, 1997.

[7] H. Feldmann, *Masking of tinnitus - Historical remarks.* In Proceedings III International Tinnitus Seminar, pp. 210-213, 1987.

[8] Widex Inc.*http://www.widex.com/*, Lynge, Denmark, 2013.

[9] Siemens Hearing Instruments, *Introduction to tinnitus management*, http://hearing.siemens.com, 2008.

[10] R.M Rodriguez-Osorio, A.D.C. Urbina, L.H. Ariet, M.C. Ramon and M.G Sanchez. *A DSP-based impulsive noise generator for QoS and EMC tests in wireless systems*, 60th IEEE Conference: Vehicular Technology Conference, 2004, Volume: 6, 2004.

[11] M. Sadasivam, J. Wagholikar and H. Sangjin, *A look-up based low-complexity parallel noise generator for particle filter processing.* In Proc. Of International Symposium on Signals, Circuits and Systems 2003 (SCS 2003). Vol. 2. pp. 621–624. 2003.

[12] Microchip Inc., *dsPIC33EP512MU810 Data Sheet, High Performance 16-bit Digital Signal Controllers.* http://www.microchip.com/, 2011.

[13] H. Fletcher, *Auditory Patterns*, Modern Physics Vol. 12, No. 1, pp. 47–65, 1940.

[14] T. Painter and A. Spanias, *Perceptual coding of digital audio.* In Proceedings of the IEEE, Vol. 88, No. 4, pp. 451–513, 2000.

[15] G. Marsaglia and W.W. Tsang, *Some difficult-to-pass tests of randomness*, Journal Statistical Software, Vol. 7(3). 2002.

[16] Texas Instruments,*PCM3052A Datasheet*, http://www.ti.com/, 2005.

[17] A.J. Uriz, P.D. Agüero, J. Castiñeira Moreira, J.C. Tulli, E.L. González, G. Moscardi and E.E. Sajama. *A development and implementation of a tinnitus treatment method.* Journal of Physics: Conference Series (JPCS). Londres: IOP SCIENCE. 2013 . ISSN: 1742-6596

[18] A.J. Uriz, P.D. Agüero, J. Castiñeira Moreira, J.C. Tulli, E.L. González and F. Denk. *Método y Dispositivo de Asistencia auditiva.* Patente de Invención - INPI:20130103576 . Buenos Aires, Argentina. 2013.

[19] A.J. Uriz, P.D. Agüero, J. Castiñeira Moreira, J.C. Tulli, E.L. González and R.M Hidalgo. *Low complexity noise power estimator for speech enhancement implemented on a dsPIC.* Anales del CASE 2014. Buenos Aires, Argentina. Agosto 2014.

[20] G. De Sebastian, J.J. Badaroco and D.G. Postan, *Audiología Práctica*, Ed. Médica Panamericana, 1992.

[21] G. Marsaglia, *The Marsaglia Random Number CDROM, with The Diehard Battery of Tests of Randomness.* Disponible en: www.stat.fsu.edu/pub/diehard. 1995.

[22] Microchip Inc. 16-Bit Lenguage Tools Libraries. http://www.microchip.com/, 2005.

[23] E.L. González. *User manual Probability Density Function meter.* http://www3.fi.mdp.edu.ar/mediciones/apuntes/Manual_Pdfimetro.pdf

Journal of Physics: Conference Series **705** (2016) 012006 doi:10.1088/1742-6596/705/1/012006

Skin-electrode impedance measurement during ECG acquisition: method's validation

Leonardo Casal[1] **and Guillermo La Mura**
Universidad Nacional de San Martín, Escuela de Ciencia y Tecnología, San Martín, Buenos Aires, Argentina

E-mail: leoncasal@gmail.com

Abstract. Skin-electrode impedance measurement can provide valuable information prior, during and post electrocardiographic (ECG) or electroencephalographic (EEG) acquisitions. In this work we validate a method for skin-electrode impedance measurement using test circuits with known resistance and capacitor values, at different frequencies for injected excitation current. Finally the method is successfully used for impedance measurement during ECG acquisition on a subject using 125 Hz and 6 nA square wave excitation signal at instrumentation amplifier input. The method can be used for many electrodes configuration.

1. Introduction

For bioelectric recordings such as electrocardiography (ECG) and electroencephalography (EEG), measurement of skin-electrode impedance can provide valuable information:

1- a mismatch in electrode-skin impedance reduces the common mode rejection ratio (CMRR) of the amplifier and leads to excessive power line interference [1];

2- impedance measurement allows determining if skin-electrode interface is good enough, avoiding skin abrasion prior to a recording [2];

3- continuous electrode impedance monitoring can be used for motion artefact estimation [3];

4- high impedance can lead to signal attenuation and distortion [4].

Because of the capacitive components corresponding to skin-electrode interface, alternate current (AC) measurement is needed to estimate the skin-electrode impedance. In particular, for dry contact electrodes, which capacitance is considerably high than wet electrodes [1-4].

The aim of this work is to validate a method capable of measuring skin-electrode impedance during ECG acquisition. The paper is organized as follows. We first describe the hardware used for signal acquisition and the implemented algorithm for measuring voltage drop due to injected current. Then we describe the electrical circuits used for method's validation based on known resistance R and capacitor C. We show results for measured impedances at 7.8 Hz, 31.2 Hz and 125 Hz and compare to model impedances. Finally, the method was successfully used for impedance measurement during an ECG acquisition on a subject.

2. Methods

2.1. Hardware description

[1] To whom any correspondence.

We implemented a biopotential measurement system based on Texas Instruments ADS1299 integrated circuit (IC) [5-6]. The IC provides an option for lead-off detection: it injects a square wave current at chosen instrumentation amplifier input with selectable amplitude and a frequency fixed at 7.8 Hz, 31.2 Hz or 125 Hz. We used 6 nA current peak amplitude (the smallest possible) in order to not disturb ECG signal in future recordings. The 7.8 Hz, 31.2 Hz and 125 Hz frequencies were used for method's validation, but only the 125 Hz frequency can be used for simultaneous ECG recording and electrode's impedance measurement. If necessary the 125 Hz frequency can be increased in multiples of 2. The voltage on instrumentation amplifier measures both ECG and excitation signal simultaneously. The data was sampled at 500 sps during 10 seconds for each frequency. When enabling lead off detection, the IC input impedance at direct current (DC) reduces from 1000 Mohm to 500 Mohm, based on its datasheet [5]. The instrumentation amplifier includes a patient protection resistor and an anti aliasing capacitor, as shown in figure 1.

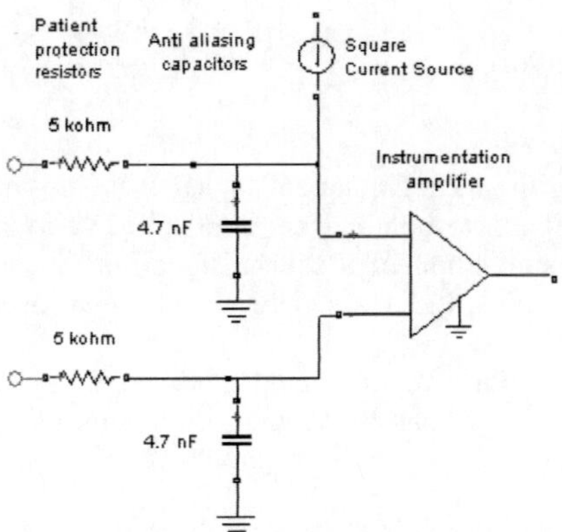

Figure 1. Implemented device's instrumentation amplifier with patient protection resistor, anti aliasing capacitor and square current source for skin-electrode impedance measurement.

2.2. Algorithm description

In order to measure skin-electrode impedance, we apply Ohm's law to injected current and measured voltage at instrumentation amplifier:

$$Z = \frac{V^{BP}{}_{RMS} \cdot \sqrt{2} \cdot \pi / 4}{I_{peak}}$$

where $V^{BP}{}_{RMS}$ is the root mean square (RMS) of the band-passed measured voltage and I_{peak} is the peak amplitude of the excitation current signal, i.e. 6 nA. Next we explain in detail this equation.

The first step is to separate the excitation signal from the ECG and 50 Hz power line noise using a bandpass filter, figure 2. We implemented a finite impulse response (FIR) filter, 300 order, Hamming window with centre frequency at corresponding excitation frequency. In this way we obtain the first harmonic of the square wave excitation signal, V^{BP}.

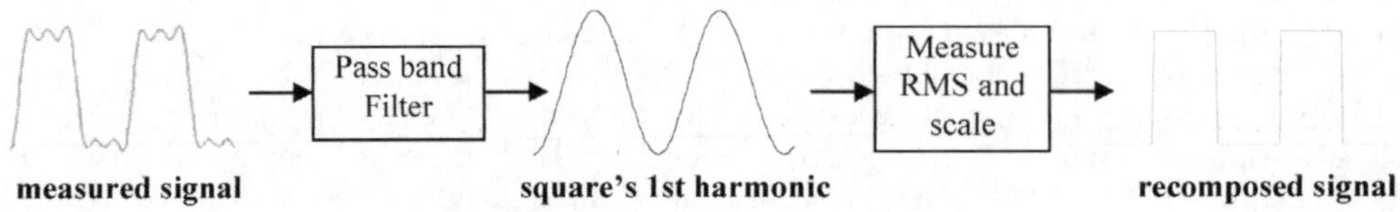

Figure 2. Signal processing of measured voltage due to injected current.

Next, we calculate RMS amplitude of the filtered signal, $V^{BP}{}_{RMS}$ and multiply it by √2 to obtain first harmonic peak amplitude. Then we multiply by the scale factor π/4 to get the corresponding voltage square peak amplitude, according to Fourier Series Coefficients when n = 1:

$$square(t) = \frac{4}{\pi} \sum_{n=1,3,5\ldots}^{\infty} \frac{1}{n} \sin\left(\frac{n \cdot 2 \cdot \pi \cdot t}{T}\right) \tag{1}$$

As we sampled at 500 sps, the 125 Hz square signal was subsampled and the measured first harmonic peak is 90.21 % of its corresponding amplitude, so we must correct for this factor. This value was obtained running a simulation of a square signal and measuring peak value of the first harmonic sampled at 500 sps. For a 125 Hz square signal we measured a 9.79 % peak reduction compared to expected value.

For 7.8 Hz and 31.2 Hz excitation, we subtracted ECG energy at 7.8 Hz and 31.2 Hz (measured without current injection) from band passed voltage, although this value was almost negligible compared to square signal energy.

2.3. Skin-electrode interface model

The skin-electrode interface can be modelled as a non-linear second order filter, described by a 6 time-varying parameters [4]. In our case we assumed only the predominant parameters: the skin-electrode interface was described by a parallel RC circuit, as shown in figure 3.

2.4. Method's validation: R and C known values

In order to validate the method, we measured the impedance of different load circuits, using 5 % tolerance R and C components. Configuration n1 consists of one resistor. Configuration n2 consists of two RC circuits to simulate the electrodes. These configurations, the measurement circuit and equivalent circuit are shown in figure 3 together with its corresponding impedance equations. The comparison between impedance model equation (using known R and C values) and measured impedance values allowed us to validate the method and determine the inputs stray capacitance due to cables, PCB and circuit implementation.

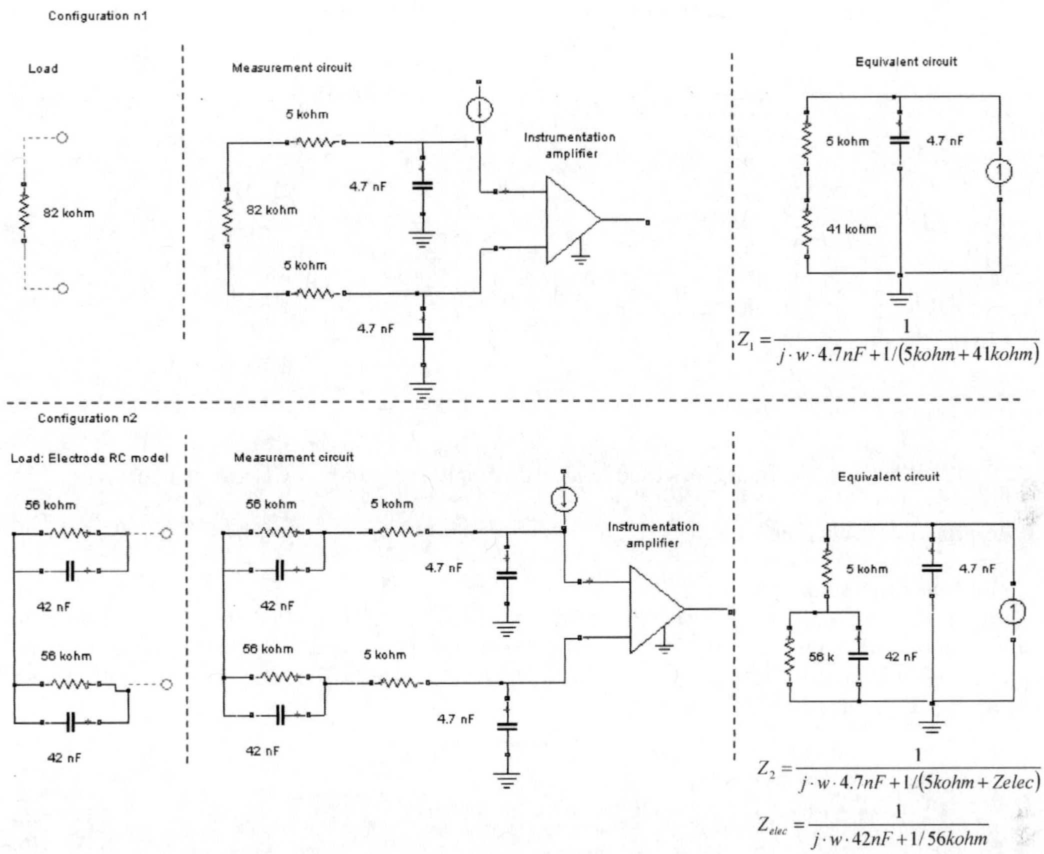

Figure 3. Configurations used to measure impedance and validate the method.

2.5. Method's application: simultaneous ECG acquisition

The method was applied to measure electrode impedance during ECG acquisition at different circuits, shown in figure 4. Configuration n3 includes an ECG patient simulator using the proposed RC model as electrodes. Configuration n4 consist of a real ECG measured on a subject using wet electrodes and right leg drive.

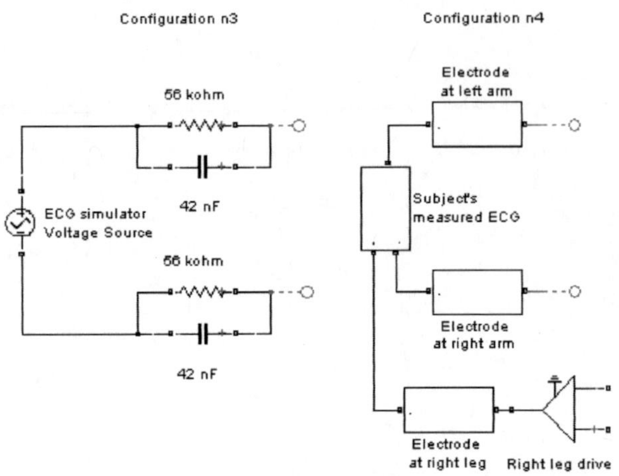

Figure 4. Configurations used to measure impedance during ECG acquisition.

3. Results and Discussion

3.1. Method's validation

Figure 5 shows measured impedance module at 7.8 Hz, 31.2 Hz and 125 Hz for different configurations n1 and n2 (with 20 % input current error bar according to IC datasheet) and corresponding impedance model including input stray capacitance. The total stray capacitance was estimated with a value of 10 nF.

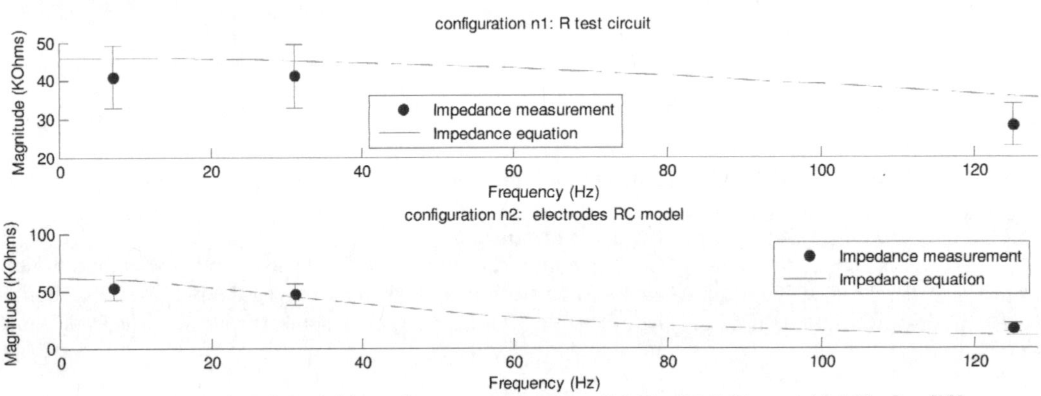

Figure 5. Measured and model impedance magnitude at 7.8 Hz, 31.2 Hz and 125 Hz for different configurations.

3.2. Method's application

Figure 6 shows time and frequency measured signals for configuration n3 (ECG simulator and electrodes simulator) for 125 Hz frequency excitation. We can see the corresponding peak at 125 Hz.

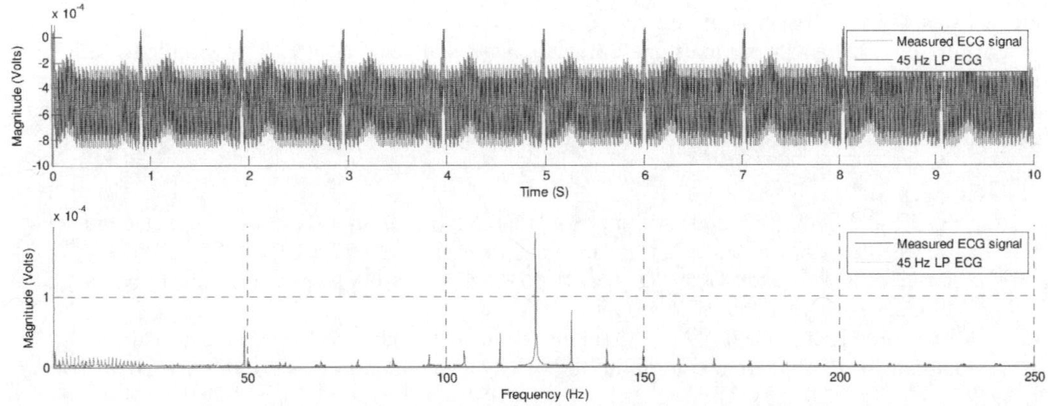

Figure 6. Time and frequency signals for configuration n3 (ECG simulator and electrodes simulator) for 125 Hz frequency excitation.

Figure 7 shows time and frequency signals for configuration n4 (ECG measured on a subject using wet electrodes) for 125 Hz frequency excitation. The 125 Hz excitation signal was easily filtered from the ECG signal using a low pass FIR filter, 45 Hz cut-off frequency, order 100.

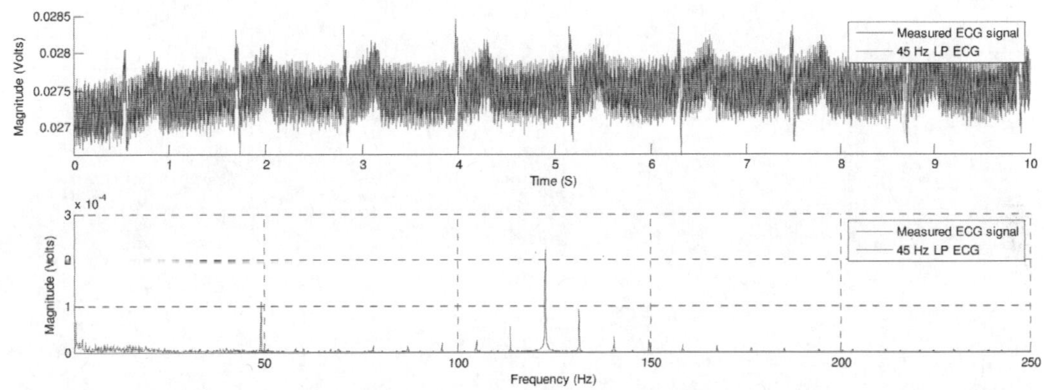

Figure 7. Time and frequency signals for configuration n4 (ECG measured on a subject using wet electrodes) for 125 Hz frequency excitation.

The electrode model parameters for configuration n4 (ECG measured on subject) are R = 56 KOhm and C = 42 nF for each electrode. This measurement can be repeated over time to evaluate impedance evolution during ECG acquisition. If more electrodes are used, excitation current can be injected on each channel to determine each electrode R and C value. The ECG signal shows low Signal to Noise Ratio (SNR) possibly due to elevated high electrode impedance.

4. Conclusions

A method for skin-electrode interface impedance measurement during ECG was proposed and validated using test circuits at different excitation frequencies. The chosen parameters for the excitation signal allowed us to use a simple pass band filter to separate ECG and excitation signal. The measured impedance showed good agreement with corresponding test circuit, thus validating the method. The implemented device's input stray capacitance was estimated to 10 nF. Finally, the method was successfully used during an ECG acquisition on a subject.

For future work we propose to estimate R and C values impedance at 125 Hz sample by sample by an adaptive algorithm. Besides, the method should be tested on many electrodes configuration to determine each electrode R and C value. For special applications that need higher ECG frequencies, excitation and sampling frequencies must be increased and method's validation should be repeated.

5. References

[1] Degen T and Loeliger T 2007 An improved method to continuously monitor the electrode-skin impedance during bioelectric measurements *Proc IEEE EMBS* 6294–6297

[2] Ferree T, Luu P, Russell G and Tucker D 2001 Scalp electrode impedance, infection risk, and EEG data quality *Clinical Neurophysiology* 536–544

[3] Kim S, Firat R et al 2010 A 2.4 uA continuos-time electrode-skin impedance measurement circuit for motion artefact monitoring in ECG acquisition system *Symp on VLSI IEEE* 219–220

[4] Assambo C, Baba A, Dozio R, Burke M 2007 Determination of the parameters of the skin-electrode impedance model for ECG measurement *Proc. Int. Conf. on Elect. HW* 90–95

[5] ADS1299 Datasheet, Texas Instruments, http://www.ti.com/lit/ds/symlink/ads1299.pdf

[6] ADS1299EEG-FE User's Guide, Texas Instruments, http://www.ti.com/lit/ug/slau443a/slau443a.pdf

Automatic digital-analog impedance plethysmograph

C B Goy[1,2,3], K A Mauro[1], L M Yanicelli[1,3], N F Parodi[1,3] M A Gómez López[2] and M C Herrera[1,3]

[1] Departamento de Bioingeniería, Facultad de Ciencias Exactas y Tecnología, Universidad Nacional de Tucumán, Av. Independencia 1900, Tucumán, Argentina.

[2] Departamento de Ing. Eléctrica, Electrónica y Computación, Facultad de Ciencias Exactas y Tecnología, Universidad Nacional de Tucumán, Av. Independencia 1900, Tucumán, Argentina.

[3] Instituto Superior de Investigaciones Biológicas (INSIBIO), Consejo Nacional de Investigaciones Científicas y Técnicas (CONICET), Chacabuco 461, Tucumán, Argentina.

E-mail: carlabgoy@gmail.com

Abstract. Venous occlusion plethysmography (VOP) is a traditional method widely used to assess limb blood circulation. One common mode to record VOP is by means of evaluating limb volume changes using impedance plethysmography (IP). In this paper the design and implementation of an automatic digital-analog impedance plethysmograph (ADAIP) for VOP is presented. The system is tested using precision resistances in order to calculate its repeatability. Then its global performance is assessed by means of VOP recordings on the upper and the lower limb of a healthy volunteer. The obtained repeatability was very high (95%), and the VOP recordings where the expected ones. It can be concluded that the whole system performs well and that it is suitable for automatic VOP recording.

1. Introduction

Venous occlusion plethysmography (VOP) is a traditional method widely used to assess limb blood circulation [1]. Venous occlusion plethysmography can be equally well applied to the upper or lower limb. Ward *et al.* proposed the use of VOP applied to the arm for the measurement of central venous pressure (CVP) [2]. This technique has been also applied to the lower limb to assess vascular parameters such as calf venous compliance, capillary filtration coefficient, venous capacitance response, fluid filtration and vascular resistance among others [3]. VOP is useful in a wide variety of medical conditions such as venous insufficiency, venous thrombosis or cardiac heart failure (CHF) [4-5], it can also be used to assess the effect of new vasoactive drugs and hormones in humans [1], to measure calf blood flow during exercise [6] or reactive hyperemia [7].

The underlying principle of venous occlusion plethysmography is simple: when the limb venous drainage is briefly interrupted, which is achieved by inflating a cuff placed above the knee, arterial inflow is unaltered and blood can enter the extremity but cannot escape. Before venous occlusion occurs the limb volume remains constant at an initial value (V_0). When the occlusion takes place, a linear growth of the limb blood volume over time, that is proportional to arterial blood inflow, is observed until venous pressure reaches the occluding pressure value and a plateau is established. Once the pressure is released, the venous is drained and venous outflow can be recorded.

One common mode to record VOP is by means of evaluating limb volume changes using impedance plethysmography (IP). This is based on the fact that the limb impedance reflects the filling state of the blood vessels contained [8]. The limb impedance can be known by means of a potential difference recording when a known constant amplitude alternating current (CAAC) is injected into the limb. The alternating current frequency range used for VOP measurement goes from 20 to 100KHz [9].

When IP is used for venous occlusion recording, the voltage measured before the occlusion (V_{bo}) reflects the initial limb volume (V_0). Then, once the occluding procedure takes place, the limb volume varies and, therefore, the instantaneous measured voltage (V_i) also varies. So, after the occlusion, V_i can be divided into V_{bo} -that represents the initial limb impedance (impedance without venous occlusion)- and a variable voltage (V_v) that is a consequence of the body segment impedance changes that occur when the occlusion takes place.

In this paper, the design and implementation of an automatic digital-analog impedance plethysmograph (ADAIP) for VOP is presented. The system is tested using precision resistance in order to calculate its repeatability when a known and constant resistance is measured. Then its global performance is assessed by means of VOP recordings on the upper and the lower limb.

2. Methods

The ADAIP consist of hardware, software and an acquisition device -LabJack U3-HV (Lakewood, Colorado, USA)- that is used to connect the hardware and software between them. The system was designed to work with a tetrapolar electrode configuration which reduces errors due to the nonlinear potential differences generated in the area near to the injecting-electrodes. These errors are observed when a bipolar configuration is used, as a consequence of the electrode-tissue interface polarization impedance [9]. A scheme of the complete system is presented in Figure 1.

2.1 Hardware

The custom-made especially designed hardware is responsible for generating and injecting a constant amplitude alternating current into a body segment, sensing the generated potential difference and raising, maintaining and releasing the cuff pressure as required for VOP recording. The design includes a variable gain implemented with a precision potentiometer and an external dial. The hardware is supplied with a ±15V/0.5A voltage source and with a 6V/4Ah battery used only for the air pump.

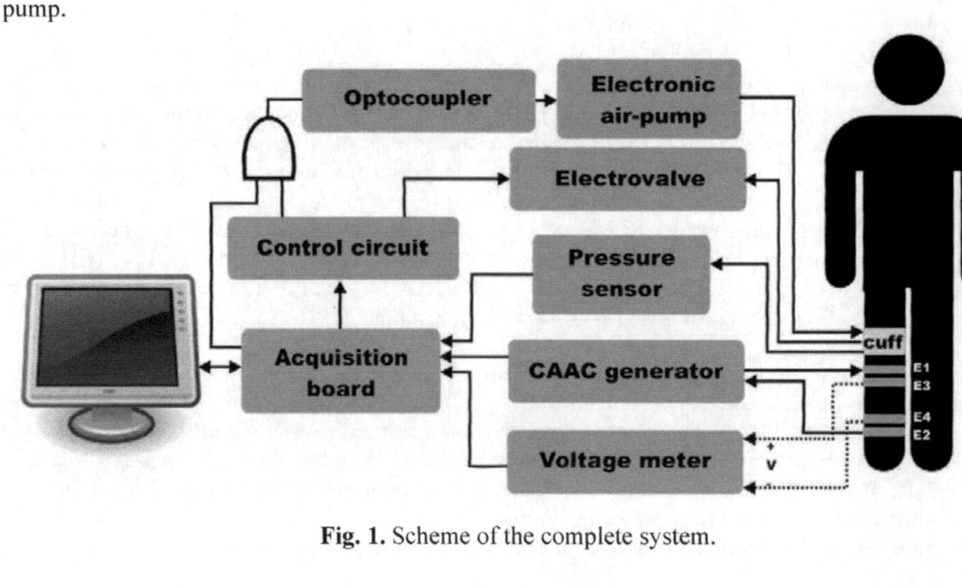

Fig. 1. Scheme of the complete system.

2.1.1. Constant amplitude alternating current (CAAC) generator. This block is the responsible for injecting a 23kHz alternating current into the body segment by means of two electrodes, E1 and E2. This block is composed of a Wien bridge oscillator, a second-order passband active filter (passband: 10-30KHz) and a voltage to current converter. The generated current has a 3mA constant amplitude.

2.1.2. Voltage meter. This block records the instantaneous voltage that is generated in the body segment when the current is injected ($V_i=V_{bo}+V_v$). This voltage is picked up by two electrodes, E3 and E4, placed between electrodes E1 and E2. As the voltage is proportional to the injected current, it is an alternating signal whose amplitude is modulated by the limb impedance changes. The voltage meter can be divided into three parts: an input stage, a demodulator circuit and a subtraction circuit (Figure 2).

The input stage consist of an instrumentation amplifier (voltage gain= 6.9), a second non inverter amplifier with variable gain and a second-order high pass active filter (cut frequency= 16KHz). The demodulator circuit is responsible for providing the amplitude of the voltage V_i which is directly proportional to the electrical impedance of the body segment placed between the electrodes E3 and E4; this is achieved with a full wave rectifier, an envelope detector and a second order low-pass active filter. The subtraction circuit receives two voltage signals: V_i ($V_i=V_{bo}+V_v$) and V_{bo} (V_{bo} comes from the output DAC0 of the acquisition device); it subtracts the signals, amplifies the resultant voltage V_v ($V_v= V_i-V_{bo}$) and sends it to the input AIN2 of the acquisition device. This last circuit consist of an instrumentation amplifier (voltage gain= 20).

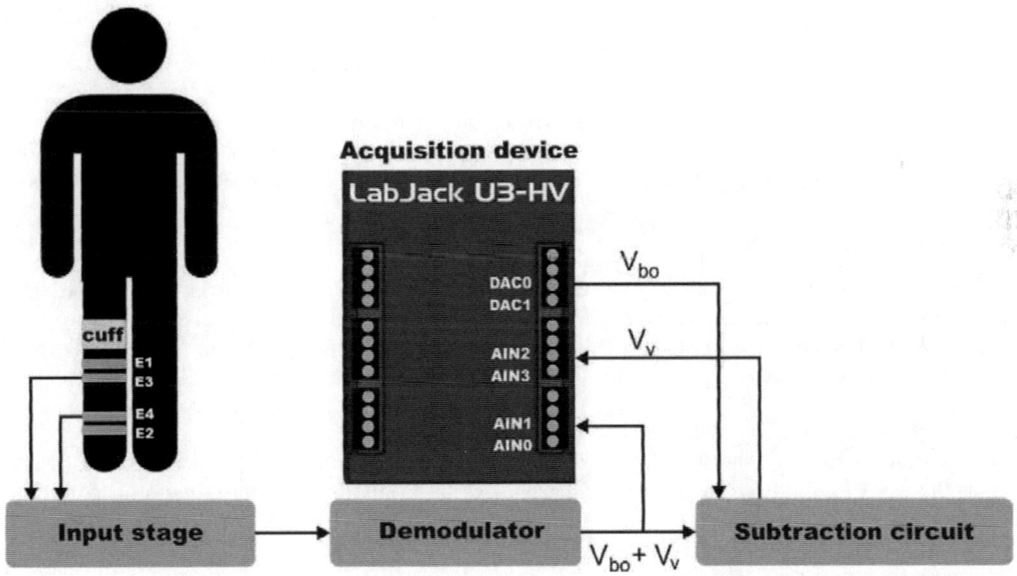

Fig. 1. Stages of the voltage meter module and their connections whit the acquisition device.

2.1.3. Pressure sensor. In this work, the pressure sensor MPXV5050GP from Freescale Inc. (Austin, TX) is used for cuff pressure recording. It consist in a piezoresistive transducer based on monolithic silicon that provides an accurate, high level analog output signal that is proportional to the applied pressure. Its principal characteristics are presented in table 1. Its transfer function is supplied by the manufacturer (equation 1). The supply voltage of the sensor is provided by the output V_s of the acquisition device.

$$V_{sp} = V_s\,(P \times 0.018+0.04) \pm ERROR \tag{1}$$

Where V_{sp} is the sensor output voltage and ERROR is function of the temperature and V_s which is specified in the datasheet of the sensor.

Table 1. Principal Characteristics of the pressure sensor MPXV5050GP

Characteristic	Value	Unit
Pressure range	0 - 50	kPa
Supply voltage	4.75 – 5.25	V
Accuracy	$\pm$ 2.5	%V_{FSS}(*)
Sensitivity	90	mV/kPa
Operating temperature	-40 a 125	°C

(*)Full scale span (V_{FSS}) is defined as the algebraic difference between the output voltage at full rated pressure and the output voltage at the minimum rated pressure.

2.1.4. Electrocoupler and electronic air pump. The electronic air pump is used to increase the cuff pressure; it works with 6VDC/30mA and it is controlled by the output FIO6 of the acquisition device and the control circuit through an AND gate (Figure 1). The air pump is only activated when the gate output is high. An optocoupler is connected between the AND gate and the air pump to provide an electrical insulation to the air pump from the rest of the circuit. The air pump is supplied with a 6V/4Ah battery.

2.1.5. Control circuit. This circuit is used to have an accurate control of the cuff pressure, it consist of a comparator circuit and an anti-rebound circuit. The comparator circuit receives two signals: V_{sp} and a reference voltage (V_r) equal to the pressure sensor output when the applied pressure is the desired cuff pressure; V_r is supplied by the output DAC1 of the acquisition device. The anti-rebound circuit is used to remove spurious pulses that can be present at the comparator circuit output when the desired pressure is reached.

2.1.6. Electrovalve. In order to maintain or release the cuff pressure as required, a solenoid electrovalve is used. It is controlled by the output FIO7 of the acquisition device. As the electrovalve works with 6VDC/20mA, and the acquisition device can deliver only 3.16VDC/18mA, a current and voltage amplifier is necessary (not shown) which is implemented using an operational amplifier and a bipolar junction transistor.

2.2 Acquisition device (Labjack U3-HV)

This device connects the hardware and software via an USB connection. It receives and delivers signals that are necessary for the proper performance of the ADAIP. The inputs and outputs used are listed below:

- AIN0: Signal from the pressure sensor.
- AIN1: Signal from the demodulator circuit.
- AIN2: Signal from the subtraction circuit.
- DAC0: Signal to the subtraction circuit.
- DAC1: Reference voltage.
- FIO6: Signal for air pump control.
- FIO7: Signal for electrovalve control.
- VS: 5V supply voltage.

2.3 Custom-made software
The software has been implemented using Labview 2009 (National Instruments, Austin, TX) and a special library from Labjack Corporation (Lakewood, Colorado, USA). It receives, processes and displays signals from the acquisition device (Labjack U3-HV (Labjack Corporation)) and controls the cuff pressure. When a VOP recording is carried out, the software manages, for cuff pressure control, two digital outputs of the acquisition board -FIO6 and FIO7- that deliver -when high- a maximum voltage of 3.16V and a maximum current of 18mA.

The received signals are V_i, V_v and Vs (voltage from the pressure sensor) which are presented at AIN1, AIN2 and AIN0 respectively. They are sampled at a 25Hz frequency and are sent to the computer into 25 samples packets. The user interface has a START button to initiate the VOP measurement procedure, a STOP button used to cancel the cuff pressure raise, two displays that show the signals V_{sp} (from the pressure sensor) and V_{bo} or V_i (from the demodulator or subtraction circuit respectively) -which of these last two signals is visualized is determined by a button named AC/DC- and a "SAVE" button that allows to store the data for subsequent analysis using ".tdms" files.

When the software is started the AC/DC button is on the position DC and the signal at the input AIN1 is visualized. Once the electrodes are placed on the limb, if no pressure is applied, the software displays the voltage level that represents the initial limb impedance value (V_{bo}) which is presented at AIN1. At this moment, the user press AC/DC button which changes to AC and the voltage V_{bo} is recorded, stored and presented at DAC0 and the signal at the input AIN2 is visualized. When V_{bo} is at DAC0 the subtraction circuit delivers a cero value at AIN2 (V_v=0) until venous occlusion starts and, therefore, the visualized voltage (V_i) changes. This is important since venous occlusion leads to voltage variations much smaller than V_{bo} and, in order to visualize them correctly, V_{bo} must be removed.

3. Results
The ADAIP has been tested using precision resistance (1% tolerance) in order to calculate its repeatability when a known and constant resistance is measured. To calculate the repeatability of measurements from recapture data an extending of the statistical tests in [10] is used as in [11]. Two precision resistances have been measured two times per day in two consecutive days for three different gains. The number of measurements and days has been selected as in [12]. The results of the measurements are presented in table 2. The first step for repeatability calculation is to compute a one way ANOVA (analysis of variance; prior to ANOVA analysis variances homogeneity and normality of residuals were tested). Then, the repeatability can be calculated using equation (2) where B and W are the between individuals and within individuals variances respectively, and N is the sample size (N=4 in this case). In table 3 the results from the ANOVA are presented. The repeatability obtained by replacing B, W –from table 3- and N in (2) is 95%.

$$R=(B-W)/[B+(N-1)W] \tag{2}$$

Table 2. Voltage values obtained for the measurement of two resistances in two consecutive days with three different gains.

R	*Gain*	*day 1*		*day 2*	
	1	0,5	0,6	0,45	0,4
150	**2**	3,8	3,5	3,8	3,6
	4	6,6	7,5	7,6	7
	1	0,4	0,4	0	0
39	**4**	4,8	4,8	4	4
	8	7,8	8	6,8	7

Table 3. Results from the ANOVA analysis.

Source of variation	*Sum of squares*	*degrees of freedom*	*F*	*p*
Between individuals	194,4430208	5	271,64487	2,7143E-16
Within individuals	2,576875	18		
Total	197,0198958	23		

The global performance of the ADAIP has been assessed by means of VOP recordings on the upper and the lower limb of a healthy volunteer. The measurement procedure has been approved by the Ethics Committee of the Universidad Nacional de Tucumán (Tucumán, Argentina) and prior to volunteer participation in the trial, a written informed consent form was signed according to the Declaration of Helsinki. The data resultant from the measurements is plotted and presented in Figure 3. It is important to notice that the record of VOP from the lower limb is longer than the one obtained from the arm, it is necessary in order to be able to calculate certain vascular parameters as described by Skoog *et al.* [3]. In Figure 3 it can be observed that once the occlusion take place –point 1- the volume of the limb changes until a plateau is established, then when the occlusion is released –point 2- the volume changes again in the opposite direction. This is the expected waveform of a VOP recording. Finally, a photograph of the equipment and the software user interface is presented in figure 4.

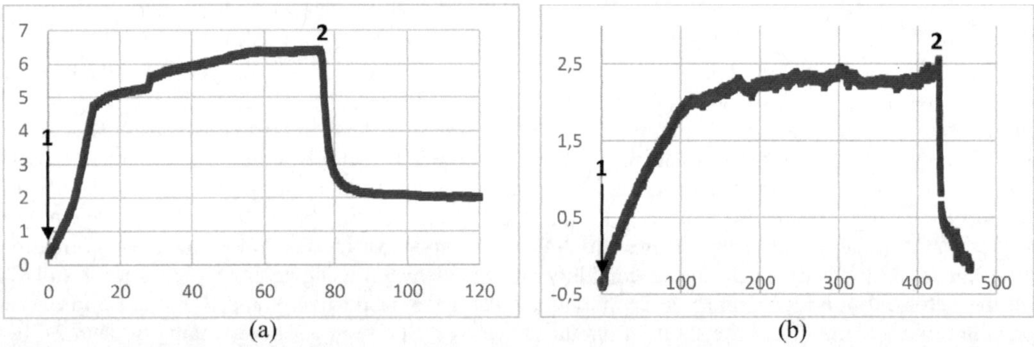

Fig. 3. VOP recordings obtained from (a) the upper and (b) the lower limb of a healthy volunteer using the ADAIP.

4. Conclussion and discussions

In this paper the design and implementation of an automatic digital-analogue impedance plethysmograph has been described. The system present a very high repeatability when a single resistance is evaluated. More tests need to be done in order to obtain the repeatability of the system when the measurements are carried out on humans. The global performance of the system has been assessed through VOP recordings on the upper and the lower limb, the obtained waveform where the expected ones, so it can be concluded that the whole system performs well and that it is suitable for automatic VOP recording.

In the future the system will be used to evaluate the vascular system of patients with certain pathologies, the intention is to compare VOP waveforms obtained from different stages of the disease in order to determine if there are any significant change that could help physicians with the management of the patient. As the intention is to compare waveforms and not absolute values the calibration of the system would not be necessary, but it could increase its applicability.

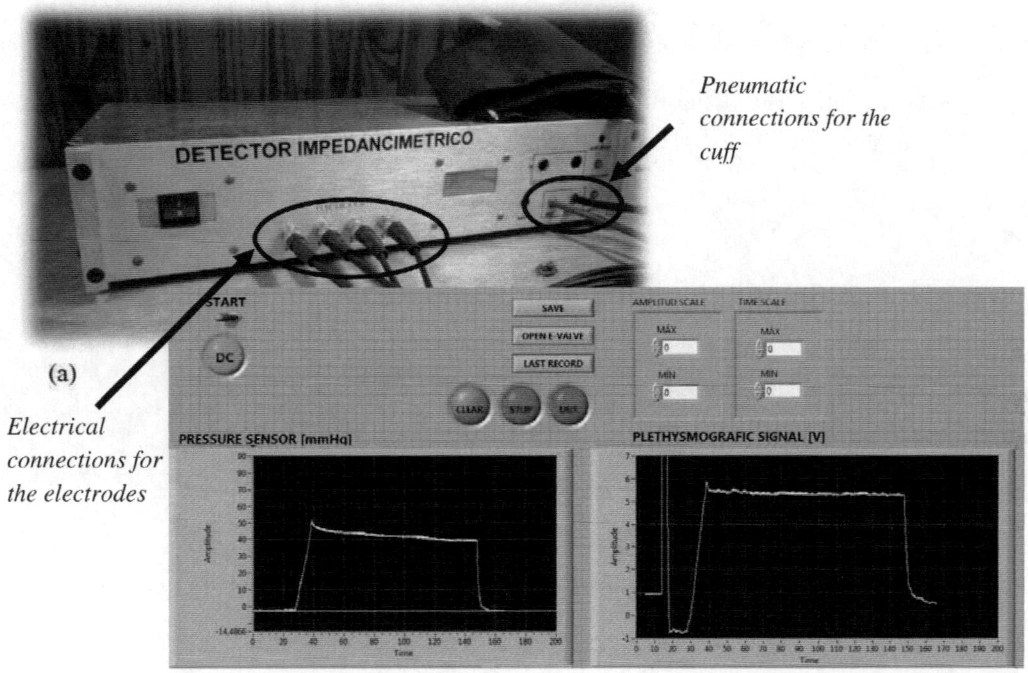

(b)

Fig. 4. (a) Photograph of the front face of the hardware with the cuff above it. The pneumatic connections -for the cuff- and the electrical connections for the four electrodes can be seen. (b) Photograph of the user interface of the software.

References

[1] Wilkinson I and Webb D 2001 Venous occlusion plethysmography in cardiovascular research: methodology nd clinical applications *Br. J. of Clin. Pharm.* **52** 631-46

[2] Ward K R, Hakam T M, Draucker G T, Proffitt K E, Barbee R W, Gunnerson K J, Reynolds P S and Spiess B D 2010 A Novel noninvasive impedance based technique for central venous pressure measurement *Shock* **33**(3) 269-73

[3] Skoog J, Zachrisson H, Lindenberger M, Ekman M, Ewerman L and Länne T 2015 Calf venous compliance measured by venous occlusion plethysmography: methodological aspects *Eur J Appl Physiol* **115** 245-46

[4] Straden E 2011 Edema en venous insufficiency *Phlebology* **18**:1, 3-14

[5] Patterson J, Selig S, Toia D, Bamroongsuk V and Hare D 2007 Foream blood flow in individuals with CHF and age-matched healthy volunteers: a study and historical review *J. Exercise Physiology.* JEP10(2007)005

[6] Joyner M J, Lennon R L, Wedel D J, Rose S H and Shepherd J T 1990 Blood flow to contracting human muscles: influence of increased sympathetic activity *J. Appl. Physiol.* **68** 1453–7

[7] Hayoz D, Drexler H, Munzel T, Hornig B, Zeiher A, Just H, Brunner H and Zelis R 1993 Flow Mediated arterial dilation is abnormal in congestive heart failure *Circulation* VII-92-VII-96

[8] Geddes L A and Baker L E 1989 *Principles of Applied Biomedical Instrumentation* (New York:Wiley) pp 473-5.

[9] Webster J G 1992 *Medical Instrumentation, Application and Design* (Boston: Houghton Mifflin Company) pp 442-4
[10] Fowler J and Cohen L 1986 *Statistics for Ornithologists* (Tring: British Trust for Ornithology, Tring)
[11] Harper D G C 1994 Some comments on the repeatability of measurements *Ringing & Migration* **15**:2 84-90
[12] Rankin G and Stokes M 1998 Reliability of assessment tools in rehabilitation: an illustration of appropriate statistical analyses *Clinical Rehabilitation* **12**: 187–99

Open Source Quartz Crystal Microbalance with dissipation monitoring

C Mista, M Zalazar, A Peñalva, M Martina, J M Reta
Laboratorio de Prototipado Electrónico e Impresión 3D. Universidad Nacional de Entre Ríos, Oro Verde, Entre Ríos, Argentina.

E-mail: cmista@bioingenieria.edu.ar, martinzalazar@bioingenieria.edu.ar, prototipado@bioingenieria.edu.ar

Abstract. The dissipation factor and subsequently the characterization of the viscoelasticity of deposition films have become crucial for the study of biomolecular adsorption. Most of the commercial quartz crystal microbalance (QCM) systems offer this feature, but it has not been incorporated in open source systems. This article describes the design, construction, and simulation of an open source QCM module for measuring dissipation factor. The module includes two blocks: switch and envelope detector. The switch rapidly disrupts the excitation of the crystal, and connects the output to the envelope detector which demodulates the amplitude of the signal. Damped sinusoidal exponential signals with different time constant were used for simulating viscosity interfaces. The incorporation of few elements facilitated a double-faced PCB design with reduced dimensions. The results from simulation show that the system has a good performance in the range of the biomolecular processes; greater relative error are observed for time constant lower than 1 µs. In conclusion, a dissipation module has been developed for calculate dissipation factor using QCM, which is compact and shows great performance for use in biomolecular adsorption.

Keywords: Dissipation, QCM, viscoelastic, envelope.

1. Introduction

Nowadays, new technologies are contributing to facilitate accurate and fast measure of small amount of organic and non-organic substances. The progress of sensor technologies has been feasible due to innovations in different fields including materials and electronic systems. In addition to the development of new tools, applying known technologies to diverse fields has shown to be a great alternative to obtain novel results. In this sense, Open source technologies have shown to have a great impact in the social [1] and the research community [2], due to their easy access and simplicity. Moreover, open source tools rapidly expand to different research fields due to their capability to adapt and satisfy different requirements [3].

In order to determine mass ranging from nanograms to micrograms, several devices has been developed using different technology to measure mass, including cantilever [4], surface acoustic wave[5], and quartz crystal microbalance (QCM) [6,7]. In particular, QCM is attractive since quartz crystals are massively produced by the electronic industry nowadays.

During last decades, quartz crystal microbalance (QCM) becomes the most utilized mass sensor. QCM is highly sensible to small mass variations due to changes in quartz crystal resonance frequency. Some QCMs reach a resolution up to 0,05ng [8]. QCMs have mayor sensibility than traditional analytical microbalances used nowadays in laboratories. The QCM is based on the piezoelectric effect, when mechanical forces are applied to the sensor producing deformation of the crystalline structure. This results in a dipole moment, and subsequently in currents that can flow across external circuits. On the other hand, applying alternating voltages to the crystal induce mechanical oscillations of the crystal lattice. The resonant frequency of the quartz crystal is established by the distance between the electrodes, and changing the thickness of the resonator (e.g. by thin film mass depositions) results in a shift of the resonant frequency.

Interestingly, the QCM have shown to operate in liquid phase [9], and this feature has been ideal for a vast applications on different fields including chemistry, biology, etc. The real-time quantification of the frequency shifts, due to mass deposition, has facilitated the study of biomolecules absorption or binding such as antibodies[10], proteins[11], and even cells[12]. However, the equations associated with the shift of the resonant frequency and the mass loading assume that the mass must be rigidly adsorbed [13,14]. Therefore, since biological interfaces in the QCM present viscoelastic characteristics, a more accurate index is required. The alternative solution is to measure the energy dissipation due to frictional dissipative losses. This feature often offered in commercial instrument [15] has not been introduced in open source devices.

The present article describes the design, simulation, and fabrication of a dissipation module integrated to a conventional QCM system in order to address dissipation due to viscoelastic film depositions.

2. Materials and methods

2.1. Overview

The dissipation module has been design to be integrated with an open source microbalance system (OpenQCM, http://openqcm.com/). The operating principle of the dissipation module is based on the rate of decay of oscillations generated by a crystal, after a rapid disruption of the excitation voltage. The generated decay voltage can be modeled as an exponential damped sinusoidal, so resonant frequency and dissipation factor can be assessed simultaneously [14,16].

The system consists of a microcontroller interacting to a computer through USB communication (Fig. 1). The system includes two major parts: digital and analog. The digital part not only records the frequency of the input signal, but also commands the switching of the power and the signal, i.e., deactivating the excitation of the crystal and connecting the crystal output to the analog part of the circuit. When signal are connected to the analog circuit, the envelope of the damped sinusoidal is processed by the envelope detector, and then converted to digital by the microcontroller ADC. The switching signal is commanded by the computer.

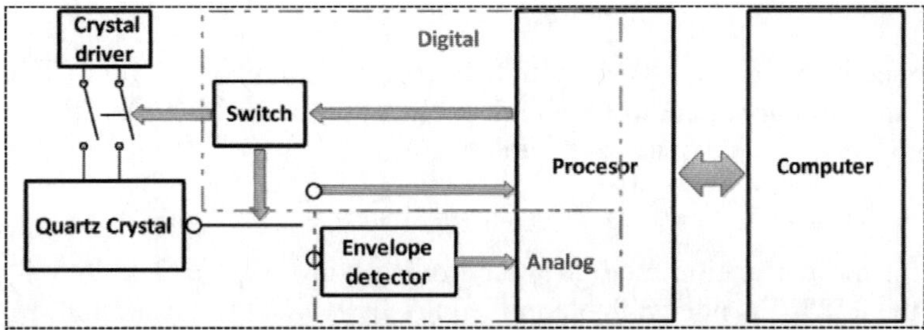

Figure 1. Block diagram of the QCM system with the dissipation module

2.2. Hardware

The dissipation module has been adapted to operate with an Arduino micro since it is the core of the original OpenQCM. One GIO pin of the microcontroller control the switching block. The switching is implemented by a digital controlled analog switch using the IC 74HCT4053. When the controlling pin changes the logic level, the switch disconnects the crystal and simultaneously connects the crystal output to the envelope detector circuit. The envelope detector circuit consists of an AM demodulator LTC5507. This IC can operate with carriers ranging from 100 kHz to 1 GHz. The schematic of the system (Fig. 2) was performed in a free PCB designing program (Ealge v7.2, CadSoft).

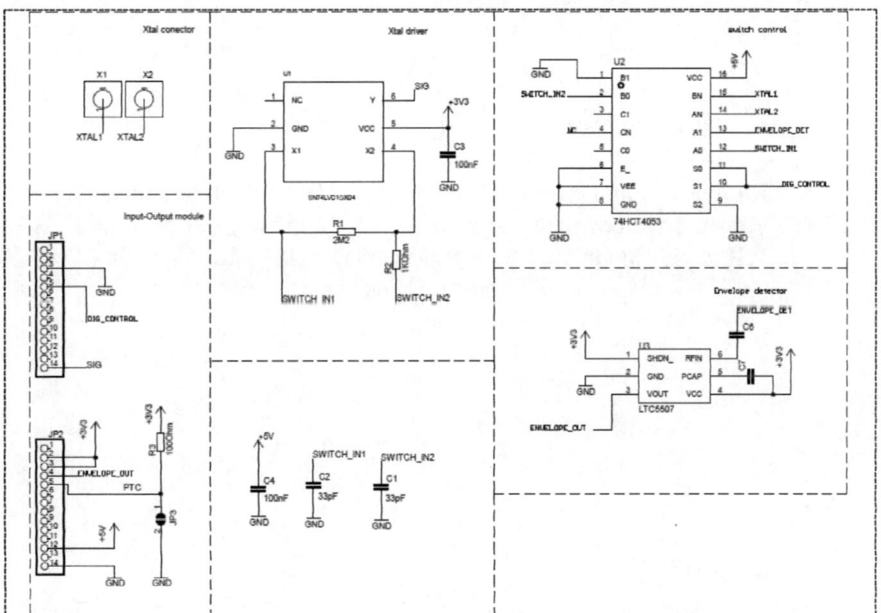

Figure 2. Schematic of the QCM system with the dissipation module. The input-output module facilitates the connection with the arduino board. The switch control deactivates the excitation of the crystal and connects the crystal output to the envelope detector, according to DIG_CONTROL status. The envelope detector quantifies the envelope of the damped sinusoidal and the output is digitalized by the ADC of the microcontroller to estimate the dissipation factor.

2.3. Software

The original user interface to the QCM system has been developed in Java JDK8. The GUI was modified in order to introduce a control switching block button (DIG_CONTROL). In addition, the GUI could be compiled to a standalone executable file.

2.4. Simulation

The performance of the dissipation module was simulated in LTspice. The ltc5505 was used in the simulation, since the ltc5507 is not available in the current version of the simulation program. This IC presents similar operation characteristic with the ltc5507, despites the difference in the frequency working range (100 kHz-1 GHz for the ltc5507 and 300 MHz to 3 GHz for the ltc5505). The sinusoidal damped signals generated by the crystal were simulated using a 1 GHz sinusoidal signal. The curricr signal was modulated with different exponentially decreasing function in order to simulate different viscoelastic interfaces; lower time constant (τ) were associated with more viscoelastic interfaces. Tau values used in the simulation were 0.1, 1, 3, 5, and 7 microseconds. The schematic of the simulation is shown in figure 3.

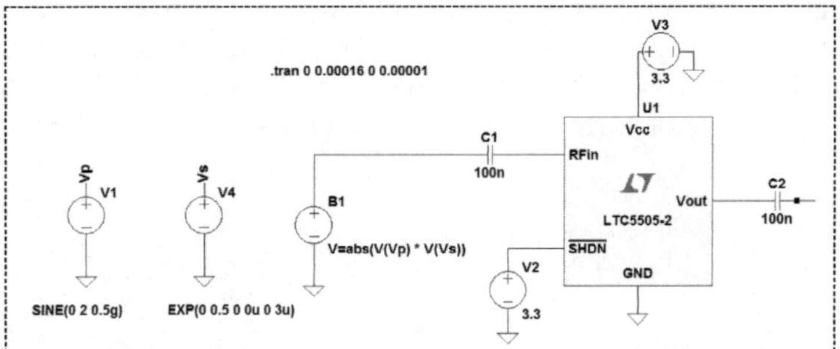

Figure 3. Simulation in LTspice to explore the performance of the envelope detector module. The frequency currier is 1GHz with modulated with exponential signal with different fall constant time (τ).

3. Results

The incorporation of few SMD elements facilitated a double-faced, PTH (plated-through-holes), PCB design with reduced dimensions (Fig. 4). PCB was produced using the ProtoMatS103 PCB milling machine from LPKF. The ProtoMat works straight from CAD data to process boards. This technology provides precise, quick, and simple PCB prototyping, achieving tracks width and clearance of 100 um (Fig. 5). All the components of the OpenQCM and the integration with the designed dissipation module are shown in Fig. 6.

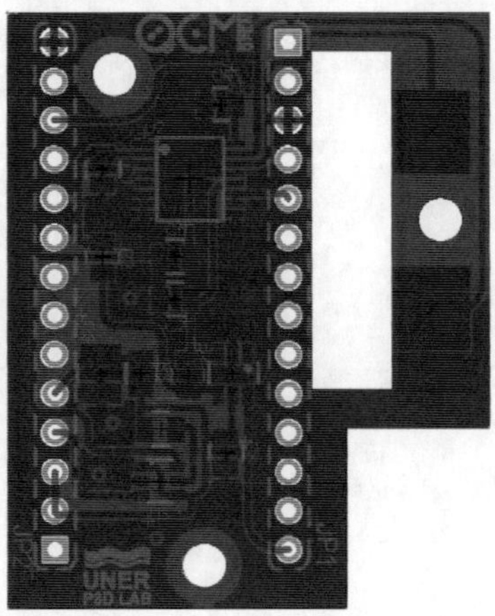

Figure 4. PCB design for the QCM with the dissipation module.

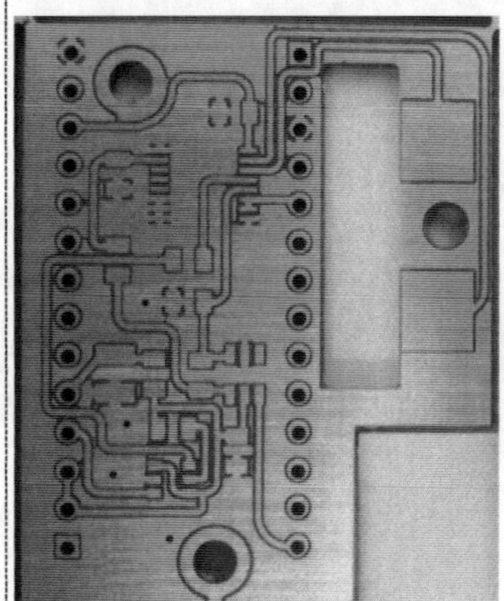

Figure 5. Board manufactured using a PCB milling system (Protomat, LPKF).

Figure 6. OpenQCM and integration of the dissipation module. The case was printed using a PolyJet 3D printer (Objet Eden260VS®, Stratasys)

The output of the dissipation module due to different input signals is observed in Fig. 7. The response to the selected input of τ with the relative error for each measure is shown in table 1.

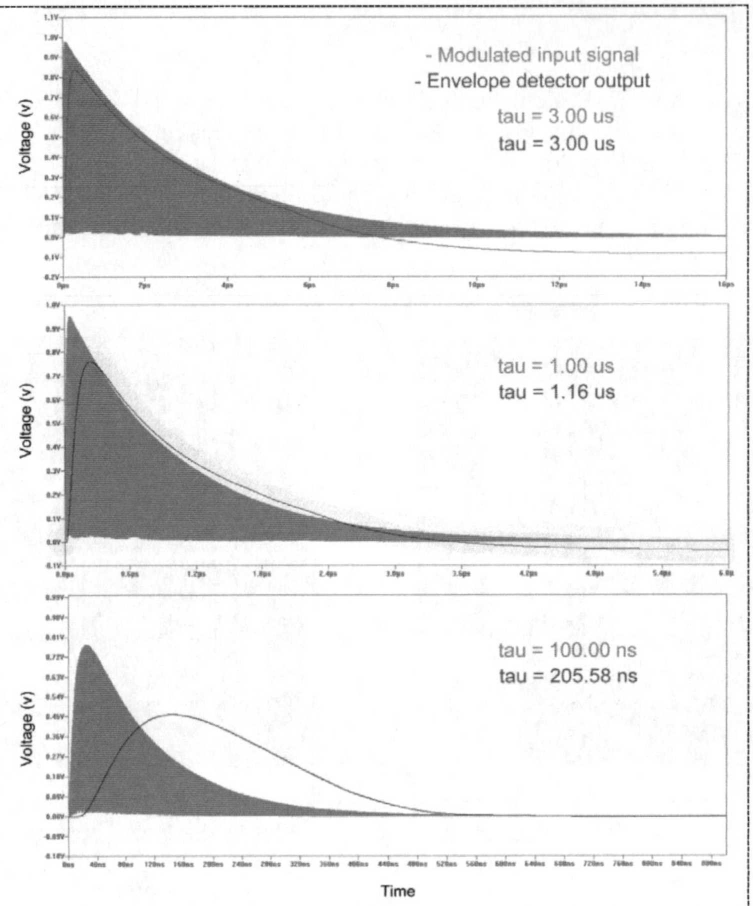

Figure 7. LTspice simulation used to test the envelope detector module. The frequency curried is 1GHz with modulated with exponential signal with different fall constant time (τ).

Table 1. Time constant (τ) estimated (Out τ) at the dissipation module for different exponential damped sinusoidal signal (input τ).

Input τ (us)	Out τ (us)	Error %
0.1	0.2	100
1	1.16	16
3	3	0
5	4.82	3.6
7	7.25	3.57

4. Discussion

The dissipation module is an important enhancement in the open source QCM system. In particular, the reduced number of electronics pieces introduced can be easily incorporated to any system. The present work has described how the module could be successfully integrated to the OpenQCM, showing efficient design.

The quantification of the dissipation phenomenon becomes significant in the study of biomolecular process. For example, during biomolecular adhesion, shifts of frequency without significant dissipation changes are found when the biomolecular mass is rigidly absorbed [14]. Thus, the biomolecules follow the oscillation of the crystal. However, when the absorbed mass is viscoelastic and not hard enough, the biomolecules no longer can follow the oscillation of the crystal, presenting deformation and resulting in energy losses [14]. A previous study showed how two molecules, human serum albumin and antibody from serum albumin, impact differently in the frequency shift and the dissipation factor [13]. The adsorption of the albumin results in shift of the frequency without changes in the dissipation, indicating a rigid characteristic of the serum albumin film. On the other hand, adsorption of antibody yields to large frequency and dissipation changes, reflecting the viscoelasticity characteristic of the biomolecular film.

Interestingly, the simulation shows that the dissipation module has low relative error for modulated exponential signals with time constant higher than 1 us. The time constant is inversely proportional to the dissipation factor ($D = 1 / f\pi\tau$) [14]. The typical time constant decay for a 10 MHz QCM in air has estimated to be around 2-3 ms. Likewise, the QCM with deionized water presents a decay constant time around $x10^{-3}$s, and around $x10^{-4}$s for proteins such as protein A and fibronectin [17]. Most of the dissipation factor for biomolecular adsorption are in the range of single to tens ($\times$ 10–6) of dissipation units [13]. Therefore, the present module is able to characterize biomolecular viscoelastic films, because it is capable of measuring faster decay time constant. This feature could be used for characterization of cell adhesion on the surface of implants, which is critical to assess biocompatibility (although additional investigation are required to reduce the size of the sensor), or it can be include into label-free sensors.

It has to be mention that most of the commercially available peak voltage detector, including the LTC5507, presents non-linearity for small signals [18]. This effect can be noted at the end of the decay exponential in Fig 7, although low-level voltages are not used for quantifying the time constant decay in the case of the QCM. Nevertheless, this non-linearity observed in commercial peak-voltage detector IC has been associated with diode performance, and can be sort out by using circuits without diodes or other nonlinear components, enhancing significantly the linearity of the system [19]. In addition, further analysis is needed in order to calculate the resolution of the system. The resolution to estimate the dissipation factor can be increased by fitting an exponentially damped sinusoidal. Thus, works in this direction are being developed by the authors.

In conclusion, a dissipation module has been developed for quantifying dissipation factor using QCM. The system is compact and can be easily incorporated to any QCM system. In addition, the module exhibit satisfactory performance for measuring viscoelasticity on the interface, addressing on real-time the quantification of time decay constants usually observed in biomedical applications. In particular, the module results a desirable system for studying biomolecular adsorption.

Reference

[1] Lim H, Park J, Lee D and Kim H J 2012 Build Your Own Quadrotor: Open-Source Projects on Unmanned Aerial Vehicles *Robot. Autom. Mag. IEEE* **19** 33–45

[2] Colorimeter O 2013 Open-Source Colorimeter 5338–46

[3] Pearce J M 2013 Building Research Equipment with Free, Open-Source Hardware *Science (80-.).* **337** 1303–4

[4] Savran C a., Knudsen S M, Ellington A D and Manalis S R 2004 Micromechanical detection of proteins using aptamer-based receptor molecules *Anal. Chem.* **76** 3194–8

[5] Gronewold T M a, Glass S, Quandt E and Famulok M 2005 Monitoring complex formation in the blood-coagulation cascade using aptamer-coated SAW sensors *Biosens. Bioelectron.* **20** 2044–52

[6] Heitmann V, Reiß B and Wegener J 2007 The Quartz Crystal Microbalance in Cell Biology: Basics and Applications *Piezoelectric Sensors* **5** 303–38

[7] Lu C-S 1972 Investigation of film-thickness determination by oscillating quartz resonators with large mass load *J. Appl. Phys.* **43** 4385

[8] Sullivan C K O and Guilbault G G 1999 Commercial quartz crystal microbalances – theory and applications **14** 663–70

[9] Nomura T, Watanabe M and West T S 1985 Behaviour of piezoelectric quartz crystals in solutions with application to the determination of iodide *Anal. Chim. Acta* **175** 107–16

[10] Schaible U, Liss M, Prohaska E, Decker J, Stadtherr K and Wolf H 2004 by a Quartz Crystal Microbalance (QCM) Biosensor **94**

[11] Jalit Y, Gutierrez F a., Dubacheva G, Goyer C, Coche-Guerente L, Defrancq E, Labbé P, Rivas G a. and Rodríguez M C 2013 Characterization of a modified gold platform for the development of a label-free anti-thrombin aptasensor *Biosens. Bioelectron.* **41** 424–9

[12] Wegener J, Janshoff a and Steinem C 2001 The quartz crystal microbalance as a novel means to study cell-substrate interactions in situ. *Cell Biochem. Biophys.* **34** 121–51

[13] Dixon M C 2008 Quartz crystal microbalance with dissipation monitoring: Enabling real-time characterization of biological materials and their interactions *J. Biomol. Tech.* **19** 151–8

[14] Rodahl M, Höök F, Fredriksson C, Keller C a, Krozer a, Brzezinski P, Voinova M and Kasemo B 1997 Simultaneous frequency and dissipation factor QCM measurements of biomolecular adsorption and cell adhesion. *Faraday Discuss.* 229–46

[15] Q-sense 2011 Quartz crystal microbalance with dissipation *Langmuir* **27** 2

[16] Edvardsson M, Rodahl M, Kasemo B and Höök F 2005 A dual-frequency QCM-D setup operating at elevated oscillation amplitudes *Anal. Chem.* **77** 4918–26

[17] Marxer C G, Coen M C and Schlapbach U F G L 2003 Simultaneous measurement of the maximum oscillation amplitude and the transient decay time constant of the QCM reveals stiffness changes of the adlayer 570–7

[18] Linear Technology LTC5507 - 100kHz to 1GHz RF Power Detector (datasheet) 1–8

[19] Asfour A, Zidi M and Yonnet J-P 2014 High Frequency Amplitude Detector for GMI Magnetic Sensors *Sensors* **14** 24502–22

Wearable System for Acquisition and Monitoring of Biological Signals

D J Piccinini[1], N B Andino[1], S D Ponce[1], MA Roberti[1] y N López[2]

[1] Analysis, Development and Biomedical Research Group, Facultad Regional San Nicolás, Universidad Tecnológica Nacional, Argentina.
[2] Medical TechnologyCabinet, Universidad Nacional de San Juan, Argentina.

Abstract. This paper presents a modular, wearable system for acquisition and wireless transmission of biological signals. Configurable slaves for different signals (such as ECG, EMG, inertial sensors, and temperature) based in the ADS1294 Medical Analog Front End are connected to a Master, based in the CC3200 microcontroller, both from Texas Instruments. The slaves are configurable according to the specific application, providing versatility to the wearable system. The battery consumption is reduced, through a couple of Li-ion batteries and the circuit has also a battery charger. A custom made box was designed and fabricated in a 3D printer, preserving the requirements of low cost, low weight and safety recommendations.

1. Introduction

Medical technology is a key element in global problems concerning health, as it can improve the quality of life or extend it through the appropriate use and prevention. Its value is calculated not only for its economic cost but also for its impact in practice and the success of medical treatments. Due to medicine improvements, life expectancy has increased.Along with this, the caring for the elderly and people with chronic diseases, havealso improved emphasizingthe necessity of preventive medicine and monitoring of this patients [1, 2].

In this context, what is proposed is the development of a platform of wearable sensors for the monitoring of the health and well-being of the individuals, in a direct or remote way. The incorporation of this technology improves the quality of the elderly's life, people with cognitive or memory disorders, disabled or with motor disability, obese patients, or any other pathology that may place the user in a vulnerable situation.It also allows a continuous and non-invasive monitoring during sport practices, exposition to adverse environmental situations, and other applications in healthy individuals [3-5].In recent years, the use of these technologies has been incorporated in entertaining and rehabilitationvideo games and also in augmented reality systems.

A wearable system should meet certain design conditions, such as data storage, wireless connection, efficiency in the supply, portability, versatility and capacity of pairingseveral sensors, among others [6].

The proposed system consists of a master module in charge of data reception andcollection, synchronization and wireless transmission. This module is connected to sensors or slaves which acquire a biological signal and then process and formatitadequately to send it to the master. It is important to remark that the communication between the disposals should respect the same protocol, independently of the nature of the acquired signal.

The data transmitted from the master module to the PC can be seen, processed and exported to specific softwares according to the user. The software should be modular to suit the sensors chosen for each application, which give versatility to the system, functioning in an interchangeable way. The proposed biologic variables to be sensed, each with its own specific acquisition and preprocessing module:

- Temperature and humidity module: it registerstwo temperature channels, which can be put in two body segments or in one external segment to determine the metabolic answer to adverse environmental conditions. The humidity is an important variable in the control of the body temperature, and its sensing lets us know the user's adaptation to external conditions.
- Heart rate module and pulse oximetry: with only one module both variables are registered through a LED emitter of specific frequency to the absorption of red cells or red corpuscles, which indicates the degree ofoxygenation. Moreover, it allows inferring the heart rate. On one hand, this module is of vital importance in the remote monitoring of cardiac, respiratory and elderly patients. On the other hand, the module is a body activityindicator for sedentary and obese users. Finally, it is crucial for the use in sports. The placement of these sensors should be done directly on the skin, because of the operating mechanism above-named.
- Kinetics measurements module: this module has an inertial sensor IMU (Inertial Measurement Unit) that contains an accelerometer and a gyroscope. It is necessary to place the sensor on the body segment wanted to be studied, or to put an IMU in each segment. The information of each IMU module is processed to avoid singularities and both sources (accelerometer and a gyroscope) are merged through Kalman filter.It is possible to determine the established angles between them with two or more segments, what is a true articular anglestimulator. Only one IMU provides information related to the user's activity, inactivity level, waste of energy (with reference with the temperature).
- Magnetometer / GPS: this sensor is incorporated into the above-named IMU, but it will be used strictly to monitor and locate the patient. This module is particularly useful for patients with cognitive impairment, Alzheimer's disease or similar pathologies. In the case of risk patients it is vitally important if any alarm sounds, for example the heart rate. Furthermore it is used when the continuous monitoring of a user is done, to study the behavior in everyday activities. The location allows knowing the activity and surroundings that are related to each event.
- Surface electromyography module (EMG): in recent years, it has been a progress in the study of muscular behavior in rehabilitation patients through the non-invasive EMG. It is possible to use these technologies to quantitative evaluations of ergonomics, fatigue and muscular effort in repetitive and forced labor, where musculoskeletal system injuries are the main cause of disability.

The modular design and the configuration of variables that are proposed in this design allow the configuration of active sensors (minimizing the battery consumption), the selective data storage and the continuous transmission of monitoring parameters depending on the application.

In this paper, de development of the master system is presented together with one slave module to acquire signals; in this case they were used electrocardiographic (ECG) and electromyography (EMG) signals as examples.

2. Materials and Methods

The system consists of individual sensors (Slaves) that includes, in situ, a stage of preparation of the biologic signal, another of regulation and feeding filtering and an analogue digital converter. These sensors are connected to a Master board that is in charge of: controlling the communication with the Slaves, the wireless transmission to a PC and the feeding of the whole system, including the batteries charging. The connection between Slaves and Master is intended to be in a properly wired article of clothing. The data transmission from the Master is carried out through Wi-Fi using TCP/IP protocol to a personal computer in order to process and extract characteristics (Figure 1). The system gives the possibility of having the data in flat file format, facilitating the analysis and application of algorithms with any programming languageor even in a simple commercial or open spreadsheet.

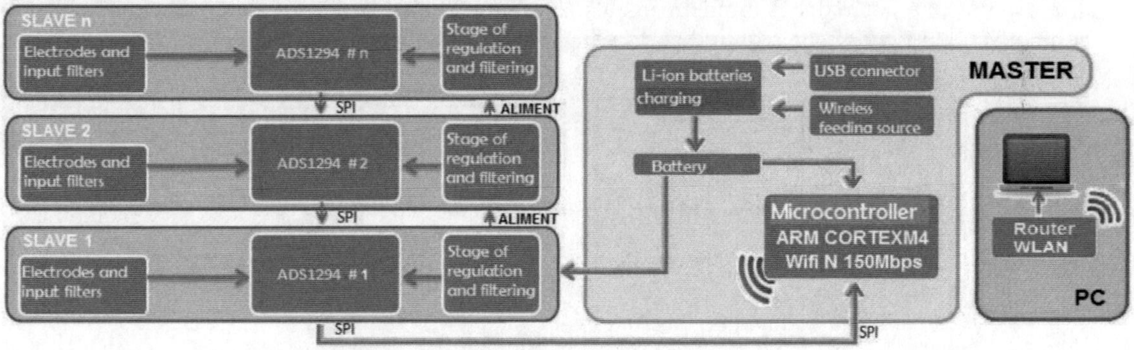

Figure 1. System block diagram

2.1. Slave Units

As it has been said before, these Slave units (color blue in figure 1) consists of one stage of low noise and high 115dB CMRRamplification of programmable gain, a multichannel delta-sigma A/D converter of 24 bits and sampling rate up to 32kSPS and low noise internal tension reference, possibility of common mode feedback through the RLD (Right-Leg Drive) circuit and SPI (Serial Peripheral Interface) digital output compatible with cascade connection of multiple disposals (Daisy Chain). All this is given for the ADS129X integrated circuit from Texas Instruments. In our case, that is, Slave module for ECG/EMG the ADS1294 model was used, which has 4 differential channels and internal clock for its digital section and delta-sigma converters. It has to be emphasized that all the family ADS129X satisfies the IEC60601 standard [7].

To feed this unit, a +3.3v unipolar unit is proposed to the digital part and a +5v unipolar tension to the analogic part. To do that, ultra-low noise linear regulators TPS7A49 from Texas Instruments are used; all recommended to analogic applications of high accuracy and precision.

The SPI bus functions at 4MHz speed (within a maximum of 20 MHz) and can add up to 41 Slavesthat are directly placed on the disposable electrodes. The maximum quantity indicated is according to the SPI bus, but also limitations can exist because when the Wi-Firouter is fur away from our position, the communication bandwidth is dropped.

The following equation corresponds to a calculation of the estimation of the number of Slaves connected through the SPI bus:

$$N_{DEVICES} = \frac{f_{SCLK}}{f_{DR}\,(N_{BITS})(N_{CHANNELS}) + 24}$$

f_{SCLK}stands for the frequency of the ISP used, $f_{DR}$is the frequency of the ADC sampling, N_{BITS} is the quantity of sample bits, in this case there are 24 bits and $N_{CHANNELS}$ corresponds to the four ADS1294 channels. The chosen configuration is 1000SPS (Samples per Second). It is possibleto reach to 208 connected Slaves if using an SPI speed of 20 MHz, since the microcontroller and the ADS128X allow it, but this can result in problems in the communication originated from the length of the wires. Even though, the necessary precautions were taken into account in the moment of design the board if this or other application is required.

The Daisy Chain configuration was selected to interconnect the Slaves. The Daisy Chain share the SCLK, DIN and CS bus signals, going through each ADS1294. The DOUT line corresponding to the output data of each ADSgoes intothe DAISY_INentry of the previous module, and so on, creating a continuous chain. That is to say, the minimum number of wires needed to make the communication in this way, is 5 and it is always the same. There is a START reference indicating the beginning of the conversion, however, it was decided to replace the physical signal to an OPERATION COMMAND through the SPI to avoid the use of another wire. The internal clock is also used for the internal synchronism signals, avoiding another extra wire and additional electronic components. In figure 2 the required wiring diagram is appreciated.

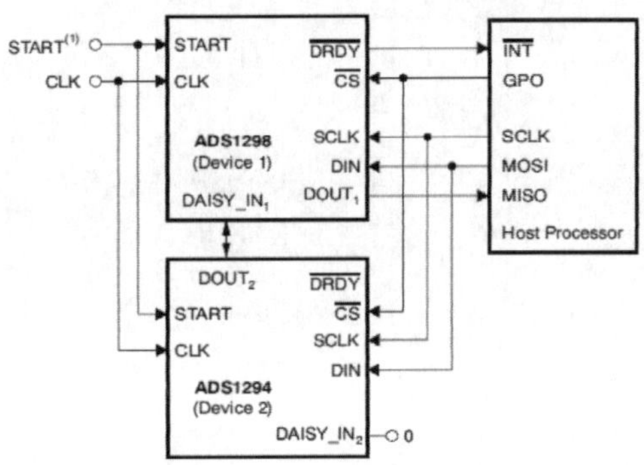

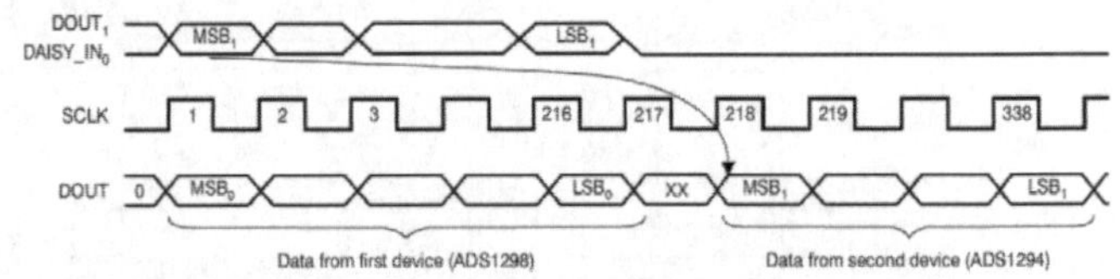

Figure 2. Multi-configuration used.**Figure 3.**Diagram of data synchronization for figure 2.

The electromagnetic interference would largely affect the system, through common mode tensions produced by the magnetic field couplingthat comes from the feeding line and other sources, including fluorescent lights. For this, it is made use of a RLD circuit that senses the common mode of a set of determined electrodes and originates a negative feedback injecting the patient with an inverted common mode signal in phase [8, 9]. The reference voltage applied in this system corresponds to +2.5V, that is to say (AVDD + AVSS)/2, generated in an internal manner.

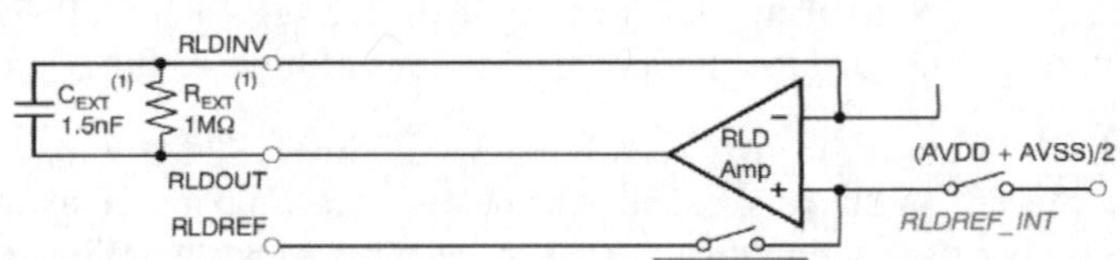

Figure 4. RLD internal block.

The producer of the front-endADS1294 recommends the following wiring diagram for the RLD when multiple interconnected disposals are used, providing feedback through a single Slave but with all of the Slaves taking part in the average, bearing in mind the body as unipotential.

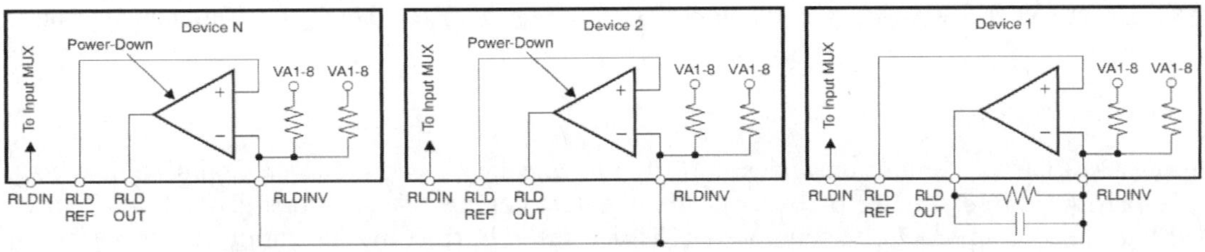

Figure 5. Connection diagram of RLD for multiple disposals.

2.2 Master Unit

The Slaves of this system are connected to a central unit or master. The microcontroller responsible for performing all the control tasks and transmission is the CC3200 of the Texas Instruments. This microcontroller corresponds to one SOC of dual core ARM CortexM4 of 32bits with floating-point, running at 80MHz and an external memory for the user program of 8 Mbit of capacity. As a remarkable characteristic of this model is a Wi-Fi connection embedded in the same module ready to be used, coordinated by a sub-processor network.It has a capacity for make transmissions of 150 Mbps under N norm and allows the user to have the possibility to use data encryption, such as AES, DES, SHA, TLS/SSL or MD5.It also has common peripherals as GPIO, PWMs, UARTs, SPI, IIC, etc. Another important characteristic of this integrated circuit itthat it has certifications (FCC, IC, CE, Wi-Fi CERTIFIED) that enables the streamline of the development, economize its commercial implementation and reduce installation timings in the market. It can be appreciated in greater detail, in figure 6 the internal peripherals of the module.

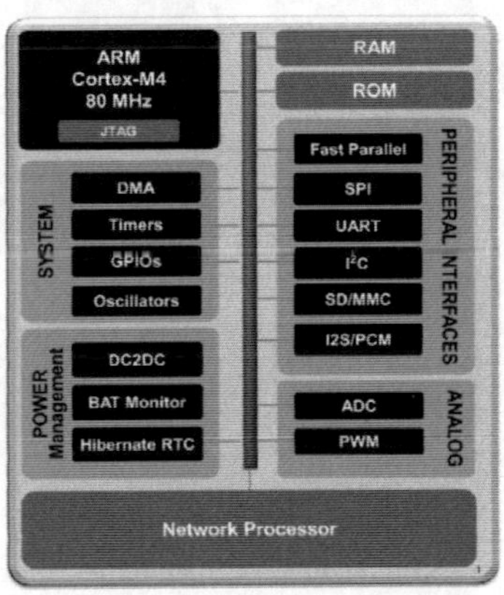

Figure 6. Peripherals conforming the CC3200 module.

Through this plaque, two media are controlled, on one hand a Wi-Fi communication, on the other hand, a SPI bus. The Master sends wirelessly the data of the acquisition through the TCP/IP protocol. The serial peripheral interface or SPI bus is a communication standard to control the transference of information between diverse integral circuits, in this case, between the slave sensors and the master unit.

The complete system functions with a pair of Li-ion batteries, for electric reasons, this is the only way of operating. These batteries are changed using the MAX1555 integral circuit of Maxim Integrated, which allows the charge with energy that comes from a USB connector or a DC source. As an innovative approach, it was used a wireless feeding source suitable to the Qi specification, BQ541013B integrated circuit of Texas Instruments, designed for portable systems. This stage consists of a transmitterand a receiver that through the beginning of electromagnetic induction allows charging the batteries mentioned above. This source gives a capacity of 5W to 5 volts output power.

It can be seen in figure 1 that the Master part is represented by an orange color, consisting on the stages named above. Moreover, it can be observed in figure 10, the physical implementation of the system.

2.2. Firmware

The firmware of the Master unit was programmed in C language, using the Code Composer Studio software distributed freely by the producer of the integrated circuit, Texas Instruments. It has been decided that the system would be composed of varied tasks in the same operating system in real time, TI RTOS, running in the microcontroller of the module. Even though, ifthe producer does not give examples of the Wi-Fi interface applications under this operating system, the Wi-Fi communication has been manipulated and all the other segments that make the communication, as sockets TCP/IP, Smart Link, etc. In figure 7, the firmware diagram for the Master is presented in detail.

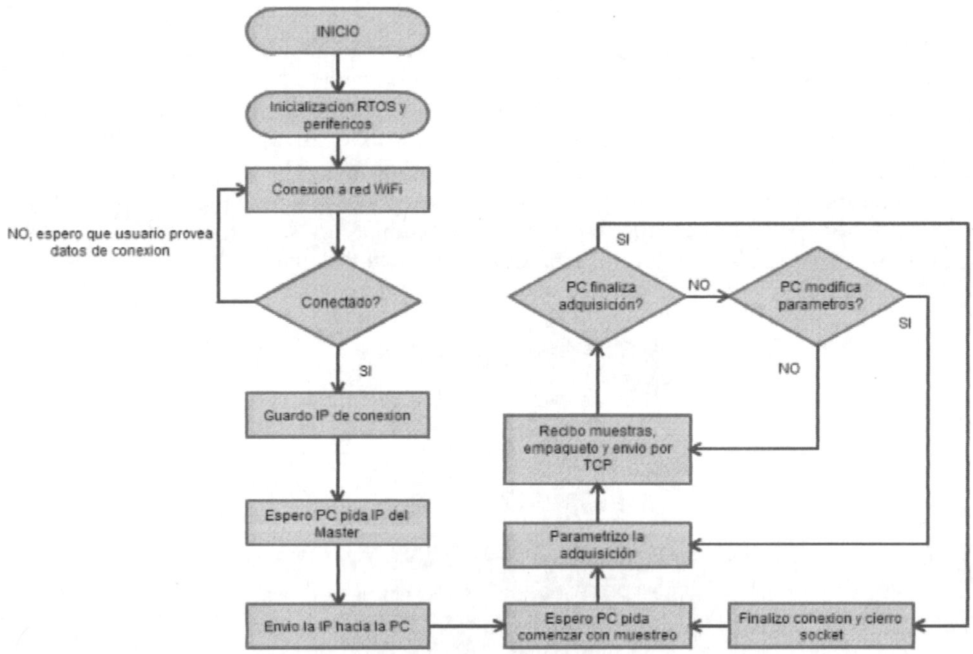

Figure 7. Diagram of the Master state.

To use theWi-Fi connection, the bookshops provided by the producer were used, configuring it as a client in the Wi-Fi connection. The module has the ability of storing profiles for different Wi-Fi networks. When turned on, it will search for nearby networks and will connect to that of the strongest signal, used and stored in a previous occasion. In case of not having stored networks, or if the device is been used for the first time, or with the presence of networks having been detected, any of them has been connected preciously, the module will be waiting for the user's indication of what network choose. This can be done with a Smartphone through the application "Wi-Fi Starter" that is distributed freely by the Texas Instruments.

2.4. Data Reception

To begin acquiring the samples, first of all the PC, through Wi-Fi, will indicate the Master how to set-up the ADC ADS1294 of each Slave. However, it should previously know the IP address that the Wi-Fi network provided to the Master before. This can be solved assigning a fixed IP address in the Master, which is not recommended because it will only work with that address and Wi-Fi

Journal of Physics: Conference Series **705** (2016) 012009 doi:10.1088/1742-6596/705/1/012009

connection, being a bit inflexible. The following may be the possible solution: to the PC sends an UDP (*User Datagram Protocol*) packetby the Wi-Fi connection directed to a well-known port of each Master, with the word "Discovery" in a Broadcast way, that is, it sends to all of the devices connected to the network.

The Master unit receives this packet, together with an IP address of the PC that is interested in start a connection, and responds to this packet with another one having its IP address. In this way, when this answer returns, the PC would have been informed of which IP address corresponds to the Master unit and the communication starts. After that, the ADS1294 should be set with the parameters as sampling frequency, internal reference tension, the status of each channel, programmable gain of each channel, RLD uses, etc. the sample reception from the ADC is done without interruptions to the sample frequency established by the user. Each sample is received by the Master of each Slave, and is sent to thePC through TCP (*Transmission Control Protocol*)to guarantee the order and delivery of the packets.

2.5. Hardware design

The printed circuit boards are designed with free or lite tools. In our case, the software KiCAD was used for the Slaves (figure 8), and in its Lite version, the software Eagle for the Master (figure 9). The designed technology adopted was SMD. For welding the chips CC3200 and BQ51013B it was needed to turn to companies of surface mount. In figure 10, it can be seen all the boards finished with the main components welded, ready for the first tests.

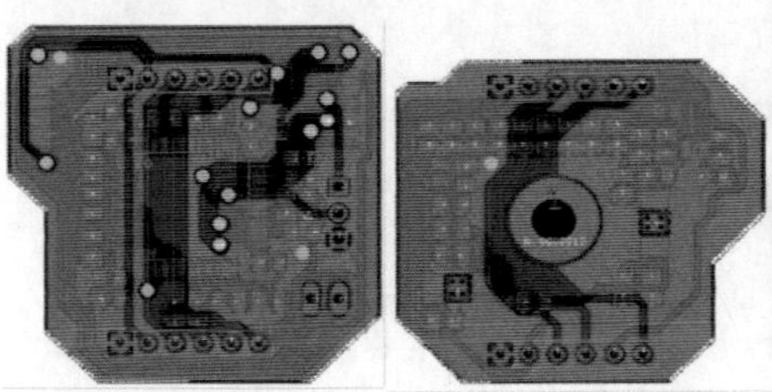

Figure 8. Design and SLAVE boards manufacture.

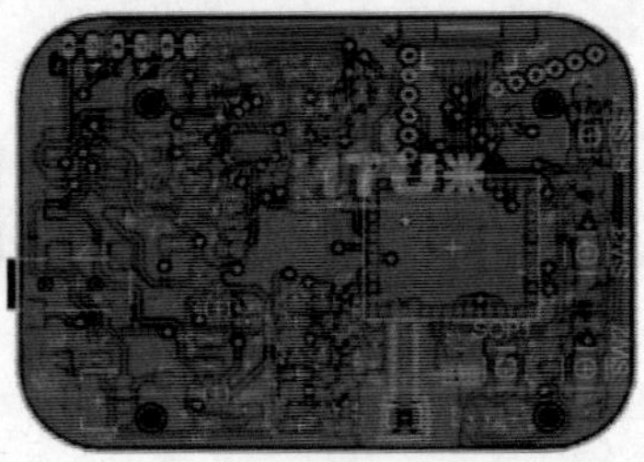

Figure 9. Design showing both layers of the MASTER board

.

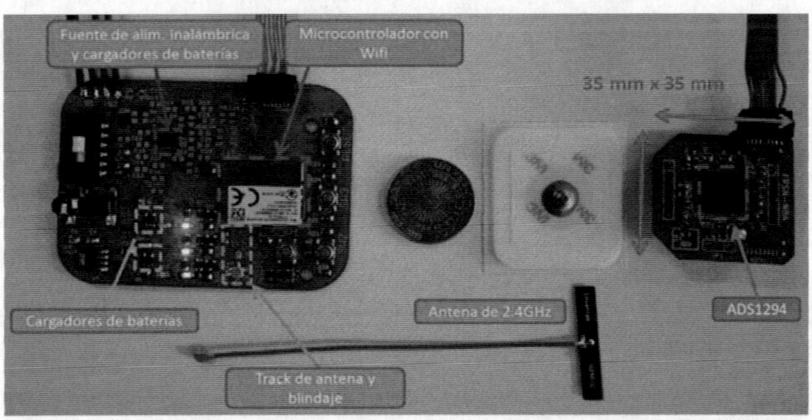

Figure 10. Finished base kit ready for the first tests.

As a final comment regarding the boards, the design has been thought to be done in two layers with PTH in FR-4 material with 1mm thickness. The most critical part of the Master board has been the design of the connection to the antenna, which must meet requirements of impedance adaptation; the calculation was verified with KiCAD (Figure 11). In figure 10, it can be seen also the additional shield with PTH holes around the course to the connector of the antenna.

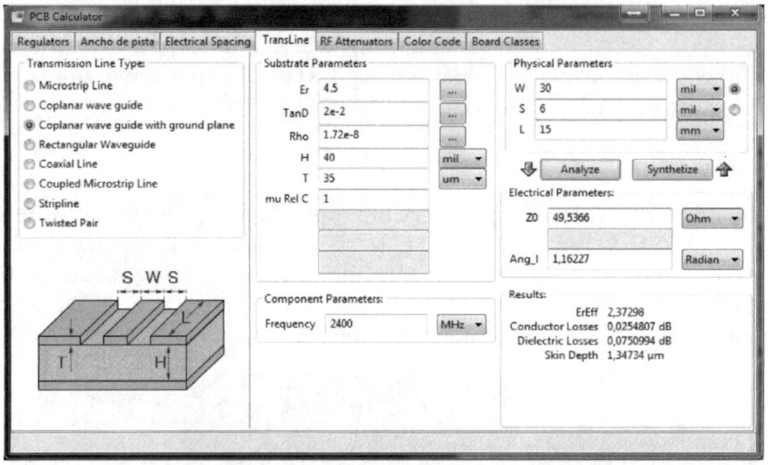

Figure 11. Calculation of the course for the antenna connector.

Lastly, the cabinet design has been performed with the use of a 3D software (figure 12), which is currently in the stage of production. A 3D printer or another conventional method of the industry dedicated to this subjectis needed.

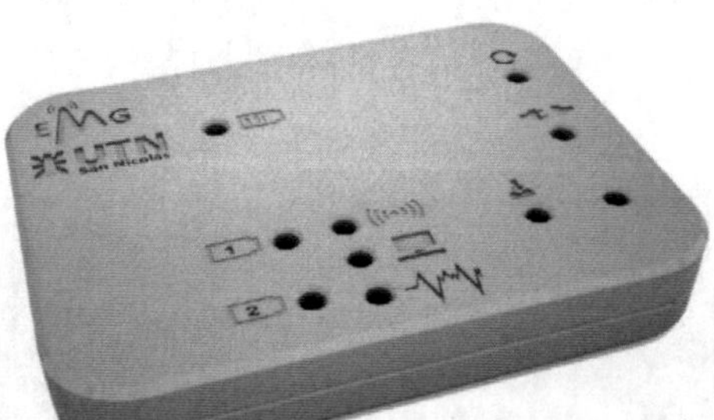

Figure 12. Prototype 3D model of the Master custom made box.

3. Results

3.1. Equipment evaluation

Before connecting the circuit with volunteers, each block was proved separately and then the complete system of ECG/EMG was proved making an acquisition of a real electrocardiogram. The performance was evaluated through the measure of its parameters and its practical use in the laboratory. A user interface that allows visualizing the signal in real time and the possibility of exporting the data for a later analysis was developed. In figure 13, it is showed the acquisition of the ECG signal.

The noise voltage equivalent to the entry was measured. If the entries are shorted out, an interpretation of the noise voltage of the ADS1294 is obtained. To a set gain configuration in 3, an ECG signal is injected. Its bandwidth is of 100Hz and is a result of a reading that enables us to establish that all the noise to the entry is low, of some $2uV_{RMS}$.

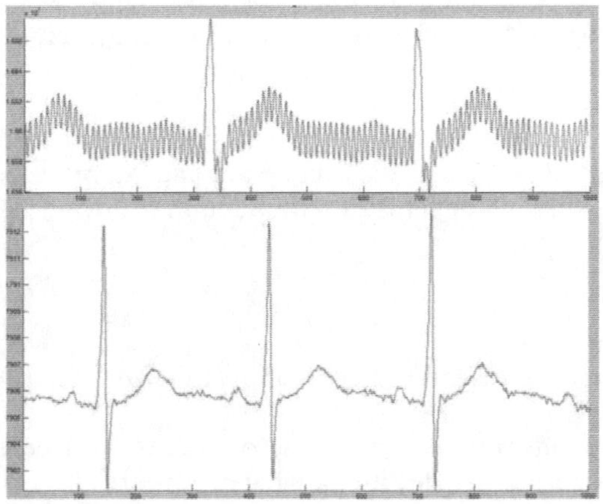

Figure 13. Capture of an analogic acquisition of the ACG signal. RLD (PD) verification.

Figure 14. Interpretation of noise voltage in the entries during 10 seconds.

In the area of wireless communication with the PC it is very important that the bandwidth must be the as much as possible to guarantee the proper functioning of all the system. For that reason, two designs of implementation were compared:

- Board A: Evaluation board of the CC3200 microcontroller, commercialized by Texas Instruments. It uses an antenna mounted on board like a chip model 316M245001, which manufacturer is TaiyoYuden.
- Board B: Own designed board, commented through this entire document, and with an antenna that is not mounted on board, using a U.FL connector for radiofrequencies.

The highest bandwidths for each Board are compared, placing these 18 meter of distance from the Wi-Fi router. The analysis is done with the open **iPerf3**application.

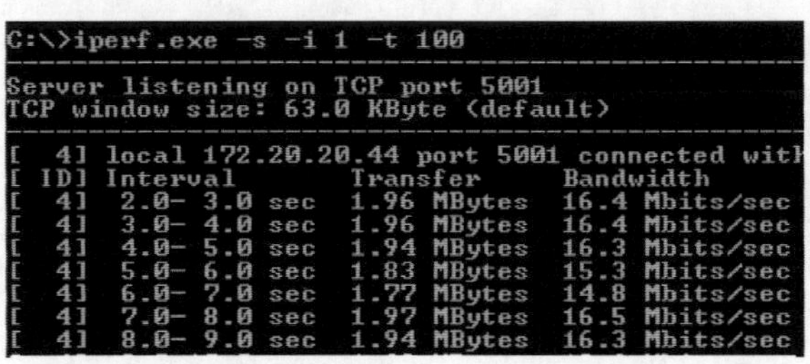

```
C:\>iperf.exe -s -i 1 -t 100
------------------------------------------------------------
Server listening on TCP port 5001
TCP window size: 63.0 KByte (default)
------------------------------------------------------------
[  4] local 172.20.20.44 port 5001 connected wit
[ ID] Interval       Transfer     Bandwidth
[  4]  2.0- 3.0 sec  1.96 MBytes  16.4 Mbits/sec
[  4]  3.0- 4.0 sec  1.96 MBytes  16.4 Mbits/sec
[  4]  4.0- 5.0 sec  1.94 MBytes  16.3 Mbits/sec
[  4]  5.0- 6.0 sec  1.83 MBytes  15.3 Mbits/sec
[  4]  6.0- 7.0 sec  1.77 MBytes  14.8 Mbits/sec
[  4]  7.0- 8.0 sec  1.97 MBytes  16.5 Mbits/sec
[  4]  8.0- 9.0 sec  1.94 MBytes  16.3 Mbits/sec
```

Figure 15. Measurement of bandwidth for board B.

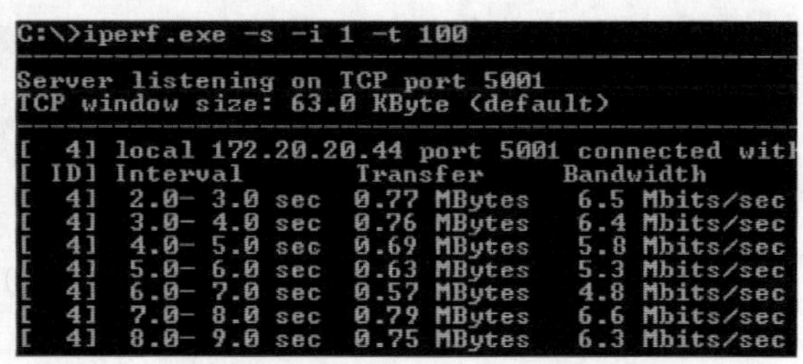

```
C:\>iperf.exe -s -i 1 -t 100
------------------------------------------------------------
Server listening on TCP port 5001
TCP window size: 63.0 KByte (default)
------------------------------------------------------------
[  4] local 172.20.20.44 port 5001 connected wit
[ ID] Interval       Transfer     Bandwidth
[  4]  2.0- 3.0 sec  0.77 MBytes  6.5 Mbits/sec
[  4]  3.0- 4.0 sec  0.76 MBytes  6.4 Mbits/sec
[  4]  4.0- 5.0 sec  0.69 MBytes  5.8 Mbits/sec
[  4]  5.0- 6.0 sec  0.63 MBytes  5.3 Mbits/sec
[  4]  6.0- 7.0 sec  0.57 MBytes  4.8 Mbits/sec
[  4]  7.0- 8.0 sec  0.79 MBytes  6.6 Mbits/sec
[  4]  8.0- 9.0 sec  0.75 MBytes  6.3 Mbits/sec
```

Figure 16. Measurement of bandwidth for board A.

As a final comment, it can be said that all the entire system is small and with low weight, making it wearable and portable. With an approximate consumption of 270mA when transmitting connected to a Slave and all the leds on. This consumption allows an autonomy of a bit more than 3,5 hours with Li-ion batteries of 1000 mAh. Previously these batteries are fully charged in 1,3 hours, loosing efficiency because of the use in the last years in other projects.

4. Conclusions

The design implemented in Master/Slave enables the reduction of the size in the stage of acquisition of biological signals, what allows a projection of a future system integrated to the clothing of the user.

The preliminary obtained results are promising in terms of similar size to a humid electrode, robustness in the wireless transmission and reliability in data reception and processing. Now, the design of dry electrodes for the acquisition is being developed with the aim of lower the noise and improves the system portability.

5. References

1. Muennig PA, Glied SA: What changes in survival rates tell us about US health care? Health Affair 2010, 29:2105-2113.
2. Gulley S, Rasch E, Chan L: If we build it, who will come? Working-age adults with chronic health care needs and the medical home. Medical Care 2011, 49:149-155.
3. Gulley SP, Rasch EK, Chan L: Ongoing coverage for ongoing care: access, utilization, and out-of-pocket spending among uninsured working-aged adults with chronic health care needs. Am J PublicHealth 2011, 101:368-375.
4. Teng X-F, Zhang Y-T, Poon CCY, Bonato P: Wearable medical systems for p-Health. IEEE Reviews in BiomedicalEngineering 2008, 1:62-74.
5. Bonato P: Wearable sensors and systems. From enabling technology to clinical applications. IEEE EngMedBiolMag 2010, 29:25-36.
6. Patel S, Park H, Bonato P, Chan L, Rodgers M: A review of wearable sensors and systems with application in rehabilitation. Journal of NeuroEngineering and Rehabilitation 2012, 9:21 doi:10.1186/1743-0003-9-21
7. ADS129x Datasheet. Texas Instruments. http://www.ti.com/lit/ds/symlink/ads1294.pdf
8. Application Report: Understanding Lead-Off Detection in ECG. Texas Instruments. http://www.ti.com/lit/an/sbaa196a/sbaa196a.pdf
9. Application Report: Improving Common-Mode Rejection Using the Right-Leg Drive Amplifier. Texas Instruments. http://www.ti.com/lit/an/sbaa188/sbaa188.pdf

From 3D Bioprinters to a fully integrated Organ Biofabrication Line

V E Passamai, J A Dernowsek, J Nogueira, V Lara, F Vilalba, V A Mironov, R A Rezende and J V da Silva

Divisão de Tecnologias Tridimensionais (DT3D)/Centro de Tecnologia da Informação Renato Archer (CTI). Rodovia Dom Pedro I (SP-65), km 143,6 - Amarais - Campinas, SP, Brazil.

E-mails: rodrigo.rezende@cti.gov.br; veronica.passamai@cti.gov.br

Abstract. About 30 years ago, the 3D printing technique appeared. From that time on, engineers in medical science field started to look at 3D printing as a partner. Firstly, biocompatible and biodegradable 3D structures for cell seeding called "scaffolds" were fabricated for *in vitro* and *in vivo* animal trials. The advances proved to be of great importance, but, the use of scaffolds faces some limitations, such as low homogeneity and low density of cell aggregates. In the last decade, 3D bioprinting technology emerged as a promising approach to overcome these limitations and as one potential solution to the challenge of organ fabrication, to obtain very similar 3D human tissues, not only for transplantation, but also for drug discovery, disease research and to decrease the usage of animals in laboratory experimentation. 3D bioprinting allowed the fabrication of 3D alive structures with higher and controllable cell density and homogeneity. Other advantage of biofabrication is that the tissue constructs are solid scaffold-free. This paper presents the 3D bioprinting technology; equipment development, stages and components of a complex Organ Bioprinting Line (OBL) and the importance of developing a Virtual OBL.

1. Introduction

3D bioprinting or organ printing can be defined as the use of material transfer processes for patterning and assembling biologically relevant materials (molecules, cells, tissues, and biodegradable biomaterials) with a prescribed organization to accomplish one or more biological functions [1].

Since the inception of organ printing concept as a computer-aided, jet-based 3D tissue-engineering of living human organs [2], this technique has undergone in a progressive development [3-5]. It gradually gained recognition as a reasonable bottom-up solid scaffold-free alternative to the traditional scaffold fabrication approach in tissue engineering (TE), [6].

3D bioprinting has already been used for the generation of several tissues, including multilayered skin, bone, vascular grafts, tracheal splints, heart tissue and cartilaginous structures. Applications of these constructs include the developing of high-throughput 3D bioprinted tissue models for research, drug discovery and toxicology [7]. The field is moving forward, there is a plethora of research being done on bioprinting technology and its potential as a source for tissue grafts and full organ transplants [8]. The increasing number and broadening geography of participants, the emergence of new exciting bioprinting technologies, and the attraction of young investigators indicates the strong growth potential of this emerging field [9].

Research groups in several countries as China, United State, South Korea, The Netherlands and Singapore, besides others, have been investing strongly in the bioprinting area. Singapore, for example, created the Singapore Centre for 3D Printing at the NTU (Nanyang Technological University), which is applying 3D printing for living solutions, among other applications [10]. Researchers from the Wyss Institute for Biologically Inspired Engineering at Harvard University are searching in bioprinting area, for example, they have enabled the creation of tissue constructs with small blood vessels and multiple cell types, marking important progress toward the printing of living tissues [11]. The University of California, San Diego, has the Chen Laboratory, were the Chen group is interested in developing 3D bioprinting techniques with a micro or nanoscale printing resolution [12]. Concerning Brazil, in 2008 the Brazilian Institute of Biofabrication (INCT-BIOFABRIS) was created by CNPq (National Counsel of Technological and Scientific Development) with the affiliation of many research groups, including UNICAMP (State University of Campinas), USP (University of Sao Paulo), INT (National Technological Institute), UFRGS (Federal University of Rio Grande do Sul) and CTI (Information Technology Center Renato Archer). CTI created, among other groups, the bioprinting research group that develop professional workforce in this emerging field of medical science and provides strong impulse for the ascension of Brazil as a recognized player in bioprinting research.

2. Developments in 3D bioprinting area

There are several companies producing 3D tissues, biomaterial constructs and developing research in the bioprinting area, as shown in table 1.

Table 1. Companies that produce 3D tissues and biomaterials constructs related to 3D bioprinting.

Company/Location	Product
Insphero/Schlieren, Switzerland	3D microtissue models: liver, pancreatic, tumor. 3D cell culture products and assay kits.
3D BioMatrix/Michigan, USA	Perfecta3D Hanging Drop Plates
Rainbow Biosciences and 3D Nano Biosciences/Florida & Texas, USA	BiO Assay, spheroid formation magnetic levitation printer system.
ACEA Biosciences/San Diego, USA	RTCA plates, Impedance-based systems for cell-based assays.
n3D Biosciences/Houston, USA	Plates and Bioprinting Kit for magnetized spheroids.

The production and commercialization of 3D bioprinting technology and organ bioprinting is already an ongoing process. There are several companies producing 3D bioprinters, some of them are: EnvisionTEC, Germany; RegenHu, Switzerland; Sciperio/nScript, USA; Organovo, USA; 3D Bioprinting Solutions Laboratory, Russia; Cyfuse Biomedical K.K., Japan; Bio 3D Technologies, Singapore; GeSiM, Germany; OxSyBio, UK; Aspect Biosystems, Canada; DigiLab, USA, Advanced Solutions Life Sciences, USA; MicroFab Technologies, USA; Seraph Robotics, USA [13].

Particularly, NovoGen MMX Bioprinter (Organovo) has two robotically controlled precision print heads: one for placing the bioink, material containing thousands of cells, other for placing hydrogel scaffolds, or a support matrix (figure 1. a) [14]. Regenova Bioprinter (Cyfuse Biomedical K.K.) has a distinct technique for obtaining 3D tissue constructs; it assembles spheroids containing cell aggregates into a three-dimensional shape by placing them in needle arrays according to the pre-designed 3D data (figure 1. b) [15]. 3D Bioprinting Solutions Laboratory has the Fabion Bioprinter, which is supplied with its own software and has a regular cartesian design allowing it to move in all directions. It also has five nozzles that can pump spheroids and extrude hydrogel and biogel in successive layers with a very high resolution (figure 1. c) [16]. BioFactory (RegenHu) is a high end, versatile and cell friendly three dimensional bio-manufacturing device that allows researchers to pattern cells, biomolecules and a range of soft and rigid materials in desirable 3D composite structures, mimicking tissue models (figure 1. d) [17].

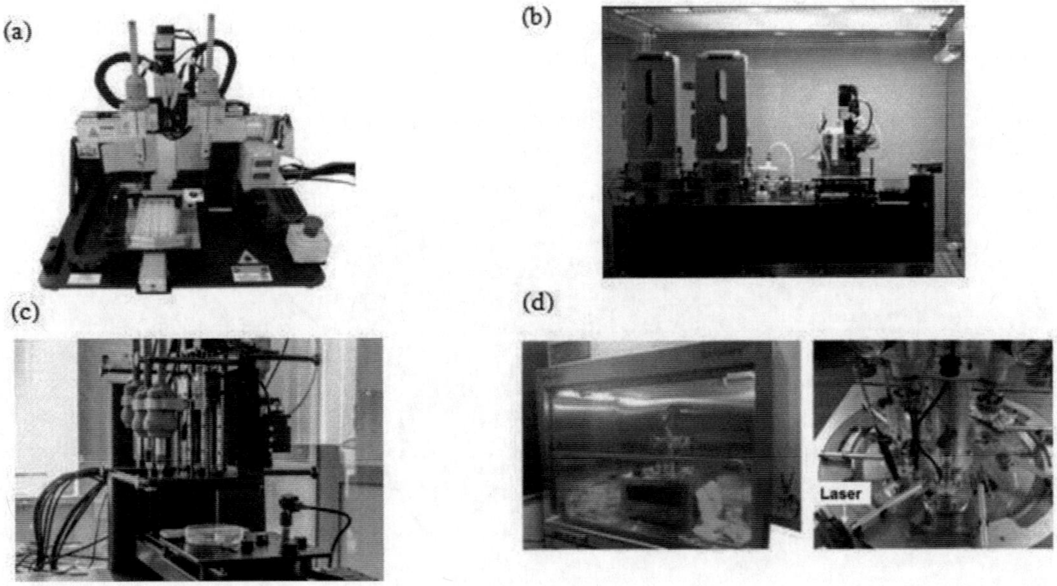

Figure 1. (a) NovoGen MMX bioprinter (Organovo, CA, USA). (b) Regenova bioprinter (Cyfuse Biomedical, Tokio, Japan). (c) Fabion bioprinter (3D Bioprinting Solutions Laboratory, Moscow, Russia). (d) BioFactory bioprinter (RegenHU, Switzerland).

3. Organ Biofabrication Line

Nowadays there are many researches and developments on robotic bioprinters, focusing in precision placement of spheroids, new biomaterials, cell-laden biomaterials and rapid prototyping technologies for 3D tissue-like structures with a resulting high density [8]. From an engineering point of view, the successful clinical translation of organ printing technology will depend not only on the development of a single tool or device, like a robotic bioprinter, but rather on the entire complex of related technologies and their seamless integration, i.e., an Organ Biofabrication Line (OBL) [18]. The design and development of a fully integrated OBL, or development of a series of integrated automated robotic tools, is imperative for the commercial translation of the organ printing technology [18]. Figure 2 shows a virtual design of an OBL that was designed at Three-dimensional Technologies Division (DT3D) – CTI [19].

The additive manufacturing OBL requires the development or employing of special equipment: clinical cell sorter, stem cell propagation bioreactor, cell differentiator, tissue spheroid bio-fabricator, tissue spheroids encapsulator, robotic bioprinter, and perfusion bioreactor [19].

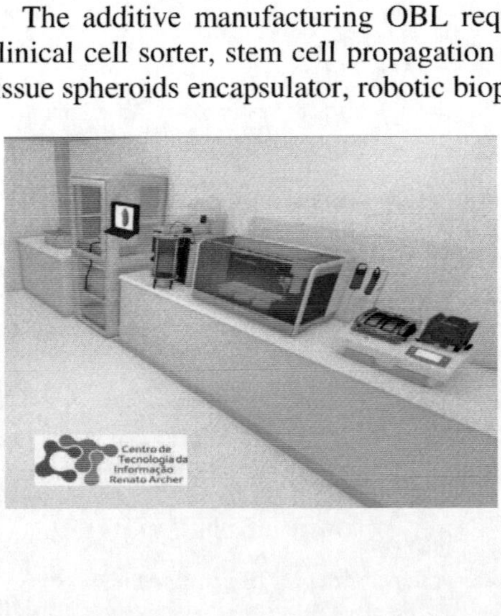

Figure 2. Organ Biofabrication Line (OBL). Virtual manufacturing is an important technological advance to predict the development of complex products from many areas, as it is a human organ [19].

The development of this integrated line of automated robotic tools at industrial scale requires the interaction of three principal research fields: Engineering, Biology and Material Science (figure 3), including a complex multidisciplinary approach and close research and development collaboration of mechanical engineers, experts in rapid prototyping technology, computers scientists, biomedical engineers, chemical engineers, material scientists with biologists and tissue engineers [20].

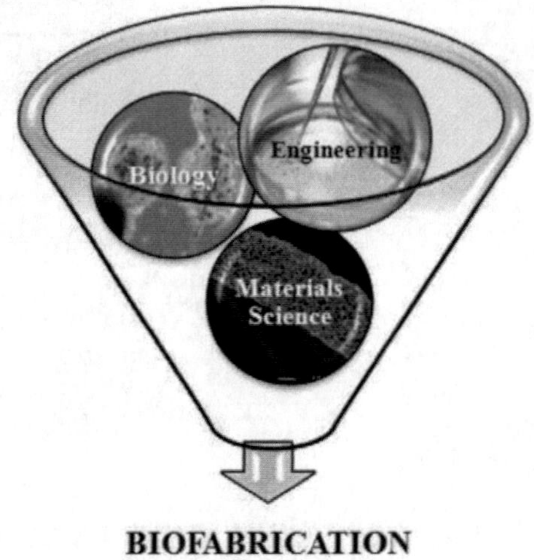

Figure 3. Research fields of the biofabrication area. Adapted from [20].

3.1 Bioprinting stages

Bioprinting is normally splitted into three parts: pre-processing, processing and post-processing, as can be seen in figure 4 [18].

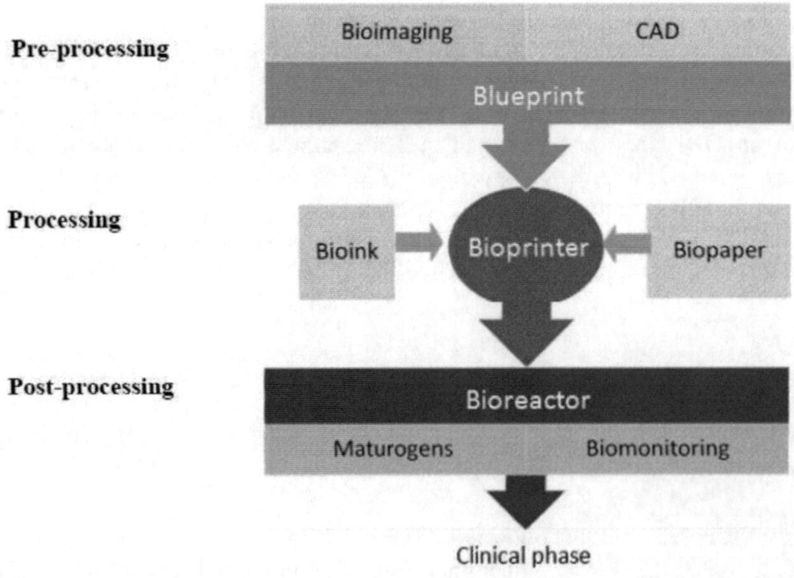

Figure 4. Three main steps in organ printing technology. Adapted from [18].

3.1.1 Pre-processing. It is the design stage. It involves the development of computer-aided design (CAD) or a "blueprint" of a 3D human tissue or organ, based on the use of clinical images and special additive manufacturing. This stage relates to the transformation of virtual reality into physical reality. Magnetic resonance imaging (MRI), computed tomography (CT) and micro-CT are already approaching the desirable level of resolution. Computer-aided reconstructions of human microanatomy and serial histological sections are highly necessary to determinate specific depositions of tissue spheroids and to obtain precise microstructures [21].

InVesalius is an example of a 3D anatomy reconstruction software, developed in CTI (http://svn.softwarepublico.gov.br/trac/invesalius). It uses CT or MRI data for the virtual representation of 3D human body parts (figure 5. a). Another method to develop a blueprint of tissue structures has been developed by the company Uformia from Norway (http://www.uformia.no), a CTI's partner, using mathematical functions to generate a digital representation (figure 5. b).

(a)

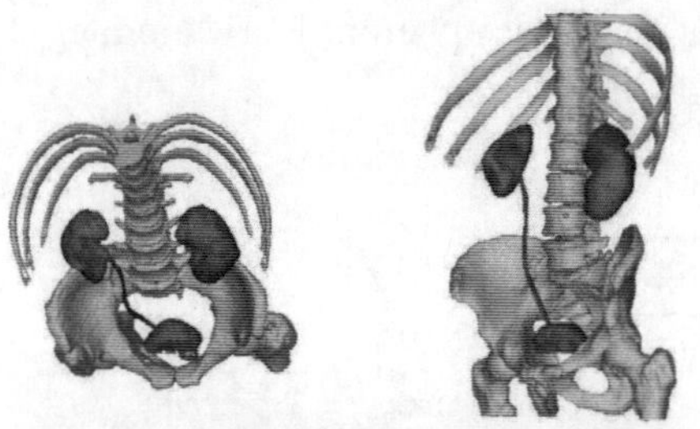

(b)

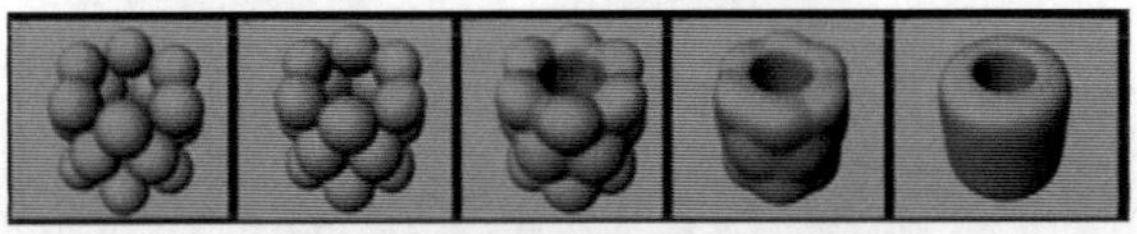

Figure 5. Blueprint examples. (a) Virtual representation of kidneys and a part of skeletal human structure (InVesalius public software for 3D medical imaging manipulation). (b) Generation of a digital representation of tissue spheroids fusion. (Courtesy of Uformia, Norway).

Mathematical modeling of anatomical and histological structures for the design of a blueprint is a powerful tool on the pre-processing stage. One of the most important principles of developmental biology, which inspired numerous discrete cell models, is the differential adhesion hypothesis (DAH). This principle leads to a close analogy between true liquids and living tissues made of adhesive and motile cells, such as most embryonic and some artificial tissues [22]. Also, there is the Monte Carlo Method, a convenient way for studying energetically driven conformational changes of a system. It is a large collection of computational algorithms that involves the usage of random numbers [23]. This and other methods are used in computational modeling, supporting the design of several software to create simulations of biocomplexity problems. Some examples are: CompuCell3D (http:// www.compucell3d.org), CellSys (http//ms.izbi.uni-leipzig.de/index.php/software/cellsys2010) and TSim - Tissue Simulator (http://www.tsimsoftware.com).

The required clinical image information is not a problem to the creation of CAD human organs, the challenge is the way to transform this accumulated knowledge of human anatomy and histology into a viable blueprint with necessary and sufficient instructions for a robotic printer to be able to print a human organ [24].

3.1.2 Processing. It is the production stage. The computer-aided robotic bioprinting process includes the preparation of a "bioink" or self-assembled tissue spheroids, development of a "biopaper" or processable and biocompatible hydrogel and the usage of a "robotic bioprinter" or computer controlled robotic precise dispenser [21].

The bioink is a material made from living cells that behaves much like a liquid; the bioprinter will dispense it, in order to create the desired shape [25]. The search of bioinks for usage in tissue printing applications should have a good biocompatibility and tailored viscosity transitions for effective printing and continuous handling. For an extrusion-based process, the bioink should show shear thinning to allow extrusion through a needle and also immediate cessation of flow, upon deposition on the substrate, to retain its plotted shape. Such material would allow the fabrication of complex structures with high resolution and the final mechanical properties of the printed construct should also be sufficient for manipulation and transportation to the next stage [26, 27].

Simple calculations based on the size and volume of tissue constructs and human organs indicates that in order to bioprint a human organ of desirable size, it will be necessary to develop technology for a scalable biofabrication of millions of tissue spheroids. Thus, scalable tissue spheroid biofabrication must be considered as one of critically important technologies enabling organ printing. There are many different technologies for tissue spheroid fabrication, eg. hanging drop method, use of external forces, microarrays, etc (figure 6) [27].

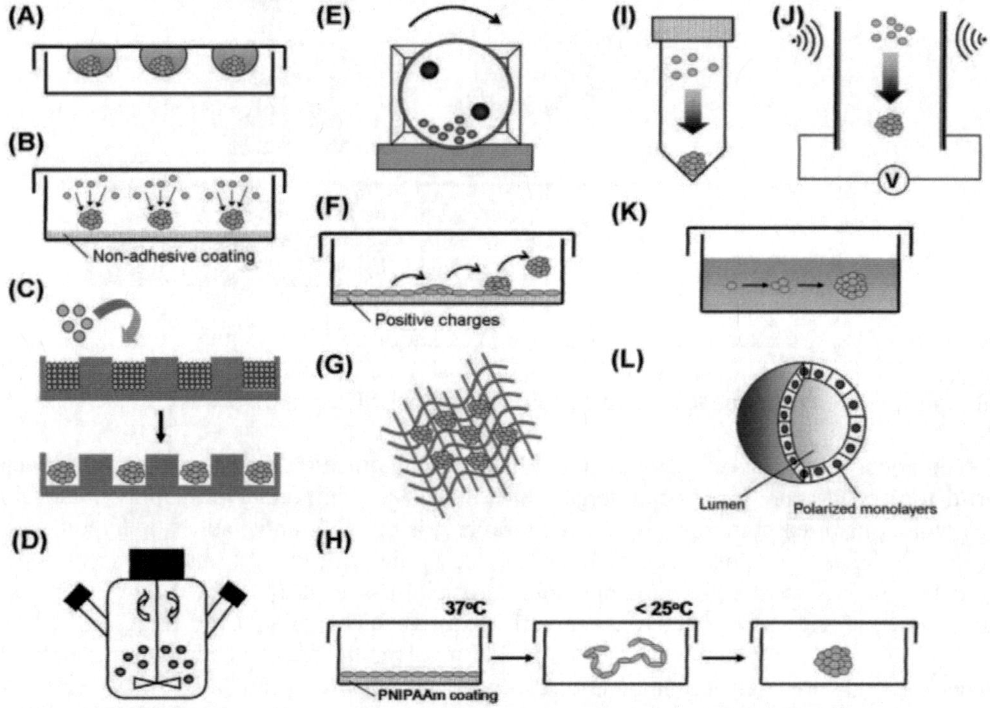

Figure 6. Spheroids fabrication methods [27]. Methods for multicellular spheroids generation. (A) Hanging-drop culture. (B) Single cell culture on nonadhesive surface. (C) Micromolding techniques. (D) Spinner flask culture. (E) Rotary cell culture systems. (F) Hepatocyte self-assembly on Primaria dishes. (G) Porous 3-D scaffolds. (H) The use of Poly(N-isopropylacrylamide)-based cell sheets. (I) Centrifugation pellet culture. (J) Electric, magnetic or acoustic force cell aggregation enhancement. (K) Monoclonal growth of tumor spheroids. (L) Polarized epithelial cysts.

In the research [28], a comparison of 600 tissue spheroids fabricated by the micromolded method with 600 spheroid generated by the conventional hanging drop method was developed. The analyses showed that the use of micromolded recessions in a non-adhesive hydrogel, combined with automated cell seeding, is a reliable method for scalable robotic fabrication of uniform-sized tissue spheroids, but uniform-sized is one of several issues to be considered [28]. To choose the best method for scalable tissue spheroid fabrication there are several aspects to be evaluated: well-controlled spheroid size, spheroid velocity formation, co-culture of different cell types, easy to scale up, good preservation of viability, rapid cell aggregation stimuli, high level of cell density, etc. An important aspect is that spheroids have to be capable of being placed in special reservoirs suitable for be used as cartridges. Taking into account all these factors, it must be chosen the most appropriate method and make adaptations to obtain an efficient automated process for scalable tissue spheroid fabrication [27, 28].
After obtaining the microtissues with a high cell density level, it is important the encapsulation of these structures. In this situation, encapsulate tissue spheroids is a way to prevent preliminary and undesirable tissue spheroid fusion before the effective printing. Beside this, the encapsulation process turns them into reinforced structures, protecting the internal live environment. Biomaterials or hydrogels used should provide a lubrication effect for the deposition stage and be permissible for tissue fusion during the post-processing [29].

Once spheroids are obtained and encapsulated, they have to be transported and loaded into adequate "biocartridges" [30]. Then, using blueprint information, the bioprinter will deposit the bioink in well-defined topological patterns into biopaper sheets [31]. In the next step, the construct will be transferred into a bioreactor and spheroids will fuse. The biopaper, a hydrogel, can be removed after construction if required; in this context, it only provides a temporary support for the deposited bioink particles. Consequently, the biopaper is clearly different from solid scaffolds used in classical scaffold-based tissue engineering and in most cases it is desired to be biodegradable, to remove it more easily [32].

3.1.3 Post-processing: The bioprinted outcome is not a functional tissue construct immediately suitable for human implantation [19]. To transform the 3D tissues and organs into functional structures they must undergo accelerated tissue maturation inside a bioreactor [8]. Being a bioreactor an essential component of classic tissue engineering, they were initially used as tools for enhancing cell seeding on solid scaffolds and for providing mechanical conditioning of tubular tissue engineered constructs. The bioreactor is also a container for keeping tissue constructs in a wet and controlled environment and operating conditions (pH, temperature, gas concentrations, nutrient supply and waste removal, physical stimuli supply), and thus maintaining their viability [24]. Those chemical and physical factors that enable accelerated tissue maturation, can be defined as maturogenic factors or simply maturogens. Also, the bioreactor serves as a packing and transportation device for mature tissue engineered constructs. Computational simulations and in situ assays are crucial in this stage. Computational fluid dynamic software packets have been increasingly developed during the past decade and are powerful tools to calculate flow fields, shear stresses and mass transport within and around 3D constructs, including the bioreactor environment [32].

Post-processing is probably the most crucial step in organ printing procedures. An effective tissue maturation will require the development of new bioreactors. Further discovery and experimental validation of potential maturogenic factors are important tasks in the developing of equipment for acceleration of tissue maturation [18].

Strategies for non-invasive and non-destructive biomonitoring, of bioprinted tissue maturation, are still not properly addressed questions in the organ printing technique. It is safe to predict that with a progress in the development of bioprinting equipment, the role of non-destructive and non-invasive methods for estimation of tissue maturation and functionality will only increase. Without the development of effective methods of pre-implantation and post-implantation "quality control" of bioprinted organs functionality, maturation, and integrity, their approval for clinical use and implantation by regulatory agencies will be practically impossible [21].

The research [33] presented three perfusion circuits in a type of a perfusion bioreactor (figure 7): one perfusion system provides wet environment around the printed constructs; second perfusion system is designed for intravascular perfusion of maturated build in vascular tree; and, finally, the third perfusion circuit is designed for enabling the temporal interstitial flow through removable temporal porous minitubes (needles). These removable porous tubes also provide temporal support and serve as some sort of non-biodegradable but removable supporting structure. The distance between these tubes as well as their porosity, must be designed based on mathematical modeling and computer simulation [33].

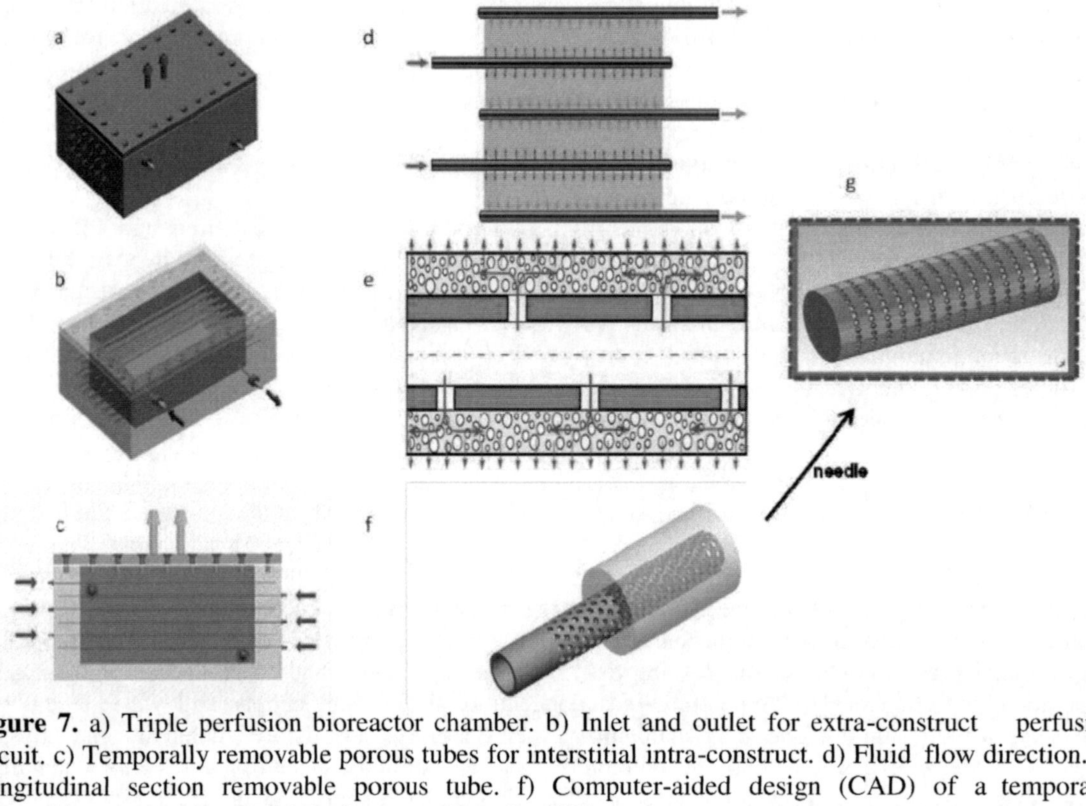

Figure 7. a) Triple perfusion bioreactor chamber. b) Inlet and outlet for extra-construct perfusion circuit. c) Temporally removable porous tubes for interstitial intra-construct. d) Fluid flow direction. e) Longitudinal section removable porous tube. f) Computer-aided design (CAD) of a temporally removable porous tube. g) CAD of a general needle. [33]

4. Conclusions and future perspectives of an OBL

Organ printing is a rapidly emerging technology that promises to transform tissue engineering into a commercially successful biomedical industry. It is an interdisciplinary field involving many science areas, among biology, computer, physics, materials and engineering. This computer-aided layer-by-layer additive biofabrication uses aggregates of thousands pre-sorted cells, microtissues called spheroids, as building blocks for originating a 3D bioprinted living structure.

It is obvious that organ printing cannot be only produced with just a robotic bioprinter, it would be like assuming that intrinsic biological structures fabrication is a very simple process, whereas the reality is quite the opposite. It requires a set of specific equipments and a controlled aseptic area. A future OBL has to be specifically well designed and studied, being that the reason why design and computer simulation of a previous virtual biofabrication line will simplify and improve the process. It must combine all possible visual information of machines, devices, human resources and processes, as well as it must include *in silico* assays about tissue spheroids formation, spheroids fusion, cell viability, tissue and organ development in each corresponding stage. Virtual OBL must integrate essential tools like CAD, computer simulation and modelling, complex mathematical modelling, new specific software for biological simulations, special equipments and information technologies in general.

5. Acknowledgements

To São Paulo Research Foundation (FAPESP), Brazil; The Nacional Council for Scientific Technological Development (CNPq), Brazil, and The Brazilian Institute of Biofabrication (INCT-Biofabris).

6. References

[1] Mironov V A, Reis N and Derby B. 2006. Bioprinting: A Begining. *Tissue Engineering*, **12**, Number 4, pp. 631.

[2] Mironov V A, Boland T, Trusk T, Forgacs G and Markwald R. 2003. Organ printing: computer-aided jet-based 3D tissue engineering. *Trends Biotechnol.*, **21**, pp. 157–161.

[3] Boland T, Xu T, Damon B and Cui X. 2006. Application of inkjet printing to tissue engineering. *Biotechnol J*, **1**:910-917

[4] Campbell PG and Weiss LE. 2007. Tissue engineering with the aid of inkjet printers. *Expert Opin Biol Ther.*,**7**:1123-1127

[5] Jakab K, Norotte C, Marga F, Murphy K, Vunjak-Novakovic G and Forgacs G. 2010. Tissue engineering by selfassembly and bio-printing of living cells. *Biofabrication*, **2**, 022001.

[6] Nichol J W and Khademhosseini A. 2009. Modular tissue engineering: engineering biological tissues from the bottom up. *Soft Matter,* **5**:1312-1319.

[7] Murphy S V and Atala A. 2014. 3D bioprinting of tissues and organs. *Nature Biotechnology* **32**, pp. 773–785.

[8] Dababneh A B and Ozbolat I T. 2014. Bioprinting Technology: A Current State-of-the-Art Review. *ASME J. Manuf. Sci. Eng.*; 136(6): 061016.

[9] Guillemot F, Mironov V A and Nakamura M. 2010. Bioprinting is coming of age: report from the International Conference on Bioprinting and Biofabrication in Bordeaux (3B'09). *Biofabrication*, **2**, p. 010201.

[10] Singapore Centre for 3D Printing - Nanyang Technological University. Singapore. Site: http://sc3dp.ntu.edu.sg/Pages/Home.aspx. Accessed in June 2015.

[11] Kolesky D B, Truby R L, Gladman A S, Busbee T A, Homan K A and Lewis J A. 2014. 3D Bioprinting of Vascularized, Heterogeneous Cell-Laden Tissue Constructs. *Adv. Mater.*, **26**, pp. 3124–3130.

[12]Chen Laboratory. Nanobiomaterials, Bioprinting and Tissue Engineering. Site: http://schen.ucsd.edu/lab/. Accessed in June 2015.

[13] Hockaday L. 3D Bioprinting, a deliverate bussines. Genetic Engineering & Biotechnology News. http://www.genengnews.com. January 1, 2015. Pp. 14-23.

[14] Robbins J B, Gorgen V, Min P, Shepherd B R and Presnell S C. 2013. A novel in vitro three-dimensional bioprinted liver tissue system for drug development. *The FASEB Journal.* **27**:872.12

[15] Cyfuse Biomedical K.K. University of Tokyo Entrepreneur Plaza. Tokyo, Japan. http://www.cyfusebio.com. Accessed in May/2015.

[16] 3D Bioprinting Solutions. Laboratory for biotechnological research. Moscow, Russia. http://www.bioprinting.ru. Accessed in May/2015.

[17] RegenHU Ltd. Villaz-St-Pierre, Switzerland. http://www.regenhu.com. Acceced in May/2015.

[18] Mironov V A, Kasyanov V and Markwald R R. 2011. Organ printing: from bioprinter to organ biofabrication line. *Current Opinion in Biotechnology*, **22**, pp. 1-7.

[19] Mironov V A, Rezende R A, Kasyanov and Da Silva JVL. 2013. Virtual Biofabrication Line. In: 6th IFAC Conf. on Management and Control of Production and Logistics. vol 6, pp. 289-294.

[20] Mironov V A, Trusk T, Kasyanov V, Little S, Swaja S and Markwald R. 2009a. Biofabrication: a 21st century manufacturing paradigm. *Biofabrication*, **1**, Num. 2, 022001.

[21] Rezende R A, Pereira F, Kemmoku D, Mironov V A, Kasyanov V, Vilbrandt T, da Silva J V L. 2012. *Innovative developments in virtual and physical prototyping.* Proc. of the 5th Int. Conf. on advanced research and Rapid Protot., Leiria, Portugal, 28 Sept-1Oct., 2011. EnablingTechnologies for Robotic Organ Printing. Bártolo *et al.* (eds) , published by Taylor & Francis. pp. 121-129.

[22] Kosztin I, Jakab K, Barz B, Neagu M, Jamison R and Forgacs G. 2006. Computational Modeling of tissue self-assembly. *Mod. Phys. Lett. B* **20**, 1217.

[23] Sun Y, Yang X and Wang Q. 2014. In-silico analysis on biofabricating vascular networks using kinetic Monte Carlo simulations. *Biofabrication* **6**(1):015008.

[24] Rezende R A, Mironova V A and da Silva, J V L. 2011. Modeling on the Three-dimensional Printing of Human Organs. *Mecánica Computacional* Vol XXX, . O Möller, J W Signorelli and M A Storti (Eds.), Rosario, Argentina, pp. 3449-3463

[25] Mironov V A, Visconti R, Kasyanov V, Forgacs G, Drake C and Markwald R. 2009b. Organ printing: tissue spheroids as building blocks. *Biomaterials*, **30**, pp. 2164–2174.

[26] Kesti M, Müller M, Becher J, Schnabelrauch M, D'Este M, Eglin D and Zenobi-Wong M. 2015. A versatile bioink for three-dimensional printing of cellular scaffolds based on thermally and photo-triggered tandem gelation. *Acta Biomater*;**11**:162-72.

[27] Lin R Z and Chang H Y. 2008. Recent advances in three-dimensional multicellular spheroid culture for biomedical research. *Biotechnology*, **3**, 1172-1184.

[28] Mehesz A N, Brown J, Hajdu Z, Beaver W, da Silva J V L, Visconti R P, Markwald R R and Mironov V A. 2011. Scalable robotic biofabrication of tissue spheroids. *Biofabrication*; **3**(2):025002.

[29] Rezende R A, Pereira F D A S, Kasyanov V, Kemmoku D T, Maia I, Da Silva J V L and Mironov V A. 2013a. Scalable Biofabrication of Tissue Spheroids for Organ Printing. *Procedia CIRP* **5**, pp. 276–281.

[30] Jakab K, Damon B, Neagu A, Kachurin A, Forgacs G. 2006. Three-dimensional tissue constructs built by bioprinting.. *Biorheology*; **43**(3-4):509-13.

[31] Billiet T, Vandenhaute M, Schelfhout J, Vlierberghe S V and Dubrue P. 2012. A review of trends and limitations in hydrogel-rapid prototyping for tissue engineering. *Biomaterials*, **33**, pp. 6020-6041.

[32] Rezende R A, Laureti C A R, J V L da Silva, Mironov V A, V Kasyanov and R M Filho. 2012. Innovative developments in virtual and physical prototyping. Proc. of the 5th Int. Conf. on advanced research and Rapid Protot., Leiria, Portugal, 28 Sept-1Oct., 2011. Towards simulation of a bioreactor environment for biofabricated tissue maturation . Bártolo et al. (eds) , published by Taylor & Francis, pp. 691-698.

[33] Rezende R A, Nogueira J, Lara V, Mironov V A, Maciel Filho R, Da Silva J V L. 2013b. Development of a Bioreactor by Computer Fluid Dynamics Simulations for the Maturation of 3D Printed Organs by Rapid Prototyping. *Chemical Engineering Transactions*, **32**, pp. 1153-1158.

Obtaining liver tridimensional scaffold through the decellularization of rabbit whole liver in 24 hours

Schliamser Federico[1]; Rinaldi Ayelen[1]; Comin Romina[1,2], Borchert Alba Nelly[5] Nari Gustavo Adrian[3,4], Salvatierra Nancy Alicia[1,2], Cid Mariana Paula[1,2]

1- Dpto. de Química, Ingeniería Biomédica, Facultad de Ciencias Exactas Físicas y Naturales - Universidad Nacional de Córdoba.

2- IIBYT (CONICET-UNC)

3- Servicio de Cirugía Hospital Florencio Díaz

4- Cátedra de Cirugía I UHC 4 - UNC.FCEFyN-UNC; IIByT (CONICET-UNC);Type the author addresses here

5- H.U.M.N-Facultad de Ciencias Médicas.

cidmarianapaula@gmail.com

Abstract. In the present work, we development a new protocol for liver decellularization in which the hole decellularization was reached over 24 h.
Introduction: the availability of transplantable livers is not sufficient to fulfill the current demand for grafts, with the search for therapeutic alternatives having generated different lines of research, one of which is the use of decellularized three-dimensional biological matrices and subsequent cell seeding to obtain a functional organ.
Objective: to produce a decellularization protocol from rabbit liver to generate a three-dimensional matrixin which the time period involved didn't pass 24 h.
Methods: The decellularization is obtained through the use of water and SDS (0,1-0,3 %), after freezing at -80 degrees, is the best alternative of different physical and/or chemical mechanisms to break down organ cells and leave only the extracellular matriz. After 24 h of retrograde perfusion, a decellularized translucent matrix was generated. To evaluate if the decellularization protocol was successful, with the extracellular matrix being preserved, we carried out histological (light microscopy) and biochemical (DNA quantification) studies.
Results: the decellularization process was verified by macroscopic observation of the organ using microscopic observation corroborated the macroscopic results, with the hematoxylin-eosin and Masson staining showing no cells or nuclear material. In addition, the DNA quantification was less than 10 % in the decellularized liver compared to control. Finally,the time taken to develop the decellularization protocol was less than 24 hours.

Introduction

Liver transplantation is currently the only therapeutic alternative for acute liver failure, terminal liver disease or metabolic disorders originated in the liver. The complexity of the function of the liver makes it impossible to use artificial systems to provide temporary hepatic support as for example in patients with renal insufficiency [1, 2]. Biologic scaffolds composed of extracellular matrix (ECM) are commonly used in reconstructive surgery and it is advancing its development for application in regenerative medicine to replace tissues and organs. The ECM represent the secreted products of resident cells of each tissue and organ, and provide the molecular keys necessary for cellular mechanisms of adhesion, migration and proliferation [3]. The ECM is currently used to create scaffolds of many tissues including skin [4], intestinal submucosa [5], urinary urethra [6], blood vessels [7] and heart valves [8], among others.

Recently the decellularization of organs such as heart [9], [10] and lung [11, 12], generated acellular natural three-dimensional scaffold which can be subsequently seeded with functional parenchymal cells or progenitor cells of these populations to generate a complete organ suitable for transplant.

The decellularization is obtained through the use of physical and/or chemical agents that disrupt by different mechanisms the organ cells leaving only the ECM, which then provide support to posterior cell seeding [13]. The techniques and materials used for this process are constantly being reviewed in order to achieve a greater cell lysis, which preserves the ECM and decreases the time required to carry it out. To date, liver decellularization protocols have involved perfusion of a combination of detergents: Triton X-100 and SDS were used in the portal or caval vein. Shupe and his colleagues [14] used increasing concentrations of Triton X-100 followed by SDS 0.1 % in a rat liver, while Uygun et al. [15] perfused 0.1 % SDS. Moreover, Soto-Gutierrez et al. [16] first used an enzymatic agent: trypsin and then 3 % Triton X-100. In our group, we were able decellularize one rabbit liver in approximately 72 hours with a combination of physical (freeze-thaw) and chemicals agents [17]. The new protocol described in the present investigation involved an initial treatment with trypsin, which allowed a better detergent transference to the tissue, then perfusion was carried out with 1% SDS and finally with Triton X-100.

In this work the decellularization technique previously published by our group [17] was improved. He recently began to discuss the process of decellularization; this is because the proper preservation of the matrix and its conservation are essential to prevent failure of the recellularizationsubsequent step [18, 19]. While the protocol described [17] was verified by optical microscopy and scanning electron microscopy, vasculature, biliary ducts and Glisson capsule preserved; it was considered necessary to reduce the time required for complete organ decellularization.

Materials and methods

2.1. Animals: Livers of male New Zealand rabbits were used 2,100 ± 120 g.
Total hepatectomy was performed with preservation of the vasculature. Subsequently, the extracted livers are decellularized by the following protocol. As a control were used rabbit livers which not were decellularized. All procedures were conducted in accordance with the NIH Guide for the Care and Use of Laboratory Animals, as approved by the Animal Care and Use Committee of the Universidad Nacional de Córdoba, Argentina, and efforts were made to minimize animal suffering and the number of animals used.

2.2. Surgical technique**:** Anesthesia was performed by injecting with 35 mg/kg ketamine and 5 mg/kg Xylazine intramuscularly (IM) as asingle dose.The rabbits were placed in the supine position with their legs fixed. Was performed median incision which was later extended to the right hemithorax. Mesenteric vein was identified and immediately 1000 IU of heparin was injected intravenously. Identification of the portal triad was performed andattached by a ligature, identification and dissection of the infra-hepatic inferior caval vein was carried out. The adhesions of the liver were removed and the incision was extending into the thorax. Identification, dissection, ligature andcutting of the supra-hepatic inferior caval vein were performed along with a total hepatectomy. In order to start with the protocol decellularization it was introduced a catheter or Teflon cannula in the suprahepatic or infra-hepatic inferior caval vein respectively, which was fixed with a prolene 4/0 stitch. It was perfused using a syringe with a PBS wash solution and then perfused with a peristaltic pump (15-20 ml / min) for about three hours with deionized water; subsequently it is frozen at -77 ° C.

2.3. Decellularization protocol: as already described after hepatectomy, the liver be infused for about three hours with deionized water and then frozen at -77 ° C. After 24 h at -77 ° C the liver was thawed at room temperature to aid cell lysis. The catheter or teflon cannula inserted into the inferior caval vein was connected to a peristaltic pump and retrograde perfusion cycles was started at a rate of between 15 and 20 ml/min at 4°C as follows: it perfused with 0.3% SDS in deionized water, alternating with deionized water. Once the process of decellularization (which is checked macroscopically) completed, successive washes were performed with deionized water and PBS to drag cell debris and detergent (1h of deionized water, 2 washes 1 h with PBS, 30 min of deionized water, 2 washes of 30 min with PBS). Later, the organ was perfused for 1 hour with 4 % ethanol in PBS to disinfect the organ and finally, 1h perfused with 10,000 units / ml penicillin-10,000 ug / ml streptomycin in PBS. Macroscopically decellularization evidenced by a MEC translucent.Staining of the biliary tree on two rabbits was performedwith methylene blue through the gallbladder puncture,which revealed the existence of the bile ducts. A mixture of methylene blue with agarose was perfusedthrough the inferior caval vein in two other rabbitsand a good vascular staining was observed.

2.4. Microscopic morphological evaluation: for evaluation of the effectiveness of the decellularization, the two histological staining techniques, hematoxylin-eosin (HE), and Masson's trichrome staining were used. These stainings were performed on the decellularized and control livers.

2.5. DNA quantification: The amount of DNA was quantified using a method described by Laird et al. [20]. Briefly, 25 mg of tissue from the control and decellularized liver were incubated with a solution containing 2 % SDS, 5 mM EDTA, 200 mMNaCl and 100 mM TRIS-HCl, pH 8.5 for 48 h at 55 °C. The DNA extraction was carried out in isopropanol and later dissolved in a solution of 10 mMTris- HCl, 0.1 mM EDTA, pH 7.5. The amount of DNA was determined spectrophotometrically at 260 nm.

1. Results and discussion

The objective of this work was reduce the time to obtain a rabbit decellularized liver [17, 19]. We assayed different chemical combinations, different detergent concentrations and other chemical agents. Also it was changed the order that these agents were used. It was established that concentrations of SDS bellow 0.3% were the ones which obtained best results. On the other hand, for good detergent penetration, it was necessary (this is the step was shown to be determinant), perfusing the fresh organ with cold water before being frozen. The water was perfused until the liquid perfused stopped go out colored. This perfusion was fundamental so then the detergents can penetrate the organ. Generally, the water was circulated for a period of about three hours. It was evident that is complicated establish duration time and perfusion volume, each organ has different weight, morphology and anatomy. In some cases, we can observe internal striations in the lobes that difficult the decellularization because they produced an asymmetric perfusion of the detergents. Finally, the organ was perfused while submerged. We tried organ perfusion with a container mesh, suspending and submerging the organ; also alternating suspension, submerging and the mesh. Finally, we can observe a better preservation of the elasticity and the macroscopic aspect of the MEC, when the organ was submerged during the decellularization and in the mesh during the washes. Frequently, the organ suspension causes more difficult for detergents working because when we suspend the organ, the perfusion diverts to the lobe that has more cells. But the suspension produces seriously damages to the vasculature and Glisson´s capsule, and this affects the reproducibility of the process and the protocol success. The decellularizated mesh of the organ is very frail, which obstructs to suture it in cases of vasculature breaking. Moreover, when MEC obtained with Nari et al. [17] protocol was incubated with epithelial cells to study the citotoxicity, we observed bacterial contamination, and also shelling of the MEC. Although they were preserved the tract biles and the tract blood; those structures did not support incubation at 37 °C during 72 hours. To try to improve the culture resistance at 37°C, we worked with at 4°C with different procedures of washes and we aggregated different antibiotics. With the improvements introduced in the protocol, we have reduced the decellularization time and increased the reproducibility and success of the protocol. In the next scheme, we show a summary of the new protocol of liver decellularization:

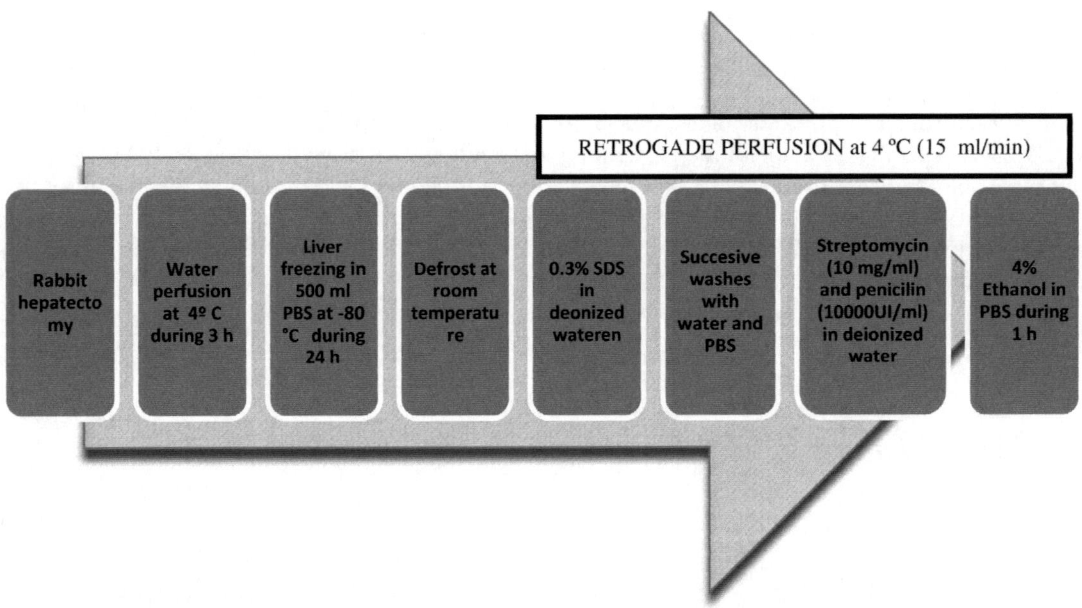

In the figure 1, we can observe the process of the liver descellularization, over the course of 24 hours while the organ is perfused. In the pictures, we can see that the MEC obtained is totally translucent and also we can see that the vasculature of the organ is preserved. The perfusion rate is 15-20 ml/min.

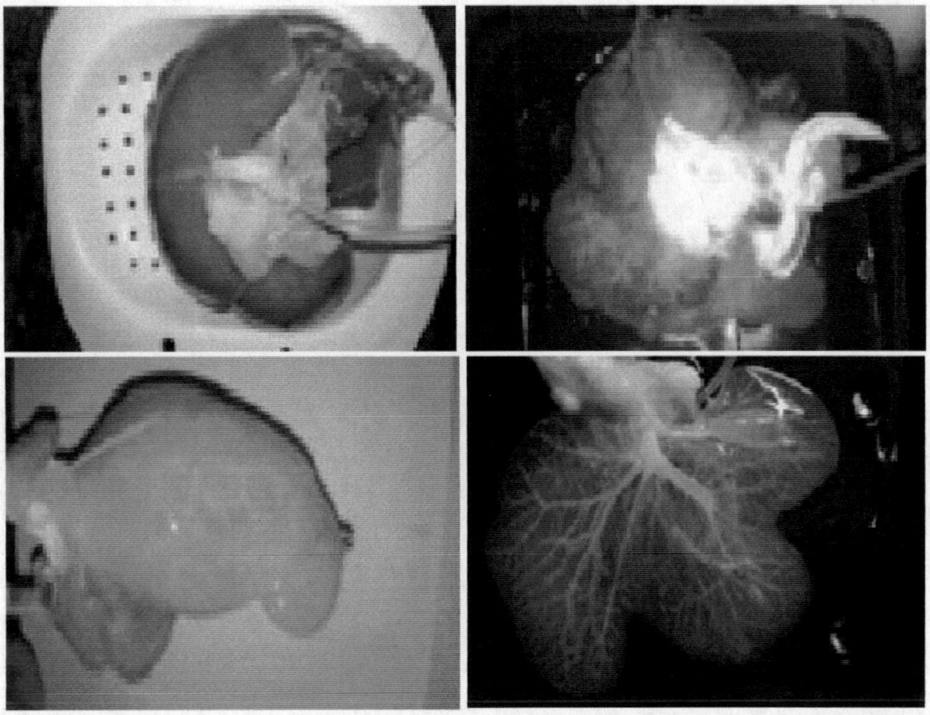

Figura 1:Macroscopic observation of decellularization procedure. A) Liver thawed (perfused with water, before freezing to -70°C); B) Liver perfused after 8 hours; C) y D) MEC obtained after 24 hours of perfusion.

Once the decellularization protocol is established, its efficiency was verified by microscopic techniques. For this, we take tissues cuts of control liver and of the MEC obtained and fixed it in glutaraldehyde at 4% in PBS, that then we process for perform the histologic stains of Hematoxylin-Eosin and Trichromic staining of Masson. In the figure 2, with the Hematoxylin-Eosin stain, we can observe de totally absence of cores or nuclear debris in the MEC. Washing is a critical step of the process of decellularization, because the heavier debris is more difficult to eliminate of the tissue. The microscopic techniques showed that there was not nuclear debris. In the figure 3, with the Masson trichromic staining, we verified the presence of collagen fibers and elastic fibers.This staining uses three dyes: hematoxylin, fuchsin and green light. This staining is very useful to show the collagen fibers and the connective tissue.

Journal of Physics: Conference Series **705** (2016) 012011 doi:10.1088/1742-6596/705/1/012011

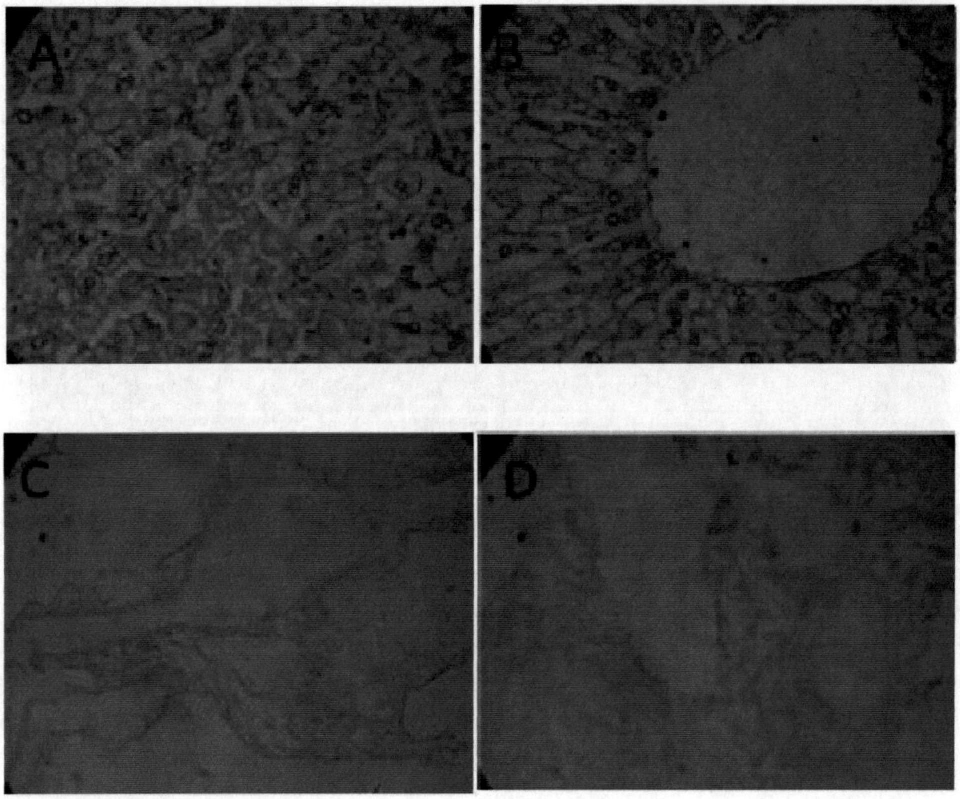

Figura 2: Hematoxylin-Eosin staining. A) and B) are histological sections of control´s livers. C) and D) are histological sections of MEC. The pictures are representative of respectives groups.(control: n=3, MEC: n=4).

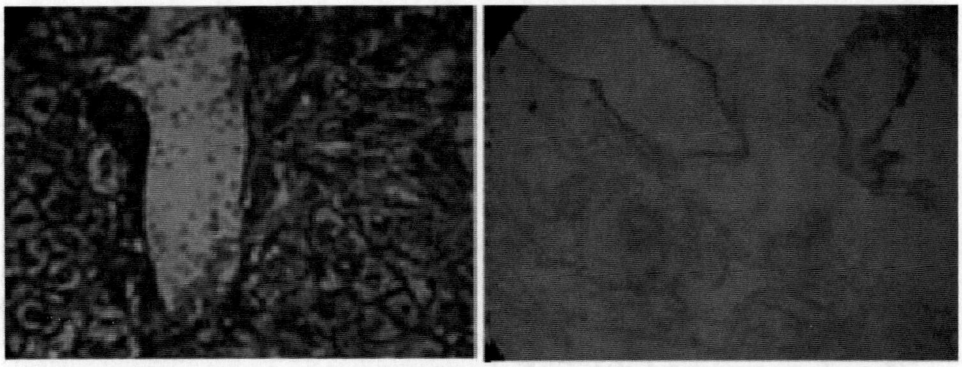

Figura 3: Masson trichromic staining of histological cuts. Left: correspond to control´s liver. Right: correspond to MEC. The pictures are representativeof the respective groups (Control: n=2; MEC:n=4).

On the other hand, DNA amount was 100 times lower (two orders of differences of magnitude) in the tissue from the MEC, than in the tissue from the control liver (Figure 4). These results are accord with the optical microscopic studies, where we showed the absence of cores and nuclear debris.

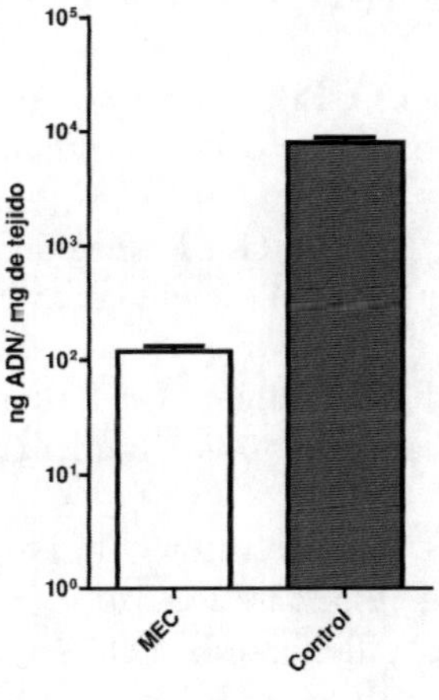

Figure 4: Quantification of DNA. Bars represent mean ± S.E.M.. Control Group: 10350 ± 1750 ng DNA/mg of tissue (n=3), MEC group: 264 ± 37ng DNA/mg of tissue(n=6).

Conclusion

The decellularization protocol allowed to obtain a preserved and translucent matrix without remains of cells and DNA after 24 hours. The protocol was developed at 4 ° C and involved an initial perfusion of water during 3 hours, freezing the organ to -77 ºC, liver defrost at ambience temperature and a perfusion of 24 hours with 0.3 % SDS intercalated with washing water at a rate of 15-20 ml / min.

References

[1] Carpentier B, Gautier A, Legallais C. Artificial and bioartificial liver devices: present and future. Gut.2009; 58:1690.

[2] Stutchfield BM, Simpson K, Wigmore SJ.Systematic review and meta-analysis of survival following extracorporeal liver support.The British journal of surgery.2011; 98:623–631.

[3] Vorotnikova E, McIntosh D, Dewilde A, Zhang J, Reing JE, Zhang L, Cordero K, Bedelbaeva K, Gourevitch D, Heber-Katz E, Badylak SF, Braunhut SJ. Extracellular matrix-derived products modulate endothelial and progenitor cell migration and proliferation in vitro and stimulate regenerative healing in vivo. Matrix biology: journal of the International Society for Matrix Biology. 2010; 29:690–700.

[4] Chen RN, Ho HO, Tsai Y T, Sheu M T. Process development of an acellular dermal matrix (ADM) for biomedical applications. Biomaterials. 2004; 25:2679–2686.

[5] Badylak SF. Xenogeneic extracellular matrix as a scaffold for tissue reconstruction.Transplant immunology.2004; 12:367–377.

[6] Chen F, Yoo JJ, Atala A. Acellular collagen matrix as a possible "off the shelf" biomaterial for urethral repair.Urology.1999, 54:407–10.

[7] Dahl SL, Koh J, Prabhakar V, NiklasonLE.Decellularized native and engineered arterial scaffolds for transplantation. Cell Transplantation.2003; 12:659–666.

[8] Kasimir MT, Rieder E, Seebacher G, Silberhumer G, Wolner E, Weigel G. Simon P. Comparison of different decellularization procedures of porcine heart valves. The International Journal of Artificial Organs. 2003; 26:421–427

[9] Ott HC, Matthiesen TS, Goh SK, Black LD, Kren SM, Netoff TI, Taylor DA. Perfusion-decellularized matrix: using nature's platform to engineer a bioartificial heart. Nature Medicine.2008; 14:213–221.

[10] Baptista PM, Siddiqui MM, Lozier G, Rodriguez SR, Atala A, Soker S. The use of whole organ decellularization for the generation of a vascularized liver organoid.Hepatology.2011; 53:604–617.

[11] Petersen TH, Calle EA, Zhao L, Lee EJ, Gui L, Raredon MB, Gavrilov K, Yi T, Zhuang ZW, Breuer C, Herzog E, NiklasonLE.Tissue-engineered lungs for in vivo implantation. Science 2010; 329:538–541.

[12] Ott HC, Clippinger B, Conrad C, Schuetz C, Pomerantseva I, Ikonomou L, Kotton D, VacantiJP.Regeneration and o rthotopic transplantation of a bioartificiallung.Nature Medicine. 2010; 16:927–933.

[13] Crapo P, Gilbert T, Badylak S. An overview of tissue and whole organ decellularization processes Biomaterials. 2011; 32:3233-3243.

[14] Shupe T, Williams M, Brown A, Willenberg B, Petersen BE.Method for the decellularization of intact rat liver.Organogenesis.2010; 6:134–136.

[15] Uygun BE, Soto-Gutierrez A, Yagi H, Izamis ML, Guzzardi MA, Shulman C, Milwid J, Kobayashi N, Tilles A, Berthiaume F, Hertl M, Nahmias Y, Yarmush ML, Uygun K. Organ reengineering through development of a transplantable recellularized liver graft using decellularized liver matrix. Nature Medicine.2010; 16:814–820.

[16] Soto-Gutierrez A, Zhang L, Medberry C, Fukumitsu K, Faulk D, Jiang H, Reing J, Gramignoli R, Komori J, Ross M, Nagaya M, Lagasse E, Stolz D, Strom SC, Fox IJ, Badylak SF.A whole-organ regenerative medicine approach for liver replacement. Tissue Engineering Part C Methods. 2011; 17:677-686.

[17] Nari GA, Cid M, Comin R, Reyna Musso L, Juri G, Taborda R, SalvatierraNA."Preparation of a three-dimensional extracelular matrix by decellularization of rabbit livers".(2013). Revista Española Enfermedades Digestivas;105:138-143.

[18] Badylak SF, Taylor D, Uygun K. Whole-Organ Tissue Engineering: Decellularization and

Recellularization of Three-Dimensional Matrix Scaffolds. Annual review of biomedical engineering.2011; 13:27–53.
[19] Scarritt ME, Pashos NC, Bunnell BA.A review of cellularization strategies for tissue engineering of whole organs.FrontBioengBiotechnol. 2015 doi: 10.3389/fbioe.2015.2015.
[20] Laird P, Zijderveld A, Linders K, Rudnicki M jaenisch R et al. Simplified mammalian DNA isolation procedure. Nucleic Acids Research 1991;19:4293-5.

Theoretical study of the adsorption of histidine amino acid on graphene

S J Rodríguez[1], **L Makinistian**[2] **and E Albanesi**[1]

[1]IFIS-CONICET, Güemes 3450, 3000 Santa Fe, Argentina and Facultad de Ingeniería, Universidad Nacional de Entre Ríos, 3101 Oro Verde (ER), Argentina
[2]Departamento de Física, e Instituto de Física Aplicada (INFAP), Universidad Nacional de San Luis-CONICET, Ejército de los Andes 950, D5700BWS San Luis, Argentina

E-mail: `sindy.rodriguez@santafe-conicet.gov.ar`

Abstract. Previous studies have demonstrated how the interactions between biomolecules and graphene play a crucial role in the characterization and functionalization of biosensors. In this paper we present a theoretical study of the adsorption of histidine on graphene using density functional theory (DFT). In order to evaluate the relevance of including the carboxyl (-COOH) and amino (-NH_2) groups in the calculations, we considered i) the histidine complete (i.e., with its carboxyl and its amino groups included), and ii) the histidine's imidazole ring alone. We calculated the density of states for the two systems before and after adsorption. Furthermore, we compared the results of three approximations of the exchange and correlation interactions: local density (LDA), the generalized gradients by Perdew, Burke and Ernzerhof (GGA-PBE), and one including van der Waals forces (DFT-D2). We found that the adsorption energy calculated by DFT-D2 is higher than the other two: $E_{ads-DFT-D2} > E_{ads-LDA} > E_{ads-GGA}$. We report the existence of charge transfer from graphene to the molecule when the adsorption occurs; this charge transfer turns up to be greater for the complete histidine than for the imidazole ring alone. Our results revealed that including the carboxyl and amino groups generates a shift in the states of imidazole ring towards E_F.

1. Introduction

Graphene is a two-dimensional crystal with special chemical and physical properties [1, 2]. Consequently, since its discovery, it has been used in fields such as electronics, biomedicine, biotechnology and pharmacology, among others [3, 4, 5]. For instance, Wang *et al.* [6], in the area of drug delivery devices and molecular transport, reported that graphene could be used as a good vehicle of DNA aptamers into living cells, because it provides effective protection of the oligonucleotides from enzymatic cleavage. Furthermore, Hess *et al.* [7, 8] developed graphene transistor arrays for recording cellular action potentials, which have applications in neural prosthesis. The study suggests the simple integration of graphene electronics with flexible technologies (an important aspect of biomedical implants design), because they prevent severe tissue damage [7].

In the area of molecular biology and nanomaterials, different studies have been developed on the adsorption of biological molecules on graphene sheets and carbon nanotubes [9, 10, 11, 12]. Graphene sensors can provide high sensitivity, selectivity and low intrinsic electronic noise, which is a great advantage over other materials [3]. Chemical sensors of carbon nanotubes for small molecules such as NO_2, NH_3, O_2 and H_2 have been tested experimentally and the graphene electronic structure shows high sensitivity when different molecules are adsorbed [13, 14].

Theoretical works use first-principles calculations to analyze the nature of the interactions between biomolecules and graphene [15, 16, 17]. For example, Gowtham *et al.* [11, 18], using DFT-LDA and *Møller – Plesset* second-order perturbation theory (MP2), investigated the interaction of the nucleobases adenine (A), cytosine (C), guanine (G), thymine (T), and uracil (U) with graphene, demonstrating that they exhibit significantly different interaction strengths when physisorbed on graphene. With a similiar approach, Rajesh *et al.* [9] reported significant differences in the binding strength between aromatic rings of amino acids phenylalanine, histidine, tyrosine and tryptophan with graphene (and also with single-walled carbon nanotubes), suggesting once again the potential use of graphene as a selective biosensor. As Gowtham *et al.* [11] and Rajesh *et al.* [9] mentioned, the details of the interaction between DNA, proteins and amino acids with graphene —and carbon nanotubes— have not been fully understood. In this work, theoretical results based on DFT for the adsorption of histidine on graphene are presented. Two adsorbates were considered (both on an infinite graphene sheet): histidine with its -COOH and -NH_2 groups, and the imidazole ring alone. The discussion focused on: i) comparing the results of three approximations to the exchange and correlation interactions —LDA, GGA-PBE and DFT-D2—, ii) evaluating the relevance of including the carboxyl and amino groups in the calculations (relevance that, to the best of our knowledge, has not been reported in the literature). Equilibrium configurations for each system, energies and adsorption distances, densities of states and net charge transfer are presented.

2. Computational Details

The *ab initio* calculations were developed in the formalism of DFT using the OpenMX package [19]. Numerical pseudo-atomic orbitals (PAOs) were used as basis function to expand one-particle Kohn-Sham wave functions. The PAO were specified by H5.0-s1, C5.0-s2p2d1, O6.0-s2p2d1, N6.0-s2p2d1, where H, C, O and N are the atomic symbols, 5.0 and 6.0 represent the cutoff radius (Bohr) in the generation by the confinement scheme [20, 21], and s2p2d1, for example, indicates the employment of two primitive orbitals for each of the s and p components and one primitive orbital for the d component.

In studies on the interaction between simple benzene derivatives with graphene, graphite and carbon nanotubes, the LDA approximation correctly described the electronic structure and adsorption geometry (comparison between adsorption-energy DFT descriptions of graphite and the experimental thermal desorption energy) [17]. Furthermore, adsorption calculations of DNA nucleobases on graphene by Lee *et al.* [15], and of adenine on graphite by Ortmann *et al.* [16] suggest that the GGA-PBE approximation does not accurately represent the $\pi - \pi$ bonds in the graphene molecule interface, resulting in low energy bonds and high adsorption distances. LDA, on the other hand, improves adsorption distances, approximating better to experimental values and DFT corrections including van der Waals interactions [11]. With the aim of contributing in this aspects, we studied the adsorption of histidine on graphene using three approximations: local LDA, generalized gradients GGA-PBE [22], and van der Waals DFT-D2 [23].

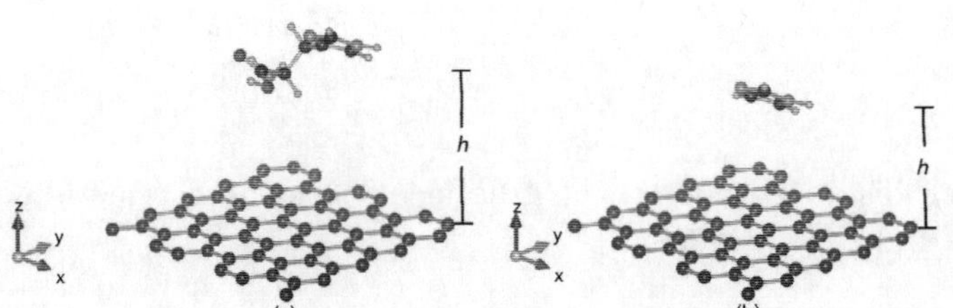

Figure 1. (a) Histidine with -COOH and -NH_2 groups on graphene. (b) Histidine imidazole ring on graphene.

The system was modelled with an orthorhombic supercell of 12.78 × 12.29 × 25.00 $Å^3$ (these dimensions are sufficient to uncouple the graphene sheets and avoid interactions between molecules of adjacent cells); an energy cut-off of 180 Ry and a k-mesh of 5 × 5 × 1 for the SCF, with a refinement up to 50 × 50 × 1 for DOS calculations; and, for geometry optimization, the convergence criterion was set to 0.02 eV/Å for histidine with -COOH and -NH_2, and 0.015 eV/Å for the imidazole ring. The graphene sheet was located at the bottom of the cell as shown in Fig. 1.

Initially we located the molecule at distance h from graphene, far enough from it, so a reference energy decoupled (E_{ref}) could be determined for the latter calculation of the interaction energy (attraction, or rejection if too close): $E_{int}(h) = E_{tot}(h)-E_{ref}$, where $E_{ref} = E_{tot}(h{=}12.5$ Å). The adsorption energy E_{ads} is that of the interaction at the global minimum of the curves $E_{int}(h)$ in Fig. 2. In order to determine the adsorption energy and the mean final distance of the biomolecule to the graphene sheet, two relaxations were performed. In the first relaxation we considered the molecule alone. Subsequently, the relaxed molecule was placed at several heights from the graphene sheet. For each height considered, a second relaxation for all atoms (histidine and graphene sheet) was performed. Once the adsorbate-substrate system was relaxed, the final average distance (measured from the geometric center of the biomolecule to the graphene sheet) and the interaction energy were obtained.

3. Results and Discussion

We begin by describing the equilibrium structures for the two systems considered (for LDA, GGA-PBE and DFT-D2). These results are shown in Fig. 2, from where it is clear that the interaction is negligible for distances greater than 6 Å . Table 1 displays the vertical distances between the molecule and the graphene sheet, and the corresponding adsorption energies, for each of the three exchange and correlation schemes used. For the two systems, we find that $E_{ads-DFT-D2} > E_{ads-LDA} > E_{ads-GGA}$. The GGA-PBE approximation resulted in very low adsorption energies and relatively long distances between substrate-adsorbate, which clearly indicates that it does not correctly describe the π-π interactions. The results for LDA and DFT-D2 are relatively close to each other. There is a strong physisorption between histidine and graphene, when considering van der Waals interactions. The adsorption energies are improved in the systems studied when the DFT-D2 scheme is performed—because it includes long-range forces—. This coincides with results for adsorption of nucleotides [15].

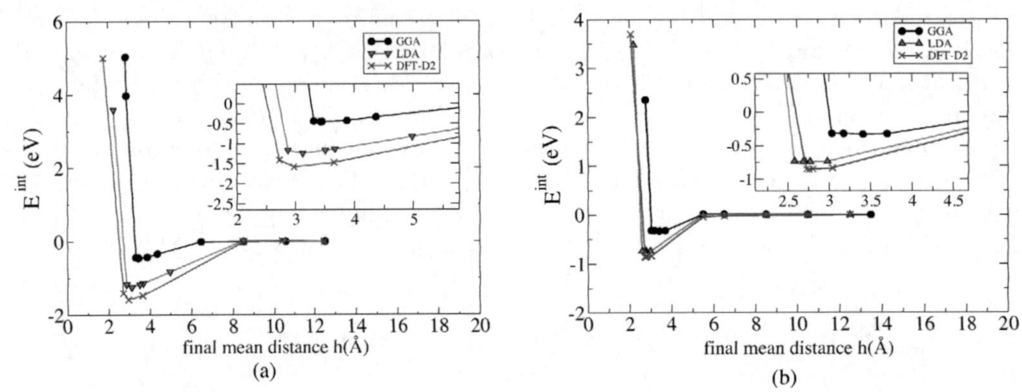

Figure 2. Interaction energy vs. mean distance between the graphene sheet and the histidine with (a), and without (b) the -COOH and -NH_2 groups.

Table 1. Theoretical results of adsorption energy for exchange and correlation approaches LDA, GGA-PBE and DFT-D2. d_{ads}^{His-Gr} is the distance measured between the geometric center of the biomolecule and graphene sheet. In Ref. [24]. hybrid meta exchange- correlation functionals called M06- 2X/6-311G (Gaussian 09). In Ref. [25] hybrid meta exchange-correlation functional, called M05-2X (Gaussian 09). No experimental results are available for these systems.

		This Work		Other work	
Approx.	System	E_{ads} (eV)	d_{ads}^{His-Gr} Å	E_{ads} (eV)	d_{cell} Å
LDA	Histidine	1.24	3.13	—	—
	Imidazole ring	0.73	2.77	—	—
GGA-PBE	Histidine	0.45	3.44	—	—
	Imidazole ring	0.32	3.41	0.21(0.55) [9]	3.21 [9]
DFT-D2	Histidine	1.58	2.97	—	—
	Imidazole ring	0.85	2.90	—	—
Other	Histidine	—	—	0.77 [24]	3.195 [24]
	Imidazole ring	—	—	0.14 [25]	3.38 [25]

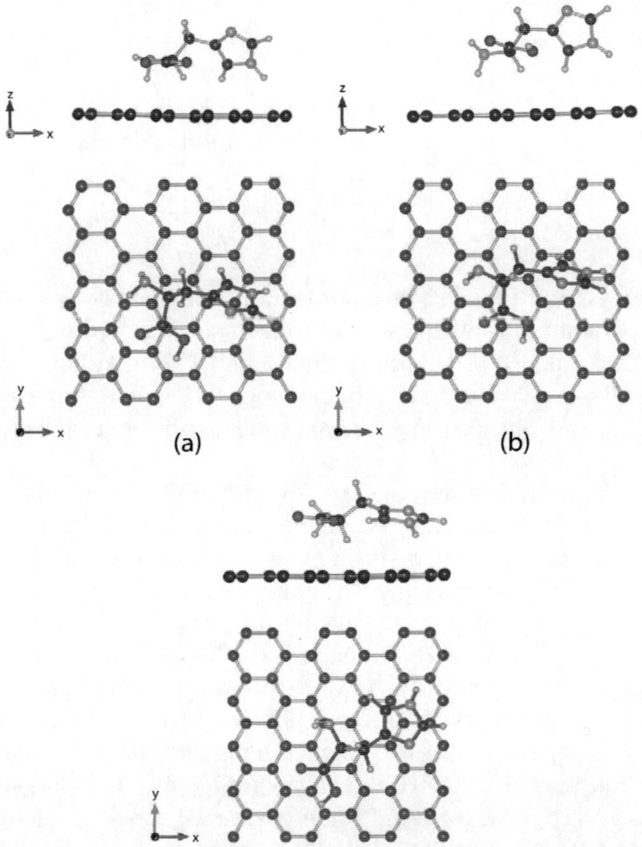

Figure 3. Equilibrium geometry for histidine with -COOH and -NH_2 groups on graphene. (a) LDA (b) GGA-PBE (c) DFT-D2.

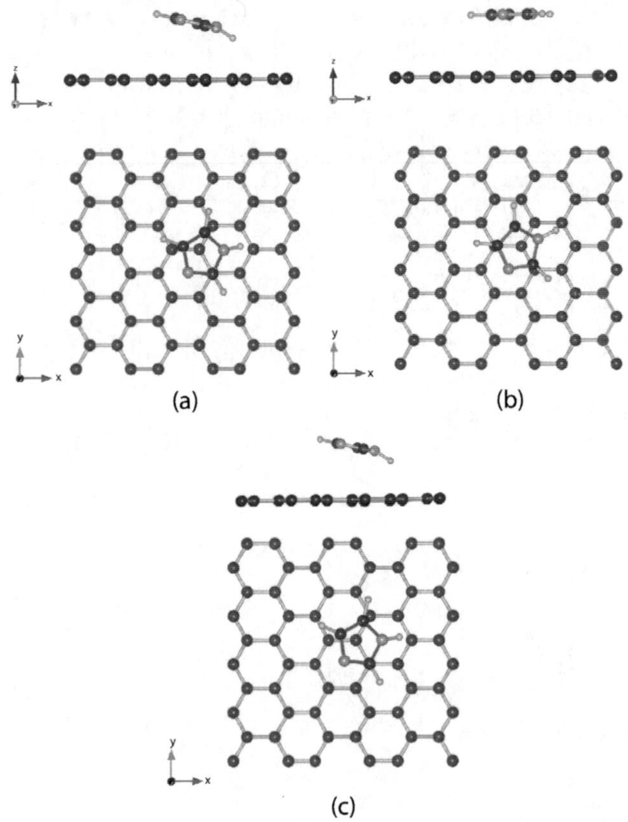

Figure 4. Equilibrium geometry for imidazole ring on graphene. (a) LDA (b) GGA-PBE (c) DFT-D2.

Equilibrium geometries are shown in Fig. 3 and Fig. 4. The equilibrium configurations for the imidazole ring show that the molecule is oriented parallel to the graphene sheet in GGA-PBE, while the ring has a slight angle in the LDA and DFT-D2 approximations. The histidine with the carboxyl and amino groups did not displayed a parallel orientation of the ring, however, when including van der Waals interactions, the ring of the molecule tends to be more parallel to the graphene sheet, see Fig. 3(c).

In order to study the possibility of using graphene as a sensor for histidine, we calculated the DOS for: histidine alone —with and without carboxyl and amine groups—, graphene alone, histidine on graphene, imidazole ring on graphene, and histidine on graphene including the individual contribution of the -COOH and -NH_2 groups, and the imidazole ring (see Fig. 5). Figure 5(a) displays the well known zero gap DOS of pure graphene. Figures 5(b)(c) and (d) show a typical molecular DOS with many very sharp peaks (as expected, there are more peaks for histidine with carboxyl and amine groups). Figures 5(e)(f) and (g) clearly show how the presence of the carboxyl and the amino groups produces significant differences. In particular, it is to be remarked that the peaks that lay closer to E_F (in the valence band) are indeed due to the imidazole ring, see Fig. 5(h). However when the -COOH and -NH_2 groups are included the states are shifted closer to E_F —first peak in the valence band, blue curve, see arrow—. If van der Waals forces are considered the shifting of that first peak is 0.29 eV towards E_F.

With regards to the three approaches: GGA-PBE and DFT-D2 show slight differences, while LDA is sensibly different from the other two (the positions and the heights of the peaks). The peaks marked with arrows in Figs. 5(e)(f) and (g) for histidine and imidazole ring respectively, lay at, LDA: -0.76 eV and -1.67 eV; GGA-PBE: -1.33 eV and -1.89 eV; DFT-D2: -1.24 eV and

-1.53 eV.

Including the carboxyl and amino groups should be of great importance in studies of electronic transport in graphene, since their presence implies more states available near the Fermi energy that can move from the valence band to the conduction band. Furthermore, the shift of states of the imidazole ring towards E_F that appears when the carboxyl and amino groups are included, could improve the sensitivity of a graphene-based sensor upon the application of a bias voltage.

In all systems studied, no gap opening is observed after the adsorption, so that the molecule-graphene system retains the semi-metallic character of pure graphene.

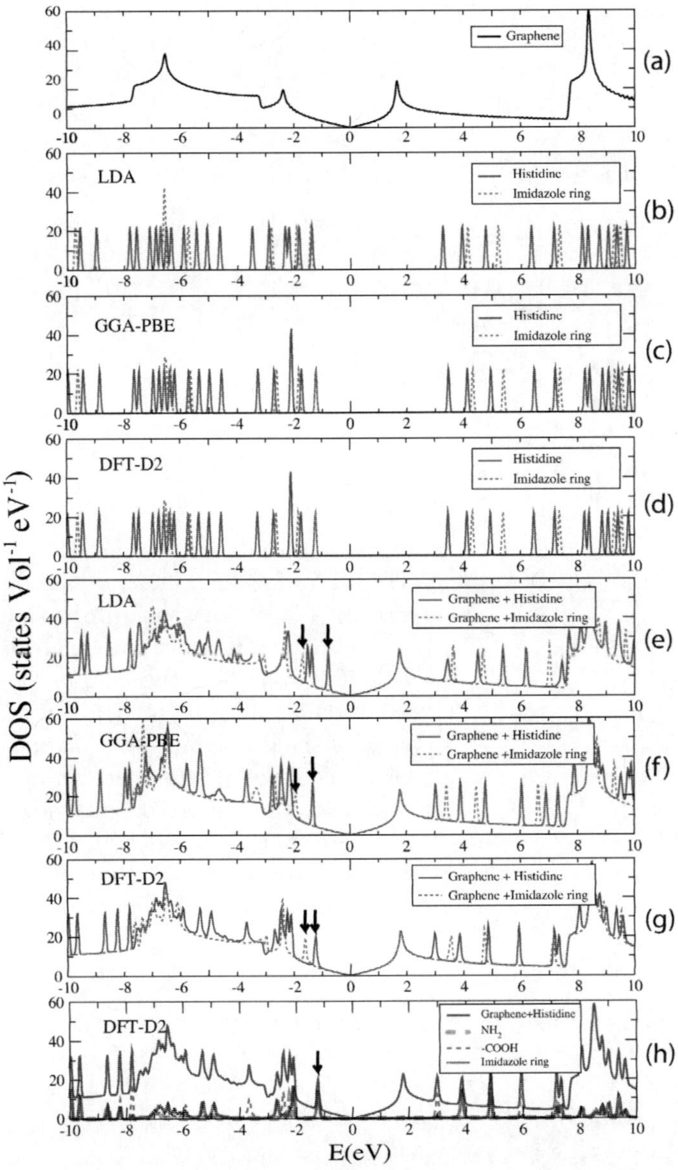

Figure 5. Total DOS of (a) pristine graphene, (b)(c)(d) histidine and imidazole ring, (e)(f)(g) histidine on graphene —with and without carboxyl and amine groups—, and (h) histidine on graphene, including DOS for -COOH, -NH_2 and imidazole ring separately.

The charge density of graphene plus histidine with carboxyl and amine groups, gives a qualitative measure of the adsorption-induced charge redistribution. In Fig. 6 we present the charge density for two planes in the direction [001], in (a) the adsorbate-substrate interaction is negligible and the charge distribution is uniform for each carbon atom of the graphene sheet. In Fig. 6(b) the charge density for graphene after the adsorption process is displayed: changes of charge distribution are more remarkable to the carbon atoms that are under the molecule, see Fig. 6(c). The analysis of the Mulliken charges in the calculations shows a very small charge transfer. We report the existence of a charge transfer from the graphene sheet to the molecule, see Table 2. For the three approaches, the charge transfer is higher when the groups -COOH and -NH_2 are considered. Again, we find that including them in the calculations is a relevant aspect of the functionalization and characterization of graphene as a biosensor: a greater charge transfer could be significant regarding electronic transport through the histidine-graphene system.

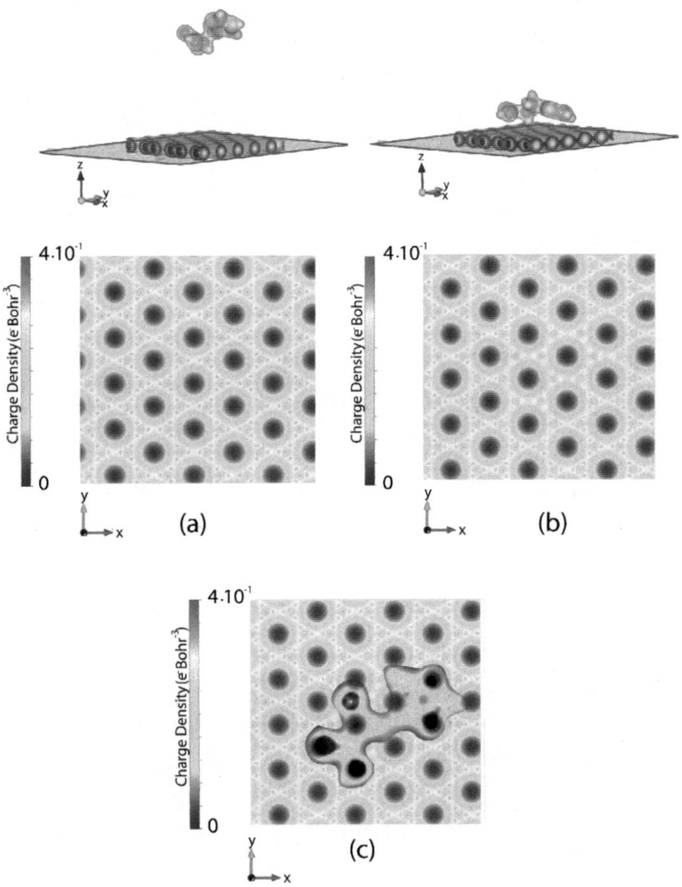

Figure 6. The charge density (DFT-D2) for the planes in the direction [001]. (a) the interaction between the adsorbate-substrate is negligible h=12Å , the charge distribution is uniform for each carbon atom of the graphene sheet. (b) adsorption between the molecule and graphene, the charge density show there are changes in distribution for graphene (c) overlap of charge density —molecule on graphene—.

Table 2. Charge transfer in the system graphene-histidine.

Approx	System	Net charge transferred to the molecule
LDA	Histidine	0.178 e
	Imidazole ring	0.076 e
GGA-PBE	Histidine	0.118 e
	Imidazole ring	0.023 e
DFT-D2	Histidine	0.188 e
	Imidazole ring	0.079 e

4. Conclusions

We studied the adsorption of histidine and imidazole ring on graphene using three approaches to the exchange and correlation interactions: LDA, GGA-PBE and DFT-D2. The calculations demonstrate that the adsorption energies satisfy the relation $E_{ads-DFT-D2} > E_{ads-LDA} > E_{ads-GGA-PBE}$. The adsorption distance for histidine meets the relation $d_{ads-DFT-D2} < d_{ads-LDA} < d_{ads-GGA-PBE}$, while for imidazole ring $d_{ads-LDA} < d_{ads-DFT-D2} < d_{ads-GGA-PBE}$. We reported the underestimation in adsorption energy and overestimation in adsorption distance in the approach GGA-PBE. We found a strong physisorption between histidine and graphene, when considering the van der Waals forces.

The interaction of histidine and the imidazole ring with graphene, showed signicant differences in the adsorption process when the carboxyl and amino groups were included, since their presence produces an important shift of the imidazole ring edge states towards E_F, improving eventual transport effects. For the three exchange and correlation schemes used, the adsorption energy was always lower in the imidazole ring system.

Also, there was a greater charge transfer from the graphene to histidine when the carboxyl and amino groups were included —0.18 e vs 0.07 e for the imidazole ring, from the approach DFT-D2—. All the studied substrate-adsorbate systems maintained the semi-metallic character (zero gap) of pure graphene.

The results suggests that graphene may be more sensitive when the adsorbed molecule includes carboxyl and amino groups. This information may be a useful guidance to develop novel graphene-based technology for the detection of amino acids.

Acknowledgements The authors acknowledge financial support from the Consejo Nacional de Investigaciones Científicas y Técnicas (CONICET), the Universidad Nacional de Entre Ríos (UNER), Argentina, and the Universidad Nacional de San Luis (Argentina). Also, we thank the work of the developers of Xcrysden [26] and VESTA codes [27], utilized in this work.

References

[1] Neto A C, Guinea F, Peres N, Novoselov K and Geim A 2009 *Rev. Mod. Phys.* **81** 109–162
[2] Novoselov K, Geim A, Morozov S, Jiang D, Zhang Y, Dubonos S and Firsov A 2004 *Science* **306** 666–669
[3] Ferrari A 2014 *Nanoscale* **1** 1–343
[4] Yang Y, Asiri A, Tang Z, Du D and Lin Y 2013 *Materials Today* **16** 365–373
[5] Bellido E and Seminario J 2012 *J. Phys. Chem. C* **116** 8409–8416
[6] Wang Y, Zhaohui L, Dehong H, Chiann-Tso L, Jinghong L and Yuehe L 2010 *J. Am. Chem. .Soc.* **132** 9274–9276
[7] Hess L, Jansen M, Maybeck V, Hauf M, Seifert M, Stutzmann M, Sharp I, Offenhusser A and Garrido J 2011 *Advanced Materials* **23** 5045–5049
[8] Bendali A, Hess L, Seifert M, Forster V, Stephan A, Garrido J and Picaud S 2013 *Advanced Healthcare Materials* **2** 929–933
[9] Rajesh C, Majumder C, Mizuseki H and Kawazoe Y 2009 *J, Chem. Phys.* **76** 033401
[10] Lee E 2012 *Appl. Phys. Lett.* **100** 153117
[11] Gowtham S, Scheicher R, Ahuja R, RPandey and Karna S 2007 *Phys. Rev. B.* **76** 033401
[12] Berland K, Chakarova S, Cooper V, Langreth D and Schroder E 2010 *J. Phys.: Condens. Matter* **23** 135001
[13] Leenaerts O, Partoens B and Peerters F 2008 *Phys. Rev. B* **77** 125416
[14] Schedin F, AKGeim, Morozov S, Hill E, Blake P, Katsnelson M and Novoselov K 2007 *nature materials* **6** 652–655
[15] Lee J, Choi Y, Kim H, Scheicher R and Cho J 2013 *J. Phys. Chem. C* **117** 13435–13441
[16] Ortmann F, Schmidt W and Bechstedt F 2005 *Phys. Rev. Lett.* **95** 186101
[17] Woods L, Bdescu S and Reinecke T 2007 *Phys. Rev. B* **75** 155415
[18] Gowtham S, Scheicher R, RPandey, Karna S and Ahuja R 2008 *Nanotechnology* **19** 125701
[19] Ozaki T, Nishio K and Kino H 2010 *Phys. Rev. B* **81**
[20] Ozaki T 2003 *Phys. Rev. B.* **67** 155108
[21] Ozaki T and Kino H 2004 *Phys. Rev. B.* **69** 195113
[22] Perdew J P, Burke K and Ernzerhof M 1998 *Phys. Rev. Lett.* **80** 891

[23] Grimme S 2006 *J. Comp. Chem.* **27** 1787–1799
[24] Umadevi D and Sastry G 2013 *ChemPhysChem* **14** 2570–2578
[25] Vovusha H, Sanyal S and Sanyal B 2013 *J. Phys. Chem. Lett.* **4** 3710–3718
[26] Kokalj A 2003 *Comput. Mater. Sci.* **28** 155
[27] Momma K and Izumi F 2011 *J. Appl. Crystallogr.* **44** 1272

Metalworking defects in surgery screws as a possible cause of post-surgical infections

Spector Mario[1], Peretti Leandro E[2], Romero Gustavo[1]

[1]Universidad Tecnológica Nacional, Facultad Regional Paraná. Almafuerte 1033 Paraná, E. Ríos, Argentina.

[2]Instituto de Desarrollo para la Industria Química, INTEC (Universidad Nacional del Litoral/CONICET), Güemes 3450, Santa Fe, Argentina.

E-mail: spectormario@gmail.com, lperetti@santafe-conicet.gov.ar

In the first phase of this work, surface defects (metalworking) in stainless steel implantable prostheses and their possible relation to infections that can be generated after surgery was studied. In a second phase, the results obtained in the aforementioned stage were applied to knee cruciate ligaments surgery screws, considering the fact that a substantial number of Mucormycetes infections have been reported after arthroscopic surgery in Argentina since the year 2005. Two types of screws, transverse and interference screws, were analyzed. The Allen heads presented defects such as burrs and metalworking bending as a result of the machining process. These defects allow the accumulation of machining oil, which could be contaminated with fungal spores. When this is the case, the gaseous sterilization by ethylene oxide may be jeopardized. Cortical screws were also analyzed and were found to present serious metalworking defects inside their heads. To reduce the risk of infection in surgery, the use of screws with metalworking defects on the outer surface, analyzed with stereomicroscope and considering the inside part of the Allen as an outer surface, should be avoided altogether.

1. Introduction

There are many cases of bacterial and fungal infections which come as a result of contaminated surgical implants in clinics and hospitals. Infectious agents in steel prostheses are particularly difficult to eliminate, especially when the bacteria have completely colonized the piece [1]. In some cases, patients have to undergo a new surgery to remove the prosthesis because the effect of antibiotic (ATB) disinfection is not effective enough. Not only do these infections bring economic losses to health institutions but they may place a patient's life at risk of death, especially when the individual is an elderly person.
Since 2005, in Argentina there have been over 40 cases of mucormycosis fungal infections in cruciate ligaments surgery patients [2], many of which were detected in the city of Paraná (Entre Ríos, Argentina), with serious health implications.

The clinical forms of mucormycosis are varied and severe in all the cases. The one isolated in the aforementioned context was the *Rhizopus microsporus var. rhizopodiformis*. This fungus is an environmental opportunistic human pathogenic fungus and as such it requires special conditions to produce an infection. One of these conditions is its accidental inoculation in traumatological surgery owing to the use of contaminated material. Bone mucormycosis produces catastrophic effects on bones and the affected surrounding tissue, causing necrosis manifested by severe pain, need for surgical bone resection, prosthetic bone replacement, temporary or permanent disability and severe death cases.
Several authors have delved into bacterial adherence to biomaterials and its characteristics. Na and Friedman [3] studied the physical and chemical characteristics of adhesion. The importance of a smooth surface finish of stainless steel pieces was reported by Arnold and Bailey [4] in the use of pipes for the food industry. Katsikogianni and Missirlis [5] studied the mechanisms of bacterial adhesion to biomaterial surfaces, factors affecting adherence, techniques used in estimating bacteria-material interactions and the models that have been developed to predict the adherence.
The first part of this work was performed on stainless steel prostheses with and without metal mechanical defects. Colonization of the surface by microorganisms and subsequent disinfection was studied. Even though some of this results have been reported elsewhere [6], it was thought of utmost importance to present them here as well to provide the basis for the hypothesis proposed for the second part of the present work.
Two titanium screws, one transverse and another one interferential, used in knee cruciate ligaments surgery and later on drawn from the patient were analyzed in the second phase of the study. The analysis consisted in finding metalworking defects that may foster microorganism adherence. Similarly, a cortical screw used in osteosynthesis was studied. Tests were also performed on machining oil employed in the manufacture of screws in order to delve into and its properties as means to allow bacteria and fungi growth [7, 8].

2. Materials and methods

2.1 First phase: steel prostheses

In the first part of the work, the specimens studied were pieces of steel prostheses that had been used and then removed from patients. These pieces were cut so that they could fit in a test tube. Pieces with some fold defects, produced by the placement of the prosthesis in the patient, and similar pieces without defects were selected. In some cases flawed specimens were polished to remove imperfections completely. Depending on the trial, the specimens were of the same surface and the same shape or they presented different surface or shape.
Once the pieces were carefully chosen, the test samples were cleaned and sterilized by dry heat in a test tube at 270 °C for 2 hours. The next step was the inoculation of the samples, which entails the colonization of the metallic sample with a microorganism.
Commercial yeasts (*Saccharomyces* spp) and the *Staphylococcus aureus* bacterium were used in this study, following the next procedure. A sterile atmosphere was guaranteed by working between burners and using glassware sterilized by dry heat for 120 minutes at 270 °C. Four grams of commercial yeast were placed in an Erlenmeyer flask and 200 mL of culture medium (pasteurized grape juice) were added. Previously sterilized specimens were placed in an oven and incubated at 28 °C for 96 hours. The specimens were then removed and rinsed in sterile distilled water and allowed to dry under sterile conditions. Then the specimens were disinfected by introducing them into a commercial disinfectant solution (composition: 5.91% Triethylene, 0.29% alkyl dimethyl benzyl ammonium chloride saccharinate) for 1 hour, rinsed in sterile distilled water for 2 hours to remove residual disinfectant and finally each sample placed in a test tube with sterile culture medium, incubated at 28 °C to observe occurrence of development.

All assays were performed by incubating a contaminated specimen that was not disinfected and a negative control tube with culture medium only.

2.1.1 Tests with Staphylococcus aureus

S. aureus was used because it has the following advantages: i) unlike yeasts, its smaller size allows it to access and colonize smaller cavities, and ii) most hospital-acquired infections are caused by this bacterium, which highlights the importance of the present research. On the other hand, it is difficult to see *S. aureus* in the metallographic microscope unless it is dyed. Also, more biosecurity measures must be taken since it is a potentially pathogenic microorganism.

Some colonies were taken from a Petri dish, in which the bacteria had developed in Tryptic Soy Agar (Britania), and placed in a test tube containing 1 mL tryptic soy broth with 1 % of glucose, to promote biofilm formation. The bacteria were incubated in a culture oven for 24 hours at 37 °C.

Once the inoculum developed, it was poured into 100 mL of tryptic soy broth with 1 % of glucose and sterile prostheses (with defects and perfectly polished samples) were introduced. They were incubated in a culture oven for 24 hours at 37 °C.

After 24 hours of incubation, the culture medium became turbid, indicating that there was bacteria development. The samples were taken and rinsed in sterile distilled water for 10 minutes with mild agitation. Washing was repeated 3 times and the samples were allowed to dry in a sterile atmosphere. Specimens were placed in individual Erlenmeyers containing the antibiotic cefazolin at a concentration of 1.28 mg of antibiotic per milliliter of sterile saline and allowed to react for 6 hours. After this time period, the specimens were rinsed in sterile distilled water for 15 minutes in individual Erlenmeyers to remove the antibiotic and allowed to dry.

The specimens were placed in test tubes containing 3 mL of sterile Tryptic Soy Broth (Britania) and incubated in a culture oven at 37 °C. The incubation was followed by means of a camera connected to a computer which took pictures every 20 minutes during the period of the trial.

2.1.2 External validation

The results obtained in this study were validated at the Laboratory of Microbiology, Facultad de Bioquímica y Ciencias Biológicas, Universidad Nacional del Litoral by means of additional tests.

In order to evaluate the prostheses two samples, with and without metalworking defects were used and inoculated with *S. aureus*. In the same way it was described above. Upon inoculation they were kept for 7 hours in an antibiotic solution of cefazoline at the same concentration the antibiotic would have in the in an infected patient's bloodstream (32 μg/mL). Following this procedure, the samples were removed and rinsed three times. The bacteria developed were counted in the residue of each rinse and reported as colony forming units (CFU) per mL.

2.2 Second phase: surgical screws

In a second phase of the study, two set screws (Figure 1) extracted from infected patients who had undergone knee cruciate ligaments surgery were used. Then, four sets of new screws available on the market were analyzed. Cortical screws were also evaluated.

2.2.1 Eye inspection of screws

An inspection was carried out with a magnifying glass at different magnifications with a maximum of x80 to determine the metalworking conditions of screws, both in terms of their external and internal structure.

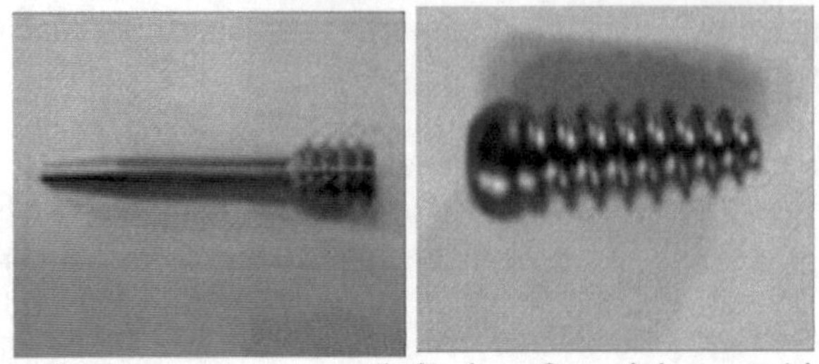

Figure1. Transverse screw (left); interferential screw (right).

2.2.2 Feasibility of machining oil as a source of infection

Sterile metalworking oil was inoculated with different amounts of *R. microsporus* var. *rhizopodiformis* spores (reaching final concentrations of 10^2, 10^3, 10^4, 10^5 and 10^6 spores mL oil^{-1}). The inoculated oil samples were left at room temperature for 24 hours and then covered with Potato Dextrose Agar (PDA, Sigma Aldrich). The plates were incubated at 35ºC for 7 days and observed daily.

3. Results and discussion

3.1 First phase: steel prostheses

3.1.1 Tests with yeast on stainless steel samples

In the first phase of the work, ten assays were performed, evaluating a different number of samples in each one. The obtained results are shown in Table 1. In 60% of the trials, yeast growth was observed in at least one sample, indicating that the disinfectant could not remove the total yeast inoculated (Figs. 2-3). Despite the disinfectant effect, there was colonization of samples in areas protected by mechanical defects. Where there were samples without growth inoculation, it could be attributed to yeasts not being able to enter protected sites (Tests 2, 4, 6 and 8). Therefore, if the stainless steel prosthesis presents some of the mentioned defects, there is a potential risk that a pathogen finds a place not reached by the antibiotic or antifungal and the patient's infection cannot be eradicated.

Table 1. Conduction tests with yeast

Test	*Number of Samples*	*Samples with growth*	*Time (days)*
1	6	1	10
2	7	0	-
3	9	1	5
4	9	0	-
5	9	3	5
6	7	0	-
7	7	1	8
8	7	0	-
9	7	1	6
10	7	1	5

Journal of Physics: Conference Series **705** (2016) 012013 doi:10.1088/1742-6596/705/1/012013

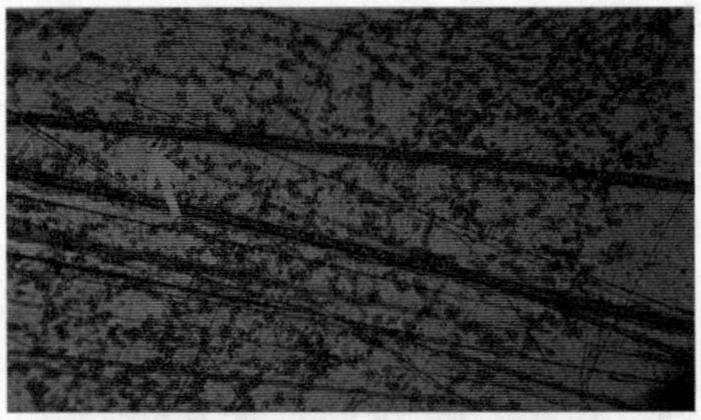

Figure 2. Striped test piece and inoculated with yeast. Notice that the thicker stripes permit yeast housing.

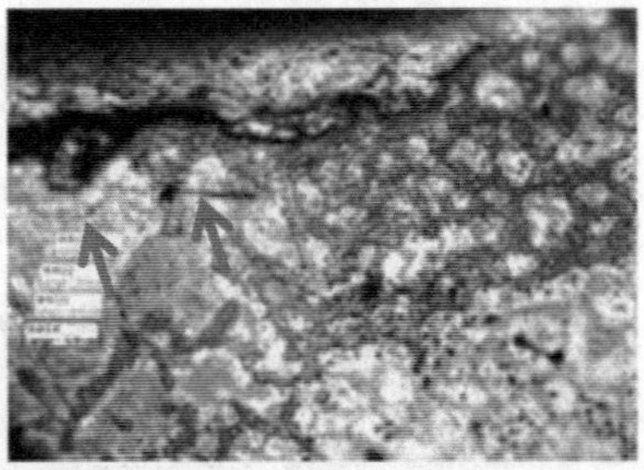

Figure 3. Sample with metallurgical defect possibly induced during surgery. Yeasts survived disinfection and developed into a new culture medium.

3.1.2 Tests with Staphylococcus aureus

Table 2 shows the results of tests with *S. aureus*. In 100% of the cases, bacterial growth was observed in the samples with defects. These results suggest that defects protected the bacteria from the action of the ATB. Note that in colonized perfectly polished samples, the antibiotic acted without impediments and removed all bacteria (in 6 out of 7 tests) or significantly delayed its development (see test number 3). Figure 4 shows an illustrative image of one of the tests, where positive (turbid) and negative (crystalline) tubes can be observed.

Table 2. Conduction tests with *Staphylococcus aureus*

Test	*Samples with defects*				*Samples without defects*		*Positive control*
	1	2	3	4	5	6	7
I	+	+	+	-	-	-	+
II	+	+	+	+	-	-	+
III	+	+	+	+	+	+	+
IV	+	+	+	+	-	-	+
V	+	+	+	+	-	-	+
VI	+	+	+	+	-	-	+
VII	+	+	+	+	-	-	+

Symbol "+" indicates development and "-" indicates the absence of development.

Figure 4. Conduction tests with *Staphylococcus aureus*. Tubes with development (turbid) contain samples with metalworking defects and tube with no development (crystalline) contains a sample without defects.

3.1.3 External Validation

The results of the validation tests carried out at Universidad Nacional del Litoral are presented in Figure 5. After a 7 hour contact of the samples with a solution of 32 μg/mL of the cefazolin ATB *S. aureus* still persisted in spite of the different washing sessions carried out.

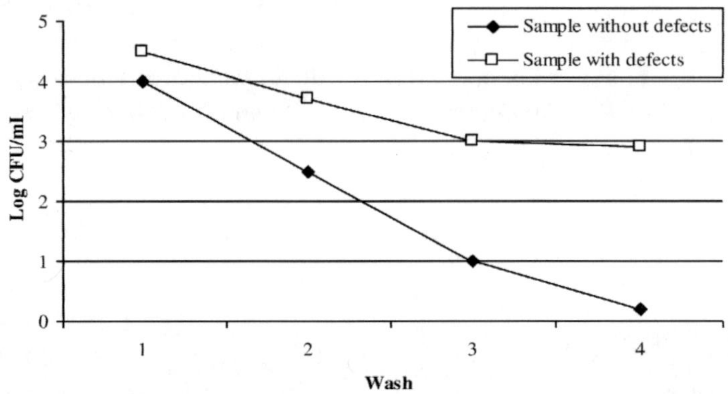

Figure 5. Persistence of *S. aureus* in samples with and without defects

By means of these validation tests previously obtained results were replicated. When a sample with defects was colonized by bacteria the ATB effect was reduced in comparison to a colonized sample without defects. In the case of the sample with defects, the persistence of viable cells of *S. aureus* reached a constant value and it was not possible to complete elimination, which was achieved in the case of the sample without defect after 4 washing sessions.

The hypothesis of this work suggests that the samples or prostheses containing manufacture related or surgery induced defects could provide sites of bacteria protection if there is prosthesis infection. According to results, it can be confirmed that defects in prostheses will protect bacteria and antibiotic treatment will yield poor results [8]. However, if there is an infection in a perfectly polished prosthesis, free of defects, antibiotic treatment will be considerably more effective.

3.2 Second phase: surgical screws

3.2.1 Eye inspection on screws infected by Rhizopus microsporus var. rhizopodiformis
After visual inspection of the screws at x60 magnification, it was found that the external surfaces of screws showed no deformities. Conversely, metalworking defects were found within the screw head (Allen hexagon, Figure 6). Using a metallographic microscope, the metalworking defects were more evident (Figure 7).

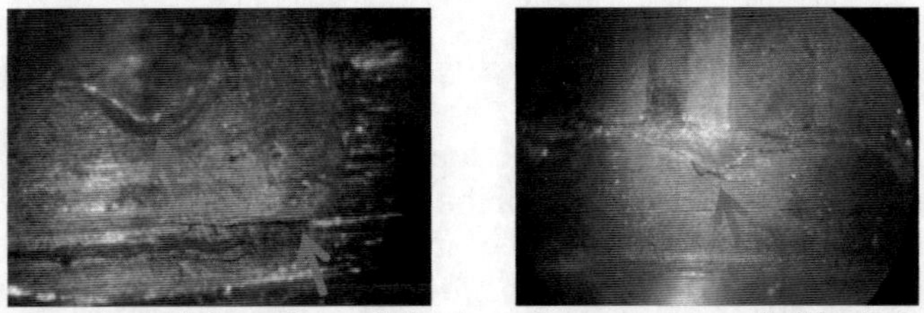

Figure 6. Axial section of the interference screw (left). Axial cut of the hexagon corner (right). Arrows indicate metalworking deformations.

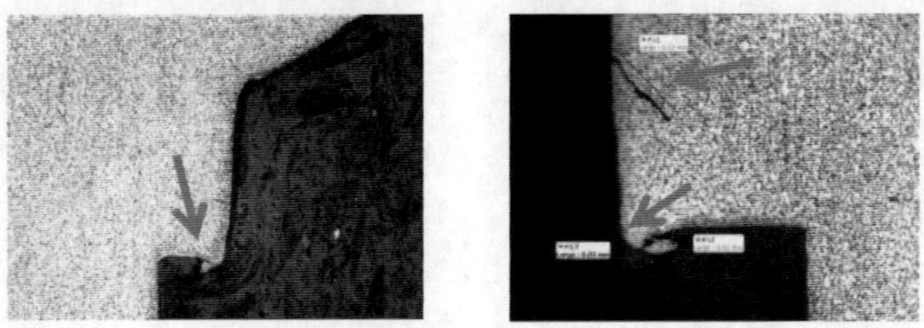

Figure 7. Metallography of a longitudinal cut of the screw heads (left); metalworking defects, bending and cracking (right).

Sections 3.1.1, 3.1.2 and 3.1.3 which report the results in the first phase of this study show that the metalworking defects such as folds, burrs and embossing become areas in which bacteria can develop and survive the effect of antibiotics, even when high concentrations are used. The application of these concepts to the occurrences on the screw heads infected with *Rhizopus microsporus* var. *Rhizopodiformis* leads to a new hypothesis: "if these metalworking defects on prostheses can protect bacteria from the antibiotic effect, they could also harbor machining oil used in the manufacture of screws. The oil could protect fungal spores from the sterilizing effect of the ethylene oxide".

3.2.2 Eye inspection of cortical screws
If the aforementioned concept of metalworking defects as a bacteria harboring area is extended to the cortical screws, it could be assumed that the defects observed in the latter may also protect bacteria from the ATB effect.
In Figure 8 metalworking defects are observed at the bottom of the screw head, which may perfectly harbor bacteria and protect them from the ATB effect.

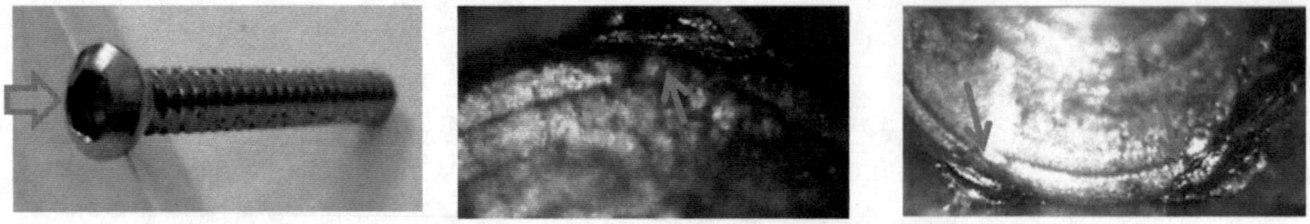

Figure 8. Bends and burrs on the bottom of the hexagon of a cortical screw.

3.2.3 Feasibility of machining oil as the medium for Rhizopus microsporus var. rhisopodiformis to survive in the screws

When the machining oil was studied as a possible source of screw contamination, it was found that the fungicidal capacity of the oil is low. It can inhibit growth of less than 10^3 spores per mL of oil (Figure 9). These results demonstrate that the metalworking defects could act as a suitable environment for the *R. microsporum* var. *rhizopodiformis* to develop in the machining oil accumulated in the burrs and cracks.

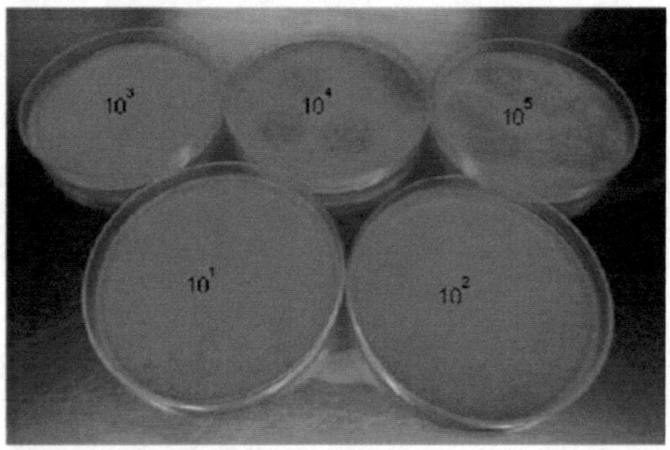

Figure 9: PDA Petri dishes where contaminated machining oil was contaminated with known inocula. Oil can inhibit growth of less than 10^3 spores per mL of oil.

4. Conclusions

This experimental work delved into the influence of surface defects (metalworking) in stainless steel implantable prostheses and their relationship with infections that are generated after surgery and generally lead to their removal.

In the first part of this work, the experiments were carried out according to the protocol here proposed and validated externally (sections 2.1, 2.1.1 and 3.1.3). Since the tests coincided with the hypothesis proposed, especially when it is a random event such as the colonization of bacteria and the place of the test piece which would be inoculated, it can be regarded as a scientific fact that if prostheses with metal mechanical defects become infected for any reason, they will carry bacteria, which places an implanted person's life at risk.

The concepts of these results were applied retrospectively, in a second phase, to knee cruciate ligaments surgery screws, in order to elaborate on the most probable theory that produced serious postsurgical fungal infections. Two types of screws, transverse and interference screws, were analyzed and they were found to contain defects such as metalworking burrs and bending inside the Allen

heads. These defects come as a result of the machining process and are invisible to the naked eye. In this case, metalworking defects would house machining oil that may be contaminated with fungal spores, and the action of sterilizing (ethylene oxide) gas would be affected. As well as this, tests on machining oil properties as a medium capable of allowing the viability of fungi were positive.
Consistent with these results, two practical measures during surgery are necessary in order to reduce the risk of infection:
1) Do not allow the use of screws with metalworking defects on the outer surface, analyzed with a stereomicroscope, with the inside part of the Allen screw regarded as part of the exterior surface.
2) Use screws that were sterilized by radiation only, as opposed to the gaseous sterilization by ethylene oxide.
Upon the application of these preventive measures in nearly 200 interventions, the percentage of infections in the context aforementioned has fallen to 0 % to date.
This research project points to the need to improve quality control standards in relation to prostheses, especially as far as screws are concerned, requiring that surface control is increases up to 80 and that the inside part of the screw head is considered as an outer surface.

Acknowledgements

The authors would like to acknowledge the Government of the Province of Entre Rios for funding this project, Universidad Tecnológica Nacional Regional Paraná for its logistics support and infrastructure, to Dr. Maria Cristina Lurá for providing the strain of *S. aureus* and to Prof. Graciela Yugdar for revise the manuscript.

References

[1] C. Bergallo. "Infección de prótesis de cadera: paradigma de las infecciones de prótesis articulares", Revista Chilena de Infectología, Vol. 17 (2) (2000) p. 87-91
[2] Makino A, Carbo L, Muscolo L, M Ayerza, Costa Paz M, Astoul Bonorino J, Aponte L: fungal osteomyelitis after ACL reconstruction. Revista Argentina de Artroscopía 2008; 15: 41-45.
[3] Y.H. Na y R.J. Friedman, "Concise review of mechanisms of bacterial adhesion to biomaterial surface", Journal of Biomedical Materials Research, Vol. 43 (1998), p. 338-348.
[4] J.W. Arnold and G.W. Bailey, "Surface finishes on stainless steel reduce bacterial attachment and early biofilm formation: scanning electron and atomic force microscopy study", Poultry Science, Vol. 79 (2000), p. 1839-.1845
[5] M. Katsikogianni y Y.F. Missirlis, "Concise review of mechanisms of adhesion to biomaterials and of techniques used in estimating bacteria-material interactions", European Cells and Materials, Vol. 8. (2004) p. 37-57.
[6] Spector, M. Peretti L. Salas, F, Romero, G, Iglesias, L. Bacterial Conduction in Prostheses. Procedia Materials Science, Volume 8, 2015, Pages 351–357.
[7] Cheng VC, Chan JF, Ngan AH, To KK, Leung SY, HW Tsoi, Yam WC, Tai JW, Wong SS, Tse H, Li IW, Lau SK, Woo PC, Leung AY, Lie AK, Liang RH , Que TL, Ho PL, Yuen KY: Outbreak of intestinal infection due to Rhizopus microsporus. J Clin Microbiol 2009; 47: 2834-2843.
[8] Gomes MZ, Lewis RE, Kontoyiannis DP: Caused by mucormycosis mucormycetes unusual, non-Rhizopus, -Mucor, and -Lichtheimia species. Clin Microbiol Rev 2011; 24: 411-445.

Optimization of a surface wave elastography method through diffraction and guided waves effects characterization

G A Grinspan[1], S Aguiar[2], N Benech[3]

[1]Sección Biofísica, Facultad de Ciencias, Universidad de la República, Montevideo, Uruguay
[2]Instituto de Ensayo de Materiales, Facultad de Ingeniería, Universidad de la República, Montevideo, Uruguay
[3]Laboratorio de Acústica Ultrasonora, Facultad de Ciencias, Universidad de la República, Montevideo, Uruguay

E-mail: gustavogrinspan@hotmail.com

Abstract. Soft biological tissue elasticity is a parameter whose reliable measure is relevant to many applications in fields as diverse as medicine and the agrifood industry. The ultrasonic elastography methods are often unviable to be applied to provide such solutions. In this way, the surface wave elastography (SWE) appears as a viable alternative, due its low cost, easy to use, non-invasive-destructive character as well as its ability to provide *in vivo* estimates. Previous studies have described a good correlation between the overall behavior of ultrasonic elastography and SWE, although the latter overestimates the elasticity values compared to the first. It has been suggested that this is due to the influence of certain physical effects related to the exclusive use of low frequency waves, as well as by characteristics of the experimental setup and/or medium. In this work we confirm the influence of such effects and discuss different strategies to make independent the estimations thereof. This allows achieving a good agreement between the ultrasonic reference method and SWE. Thus, SWE becomes a reliable method to estimate soft biological tissue elasticity.

1. Introduction

Elasticity is an important parameter to assess in those areas where it is relevant to infer the intrinsic mechanical state of soft solids. Some examples of soft solids include rubber, gels, agrifoods and soft biological tissues. The interdisciplinary study of the mechanical properties of this type of medium, allows addressing complex problems of different areas of knowledge, with potential applications in different fields. For example, the elasticity of skeletal muscle is a parameter closely related to relevant factors both for the meat industry as well as for the rehabilitative medicine, such as the tenderness of meat and muscle tone, respectively.

The elastography research field has contributed greatly in recent years, providing numerous studies and methods to evaluate the Young modulus (Y) of this type of medium in a non-destructive and non-invasive way. Particularly, the transient elastography (TE) has positioned as a reference in the field of elastography. Thus, its applications have served both to determine the elasticity of soft biological tissues *in vivo* as skeletal muscle ([1], [2], [3]), or *in vitro* as agar-gelatin phantoms and meat ([4], [5]). However, the dependence between its functioning and the presence of multiple diffusers into the medium (which originates the speckle acoustic signals), coupled with the high costs of implementation due to the requirements of its electronics, implies that the use of the ET is unviable within certain specific areas.

An alternative to ultrasonic methods has been the exclusive use of low-frequency waves. In this sense, it is possible to find works that refer to the use of surface waves (or Rayleigh waves) to determine the mechanical properties of soft solids ([6], [7], [8], [9], [10], [11], [12]). Particularly, an SWE method composed of a linear array of piezoelectric sensors, was proposed by the authors of [13] to measure the elasticity of the human brachial biceps *in vivo*. In this work are shown dispersion curves that describe how at frequencies in the order of 100 Hz and higher, the behavior of the method have a good agreement with ET, although it overestimates the elasticity values with respect to the latter. It is suggested that this could be due to diffraction and guided waves effects, being then necessary confirm the presence of such effects and to develop strategies that allow obtaining accurate estimations of elasticity.

Therefore, the objective of this work is to study in detail the impact of diffraction and guided waves effects on the SWE method proposed in [13], as well as alternatives are proposed to make independent the estimations of such effects.

2. Materials and methods

2.1. Surface wave elastography method (SWE)

The SWE involves the propagation of waves across the surface of medium, to estimate its elasticity through the study of such vibrations. In this sense, the experimental setup that composes this method is an external source (mechanical vibrator), whose frequency range is determined by the particular properties of the medium of interest. Also, depending on the specific vibrator used, this may have, or not, an attached rod to transmit the vibration to the medium. In this sense, we used a speaker with an attached piston, which generates a cycle of sinusoid with a central frequency between 50 and 250Hz. The vibration is normal to the surface, so excites mainly the vertical component of the surface waves. The wave thus generated is recorded by a linear arrangement of four piezoelectric vibration sensors, which are disposed on the surface of medium and record the vertical component of the vibration. Subsequently, an 8-channel data acquisition board (converting A/D) acquires the recorded signals and transfers them to a computer to the data processing (figure 1).

2.2. Signal processing

As we have seen previously, the method is composed by 4 sensors aligned with the exciting source. These sensors capture the acoustic vibration producing a potential difference between its terminals that is digitalized by the data acquisition board (figure 1).

From the digitized signals, the main objective of the method is to estimate the temporal delay between the signals recorded by the sensors of the linear array (figure 2a). Once the delay is estimated, as the distance d between sensors is fixed, the speed of the vibration in the medium can be obtained. The first step in the signal processing consists of applying a bandpass filter centered on the emitter frequency and a bandwidth of 30%. This procedure allows eliminating the unwanted frequencies in the signal that may affect the estimation of the surface wave speed.

The second step is to estimate the delay between consecutive signals. To do this, we used an correlation algorithm of signals based on normalized covariance between them:

$$R(j) = \frac{\sum_i (x(i)-\bar{x})(y(i+j)-\bar{y})}{\sqrt{\sum(x(i)-\bar{x})^2 \sum(y(i)-\bar{y})^2}} \quad (1)$$

In the above expression x and y represent two acoustic signals recorded by two consecutive sensors, and the values of indexes i, j represent the time (discrete because the signal is sampled at a given frequency fs). The value of j, which maximizes the value of R is the first step to estimate the shift δt between two consecutive signals: $\delta t = j/fs$. In this way the value of temporary delay that can be measured is given by discrete steps limited by the sampling frequency fs.

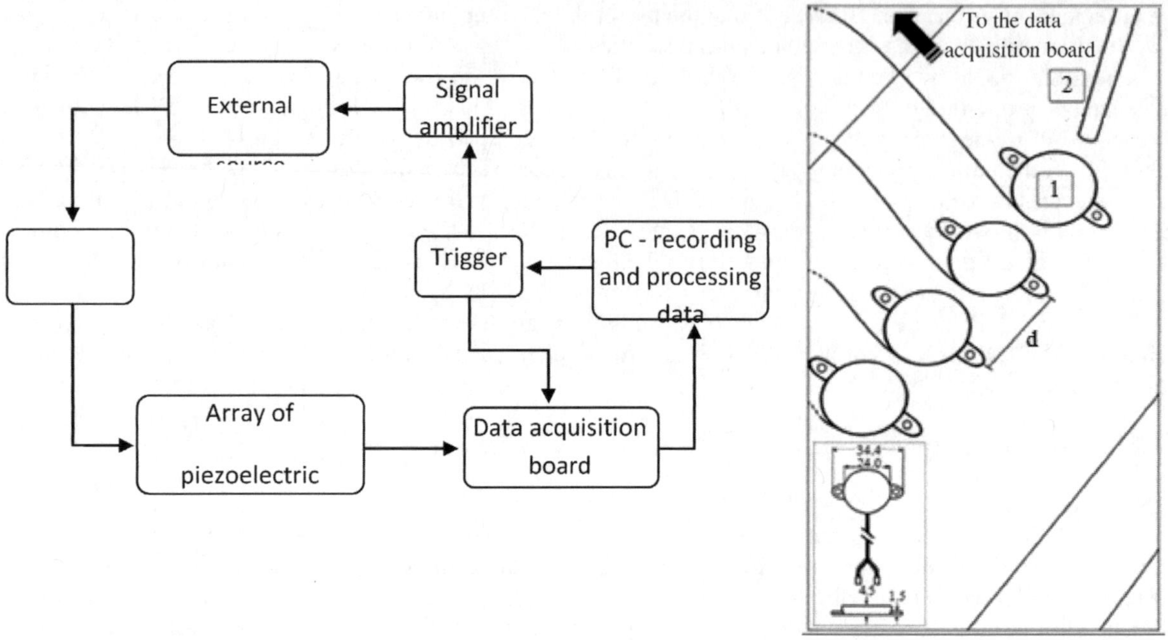

Figure 1. Flowchart of the experimental setup of SWE method and linear array of piezoelectric vibration sensors (1) aligned with the rod attached to the mechanical vibrator (2).

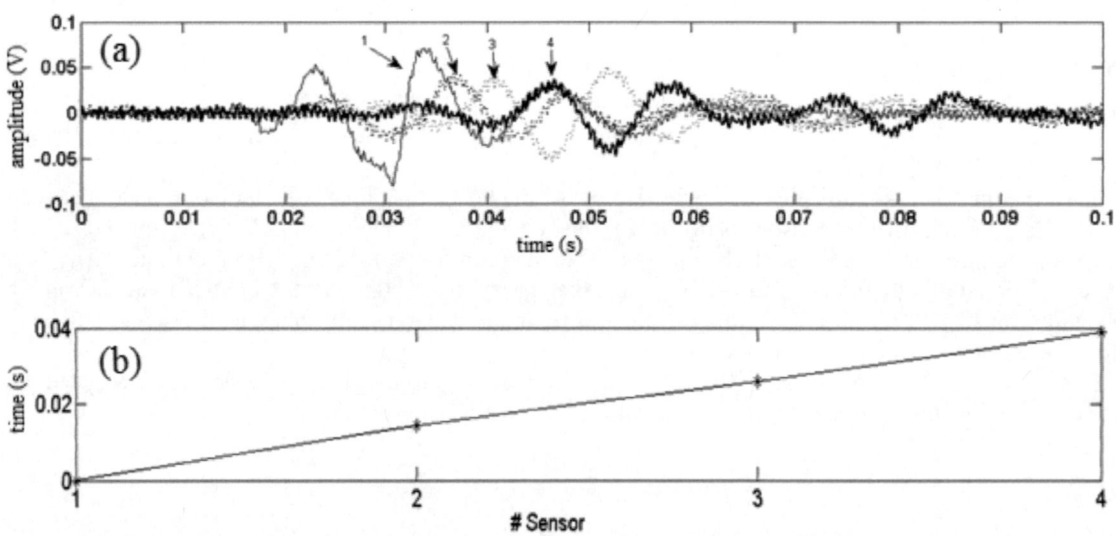

Figure 2. (a) typical signals recorded by the vibration sensors. (b) accumulative sum of the time-shift between sensors, which determines the time of arrival of surface wave to each.

To overcome this limitation, the third step in the signal processing consists of making a parabolic interpolation of the correlation coefficient R around the maximum. We chose the parabolic interpolation because it is computationally efficient. Assuming for the first sensor time 0, the final result is a vector of 4 inputs, each one with the temporary shift of the signal between sensors. The accumulative sum of these times provides the arrival time of the surface wave to each sensor (taking as origin the arrival time to the first sensor). Finally, we get a $\delta t(r)$ curve, where r represents the position (known) of each sensor (figure 4b). The next step in the signal processing is to perform a linear fit of the $\delta t(r)$ curve. The inverse of the slope of this fit is the Rayleigh surface wave speed. There is a relationship between the Rayleigh waves speed and the shear waves c_S given by the expression known as Viktorov formula [14]:

$$\frac{c_R}{c_S} = \frac{0{,}718-(c_S/c_P)^2}{0{,}75-(c_S/c_P)^2} \tag{2}$$

where C_p is the compression waves speed. Since in soft solids the elastic modulus of compression (λ) is much greater than the elastic modulus of shear (μ), we have that $C_p \gg C_s$. As a result we can obtain an approximate expression that relates the Rayleigh wave speed with the shear wave speed:

$$\frac{c_R}{c_S} \approx \frac{0{,}718}{0{,}75} = 0{,}96 \Rightarrow c_R = 0{,}96c_S \tag{3}$$

Therefore, and according to the fact that in soft solids the speed of shear wave is related to Young's modulus by $Y = 3\rho {c_S}^2$ (where ρ is the density of the medium), by measuring the speed of Rayleigh waves through the procedure described, we can finally estimate Y by:

$$Y = 3\rho(1{,}04c_R)^2 \tag{4}$$

2.3. Experimental setup

It was used the SWE method in different experimental conditions, with the objective of characterizing the diffraction and guided waves effects, as well as to analyze different strategies to make the estimations independent of such effects.

We perform measurements in agar-gelatin phantoms, which simulate the mechanical properties of soft biological tissues. Two phantoms were manufactured: N°1: 2.5% of gelatin, 1% agar, dimensions 12.0 x 12.0 x 4.0 cm^3 (length, width and height); N ° 2: 3.5% of gelatin, 1.5% agar, dimensions 12.0 x 12.0 x 4.0 cm^3. Each of these phantoms were measured intact, as well as after cutting them at the middle of height, being then its dimensions 12.0 x 12.0 x 2.0 cm^3. Furthermore, the phantoms were measured with different values of frequency, varying within the range of 50-160Hz.

3. Results and discussion

3.1. Experimental confirmation of the incidence of diffraction and guided wave effects

Results presented in [13] denote that the SWE is a useful elastography method to evaluate changes in the mechanical properties of soft biological tissues. However, although this version has shown characterize analogously the changes in the elasticity of brachial biceps under different situations of static load, it is appreciated that overestimates the values of C_R with respect to the real values provided by the ET (figure 7). It is suggested that this may be due to the influence of diffraction as well as geometric dispersion effects associated with guided waves.

The diffraction effect occurs when the separation between the source and the receiver is in the order of the wavelength and the source has comparable dimensions to the latter. In addition to the overestimation of velocity values, this produces a characteristic dispersive effect whereby the value of C_R decreases as increases the frequency of the wave [15]. In this regard, figure 3 clearly reflects the impact of the diffraction effect on the experiences made with phantoms N°1 and N°2, coinciding with the frequency analysis displayed in [13] obtained from the estimates in braquial biceps. The characteristics of the experimental setup used in both cases allow us to assume a punctual source, since your radius is much less than the wavelength in soft solids at the typical frequencies (between 10 and 60 mm, depending on the speed). However, the size of the piezoelectric sensors (10mm radius) is not depreciable relative to the wavelength and the separation between them is small compared to the same. As a result, and given the reciprocity emitter/receiver of the acoustic field generated, we must consider the diffraction effects associated with the reception of waves.

The guided waves effect occurs when the height of the sample is comparable to or less than wavelength. From the analytical point of view, the effect of guided waves is more complex than the diffraction, since there are two types of modes of guided waves, and each mode has in principle several possible orders of propagation. Fortunately, in recent years it has been reported that in soft solids the predominant mode (or the only possible depending on the experimental conditions) at low frequencies is known as antisymmetric mode of order 0 or *A0* [12]. This mode has an inverse dispersion condition with respect to the diffraction effect, since by increasing the frequency and/or the height of the sample, the apparent speed is higher and closer to the real speed (Figure 4). This is reflected in the results obtained with phantoms N°1 and N°2 halved by a horizontal cut (Figure 5). Compared with figures 3a and 3b, we can see the maintenance of the downward trend of the speed according to the frequency, but numerical values are lower. The downward trend is maintained because the effect of diffraction is still present and it is dominant with respect to guided waves, but the decrease of numerical values is attributable to this effect since the dispersive pattern operates in the opposite direction. From previous experiences, it can be concluded that to avoid the effect of guided waves, it is suitable to work with high-altitude samples (compared with the wavelength) or in equivalent form, increase the frequency f_0 of the surface wave. For the applications of this method, it is not feasible to modify at will the height h of the samples. Therefore, it is necessary to choose appropriately the optimal operating frequency according to the application itself.

3.2. Correction of diffraction from analytical expressions

In some particular cases it is possible to obtain an analytic expression for the diffracted wave field. For surface waves in elastic media, the authors of [16] obtained an expression for the radiation at frequency ω of a circular piston acting normally to the free surface of solid:

$$u_z(r) = \frac{ae^{i\omega t}}{\mu}\int_0^{\infty} \frac{J_1(aq\zeta)\sqrt{\zeta^2-1}}{F_0(\zeta)}[2\zeta^2+\eta^2-2\zeta^2]J_0(\zeta rq)\,d\zeta \tag{5}$$

In the above expression $u_z(r)$ is the normal component of the field at distance r from the source, a is the radius of the circular piston, q is the number of wave of compression waves, η is the number of wave of shear wave normalized with respect to q, J_n is the Bessel function of order n, ζ is the variable of integration with respect to q and F_0 is the Rayleigh function:

$$F_0(\zeta) = (2\zeta^2-\eta^2)^2 - 4\zeta^2\sqrt{\zeta^2-\eta^2}\sqrt{\zeta^2-1} \tag{6}$$

Given the reciprocity emitter/receiver of the acoustic field generated, the expression (5) allows us to compute the module and the phase of the field emitted by a punctual source and received by a circular sensor of radius a, as in our experimental arrangement. Because the integral expressed in (5) has no known analytical solution in the near field, we must resort to numerical solutions to know the module and phase of the field in each sensor of the linear array at a given frequency ω. This calculation allows us to compute the apparent speed c_a according to the frequency through:

$$c_a = \frac{\omega}{k} = \omega\left(\frac{\Delta\varphi(\omega)}{\delta}\right)^{-1} \tag{7}$$

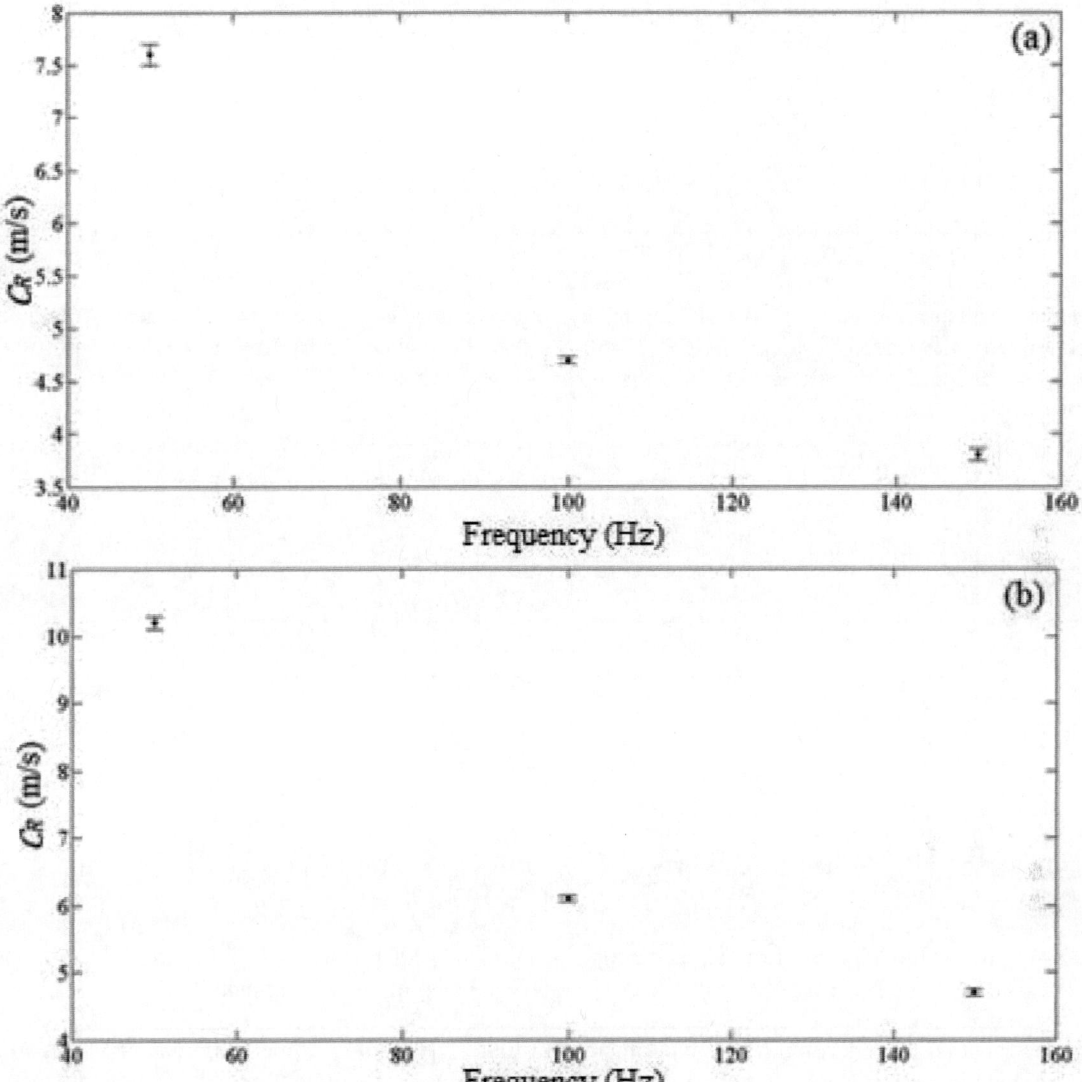

Figure 3. Dispersion curve of Rayleigh wave speed obtained in phantoms N°1 (a) and N°2 (b) at 50, 100 and 150Hz. Error bars correspond to the standard deviation on 15 measures.

where k represents the apparent number of wave, $\Delta\varphi(\omega)$ is the phase difference between the two sensors to the frequency ω and δ is the distance between them. We should note that to obtain numerically the phase difference $\Delta\varphi(\omega)$ from the expression (5), it is necessary to know the speed of the shear waves (because the shear modulus μ is present in the expression). But this is precisely the purpose of the SWE method. To find it, we propose to use an inversion algorithm based on the model of diffraction for an only experimental measure. The range of speed of the shear waves introduced into the algorithm is based on certain prior knowledge (in the case of agar-gelatin phantoms we set the range in 1-6 m/s). In this way, using the model of [16] we obtain numerically a curve $c_a(c_s)$ at a given frequency (we should note that all other experimental parameters remain fixed for a given configuration of the source and sensors). From this curve, we can then introduce the experimental result of the apparent speed and obtain the corresponding value of shear waves as shown in figure 6.

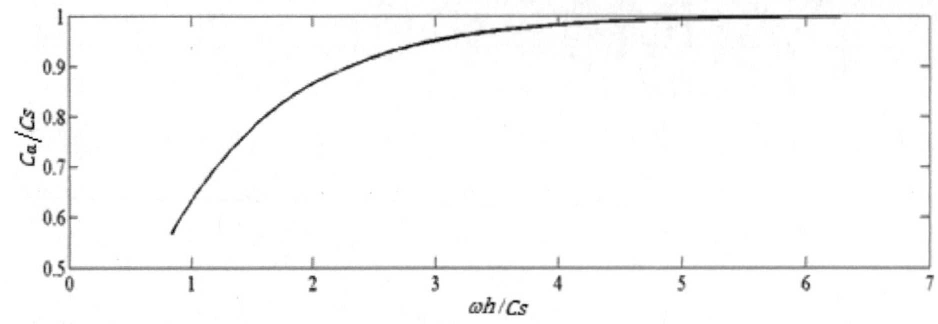

Figure 4. Dispersion curve to the *A0* mode. c_a: apparent speed; c_s: shear waves speed; ω: angular frequency of the wave; h: sample height. It follows that the higher the variable $x = \omega h/c_s$, nearest is the apparent speed to the real speed.

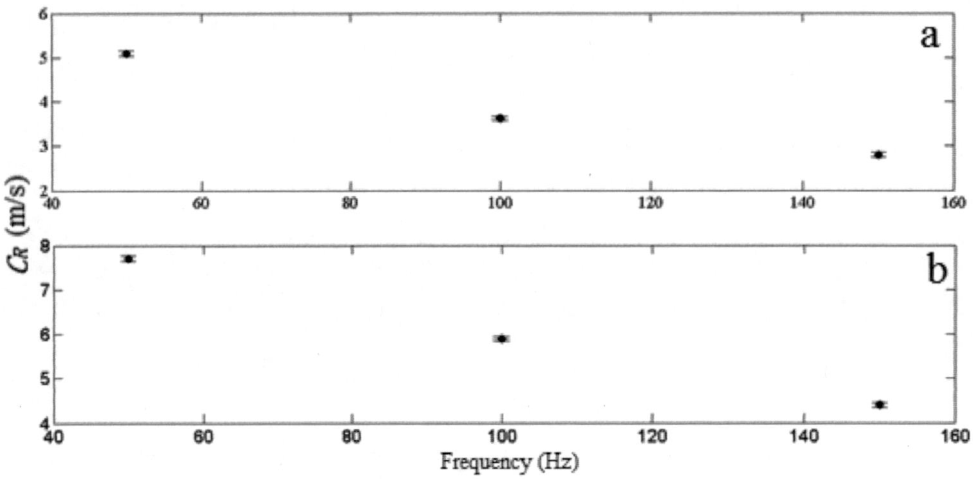

Figure 5. Rayleigh wave speed in function of frequency for phantoms N°1 (a) and N°2 (b) with h = 2cm. Error bars correspond to the standard deviation on 15 measures.

We perform this procedure with the experimental results of [13] for the measurement of elasticity in brachial biceps under different load conditions. The results shows a significant reduction in the diffraction effect, whereby the SWE already not overestimates the values of c_s, but on the contrary the results fit very well to the values provided by the ET (figure 7). For its part, after to perform this procedure with the experimental results obtained with gels N°1 and N°2, it is observed that our inversion algorithm retrieves the value of c_s for all frequencies in both gels (figures 8a and 8b). In this way, beyond the experimental uncertainty, the velocity values already not evidence the dispersive pattern of diffraction effect, as may be expected in an elastic solid. Likewise, if we apply the inversion algorithm to the results obtained with the phantoms N°1 and N°2 halved by a horizontal cut, we see that trend is modified with respect to figure 5, eliminating the effect of diffraction since now the speed of the Rayleigh wave increases in accordance with the increase of frequency (figures 9a and 9b). This is in agreement with the theory for guided waves taking into account the experimental conditions (h = 2cm). In addition, we can see that the value obtained at 150 Hz retrieves the speed value shown in figures 8a and 8b (h = 4cm) to the same frequency. The latter has important connotations related with avoid the effect of guided waves, since we can conclude that if we put our experimental condition for values of $x > 3$ (figure 4) it is possible to depreciate the incidence of this effect.

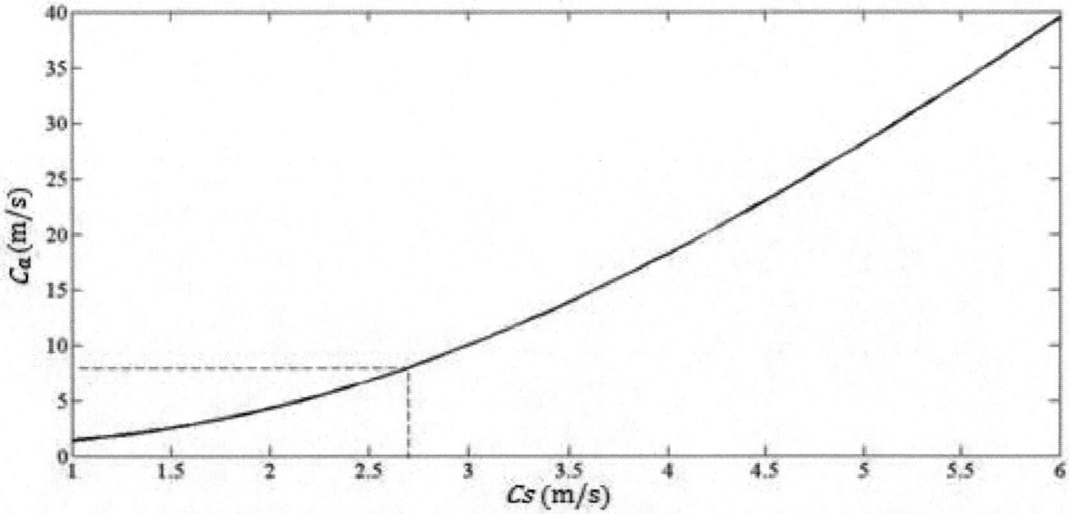

Figure 6. Apparent speed (c_a) in function of shear wave speed (c_s) for a fixed frequency (50Hz), from which we can estimate c_s knowing the c_a measured experimentally. Here we display the estimation for phantom N°1 at 50 Hz, whose estimated value is $c_s = 2.72\, m/s$.

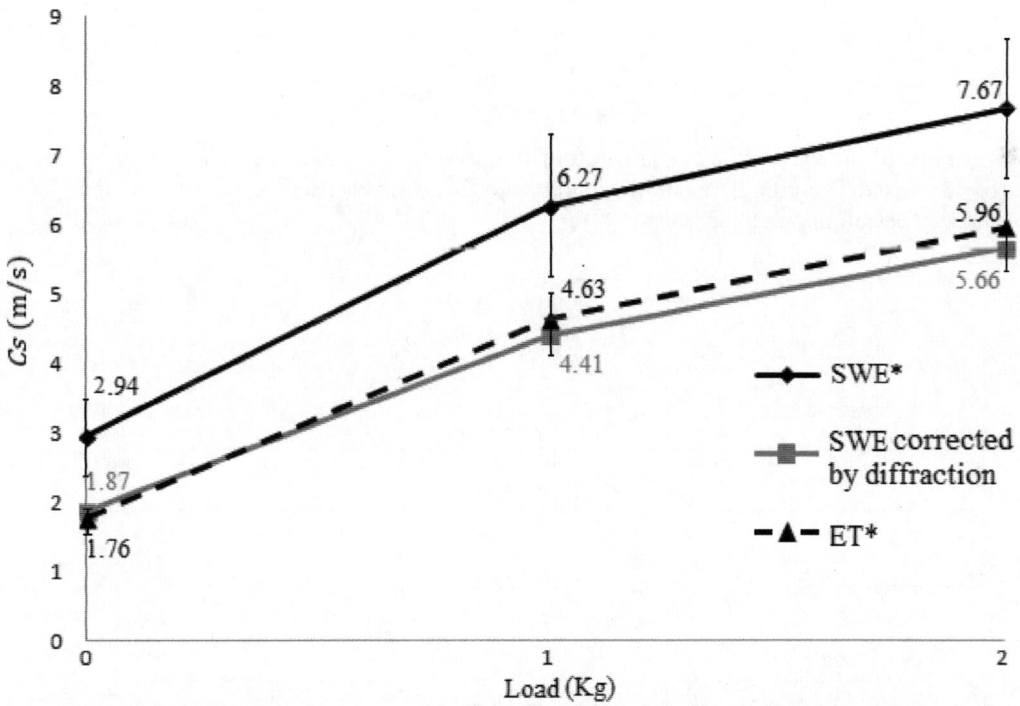

Figure 7. Numerical correction of diffraction effect applied to measurements of muscular elasticity of [13] (*).

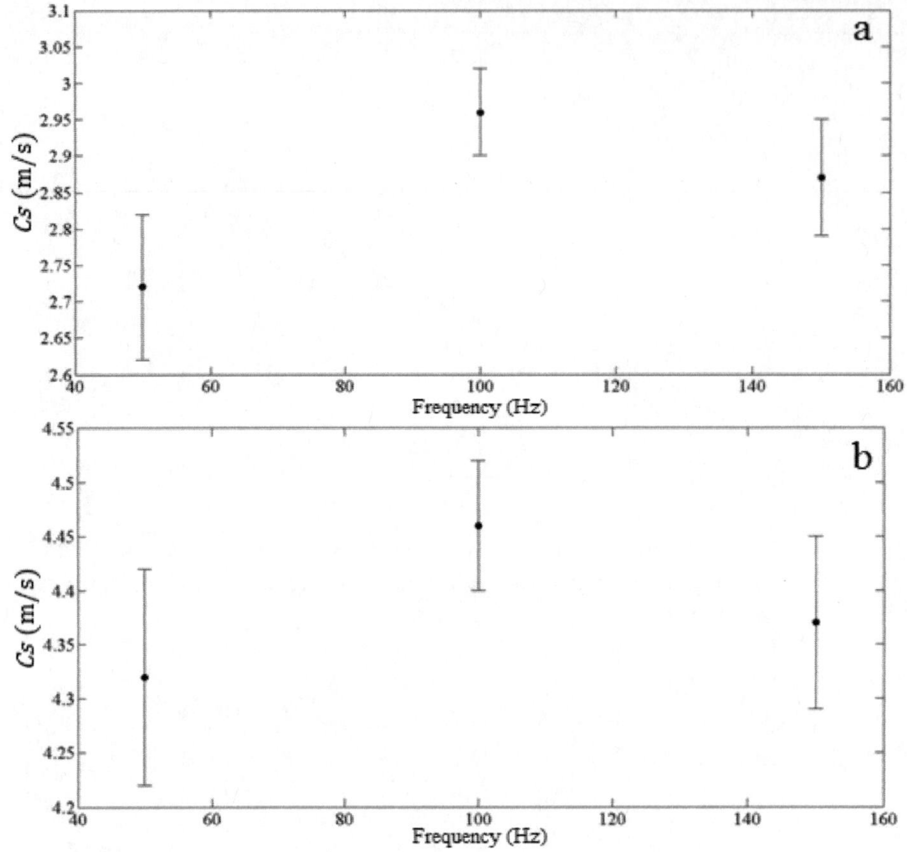

Figure 8. Shear wave speed in function of frequency for phantoms N°1 (a) and N°2 (b), corrected according to our inversion algorithm. Error bars correspond to the standard deviation on 15 measures.

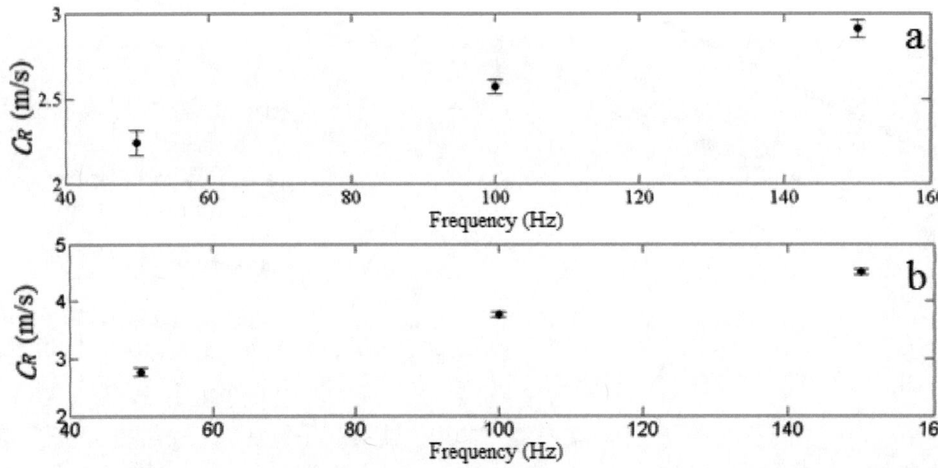

Figure 9. Rayleigh wave speed in function of frequency for phantoms N°1 (a) and N°2 (b) with h = 2cm, corrected according to our inversion algorithm. Error bars correspond to the standard deviation on 15 measures.

4. Conclusions

The present work shows that the SWE is a viable alternative to characterize the biomechanical properties of soft biological tissues. It have been confirmed the hypotheses proposed in [13], about that the exclusive use of low-frequency waves induces diffraction and guided waves effects. This results in an overestimation of the elasticity in relation to the ultrasonic reference methods. In this sense, the inversion algorithm proposed to correct numerically the diffraction and the criterion to depreciate the guided waves effects, have allowed achieving the independence of such effects and adjusting very well the results to the ET. However, beyond these strategies of correction are effective, are not sufficiently efficient for the applications of interest since they require a further processing of data and do not allow obtaining real-time estimates.

The non-invasive character, in vivo applicability, easy use and low cost, are features of the SWE that open up a wide spectrum in terms of their potential applications. This work represents a significant advance in what refers to the understanding of these methods, its limitations and how to overcome them. Developing strategies that allow to correct in real time the incidence of the described effects, is the key step to achieve that the SWE can offer the effective solutions that are necessary in those fields that do not have objective methods to quantifying the soft biological tissue elasticity. Rehabilitative medicine and agrifood industry (e.g. meat industry) are clear examples of the latter.

5. References

[1] Gennisson J L, Cornu C, Catheline S, Fink M and Portero P 2005 *J. Biomech.* **38** 1543
[2] Gennisson J L, Catheline S, Chaffaï S and Fink M 2003 *J. Acoust. Soc. Am.* **114** 536
[3] Nordez A, Gennisson J L, Casari P, Catheline S and Cornu C 2008 *J. Biomech.* **41** 2305
[4] Benech N, Negreira C and Brito G 2012 *Elastografía ultrasónica para evaluación de la terneza de carne vacuna* (Montevideo, PROCISUR publicaciones).
[5] Catheline S, Gennisson J L , Delon G, Fink M, Sinkus R, Abouelkaram S and Culioli J 2004 *J. Acoust. Soc. Am.* **116** 3734
[6] Kazarov V V and Klochkov B N 1989 *Biophysics* **34** 742
[7] Sarvazyan A, Ponomarjev V, Vucelic D, Popovic G, Vexler A 1992 *EP Patent* **0329817** B1
[8] Courage W 2003 *US Patent* **6619423** B2
[9] Sabra K G, Conti S, Roux P, Kuperman W A 2007 *Appl. Phys. Lett.* **90** 194101
[10] Salman M and Sabra K G 2013 *J. Acoust. Soc. Am.* **133** 1245
[11] Zhang X, Qiang B and Greenleaf J 2011 *Ultrasonics* **51** 157
[12] Brum J, Gennisson J L, Nguyen T M, Benech N, Fink M, Tanter M and Negreira C 2012 *IEEE Trans. Ultras. Ferroelec. Freq. Contr.* **59** 703
[13] Benech N, Aguiar S, Grinspan G A, Brum J and Negreira C 2012 *Proc. IEEE Ultras. Symp.* **2012** 2571
[14] Diuelesaint E and Royer 1980 D *Elastic waves in solids* (New York: Wiley)
[15] Ruiz A and Nagy P B 2002 *J. Acoust. Soc. Am.* **112** 835
[16] Miller G and Pursey H 1954 *Proc. R. Soc. A.* **223** 521

Effects of walking speed and age on the muscle forces of unimpaired gait subjects

Carlos G. Fliger[1], Marcos J. Crespo[2], Ariel A. Braidot[1], Emiliano P. Ravera[1,3]

[1]Laboratory of Biomechanics, School of Engineering, National University of Entre Ríos, 3100, Argentina.
[2]Gait and Motion Analysis Laboratory, FLENI Institute for Neurological Research, B1625XAF, Argentina.
[3]National Council of Scientific and Technical Research (CONICET), Argentina.

E-mail: emilianoravera@bioingenieria.edu.ar

Abstract. Clinical gait analysis provides great contributions to the understanding of gait disorders and also provides a mean for a more comprehensive treatment plan. However, direct measures of muscle forces are difficult to obtain in clinical settings because it generally requires invasive techniques. Techniques of musculoskeletal modeling have been used for several decades to improve the benefits of clinical gait analysis, but many of the previous studies were focused on analyzing separately the muscle forces distribution of children or adult subjects with only one condition of walking speed. For these reason, the present study aims to enhance the current literature by describing the age and speed gait effects on muscle forces during walking.

We used a musculoskeletal model with 23 degrees of freedom and 92 musculotendon actuators to represent 76 muscles in the lower extremities and torso. The computed muscle control algorithm was used to estimate the muscle forces from the kinematics and to adjust the model obtained in the residual reduction algorithm.

We find that hamstrings has an important peak in the mid-stance phase in the adult group but this peak disappears in the children group with the same walking speed condition. Furthermore, the rectus femoris presents an increase in the muscle force during the pre- and mid-swing in concordance with the increment in the walking speed of subjects. This behavior could be associated with the role that the rectus femoris has in the acceleration of the knee joint. Finally, we show that the soleus is the muscle that perform the major force throughout the gait cycle regardless of age and walking speed.

1. Introduction

Clinical gait analysis provides great contributions to the understanding of gait disorders and also provides a mean for a more comprehensive treatment plan [1, 2]. However, a complete distribution of dynamic muscle forces while walking is a challenge for many researchers [3, 4]. Direct measures of muscle forces are difficult to obtain in clinical settings because it generally requires invasive techniques. Computational models that represent the human locomotor system are being currently proposed to sort out those limitations [3].

A musculoskeletal model represents a numeric set of anatomical parameters to quantify their interaction. Hence, the muscles are described as a simple line between patches of origin and insertion, while the joints are represented as fixed centers of rotation [5,6]. Estimation of muscle forces using musculoskeletal models usually requires solving an optimization problem regardless of the method used to solve the equations that describe the dynamics musculoskeletal system, inverse or forward dynamics.

Techniques of musculoskeletal modeling have been used for several decades to improve the benefits of clinical gait analysis. Just in gait, *Seireg and Arvikar (1975)*, *Crowninshield et al. (1978)*, *Rohrer et al. (1984)* minimizing the sum of muscle forces to analyze the relationship between the net joints moments and muscle forces of the lower limb. *Patriarco et al. (1981)* found the muscle chemical-mechanical power, in addition to minimizing the muscle forces. *Crowninshield and Brand (1981)*, *Brand et al. (1986)*, *Glitsch and Baumann (1997)* and *Pedersen et al. (1997)* maximize the endurance by minimizing the muscle stress cubed [3]. Recent works like *van der Krogt et al. (2012)* [7] used the computed muscle control algorithm developed by *Thelen et al. (2003)* [8] to analyze the lower limb muscles weakness throughout the gait cycle in a group of adult subjects that walking at self-selected comfortable speed. *Ravera et al. (2014)* [9], developed a musculoskeletal model of the lower limb that simulate the smoothed EMG data through the sum of different Gaussian bells to analyze the muscle forces throughout the gait cycle in a group of children that walking at self-selected comfortable speed.

Many of the previous studies were focused on analyzing separately the muscle forces distribution on children or adults subjects, and generally in groups that walk at self-selected comfortable walking speed. For these reasons, the present study aims to enhance the current literature by describing speed gait effects on muscle forces of the lower limbs for a group of children and adult subjects that typically develop an unimpaired walking at three self-selected walking speeds. The aim of our study is to propose a single comprehensive source for muscle forces at a wide range of speeds and ages. We intend not to explain the underlying biomechanics of the speed and age variations, but rather to provide a thorough report of the data in a single source.

2. Methods

2.1. Participants

Two groups participated in this study (Table 1) to obtain a reliable data sample for a wide range of walking conditions. The first group included five children (7-14 years of age, 1.24-1.67 m in height and 22-53 kg in mass) that walk at a self-selected, comfortable walking speed. For operative reasons, the children were measured only once. The second group included eight adult subjects (23-42 years of age, 1.61-1.83 m in height and 63-83 kg in mass) that walk at three self-selected walking speeds (comfortable or free, slow and fast). After sufficient practice attempts, six successful trials were recorded for each speed condition. This resulted in a total of forty-eight trials per speed condition for the adults and ten trials for the children.

Walking speeds were rendered dimensionless using leg length (L_{leg}) following the scheme proposed by *Hof (1996)* [10]:

$$v^* = \frac{v}{\sqrt{gL_{leg}}} \ . \tag{1}$$

Both groups were examined by the Gait and Movement Laboratory at FLENI Institute for Neurological Research (Escobar, Argentina) team, and they exhibited normal gait patterns. This study was reviewed and approved by the research Ethics Committee. The protocol was explained to each subject, and either the participants or their caregivers signed informed consents.

Kinematic data were recorded using an Elite 2002 motion capture system (BTS Bioengineering, Italy) with 8 cameras ($100Hz$) and two force plates (Kistler 9281E, Kistler Group, Switzerland). Twenty-two retro-reflective skin markers were placed over bony landmarks (as indicated in the protocol of *Davis et al. (1991)* [11]). Electrical muscle activity data was recorded from the rectus femoris, semimembranosus, gastrocnemius and tibialis anterior muscle

Table 1: Subject characteristics and walking speeds.

Subject	Gender	Age (years)	Mass (Kg)	Leg length (m)	Slow speed (m/s) (non-dimensional)	Free Speed (m/s) (non-dimensional)	Fast speed (m/s) (non-dimensional)
Children							
1	F	11	40	0.79	-	1.32 (0.47)	-
2	M	14	53	0.89	-	0.98 (0.33)	-
3	F	7	22	0.65	-	1.45 (0.57)	-
4	F	8	38	0.69	-	1.34 (0.51)	-
5	F	9	30	0.72	-	1.02 (0.39)	-
Mean	-	9.80	36.60	0.75	-	1.27 (0.48)	-
SD	-	2.77	11.61	0.09	-	0.19 (0.09)	-
Adult							
1	M	28	78	0.95	1.04 (0.34)	1.27 (0.42)	1.55 (0.51)
2	M	42	79	0.95	1.08 (0.35)	1.11 (0.36)	1.80 (0.59)
3	M	29	75	0.94	0.93 (0.30)	1.37 (0.45)	1.44 (0.47)
4	M	27	64	0.86	0.93 (0.32)	1.08 (0.37)	1.15 (0.39)
5	F	30	63	0.81	0.92 (0.33)	1.12 (0.40)	1.55 (0.55)
6	M	29	83	1.00	1.01 (0.32)	0.99 (0.32)	1.03 (0.33)
7	F	23	68	0.92	0.58 (0.19)	1.09 (0.36)	1.24 (0.41)
8	M	27	77	0.91	0.94 (0.31)	1.06 (0.36)	1.34 (0.45)
Mean	-	29.37	73.37	0.92	0.93 (0.31)	1.14 (0.38)	1.39 (0.46)
SD	-	5.53	7.42	0.06	0.15 (0.05)	0.12 (0.04)	0.25 (0.08)

using a Teleemg surface dynamic electromyograph (BTS Bioengineering, Italy) with a sampling frequency of 2000 Hz [12].

2.2. Musculoskeletal Model

We used a musculoskeletal model, available in OpenSim (Fig. 1), with 23 degrees of freedom and 92 musculotendon actuators to represent 76 muscles in the lower extremities and torso. The degrees of freedom in this model included three translations and three rotations of the pelvis; three ball and socket joints, one located at the third lumbar vertebrae to model the interaction between pelvis and trunk and two at each hip joints; a custom joint with coupled translations and rotations at each knee; and a revolute joint at each ankle [13, 14].

OpenSim is an open-source platform for modeling, simulating, and analyzing the neuromusculoskeletal system. It includes low-level computational tools that allows one to derive equations of motion for dynamical systems, perform numerical integration, and solve constrained non-linear optimization problems. It offers access to control algorithms, actuators, and analyses; by integrating these components into a modeling and simulation platform [13].

In addition, SimTrack (one of the OpenSim tools) enables researchers to generate dynamic simulations of movement from motion capture data. This guides users through four steps to create a dynamic simulation: 1) a dynamic musculoskeletal model is scaled to match the anthropometry of an individual subject, 2) an inverse kinematics (IK) problem is solved, 3) a residual reduction algorithm (RRA) is applied and 4) the computed muscle control (CMC) is used to generate a set of muscle excitations.

Inverse kinematics and inverse dynamics, from the experimental marker trajectories and ground reaction forces, were used to calculate joint angles and moments. A dynamic simulation of one gait cycle was generated for each trial of each subject. The RRA was used to reduce residuals at the pelvis. This was proposed to minimize the effects of modeling and marker data processing errors that aggregate and lead to large nonphysical compensatory forces called residuals. Specifically, it altered the torso mass center of a subject-specific model and adjusted the kinematics of the model from inverse kinematics in order to be dynamically consistent with

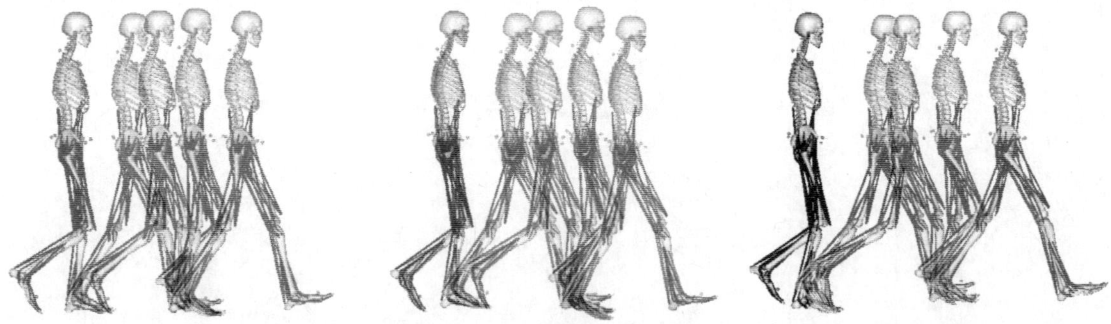

Figure 1: Musculoskeletal model used to generate three-dimensional simulations throughout the gait cycle. The musculotendon actuator colors indicate the level of activation on a scale from dark blue (no activation) to bright red (full activation).

the ground reaction force plate data. The CMC algorithm [8] was used to estimate the muscle forces from the kinematics and to adjust the model obtained in RRA. CMC used the static optimization approach along with feedforward and feedback controls to drive the kinematic trajectory of the musculoskeletal model toward a set of desired kinematics. In particular, the optimization criteria or cost function (2) was performed to be the sum of squared muscle activations [15, 16]:

$$J = \sum_{m=1}^{76} (a_m(t))^2 \ , \tag{2}$$

where $a_m(t)$ is the steady-state activation of muscle m. All steps of our musculoskeletal simulations were implemented in concordance with the "best practices to verification and validation of musculoskeletal models" discussed by *Hicks et al. (2015)* [17].

3. Results

Figure 2 shows the mean and standard deviation of the sagittal, coronal, and transverse plane kinematics. It shows the same speed-related effects in the joint angles that were presented by *Schwartz et al. (2008)* [18]. In particular, we highlight a scaled nearly linear rise in hip flexion angle at initial contact and a plateau in ankle dorsiflexion angle of adult subject. Also, an increase in the platarflexion angle, hip flexion angle and hip rotation angle was found in children throughout the gait cycle in contrast with adult subjects.

The mean and standard deviation of net joint moments of hip, knee and ankle are shown in Figure 3. Similar morphological patterns are found in both age groups, in particular a slight rise was found in the joint moment peaks when adult subjects increase their walking speed. Abduction/adduction hip joint moments present two main peaks at 10% and 50% of the gait cycle respectively. Only the first peak shows a significant increase when the subjects walk faster.

Prior to analyzing the results of the musculoskeletal simulations for each participant through OpenSim, we compared the muscle activations from CMC with the EMG signals recorded for each individual trial, which were qualitatively similar (i.e. present their peaks in sames areas of the gait cycle and are morphologicaly similar), as shown in Figure 4.

Although the simulations performed in OpenSim includ a complete set of lower limb muscles, as was described in section 2.2, for the later analysis we have considered the behaviors of the Gluteus Maximus (GM), Gluteus Medius (Gm), Hamstrings (Ham), Vastus (Vas), Rectus Femoris (RF), Gastrocnemius (Gas), Soleus (Sol) and Tibialis Anterior (TA) as the most

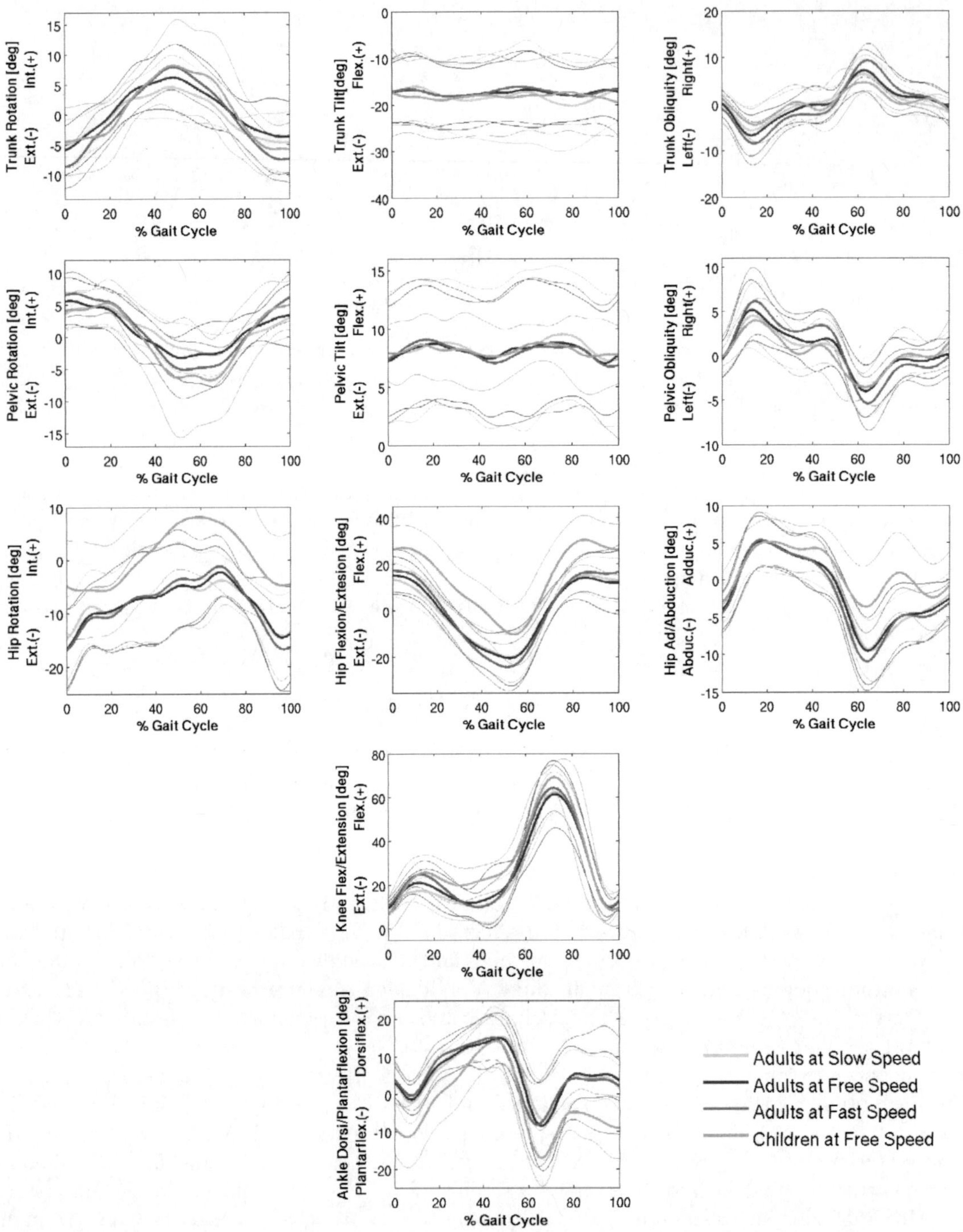

Figure 2: Joint angles for children and adult groups across walking speed. Each bold line represents the average of trials and the thin lines represent the standard deviation of trials within the corresponding group.

representative muscle groups in vertical and fore-aft accelerations during gait [19].

Muscle forces estimated by CMC algorithm are shown in Figure 5. Hamstrings has an

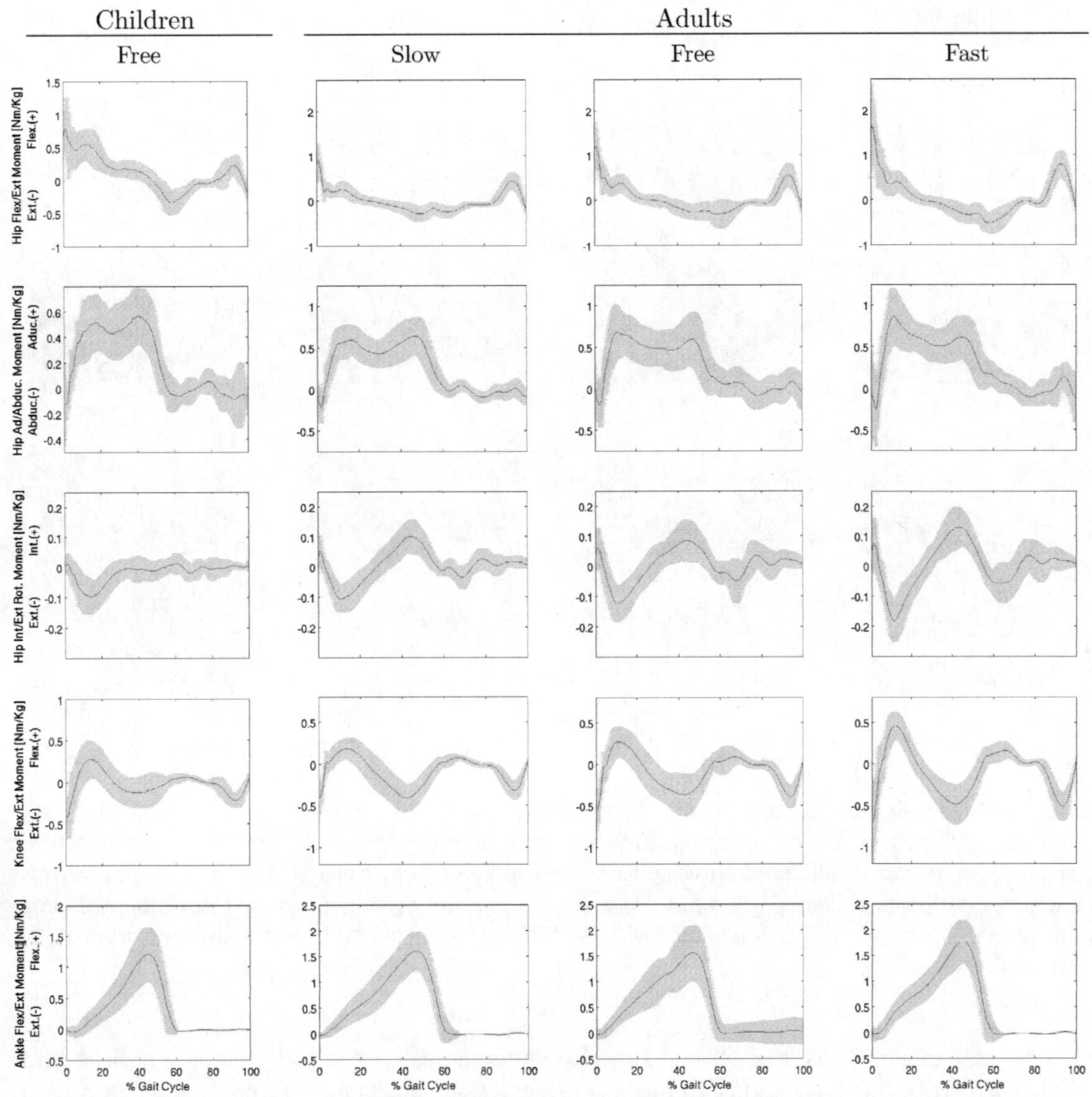

Figure 3: Mean (blue lines) and standard deviation (shaded area) of the joint moments for children that walking at self-selected comfortable speed, and adults that walking at three self-selected speeds (slow, comfortable or free and fast).

important peak in the mid-stance phase in the adult group but this peak disappears in the children group with the same walking speed condition. Furthermore, the rectus femoris presents an increase in the muscle force during the pre- and mid-swing in concordance with the increment in the walking speed of subjects. This behavior could be associated with the role that the rectus femoris has in the acceleration of the knee joint [20]. In addition, in the children group the gluteus maximus has a great peak of muscle force during mid-stance, whereas the same peak is only showed weakly in adults. On the other hand, the gastrocnemius, soleus and tibialis aterior present their main variations in the stance phase, in accordance with the role that they have in the plantar/dorsiflexion joint moment [21, 22]. In particular, we show that tibialis anterior has a marked peak in the initial contact of the gait cycle in accordance with the dorsiflexion peak

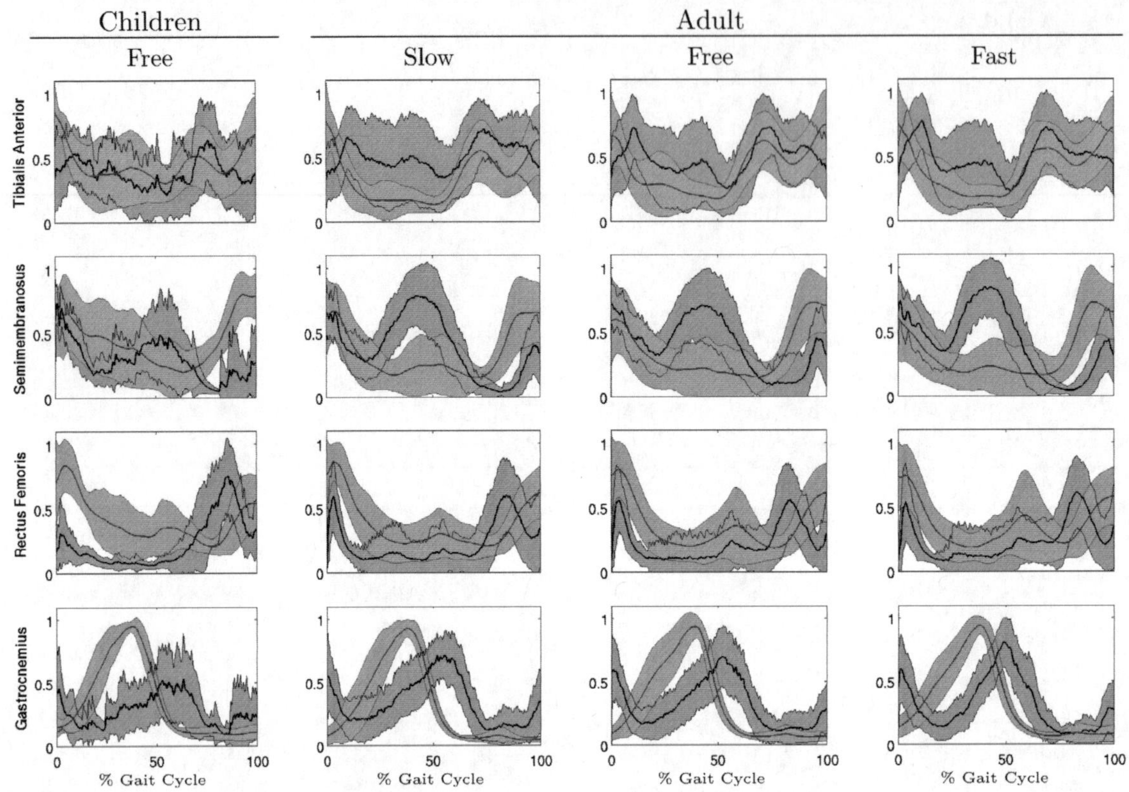

Figure 4: Mean simulated muscle activations (black lines) and mean experimental EMG (blue lines) throughout the gait cycle for children that walking at a comfortable (or free) self-selected speed and adults that walking at three self-selected speeds (slow, free and fast) for all four muscles for which EMG was collected. Shaded areas show the standard deviations. Experimental EMG data are rectified and bi-directionally low-pass filtered at 6 Hz. All signals were normalized by their peak value.

of ankle joint moment (see Fig. 3) and this muscle force peak appears to increase with walking speed. Finally, the gastrocnemius and soleus present their peaks at the end of stance phase by acting in antagonistic form with tibialis anterior.

In the end, Figure 6 shows the average values over the gait cycle of the mean and standard deviation of each muscle force in order to find which muscles perform more force throughout the gait cycle. Hence, we conclude that the soleus is the muscle that provides the major force during walk in concordance with the study of *Zajac (2002)* [22], and the gluteus maximus (a big muscle) is the muscle that provides the smaller force throughout the gait cycle.

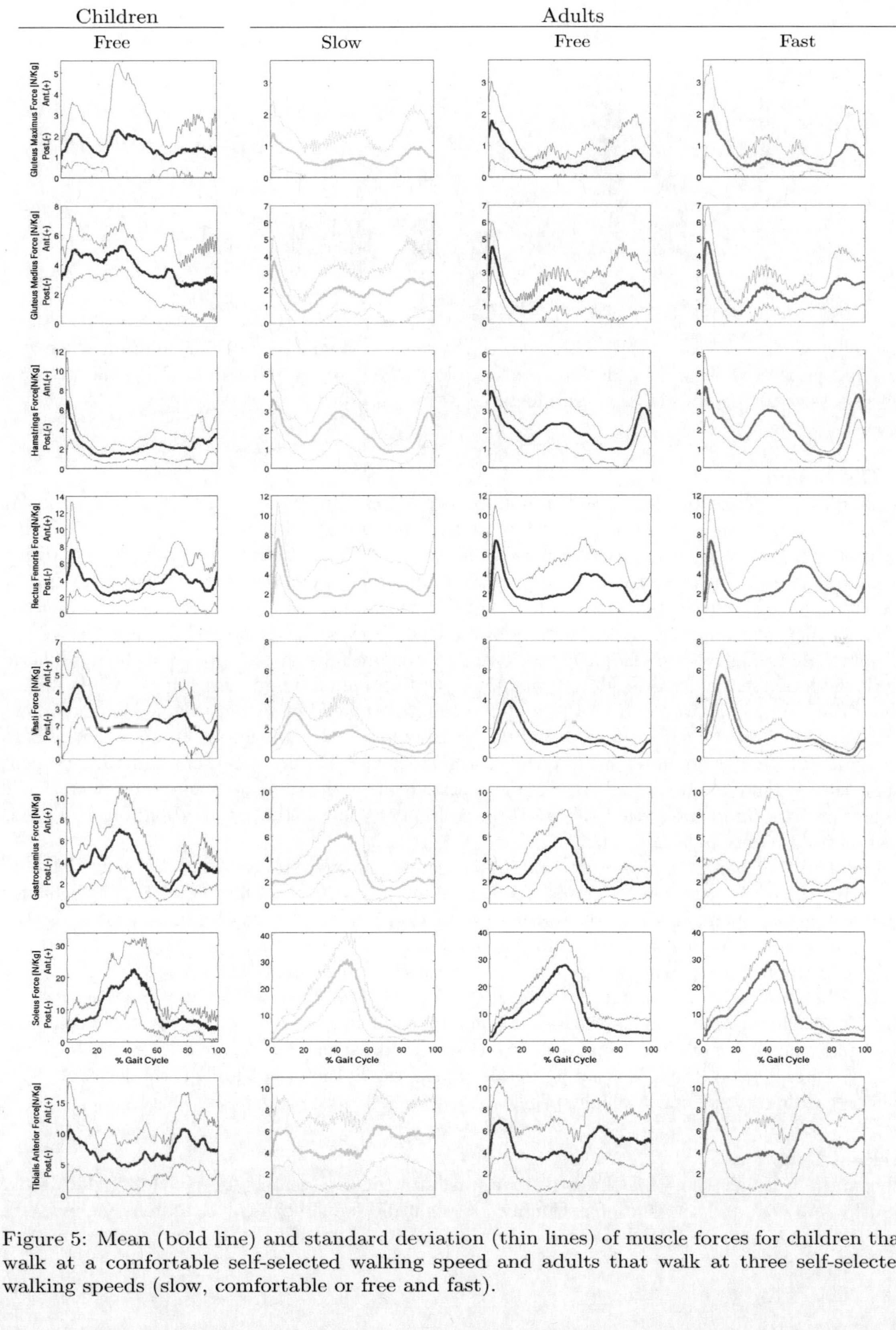

Figure 5: Mean (bold line) and standard deviation (thin lines) of muscle forces for children that walk at a comfortable self-selected walking speed and adults that walk at three self-selected walking speeds (slow, comfortable or free and fast).

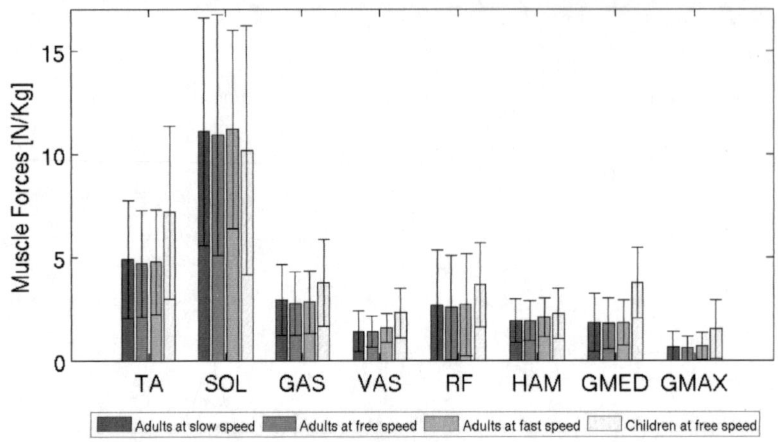

Figure 6: Average values over the gait cycle of the mean and standard deviation of muscle forces for children that walk at self-selected comfortable walking speed and adults that walk at three self-selected walking speeds (slow, comfortable or free and fast).

4. Discussion

Techniques of musculoskeletal modeling have been used by several decades to improve the benefits of clinical gait analysis, but many of the previous studies were focused on analyzing separately the muscle forces distribution of children or adult subjects with only one condition of walking speed. For these reason, the present study aims to enhance the current literature by describing speed gait effects on muscle forces of the lower limbs for a group of children and adult subjects that typically develop an unimpaired gait at three self-selected walking speeds.

When analyzing the hip flexion/extention moment joint (Fig. 3), it appears to be modulated by the action of rectus femoris, gluteus maximus, gluteus medialis and hamstrings mainly during the initial contact. However, from the mid-stance until the end of swing, this net joint moment tends to zero while the antagonistic activity of hamstrings and rectus femoris is shown to rise. In the abduction/adduction hip joint moment, the gluteus medialis produces an abduction moment while the gluteus maximus and hamstrings perform an adduction moment. The decrease in abduction joint moment during the mid-stance happens followed by a modulation of gluteus medialis and hamstrings.

On the other hand, the knee extensor joint moment in mid-stance (Fig. 3) occurs because an activation of the gastrocnemius together with the hamstrings muscles (Fig. 5). Its behavior produces a force peak at the end of swing phase that coincides with the peak present in the hamstrings force. Also, when the minimum value of knee flexion joint moment occurs, around the pre-swing or initial swing, an increase in the muscle forces of vastus and rectus femoris was found; they appear to be the main responsibles of the second knee net flexion joint moment.

Finally, the ankle flexion/extension net joint moment is produced mainly by the soleus and the gastrocnemius while the tibiales anterior acts modulating the net joint moment behavior (Fig. 5). In addition, the soleus is the muscle that perform the major force throughout the gait cycle regardless of age and walking speed.

Acknowledgments

The authors thank the FLENI Institute for Neurological Research for providing data from healthy subjects and the National Council of Scientific and Technical Research (CONICET) and PICTO 222-2009 (AGENCIA) for providing the funds needed for this research.

References

[1] Wren T a L, Otsuka N Y, Bowen R E, Scaduto A a, Chan L S, Sheng M, Hara R and Kay R M 2011 *Gait & Posture* **34** 364–9 ISSN 1879-2219 URL http://www.ncbi.nlm.nih.gov/pubmed/21723131
[2] de Morais Filho M C, Kawamura C M, dos Santos C A and Mattar R 2012 *Gait & Posture* **36** 201–4 ISSN 1879-2219 URL http://www.ncbi.nlm.nih.gov/pubmed/22425638
[3] Erdemir A, McLean S, Herzog W and van den Bogert A J 2007 *Clinical Biomechanics (Bristol, Avon)* **22** 131–54 ISSN 0268-0033 URL http://www.ncbi.nlm.nih.gov/pubmed/17070969
[4] Amarantini D, Rao G and Berton E 2010 *Journal of Biomechanics* **43** 1827–30 ISSN 1873-2380 URL http://www.ncbi.nlm.nih.gov/pubmed/20206935
[5] Kaufman K R, An K N W, Litchy W J and Chao E Y 1991 *Neuroscience* **40** 781–92 ISSN 0306-4522 URL http://www.ncbi.nlm.nih.gov/pubmed/2062441
[6] Arnold E M, Ward S R, Lieber R L and Delp S L 2010 *Annals of Biomedical Engineering* **38** 269–79 ISSN 1521-6047 URL http://www.ncbi.nlm.nih.gov/pubmed/19957039
[7] van der Krogt M M, Delp S L and Schwartz M H 2012 *Gait & Posture* **36** 113–9 ISSN 1879-2219 URL http://www.ncbi.nlm.nih.gov/pubmed/22386624
[8] Thelen D G, Anderson F C and Delp S L 2003 *Journal of Biomechanics* **36** 321–328 ISSN 00219290 URL http://www.ncbi.nlm.nih.gov/pubmed/12594980
[9] Ravera E P, Crespo M J and Braidot A A A 2014 *Computer Methods in Biomechanics and Biomedical Engineering* 1–12 ISSN 1476-8259 URL http://www.ncbi.nlm.nih.gov/pubmed/25408069
[10] Hof A 1996 *Gait & Posture* **4** 222–3 URL http://www.sciencedirect.com/science/article/pii/0966636295010572
[11] Davis R B, Õunpuu S, Tyburski D and Gage J R 1991 *Human Movement Science* **10** 575–587 ISSN 01679457 URL http://www.sciencedirect.com/science/article/pii/016794579190046Z
[12] Stegeman D F and Hermens H J 1999 Standards for surface electromyography: the European project "Surface EMG for non-invasive assessment of muscles (SENIAM) Tech. rep. URL www.seniam.org
[13] Delp S L, Anderson F C, Arnold A S, Loan P, Habib A, John C T, Guendelman E and Thelen D G 2007 *IEEE Transactions on Bio-medical Engineering* **54** 1940–50 ISSN 0018-9294 URL http://www.ncbi.nlm.nih.gov/pubmed/18018689
[14] Anderson F C and Pandy M G 1999 *Computer Methods in Biomechanics and Biomedical Engineering* **2** 201–231 ISSN 1476-8259 URL http://www.ncbi.nlm.nih.gov/pubmed/11264828
[15] Anderson F C and Pandy M G 2001 *Journal of Biomechanics* **34** 153–61 ISSN 0021-9290 URL http://www.ncbi.nlm.nih.gov/pubmed/11165278
[16] Thelen D G and Anderson F C 2006 *Journal of Biomechanics* **39** 1107–15 ISSN 0021-9290 URL http://www.ncbi.nlm.nih.gov/pubmed/16023125
[17] Hicks J L, Uchida T K, Seth A, Rajagopal A and Delp S 2015 *Journal of Biomechanical Engineering* **137** 1–24 ISSN 0148-0731 URL http://www.ncbi.nlm.nih.gov/pubmed/25474098
[18] Schwartz M H, Rozumalski A and Trost J P 2008 *Journal of Biomechanics* **41** 1639–50 ISSN 0021-9290 URL http://www.ncbi.nlm.nih.gov/pubmed/18466909
[19] Liu M Q, Anderson F C, Pandy M G and Delp S L 2006 *Journal of Biomechanics* **39** 2623–30 ISSN 0021-9290 URL http://www.ncbi.nlm.nih.gov/pubmed/16216251
[20] Arnold E M, Hamner S R, Seth A, Millard M and Delp S L 2013 *The Journal of Experimental Biology* **216** 2150–60 ISSN 1477-9145 URL http://www.ncbi.nlm.nih.gov/pubmed/23470656
[21] Neptune R R, Kautz S a and Zajac F E 2001 *Journal of Biomechanics* **34** 1387–98 ISSN 0021-9290 URL http://www.ncbi.nlm.nih.gov/pubmed/11672713
[22] Zajac F E, Neptune R R and Kautz S a 2002 *Gait & Posture* **16** 215–32 ISSN 0966-6362 URL http://www.ncbi.nlm.nih.gov/pubmed/12443946

Video-analysis Interface for Angular Joint Measurement

M Mondani[1,3]; I Ghersi[1,2]; M T Miralles[1,2]

[1]Pontificia Universidad Católica Argentina (LaBIS-Facultad de Ciencias Fisicomatemáticas e Ingeniería-UCA), Av. A. Moreau de Justo 1500, CABA, Argentina.
[2] Universidad de Buenos Aires (CIDI-Facultad de Arquitectura Diseño y Urbanismo-UBA), Intendente Guiraldes 2160, Pabellón III, Cuidad Universitaria, CABA, Argentina.

e-mail: mmondani@gmail.com

Abstract. Real-time quantification of joint articular movement is instrumental in the comprehensive assessment of significant biomechanical gestures. The development of an interface, based on an automatic algorithm for 3D-motion analysis, is presented in this work. The graphical interface uses open-source libraries for video processing, and its use is intuitive. The proposed method is low-cost, of acceptable precision ($|\varepsilon_\theta| < 1°$), and minimally invasive. It allows to obtain angular movement of joints in different planes, synchronized with the video of the gesture, as well as to make comparisons and calculate parameters of interest from the acquired angular kinematics.

1. Introduction

The systematic study of different stages found in normal and dysfunctional gait involves relevant kinematic variables, concerning the angular functions of hip, knee and ankle joints, throughout the full gait cycle [1]. The need to supplement previous studies based on 3D-accelerometry in adult patients with a higher risk of falling [2-4] motivated the development of a system that would allow the accurate assessment of joint movements.

Over the past years, there has been an enormous production of different, alternative methods for the evaluation of joint angles [5-7], with relative advantages related to performance, cost and ease of use. The measurement of angles between body segments consists in measuring the spatial orientation of one body segment with respect to another adjacent to it, separated from the first by the studied joint, or in relation to a constant reference, like the vertical line. Measuring of lower-limb joints is essential in the different stages of walking, as in any other gesture [8-9]. This knowledge completes the evaluation of the active and passive mobility, the characterization of axial deviations and, finally, the search of abnormal movements. It is also one of the parameters used for the complete evaluation of a running cycle.

This paper presents a method for measuring joint movements during any gesture, in particular walking, that is simple, low-cost and with excellent results. It is based on a video-analysis algorithm, which isolates active linear markers that are fixed to the moving segments, and acquires information

[3] Corresponding author

concerning the dynamics of such joint angle over a gesture. The results are displayed in a dedicated graphical interface that was developed for that purpose.

2. Materials and Methods.

2.1. Light-emitting linear markers

The procedure for measuring angles between adjacent body segments in this work relies on a specially-developed software that identifies and measures the angle between light-emitting linear markers during joint movements for body segments and gestures of interest.

The most common markers are spherical markers that either reflect light from a separate lighting system, or emit visible light as well as in the infrared range (which allows an easier detection, when using the appropriate hardware). Self-adhesive markers, which have a printed pattern, are also available in the market. This pattern is looked upon on each frame of the video.

Markers of low cost using plastic, cylindrical, flexible rods were built for this work. As an example, this paper presents the markers developed for the study of knee joint movement. The rod dimensions were 15 cm long and 1 cm in diameter (see figure 1), being adaptable to any other size, if necessary. Light emission is obtained by adhering a strip of light-emitting diodes (LEDs) in front of which a diffusing paper is placed to generate uniform light. The shape of the markers makes the system more resistant to partial occlusions and is easier to associate with a straight line. A 12 Volt DC source provides power to the system (with an approximate consumption of 45 mA per marker). Strips are fixed to the concerned corporal segments by hook and loop (Velcro) elastic clamps (see figure 2), facilitating the free movement of the segments. The markers must be attached parallel to the corporal segments to ensure the correct measurement of the joint angle. Two high speed cameras were used for video recording (Casio EXFH25) capturing images at 120 fps, at the same time recording the sagittal and frontal planes of the moving subject, respectively.

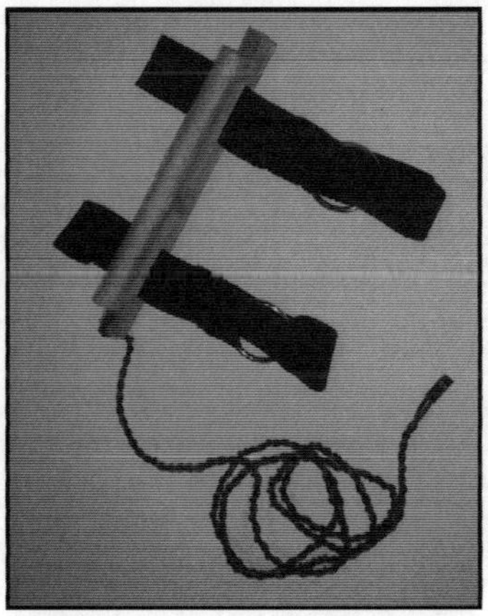

Figure 1. Linear cylindrical optical marker with two elastic clamps and power cable.

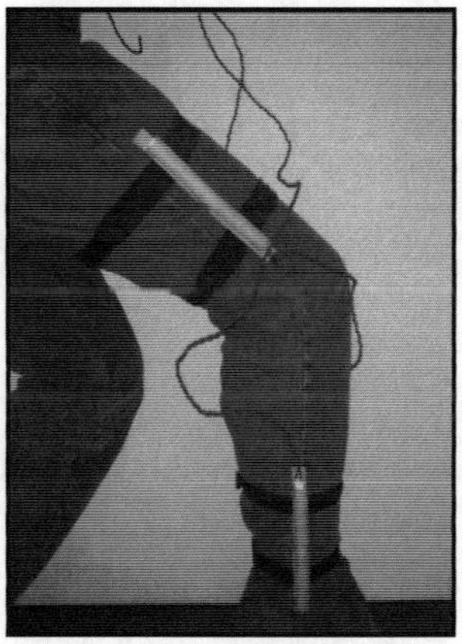

Figure 2. Markers located in the lower limb to study the angular dynamics of the movement of the knee.

2.2 Algorithm for acquisition and calculation of angles

In order to calculate the angle between the segments that are instrumented with the markers, an algorithm that associates a straight line to each marker in each video frame was developed. The algorithm further calculates the angle of inclination and the centroid of each of the markers, and the angle between relevant adjacent segments.

OpenCV computer-vision libraries and tools, developed by Intel and freely distributed under BSD license, were used.

The stages of the algorithm were as follows:

- *Pre-processing stage*: Each video frame is smoothed with a Gaussian blur to eliminate details that could be wrongly detected as markers (false detection), thus the borders of the markers are also smoothed. This allows calculating the marker's tilt more accurately. Then, a binary threshold is applied (luminance level), resulting in a black and white image. See figures 3 and 4 for an example of this stage.

Figure 3. Video frame of a subject under study, with markers in lower limb, from which the pre-processing stage begins.

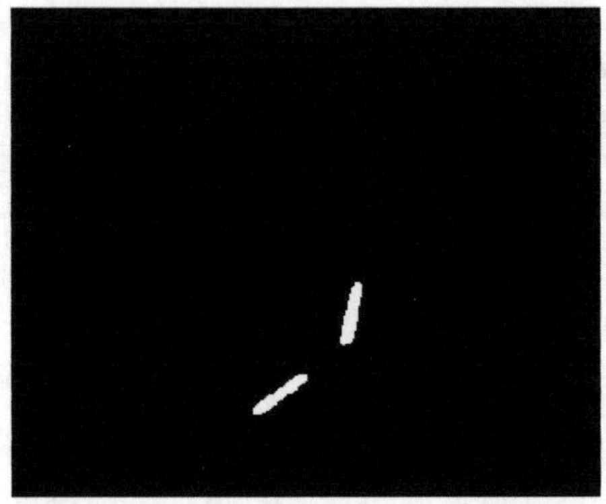

Figure 4. Black-and-white image of markers obtained by processing the image with a binary threshold.

- *Valid contour detection stage*: on the resulting image of the pre-processing stage (see figure 4) all contours are acquired, followed by a Hough-transform process. This transform can be used to find parametric curves in an image, which allows to find all the lines whose length (in pixels) is greater than a specified minimum. Once these lines are found, an analysis of which of the obtained contours contain, at least, one of these lines, is carried out. Therefore, the contours that contain a line are the ones that have the appropriate length (the linear marker length) and any other contour found in the figure, not eliminated by the binary threshold, is removed.

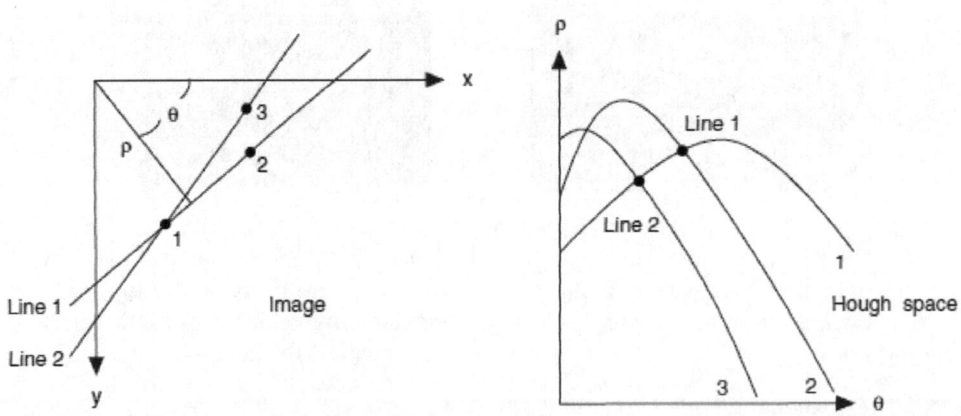

Figure 5. Illustrative diagram of stage 2: a) Hough-transform parameters (θ and ρ) of the intersection of example straight lines; b) Shape graph illustrating how the Hough transform operates on these lines.

Figure 5 shows how the Hough transform works for this algorithm. A series of straight lines of different angles are crossed through each point of interest of the image under analysis in order to measure the normal distance between the origin and each of these lines. Subsequently, a point in space (x, y) of the image leads to a sinusoid in the Hough space (ρ, θ). The number of lines running through each point will depend on the discretization of the parameter θ. The straight line joining two points in the space (x, y) will be found by the intersection of the sine waves produced by those two points in the Hough space. This point will indicate the value of ρ and θ that describe the selected straight line.

- *Segment calculation stage:* The theory of image moments is applied once contours are correctly identified in order to find the centroid and the inclination of each of these contours (thus of each body segment) as detailed in equation (1) [10]:

$$m_{p,q} = \iint x^p y^q f(x,y)\, dx\, dy \tag{1}$$

being:

$$f(x,y) \rightarrow b(x,y) = \begin{pmatrix} 1 \\ \\ 0 \end{pmatrix} \tag{2}$$

the moment orders determined by parameter $p+q$. The orders of interest applied for this work were:

- $m_{0,0}$ → Related to the contour area
- $m_{1,0}$ / $m_{0,0}$ → x coordinate of the contour centroid
- $m_{0,1}$ / $m_{0,0}$ → y coordinate of the contour centroid

The central moments are obtained employing definition (1), as shown in equation (3)

$$\mu_{p,q} = \iint (x - x_c)^p (y - y_c)^q f(x,y)\, dx\, dy \tag{3}$$

where x_c and y_c are the coordinates of each contour centroid. These coordinates are obtained on the basis of the above mentioned moments. The inclination of the contour can be calculated using equation (4):

$$\phi = \frac{1}{2}\tan^{-1}\left(\frac{2\mu_{1,1}}{\mu_{2,0} - \mu_{0,2}}\right) \tag{4}$$

- *Search of line intersection stage*: For each line "i" of the N lines found in the previous point, the intersection with the remaining lines (N - 1) and the points of intersection are stored in a vector. The point being at the shortest distance to line "i" is selected, which will correspond to the intersection of line "i" with the closest line in the image. As an example, a case of three straight lines is shown in figure 6. The outline found in the image is shown in white and the line representing the contour appears in green (with a red dot indicating the position of the centroid). The distance between each of the lines is shown in yellow.
  Since the presence of intersection points is not constant in all frames, a tracking of each of the intersection points between the frames was carried out.

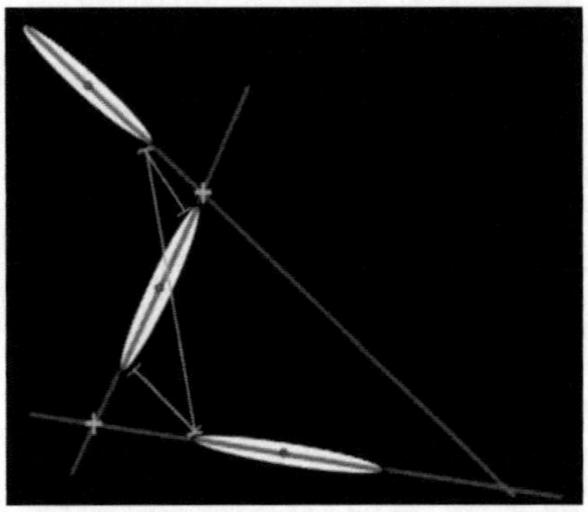

Figure 6. Example of line "i" intersection with each of the remaining N-1. The closest intersection with line "i" is obtained.

Figure 7. Example of the calculation of an angle between adjacent segments using the developed algorithm.

- *Calculation of angle between adjacent segments stage*: The angle between the vectors ranging from the intersection point to the farthest end of each intersecting line is calculated (see figure 7).

- *Three-dimensional representation of each segment stage:* This algorithm for searching segments and angular calculation is equally applied to the frames of each of the two videos (sagittal and frontal). Each pair of frames is analyzed in order to relate the obtained information to the three-dimensional representation of each segment. Thus, 3D graphs of the joint segments under study are achieved.

2.3 User interface
Figure 8 shows the graphical interface that was developed for this work. Different controls are shown on the left. The dedicated panel allows to:

- Control video.
- Set-up pre-processing parameters of each frame: smoothing level, binarization threshold.
- Determine the size of the markers to be detected, based on the specification of a minimum length (in pixels) for the lines obtained by the Hough transform.
- Set-up data visualization: Establish a reference level (for example 180°) from which measurements are plotted. The number of samples to be considered in a moving average can also be set.

Figure 8 shows video images of the moving subject during a normal walk, with markers placed on the leg and thigh –the top section shows the sagittal plane, while the bottom shows the frontal plane. In each video, the areas under which the analysis should be performed were established using vertical controls, shown on the left side of each video. Thus, a reduction of the processing time for each frame is achieved.

The controls that set the video processing parameters mentioned above are shown on the left side of the graphical interface.

On the right side, the found segments are displayed as three-dimensional shapes, through three views. The first one matches the frontal plane. The second is a free camera which can be rotated while the video is playing, simplifying the view of certain movements. The third one matches the sagittal plane.

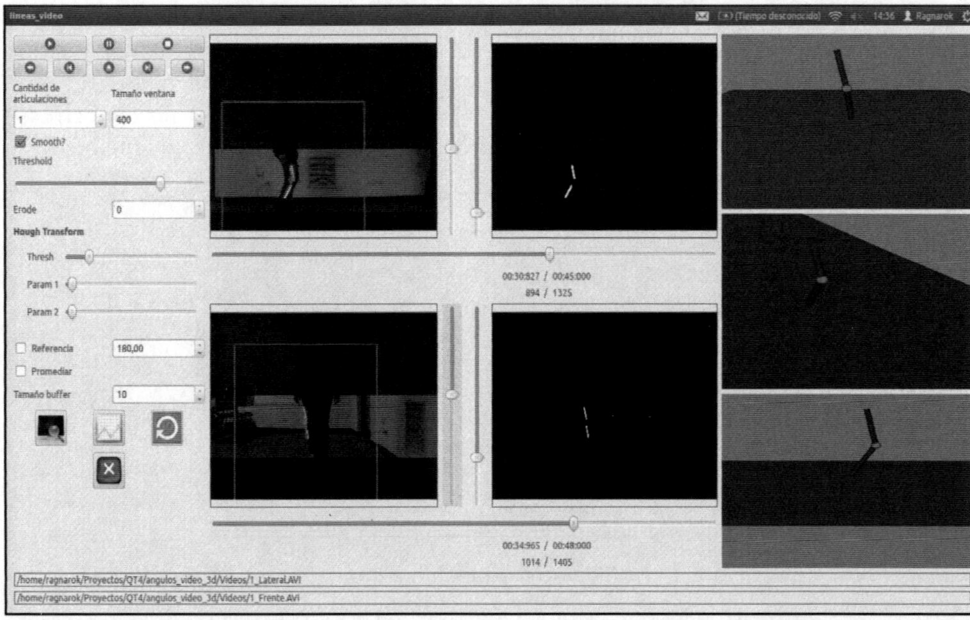

Figure 8. Graphical interface that allows the simultaneous view of both videos and the three-dimensional representation of the found segments.

Synchronization between both videos is essential for this method. An easy way to achieve synchronization is through the adequate starting of the markers, by initiating the recording on both planes with the markers turned off, and powering them when both cameras are running. The graphical interface has an option that finds the first frame in which the number of requested joints appears. Once the corresponding frames in each video are found, the video processing strategy can take place.

3. Results

Figure 9 shows the angular motion results from a test video, for the frontal plane (bottom), as well as sagittal (top).

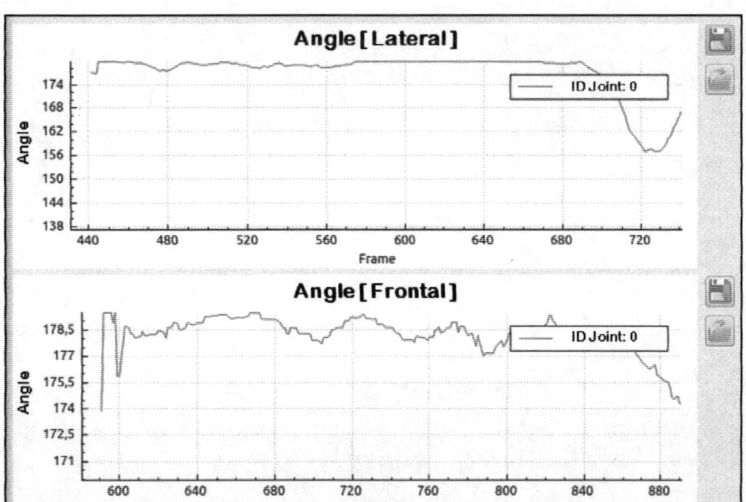

Figure 9. Time-graphs from test recording in the sagittal plane (upper image) and frontal plane (lower image).

Figure 10 illustrates the absolute error of the adjacent segments angle in the range of 1° to 150 °. To measure the error, the algorithm described in this paper was compared with an analog goniometer. The absolute error of this algorithm was obtained using the goniometer's measurement as a reference value (universal goniometer, see [11]). The displayed error (ε_θ) is less than 0.8°, which indicates that this measuring method is within the order of magnitude of many commercial electrogoniometers.

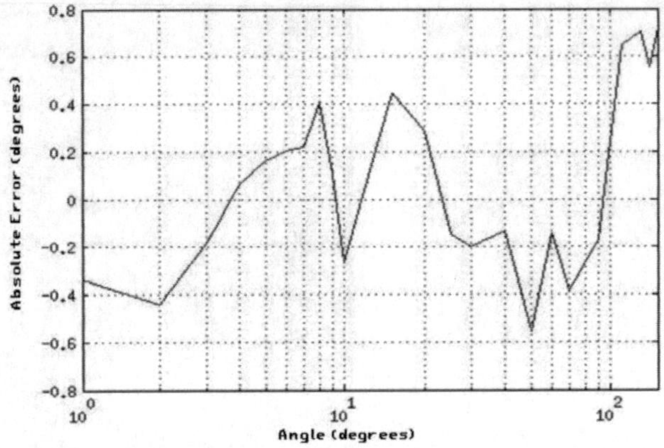

Figure 10 Absolute error achieved by the developed method.

From the acquired gesture graphs, a full study of the angular movement during any gesture can be performed. As an example, Figure 11 shows an analysis of the knee-joint angle over the sagittal plane, with the user performing a barefoot normal walk (blue line), and while performing the same gesture, wearing tennis shoes featuring a special sole, with a modified support surface in the heel (red line). Stages of initial contact, loading response, pre-swing and swing [12] are observed:

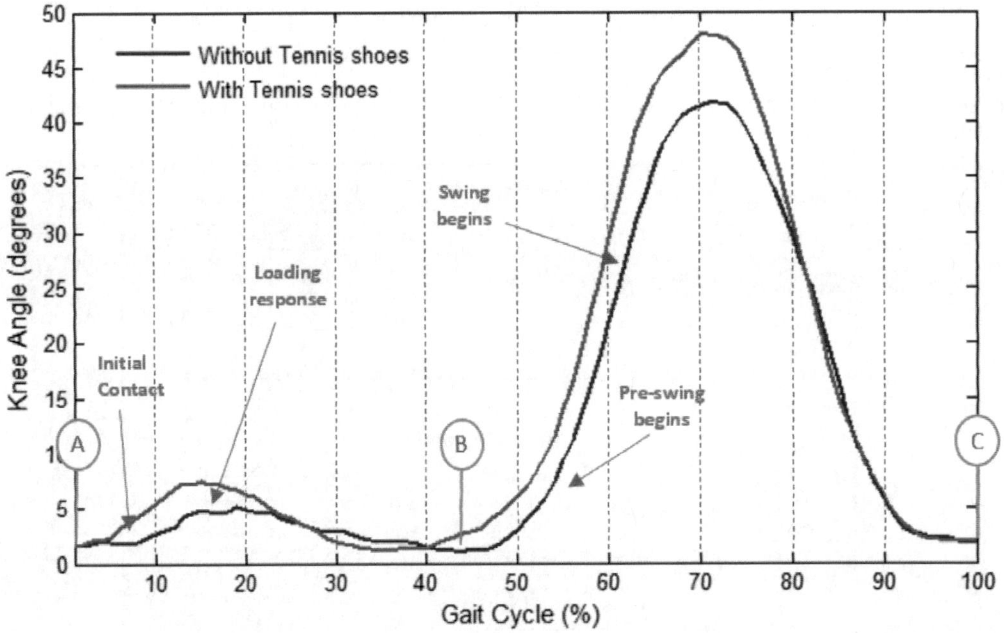

Figure 11. Gait cycle graph of knee joint movement in sagittal plane with (red) and without (blue) special tennis shoes

4. Conclusions

This work presented the development of a low-cost, acceptably precise ($|\varepsilon_\theta| < 1°$) and minimally invasive method for quantifying joint movements. The graphical interface that was developed for this work uses open-source libraries and its use is intuitive. The way in which the algorithm was designed allows extending the number of markers for different tests, thus allowing the potential simultaneous study of two or more joints. The inclusion of two cameras prevents marker overlapping in the recording. The current logic for finding and tracking the markers in each video frame requires a proper analysis to improve its efficiency. Potential improvements for this system include the optimization of algorithms as well as the introduction of new functions, allowing to quantify and compare joint angular movement, in real time, with other measurements of the same subject as well as with those considered as reference.

References

[1] Gordon D, Robertson E et al. 2004 *Research methods in biomechanics*, London, Human Kinetics.

[2] Vecchio V, Miralles, M, Ghersi I, Val G Capalbo I and Rodríguez H 2012*:* Portable and programmable wireless system for evaluation of balance and gait based on MEMS capacitive triaxial accelerometer, *Proc. Argencon 2012*, Córdoba, Argentina.

[3] Miralles M T, Ghersi I, Vecchio R, Paterson R, Perez Akly M, Ferrando M, Paterson A and Alvarez F 2013 Comprehensive feature extraction for objective dynamic gait index assessment of risk of falls in the elderly *J. Phys.: Conf. Series.* **477** 012001.

[4] Ghersi I; Alvarez F; Miralles M T 2015 Classification of Performance in risk-of-falls assessment based on accelerometer data and feature boosting, *IFMBE Proceedings* **49**, 222-225.

[5] Djuric-Jovicic M D et al 2011 Kinematics of Gait: New Method for Angle Estimation Based on Accelerometers, *Sensors* **11**, 10571-10585

[6] Rowe P J, Myles C M, Hillmann S J and Hazlewood M E, 2001, Validation of Flexible Electrogoniometry as a Measure of Joint Kinematics, *Physiotherapy*, **87** issue 9, pp 479-488.

[7] Nasseri N. Hadian M R et al 2007 Reliability and accuracy of joint position sense measurement in the laboratory and clinic; utilizing a new system; *Acta Medica Iranica* **45**, 5.

[8] Whittle M 2007 *Gait Analysis: An Introduction*, Butterworth-Heinemann.

[9] Stergiou N 2004 *Innovative Analysis of Human Movement*, Human Kinetics.

[10] Teagle R 1980 Image Analysis via the General Theory of Moments; *Journal of the Optical Society Of America*, **70,** 8.

[11] Norkin C C, White D J 2009 *Measurement of Joint Motion: A Guide to Goniometry*, F. A. Davis Company, pp 21-25.

[12] Öunpuu S 1994 The biomechanics of walking and running, *Clinics in Sports Medicine* **13**, 4.

Acknowledgments

This work was carried out under -and supported in part by- Project 20621201000011BA UBACyT 2013-2016, Facultad de Arquitectura, Diseño y Urbanismo, Universidad de Buenos Aires (FADU-UBA), and Facultad de Ciencias Fisicomatemáticas e Ingeniería, Pontificia Universidad Católica Argentina (UCA).

Two-dimensional posture evaluation in Parkinson's disease: effect of loads on the spinal angle during gait.

Paula Celoria[1], Federico Nanni[1], Flavia Pastore[1], Sebastian Pulenta[1], Matias Tajerian[1], Lucio Pantazis[1,2], Marcela Moscoso-Vasquez[1,2], Daniel Cerquetti [3], Marcelo Merello [2,3] and Marcelo Risk[1,2]

[1] Departamento de Bioingeniería, Instituto Tecnológico de Buenos Aires (ITBA), Av. Eduardo Madero 399, C1106ACD, Buenos Aires, Argentina.
[2] Consejo Nacional de Investigaciones Científicas y Técnicas (CONICET), Argentina.
[3] Fundación para la Lucha contra las Enfermedades Neurológicas de la Infancia (FLENI), Argentina.

E-mail: mrisk@itba.edu.ar

Abstract.
Parkinson's Disease patients present diminished coordination caused by neural degeneration. This leads to large motor difficulties during gait such as balance loss and pronounced forward inclination of the upper body. This work assessed the spinal sagittal plane angle alterations in two groups: six parkinsonian patients and six control healthy subjects. This parameter was analyzed during gait under three conditions: without external loads and with external loads applied either on the chest or on the lower back area. Results were statistically compared by means of t-test of paired samples in both groups. For patients, a significant effect was found when loads were applied on the chest. On the other hand, healthy subjects showed no significant differences in either case.

1. Introduction

The arthro-osteo-muscular and nervous systems are closely interrelated in the complex dynamics and patterns present in the human gait. Therefore, neurodegenerative diseases, such as Parkinson's disease (PD), often cause movement disorders that could eventually affect the patient's gait. PD common symptoms are stiffness, tremor and bradykinesia (slowing down of voluntary movements).

In PD patients, it is observed a progressive loss of automatic movement patterns and postural reflexes, which leads to imbalance, diminished proprioception and festination (sequential effect of reducing consecutive footsteps length with accompanying reduction in gait speed).

Another motor sign present in PD patients is the tendency to tilt the torso forwards. The result is a bent posture and a displacement of the center of mass towards the front, ending up ahead of the ordinary center of pressure. These disorders not only affect the quality of life of PD patients, but also are often responsible for tripping and serious trauma [1, 2, 3, 4].

Moreover, some studies suggest that physical activities and muscular strengthening can help in the rehabilitation process [5]. In this way, PD patients who train postural maintenance related muscles tend to develop more control over their balance and gait.

The present work assessed the influence of a load over the posture of PD patients during gait. The chosen parameter to assess this influence was the Sagittal Plane Angle (SPA) described by the spine and the vertical axis, as shown in 1.

The proposed two-dimensional model is intended to compare the antero-posterior trunk inclination under different experimental conditions, which is the main cause of the forward displacement of the center of mass.

2. Materials and Methods

This study was approved by the bioethics committee of *Fundación para la Lucha contra las Enfermedades Neurológicas de la Infancia* (FLENI).

In order to evaluate whether the SPA varies significantly when the subjects walk carrying a standardized load, a paired two-sample t-test was calculated between PD patients and a control group. The sample size, experimental settings and numerical methods are described below.

2.1. Subjects

Two groups were evaluated: the control group was a sample of six healthy subjects between 22 and 75 years old, and the PD group consisted of six patients, between 50 to 75 years old, in third-stage Parkinson's disease according to the Hoehn and Yahr scale [6], diagnosed at the Rehabilitation and Educational Therapy Institute at FLENI.

All subjects signed informed consents, and patients were recruited from the out-patient clinic at the Movement Disorders section, FLENI.

Demographic data of both groups are presented in Table 1.

Table 1. Demographic data of healthy subjects (C) and PD patients.

Subject #	Gender	Group	Weight (Kg)	Height (m)	Body Mass Index (Kg/m^2)	Load Applied (Kg)
1	M	C	78	1.7	27	4.7
2	M	C	101	1.75	28	6.1
3	F	C	57	1.63	22	3.4
4	F	C	57	1.7	21	3.4
5	F	C	61	1.71	21	3.7
6	F	C	80	1.64	29	4.8
7	M	PD	92	1.66	31	5.5
8	M	PD	94	1.76	29	5.6
9	M	PD	100	1.7	33	6.0
10	F	PD	54	1.51	27	3.2
11	M	PD	82	1.72	24	4.9
12	M	PD	81	1.7	29	4.9

2.2. Experimental measurements

The gait assessments were performed at the FLENI Gait and Motion Laboratory. By using 17 sensors placed on the body of each subject (according to STCISB standards), the space-time coordinates were registered through a system of six ELITE optoelectronic cameras (BTS Bioengineering, Milan, Italy) with a sampling frequency of 100 Hz [7].

The subjects walked several times a distance of 10 meters in two different conditions:

(i) No load.

(ii) With load applied on the chest.

Loads were applied by means of a portable and noninvasive weighted backpack device capable of shifting the subjects' center off mass. Each load was calibrated to be 6% of the body weight of each corresponding subject [8] (see Table 1).

2.3. Statistical Analysis

In order to model the spinal angle, a straight line intersecting two dots was projected. The upper dot was located at the 7th cervical vertebrae and the other one at the sacrum bone. This is shown on figure 1.

The angle between this line and the vertical axis on the sagittal plane was the measurement used to evaluate the inclination. This SPA was calculated for each space-time stamp obtained by the camera along the 10 meters walk.

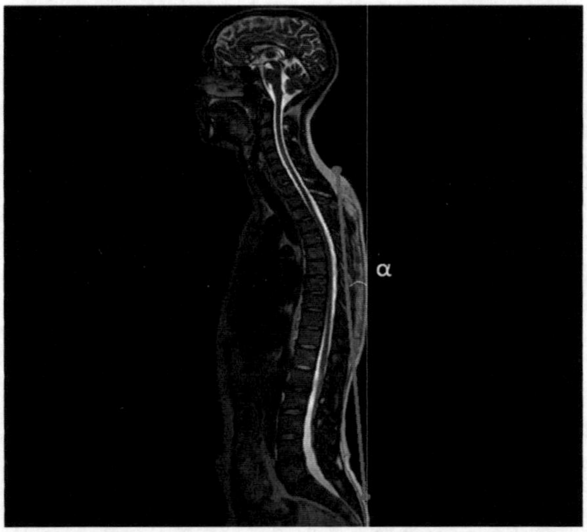

Figure 1. Spinal angle model (in red) and the SPA, (α). Drawn in white is the vertical axis reference.

SPA histograms for each subject were plotted . Figures 3 and 2 show plots for patient 12 and 8 respectively. It can be observed that mean SPA values under condition 2 were lower than under condition 1. However, this contrast was not observed in control subjects.

Normality test was performed by the qq-plot method [9, 10], shown in figure 4.

A paired sample t-test was run, comparing results for both conditions 1 and 2. Significance threshold was arbitrarily defined for $P < 0.05$, 2-tailed.

To take into consideration angular variability, the following variables were computed:

- $\bar{X}_i$ SPA average along the gait of the ith individual without load.
- $\bar{Y}_i$ SPA average along the gait of the ith individual with load.
- n_i^x Amount of measurements along the gait of the ith individual without load.

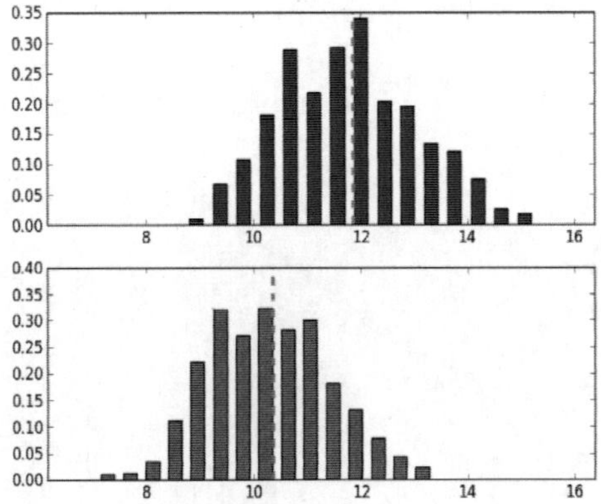

Figure 2. Histogram PD patient 12, upper panel without load and lower panel with load, red dotted lines are the means.

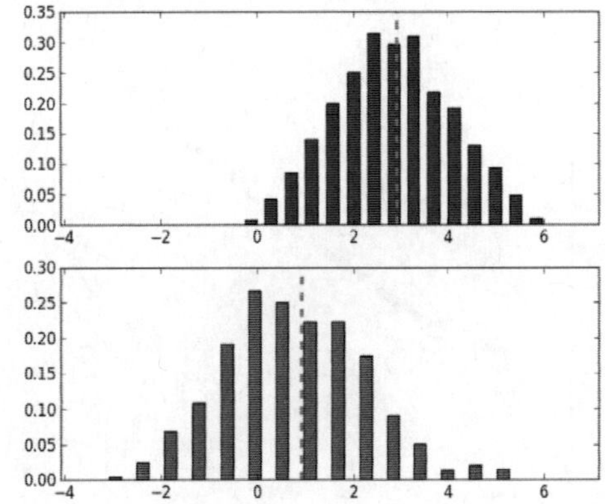

Figure 3. Histogram of PD patient 8, upper panel without load and lower panel with load, red dotted lines are the means.

- n_i^y Amount of measurements along the gait of the ith individual with load.
- σ_i^x Standard deviation of the SPA along the gait of the ith individual without load.
- σ_i^y Standard deviation of the SPA along the gait of the ith individual with load.

Statistical analysis was performed on the transformed variables according to the following expressions:

- $\widetilde{X_i} = \dfrac{\bar{X_i}}{\sqrt{\frac{(\sigma_i^x)^2}{n_i^x} + \frac{(\sigma_i^y)^2}{n_i^y}}}$
- $\widetilde{Y_i} = \dfrac{Y_i}{\sqrt{\frac{(\sigma_i^x)^2}{n_i^x} + \frac{(\sigma_i^y)^2}{n_i^y}}}$

Software processing was calculated using packages NumPy and SciPy, with Python Language v2.7 [11], and statistical analysis was performed with R language [12].

3. Results

Statistical analysis showed a significant decrease on SPA measurements in condition 2 compared to condition 1, corresponding to straighter postures when subjects were loaded.

Despite this decrease, it was found no correlation between those variations and their initial or basal SPA (condition 1). The results are summarized in Table 2.

4. Discusion

The two-dimensional model of the spine projection in the sagittal plane (SPA) is a simplfied definition implemented to assess a specific parameter.

Nevertheless, the same model could be used in a further study to observe the effect of loads in lateral sways of the spine during gait.

Considering that postural instabilities or falls could have predominance in either direction, it might be useful to analyze whether loads placed on the anterior-posterior plane, could

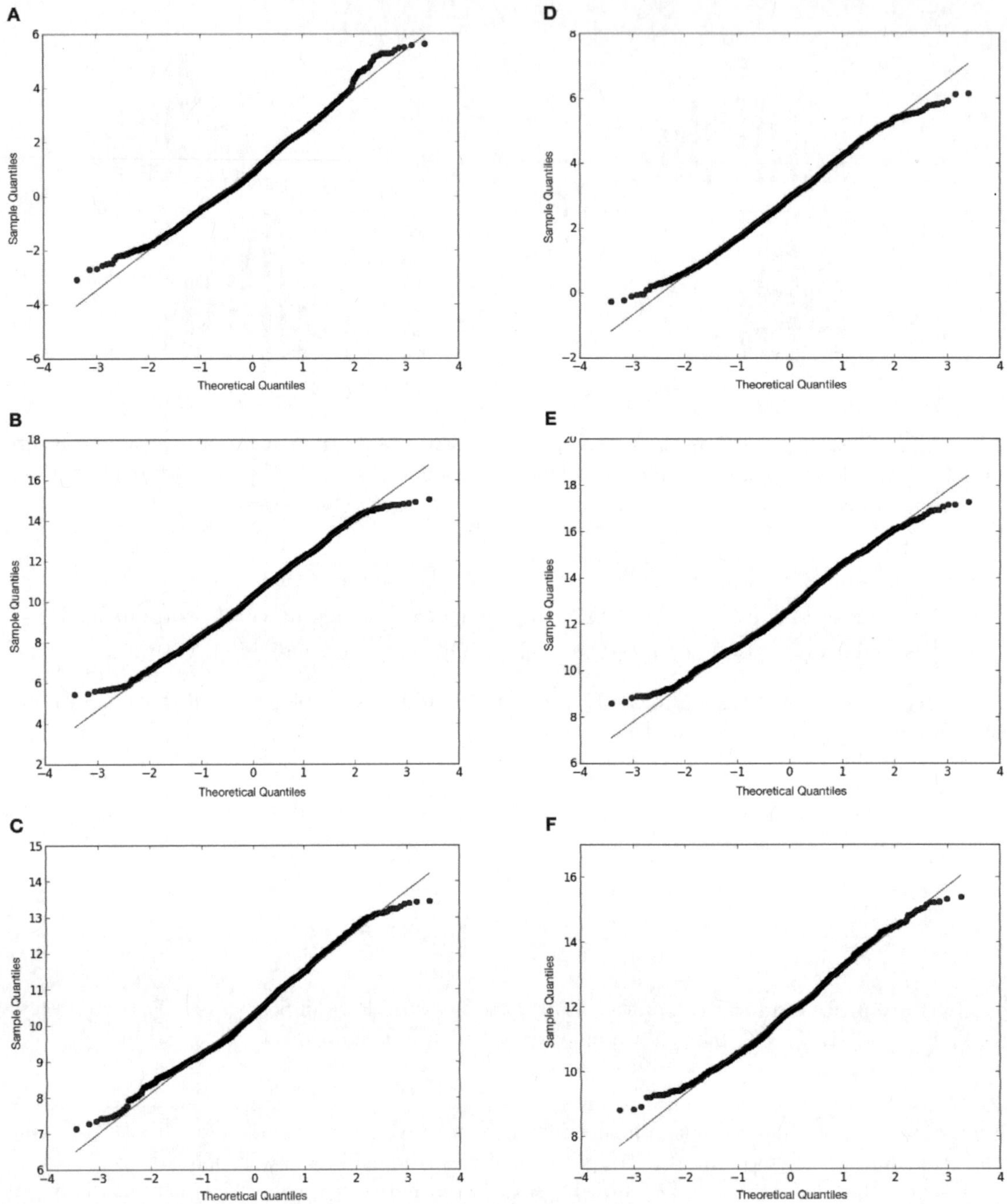

Figure 4. Q-Q plots of PD patients 8 (A and D), 9 (B and E) and 12 (C and F), where panels on the left are for condition 2, and panels on the right are for condition 1.

Table 2. Statistical results for each group

Group	Difference of means	Statistic t	df	P
PD	1.74°	8.86	6	0.0001
C	4.36°	1.86	6	0.11

alter lateral components of gait patterns that might have a deleterious effect on overall gait performance.

This study was limited due to age variability in control subjects and sample sizes. Further research is needed in order to support these preliminary results.

The results suggest that the incorporation of loads in the treatment of the instability and posture in patients in the third stage of Parkinson Disease could influence the clinical results.

Several researches have determined that therapeutic techniques could lead to determine the real impact of the rehabilitation exercises with loads applied, and eventually include them in the corresponding protocols.

5. Acknowledgments

The authors would like to thank Marcos Crespo (FLENI) and Franka van Velthoven (University of Twente), for their assistance in the assembly of all the experimental settings.

References

[1] Wood B H, Bilclough J A, Bowron A and Walker R W 2002 *J Neurol Neurosurg Psychiatry* **72** 721–725
[2] Iansek R, Huxham F and McGinley J 2006 *Movement Disorders* **21** 1419–1424
[3] Morris M, Iansek R and Galna B 2008 *Movement Disorders* **23** 451–460
[4] Merello M, Ballesteros D, Rossi M, Arena J, Crespo M, Cervio A, Rivero A, Cerquetti D, Risk M and Balej J 2012 *Functional Neurology* **27** 217–224
[5] Ergun Y U, Doerschug K C, Magnotta V, Dawson J D, Thomsen T R, Kline J N, Rizzo M, Newman S R, Mehta S, Grabowski T J, Bruss J, Blanchette D R, Anderson S W, Voss M W, Kramer A F and Darling W G 2014 *Neurology* **83** 413–425
[6] Bloem B, Hausdorff J, Visser J and Giladi N 2004 *Movement Disord* **19** 192–195
[7] Bloem B, Grimbergen Y, Cramer M, Willemsen M and Zwinderman A 2001 *J Neurol* **248** 950–958
[8] Giladi N, McDermott M P, Fahn S, Przedborski S, Jankovic J, Stern M, Tanner C and Group T P S 2001 *Neurology* **56** 1712–1721
[9] Wilk M and Gnanadesikan R 1968 *Biometrika* **55** 1–17
[10] Becker R, Chambers J and Wilks A 1988 *The New S Language* (Wadsworth & Brooks/Cole)
[11] Jones E, Oliphant T, Peterson P *et al.* 2001– SciPy: Open source scientific tools for Python [Online; accessed 2015-06-09] URL `http://www.scipy.org/`
[12] R Core Team 2014 *R: A Language and Environment for Statistical Computing* R Foundation for Statistical Computing Vienna, Austria URL `http://www.R-project.org/`

Biomechanics and Biotensegrity: Study Method and Frequency Response of the Simplex and 3-bar-SVD Tensegrity Configurations

C Castro Arenas[1,2,4], I Ghersi[2,3], M T Miralles[2,3,4]

[1] Centro Laboratorio de Morfología, Instituto de la Espacialidad Humana, Facultad de Arquitectura Diseño y Urbanismo, Universidad de Buenos Aires, Intendente Guiraldes 2160, Pabellón III, Cuidad Universitaria, CABA, Buenos Aires, Argentina.
[2] Centro de Investigación en Diseño Industrial de Productos Complejos, Facultad de Arquitectura Diseño y Urbanismo, Universidad de Buenos Aires, Intendente Guiraldes 2160, Pabellón III, Cuidad Universitaria, CABA, Buenos Aires, Argentina.
[3] Pontificia Universidad Católica Argentina (LaBIS-Facultad de Ciencias Fisicomatemáticas e Ingeniería-UCA), Av. A. Moreau de Justo 1500, CABA, Buenos Aires, Argentina.

e-mail: ccastroarenas@gmail.com, mmiralles@gmail.com

Abstract. The purpose of this work is to study the frequency response of 3D tensegrity structures. These are structures that have been used, since the 80's, to model biological systems of different scales. This fact led to the origin of the field of biotensegrity, which includes biomechanics as a natural field of application. In this work: a) A simple method for the analysis of frequency response of different nodes in 3D tensegrity structures was set up and tuned. This method is based on a video-analysis algorithm, which was applied to the structures, as they were vibrated along their axis of symmetry, at frequencies from 1 Hz to 60 Hz. b) Frequency-response analyses were performed, for the simplest 3D structure, the Simplex module, as well as for two towers, formed by stacking two and three Simplex modules, respectively. Resonant frequencies were detected for the Simplex module at (19.2± 0.1) Hz and (50.2±0.1) Hz (the latter being an average of frequencies between homologous nodes). For the towers with two and three modules, each selected node presented a characteristic frequency response, modulated by their spatial placement in each model. Resonances for the two-stage tower were found at: (12 ±0.1) Hz; (16.2±0.1) Hz; (29.4±0.1) Hz and (37.2 ± 0.1) Hz. For the tower with three Simplex modules, the main resonant frequencies were found at (12.0± 0.1) Hz and (21.0± 0.1) Hz. Results show that the proposed method is adequate for the study (2D) of any 3D tensegrity structure, with the potential of being generalized to the study of oscillations in three dimensions. A growing complexity and variability in the frequency response of the nodes was observed, as modules were added to the structures. These findings were compared to those found in the available literature.

1. Introduction

This paper aims to explore the applications that biotensegrity may have in areas of human biomechanics [1], both for modeling the behavior of well-known anatomical structures, as well as for

[4] Corresponding author

initiating preliminary studies on the still little explored dynamic behavior of these structures. The term "biotensegrity" was created by S. M. Levin, refers to the application of tensegrity principles to biological forms, with a focus on their structural and mechanical performance [2, 3]. This paper uses the term biotensegrity in a narrow sense, on one hand to denote the analogy between the biological structures bio-mechanical behavior, and on the other hand, the dynamic behavior of tensegrity structures.

Tensegrity defines class-specific structural systems in stable equilibrium known since the 50's [4]. These systems constitute spatial networks formed by independent interconnected tensors of traction and compression bars.

The interest in this type of study lies on the importance of geometry in relation to the mechanical properties of materials and, as well as on the need to have the quantity and quality of mass in specific parts of the structures, to support and pass the preset loads.

Specifically, tensegrity structures are material arrangements with low mass that adds the ability to replace other self-similar element of tensegrity, to achieve greater efficiency in their behavior. Dynamic knowledge, in particular frequency response, allows improving current theoretical models of control to implement the tensors with different types of sensors and replace the bars with actuators or active mechanical elements (smart tensegrity structures).D.E. Ingber argued that tensegrity is the fundamental architecture of life, based on his observations in cells [5].Tensegrity structures possess relevant performance qualities such as flexibility, compactness, material optimization and resilience, extending its applications to different scales, manufacturing processes and materials. However, little is known about its dynamic properties.

There are numerous works on classification and mechanical behavior of the tensegrity simplest module, known either as "module Simplex"[4]. All members of a tensegrity structure are axially loaded. The overall structure is flexed with external static loads while no elements of the tensegrity structure flex in themselves.

The choice of the geometric parameters determines the mechanical properties having a great influence on the resistance of the structure. One of the most outstanding properties of these structures, of interest to biomechanics, is the increase in stiffness by load increase (unless a tensor is loosen) and to the fact that the shape of the structure can be changed with a minimum application of energy.

This paper presents the development of a simple method for studying response in frequency and resonance forms of different nodes of tensegrity vibration structures (up to 60 Hz), for a: a) Simplex module, b) 3-bar two stages SVD tensegrity, and, c) 3-bar three stages SVD tensegrity. The aim is to relate the findings with future models of biomechanics of lumbar spine for which two adjacent vertebrae and their interposed inter-vertebrae disk form a motion segment of the spine [6].

2. Methodology

2.1. Preparation of models

The Simplex tensegrity module is the minimal and simplest 3D tensegrity structure. It is composed by three bars and nine strings and is the basic module of 3-bar SVD tensegrity configurations. The *3-bar-SVD* tensegrity (class 1), is comprised of three bars held together in space by strings. The strings supporting the next stage are known as the "saddle strings" (S). The strings connecting the top of the bars of one stage to the top of bars of the adjacent stages, or the bottom of the bars of one stage to the bottom of bars of the adjacent stages, are knows as the "diagonal strings" (D). Strings connecting the top of the bars of one stage to the bottom of the bars of the same stage are known as "vertical strings" (V). It is a pre-tensed, self-supporting structure holding the position or stability in its vertices by the pre-stress (pre-existing stress tensor or isometric tension) of its own structure. From the biomechanics point of view, bones are analogous to bars that resist muscle traction, and those of tendon and ligament traction are related to tensors. Muscle tone (pre-stress) modulates the stability and stiffness of the body structure.

For this work, three models corresponding to a Simplex module (Figure 1a), a 3-bar two stages SVD tensegrity (Figure 1b) and a 3-bar three stages SVD tensegrity (Figure 1c) were built. The compression bars of the scale models were built in pieces of hollow aluminum bars (0.8 mm wall thickness) of (0.5 ± 0.1) cm external diameter and (20 ± 0.1) cm length. Commercial traction springs in three different lengths (at rest) and 4 different lengths (under tension) were used as links. The interconnection between springs and bars were made from metal anchors located at the ends of the bars.

2.2. Frequency Response and modes of vibration of the nodes in tensegrity structures
The models were placed on an acrylic square platform perforated with small in line holes, to lose weight and to allow the correct fixation of the structures to the uni-axial mechanical oscillator (Pasco Mech. Wave Driver SF-9324).The nodes of the studied structures were numbered consecutively and instrumented with Light Emitting Diodes (LEDs) in order to be properly identified, as shown in figure 1. The input signals were produced by a sine wave generator.

Given that the mechanical actuator has its own transfer function, for all cases a reference node for the models was defined (node), allowing to record the properties of base displacement. The marker movements were analyzed with respect to the position of this base marker, to its own spatial position in the structure and to the extent of effective displacement in the image plane (x-y plane).

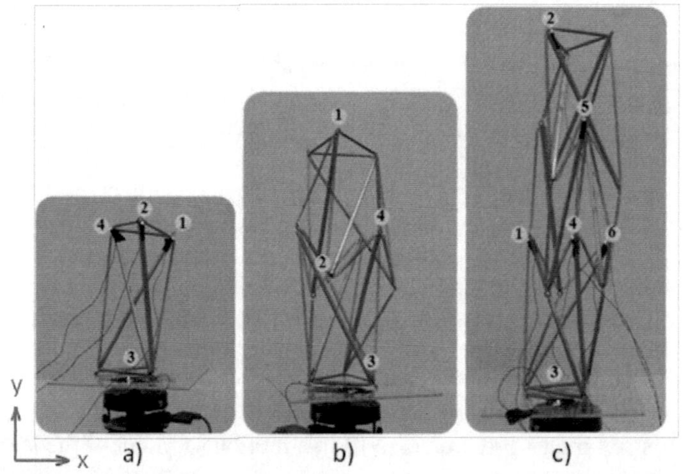

Figure 1. Models implemented with LEDs in nodes of interest. Shown: a) Simplex module; b) 3-bar two stages SVD tensegrity; and, c) 3-bar three stages SVD tensegrity. Numbers were independently assigned in each of the structures to denote nodes of interest in each case.

A sine wave signal generator (Pasco WA-9867) was used to power the mechanical oscillator. Its output was variable in frequency (0.1 Hz - 1000.0 Hz), and of controllable amplitude (0 V – 10 V). During the tests, inputs to the oscillator were kept constant in amplitude (50 %, 75 %), while their frequency was swept with the signal generator at (1.0 ± 0.1) Hz steps. With a sampling rate of 120 fps for the camera, a frequency ramp was implemented, from 1.0 Hz to 60.0 Hz. The position of the markers at each frequency level was recorded by a high speed Casio EXFH25 camera, for 3 seconds.

The coordinates -in pixels- of the estimated centroid of each LED marker were acquired as a two-dimensional vector (x-axis: horizontal; y axis: vertical) for each table. The estimated path travelled by each marker was obtained from the timed succession of these coordinates. Figures 2 and 3 illustrate the above mentioned analysis for a three stages module that served to adjust the method. The sampling frequency is high enough for maintaining an acceptable work resolution (640 x 480 pixels). The videos are stored in the memory of the camera, later retrieved and processed by a specific application for these studies.

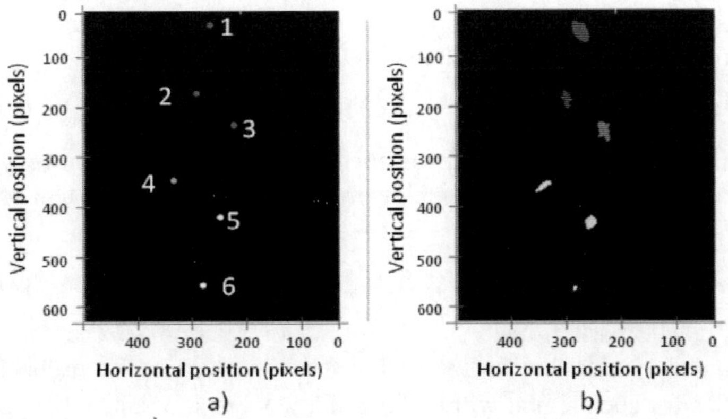

Figure 2. a) Test 3-bar three stages SVD tensegrity instrumented with LEDs in each of the six nodes,
b) Image of the LEDs vibrating at its resonance frequency in the x-y plane.

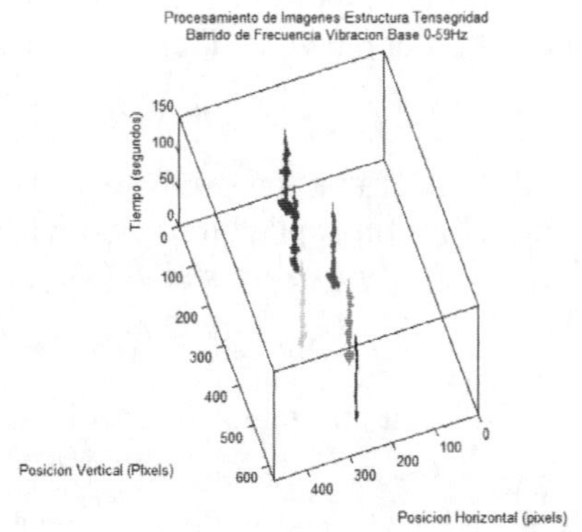

Figure 3. Temporary paths of 6 markers on the time axis of LEDs shown in Figure 2.

2.3. Frequency Response and modes of vibration of the nodes

The application developed for this work allows opening the video file and to process it frame by frame, segmenting markers and obtaining centroids for the "spots" that represent the position of each marker in each video frame (see Figure 2b). Segmentation is achieved after a smoothing process, with a threshold for the intensity and/or color-features of the pixels in the image. The remaining, isolated areas represent each LED marker in a black and white image, and the centroid coordinates for each of these areas are calculated. For greater robustness of the solution, it is possible and necessary to alter the weights within the segmentation algorithm, according to the lighting conditions in which the video was obtained (intensity threshold for the illumination corresponding to the markers). As a result of executing this application, a matrix by the components "x" and "y" of each centroid is obtained (one corresponding to each node, in each picture of the high speed video). This matrix has the information of the relative position of the nodes along the experience. The analysis of the frequency response offers direct and intuitive information for the response of a dynamic system facing an oscillatory input.

A sliding window strategy was implemented for each marker (m_k), and for each frequency level between (1.0 Hz-60.0 Hz), to analyze re sponse (displacements in the x-y plane) with regard to those of the base node (m_b). The width of this window was defined in 1 second (120 samples), the output (m_k)/input (m_b) relation was analyzed for each window displacement on the signal.

For each sliding window displacement (n), an oscillation frequency of the structure base was estimated ($f_{oscillator}(n)$) by searching the maximum level shown by the Fast Fourier Transform (FFT) of the base marker vertical displacement coordinate (y_{m_b}) in that window ($y_{m_b}(n\colon n+120)$. This expression is detailed in (1):

$$f_{oscillator}(n) \approx \max(|FFT(y_{m_b}(n\colon n+120))|) \quad (1)$$

Given the waveform of the input sign, this estimation of the oscillation frequency is indicative of the oscillation frequency assigned to the structure. In case that the dynamics of the structure affects significantly the action of the base level, and consequently the form of its FFT, the estimation of the mentioned frequency will allow to detect it as a frequency of particular interest for further analysis. The estimated frequencies are, in turn, rounded off to entire values for further utilization. Once the oscillation frequency of the structure is identified, the response frequency for each marker of interest is obtained, in any of the two axes. To obtain a relative response to the structure base level it is necessary to subtract the respective displacement of the base marker to the vertical coordinate of each marker in the window. Since the extent of the vertical displacement in the base level changes with the frequency (for the attenuation of the actuator response, and the potential dynamic impact of the structure on itself), the response of each marker was normalized in the vertical axis, appealing to the standard deviation of the vertical displacement registered in the base, as shown in the equation (2):

$$y_{m_k}{}^N(n{:}\,n+120) = (y_{m_k}(n{:}\,n+120) - y_{m_b}(n{:}\,n+120))/std(y_{m_b}(n{:}\,n+120)) \quad (2)$$

Provided that the oscillatory action analyzed on the structures occur on the vertical axis, the only normalization applied to the horizontal coordinates consists in an adjustment for the effective extent of the input action expressed in (3):

$$x_{m_k}{}^N(n{:}\,n+120) = (x_{m_k}(n{:}\,n+120))/std(y_{m_b}(n{:}\,n+120)) \quad (3)$$

With the normalized coordinates of each marker inside the window in relation to the base $(x_{m_k}{}^N(n{:}\,n+120), y_{m_k}{}^N(n{:}\,n+120)$, an absolute value for the FFT value was obtained for each of these signals. Prior to the FFT, mean values for each signal within the window were subtracted. These frequency responses (FR_{xk}, FR_{yk}), that provide information between 0 Hz and 60 Hz, were assigned to each frequency of the oscillator detected by the sliding window in that point:

$$FR_{yk}(\mathrm{f}, \mathrm{f}_{\mathrm{oscillator}}(\mathrm{n})) = |FFT(y_{m_k}{}^N(i_v{:}\,j_v))| \quad (4)$$

$$FR_{xk}(\mathrm{f}, \mathrm{f}_{\mathrm{oscillator}}(\mathrm{n})) = |FFT(x_{m_k}{}^N(i_v{:}\,j_v))| \quad (5)$$

Since window displacements generated multiple instances of $FR_{xk}yFR_{yk}$, the signals with the same associated frequency were averaged for each marker, improving overall estimations and mitigating errors in the estimation of the frequency of oscillation. With this information, it is possible to obtain two surfaces, representing the complete frequency response of each marker in the structure, facing all the assigned input frequencies. By the proposed strategy of analysis the points of significant oscillation, as well as the vibration modes that demonstrated minimal movements, were automatically, and by graphical means, identified for each structure of interest.

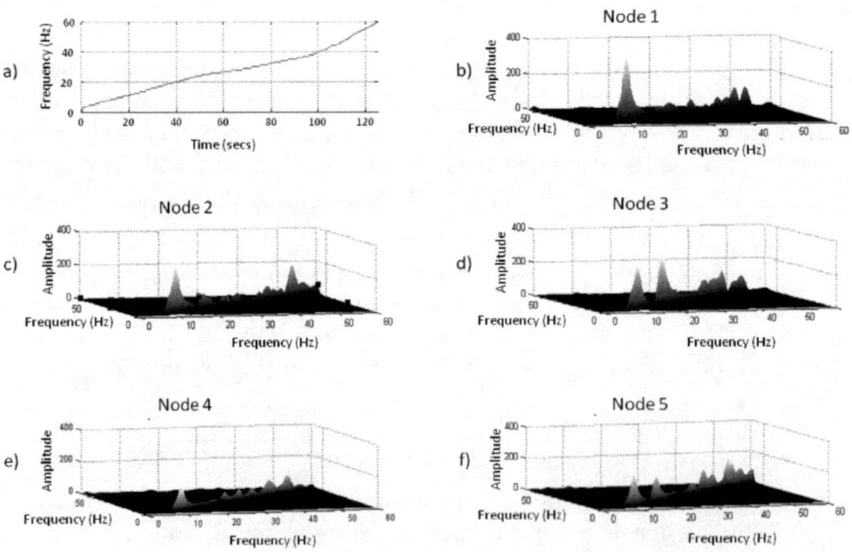

Figure 4. a) Frequency sweep, b)-f) Frequency-response of the structure in Figure 2, over the vertical axis, for nodes 1 through 5, respectively, compared to the base marker (node 6).

3. Results

3.1 Frequency response of Simplex module

The results of the study of the frequency response of the Simplex module are the following. The estimation curves of the frequency of oscillation offered results compatible with the scale of input frequencies between 1.0 Hz and 60.0 Hz. Local maximum points from the surfaces, derived from the signal processing strategy, are associated to response peaks -structure resonances- whereas the minimums indicate fixed positions. As expected, the three surfaces obtained from the markers offer similar overall results, given the symmetry of the structure. The markers have a spatial distribution around the central axis of the module (actuator oscillation axis) in a radius of (10.0 ± 0.1) cm (see figure 1a). For the three markers, a general resonance frequency is found at (19.2 ± 0.1), concentrated on the horizontal axis, and to a lesser degree at (50.2 ± 0.1) Hz on the vertical axis. Table 1 summarizes relevant quantitative results derived from the experimental setup on this structure.

Table 1. Frequency-response for nodes 1,2 and 4 of the Simplex module

Node	*Peak Frequency (± 0.1)Hz*		*Amplitude (adimensional)*	
	x-axis	y-axis	x-axis	y-axis
1	19.2	-	49	-
	-	49.2	-	91
2	19.2	19.2	108	29
	-	51.0	-	76
4	19.8	19.2	45	46
	48.0	52.8	23	66

General oscillation amplitudes vary between nodes. Observable differences, like the response of node 1, for which there was no significant peak at 19.2 Hz in the vertical axis (contrary to results from analogous nodes 2 and 4), reflect differences in pre-tension states surrounding that node. The value for the second vibration mode (50.2 ± 0.1) Hz was obtained as an average of the respective detected values among all homologous nodes (see Table 1).

Figure 5, from left to right, shows a summarized response to the input, over the horizontal axis, for nodes 1, 2 and 4 respectively. These surfaces hold similar characteristics, while their general magnitude levels and peak values are variable. Nodes 1 and 4 hold comparable response levels, while node 2 approximately doubles their amplitude. Horizontal displacements for these nodes are mostly modulated, according to the axial symmetry of rotation that defines the structure, and are therefore variably projected over the depth axis, which is not registered by the frontal camera. The standout magnitude for the response of node 2 can be explained by this fact, while other minor differences can be attributed to material flaws, as well as to involuntary assembly imperfections. Additional local maximums that were found in the horizontal axis (see Figure 5a) were not included in the table, while the selected values are a match among homologous nodes.

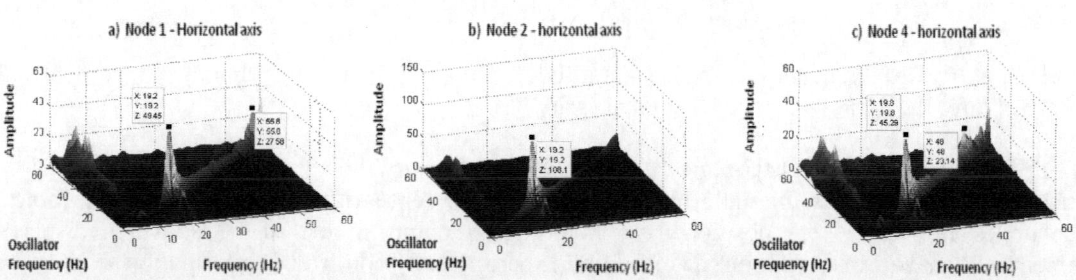

Figure 5. Simplex Module: a), b) and c) Frequency response in the horizontal axis, for nodes 1, 2 and 4, respectively.

3.2 Frequency response of the 3-bar two stages SVD tensegrity
This tower was built by piling two Simplex modules as shown in Figure 1b).The spatial orientations are opposite, being the lower anticlockwise and the superior clockwise. Once again, the frequency response of the different nodes for the Simplex module was analyzed, as described in section 2.3. Among all possible nodes, those equivalent and non-equivalent between structures can be selected to carry out comparisons. The results are shown in Table 2.

The structure presents general vibration modes, of varying magnitude, for these three nodes, at (16.2 ± 0.1) Hz, (29.4± 0.1) Hz, and (37.2 ± 0.1) Hz. Figure 6 shows the magnitude of the frequency response from node 1, over the horizontal and vertical axis, respectively, for this two-stage structure. Node 1 shows a slightly different, more focused resonant response over its horizontal axis than the remaining nodes, with no significant peaks at 12.0 Hz, while the remaining nodes (2 and 4) display another standout point at that value. This difference is consistent with the relative placement of node 1, compared to 2 and 4, in the structure (see Figure 1). Responses show an overall increased complexity, with a larger number of feature points, all remaining within a diagonal (visible from the xy-view of the surface), representing a 1:1 ratio between oscillator frequency and peak response.

Table 2. Frequency response for nodes 1,2 and 4 of the 3-bar two stages SVD tensegrity

Node	*Peak Frequency (± 0.1)Hz*		*Amplitude (adimensional)*	
	x-axis	y-axis	x-axis	y-axis
1	16.2	16.2	389	652
	31.8	29.4	155	188
	34.8	-	202	-
	37.2	-	323	-
2	12.0	12.0	148	90
	18.0	16.2	470	236
	29.4	31.8	114	147
	37.2	37.2	441	298
4	12.0	12.0	174	79
	16.8	16.2	246	189
	29.4	29.4	102	48
	37.2	37.2	271	181

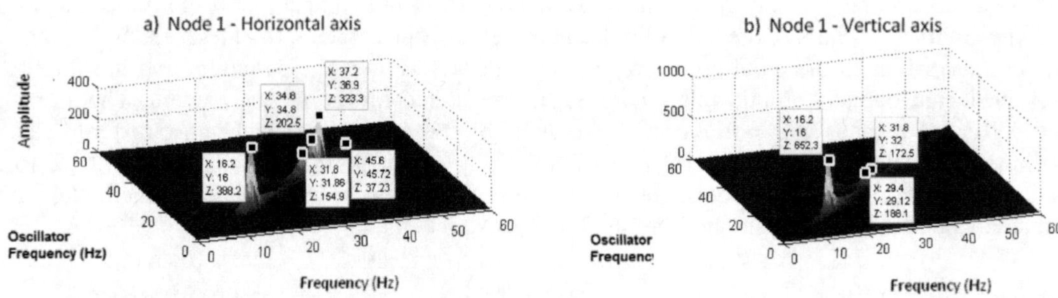

Figure 6. a) and b) Frequency response amplitude of node 1 in the horizontal and vertical axis, respectively, for the 3-bar two stages SVD tensegrity.

3.3.1 Frequency response of the 3-bar three stages SVD tensegrity.
Frequency-response results for the 3-bar three stages SVD tensegrity tower are shown in Table 3. Since nodes 1, 4 and 6 are positioned on homologous points in the structure, which display a symmetrical behavior that was both deduced and experimentally confirmed, information from two of these nodes is not included in the Table. Significant general oscillation frequencies for this structure,

shared by these nodes, occur at (12.0± 0.1) Hz and (21.0 ± 0.1) Hz, with other, less significant peaks detected at (40.8 ± 0.1) Hz, (43.8 ± 0.1) Hz and (51.0 ± 0.1) Hz.

Once again, the measured response characteristic, for both axes, is much more complex in its composition than those found in the previous sections, displaying peaks outside the 1:1 diagonal, for example at (10.8 ± 0.1) Hz, when the input, oscillator frequency was of (21 ± 0.1) Hz, a response at approximately one half of the frequency of the input excitation. One remaining peak of (1.8 ± 0.1) Hz was found for (12 ± 0.1) Hz inputs, resulting in an elaborate response of the structure to 12 Hz vibrations. Frequencies shown in Table 3 remain within the 1:1 ratio to the input signal.

Figure 7 shows four surfaces, representing the frequency response from nodes 1 and 5 of the tower. These graphs display the great variability between response characteristics, both among different nodes, as well as in the directionality of each response, for any node, which is also visible in figure 6.

Table 3. Frequency response for nodes 1,2 and 5 of the three-stage, *3-bar SVD* structure

Node	*Peak Frequency (± 0.1)Hz*		*Amplitude (adimensional)*	
	x-axis	y-axis	x-axis	y-axis
1	12.0	12.0	290	256
	21.0	21.0	290	45
	-	28.2	-	86
	43.8	43.8	130	52
	51.0	51.0	55	50
2	-	12.0	-	622
	21.0	21.0	282	175
	33.0	34.2	88	67
	40.8	40.8	105	58
	51.0	51.0	60	51
5	12.6	12.0	104	492
	21.0	21.0	93	468
	35.4	33.0	73	61
	40.8	-	113	-
	-	43.8	-	94
	-	51.0	-	164

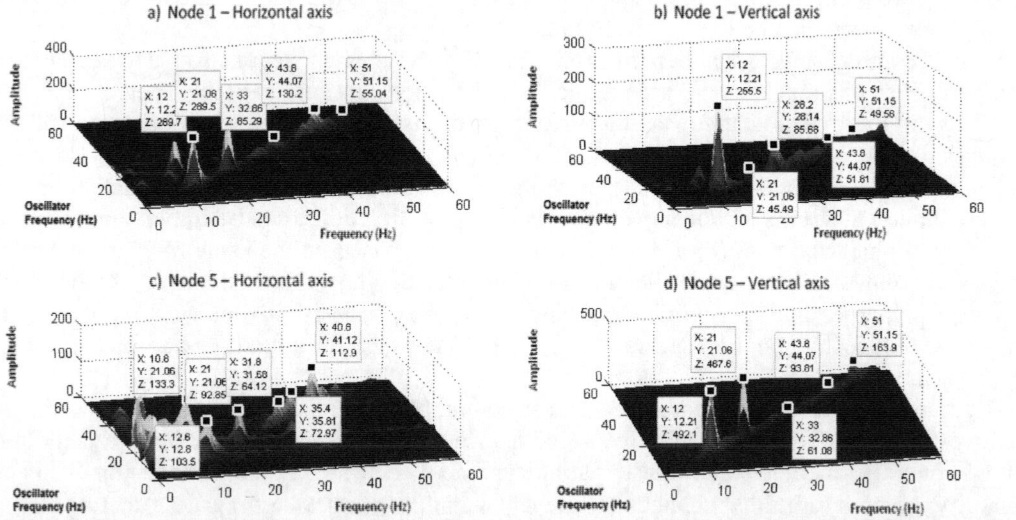

Figure 7. Frequency response of the 3-bar three stages SVD tensegrity (nodes 1 and 5, in both axes): Significant differences in response characteristics, between axes and nodes, are detected.

4. Conclusions

In this work, a simple method for analyzing the frequency response of 3D tensegrity structure nodes (structure vibration modes) was developed, based on light emitting markers (LEDs) and a signal processing algorithm to obtain information from video images taken by a high speed camera. Each node has a characteristic and well determined resonant frequency pattern. Limitations to this method are related to its frequency range (scanning between 0 Hz and 60 Hz, with ±0.1 Hz error), and to the fact that only oscillations parallel to the structure axis were considered. Node displacements were measured in each frame over the plane of each image.

The frequency response for a Simplex 3-bar SVD tensegrity module was measured, resulting in resonance or peak responses at (19.2 ± 0.1) Hz and (50.2 ± 0.1) Hz. In particular, results from this module hint at the variability of the response, which proved to be very sensitive to the material of the compressive elements. As an example, when changing aluminum to wood, for the Simplex module, the first resonant or peak frequency rises to 30 Hz. These graphs were not presented due to a lack of space.

Similar frequency responses for towers formed by two and three stages of modules were also assessed. When the modules are superposed to form towers, the number of nodes increases, but for the study, those equivalent and non-equivalent between structures were selected to carry out comparisons. These were the instrumental nodes, and some of them were compared between three structures. These results displayed an increasing complexity for the response characteristics, which in turn, and in conjunction with their sensitivity to constructive properties, highlights the value of counting with an experimental method for the assessment of real structures and prototypes, in particular, associated to more elaborate biomechanical models.

Additionally, the flexibility displayed by the response of these structures opens the possibility to "tune" them towards desirable properties for product design. These properties can, in turn, be validated through the method shown in this work.

The comparison of results to bibliography can only be accomplished with information found for the three module tower [7]. A similar tower, with a scale ratio of 2.3, was studied using 3D accelerometry, between 0 Hz and 50 Hz, and using guitar strings as tensors. As in this study, different vibration modes are found, with frequencies in the order of those found in this study. Based on a finite element model, authors associate modes to torsion and bending, and indicate the relationship in the variation of these frequencies to the pre-tension state of the structure. In the present analysis, no significant responses were found to oscillator frequencies below 12 Hz for the three-stage tower, and no evidence of "soft mode", as indicated in [7], was found.

Future research stages can benefit from further advances in acquisition and signal-processing strategies. For example, image-processing can be used to automatically detect the oscillation frequency of the power source, taking advantage of its digital display (through optical character recognition). The addition of a second, orthogonal camera to the system can also be directly implemented as a method for 3D node tracking. Finally, alternative mechanical actuators, accepting multiple input waveforms and/or movements in different directions, can be implemented, giving place to alternative characterizations for these dynamic systems, based on the same fundamental principle of analysis presented in this work. Some examples include the precise detection dedicated to low-frequency responses and to significant changes in spatial orientation (deformations), as well as the analysis of transient responses, and fault and limit conditions, for structures and prototypes based on tensegrity models.

To summarize, we are in the presence of a structural block, with a frequency response that is highly sensitive to qualitative and quantitative variables. In turn, once these variables are determined, this response can be modeled and these parameters applied with absolute precision. This adaptive versatility, given by minimal changes to relevant variables in these tensegrity structures, is a typical feature of biological structures, at every scale.

References

[1] Randel L, Swanson I 2013 Biotensegrity: A unifying theory of biological architecture with applications to osteopathic practice, Education, and Research—A Review and Analysis *J Amer. Osteo. Asso.* **113** 1 34-52 https://goo.gl/tl2lUP

[2] Levin S 2002 The Tensegrity-Truss as a Model for Spine Mechanics: Biotensegrity *J. Mech. Med. Biol.* **2** 03n04 375-388

[3] Levin S 2006 Tensegrity, The New Biomechanics *Textbook of Musculoskeletal Medicine* ed Hutson M & Ellis R (Oxford: Oxford University Press) pp 69 - 80

[4] Skelton R E and de Oliveira M 2009 *Tensegrity Systems* (Springer)

[5] Ingber D E 2003 Tensegrity I: Cell Structure and hierarchical systems biology *J. Cell Sc.* **116** 1157-1173

[6] Zatsiorsky 1998 *Kinetics of human motion* (Human Kinetics) p 312

[7] Bossens F Callafon R A and Skelton R E 2004 Modal analysis of a tensegrity structure-an experimental study *Mechanical and Aerospace Engineering Dynamic Systems and Control Group* (San Diego: University of California) http://goo.gl/CkqCHL

Acknowledgements

The authors wish to acknowledge the technical assistance provided by Mr. Gonzalo Fernandez. This work was carried out under - and supported in part by - Project 20621201000011BA UBACyT 2013-2016, Centro de Investigación en Diseño Industrial de Productos Complejos, *Facultad de Arquitectura, Diseño y Urbanismo, Universidad de Buenos Aires* (CIDI-FADU-UBA), y *Facultad de Ciencias Fisicomatemáticas e Ingeniería, Pontificia Universidad Católica Argentina (*UCA).

A possible parameter for gait clinimetric evaluation in Parkinson's disease patients

C. N. Lescano[1,3], S. E. Rodrigo[1] and D. A. Christian[2]
[1] Gabinete de Tecnología Médica, Facultad de Ingeniería, Universidad Nacional de San Juan
[2] Departamento de Matemática, Facultad de Ingeniería, Universidad Nacional de San Juan
[3] CONICET: Consejo Nacional de Investigaciones Científicas y Técnicas

E-mail: clescano@gateme.unsj.edu.ar

Abstract: The strength and usefulness of a rating scale for describing disease evolution relies on the accurate determination of variations representing clinically relevant changes. In this sense, the habitually used Hoehn-Yahr (HY) Scale for Parkinson Disease (PD) in its modified version distinguishes between the 2 and 2.5 stages to explain if the bilateral involvement is or is not accompanied by body balance impairment. Nevertheless, this scaling does not allow for differentiating the symptoms and signs associated with each stage accurately. Considering this difference, this work aims at analyzing some gait parameters -stance and swing phase times and magnitude of the vertical component of ground reaction force during the gait cycle- of PD patients classified as HY=2 and HY=2.5 in contrast with healthy subjects (HY=0), with the purpose of assessing whether there is a statistically significant difference among all these HY categories. For all gait parameters evaluated, the results indicated significant differences between HY=0 and HY=2.5. However, only the magnitude of the vertical component of ground reaction force presented relevant differences between HY=2 and 2.5. As expected, therefore, these results show the potential of such parameter to clinimetrically identify the level of gait impairment/disability in PD patients on the Hoehn-Yahr Scale.

1. Introduction

Parkinson's disease (PD) is a progressive disorder of the nervous system that affects human body movement, among other functions. Worldwide, approximately 7 to 10 million people have PD [1], and in Argentina the disease is the second most common chronic neurodegenerative disorder after Alzheimer's disease, with around 70,000 individuals affected [2]. PD can occur in different age groups, including patients less than 20 years old, though its prevalence is about of 1-2% in those over 65 years. Also, it is known that men are one and a half times more likely to have PD than women [3].

Classical research indicates that PD is gradually developed as the result of a defect in dopamine production, though today the involvement of many other neurotransmitters is recognized, such as norepinephrine, acetylcholine, glutamate and GABA [1, 3]. While PD cannot be cured at this moment, early diagnosis and medical intervention with levodopa (L-dopa) therapy may markedly improve its symptoms [1-3, 4], thus displaying the importance of accurately recognizing the early manifestations of this disease, as well as the symptoms and signs characteristic of its progressive evolution.

In this sense, it is well established that in the early stages of Parkinson's disease, a patient's face may show little or no expression and his arms may not swing when he walks. In addition, his speech may become soft or slurred. Among the most common PD features that worsen over time, we find bradykinesia (i.e., slowed movement), hypokinesia (small amplitude movements), resting tremors and rigidity, though these may not all be present, which can be accompanied by postural instability and another features described as "no-motor symptoms" [3, 5, 6].

As regards clinical features of PD related to locomotion disorder development, in addition to a decrease in arm movement (hypokinesia), symptoms in the lower limbs include: temporal / spatial variability of gait patterns and shortening step length, little foot-lifting off the ground with consequent shuffling, and gait speed decrease or increase at the expense of a cadence rise greater than the step length. It is also possible that at a given moment, the patient may have difficulty starting up his gait or making turns, an episode known as freezing [7, 8].

In order to describe and classify this complexity of signs and symptoms related with PD evolution, different scales have been developed, such as the Unified Parkinson's Disease Rating Scale (UPDRS) and the Hoehn-Yahr (HY) Scale [9, 10]. The first assesses various patient features: mental status, daily life activities, motor status, complications of therapy, etc. In contrast, the HY Scale focuses on motor disorders caused by PD and their functional consequences. It was originally designed to be a simple descriptive scale divided into five-point stages (Table 1, left), providing a general estimate of clinical function in PD by combining functional deficit (disability) and objective signs (impairment) [9].

Additionally, the HY Scale is based on the two-fold concept that the severity of overall parkinsonian dysfunction relates to bilateral motor involvement and balance/gait compromise. Consequently, parkinsonian motor impairment can be charted from a unilateral (Stage 1) to a bilateral disease without balance difficulties (Stage 2), to the presence of postural instability (Stage 3), the loss of physical independence (Stage 4), and being wheelchair- or bed-bound (Stage 5) [9].

A more detailed analysis of this gradation system shows that the definition of Stage 1 as "unilateral involvement only, usually with minimal or no functional disability," combines two concepts that are not truly equivalents. Stage 2 is defined by the lack of "impairment of balance," but this wording does not use the same descriptive wording as in Stage 1. Besides, the progressive stages of the HY scale are based on two different indices of severity: unilateral versus bilateral signs, and absence or presence of gait and balance impairments. As such, the HY scale is a categorical scale describing clinical status, but each increment does not really represent a higher degree of overall motor dysfunction [9, 10].

Table 1. Comparison between the original and modified Hoehn and Yahr Scale.

Stage	Hoehn–Yahr Scale	Modified Hoehn–Yahr Scale
1	Unilateral involvement only usually with minimal or no functional disability	Unilateral involvement only
1.5	-	Unilateral and axial involvement
2	Bilateral or midline involvement without impairment of balance	Bilateral involvement without impairment of balance
2.5	-	Mild bilateral disease with recovery on pull test
3	Bilateral disease: mild to moderate disability with impaired postural reflexes; physically independent	Mild to moderate bilateral disease; some postural instability; physically independent
4	Severely disabling disease; still able to walk or stand unassisted	Severe disability; still able to walk or stand unassisted
5	Confinement to bed or wheelchair unless aided	Wheelchair bound or bedridden unless aided

Despite these weaknesses and the scale's development in the pre-L-dopa era, the use of the HY Scale has continued to be widely used for the description of large populations of PD patients. As an example, studies from both the pre-L-dopa and L-dopa eras involving large cohorts of PD patients have found similar percentages of cases assigned to the different stages of the HY Scale. In these studies, Stage 1 and Stage 5 account for the smallest number of subjects, followed by Stage 4, with the bulk of patients, ranging from 52 to 77%, falling into Stages 2 or 3 [11].

Among the gait variability measures considered for PD patients, the following temporal parameters calculated for each gait cycle were used [13]: 1− stride time or duration of the gait cycle (time from initial contact of one foot to subsequent contact of same foot); 2− swing time (amount of time one foot is in the air); 3− percentage swing time (100 x swing time/stride time); 4− double stance time (time of bilateral foot contact); 5– percentage double stance time (100 x double stance time/stride time); and 6– step time (time from initial contact of one foot to initial contact of the other foot).

While this research group has not evaluated these gait variability measures for PD patients classified in each stage of the HY Scale, further studies on the subject led these investigators to conclude that gait variability is a quantifiable feature of walking that is altered (both in terms of magnitude and dynamics) in clinically relevant syndromes, such as falling and neuro-degenerative diseases (e.g., Parkinson's and Alzheimer's disease [15, 16]. Additionally, gait instability measures [14, 17] directly related to body balance impairment signalled by a score of 2.5 on the modified HY Scale, apparently would predict falls in idiopathic elderly fallers and other populations who share an increased fall risk.

With regard to the variability analysis of gait kinetic parameters as possible indicators of PD evolution, a previous work by our research group showed that the vertical component of ground reaction force (VGRF) can be used for automatic classification of PD patients from clustering techniques based on artificial neural networks [18, 19]. In this case, the gait variability measures utilized were: 1– mean maximum value of VGRF; 2– mean standard deviation of VGRF; 3– mean coefficient of variation of VGRF and 4– mean sum of VGRF over successive stance phases. The preliminary results of our study indicated that the VGRF contains relevant information to differentiate objectively normal and PD gait patterns, thus displaying that this parameter could be a predictor of the degree of gait impairment/disability for PD.

According to all these studies, we therefore hypothesize that gait data analysis can provide sensitive and clinically relevant information in the evaluation of PD regarding morbidity, fall risk and the response to therapeutic interventions in general terms and particularly, in the assessment of significant statistical differences between patients included in the HY=2 and HY=2.5 stages.

In order to evaluate the viability of our hypothesis and based on this background, the aim of this work was to analyze gait data of PD patients classified as HY=2 and HY=2.5 with respect to normality (HY=0), with the purpose of assessing whether there are statistically significant differences between these HY categories. In the following sections, the methodology utilized to fulfill this objective is described. Firstly, the materials and methods used to evaluate gait data variability for PD patients belonging to HY=2 and HY=2.5 are detailed. Then, the results of a statistical analysis of different gait parameters among the considered Hoehn-Yahr categories are presented. Finally, a brief discussion and conclusions are provided, in the search for clues to assess the level of gait impairment/disability in PD patients on the HY Scale from a clinimetrically consistent gait parameter.

2. Materials and Methods

2.1 Human Gait Database

To evaluate whether there are significant differences of gait disorders between the analyzed PD patient groups (HY=2 and HY=2.5) and among them and the control group (HY=0), a database of 20 patients with idiopathic PD and 20 healthy control subjects of similar age was used. This database, freely available on the *Physionet* website [20], contains clinical and demographic information of the considered subjects, besides the corresponding value of the modified Hoehn-Yahr Scale (Table 2).

Table 2. Clinical and demographic data, expressed as average values and standard deviation

	Healthy subjects	**Parkinson's disease patients**	
Interest-factor level	HY=0	HY=2	HY=2.5
Age	67.4±8.6	71.8±10.5	73.8±7.38
Height [m]	1.64±0.08	1.70±0.08	1.66±0.08
Weigth [Kg]	70.1±12.52	73.7±9.79	70.4±11.86
Gait Speed [m/s]	1.18±0.14	1.01±0.15	0.96±0.23
Study made by:	Yogev *et al.*, 2005; Toledo *et al.*, 2005; Hausdorff *et al.*, 2007 [20].	Yogev *et al.*, 2005; Hausdorff *et al.*, 2007 [20]	

In contrast to normal cases (HY=0), pathological cases analyzed here correspond to 10 PD cases of both Stages HY=2 and HY=2.5 based on the modified version of the Hoehn-Yahr Scale. As was previously mentioned, these Stages are related to patients that exhibit an altered gait with bilateral symptoms, differentiated by the lack or presence of body balance impairment [9, 12].

Also, the database includes kinetic data for both feet that were previously acquired with the F-Scan insole system [21]. From an insole with 8 sensors underneath each foot, this system measures force (in Newtons, N) as a function of time, which together represent the plantar pressures distribution for every foot. The gait test was performed during about 2 minutes on level ground at a normal self-selected cadence for each analyzed subject, the data being then digitalized and recorded at 100 samples per second. Also, the database contains two signals that reflect the sum of the 8 sensor data for each foot, that is, equivalent to the VGRF for left and right feet, respectively [20]. Although this VGRF value is not exactly equivalent to the vertical component of the ground reaction force measured on force platforms [22, 23], the analysis on this value can still be done to obtain clinically relevant information.

Furthermore, for minimizing the effects of start-up and to match the length of the vector data for all samples, the first 5 seconds of data were discarded, opting finally for a vector length of 2700. Figure 1 shows the VGRF during a gait cycle for both feet, obtained from the mentioned database for a subject with a normal gait and a Parkinson's patient with HY=2.

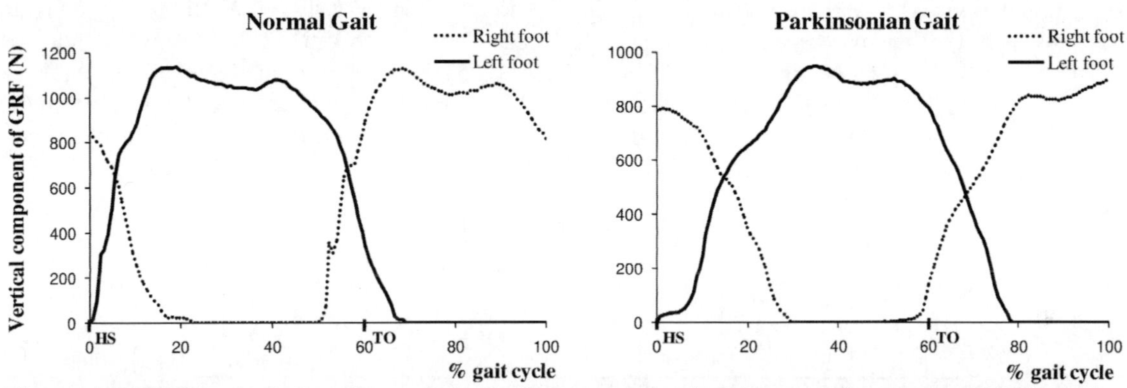

Figure 1. VGRF for both feet for a gait cycle. Left: normal gait. Right: parkinsonian gait, HY=2. For normal gait, stance phase (0-60% GC) begins at left heel strike (HS) and ends with toe off (TO). Swing phase takes between 60 and 100% GC. Notice how these percentages vary for PD gait.

2.2 Definition of Variables

The choice of variability parameters to analyze gait in PD is based on typical inter-subject differences observed in the size and shape of the curve representing the temporal variation of the VGRF. Indeed, as shown on the left in Figure 1 for a case of normal gait, the curve has two peaks. The first happens during heel strike with the ground at the beginning of the stance phase, while the second peak is caused by the upward force exerted by the ground during toe off at the end of the same phase [18-19, 24, 25]. Additionally, the stance and swing phases occur between 0-60% and 60-100% of gait cycle.

In contrast, the curve represented on the right side of Figure 1 for the case of the HY=2 PD patient, sometimes exhibits a reduction in the peak height of VGRF. Also of note is the variation of time taken for the stance and swing phases for PD patients as compared to control subjects (in this case 70% and 30% of gait cycle, respectively), so as to ensure body stability during locomotion. Later stages of this disease are characterized by a gait with small shuffling steps and a single narrow peak of VGRF, thus reducing the time required for the swing phase significantly [18-19, 26].

According to these observations and in order to select the most appropriate variables for differentiation of normal and parkinsonian gait patterns, distinct parameters were employed. Such parameters characterize the inter-subject gait variability throughout the temporal signal corresponding to consecutive gait cycles. Among them, the following were utilized: the stance phase (STPT) and swing phase (SWPT) times and the magnitude of the VGRF, as well as the respective variability obtained by calculating their corresponding coefficients of variation (CV), defined as the ratio of the standard deviation to the mean. In each case, twenty gait cycles performed by the left limb were analyzed, from which the mentioned gait parameters were statistically analyzed [27].

Furthermore, prior to analysis, the VGRF signals were filtered using a moving average filter of 5 points. Then, from the filtered signals the stance phase and swing phase times (STPT and SWPT) were determined by applying the thresholding technique [28]. For this, the stance phase was defined for signal fractions above 10N and the swing phase as the time during which the foot is not in contact with the ground and hence, the VGRF falls below 10N. Also, the STPT and SWPT data were normalized by their respective gait cycle time for each considered subject. The same procedure was applied to the magnitude of VGRF, in this case normalized in regard to the body mass of each analyzed subject.

3. Results

3.1 Analysis of Stance Phase and Swing Phase Times

Gait data (STPT and SWPT) for the samples HY=0, HY=2 and HY=2.5 yielded values of standardized asymmetry and kurtosis coefficients outside the range (-2, +2), showing that such data does not come from populations with normal distribution. Thus, the non-parametric Kruskal-Wallis test to compare medians instead of means was applied using Statgraphics Centurion XVI.II [27]. The results indicated that there is a statistically significant difference ($p < 0.05$) between the medians of the analyzed samples (Table 3). Still, when the results for STPT and SWPT were contrasted with the respective box and whisker plots (Figures 2 and 3), a significant difference was seen only between samples with HY=0 and HY=2.5, but neither between HY=0 and HY=2 nor between HY=2 and HY=2.5.

Tabla 3. Kruskal-Wallis test

		SWPT	STPT
HY	*Sample size*	*Rank*	*Rank*
0	400[a]	427,711	373,289
2	200[b]	408,668	392,332
2,5	200	337,91	463,09

Statistic= 20,4691 P Value = **0,0000359089**

[a]20 time series multiplied by 20 normal subjects.

[b]20 time series multiplied by 10 PD patients in each case (HY=2 and HY=2.5).

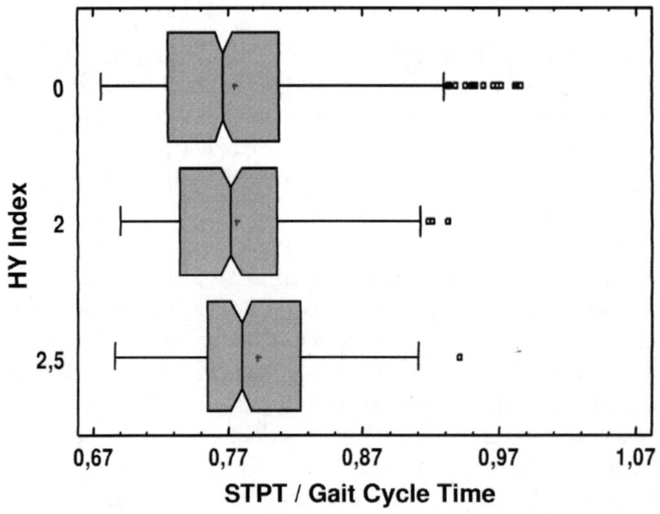

Figure 2: Box and whisker plot for stance phase time (STPT) during gait cycle for the analyzed HY categories.

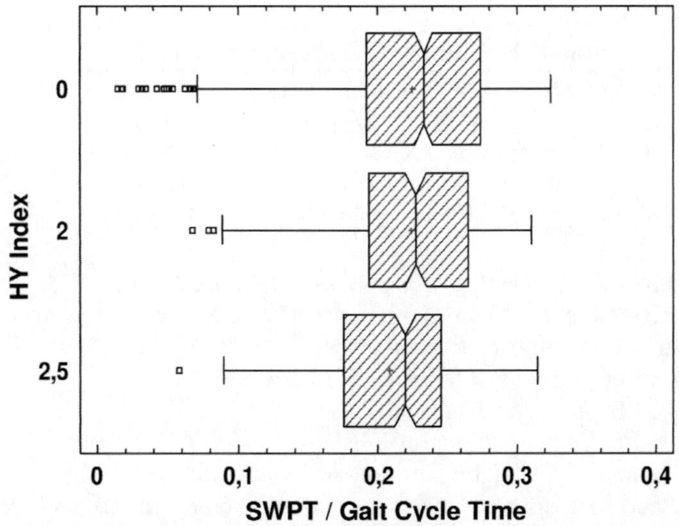

Figure 3: Box and whisker plot for swing phase time (SWPT) during gait cycle for the analyzed HY categories.

In addition, using the temporal data, the coefficients of variation (CV) of the STPT and SWPT series were calculated. According to the obtained values for standardized asymmetry and kurtosis coefficients, it is possible that the achieved transformation through CV calculation comes from normal distribution, being valid to apply statistical tests that evaluate their respective standard deviations [27]. However, when the corresponding central tendency (mean and median) and standard deviation measures for these new considered variables were compared, no significant differences between samples were evidenced. Furthermore, the Kruskal-Wallis contrast confirmed that there is no significant difference between medians with a value of $p = 0.487911$ for a confidence level of 95%. Figure 4 represents the results of this analysis for the CV(SWPT) through the box and whisker diagram.

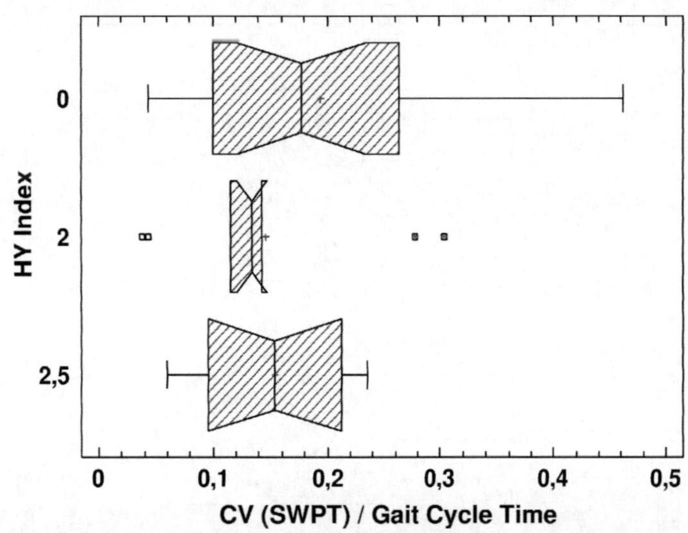

Figure 4. Box and whisker diagram for the coefficient of variation of swing phase time normalized by their respective gait cycle time for the HY categories analyzed.

Finally, Table 4 displays the mean, standard deviation, and maximum and minimum values obtained for the respective coefficients of variation (CV(SWPT)) of the analyzed HY samples.

Table 4. Statistical evaluation of CV(SWPT)

HY	*Simple size*	*Mean value*	*Standard deviation*	*Minimum*	*Maximum*
0	20	0,193536	0,113304	0,044103	0,461981
2	10	0,145837	0,0856172	0,0380525	0,304059
2,5	10	0,153885	0,064203	0,0596666	0,235273
Total	40	0,171698	0,0969254	0,0380525	0,461981

3.2 Analysis of the Magnitude of the Vertical Component of Ground Reaction Force

In order to find parameters that establish significant differences for the analyzed samples, three features of the VGRF were evaluated for each gait cycle:

-Maximum VGRF value normalized to respective body weight,
- Area under the VGRF curve during stance phase,
- Coefficient of variation of the maximum VGRF.

In the first two cases (maximum VGRF value and area value under the VGRF curve), the values of standardized asymmetry and kurtosis coefficients were out of range, and as such, the non-parametric Kruskal-Wallis test was applied [27]. In the case of the maximum VGRF, the obtained p value indicated that there is statistically significant difference between the medians of the analyzed HY samples (Table 5). Also, the box and whisker plot for the three samples exhibited differences between HY=0 and HY=2.5, and between HY=2 and HY=2.5. However, this difference is not significant when comparing the cases HY=0 and HY=2 (Figure 5).

Table 5. Kruskal-Wallis test

		Max VGRF/Body Weight
HY	*Sample size*	*Rank*
0	400	411,89
2	200	304,965
2,5	200	473,255

Statistic= 54,9801 P Value = **1,15141E-12**

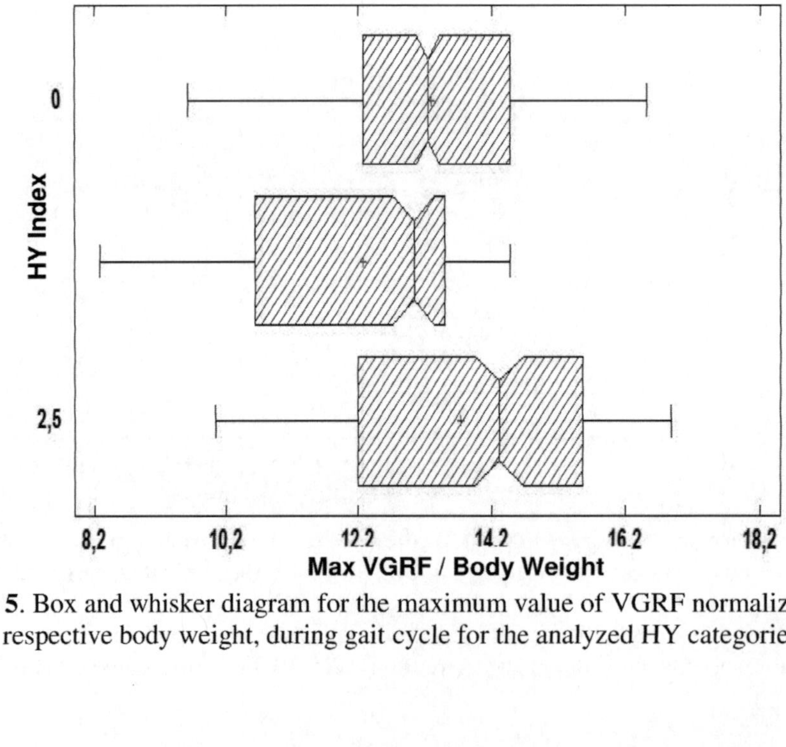

Figure 5. Box and whisker diagram for the maximum value of VGRF normalized by the respective body weight, during gait cycle for the analyzed HY categories.

Similar results were achieved when the area under the VGRF curve during stance phase were analyzed, in this case with a value p=0.03 (<0.05). Finally, in regard to maximum VGRF variability, although data are from a population with normal distribution, the central tendency measures did not show statistically significant differences for the HY index [27]. As for the comparison of standard deviation, only significant differences between the HY=0 and HY=2.5 samples were found.

4. Discussion and Conclusions

In the literature, there are various examples of previous research analyzing the stride to stride fluctuations in PD patients from spatial-temporal parameters of gait patterns [7-8, 13-17]. On the one hand, through such research, both an increase of gait variability in PD in contrast to normal gait and a correlation between the degree of gait variability and PD severity [8, 13-14] are shown. On the other hand, this research also confirms that gait variability is a quantifiable feature that, specifically for PD, exhibits alterations in terms of both its magnitude and its dynamics [15-17].

With respect to the viability of using the magnitude of VGRF registered with insole systems for differentiating gait features between PD and control groups, some researchers object to its efficacy. As examples, Barnett and collaborators [22] have signalled that the VGRF measures obtained with an instrumented insole, such as the F-Scan system used for gait database acquisition [20-21], are lower than those achieved with a force platform. This is due to a threshold established in baropometric systems in order to reduce the noise during data collection. Other studies have demonstrated that although the F-Scan insole system could be a useful device to measure the VGRF during gait, care should be taken when interpreting the force data during the initial 21% and final 10% of the stance phase of gait cycle because of a delay in data acquisition with such system [23].

In this regard, our suggestion is that the comparison of VGRF data collected by different instruments should be avoided. Taking into account that the VGRF data for the PD and control groups in this study were acquired using the same insole system [20-21], it is valid to consider that only relative differences in force values between both groups are important. Our previous results based on the analysis of the VGRF measured with this insole system, in where an automatic classification of PD patients was achieved through the application of clustering techniques based on artificial neural networks [18-19], supports the argument that this measure of VGRF contains relevant information for objectively differentiating normal and PD gait patterns.

However, in all of these studies the gait variability has not been evaluated taking into account each progression stage of PD according to both the original and modified version of the HY Scale [9-10, 12]. In this sense, our present work shows possible ways to differentiate the gait alterations specifically for each HY stage. This proposal is based on the obtained results indicating that, for all temporal variables analyzed (stance phase time and swing phase time, referred to in the text as STPT and SWPT), there were significant differences between the HY=0 and HY=2.5 groups. In addition, although these differences were not evidenced between HY=0 and HY=2, this could indicate that in the early stages of Parkinson's disease, the temporal parameters of the gait cycle do not differ with respect to those for healthy subjects.

In relation to this issue, the analysis of Figure 2 indicates that the STPT achieved for the different levels of the factor of HY interest gradually increases as the HY level rises. The opposite occurs for SWPT. Such results support the theory regarding body balance impairment with PD progression [8, 16-17]. Furthermore, the differences observed between the HY=0 and HY=2.5 groups cannot be attributed to age because both analyzed PD groups (HY=2 and HY=2.5) are age-matched, having in addition similar ages to the normal group. A challenge for future studies is to better map these temporal gait parameters and their changes in response to PD evolution.

As regards the analysis of the magnitude of the VGRF from its features evaluated (maximum VGRF value and area under the VGRF curve), the results showed statistically significant difference between the cases HY=0 and HY=2.5, but not between the cases HY=0 and HY=2. However, relevant differences were obtained between the HY=2 and HY=2.5 groups in analyzing the median of maximum VGRF (Table 5 and Figure 5), thus showing the potentiality of the VGRF as a possible predictor of the level of gait impairment/disability in PD patients. In this way, these results confirm our hypothesis. Particularly, in further research, our objective will be to explore the strength and usefulness of the VGRF to quantify the lack or presence of body balance impairment that typically differentiate PD patients classified as HY=2 and HY=2.5.

Finally, as a general conclusion, we can say that this work constitutes an application example of gait analysis in the search for clues to clinimetrically identify the level of gait impairment/disability in PD patients on the Hoehn-Yahr Scale. This type of identification could be applied in our country as a routine study in clinical decision-making taking into account the lower cost of insole systems with respect to force platforms.

5. References

[1] Parkinson's Disease Foundation. http://www.pdf.org/en/parkinson_statistics

[2] Gatto E. Parkinson: en Argentina unas 70.000 personas lo padecen. [Internet]. 2015 [cited 2015 April]. Available from: http://www.docsalud.com/articulo/4526/

[3] Davie CA. A review of Parkinson's disease. British Medical Bulletin 2008; 86(1):109-127. PMid:18398010. http://dx.doi.org/10.1093/bmb/ldn013

[4] Goetz CG, Tanner CM, Shannon KM. Progression of Parkinson's disease without levodopa. Neurology 1987;37:695–698

[5] Smithson F, Morris ME, Iansek R. Performance on clinical tests of balance in Parkinson's disease. Phys Ther 1998;78(6):577– 92

[6] Webster DD. Critical analysis of the disability in Parkinson's disease. Mod Treat 1968; 5(2):257 – 82

[7] Murray MP, Sepic SB, Gardner GM, Downs WJ. Walking patterns of men with parkinsonism. Am J Phys Med Rehabil 1978; 57:278– 94

[8] Hausdorff JM, Cudkowicz ME, Firtion R, Wei JY, Goldberger AL. Gait variability and basal ganglia disorders: stride-to-stride variations in gait cycle timing in Parkinson's disease and Huntington's disease. Movement Disorders 1998; 13(3):428-37. PMid:9613733

[9] Hoehn MM, Yahr MD. Parkinsonism: onset, progression, and mortality. Neurology 1967; 17(5):427-42. PMid:6067254

[10] Goetz CG, Poewe W, Rascol O, Sampaio C, Stebbins GT, Counsell C, Giladi N, Holloway RG, Moore CG, Wenning GK, Yahr MD. Movement Disorder Society Task Force Report on the Hoehn and Yahr Staging Scale: Status and Recommendations. Mov Disord 2004; 19(9):1020-1028

[11] Maier-Hoehn MM. Parkinsonism treated with levodopa: progression and mortality. J Neural Transm 1983; 19:253–264

[12] Jankovic JM, McDermott M, Carter J, Gauthier S, Goetz C, Golbe L, et al. Variable expression of Parkinson's disease: a base-line analysis of the DATATOP cohort. Neurology 1990; 50:1529–1534

[13] Hausdorff JM, Cudkowicz ME, Firtion R, Wei JY, Goldberger AL. Gait variability and basal ganglia disorders: stride-to-stride variations of gait cycle timing in Parkinson's disease and Huntington's disease. Mov Disord 1998; 13(3):428-437

[14] Hausdorff JM. Gait variability: methods, modeling and meaning. Journal of NeuroEngineering and Rehabilitation 2005; 2:1-9. PMid:16033650 PMCid:1185560. http://dx.doi.org/10.1186/1743-0003-2-19

[15] Hausdorff JM, Lertratanakul A, Cudkowicz ME, Peterson AM, Kaliton D, Goldberger AL. Dynamic markers of altered gait rhythm in amyotrophic lateral sclerosis. Journal of Applied Physiology 2000; 88:2045-2053. PMid:10846017. Experimental Brain Research. 2003; 149(2): 187-94
[16] Hausdorff JM: Gait dynamics in Parkinson's disease: common and distinct behaviour among stride length, gait variability and fractal-like scaling. Chaos 2009, 19,026113-1-12
[17] Schaafsma JD, Giladi N, Balash Y, Bartels AL, Gurevich T, Hausdorff JM: Gait dynamics in Parkinson's disease: relationship to Parkinsonian features, falls and response to levodopa. J Neurol Sci 2003; 212: 47-53
[18] Lescano CN, Herrera CV, Rodrigo SE. Clasificación de Patrones de Marcha Humana mediante Redes Neuronales 2011. XVIII Congreso Argentino de Bioingeniería, VII Jornadas de Ingeniería Clínica, SABI 2011. Sociedad Argentina de Bioingeniería, Mar del Plata (Argentina), 28-30 Sep
[19] Rodrigo SE, Lescano CN, Rodrigo RH. Application of Kohonen maps to kinetic analysis of human gait. Brazilian Journal of Biomedical Engineering 2012; 28(3):217-226
[20] Physiobank. Physiological signal archives for biomedical research. [Internet]. PhysioNet:MIT; Cambridge; 2009 [cited 2011 Mar]. Available from: http://www.physionet.org/physiobank/database/gaitdb/
[21] Infotronic. CDG 16 channel foot pressure measurement. [Internet]. 2005 [cited 2009 May]. Available from: http://www.infotronic.nl/industrial-medical/products/cdg.html
[22] Barnett S, Cunningham JL, West S. A comparison of vertical force and temporal parameters produced by an in-shoe pressure measuring system and a force platform. Clinical Biomechanics 2001; 16: 353-357
[23] Chen B, Bates BT. Comparison of F-Scan in-sole and AMTI forceplate system in measuring vertical ground reaction force during gait. Physiotherapy Theory and Practice 2000; 16(1):43-53.
[24] Winter DA. "Biomechanics and Motor Control of Human Movement", John Wiley and Sons Inc., Canada, 2009
[25] Winter DA. Kinematic and kinetic patterns in human gait: variability and compensating effects. Human Movement Science 1984; 3:51-76
[26] Koozekanani SH, Balmaseda MT, Fatehi MT, Lowney ED. Ground reaction forces during ambulation in Parkinsonism: pilot study. Archives of Physical Medicine and Rehabilitation 1987; 68(1): 28-30
[27] Box GEP, Hunter JS, Hunter WG. Estadística para investigadores: diseño, innovación y descubrimiento 2008. Editorial Reverté, España
[28] Elliot DF. Handbook of Digital Signal Processing: Engineering Applications 1987. Academic Press, United States of America, ISBN:0-12-237075-9

Digital image processing techniques applied to pressure analysis and morphological features extraction in footprints.

F J Buchelly[1, 2, 3], **D Mayorca**[3], **V Ballarín**[1] **and J Pastore**[1]

[1] Laboratorio de Procesamiento Digital de imágenes, ICyTE (Instituto de Investigaciones Científicas y Tecnológicas en Electrónica) Facultad de Ingeniería, Universidad Nacional de Mar del Plata, Argentina
[2] Consejo Nacional de Investigaciones Científicas y Técnicas - CONICET
[3] Grupo de Óptica y Láser, Departamento de Física, Universidad del Cauca, Colombia

E-mail: fbuchelly@gmail.com

Abstract. This paper shows the correlation between foot morphology and pressure distribution on footplant by means of a morphological parameters analysis and pressure calculation. Footprint images were acquired using an optical pedobarograph and then processed for obtaining binary masks and intensity images in gray scale. Morphological descriptors were obtained from the binary images and the Hernandez Corvo (HC) index was automatically calculated for determine the type of foot. Pressure distributions were obtained from gray scale images making a correspondence between light intensity in footprints and pressure. Pressure analysis was performed by finding the maximum pressure, the mean pressure and the ratio between them that determines the uniformity of the distribution. Finally, a high correlation was found between this ratio and the type of foot determined by HC index.

1. Introduction

Obtaining footprint features is important issue in health and sports, because it permits to study pathologies related to foot morphology [1] [2], human locomotion [2] [4], or even in detecting risks produced by diabetes [5]. These pathology diagnostics are decisive in many disciplines like preventive medicine [6], clinical medicine [7], orthopedics [8], sports medicine [9] [10], and other.

Morphological and pressure features are data of medical interest. Currently, morphology features are visually obtained on footprint images in manual way [10], and also in semiautomatic way [11]. There are different forms of obtaining a person's footprint. Lara et al. [9] show some of them. Commonly, ink based methods are used, in which skin is put in contact with a paper sheet or some film. The patients not only feel uncomfortable with this procedure, but it is possible to make mistakes obtaining footprint images because diagnoses are subjective and it´s difficult to estimate pressure values.

A different system is a parallelepiped footprint viewer, or podoscope, with a flat glass surface in which a person keeps a vertical position supporting both feet, like is used in Oller [3]. In that work, fluorescent lamps emit yellow light at both sides of the glass, showing a luminous footprint when the total internal reflection condition (TIR condition) is eliminated by contact of the skin and glass. A similar version, proposed by Buchelly et al. [2] and Mayorca [5] uses green light, instead of yellow, that improves footprint intensity, due to radiation absorption of the glass that minimizes its value for near green wavelengths, allowing simpler Digital Image Processing.

Pressure distribution measurement can be made by sensor arrays [12] [13]. Usually, their spatial resolution is poor and can lose information about foot form in places where pressure value is low. Some other works [2] study feet morphology and pressure using different devices. Bates et al. [14] shows that geometry of a footprint, done in a compliant material, is related to pressure distribution. According to the previous conclusion, using a podoscope like the ones described in [2] and [5] also allows to determine the pressure of each point in a foot plant considering that skin roughness is deformed depending on the supported pressure. This deformation influences strongly the light intensity and allows integrate both devices in one called "pedobarograph".

Following the previous ideas, in this work we propose a comparative analysis of Digital Image Processing techniques used to automatically determining morphological parameters associated to each type of foot and pressure distribution in footprint images acquired from a pedobarograph device, respect to similar previous works.

2. Foot pathologies

From footprint data it is possible to make the following pathological studies.

2.1. Footprint morphology

There are several classifications of feet morphologies. One of them is the rank from plane foot to extreme cavus foot. Figure 1 shows examples of this rank. Each one produces a particular footprint and pressure distribution, affecting the gait and making feel more or less pain when the person stands in two feet. An intermediate interval between these extreme states is considered as normal foot.

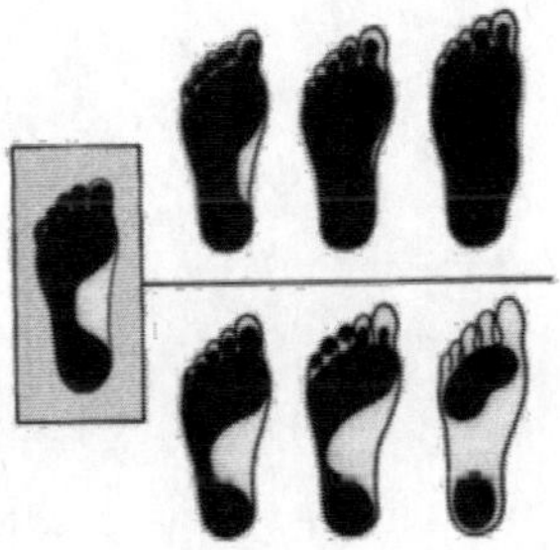

Figure 1. Example of footprints of plane foot (up) in different grades, normal foot (medium left) and cavus foot (down). Image taken from [15]

2.1.1. Plane foot

A plane foot occurs when the decreasing of the plant arc is below normal values according to the Costa-Bartani angle and the location of scaphoid bone respect to the Feiss line. The ball is configured depending on these parameters, modifying the support points and the plant morphology [16].

2.1.2. Cavus foot

The cavus foot is often characterized by a ball accentuation due to the height of the longitudinal arcs. The distribution of pressure on the floor results perturbed by the oncoming of the anterior and posterior supports as well as the reduction of the support surface, mainly at the expense of the lateral border, which serves of basis for podoscopic classification of cavus feet [17].

2.1.3. Determining type of foot

Some authors [5] [9] [10] [11] [18] use some indexes that permit assign a value to each type of foot by its respective footprint. Some of them are: Clarke's angle, Chippaux`s angle, Arc Index, and others. However, the most used is the Hernandez Corvo (HC) index, also mentioned in the previous references. In order to assess the HC index, the following steps are performed [9]

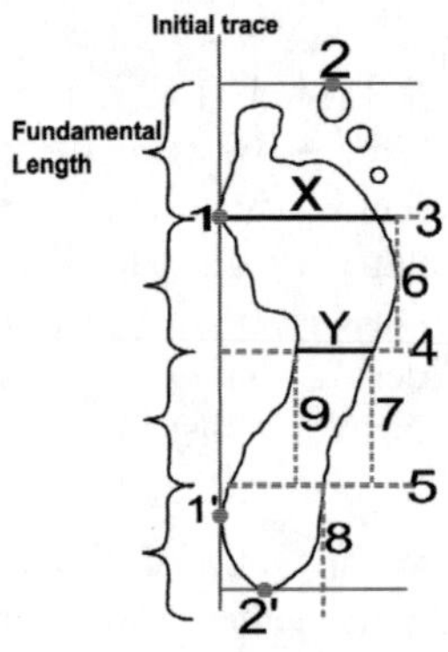

Figure 2. Hernandez Corvo procedure for footprint assessment.

- Mark two points in the inner salience (1 and 1') and do the "initial trace", joining both points
- Mark a point in the most anterior part of the footprint (including toes) and other in the most posterior point (2 and 2')
- Trace perpendicular lines to 2 and 2' respect to the initial trace. The distance between this trace and point 1 is the "fundamental length" and it has to be taken many times as possible on the initial trace (3, 4 and 5)
- Trace a line perpendicular to 3, crossing by the outer point of the footprint, and other two lines in similar way perpendicular to 4 and 5 in the outer part (6, 7 and 8, respectively). The distance between initial trace and 6 is X (metatarsial width); the distance between 9 and 7 is Y (extern arc, support surface in the middle of foot)
- Type of foot is obtained using Equation 1:

$$HCI = 100 * \frac{(X-Y)}{X} \tag{1}$$

HC index value classify foot in following percent ranges [10]

Table 1. Type of foot depending on HCI value.

IHC (%)	TYPE OF FOOT
0-34	Plane
35-39	Plane-Normal
40-54	Normal
55-59	Cavus-Normal
60-74	Cavus
75-84	High Cavus
85-100	Extreme Cavus

2.2. Pressure Distribution

The barometric parameters of foot are not independent of the morphology. The simplest example is the variation of pressure in different types of footprints determined by HC index. In extreme cavus foot, standing people feel more pain because corporal weight is supported by a smaller area. Another case in which morphology is related to pressure is Hallux Abductus Valgus (HAV) or "bunion", where not only the deformation of foot causes nuisance but also changes pressure distribution [19].

On the other hand, plant pressure analysis allows determining not only the morphologic pathologies, but also diabetic foot associated with diabetes mellitus. Among risk factors that can become in diabetic foot, the main pathology is the occurrence of abnormal pressure regions in foot [20] and redistribution of maximum pressure points. Those signs and many others may allow the generation of ulcers and even amputation [21] [22].

3. Methodology

The proposed processing system is summarized in figure 3

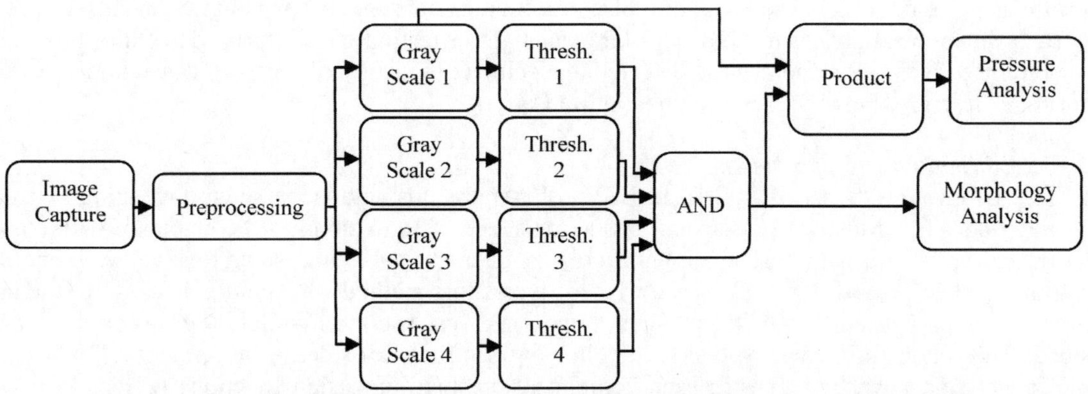

Figure 3. Flowchart that describes the general steps of the proposed methodology.

3.1. Image acquisition

In this work, a pedobarograph was fabricated to generate visible footprints, similar as the proposed in works [2] [5] [11] [23] [24] [25]. Its image formation system is described in figure 4.

Some of the mentioned devices use a film between foot skin and glass (or acrylic) with the purpose to avoid the light noise of the environment and reduce the dependency between the generated luminosity and the skin humidity. In spite of these advantages, using that material implies other kinds of noise and disadvantages because of its texture (loss of information), adherence with acrylic or glass, problems due to its thickness, and other [23]. For that reasons, this work doesn't use a transducer film, it makes footprint detection by applying different transformations to grayscale and creating binary masks by means of thresholding.

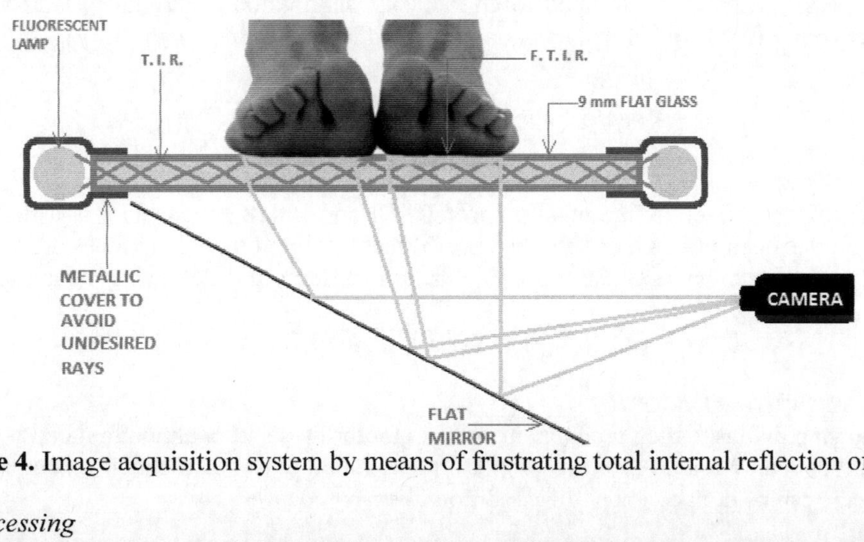

Figure 4. Image acquisition system by means of frustrating total internal reflection of light.

3.2. Preprocessing

The main objective in many Digital Image Processing applications is to measure parameters from the real world. In this case, it is necessary that the geometry of the acquired footprint image corresponds to the geometry of flat image in glass for determining morphological features (lengths, areas) correctly. The camera was calibrated to correct optical distortions (radial, tangential, center deviation) produced by the camera lens [26] [27]. The camera parameters were obtained and correction model was generated by using a chessboard pattern. Additionally, it was needed to perform a geometric transform because of the camera position and using a flat mirror introduces perspective effects.

Sometimes noise filtering is used to avoid false detection and abrupt changes in gray levels. However, low pass filters mix light from FTIR with ambient light, affecting the footprint detection. Low pass filters are used later, after footprint is detected, to facilitate the study of pressure distribution. All the algorithms described below were developed in MatLab®.

3.3. Transformation to grayscale

This step is based on the use of RGB and HSV color spaces, having accounted the meaning of green color over all other. Mayorca [5] use an opaque transducer film to simplify the detection with a gray background and a green footprint, allowing only to intensity value and saturation value define the threshold method (V and S in HSV space). In this work, that method was omitted because of the non uniformity of background. Four independent transforms were used to highlight the footprint regions, instead. These transforms are explained in following sections, considering a color pixel as a three dimensional vector whose values in each component are two dimension functions (color planes) of spatial coordinates (Equations 2 and 3) and a gray scale pixel is the value of a two dimension function of spatial coordinates, resulting of any operation with color planes. The range of gray levels are normalized (from 0 to 1) instead of the usual (0 to 255) range.

$$I_{RGB}(x,y)=(R(x,y), G(x,y), B(x,y)) \quad (2)$$

$$I_{HSV}(x,y)=(H(x,y), S(x,y), V(x,y)) \quad (3)$$

3.3.1. Subtraction of red and blue values

The first method of green highlighting is the one used in references [2] [11], defined by equation 4.

$$\mathrm{Gray}_1(x,y) = G(x,y) - 0.5R(x,y) - 0.5B(x,y) \quad (4)$$

3.3.2. Gaussian of Hue values

Hue plane of HSV space is transformed to an intensity distribution according to Gaussian function with a mean in green ($\overline{H}$=1/3), defined by equation 5, and with an adjust factor f(x,y).

$$\mathrm{Gray}_2(x,y) = \exp\left(-\frac{(H(x,y)-0.\overline{3})^2}{\sigma^2}\right).f(x,y) \quad (5)$$

3.3.3. Subtraction of mean RGB from V value

In HSV space, V is defined as the maximum of R, G, and B [28], then there is a high difference between V and the mean of R, G, and B when one of them is 1 and the other values are 0. In the other hand, the difference is low if R, G and B are similar. This difference is defined by equation 6.

$$\mathrm{Gray}_3(x,y) = V(x,y) - \mathrm{mean}(R,G,B) \quad (6)$$

3.3.4. Comparison between R, G, B

As it was shown previously, the predominant color in footprint is green, although it can be mixed with a few of blue or red. Without compare proportions of blue and red present in a footprint, a fourth grayscale image can be defined under the condition given by equation 7.

$$\mathrm{Gray}_4(x,y) = \begin{cases} G, \text{if } G > \alpha R \wedge G > \beta B \\ 0, \quad \text{otherwise} \end{cases} \quad (7)$$

In equation 7, α and β are adjust constants, fixed for all images. For this work these values are α =1.5 and β =1 and were found experimentally.

3.4. Thresholding and Morphological operations
After gray scale transformations are performed, each image is thresholded with an optimum value obtained empirically. Although every gray scale image show an intensity pattern, where the brightest pixels represent green regions in the original image, the results are not the same. The obtained binary masks usually are different and a definitive mask by means of a logical AND operation was generated. Thus, both noise and morphological differences are eliminated.

Occasionally, some isolated small regions still remain, or some borders have defects or connected objects have holes. These aspects were corrected using morphological operations. We applied an opening followed by closing to smooth borders [29], delete isolated and small undesired particles. We used a disk structuring element. Finally, an algorithm of holes filling is applied.

3.5. Foot morphology analysis
In this work, the equivalence between physical lengths on the glass surface and discrete lengths in the images was calculated (1 pixel-length=0.054cm). This conversion parameter is the same for vertical and horizontal directions in images. With this value, is possible to measure patient's feet length and width, as well as orientation of the feet, arc length, support area, although the decisive parameter to determine the type of foot was Hernandez Corvo Index, automatically calculated.

First step in determining HC index was to align the images of each foot independently along of the major axes of the feet, in approximation of the principal length direction. This line was in vertical orientation in each image. Then, X and Y were found like horizontal lines and HCI was determined using Equation 1. Nevertheless, that procedure sometimes causes errors due to the unsupervised detection of extern in the foot arc. Because of this, two alternative methods were performed for the HCI determination:

- Second method: taking the X value as the foot width and the Y value as the width of the footprint at the medium height between points 2 and 2'.
- Third method: taking the X value as the maximum horizontal width in the anterior part of foot, taking the X' value as the maximum horizontal width in the posterior part of foot and the Y value as the minimum horizontal width between X and X'.

The developed system displays HCI calculation using the three methods and the average value.

3.6. Pressure analysis
A useful intensity distribution $I_{green}(x,y)$ is obtained by multiplying the generated binary mask and a gray scale image. The best way to represent the variation of the green intensity in footprints was $Gray_1$ (See equation 4). The image $Gray_2$ may have high values even if the green intensity is low, because it is calculated using Hue. $Gray_3$ and $Gray_4$ may have abrupt intensity spatial variations that are not considered for this purpose.

From this intensity distribution it is possible to determine pressure values based in previous works. In this work, a proportional allocation was used, as proposed in [2], where the person's weight (w=m.g, product between mass and gravity acceleration) is distributed according to each pixel value by means of equation 8, proposed by Buchelly et al. [2].

$$P(x,y) = \frac{m.g}{A_{pixel}} \frac{I_{green}(x,y)}{\sum_{x=0}^{w} \sum_{y=0}^{h} I_{green}(x,y)} \quad (8)$$

Mayorca [5] shows that pressure versus intensity dependency present high correlation, so the dependency is linear. In our work, we use equation 8 because a linear dependency can be transformed to a proportional allocation performing an offset compensation.

4. Results and Discussion

In this work, the proposed system is evaluated with images of 11 different people footprints including children, adult men and adult women. The footprint image segmentation of a left foot is presented as an example in Figure 5, as well as its intensity distribution.

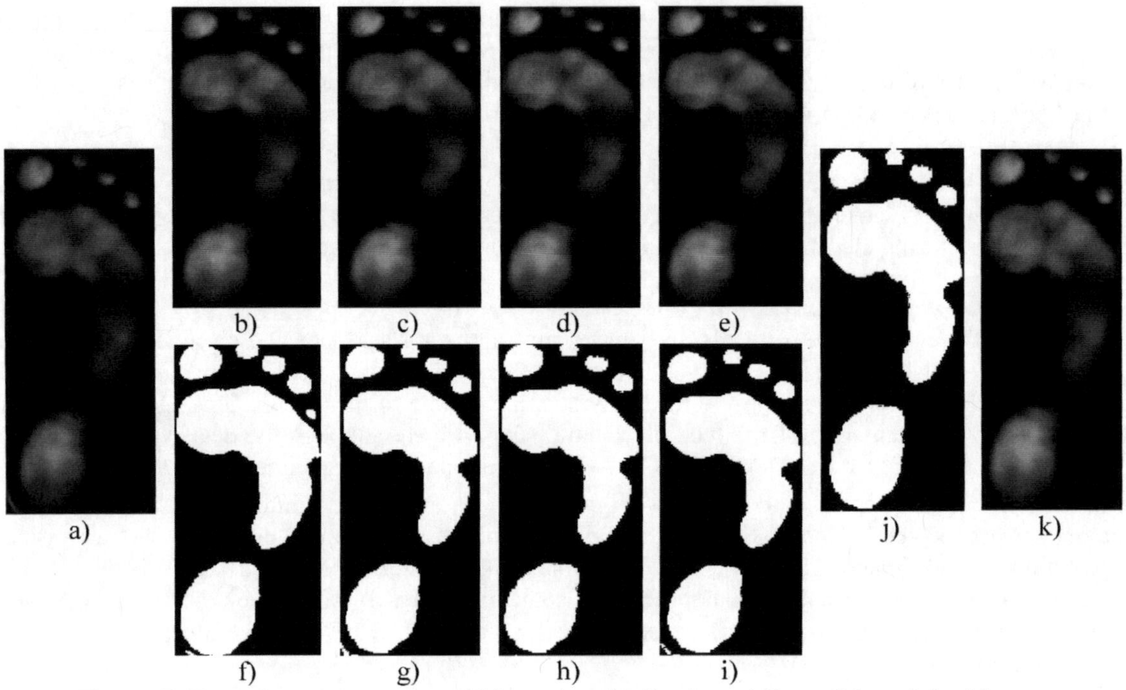

Figure 5. Footprint segmentation and intensity distribution: a) Part of the original image corresponding to a left foot, b) $Gray_1$, c) $Gray_2$, d) $Gray_3$, e) $Gray_4$, f) to i) binary images obtained by thresholding, j) AND operation between the binary images and k) intensity distribution.

The procedure applied for obtaining the final mask was simplified by thresholding only $Gray_1$ image, like the approach of [11]. However, some undesired green regions are captured (figure 5f) to 5i) near the heel) that may be large or small. In [11], those blobs were deleted after size and form analysis. In our work, they were eliminated or attenuated using the AND operation. If undesired regions remained, they were removed by means of morphological opening and closing.

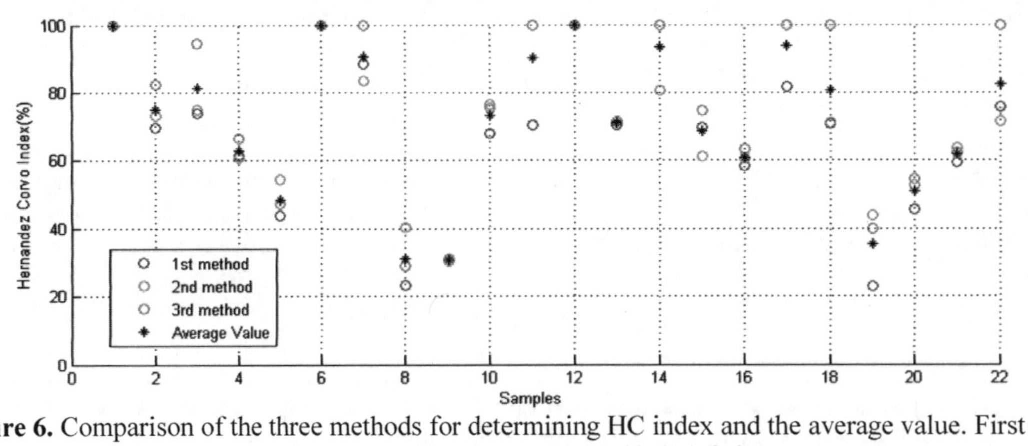

Figure 6. Comparison of the three methods for determining HC index and the average value. First half of data are right feet and second half are left feet.

In order to calculate the HC index, three mentioned methods were used as described before and then the average was calculated. They present some differences, as shown in Figure 6; however, the performance of each method is similar to the others, showing a high correlation.

In the same way, the Manual calculation of HC index is performed and the mean deviation from each method respect to the manual result is calculated, as shown in Table 2.

Table 2. Comparing performances of the three methods for HC index calculation.

	Method 1	Method 2	Method 3	Average
Square root of the mean square deviation of each method respect to the manual procedure.	9.78%	9.78%	14.98%	9.43%

On the other hand, the pressure distribution is found using equation 5. As an example, figure 7 shows pressure distributions for a plane foot, a normal foot and a cavus foot.

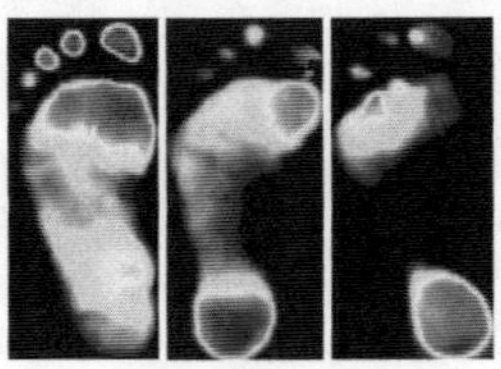

Figure 7. Pressure distribution of a plane foot (left) a normal foot (center) and a cavus foot (right).

The maximum and average values of pressure in each image varies depending on person's weight, foot size and foot type. Figure 8 shows the average pressure and the maximum pressure of each footprint.

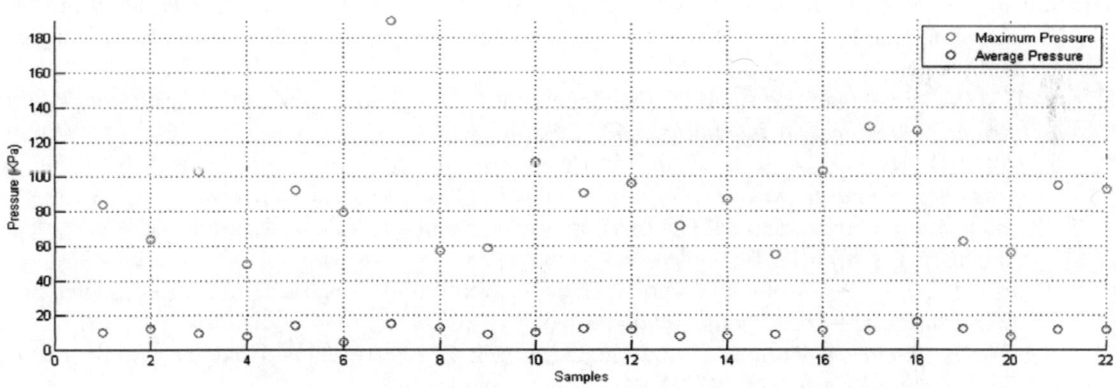

Figure 8. Mean and maximum pressure value for each sample foot.

Although the previous values are important for clinical assessments, a better diagnostic parameter may be the ratio between maximum pressure and mean pressure, because it measures the uniformity of the pressure distribution. In the analyzed samples, the minimum values of the ratio belong to the right and the left foot of the same person, and the HC index shows that both are plane. Additionally, this person is the younger child (samples 8 and 19 of figure 8). On the other hand, the maximum values of this ratio belong to extreme cavus feet (samples 6 and 17), showing high correlation with the HC index.

5. Conclusions

An automatic digital image processing system was created for detecting footprints from an optical pedobarograph for analyzing foot plant morphology and pressure distribution. Four transforms to gray scale were used for segmenting the footprint region with fixed thresholds. Definitive masks were obtained performing a logical AND operation that eliminates false detected binary objects. The remaining undesired regions were removed and borders were smoothed by means of morphological opening and closing.

Length measures can be done using a conversion factor between physical lengths in centimeters and image lengths in pixels, as well as area measurements, for footprint morphological analysis in a binary mask. Hernandez Corvo assessment was performed automatically for twenty two samples (eleven persons) and classifying each foot as plane, normal or cavus in different grades. Three methods were used for determining HC index and the average value was calculated. These values were compared with the manual procedure and square mean deviation shows that the average value is the closest.

Combining the binary mask with the intensity image allows determining the pressure distribution on foot plant. Some descriptors like maximum pressure and mean pressure were calculated for each sample. The ratio between them measures the uniformity of the distribution and is independent of foot size and person's weight. Finally, pressure distribution and morphological classification of feet were compared to show the existence of high correlation between them.

The approach was methodologically appropriate for a preliminary work. As a future work we plan to build a protocol to evaluate it with a larger number of patients in order to validate clinically this technique.

6. References

[1] Revenga C and Bulo M 2005 El pie plano valgo: evolución de la huella plantar y factores relacionados *Revista Española de Cirugía Ortopédica y Traumatología* **49** 271-280

[2] Buchelly F, Mayorca D, Ramírez A, Mayorca H and Ferrín D 2009 *Elaboración de un podobarógrafo para análisis de la huella plantar* Fundación Universitaria María Cano (Popayán, Colombia: Proyecto de la asignatura de biomecánica para el programa de fisioterapia) p 1-32

[3] Oller A 2006 *La fórmula metatarsial y su valor predictivo en los trastornos de la marcha* Universitat de Barcelona (Barcelona: Tesis doctoral del programa de Doctorado Investigació en Ciències de la Salut) p 1-130

[4] Coll M, Viladot A and Suso A 1999 Estudio evolutivo del pie plano infantil *Revista española de cirugía ortopédica y traumatología* **43** 129-136

[5] Mayorca D 2011 *Diseño y construcción de un sistema óptico de análisis de distribución de presiones plantares para el diagnóstico de pie diabético mediante técnicas de visión por computador* Universidad del Cauca (Popayán, Colombia: Trabajo de grado de la carrera de Ingeniería Física) p 1-74

[6] Vélez M 2009 "Posturología como análisis preventivo de lesiones músculo-esqueléticas." *Empresa Preven-Ergo (Quito-Ecuador)* 1-5

[7] Aguilar I, Sánchez I, Pedraza G and Guadarrama L 2012 Correlación plantar y maloclusión. Caso clínico. *Revista ADM* **49** 91-94

[8] Pinzón M *Podología Unos pies sanos son la base de un cuerpo sano* BENAmedic clínica web page: http://www.clinicabenamedic.com/podologia/ Access date: 05-06-2015

[9] Lara S, Lara A, Zagalaz M and Martínez E 2011 Análisis de los diferentes métodos de evaluación de la huella plantar *Retos. Nuevas tendencias en Educación Física, Deporte y Recreación* **19** 49-53

[10] Berdejo D, Lara A, Martínez E, Chacón J and Lara D 2013 Footprint modifications according to the physical activity practised *Rev.int.med.cienc.act.fis.deporte* **vol. 13 num. 49** 19-38

[11] Ferrín D, Magdalena X and Loaiza H 2013 Determinación semiautomática de parámetros morfológicos de la huella plantar mediante el procesamiento digital de imágenes *Revista S&T* **11(27)** 9-26

[12] Arcan M and Brull M 1976 A fundamental characteristic of the human body and foot. The foot-ground pressure pattern *J. Biomechanics* **9** 453-457

[13] Lord M 1981 Foot pressure measurement: A review of methodology *Journal of Biomedical Engineering* **3** 91-99
[14] Bates K, Savage R, Pataky T, Morse S, Webster E, Falkingham P, Ren L, Qian Z, Collins D, Bennett M, McClymont J and Crompton R 2013 Does footprint correlate with foot motion and pressure? *Interface. Journal of the royal society* **10(83)**
[15] Arce C 2015 *Arces Software* Dr. Carlos Alberto Arce Gonzales, Medio especialista en medicina de rehabilitación http://www.arcesw.com/principal.htm Acces date: 05-07-2015
[16] Salazar C 2007 Pie plano, como origen de alteraciones biomédicas en cadena ascendente *Fisioterapia* **29(2)** 80-89
[17] Curvale G and Rochwerger A 2002 *Encyclopédie Medico-Chirurgicale E-14-429* Subject title: Pie cavo. Editions Scientifiques et Médicales Elsevier SAS (París)
[18] Zurita F, Martínez A and Zurita A 2006 Influencia de la tipología del pie en la actividad físico deportiva *Fisioterapia* **29(2)** 74-79
[19] López G and Coloma S 2012 Hallux valgius y presión plantar. Revisión de la literatura *I simposio sobre biomecánica y pie durante la actividad física – Universitat de Valencia (3-4 de Mayo de 2012)*
[20] Tizón E, Dovale M, Fernández C, López M, Mouteira M, Penabad S, Rodríguez O and Vásquez R 2004 Atención de enfermería en la prevención y cuidados del pie diabético *Aten Primaria* **34(5)** 263-271
[21] Tazi O and Debure C 2011 *Encyclopédie Medico-Chirurgicale E – 98-866-B-10* Subject title: Pie diabético. Editions Scientifiques et Médicales Elsevier SAS (Paris)
[22] Gómez E, Levy A, Díaz A, Cuesta M, Montañez C and Calle A 2012 Pie diabético *Seminarios de la fundación española de Reumatología* **13(4)** 119-129
[23] Díaz C, Torres A, Ramírez J, García L and Álvarez N 2006 Descripción de un sistema para la medición de las presiones plantares por medio del procesamiento de imágenes. Fase I *Revista EIA* **6** 43-55
[24] Shah S and Patil K 2005 Processing of foot pressure images and display of an advanced clinical parameter PR in Diabetic Neuropathy *Proceedings of the 2nd international IEEE EMBS conference on Neural Engineering Arlington, Virginia March 16-19, 2005* p v-viii
[25] Periyasamy R, Mishra A, Anand S and Ammini A 2010 Static foot pressure image analysis for variation in men and women while standing. Foot Pressure image analysis *Proceedings of 2010 international conference on systems in medicine and biology 16-18 December 2010, IIT Kharagpur, India* p 386-391
[26] Szeliski R 2010 *Computer Vision: Algorithms and Applications* Springer p 327-342
[27] Ferrín C and Buchelly F 2010 *Diseño y construcción de plataformas multitoque de tipo óptico, utilizando técnicas de visión artificial en tiempo real, para facilitar la interacción de los usuarios con el computador* Universidad del Cauca (Popayán, Colombia: Trabajo de grado de la carrera de Ingeniería Física) p 28-31
[28] 2009 *Opencv 2.0 Documentation* p 313
[29] Gonzalez R and Woods R *Digital image processing (Second Edition)* Prentice Hall (Upper Saddle River, New Jersey) p 528-531

Acknowledgments
Authors acknowledge the assistance from the GOL members (Grupo de Optica y Láser) of Universidad Del Cauca, especially to engineers Leonairo Pencue Fierro and Carlos Diego Ferrín.

An evolutionary algorithm for the segmentation of muscles and bones of the lower limb.

Marco A.Lpez [1,2,3], **A. Braidot** [1,2], **Anbal Sattler** [1], **Claudia Schira** [1], **E. Uriburu.** [4]

[1] School of Life and Health Sciences, Autonomous University of Entre Ríos, Paraná, Argentina.
[2] National Council of Scientific and Technical Research (CONICET), Buenos Aires, Argentina.
[3] Laboratory of Biomechanics, School of Engineering, National University of Entre Ríos, Oro Verde, Argentina.
[4] Magnetic Resonance Entre Ros, Paran, Argentina

E-mail: lopez.marco@docentes.uader.edu.ar

Abstract. In the field of medical image segmentation, muscles segmentation is a problem that has not been fully resolved yet. This is due to the fact that the basic assumption of image segmentation, which asserts that a visual distinction should ex- ist between the different structures to be identified, is infringed. As the tissue composition of two different muscles is the same, it becomes extremely difficult to distinguish one another if they are near. We have developed an evolutionary algorithm which selects the set and the sequence of morphological operators that better segments muscles and bones from an MRI image. The achieved results shows that the developed algorithm presents average sensitivity values close to 75% in the segmentation of the different processed muscles and bones. It also presents average specificity values close to 93% for the same structures. Furthermore, the algorithm can identify muscles that are closely located through the path from their origin point to their insertions, with very low error values (below 7%) .

1. Introduction

Image segmentation has been traditionally done in two ways: one way is tracing each required structure manually in each image in the study. The results obtained with this process precisely represents the anatomy, but are inefficient due to the large amount of time required for obtaining each structure. Another method is to use computer programs that perform this process in an automatic or semi-automatic way, reducing processing times and therefore increasing the efficiency. Furthermore these methods allow reproducibility, feature that cannot be achieved with manual methods. Therefore, the objective is to make these computational methods reach anatomical representations comparable to those obtained manually. It is important to note that with exception of the manual segmentation, there is not yet a universal semiautomatic or automatic segmentation technique that can be used at any application, so it is necessary to adapt the available segmentation techniques to each specific application.

In the field of medical image segmentation, muscles segmentation is a problem that has not been fully resolved yet. This is due to the fact that the basic assumption of image segmentation, which asserts that a visual distinction should exist between the different structures to be identified, is infringed. As the tissue composition of two different muscles is the same, it becomes extremely difficult to distinguish one another if they are closely located [1].

The algorithms based on mathematical morphology are widely used in the field of digital image processing in the process of extracting information about the geometry and topology of an unknown set of pixels. This is achieved through the use of a test element whose shape and size are known. This element allows to compare and quantify if it is contained or not, within the image. This element is called Structuring Element (SE). For example this kind of algorithms have been used as an aid in the segmentation of the pectoral muscle from mammography studies [2], for the segmentation of computed tomography images [3], among others.

Evolutionary algorithms are adaptative heuristic search algorithms based on the evolutionary ideas of natural selection and genetics. These algorithms use an intelligent exploitation of a random search for solving an optimization problem. However, they are not in any way random, on the contrary they rather use historical information to direct the search to a region of better performance within the search space [4]. Evolutionary algorithms simulate the Charles Darwin principle of Survival of the fittest, i.e., through several generations the fittest individuals survive to achieve the solution of the problem. Each generation consists of a population of sequences of characters that are analogous to the chromosomes of DNA. Each individual represents a point in the search space and a possible solution to the problem. These individuals are taken through the process of evolution, where they compete with each other and the fittest will be more likely to generate offspring. The genes of the best individuals then tend to spread through the population so that two good parents produce in some combinations, offsprings with better performance than their parents [4]. Thus it is expected that the last generation would obtain individuals that carry the best solutions to the initial

In this article we have developed an evolutionary algorithm that aims to identify the set and order of application of different morphological operators, which under a defined set of evolution conditions, may achieve the best segmentation of a group of muscles and bones of the lower limb; from a set of magnetic resonance images (MRI).

2. Evolutionary Algorithm

The developed evolutionary algorithm is based on the algorithm proposed by Janc et al. [5] for the segmentation of trabecular and cortical bone. Jancs's algorithm has been modified to also achieve the segmentation of muscles. The algorithm presented in this article, aims not only to achieve the segmentation of bone estructures, but also muscle structures through morphological operators.. Noraini Geraghty [6] describes the general scheme of an evolutionary algorithm. The algorithm presented in this paper uses this scheme, and the functions contained in the Matlab's digital Image Processing Toolbox (IPT) , for the application of the different morphological operators over the set of MRI. The features of the developed evolutionary algorithm are:

2.1. Chromosome:

The chromosome is the descriptor of the characteristics that determine each one of the individuals that represent a solution to the problem to be solved. For this algorithm was defined a chromosome consisting of 3 different processing blocks located one after the other. The three blocks add up seventy five genes.

1. Gray Level Processing: This block refers to the gray level morphological processing of the image. It is represented by the first twenty-four genes of the chromosome , which in groups of four genes represent the parameters for each of the six possible morphological processes. The structure that represents each of the processes in the chromosome is shown in Figure. 1:

Figure 1. Structure of the genes which encode gray level processing.

- Gene 1: this gene determines the type of SE used. This gene is defined by a number which refers to the geometry of SE. This gene can take the values: 1. Square, 2. Disk, 3. Diamond.
- Gene 2: this gene determines the size of the SE defined by the previous gene. 120 pixels is defined as the maximum size that can take the SE.
- Gene 3: this gene determines the type of processing to be applied. These processes are: 1. Erosion, 2. Dilatation, 3. Closing, 4. Opening, 5.Tophat , 6. Bothat.
- Gene 4: this gene defines the application of the processing. It may take 2 different values: 0. If the processing is not apply, 1. If the processing is applied.

2. Black and white processing: after applying the first processing block, a binarization of the resulting image is performed. This binarization is performed using the IPT's function *im2bw*. For the determination of the binarization threshold the function *graythresh* is used, this function calculates the global threshold of an image using Otsu method. Once the image has been binarized, the same six processes defined for gray levels are applied; but this time applied to the black and white image. The genes that represent these six processes are determined by the group of genes located from the twenty-fifth gene up to the forty-eighth gene of the chromosome.
3. *bwmorph* function processing: these last processing block is applied to the resulting images of the previous black and white processing. The IPT's function *bwmorph* allows to apply various morphological operators over black and white images. From all the different operators that this function allows to apply, nine were chosen each of which is represented by three genes of the chromosome. This latter processing block is defined by the group of pixels located from the forty-ninth gene up to the seventy-fifth gene of the chromosome. The structure which represents each of these processes in the chromosome is shown in Figure 2:

Figure 2. Structure of the genes which encode *bwmorph* function processing.

- Gene 1: this first gene determines the morphological operator to be applied over the image. It can take values from: 1. Bridge, 2. Fill, 3. Majority 4. Remove, 5. Shrink, 6. Skel, 7. Spur, 8. Thicken, 9. Thin.
- Gene 2: This gene determines the number of times the operator defined by the previous gene is going to be applied over the image. It was defined to be five the greatest number of times the operator can be applied.
- Gene 3: this gene defines the application of the processing. It may take 2 different values: 0. If the processing is not apply, 1. If the processing is applied.

2.2. Initial Population:
In this block the initial population of the algorithm is defined. The seventy five genes of the chromosome for each individual in the population are selected here.. For each gene it is randomly

selected a number between 0 and the maximum value that can take this gene. It is important to clarify that all individuals generated in this block are viable.

2.3. Fitness Calculation:
In this block the fitness of each individual in the population being evaluated is calculated. As an initial step the original image is taken and each processing defined by the chromosome is applied. Then, the Dice coefficient [5] is calculated to measure the fitness of the resulting binary image. This coefficient is calculated as:

$$Fitness = \frac{2n}{a+b} \tag{1}$$

Where, n is the number of pixels with value 0 which coincide in both images, a is the number of pixels with value 0 in the image found by the genetic algorithm, and b is the number of pixels with value 0 in the gold standard.

The obtained fitness values are in the range between 0 and 1. A value of 0 is assigned to images which have no matching pixels with the Gold Standard; and a value of 1 to images which are identical to the Gold Standard. The Gold Standard used in this article is the manual segmentation performed by the medical expert.

2.4. Selection:
Once the fitness of all individuals in the population has been calculated, a selection of the group of individuals which are going to reproduce for the generation of a new population must be performed. In this article a linear ranking selection method with selective pressure control is used. This method sorts individuals regarding their fitness, assigning the first position of the ranking to the individual with the lowest value of fitness, and the N position to the one with the highest fitness value. The probability of selecting the individual i is linear and is given by the expression:

$$ProbS(i) = 2 - SP + 2(SP-1)\frac{Pos-1}{Nind-1} \tag{2}$$

Donde:

SP: Selective Presion.
Pos: Position occupied by the individual i inside the ranking.
$Nind$: Number of individuals inside the population.

The value of SP controls the probability of the individuals with the lowest fitness values being selected, increasing their chance when assigning low SP values, and decreasing their chance with highest values. The expression used in this article allows to select SP values between 1 and 2.

2.5. Reproduction
The developed algorithm divides the reproduction block in two consecutive steps: a crossover followed by a mutation

1. Crossover: In this first step the crossover between the individuals selected in the previous block is made. Figure 3 shows the crossover procedure. In this procedure three random cutting points over the chromosomes are defined , which take into account the division of the chromosome into the three blocks described above. i.e the first random number can take values between one and twenty four, the second value between twenty five and forty eight, and the third value between forty nine and seventy five. This is done in order to ensure that the offspring obtained from the crossover process are viable as their parents.

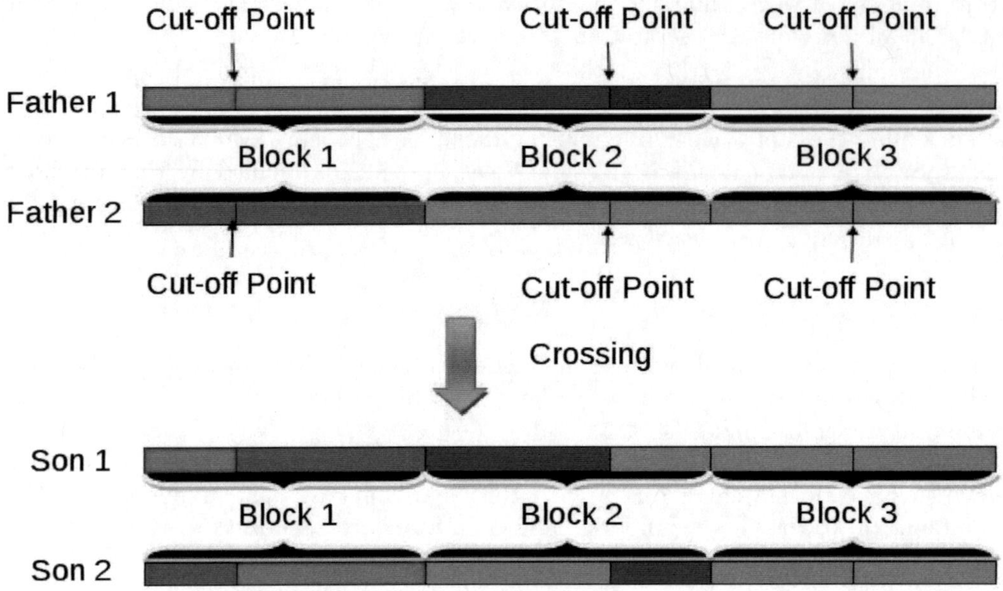

Figure 3. Crossover of the selected individuals.

2. Mutation: mutation of the children generated during the crossover depends on a parameter called mutation probability MP. If a random number between cero and one generated by the algorithm is greater than the value of the selected MP, the mutation of the offspring's chromosomes is generated. Selection of genes to be mutated is also random. The mutation of a gene is carried within the constraints determined during the generation of the initial population, this ensures that the mutated individuals remain viables. The mutation process is shown in Figure ??.

Once a new population is obtained from the reproduction of the selected individuals, the fitness values are calculated for the new individuals. If the Number of Generations (NG) determined as the stop restriction is reached, the algorithm stops; if not, the selection of the new parents to be reproduced is generated, and all the previously descripted blocks are applied once again until de NG restriction is reached.

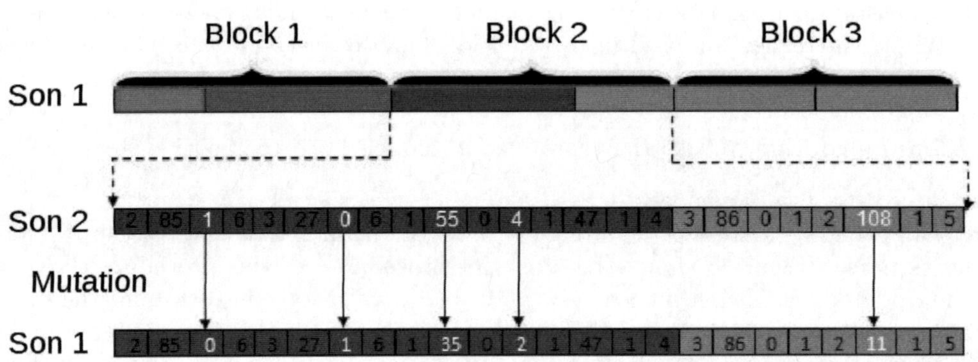

Figure 4. Children's Mutation process

Table 1. Selected working parameters for the evolutive algorithm for each segmented structure.

Estructure	PopSize	NG	MP	SP	Fitness	Dispersion
Fem	30	200	High	Variable	0.9417	0.0018
GM	50	400	Variable	Variable	0.8675	0.0098
BFCL	50	400	Variable	Variable	0.8205	0
SemT	50	400	Variable	High	0.9310	0.0021
SemM	100	200	Very Low	High	0.5936	0
Sar	50	400	Variable	High	0.8622	0
RF	50	400	Variable	Variable	0.9097	0.0012
Gra	30	400	Variable	High	0.9139	0

From the characteristics of the previously descripted algorithm, it can be identified that the performance of the evolutive algorithms is influenced by a set of working parameters which defined both the efficiency and quality of the processing results. These parameters are: the Population Size (PopSize), the NG, the MP and the SP. The values used for these parameters in this article are shown in Table 1. These values were determined after analyzing the results obtained by the developed algorithm with different combinations of these four parameters in a randomly selected image of the MRI set. The selection criteria to determine the optimal set of working parameters were: the highest average fitness value and the lowest dispersion among the individuals inside the final generation obtained with the different combination of working parameters and for each segmented structure [7].

3. Performed Tests

Different tests on this article were performed over a set of 130 T2-weigthed-spin echo images, with a Relaxation Time of $400ms$,, Echo Time $17ms$, a Field of View of $20.48 * 20.48\ cm$ (image size of $512 * 512$ pixels and a resolution of $0.4mm/pixel$ for both the x axis and the y axis), and a gap between images of $4mm$ (z axis). These images were obtained using a $1.5T$ Siemens Magnetom Vision device.

The anatomic structures segmented manually and using the developed algorithm were: the femur bone (Fem), and the muscles; Sartorius (Sar), Rectus Femoris (RecF), Gracilis (Gra), Biceps Femoris Long, Head (BFLH), Biceps Femoris Short Head (BFSH), Semitendinosus (SemT), Semimembranosus (SemM) and Gluteus Maximus (GM).

The manually segmented regions were inserted into the evolutive algorithm in order to be used as training structures for this algorithm. The developed algorithm calculates the centroid and the area of the inserted regions. With this information an automatic subsampling of the image is performed and the morphological operators contained inside the chromosome of each individual on the population are applied only to the subsampled image. The subsampled image is defined as the square region whose central point is located in the calculated centroid for the manually segmented region, and whose sides have a size of three times the calculated radius for the manually segmented region.

The quality of the results obtained with the developed algorithm for the segmented anatomical structures, were quantified using the Receiver Operating Characteristic (ROC) analysis [8]. From this analysis the algorithm's Sensibility and Specificity values were calculated. The confusion matrix for the segmented anatomical structures was also calculated. This matrix shows the algorithm's ability to distinguish different structures.

4. Results Discussions
Figure 5 shows an example of segmenting the Gra muscle using the working parameters in Table 1. As mentioned above, the segmentation process starts with an automatic subsampling of the image (Figure 5-A). This subsampling is based on the information obtained from the manual segmentation made by the medical expert (Figure 5-B). Once the subsampling of the image has been made, the developed algorithm starts the evoluting phase and if the final population is reached, the individual with the highest fitness value inside this population is selected. The information contain by the chromosome of the selected individual, are the set and the sequence of the morphological operators that better segment the muscle. Figure 5-C shows the resulting binary image of the application of the morphological operators over the subsampled image. Figure 5-D shows the region obteined for Gra muscle with the developed algorithm and the manually segmented region.

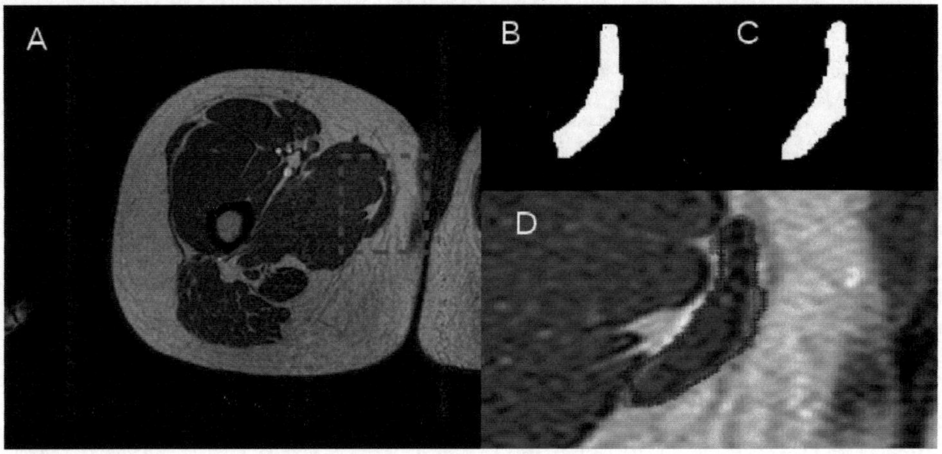

Figure 5. Gra Muscle Segmentation Process. A) Original MRI, the blue doted square shows the location of the Gra muscle in the image. B) Gra Manual Segmentation (Gold Standard). C) Gra. Binary image obtained from the morphological processing. D) Segmented Regions Comparisson, blue + symbols represents Gra gold standard segmentation and red * symbols represents Gra morphological segmentation.

Table 2 shows the mean calculated values of Sensibility (pixels's true positive fraction), Specificity (pixels's true negative fraction), pixels's False Positive Fraction (FPF) and pixels's False Negative Fraction (FNF). These values were obtained as the average of the calculated values of sensitivity, specificity, PFP and PFN for each structure in all images where it was present. From the data contained in this table, it can be seen that Sensibility values obtained for all the segmented structures are close to 75%; these values show the good capability of the algorithm on identifying those pixels that actually belongs to the muscle that has been segmented. On the other hand, the values obtained for the Specificity (close to 93% in all the segmented structures), and the low values obtained for the FPF (with average value less than 7%); show the good capability of the algorithm on identifying those pixels that actually do not belong to the muscle that has been segmented. These values were expected because the developed algorithm uses the *a priori* knowledge about the location and shape of the segmented muscles.

Despite of the good global results shown by the algorithm, the values obtained for the FPF (close to 20%) shown the tendency that the algorithm have of sub-segment different structures. It can be seen that structures with high cross-sectional area (e.g GM and RecF muscles) presents the highest values of FPF. This may be because of the initial restriction which determines that the maximum size that the SE can take is 120 pixels. This size could be smaller than the necessary one for representing the actual area of the structures in all the MRI set. The initial maximum size restriction was set based on the resulting maximum areas of the segmented structures, in a preliminary test of the developed algorithm over a randomly selected group of images which belong the set of MRI.

Table 2. Sensibility, Specificity, FPF, FNF values, calculated for each processed structure.

Estructure	Sensibility [%]	Specificity [%]	FPF [%]	FNF [%]
BFLH	76.41	94.31	5.69	23.59
BFSH	74.38	95.20	4.80	25.62
Fem	76.96	90.63	9.37	23.04
GM	72.81	86.80	13.20	27.19
Gra	82.96	98.17	1.83	17.04
RecF	74.79	95.75	4.25	25.21
Sar	81.04	98.00	2.00	18.96
SemM	72.03	87.54	12.46	27.97
SemT	83.78	98.25	1.75	16.22

Table 3 shows the confusion matrix calculated for the segmentation of the full MRI set. From this table it can be seen, in the same way as with the ROC analysis, the good performance of the developed algorithm in segmenting the processed structures (Matrix's Diagonal). Furthermore, it also can be seen the algorithm's capability to correctly segment muscles which during their path from the origin to the insertion, are closely located. This is the case of the SemM and SemT muscles. For these muscles the confusion matrix shows that the developed algorithm incorrectly classified 2.97% of the processed pixels as belonging to the SemM muscle when actually belonged to the SemT muscle; and 1.94% of the processed pixels in the opposite case. This same situation can be analysed for BFSH and BFLH muscles.

The Column Others in Table 3 refers to the fraction of pixels which the algorithm identified as belonging to one of the processed structures; but the Gold Standard identified as not belonging to any of these structures. These fractions of incorrectly segmented pixels may belong to other tissues inside the set of MRI (e.g adipose tissue), or to other muscles or bones not processed in this article

Conclusions

In this study we present an algorithm for segmentation of muscles and bones from a series of MRI. This algorithm uses the concepts of evolutionary algorithms to identify the set and order of application of different morphological operators for segmenting each of the muscles and bones. The results obtained from the application of the algorithm to the different structures are consistent with those reported by Janc *et al.* [5] for identifying spongy and trabecular bone. The results obtained in this study are the bases for the posing in future works, a Model Based segmentation algorithm. This latter algorithm could allow to perform the segmentation of a new serie of MRI using the statistical information collected from the different morphological operators that better segmented the processed structures in a set of training images series; e.g the MRI set processed in this article. In adittion, increased use of this statistical information could lead to the selection of the constraints for the algorithm (e.g the maximum size of SE)

Table 3. Results Confusion Matrix.

		Should be Classified as:									
	Estructura	**BFSH**	**BFLH**	**Fem**	**GM**	**Gra**	**RecF**	**Sar**	**SemM**	**SemT**	**Otros**
Were Classified as:	BFSH	74.38	5.91	0.11	1.33	0.00	0.00	0.00	0.00	0.00	18.27
	BFLH	1.23	76.41	0.04	1.21	0.00	0.00	0.00	0.00	5.19	15.92
	Fem	0.72	0.00	76.96	2.01	0.00	0.00	0.00	0.00	0.00	20.31
	GM	1.82	0.46	1.02	72.81	0.00	0.02	0.00	0.00	0.00	23.87
	Gra	0.00	0.00	0.00	0.00	82.96	0.00	0.00	0.14	0.00	16.90
	RecF	0.00	0.00	0.00	0.01	0.00	74.79	3.20	0.00	0.00	22.00
	Sar	0.00	0.00	0.00	0.00	0.00	0.00	81.04	0.32	0.00	18.64
	SemM	0.01	0.00	0.00	0.00	0.43	0.00	0.00	72.03	2.97	24.56
	SemT	0.00	3.31	0.00	0.34	0.00	0.00	0.00	1.94	83.78	10.63

that better suited to the actual characteristics of the segmented structures and therefore may achieved better results in segmenting muscles and bones.

Acknowledgments

The authors would like to thank to National Council of Scientific and Technical Reasearch (CONICET) and the research project PICTO 222-2009 to provide the funds needed for this research.

References

[1] Baudin P-Y, Azzabou N, Carlier P G and Paragios N 2012 Automatic skeletal muscle segmentation through random walks and graph-based seed placement 2012 9th IEEE Int. Symp. Biomed. Imaging 1036-9.

[2] Shanmugavadivu P and Sivakumar V 2013 Segmentation of pectoral muscle in mammograms using fractal method 2013 International Conference on Computer Communication and Informatics, ICCCI 2013 pp 0-5.

[3] Pastore J, Moler E and Ballarin V 2007 Operadores Morfolgicos Multiescala y Distancia Geodsica Aplicados a la Segmentacin de Imgenes de Tomografa Axial Computada IEEE Lat. Am. Trans. 5 28-31.

[4] Sivaraj R and Ravichandran T 2011 A review of selection methods in genetic algorithm Int. J. Eng. Sci. Tecnol. 3 3792-7.

[5] Janc K, Tarasiuk J, Bonnet a S and Lipinski P 2013 Genetic algorithms as a useful tool for trabecular and cortical bone segmentation. Comput. Methods Programs Biomed. 111 72-83.

[6] Noraini M and Geraghty J 2011 Genetic Algorithm Performance with Different Selection Strategies in Solving TSP Proceedings of the World Congress on Engineering vol II (London) pp 4-9.

[7] López M A, Braidot A A A, Sattler A and Schira C 2014 Analysis and Selection of Operating Parameters of an Evolutionary Algorithm for Muscle and Bone Segmentation IFMBE Proceedings: VI Latin American Congress on Biomedical Engineering CLAIB 2014 (Paran: Springer) pp 449-52.

[8] López M A and Braidot A A A 2015 Shortest Path Algorithm for Obtaining Muscular Anthropometric Data from MRI Am. J. Biomed. Eng. 5 15-23.

Tissue discrimination in magnetic resonance imaging of the rotator cuff

G J Meschino[1, 3], D S Comas[2, 4], M A González[2], C Capiel[3, 5] and V L Ballarin[2, 3]

[1] Bioengineering Lab, ICyTE, Universidad Nacional de Mar del Plata, Argentina.
[2] Digital Image Processing Lab, ICyTE, Universidad Nacional de Mar del Plata, Argentina.
[3] Group of Informatics and Health, Universidad FASTA, Mar del Plata, Argentina.
[4] Consejo Nacional de Investigaciones Científicas y Técnicas (CONICET), Argentina.
[5] Instituto Radiológico, Mar del Plata, Argentina.

E-mail: gmeschin@fi.mdp.edu.ar

Abstract. Evaluation and diagnosis of diseases of the muscles within the rotator cuff can be done using different modalities, being the Magnetic Resonance the method more widely used. There are criteria to evaluate the degree of fat infiltration and muscle atrophy, but these have low accuracy and show great variability inter and intra observer. In this paper, an analysis of the texture features of the rotator cuff muscles is performed to classify them and other tissues. A general supervised classification approach was used, combining forward-search as feature selection method with kNN as classification rule. Sections of Magnetic Resonance Images of the tissues of interest were selected by specialist doctors and they were considered as Gold Standard. Accuracies obtained were of 93% for T1-weighted images and 92% for T2-weighted images. As an immediate future work, the combination of both sequences of images will be considered, expecting to improve the results, as well as the use of other sequences of Magnetic Resonance Images. This work represents an initial point for the classification and quantification of fat infiltration and muscle atrophy degree. From this initial point, it is expected to make an accurate and objective system which will result in benefits for future research and for patients' health.

1. Introduction

Evaluation and diagnosis of diseases of the muscles within the rotator cuff can be done using different medical imaging modalities. Magnetic Resonance Imaging (MRI) is the method more widely used, though other types are also chosen, such as Computed Tomography (TC) and Ultrasound Imaging (US) [1-2]. In order to assess intramuscular fatty infiltration and muscular atrophy, there exist currently different classification criteria, which try to quantize the stage of such pathologies. These criteria have a lack of precision, since they present inter- and intra-observer variability. This is mainly because scales used to describe the stages are not sensitive to little changes in the muscle characteristics and by the other hand, scales are subjective and qualitative [3]. A wide used classification that has shown to be useful for clinical diagnosis comes from Goutallier [4], where a degree of fatty infiltration is defined, assigning 0 when no fat is found, 1 when there are only a few fat lines, 2 if there are more muscle than fat, 3 if the same amount of fat than muscle is observed, and 4 when fat exceeds muscle. Fatty infiltration correlates with the risk of failure in tendon regeneration when a shoulder arthroplasty is practiced [5].

Authors have proposed methodologies to improve the descripted stage classification with good results, but presenting the drawback of requiring a complex task for the study operator (generally the technician or the image specialist) [6]. Considering this drawback, in this work we propose a preliminary design and implementation of a software tool to contribute to the rotator cuff muscle assessment, in order to make this task relatively easy, reproducible and reliable. We propose MRI texture descriptors to discriminate tissues, including muscles having different stages of fatty infiltration.

By direct observation, the muscles of the rotator cuff present different textures in MRI. Therefore, grey level histograms of these images are not bimodal or multimodal for tissue discrimination. As a consequence, thresholding techniques and other global methods are not applicable in this case. It is necessary to consider a processing pipeline including texture feature extraction and pattern recognition in order to analyse the image in a pixel wise basis [7-9].

In this paper we present an analysis of the features that better represent the texture of the rotator cuff muscles and other tissues that will be applied in later stages of the software development. Selected texture features will be used in the ultimate method that will classify or quantify fatty infiltration and muscle atrophy. A good and reliable performance of this method strongly depends on the robustness of the selected features, so the importance of this preliminary study is well established.

We propose a typical supervised classification pipeline, where feature selection was accomplished by *forward-search* and tissue classification was performed by *k-Nearest Neighbours* (kNN) method [10-11]. In this preliminary study, we considered features extracted in T1-weighted and T2-weighted MRI separately, i.e., we do not combine features of both type of MRI simultaneously. We applied different texture features paradigms, such as grey level co-occurrence matrices, granulometry, spectral features and fractal dimensions [9]. We processed small regions of interest (samples) selected by Imaging Specialists, who marked them in the image and labelled each one. These samples were used as Gold-Standard for assessing the performance of the classification algorithm in the feature selection process. In addition, we segmented MRI images in order to visually analyse the discrimination abilities of the selected features.

2. Materials

Imaging Specialists selected 20 IRM typical studies considering T1- and T2-weighted images and other sequences generally used for diagnosis purposes (PD, TIR, TRA y SAG), with typical acquisition parameters. Different sequences were taken considering slices exactly in the same patient position, trying the patient to be as still as possible. Images were acquired following the direction of the rotator cuff muscles. In figure 1, a typical case is shown, considering T1- and T2-weighted images.

A simple software tool was developed to be used by the Imaging Specialists to make as simple as possible the regions of interest and their labelling. Elliptical regions were considered, as they naturally adapt to the intuitive concept of "region of interest" that Specialists mark when they refer to a part of an image. While selecting a region of interest in a T1-weighted image, exactly the same region was taken in the corresponding T2-weighted image and both were equally labelled. Region selection was done assuring that each one contained only one type of tissue (class). In addition, more than one region was selected for each class if it was possible. As a result, a set of small labelled images (T1 and T2) was stored. As we did not restricted the region size, we got different sized images, with width and height varying between 20 and 100 pixels, containing the useful elliptical region centred and the remaining pixels in black (zero grey level).

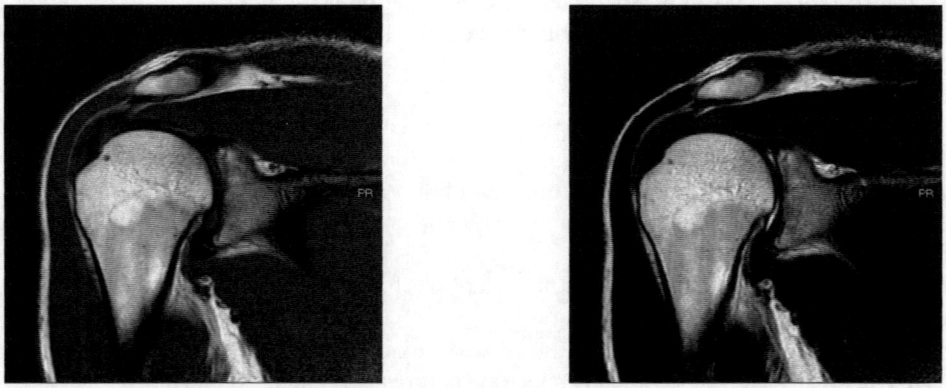

(a) (b)

Figure 1. Typical rotator cuff IRM. a) T1-weighted image. b) T2-weighted image.

These images represent the study object for the analysis we are presenting. Each one contains pixels (and their neighbourhood) corresponding only to one tissue. Imaging Specialists indicated 9 tissues, including muscles with different fatty infiltration, described in table 1. In this sense, as it can be noted, classes #6 to #9 are particularly of interest.

Table 1. Different rotator cuff tissues, selected by Imaging Specialists.

Class	Description
#1	Fat.
#2	Cortical Bone.
#3	Medullary bone.
#4	Generic fluids.
#5	Marrow bone.
#6	Muscle.
#7	Muscle with fatty infiltration.
#8	Muscle with discrete fatty infiltration.
#9	Muscle with low fatty infiltration.

Two different databases were arranged, one containing regions coming from T1-weighted images and the other regions coming from T2-weighted images. Feature selection for optimal discrimination of tissues was made independently for each database, so assessing indirectly which sequence is better in order to recognize each tissue.

In Section 3 we present details of the methods proposed for feature selection and for estimation of the generalization error in the classification.

3. Method

As we described in previous sections, texture features were preliminary studied in order to find the best ones to discriminate IRM pixels corresponding to rotator cuff tissues, including muscles with different fatty infiltration.

Two databases were achieved for different regions of interest coming from 9 tissue classes, keeping T1 and T2 images separately. The same method was used to data processing, considering the next stages: *1) Texture feature extraction, 2) Arrange dataset to be used in a supervised way (feature vectors and labels), 3) Feature selection and generalization error estimation.*

During the first stage, IRM regions of interest were processed, computing texture features considering different paradigms [7-9], analysing only pixels inside the ellipses. After that, a dataset was built, joining together the feature vectors with the tissue label assigned to the corresponding region of interest. Finally, a feature selection was done, trying to get the lower classification error, by means of a *forward-search* algorithm, and kNN for classification [10-11].

The remaining parts of this section present these steps in detail.

3.1. Texture feature extraction.

Texture recognition is one of the main approaches for computationally recognizing structures of objects in a digital image [8]. Texture is defined in a region of mutually-related pixels that follow a regular pattern throughout an image. Therefore, textures cannot be descripted only considering the grey intensity of a pixel: the grey levels of pixel neighbourhood must be considered, previously determining some type of neighbourhood.

A lot of techniques to characterize textures can be found. Detailed descriptions of different proposed approaches can be consulted in [8] and [9].

In this work, we characterized T1 and T2 MRI textures by four types of texture descriptors: grey level co-occurrence matrices, granulometry, spectral features and fractal dimensions. T1 and T2 sub-images were processed separately. Features extraction was done in the next way:

1. Each MRI sub-image was processed in a pixel wise basis. Since these sub-images are small, we considered 11x11 squared neighbourhoods surrounding each pixel being characterized. We computed features following next considerations:
   - *Grey level co-occurrence matrices:* we made 32 grey levels compression previous to the co-occurrence computing. We considered grey level transitions between each pixel and its neighbours in a -45° direction. From the co-occurrence matrices we computed 22 descriptors, which definitions can be found in [9].
   - *Granulometry:* we used 7 opening operations with a cross-shaped structure element. We computed the granulometric distribution, considering its grey levels, mean values and variance as features.
   - *Spectral features:* we analysed the entropy within 9 frequency bands computing the frequency transform for each pixel neighbourhood.
   - *Fractal dimensions:* two Hurst exponent estimations were computed, as suggested by Comas *et al.* in a previous work [8].
2. After feature extraction, only pixels inside the ellipses were considered, taking only pixels whose neighbourhood of 11x11 pixels was entirely inside the ellipse.

As mentioned before, same regions of interest were simultaneously taken from T1 and T2 images. As a consequence, both databases had the same number of MRI sections for each tissue, having both the same size and the same features were computed for each one. Considering this all, we achieved 96782 feature vectors for each database, which were linearly normalized in the [-1, 1] interval.

The number of samples for different tissues was not balanced. Relative frequencies for tissues are shown in table 2 for both databases.

Table 2. Frequency of classes in training dataset.

Class	Frequency
#1	8.55%
#2	0.08%
#3	3.20%
#4	0.01%
#5	31.01%
#6	27.25%
#7	5.03%
#8	13.50%
#9	11.38%

3.2. Dataset arrangement, to be used in a supervised way (feature vectors and labels).

Once features were computed, two labelled datasets were built, $\{X_{T1},Y_{T1}\}$ and $\{X_{T2},Y_{T2}\}$, where X_{T1} and X_{T2} are the feature vectors extracted from T1 and T2 MRI respectively; Y_{T1} and Y_{T2} are the tissues (classes) for each feature vector in X_{T1} and X_{T2}.

Formally for the T1 dataset: $X_{T1}=\{x_{T1,i}\}_{i=1,...,N}$, $Y_{T1}=\{y_{T1,i}\}_{i=1,...,N}$, $x_{T1,i}\in[-1,1]^{35}$ is the i feature vector in the dataset, $y_{T1,i}\in\{1,2,3,...,9\}$ represents the label for the feature vector i (only one for the 9 tissues considered) y N=96782. Similarly, for the T2 dataset: $X_{T2}=\{x_{T2,j}\}_{j=1,...,N}$, $Y_{T2}=\{y_{T2,j}\}_{j=1,...,N}$, $x_{T2,j}\in[-1,1]^{35}$ is the j feature vector in the dataset and $y_{T2,j}\in\{1,2,3,...,9\}$ represents its label, with N=96782.

Both dataset $\{X_{T1},Y_{T1}\}$ and $\{X_{T2},Y_{T2}\}$ were used in stage 3 for feature selection, in order to find the best features to recognize the different classes.

3.3. Feature selection and generalization error estimation.

In this stage, the labelled datasets $\{X_{T1},Y_{T1}\}$ y $\{X_{T2},Y_{T2}\}$ generated in stage 2 were separately processed. For each dataset, an optimal-feature search was performed, obtaining a feature subset that achieves the less global error for data (pixels) classification (assignment to one tissue).

The *forward-search* algorithm was used for optimal-feature search, using kNN as classification algorithm [10-11]. During the iterations of this process, we took special care in generalization abilities for the method; that is why we estimated error for each feature combination using *K-fold cross validation* (K=10) [10]. For each *K-fold* iteration step, the number of samples for different classes in each fold was balanced by a random sampling with replacement.

Different optimal feature combinations were achieved for each dataset, considering 3, 5, 7 and 9 nearest neighbours for kNN. As a result, we found two optimal feature sets, one for T1 MRI and the other one for T2 MRI.

Considering the best features, we computed the classification results using confusion matrices [8, 12], considering *10-fold cross validation* again, and the optimal number of neighbours in each case. Confusion matrices allow knowing how many pixels have been correctly labelled by the algorithm and in addition, they allow analysing how the errors are propagated for erroneously classified pixels.

The ratio of number of pixels detected for a tissue to the number of pixels corresponding actually to that tissue is computed. The biggest values are expected to be in the main diagonal. These values can be used to compute the classification Accuracy [8]:

$$Accuracy=\frac{\sum_{i=1...\lambda} c_{ii}}{\sum_{i=1...\lambda,\, j=1...\lambda} c_{ij}} \quad (1)$$

where $c_{ij}, i=1,...,\lambda, j=1,...,\lambda$ are the elements of the confusion matrix (number of pixels) and λ represents the number of classes (9 for this work). Each element c_{ij} can be interpreted as "the number of pixels that being i-class were classified as j-class."

4. Results and discussion

In this section, results obtained using optimal features for both T1 and T2 datasets are shown.

Tables 3 and 4 show the confusion matrices obtained after the classification process for T1 and T2 datasets, respectively. They allow analysing the percentage of pixels that belonged to one class that were been classified correctly, and also allow to know to what class they were wrongly assigned. Elements in the main diagonal of the confusion matrices are shaded to make easier the interpretation.

After feature selection, the best combination for T1 dataset had 25 features and the best for T2 dataset had 20 features. For both datasets a value of k = 3 for the parameter of kNN algorithm was obtained. Features selected are detailed in table 5.

We found the highest values in the main diagonal of the confusion matrices. This allows inferring that selected features and the classification algorithm proposed are both successful to solve the problem of tissue recognizing in rotator cuff MRI.

Table 3. Confusion matrix for T1 MRI applying the kNN algorithm with k =3.

	... #1	... #2	... #3	... #4	... #5	... #6	... #7	... #8	... #9	Sum
Pixels being Fat (Class #1) that were classified as...	0.85	0.00	0.01	0.00	0.14	0.00	0.00	0.00	0.00	1.00
Pixels being Cortical Bone (Class #2) that were classified as...	0.00	1.00	0.00	0.00	0.00	0.00	0.00	0.00	0.00	1.00
Pixels being Medullary Bone (Class #3) that were classified as...	0.06	0.00	0.76	0.00	0.18	0.00	0.00	0.00	0.00	1.00
Pixels being Generic fluids (Class #4) that were classified as...	0.00	0.00	0.00	1.00	0.00	0.00	0.00	0.00	0.00	1.00
Pixels being Marrow bone (Class #5) that were classified as...	0.02	0.00	0.01	0.00	0.97	0.00	0.00	0.00	0.00	1.00
Pixels being Muscle (Class #6) that were classified as...	0.00	0.00	0.00	0.00	0.00	0.99	0.00	0.01	0.00	1.00
Pixels being Muscle with fatty infiltration (Class #7) that were classified as...	0.00	0.00	0.00	0.00	0.05	0.01	0.86	0.03	0.05	1.00
Pixels being Muscle with discrete fatty infiltrations (Class #8) that were classified as...	0.00	0.00	0.00	0.00	0.00	0.04	0.01	0.89	0.06	1.00
Pixels being Muscle with low fatty infiltration (Class #9) that was classified as...	0.00	0.00	0.00	0.00	0.00	0.01	0.01	0.08	0.89	1.00

Table 4. Confusion matrix for T2 MRI applying the kNN algorithm with k =3.

	... #1	... #2	... #3	... #4	... #5	... #6	... #7	... #8	... #9	Sum
Pixels being Fat (Class #1) that were classified as...	0.87	0.00	0.01	0.00	0.12	0.00	0.00	0.00	0.00	1.00
Pixels being Cortical Bone (Class #2) that were classified as...	0.00	0.99	0.00	0.00	0.00	0.01	0.00	0.00	0.00	1.00
Pixels being Medullary Bone (Class #3) that were classified as...	0.03	0.00	0.81	0.00	0.16	0.00	0.00	0.00	0.00	1.00
Pixels being Generic fluids (Class #4) that were classified as...	0.00	0.00	0.00	1.00	0.00	0.00	0.00	0.00	0.00	1.00
Pixels being Marrow bone (Class #5) that were classified as...	0.02	0.00	0.01	0.00	0.97	0.00	0.00	0.00	0.00	1.00
Pixels being Muscle (Class #6) that were classified as...	0.00	0.00	0.00	0.00	0.00	0.96	0.00	0.04	0.00	1.00
Pixels being Muscle with fatty infiltration (Class #7) that were classified as...	0.01	0.00	0.00	0.00	0.06	0.00	0.85	0.02	0.06	1.00
Pixels being Muscle with discrete fatty infiltrations (Class #8) that were classified as...	0.00	0.00	0.00	0.00	0.00	0.14	0.01	0.80	0.06	1.00
Pixels being Muscle with low fatty infiltration (Class #9) that was classified as...	0.00	0.00	0.00	0.00	0.00	0.01	0.02	0.08	0.88	1.00

For T1 images it can be observed that Cortical Bone, Generic fluids and Muscle classes are detected with accuracy higher than 97%. Bone Marrow, Fat and muscle with different fatty infiltration classes are detected with a slightly lower accuracy, but keeping values higher than 85%. The highest error can be observed for the Medullary Bone class (accuracy of 76%), but analysing the propagation of the error to other classes, we found that most of them are assigned to Bone Marrow (18%), being this a similar tissue, and to Fat (6%). These classes are not included in the main objectives of this work.

For T2 images, Cortical Bone, Generic fluids, Bone Marrow and Muscle classes are detected with accuracy higher than 97%. It is quite similar to T1 images, adding the Bone Marrow to the best recognized tissues. Following the same analysis for T1 images, we can note that all the remaining classes are detected with accuracy higher than 80%. One of the tissues that presents the highest error is again the Medullary Bone (accuracy of 81%), keeping the same behaviour than observed in T1 images: the error propagation is to Bone Marrow class (16%) and Fat class (3%). In this case, the lower accuracy is 80% and corresponds to Muscle with discrete fatty infiltrations.

Results for T2 images seem to be in general better than for T1 images, but it depends on the types of tissues that are priority expected to be low-error detected.

Table 5. Optimal features obtained for T1 and T2 MRI.

MRI	Quantity	Description
T1	25	*Co-occurrence matrix:* Energy, Homogeneity, Entropy, Homogeneity 0.01, Mean value, Variance, Cluster Shade, Cluster Prominence, Correlation, Mean of sum, Entropy of sum, Variance of sum, Variance of difference, Correlation measure, Maximum probability, Order-2 differential inverse moment, Directivity, Dissimilarity, Mean probability. *Granulometry:* Central pixel intensity, Mean value of the granulometric distribution, Variance of the granulometric distribution. *Spectral features:* Lower frequency band entropy. *Fractal models:* both Hurst coefficient estimations.
T2	20	*Co-occurrence matrix:* Energy, Entropy, Homogeneity 0.01, Mean value, Variance, Cluster Shade, Cluster Prominence, Correlation, Mean of sum, Entropy of difference, Variance of sum, Correlation measure, Maximum probability, Order-2 differential inverse moment, Dissimilarity, Mean probability. *Granulometry:* Central pixel intensity, Mean value of the granulometric distribution, Variance of the granulometric distribution. *Spectral features:* no features. *Fractal models:* second Hurst coefficient estimation.

Values observed in the confusion matrices show good accuracy for all classes. Overall accuracy is 93% for T1 images and 92% for T2 images. Analysis of the confusion matrices is more detailed than these general values.

It is reasonably expected that combining what we obtained from both T1 and T2 datasets results would improve, which is the aim for immediate future works. The same is expected considering other MRI sequences.

In future work we will tackle the problem of computing new quantification indices for different stages of fatty infiltration.

Finally, segmentation of whole MRI images are shown for T1 and T2 in figure 2, as examples, with the only objective of visually analyse the discrimination ability of the selected features. Wrong regions can be observed, but we expect to improve these results considering T1 and T2 taken together.

For the images in figure 2, the label assignation method for kNN was modified. Considering $k = 3$, the optimal value obtained for T1 and T2, given a feature vector for labelling, a class was assigned only if the same is observed at least 2 times between its 3 nearest neighbours. If it is not so, the feature vector (and consequently the pixel it represents) are marked as "Unknown pixels."

These images are included in order to show preliminarily a possible future work and they will be analysed by Specialists. There are wrong labelled regions, mainly medullary bone, cortical bone and fat. However, muscle and its different fatty infiltrations were acceptably recognized.

5. Conclusions

In this work we presented a preliminary study of the texture features to best represent rotator cuff muscles with different fatty infiltration and other tissues, considering T1- and T2-weighted MRI.

Starting from a labelled dataset coming from elliptical regions of interest classified by Specialists, an optimal set of features was obtained to discriminate the tissues found in this type of images. At the same time, it was shown that the kNN algorithm is a suitable method for this application.

Using these features and algorithm, we obtained accuracies higher than 90% for the main tissues considered, analysing by means of confusion matrices the error cases.

This work is the starting point for future segmentation of rotator cuff MRI and for the quantification of the stage of fatty infiltration and muscle atrophy. Efficacy and performance of the future methods will depend on the robustness of the features used, which was the motivation of the present work.

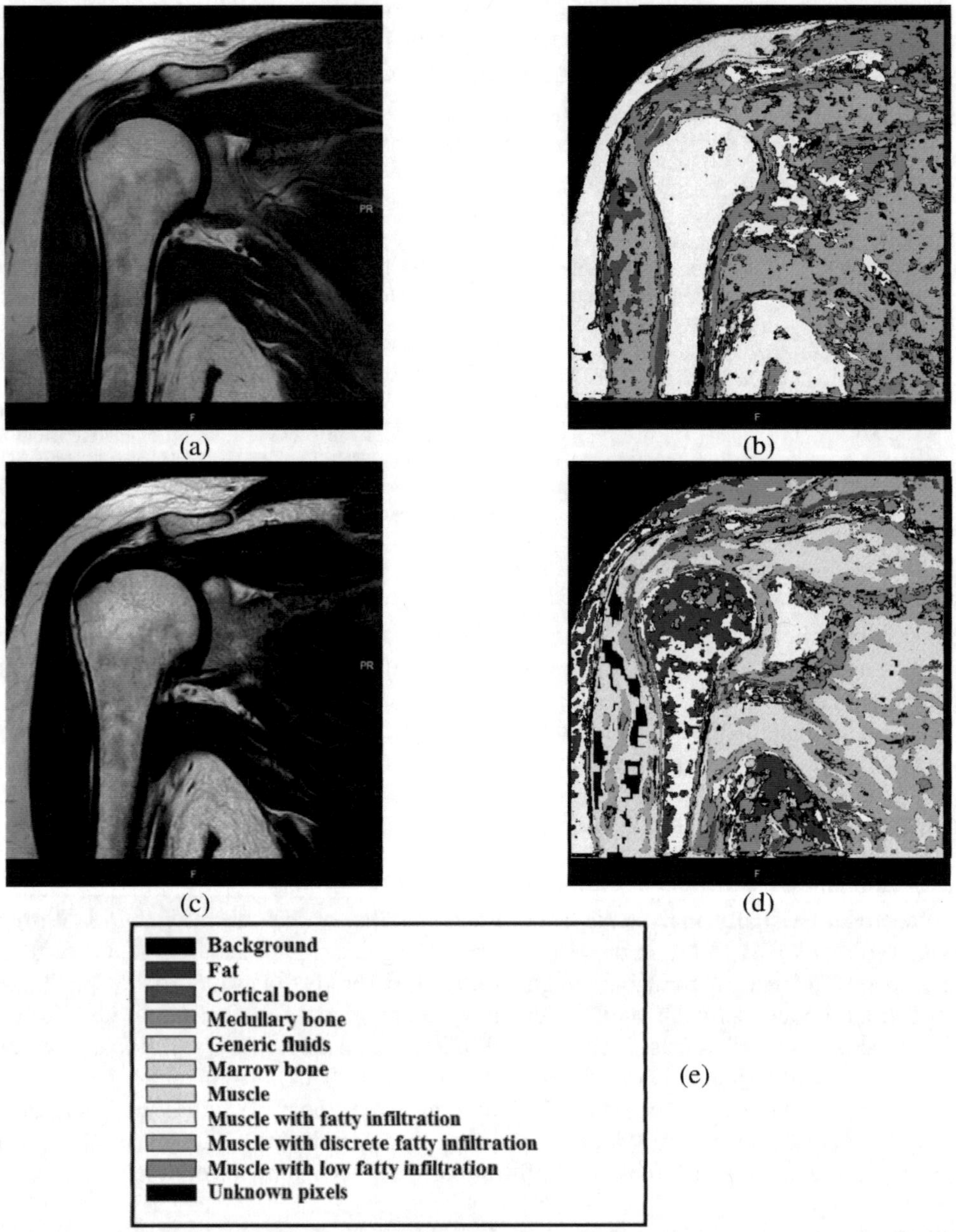

Figure 2. Preliminary segmentation for MRI obtained using the dataset coming from regions of interest as prototype feature vectors. a) Original T1-weighted MRI. b) Segmentation obtained for T1-weighted MRI, k=3. c) Original T2-weighted MRI. d) Segmentation obtained for T2-weighted MRI, k=3. e) Colour reference for tissues.

By achieving the future objective, we will accomplish an objective and easy-to-use clinical decision support system. It will be useful for assessing the grade of muscular atrophy induced by fat growing in different stages. This will result in benefits for future research and for patients' health.

Acknowledgement
D S Comas thanks the support of Consejo Nacional de Investigaciones Científicas y Técnicas (CONICET) from Argentina throughout his PhD scholarship.

References

[1] van de Sande M A, Stoel B C, Obermann W R, Tjong a Lieng J G, Rozing P M 2005 Quantitative Assessment of Fatty Degeneration in Rotator Cuff Muscles Determined With Computed Tomography *Invest. Radiol.* 40 **5** 313-9

[2] Khoury V, Cardinal E, Brassard P 2008 Atrophy and Fatty Infiltration of the Supraspinatus Muscle: Sonography Versus MRI *Am. J. Roentgenol.* 190 **4** 1105-11

[3] Spencer E, Warren R, Dunn R, Wright R, Wolf B, Spindler B, Kurt P, McCarty E, Benjamin E, Grant J, Safran M, Brian H, Kuhn J 2008 Interobserver Agreement in the Classification of Rotator Cuff Tears Using Magnetic Resonance Imaging *Am. J. Sports Med.* 36 **1** 99-103

[4] Santoro Belangero P, Ejnisman B, Arce G 2013 A Review of Rotator Cuff Classifications in Current Use in: *Proc. of the ISAKOS Upper Extremity Committees 2009-2013* pp 5-13

[5] Zanneti M, Gerber C, Hodler J 1998 Quantitative Assessment of the Muscles of the Rotator Cuff with Magnetic Resonance Imaging *Invest. Radiol.* 33 **3** 163-170

[6] Gerber C, Schneeberger A, Hoppeler H, Meyer D 2007 Correlation of atrophy and fatty infiltration on strength and integrity of rotator cuff repairs: A study in thirteen patients *J. Shoulder Elbow Surg.* 16 **6** 691-6

[7] Gonzalez R C and Woods R E 2002 *Digital image processing* (Upper Saddle River, N. J.) vol 1 (Prentice Hall)

[8] Comas D S, Meschino G J Ballarin V L 2010 Framework de segmentación y análisis de imágenes mediante reconocimiento de texturas in: *Proc. Argentinian Symposium of Technology (AST 2010)* (Buenos Aires, Argentina) pp. 1529-41

[9] Moler E G, Ballarin V L, Gonzalez M L 1999 Características Estadísticos, Espectrales y Morfológicos para Clasificación de Texturas: Un Análisis Comparativo en: *Proc. VIII RPIC Reunión de Trabajo en Procesamiento de la Información y Control* (Mar del Plata, Argentina) pp. 59-62

[10] Duda R O, Hart P E, Stork D G 2001 *Pattern Classification* (Wiley-Interscience)

[11] Rajini N II 2011 Classification of MRI brain images using k-nearest neighbor and artificial neural network in: *Proc. 2011 International Conference on Recent Trends in Information Technology*, (Chennai, India) pp. 563-568

[12] Kohavi R and Provost F 1998 Glossary of Terms *Mach. Learn.* 30 **2-3** 271-4

New Windows based Color Morphological Operators for Biomedical Image Processing

Juan Pastore[1,2,4], **Agustina Bouchet**[1,2,4], **Marcel Brun**[1,2], **Virginia Ballarin**[1,3]

1 Digital Images Processing Lab. ICyTE - UNMdP.
2 Department of Mathematics, School of Engineering, UNMdP
3 Department of Electronics, School of Engineering, UNMdP
4 CONICET

E-mail: jpastore@fi.mdp.edu.ar

Abstract. Morphological image processing is well known as an efficient methodology for image processing and computer vision. With the wide use of color in many areas, the interest on the color perception and processing has been growing rapidly. Many models have been proposed to extend morphological operators to the field of color images, dealing with some new problems not present previously in the binary and gray level contexts. These solutions usually deal with the lattice structure of the color space, or provide it with total orders, to be able to define basic operators with required properties. In this work we propose a new locally defined ordering, in the context of window based morphological operators, for the definition of erosions-like and dilation-like operators, which provides the same desired properties expected from color morphology, avoiding some of the drawbacks of the prior approaches. Experimental results show that the proposed color operators can be efficiently used for color image processing.

1. Introduction

The Mathematical Morphology (MM) field of image processing was founded by Matheron [1] and Serra [2] and became very popular in non-linear image processing. Initially, the MM had been introduced as a processing technique for binary images, which were regarded as sets; therefore, its elementary operations are based on the set theory [3]. However, the extension to gray scale images, using the umbra concept [4-5], introduced a generalization of the basic morphological operations which were subsequently used in many image processing and analysis tasks e.g. morphological filtering [6], watershed segmentation [7], etc. Gray scale morphology is based on the lattice theory, which implies a partial ordering of the data within the gray scale images. In this case, the lattice structure is not difficult to obtain, since the gray scale images are real functions, and the set of real numbers implicitly possesses a lattice structure. However, while the extension from binary to gray scale images is a natural one, the extension to color, and multi-spectral, images is not straightforward, because of the vectorial nature of the data, and the difficulty in finding a suitable ordering for it. Barnett introduced four types of vector orderings: marginal (M-order), reduced (R-order), conditional (C-order) and partial (P-order) [8]. When applied to color data, all these orderings have certain disadvantages, depending on the application. For instance, the marginal ordering introduces false colors; and the reduced and partial orderings are either relying on pre-orderings, thus lacking the anti-symmetry property, or behave like conditional orderings, generating perceptual nonlinearities [9].

Compared with gray-scale images, color images can provide richer information. With the wide use of color in medical image equipment, the interest on the color perception and processing has been growing rapidly. In this work we proposed morphological color operators, based on a new locally defined ordering, avoiding some of the drawbacks of the prior approaches. First, we apply these new operators to synthetic image, comparing the results against marginal and lexicographical ordering. Then we present two examples of application, of the new operators, to previous biomedical image processing problems: detection of exudates in ocular images and bone marrow biopsies segmentation. Experimental results show that the proposed color morphological operators can be efficiently used in color image processing.

2. Theoretical Concepts

Initially, the MM had been introduced as a processing technique for binary images, which were regarded as sets; therefore, its elementary operations are based on the set theory.

For binary images, modeled as subsets of a domain E, dilation and erosion of an image A by the structuring element B are defined, respectively, as:

$$\delta_B(A) = \bigcup_{y \in B} A_y \tag{1}$$

$$\varepsilon_B(A) = \bigcap_{y \in B} A_y \tag{2}$$

When modeling binary images by their characteristic functions, the equivalent to a set A is a function $\chi_A : E \to \{0,1\}$, defined by $\chi_A(x) = 1 \Leftrightarrow x \in A$, the dilation and erosion by a flat structuring element, defined by a set B, can also be defined by:

$$\delta_B(\chi_A)(x) = \sup_{y \in B} \{\chi_A(x-y)\} \tag{3}$$

$$\varepsilon_B(\chi_A)(x) = \inf_{y \in B} \{\chi_A(x+y)\} \tag{4}$$

Immediately after, in a natural way, were extended for gray level images. The dilation and erosion of an image $f : E \to L$, by a flat structuring element B is defined, similarly, by:

$$\delta_B(f)(x) = \sup_{y \in B} \{f(x-y)\} \tag{5}$$

$$\varepsilon_B(f)(x) = \inf_{y \in B} \{f(x+y)\} \tag{6}$$

These definitions are based on the existence of an order relationship in the space of gray levels L, where the minimum and maximum are obtained. More generally, if the range is a more complex complete lattice, the same definition can still be applied, where the minimum is replaced by the infimum and the maximum by the supremum.

Existing approaches for color mathematical morphology need to deal with the fact that usual color spaces are vector spaces, usually with three color components, like red, green and blue intensities for the classic RGB model.

A color image can be modeled as a function $f : D_f \subset \mathbb{R}^2 \to \Im \subset \mathbb{R}^3$ where $\Im$ that represents a color space. Usually the order on the space of images is derived from an order in the color space $\Im$. Let be $\Im \subset \mathbb{R}^3$ the unit cube (RGB color space) and let $\leq_\Im$ be an order in $\Im$ that provides a structure of complete lattice. The normal extension of this order to the space of images, or functions $f : D_f \subset \mathbb{R}^2 \to \Im \subset \mathbb{R}^3$, with an order $\leq$, is by defining it, for $f, g : D \subset \mathbb{R}^2 \to \Im \subset \mathbb{R}^3$, as $f \leq g$ if and only if $f(x) \leq_\Im g(x) \; \forall x \in D$ [10]. This order provides the images with a lattice structure too. Once a lattice structure is defined on the space of color images, the basic operations erosion ($\varepsilon_B^{\leq_\Im}$) and dilation ($\delta_B^{\leq_\Im}$) can be defined: for a color image $f : D_f \subset \mathbb{R}^2 \to \Im \subset \mathbb{R}^3$ its dilation and erosion by a structuring element B is defined as:

$$\varepsilon_B^{\leq_\Im}(f) = \inf_{s \in B}^{\leq_\Im} \{f \circ \tau_s\} \tag{7}$$

$$\delta_B^{\leq_\Im}(f) = \sup_{s \in B}^{\leq_\Im} \{f \circ \tau_{-s}\} \tag{8}$$

being $\tau_s : \mathbb{R}^2 \to \mathbb{R}^2$ traslation function by the element $s \in \mathbb{R}^2$, $\tau_s(x) = x + s$.

The results of erosion and dilation of a color image depends strongly on the order established in the color space. After defining the basic operators in the domain of color images, we can define the morphological opening ($\gamma_B^{\leq_\Im}$) and closing ($\phi_B^{\leq_\Im}$) filters in the same way as defined for binary images and gray levels.

$$\gamma_B^{\leq_\Im}(f) = \delta_B^{\leq_\Im}\left(\varepsilon_B^{\leq_\Im}(f)\right) \tag{9}$$

$$\phi_B^{\leq_\Im}(f) = \varepsilon_B^{\leq_\Im}\left(\delta_B^{\leq_\Im}(f)\right) \tag{10}$$

An important issue with this lattice structure for the RGB color space, inherited by the color images space, it that it results on the creation of false colors when applying the basic operator, since the infimum for a set colors may be a new color different from any of them [11]. For example, the infimum between $(0.5,1,1)$, a kind of light blue and $(1,1,0.5)$, a light yellow, is $(0.5,1,0.5)$, a light green (figure 1).

To solve this situation, many approaches were proposed in literature. Between them we can list, P-ordering and R-ordering. In C-ordering, there is a total order, obtained by sequential application of the order of the components. An example of such order is the lexicographic order. In R-ordering, the RGB vectors are mapped to a scalar (for example by its distance to a reference point), and a total order is applied. The drawback with this approach lies in the need of a reference point, and the fact that the map is not one to one [11-12].

In [11] the proposed method of lexicographic cascade depends on the distance to a reference color. Because that distance defines a partial ordering, since different colors can have the same distance to the reference color, a lexicographic cascade is applied to decide on these cases. This solution, as other related ones, suffers from the need to choose a reference color and a lexicographic order. Other solution, proposed in [12], define a total order over all the RGB colors observed in the image, solving the problem without defining a total order over the whole RGB space. This last solution deviates from the theoretical foundations of MM as operators on lattices, but erosion and dilation operators result in no false colors. Other solutions, like the one proposed in [13] provides another total ordering in the HSI color space, weighting the components using a *cost* function, and solving the ties by a lexicographic ordering. This approach solves the need for a reference color, but still depends on a lexicographic order chosen ad-hoc.

On another field of study of Mathematical Morphology, windows based operators have been studied for many years, as a way to represent locally defined morphological operators, and for their automatic design from examples [14]. In this context, translation invariant locally defined Morphological Operators can be represented as w-operators, like the case of erosions and dilations. This family of morphological operators includes and extends that subfamily of classical morphological operators. One of the advantages of designing w-operators is the freedom from the need to define complete lattices on the image space, based on the lattices defined in the color space. Any mapping from a window observation to the color space defines valid w-operators on images [15]. These advantages are used, for example, for the statistical design of morphological operators based on examples [16].

3. Proposed Approach

In this work we propose an approach, for color morphology, that deviates from the global lattice or total order used in previous works, by implementing color w-operators and defining erosion-like and dilation-like operators using a local total order on the observed windows.

Given the existence of a lattice structure in the color space, by applying the minimum (or maximum) operation on the colors observed in a symmetric window W, around a pixel x, and replacing, on the resulting image, that pixel by the computed value, it is equivalent to erode (or dilate) the color image by the structuring element B defined by W (B and W as the same subsets).

In our approach, to avoid the issues related to the need to define an ad-hoc total ordering in the color space, we define a *local minimum* and *local maximum* based on a ordering of the observed values on the window. With this local definition, equations (1) and (2) can be applied. In this case the resulting morphological w-operators are not strictly erosions and dilations, since they are not defined as infimum and supremum on complete lattices over the images, but just on a total order on the window, so they will be called erosion-like and dilation-like operators.

(a) (b) (c)

Figure 1. Example of the creation of false color: a) light blue $(0.5,1,1)$, b) light yellow $(1,1,0.5)$ and c) result of the infimum, light green $(0.5,1,0.5)$.

The key properties of the local order is that, because it does not depend on a global order, it can be made independent of ad-hoc decisions, like the election of a reference color, or the selection of a lexicographic order and, in the other hand, it can be made dependent on spatial properties of the window. The local order defined will have three main properties: a) it compares first the average intensity of the three channels; b) in case of ties, it compares the closeness, as color, to the central pixel; c) en case of ties, it compares the spatial closeness to the central pixel. The first option is based on the need to specify an order that is related to the luminosity of the pixels, as in [13]. The second option is based on that, if two candidates to the minimum (or maximum) have the same luminosity, then we want to choose the one closest, as color, to the original pixel being replaced. This would avoid color invasions in extreme situations. Finally, in the unlikely situation where two colors candidate to be the minimum have the same distance, as color, to the central pixel, then we want to avoid spatial color invasion.

Formally, the process is described in the following way. The first step consists on providing an order on the elements of B. Let say that $B=\{y_1,\ldots,y_n\}$. This order of the pixel should reflect the distance between the pixels and the central pixel. The first elements should be the closest, spatially, to the central pixel. Let x a pixel of the image where we want to compute the erosion or dilation by B. We list the observed colors in the neighbor of pixel x, as $v_1,\ldots,v_n$, where $v_i = f(x+y_i)$, and the index of the central one pixel is $k_0 = 1$ (since the spatial order should always position the central pixel first). The minimum between the values v_i is computed based on the following local total order:

$$v_i \le v_j \Leftrightarrow \begin{cases} \|v_i\| \le \|v_j\| & or \\ \|v_i\| = \|v_j\| & and \quad \|v_i - v_{k_0}\| \le \|v_j - v_{k_0}\| \\ \|v_i\| = \|v_j\| & and \quad \|v_i - v_{k_0}\| = \|v_j - v_{k_0}\| \quad and \quad i \le j \end{cases} \tag{11}$$

Finally, the erosion-like of an image by a structuring element B is obtained by computing, for each pixel x, the minimum based on the newly defined order, on the window defined by B. For dilations, since a maximum will be selected, the equation for the local order is the following:

$$v_i \ge v_j \Leftrightarrow \begin{cases} \|v_i\| \ge \|v_j\| & or \\ \|v_i\| = \|v_j\| & and \quad \|v_i - v_{k_0}\| \le \|v_j - v_{k_0}\| \\ \|v_i\| = \|v_j\| & and \quad \|v_i - v_{k_0}\| = \|v_j - v_{k_0}\| \quad and \quad i \le j \end{cases} \tag{12}$$

The dilation-like of an image by a structuring element B is obtained by computing, for each pixel x, the maximum based on the newly defined order, on the window defined by B.

This change prioritizes the closest colors and pixels by positioning them higher on the local order. It is important that the ordering of the elements of the window reflects the distance to the central pixel. Usually an indexing in spiral, from inside to outside, would provide an appropriate indexing.

As an example of application, let's assume that we are working with a 3x3 structuring element B. It defines a window W, and the index of the pixels in W is given by the following schema, where the central pixel has index 1:

7	3	6
4	1	2
8	5	9

Let's assume then that we observe an image with the windows W, and we observe the following RGB colors around the central pixel in the original image:

(10,10,10)	(10,10,10)	(10,10,10)
(10,10,10)	(5,10,11)	(9,5,11)
(5,8,12)	(10,10,10)	(10,10,10)

The list of values, based on the order on W, would be:

	Color	Intensity (square)	Distance to Central (square)
v_1	$= (5,10,11)$	8.66	0
v_2	$= (9,5,11)$	8.33	9
v_3	$= (10,10,10)$	10.0	6
v_4	$= (10,10,10)$	10.0	6
v_5	$= (10,10,10)$	10.0	6
v_6	$= (10,10,10)$	10.0	6
v_7	$= (10,10,10)$	10.0	6
v_8	$= (5,8,12)$	8.33	3
v_9	$= (10,10,10)$	10.0	6

The final order on these color values would be: $v_3 \geq v_4 \geq v_5 \geq v_6 \geq v_7 \geq v_9 \geq v_1 \geq v_2 \geq v_8$. The first six values are obtained because $(10,10,10)$ is the largest intensity value, they all have the same distance (as color) to $(5,10,11)$, so that they keep the pixel order defined in B. The following value, v_1, is selected because its intensity is larger than the rest. Finally, v_2 and v_8 have the same intensity, but v_8 is closer to v_1, as color, than for v_2, so that v_2 is the next in list ($v_8 \leq v_2$). Finally, the minimum value for this list of colors is $v_8 = (5,8,12)$, selected as the new value for the central pixel.

(a) original synthetic image

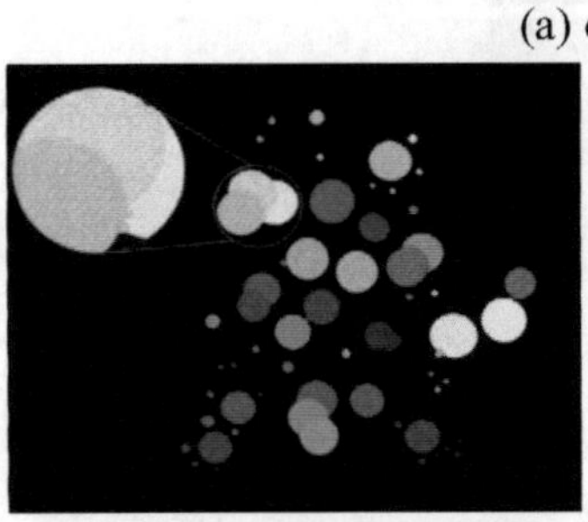

(b) erosion-like operator

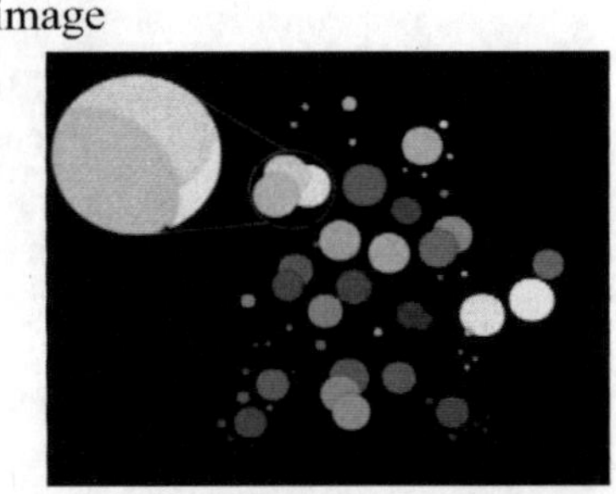

(c) M-order erosion

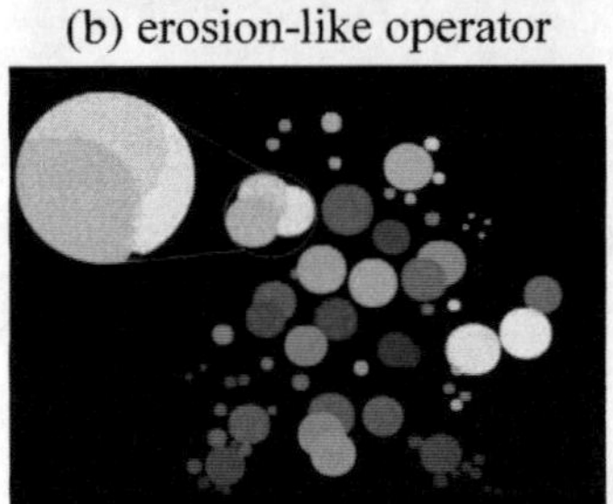

(d) dilatation-like operator

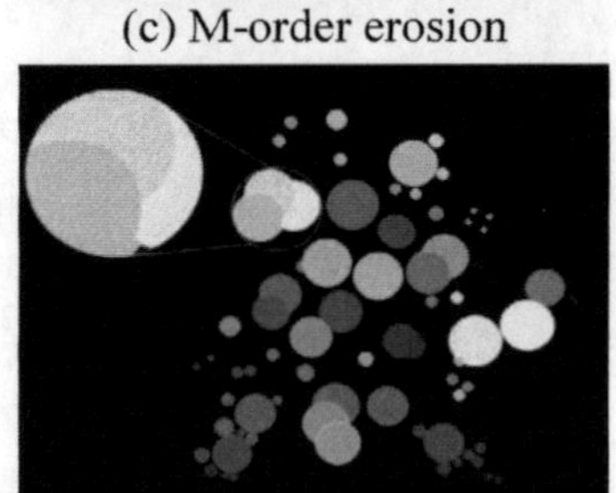

(e) M-order dilatation

Figure 2. Comparing results to M-order using a synthetic image containing colored balls

This new approach departs from the previous work in this area by moving from a total order on the RGB space to a total order on the RGB colors in the observed window. Because of this, the result of the erosion-like and dilation-like are expected to deviate less from the original pixel color, when more than one candidate (to minimum or maximum) is present, avoiding the use of ad-hoc lexicographic orders or selection of a global reference color.

To compare this approach and the previously ones, we applied basic operators: erosion and dilation based on lattice order (or M-order), erosion and dilatation based on lexicographic order, erosion-like and dilation-like operators, on a synthetic image containing colored balls. Figure 2 shows the comparison between M-order and our approach. In the figures a region of the same image was scaled in order to appreciate better the effects of the different operators. It can be shown that when applying M-order that false colors appear as it was expected unlike using the new operators.

Figure 3 shows the comparison between lexicographical order and our approach. The figure shows a region of the image in a larger scale in order to appreciate better the differences between the different operators. It can be shown that when applying lexicographic order, no false colors appear but the quality of the original image is not preserved. For erosion the border looks strongly pixelated, while for the dilation, the interior of the object looses its smooth coloring. Both of these issues don't appear for our proposed approach.

Like with standard erosions and dilation, the new erosion-like and dilation-like operators, here defined, can be combined to generate more complex operators, like opening, closing, to-hats, etc. This property allows us to port previously designed algorithms, based on gray level images, to the context of color biomedical images, as will be described in the next section.

(a) original synthetic image

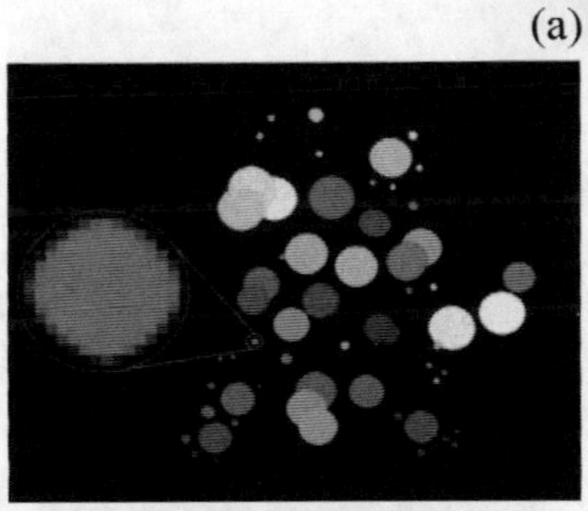

(b) erosion-like operator

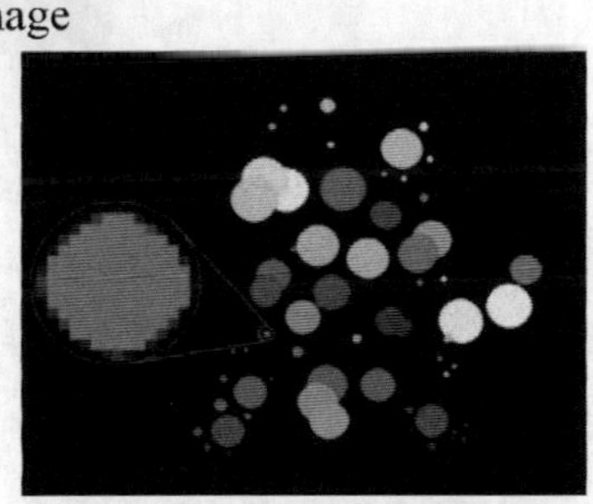

lexicographic erosion

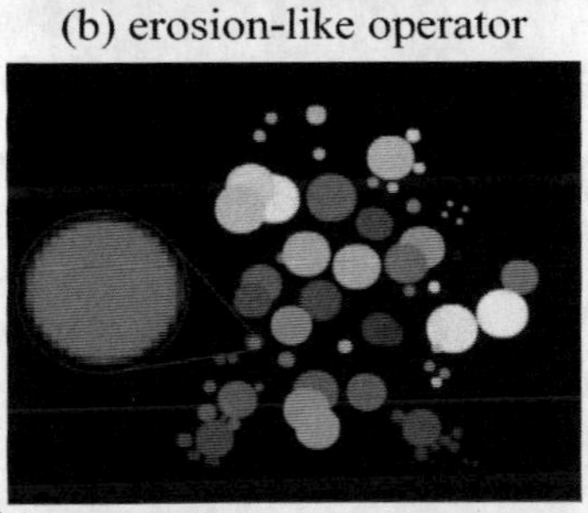

(b) dilation-like operator

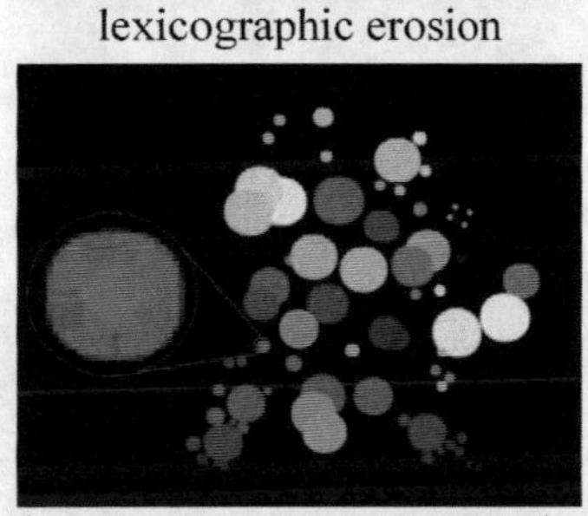

lexicographic dilation

Figure 3. Comparing results to lexicographic order using a synthetic image containing colored balls.

A. Detection of Exudates

As an example of application of the new morphological operators to previous biomedical image processing problems, we applied them to the detection of exudates in ocular images, which was done in a previous work using the M-order [17]. Hard and soft exudates are the main signs of diabetic macular edema (DME). The segmentation of both kinds of exudates generates valuable information not only for the diagnosis of DME, but also for treatment, which helps to avoid vision loss and blindness. Here we present the results obtained by applying the same method in Pastore et al. [17], but using erosion-like and dilation-like operators for the definition of opening-like and closings-like operators. The images were obtained from the Diabetic Retinopathy Database and Evaluation Protocol [18]. The algorithm presented can be summarized in the following three stages:

Step 1: We apply successive morphological opening-like with growing structuring elements (SE). This removes the objects that are smaller than the initial size of the SE. When this operator is applied again, increasing the size of the SE, the objects remained in the previous stage are removed. This process continues until only an average color background stays in the image.

Step 2: In order to highlight the bright objects, we subtract the image obtained in the previous stage to the original image.

Step 3: We binarize the resulting image using the Otsu's algorithm to segment the most relevant components [19].

Figure 4-a shows that, in retinopathy images, the exudates have a light color with respect to the background. Because of this feature, when applying the morphological opening-like, defined for color images, it removes the exudates that are smaller than the initial size SE (figure 4-b). If this operator is applied again, increasing the size of SE, the exudates remaining in the previous stages are removed. When, continuing this process *n* times, only the background of the image is retained (figure 4-c). To highlight the exudates, we subtract the image obtained in the previous stage from the original image (figure 4-d). Since the RGB image has the relevant information in the green component, this component is binarized using Otsu's method (figure 4-e). Finally, to visualize the results, a Canny Filter [20] is applied to the resulting image (figure 4-f), and the result is superimposed to the original image (figure 4-g).

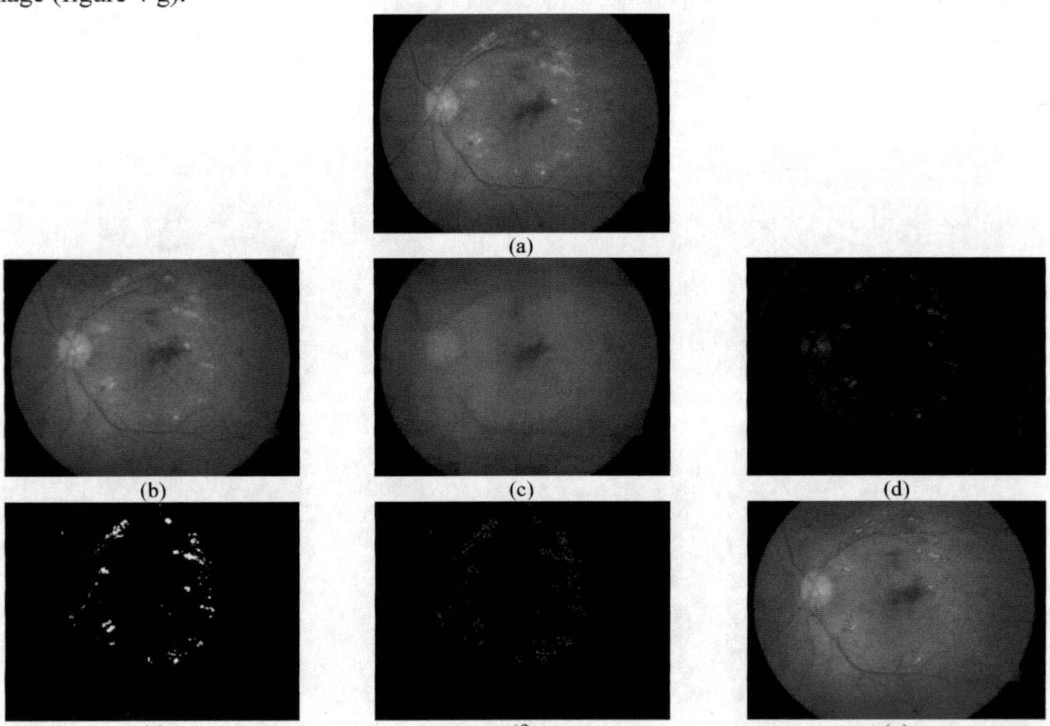

Figure 4. Steps of the proposed method. (a) Original image, (b) initial opening, (c) successive openings of the original image, (d) subtraction between the original image and the image (c), (e) binarization of the green component, (f) Canny Filter, (g) resulting image overlay to the original image.

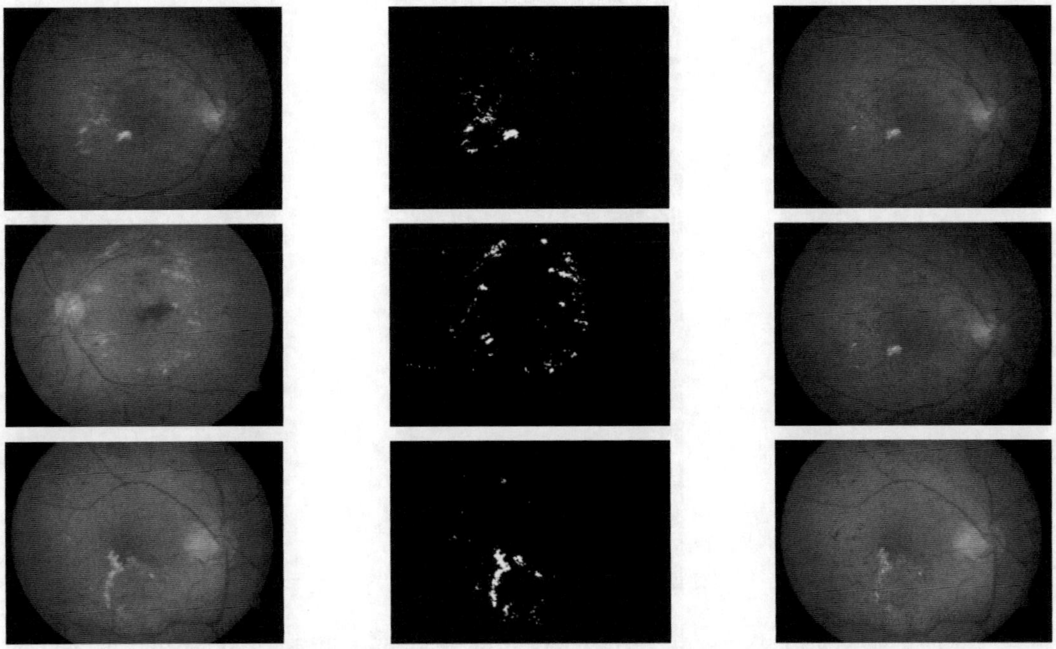

Figure 5. Results obtained by the proposed method. First column, original images. Second column, segmented images. Last column, segmented images overlaid to the original image.

Figure 5 shows the results obtained for three processed images of the Diabetic Retinopathy Database and Evaluation Protocol [18]. The first column shows the original images, the second column shows the segmentations performed by the proposed method, the third column shows the segmentation overlaid to the original image in order to assess the results.

In this example we can see how more complex operators, like openings-like operators, can be generated from the erosion-like and dilation-like operators proposed here, producing similar results than with previously defined morphological operators.

B. Segmentation of Bone Marrow Biopsies

Another example of porting a previously algorithm is the identification of trabecular structures, as done previously in [21]. The anatomic and pathologic reports of the histological cuts given results expressed in percentages, indicating the presence of the trabeculae, hematopoietic and fatty cells. This allows evaluating the existence or grade of some metabolic disorder, comparing normal values with pathologic ones [22]. To segment dark objects (trabecular tissue), we applied the same method than in the previous section, but using successive closing-like operators.

Figure 6 shows the results obtained in three bone marrow biopsies images. The first column shows the original image, the second column shows the segmentation obtained with the new color operators, and the third column shows the two of them together for a better visualization. The results of this application of like-morphological operators are similar to the ones obtained previously in [21].

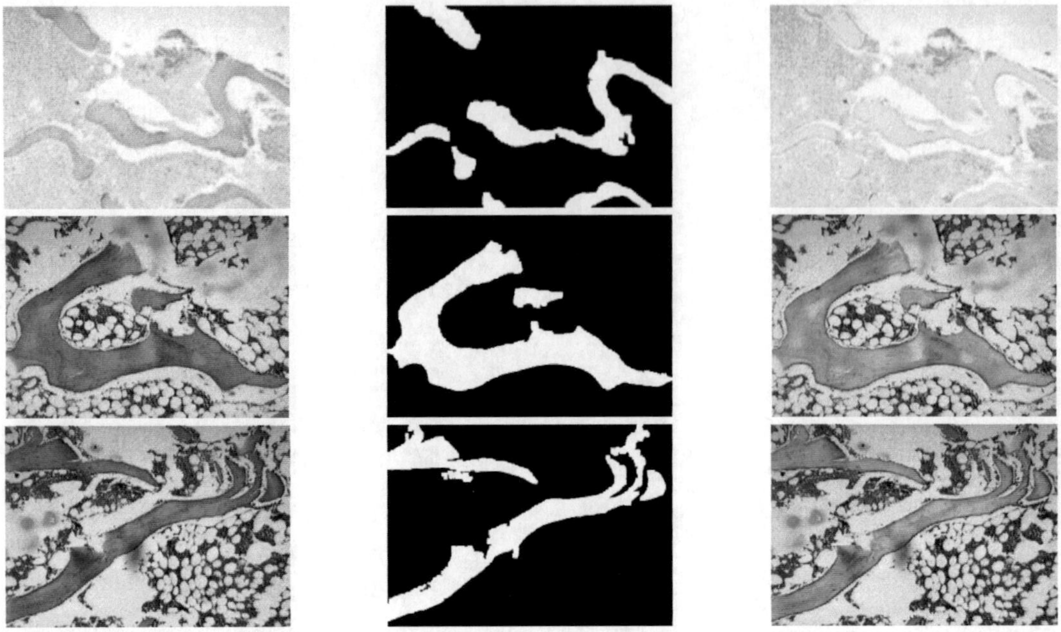

Figure 6. Results obtained by the proposed method. First column, original images. Second column, segmented images. Last column, segmented images overlaid to the original image.

4. Conclusions

In this work we presented new color morphological-like operators, named erosion-like and dilation-like operators, which can also be combined to generate more complex operators. To visualize the differences between this new approach and the previously proposed ones, we applied the basic operators: erosion and dilation based on lattice order (or M-order), erosion and dilatation based on lexicographic order, erosion-like and dilation-like operators on synthetic images. The new like-operators do not introduce false colors preserving the quality of the original image. Both of these issues are not simultaneously obtained with other approaches. Finally, as an example of its use, we show how more complex operators can be used on biomedical image processing, via operators similar to morphological opening, closing and top-hat. The new operators, part of the family of the morphological w-operators, outperform previously proposed color morphology approaches, regarding quality of the resulting image, and can be used to define more complex operators.

5. References

[1] G. Matheron, Random Sets and Integral Geometry, John Wiley and Sons, Inc., 1975.

[2] J. Serra, Image Analysis and Mathematical Morphology, vol. I, Academic Press, 1982.

[3] J. Goutsias, H.J.A.M. Heijmans, Fundamenta morphologicae mathematicae, Fundamenta Informaticae – Special Issue on Mathematical Morphology 41 (2000) 1–31.

[4] S. Sternberg, Grayscale morphology, Computer Vision, Graphics, and Image Processing 35 (1986) 333–355.

[5] R. Haralick, S. Sternberg, X. Zhuang, Image analysis using mathematical morphology, IEEE Transactions on Pattern Analysis and Machine Intelligence 9 (1987) 532–550.

[6] P. Maragos, R. Schafer, Morphological filters–part I: their set-theoretic analysis and relations to linear shift-invariant filters, IEEE T. on Acoustics, Speech and Sig. Proc. 35 (1987) 1153–1169.

[7] F. Meyer, S. Beucher, The morphological approach of segmentation: the watershed transformation, in: E. Dougherty (Ed.), Mathematical Morphology in Image Processing, Marcel Dekker, New York, 1992.

[8] V. Barnett, The ordering of multivariate data, Journal of the Royal Statistical Society 139 (1976) 318–355.

[9] A. Ledoux, N. Richard, A.S. Capelle-Laize, Limits and comparisons of orderings using colour distances, Traitement du Signal 29 (2012) 65–82.

[10] J. Serra: *Image analysis and mathematical morphology, Vol. II*. Academic Press, London, 1988.

[11] Jesus Angulo. Morphological colour operators in totally ordered lattices based on distances: Application to image filtering enhancement and analysis. Computer vision and image understanding, 2007, 107(1-2):56-73.

[12] Zhao Chunjiang, A new vectorial ordering for color morphology based on marginal ordering, The 5th Int. Conference on Computer Science & Education, Hefei, China. August 24–27, 2010.

[13] Tobar, M.C., Platero C., González P., Asensio G., Mathematical Morphology in the HSI Colour Space, J. Martín et al. (Eds.): IbPRIA 2007, Part II, LNCS 4478, pp 467-474, 2007.

[14] J. Barrera, G. P.Salas, "Set operations on closed intervals and their applications to the automatic programming of morphological machines". J. Electronic Imaging 5(3): 335-352 (1996)

[15] J. Barrera, G. Banon, and E.R. Dougherty, "Automatic design of morphological operators," in Space Structure and Randomness, 2005, pp. 257-278.

[16] Roberto Hirata Jr., Marcel Brun, Junior Barrera, and Edward R. Dougherty, "Aperture filters: theory, application, and multiresolution analysis," Advances In Nonlinear Signal And Image Processing, eds. Sephen Marshall and Giovanni L. Sicuranza, EURASIP Book Series on Signal Processing and Communication, Hindawi Publishing Corporation, 2006.

[17] Ordoñez C., Pastore J., Ballarin, V., 2012. "Segmentación de exudados en imágenes de fondo de ojo aplicando Morfología Matemática Color", 5ta Escuela y Workshop Argentino en Ciencias de las Imágenes, (ECImag-2012), pp. 27-28

[18] Kauppi, T., Kalesnykiene, V., Kämäräinen, J.-K., Lensu, L., Sorri, I., Raninen, A., Voutilainen, R., Uusitalo, H., Kälviäinen, H., Pietilä, J., 2007. DIARETDB1 diabetic retinopathy database and evaluation protocol. Medical Image Understanding and Analysis (MIUA), 61-65.

[19] Otsu, N., 1979. A threshold selection method from gray-level histogram. *IEEE Trans. Syst. Man Cybern.* 9, 62-66.

[20] Canny, J., 1986. A Computational Approach to Edge Detection. Pattern Analysis and Machine Intelligence, IEEE Transactions on PAMI-8, 679-698.

[21] Pastore J., Bouchet A., Moler E. and Ballarin V., 2013. Segmentation of Bone Marrow Biopsies by Mathematical Morphology in Color Spaces. IEEE Latin America Transactions. Vol. 11, No. 1, pp. 335-339. ISSN 1548-0992.

[22] Rubin E., Farber, E., 1992. "Patología: Fundamentos", Buenos Aires: Editorial Médica Panamericana.

Carotid wall stress calculated with continuous intima-media thickness assessment using B-mode ultrasound

A F Pascaner[1,2], D Craiem[1,2], M E Casciaro[1,2], R Danielo[3], S Graf[1,2] and E Guevara[3]

[1]Facultad de Ingeniería y Ciencias Exactas y Naturales, Universidad Favaloro. Buenos Aires, Argentina. [2]CONICET. Buenos Aires, Argentina. [3]Hospital Universitario de la Fundación Favaloro, Unidad de Ecocardiografía y Doppler Vascular. Buenos Aires, Argentina.

E-mail: apascaner@favaloro.edu.ar

Abstract. Cardiovascular risk is normally assessed using clinical risk factors but it can be refined using non-invasive infra-clinical markers. Intima-Media Thickness (IMT) is recognized as an early indicator of cardiovascular disease. Carotid Wall Stress (CWS) can be calculated using arterial pressure and carotid size (diameter and IMT). Generally, IMT is measured during diastole when it reaches its maximum value. However, it changes during the cardiac cycle and a time-dependant waveform can be obtained using B-mode ultrasound images. In this work we calculated CWS considering three different approaches for IMT assessment: (i) constant IMT (standard diastolic value), (ii) estimated IMT from diameter waveform (assuming a constant cross-sectional wall area) and (iii) continuously measured IMT. Our results showed that maximum wall stress depends on the IMT estimation method. Systolic CWS progressively increased using the three approaches ($p<0.001$). We conclude that maximum CWS is highly dependent on wall thickness and accurate IMT measures during systole should be encouraged.

1. Introduction

Several studies have aimed at detecting cardiovascular diseases in an early stage and thus improve risk prediction [1-3]. Besides traditional risk factors (such as age, smoking and hypertension), different infra-clinical factors were studied in order to refine the risk assessment of a given patient. This is the case of Intima-Media Thickness (IMT), Pulse Wave Velocity (PWV) [4-6], arterial compliance [7], mechanical stress in the arterial wall [8], to name a few. In this work we propose a new approach to calculate the arterial wall stress at the level of the common carotid artery.

The carotid wall stress depends on the blood pressure and the arterial geometry. Normally, the artery is represented as a cylinder with a thick wall. Using B-mode ultrasound it is possible to estimate arterial diameter and IMT [9][10]. During diastole, diameter variations are slow and IMT tends to reach its maximum value, thus making it the preferred moment for IMT estimation during the cardiac cycle [11]. However, using modern equipment and digital image processing techniques, we have shown that it is possible to assess diameter and IMT waveforms during the entire cardiac cycle [10]. The continuous assessment of IMT and carotid diameter allows the calculation of the time-varying arterial wall stress.

The aim of this work was to compare different maximum stress values assuming that IMT: (i) was constant and its value was estimated during diastole, (ii) was calculated assuming constant Cross-Sectional Wall Area (CSWA) from diameter measures and (iii) was continuously estimated with the proposed algorithm [10]. For this purpose, a cohort of healthy volunteers and intermediate cardiovascular risk patients was evaluated using B-mode ultrasound sequences.

2. Materials and methods

2.1. Study population

The population under study was comprised of healthy volunteers and cardiovascular patients. Volunteers were prospectively recruited under the following inclusion criteria: being less than 30 years old, non-smokers, non-diabetics and not having been diagnosed with neither hypertension nor hypercholesterolemia. Intermediate cardiovascular risk patients were retrospectively selected by the Echocardiography and Vascular Doppler Unit of the Favaloro Foundation Hospital. Table 1 shows clinical data of volunteers and patients.

Table 1. Clinical data of the subjects. Values are expressed as mean ± standard deviation or n (%). We used a T-test to compare continuous variables and a χ^2 statistic for dichotomous variables.

Parameter	Volunteers	Patients	p
Number of subjects	45	19	-
Age, years old	21.8 ± 1.6	54.4 ± 7.8	<0.01
Male, n (%)	25 (56%)	14 (74%)	n.s.
Body mass index, kg/m^2	23 ± 3	28 ± 6	<0.01
Systolic pressure, mmHg	121 ± 14	130 ± 12	<0.01
Diastolic pressure, mmHg	77 ± 8	74 ± 9	n.s.
Pulse pressure, mmHg	44 ± 12	56 ± 9	<0.01
Smokers, n (%)	0 (0%)	12 (63%)	<0.01
Hypertensive, n (%)	0 (0%)	11 (58%)	<0.01
Hypercholesterolemia, n (%)	0 (0%)	13 (68%)	<0.01
Diabetes, n (%)	0 (0%)	2 (11%)	<0.05
Heart rate, bpm	74 ± 11	68 ± 11	<0.05
Carotid plaques, n (%)	0 (0%)	9 (47%)	<0.01
Femoral plaques, n (%)	0 (0%)	11 (58%)	<0.01

2.2. Continuous diameter and IMT measurement

Each subject underwent an ultrasound examination performed by a professional with a high resolution device (Philips HD11XE) and a 10 MHz linear transducer. Five-second sequences of B-mode ultrasound images of the common carotid artery were recorded. All sequences were acquired during a single breath-hold. For volunteers, both left and right carotid arteries were registered. For patients, only the available registers in one or two carotids were processed. We developed a software that imported the sequences and measured frame-by-frame diameter and IMT, thus obtaining continuous waveforms. A detailed explanation of the software can be found in [10]. Briefly, the user selected a Region Of Interest (ROI) in the first image of every sequence. Each column of pixels within the ROI was considered as a vector of grey level intensity. Three interfaces were detected in each column: (i) near wall intima-lumen, (ii) far wall lumen-intima and (iii) far wall media-adventitia. The algorithm searched for the maximum gradient of intensity to determine their locations. The definitive interfaces were considered as the mean value of all columns in the ROI. Diameter was considered as the distance between (i) and (ii) and IMT as the distance between (ii) and (iii).

For every sequence, IMT values were smoothed with a 3-point moving-average filter. The waveforms were separated in heartbeats using extreme diameter values. For each beat, diastole and systole were considered as the points where minimum and maximum diameter occurred, respectively. Diastolic and systolic diameters and IMTs were averaged for all the heartbeats (a minimum of three) obtaining four values for each sequence: diastolic diameter (D_d), systolic diameter (D_s), diastolic IMT (IMT_d) and systolic IMT (IMT_s). Blood pressure was measured in the brachial artery with an automatic oscillometric device (Omron BP742).

2.3. Stress calculation

Normal stress in a body is defined as the extending force divided by the area of the cross-section:

$$\tau = \frac{F}{A} \tag{1}$$

where τ is normal stress, F is the force and A is the cross-sectional area. We considered the artery as a perfect cylinder with a thick wall in which only an internal pressure is acting (blood pressure). With this model, we can define medial radius R as follows:

$$R = \frac{r_e + r_i}{2} \tag{2}$$

where r_e and r_i are external and internal radius of the cylinder, respectively. Wall stress at R can be expressed in terms of external and internal radii [12]:

$$\tau = \frac{2P(r_e r_i)^2}{(r_e^2 - r_i^2)R} \tag{3}$$

where P is blood pressure. We measured continuous diameter (D) and IMT in volunteers and patients using B-mode ultrasound sequences and calculated r_e and r_i as follows:

$$r_i = \frac{D}{2} \qquad r_e = \frac{D}{2} + IMT \tag{4}$$

Diastolic wall stress was calculated using diastolic pressure, diameter and IMT values. Systolic wall stress was calculated in three different ways, using the same systolic diameter and pressure but changing IMT estimation. IMT was considered as: (i) standard IMT measured in diastole (IMT_d), (ii) estimated systolic IMT assuming a constant CSWA (IMT_{s_est}) and (iii) systolic IMT measured with the proposed algorithm (IMT_s). In the case (i), IMT_d was considered constant and the diastolic standard value was used. In the case (ii), IMT_{s_est} was calculated considering that the arterial wall is incompressible [13], i.e., its CSWA remains constant within the cardiac cycle. CSWA can be calculated as follows:

$$CSWA = \pi\, IMT(D + IMT) \tag{5}$$

Assuming a constant CSWA value during the cardiac cycle, IMT_{s_est} can be calculated with diameter and IMT_d:

$$IMT_{s\_est} = \frac{-D_s + \sqrt{D_s^2 + 4IMT_d(D_d + IMT_d)}}{2} \tag{6}$$

No longitudinal stretch was considered. However, it has been shown that the carotid arterial length changes during systole and this change affects significantly the mechanical properties of the arterial wall [10]. For this reason, we also proposed the case (iii), and we measured IMT during systole to obtain its actual value.

3. Results

3.1. Diameter and IMT measurements

Table 2 shows mean diastolic and systolic diameter and IMT values for volunteers and patients. A Student's t-test was used to compare values.

Table 2. Obtained values of diastolic and systolic diameter and the three estimations of IMT. Values are expressed as mean ± standard deviation.

Parameter	Volunteers	Patients	p
Number of carotids	90	27	-
D_d (mm)	5.16 ± 0.48	5.73 ± 0.88	<0.01
D_s (mm)	6.00 ± 0.56	6.25 ± 0.93	n.s.
IMT_d (mm)	0.45 ± 0.09	0.58 ± 0.11	<0.01
$IMT_{s\ est}$ (mm)	0.40 ± 0.08	0.54 ± 0.10	<0.01
IMT_s (mm)	0.37 ± 0.09	0.53 ± 0.12	<0.01

Both in patients and in volunteers D_d was lower than D_s ($p<0.01$) and IMT_d was higher than IMT_s ($p<0.01$) and IMT_{s_est} ($p<0.01$). IMT_{s_est} was higher than IMT_s in volunteers and in patients ($p<0.01$ and $p<0.05$, respectively).

3.2. Wall stress results

Diastolic and systolic wall stresses for volunteers and patients are shown in figure 1. From left to right: diastolic wall stress, systolic wall stress using IMT_d, systolic wall stress using IMT_{s_est} and systolic wall stress using IMT_s.

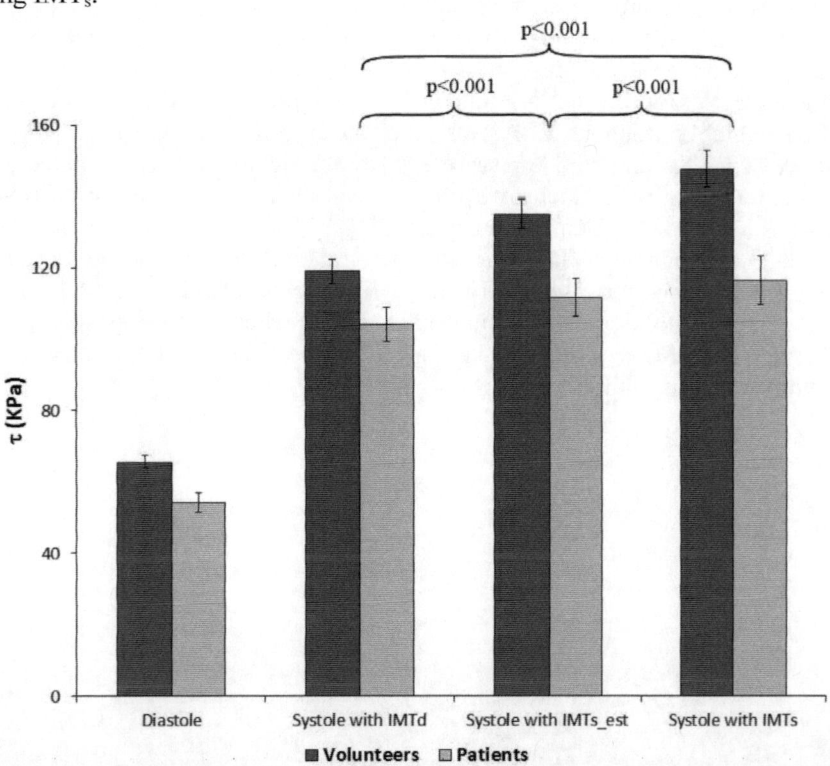

Figure 1. Minimal (diastolic) and maximal (systolic) wall stress in the carotid artery wall for volunteers and patients. Bars indicate standard error.

Wall stresses in systole for the three IMT estimations were higher than in diastole ($p<0.001$). It can be observed that as systolic IMT approaches the real, smaller, value (from left to right), wall stress increases ($p<0.001$). Finally, we observed that wall stresses were systematically higher in volunteers with respect to patients ($p<0.05$ for systolic wall stress with IMT_d and $p<0.01$ for the rest).

4. Discussion

We have found that there are significant differences in the estimation of the maximum stress exerted over the carotid artery, depending on the IMT estimation method. These differences were significant for both volunteers and patients. It can be observed in figure 2 that as IMT decreases, stress increases. Accordingly, if IMT is continuously assessed during the cardiac cycle, higher maximum stresses were found.

In figure 1, it is noteworthy that patients had a lower maximum wall stresses than volunteers, despite having a significantly higher systolic pressure. Probably, higher arterial pressure triggers an IMT increase as a protection mechanism to preserve arterial wall stress. This IMT increase reduces stress, according to equations (1) and (3). Whether this compensation mechanism was entirely responsible of the lower values of wall stresses found in patients with respect to volunteers, need to be further investigated.

This study had some limitations that need to be addressed. Firstly, it is arguable whether measuring IMT in systole is possible, since its value is near the ultrasound resolution. Nonetheless, it has been reported that technology advances allowed for accurate IMT estimation even in young subjects, where IMT values tend to be smaller [14]. The use of automated algorithms become indispensable [10]. Finally, we used the brachial pressure to calculate carotid wall stress, although central pressure should have been employed. Some studies have shown an amplification phenomenon, which makes systolic pressure in the periphery (e.g. in the brachial artery) higher than central pressure (e.g. in the carotid artery) [13]. In order to have more accurate results, central pressure could be estimated from brachial pressure and continuous Cross-Sectional Area (CSA) estimations [15]. Meinders et al proposed an exponential model and an iterative algorithm that converges into central pressure. We have tested this algorithm in a volunteer using brachial diastolic and systolic pressure and continuous carotid diameter estimations (from which continuous CSA was calculated). Figure 2 shows (on the left hand) the exponential fit between CSA and blood pressure and (on the right hand) A) estimated central pressure waveform, B) measured carotid diameter waveform, C) IMT estimations using (i) standard diastolic IMT (IMT_d), (ii) estimated IMT assuming constant CSWA (IMT_{s_est}) and (iii) continuously measured IMT (IMT_s) and D) calculated wall stress with the three different IMT assumptions. Note the incremental wall stress values for cases (i), (ii) and (iii) (also observed in figure 1). This preliminary test showed that central systolic pressure resulted 131 mmHg when brachial systolic pressure was 133 mmHg (1.5% error). As a future work, we can repeat this analysis for all subjects although pressure differences should remain negligible.

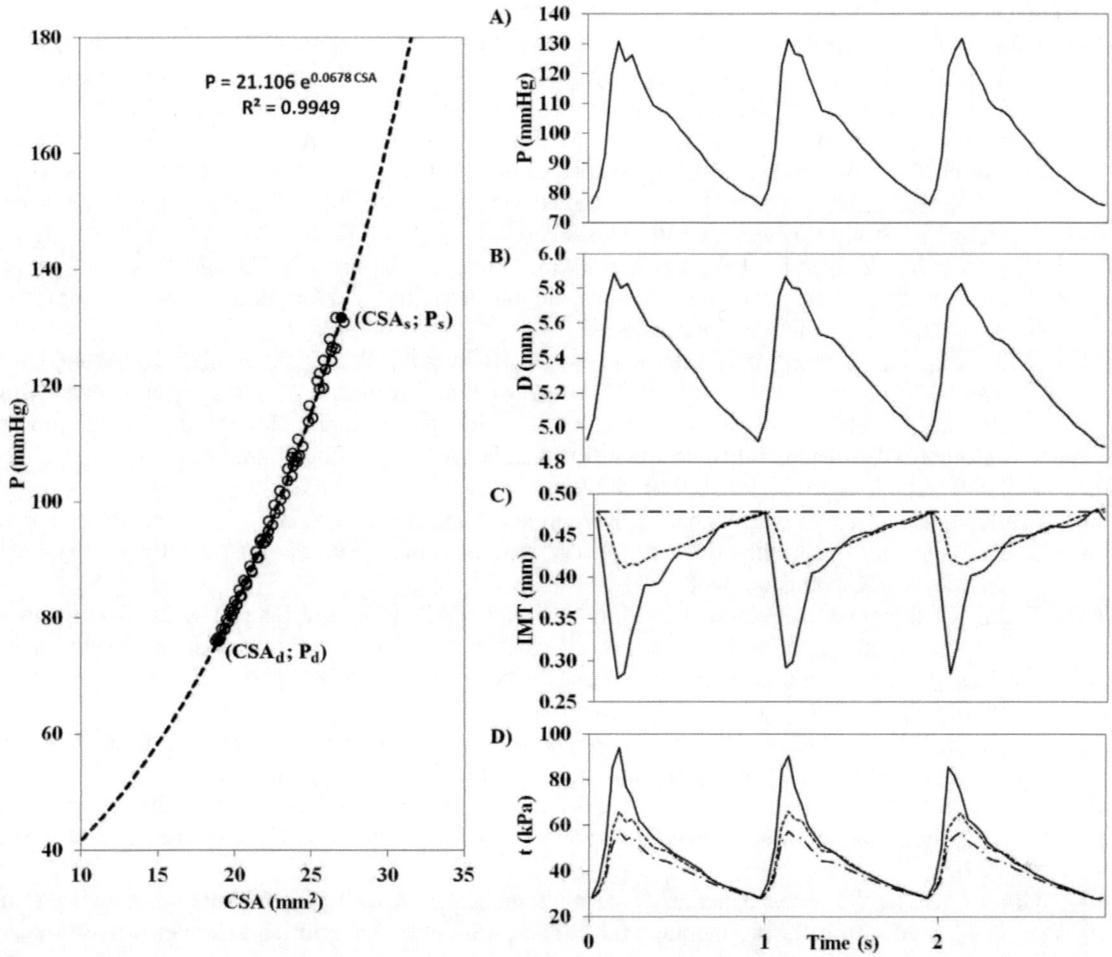

Figure 2. On the left: exponential estimation of central pressure using CSA. Dashed line is the estimated function and solid line represents the range of the representative volunteer. On the right: A) estimated pressure waveform, B) measured diameter waveform, C) IMT_d, IMT_{s_est} and IMT_s (dashed-dotted, dashed and solid lines, respectively) and D) calculated wall stress with the three approaches of IMT (same line types as in C)).

5. Conclusion

In this work we proposed a method to calculate the maximum carotid wall stress using continuous IMT measurements. We verified that IMT estimation methods can significantly modify carotid wall stress estimation. The estimation of true IMT values during systole was possible using an image processing algorithm that continuously measured simultaneous IMT and diameter values. The carotid wall stress is highly dependent on wall thickness and accurate IMT measures during systole should be encouraged.

6. Acknowledgement

This work was partly supported by the grant PICT 2012 Number 0406 of the "Ministerio de Ciencia y Tecnología" (Argentine).

7. References

[1] Laurent S, Cockcroft J, Van Bortel L, Boutouyrie P, Giannattasio C, Hayoz D, Pannier B, Vlachopoulos C, Wilkinson I and Struijker-Boudier H 2006 Expert consensus document on arterial stiffness: methodological issues and clinical applications *Eur Heart J* **27** 2588-605

[2] Xanthakis V, Enserro D M, Murabito J M, Polak J F, Wollert K C, Januzzi J L, Wang T J, Tofler G and Vasan R S 2014 Ideal cardiovascular health: associations with biomarkers and subclinical disease and impact on incidence of cardiovascular disease in the Framingham Offspring Study *Circulation* **130** 1676-83

[3] Xanthakis V, Sullivan L M, Vasan R S, Benjamin E J, Massaro J M, D'Agostino R B, Sr. and Pencina M J 2014 Assessing the incremental predictive performance of novel biomarkers over standard predictors *Stat Med* **33** 2577-84

[4] Koivistoinen T, Virtanen M, Hutri-Kahonen N, Lehtimaki T, Jula A, Juonala M, Moilanen L, Aatola H, Hyttinen J, Viikari J S, Raitakari O T and Kahonen M 2012 Arterial pulse wave velocity in relation to carotid intima-media thickness, brachial flow-mediated dilation and carotid artery distensibility: the Cardiovascular Risk in Young Finns Study and the Health 2000 Survey *Atherosclerosis* **220** 387-93

[5] Mutlu B, Tigen K, Gurel E, Ozben B, Karaahmet T and Basaran Y 2011 The predictive value of flow-mediated dilation and carotid artery intima-media thickness for occult coronary artery disease *Echocardiography* **28** 1141-7

[6] Nagai K, Shibata S, Akishita M, Sudoh N, Obara T, Toba K and Kozaki K 2013 Efficacy of combined use of three non-invasive atherosclerosis tests to predict vascular events in the elderly; carotid intima-media thickness, flow-mediated dilation of brachial artery and pulse wave velocity *Atherosclerosis* **231** 365-70

[7] Reneman R S, van Merode T, Hick P, Muytjens A M and Hoeks A P 1986 Age-related changes in carotid artery wall properties in men *Ultrasound Med Biol* **12** 465-71

[8] Makris G C, Nicolaides A N, Xu X Y and Geroulakos G 2010 Introduction to the biomechanics of carotid plaque pathogenesis and rupture: review of the clinical evidence *Br J Radiol* **83** 729-35

[9] Graf S, Gariepy J, Massonneau M, Armentano R L, Mansour S, Barra J G, Simon A and Levenson J 1999 Experimental and clinical validation of arterial diameter waveform and intimal media thickness obtained from B-mode ultrasound image processing *Ultrasound Med Biol* **25** 1353-63

[10] Pascaner A F, Craiem D, Casciaro M E, Danielo R, Graf S and Guevara E 2015 Continuous assessment of carotid intima-media thickness applied to estimate a volumetric compliance using B-mode ultrasound sequences *Physiol Meas* **36** 397-407

[11] Haller C, Schulz J, Schmidt-Trucksass A, Burkardt H, Schmitz D, Dickhuth H H and Sandrock M 2007 Sequential based analysis of Intima-Media Thickness (IMT) in common carotid artery studies *Atherosclerosis* **195** e203-9

[12] Armentano R L and Cabrera Fischer E I 1994 Biomecánica arterial. Fundamentos para su abordaje en la clínica médica

[13] Nichols W W and O'Rourke M F 1998 McDonald's Blood flow in arteries. Theoretical, experimental and clinical principles

[14] Jourdan C, Wuhl E, Litwin M, Fahr K, Trelewicz J, Jobs K, Schenk J P, Grenda R, Mehls O, Troger J and Schaefer F 2005 Normative values for intima-media thickness and distensibility of large arteries in healthy adolescents *J Hypertens* **23** 1707-15

[15] Meinders J M and Hoeks A P 2004 Simultaneous assessment of diameter and pressure waveforms in the carotid artery *Ultrasound Med Biol* **30** 147-54

Processing Pipeline of Sugarcane Spectral Response to Characterize the Fallen Plants Phenomenon

Agustín Solano[1], Alejandra Kemerer[2], Alejandro Hadad[1]

[1] Information Systems Laboratory, Faculty of Engineering, National University of Entre Ríos, [2] INTA-EEA Paraná, Natural Resources and Abiotic Factors Group, Faculty of Agronomy, National University of Entre Ríos

E-mail: asolano@bioingenieria.edu.ar, kemerer.alejandra@inta.gob.ar, ajhadad@bioingenieria.edu.ar

Abstract. Nowadays, in agronomic systems it is possible to make a variable management of inputs to improve the efficiency of agronomic industry and optimize the logistics of the harvesting process. In this way, it was proposed for sugarcane culture the use of remote sensing tools and computational methods to identify useful areas in the cultivated lands. The objective was to use these areas to make variable management of the crop. When at the moment of harvesting the sugarcane there are fallen stalks, together with them some strange material (vegetal or mineral) is collected. This strange material is not millable and when it enters onto the sugar mill it causes important looses of efficiency in the sugar extraction processes and affects its quality. Considering this issue, the spectral response of sugarcane plants in aerial multispectral images was studied. The spectral response was analyzed in different bands of the electromagnetic spectrum. Then, the aerial images were segmented to obtain homogeneous regions useful for producers to make decisions related to the use of inputs and resources according to the variability of the system (existence of fallen cane and standing cane). The obtained segmentation results were satisfactory. It was possible to identify regions with fallen cane and regions with standing cane with high precision rates.

1. Introduction

The characteristics of the argentinian agricultural production system and the necesity of the producers to achieve greater efficiency, precision and productivity are reflected in the constant demand of tools, data, information and new knowledge related to Precision Agriculture.

Precision Agriculture is a new concept characterized by inputs' management that may vary according to system variability. It is the application of technology and principles to manage the space-time variability associated to the agronomic production. It allow to explore and value the productive and economic benefits of managing the system in a site-specific way.

The implementation of precision agriculture techniques can be achieved in two levels. At strategic level there are several types of sensors which provide data and information about specific applications. At operational level the image acquisition task is increasing its importance and it allow to examine different processes on cultivated lands.

Remote sensing allow to obtain information of the earth surface without making physical contact with it, thus it is the opposite of on site observation [2]. The images of the earth surface can be satellital or aerial and they have different characteristics. The images can be integrated with others layers of spatial information in a Geographic Information System (GIS) using a Global Positioning System (GPS) for locating positions of interest. The integrated system allow to carry out management actions in the sites of interest.

The commercial emergence of high-resolution images has driven the usefulness of remote sensing for agronomic and precision agriculture purposes. The image acquisition systems which are sensitive to different bands of the electromagnetic spectrum enhance the quality and definition of the information.

Remote sensing can play an important role in monitoring the harvest and planting processes in small-scale fields where updated information is hard to find. Numerous studies [6, 9, 11, 16] show that the spectral information is associated with agronomic variables and can be used to monitor and predict crop yields.

In the present work a workflow to detect and segment homogeneous regions in aerial images of cultivated lands is proposed. The detected regions must have a size and shape of practical use for the producers to manage the farm in a site-specific way.

The crop under study is sugarcane. The cultivation of sugarcane in Argentina is found principally localized in the northwest region of the country (99 %). In Tucumán, Salta and Jujuy it represents, in agreement with [10], one of the most important productive activities. It is also found in scarce extension in the littoral region (1%), in Misiones and Santa Fe.

During the cultivation of sugarcane fields, losses of raw material occurr. The losses can be classified in two different ways: *Pre-harvest losses* and *Harvest losses*. The former is related to the canes that are fallen before harvesting, whereas the latter is related to the fact that the harvesting machines are not 100% efficient.

The presence of fallen sugarcane (CC) at the moment of harvesting brings important losses of useful product during the harvesting process and in the sugar mill. It is normal that the canes in a productive area reach an average height but they present variations of their characteristics in different sectors of the field. This might be because of the different conditions of the soil that presents different levels of fertility, humidity, compressing, etc. In addition, each plant is constituted in a different way and manifests in unique form the different external stimulus.

When the plant is young and has developed under the best conditions, it reaches certain height and begins to warp due to its own weight. It breaks if its breakage tension is surpassed, product of the wind or environmental phenomenons as hail. After that, the soil can remain exposed or new green sheets can appear , due to phototropism. These factors expose the great complexity and heterogeneity of the phenomenon under study and make it difficult to analyze and quantify.

When areas with presence of fallen canes are harvested, extraneous material without sucrose can enter the sugar mill. This extraneous material es composed of stalk tops, trash and dust. It represents an important loss of efficiency in the process of manufacture of the sugar, as assure [17]. The quality of the cane affects directly the industrial performance and the quality of the obtained sugar [8].

The quality of the raw material (cane) can be affected by changing aspects of the sugarcane agriculture. Some of these aspects are the introduction of new cultivars, climatic variations, the use of ripening chemists, changes in the cultural practices and in the systems of harvest or the appearance of illnesses or infections in the plants.

It is observed that the content of extraneous material picked up together with the cane harvested in mechanized way is appreciably lower than in the semi-mechanized way, being in the range 2 to 5,7%. At the same time, the values of fiber (plant residues) rise from 14 to about 18% due to the presence of increased amount of plant parts after a less efficient harvesting [15]. As a result, the producers are paid for a significant part of the weight of the extraneous material of the collected raw material at the same price of the cane, because the way of sampling and analysis of this parameter is generally inadequate.

Washing tables, to wash the cane, have been installed in several mills. This was done to reduce to the minimun possible the percentage of extraneous material delivered with the sugarcane. There, it is removed the present dust with big quantities of water (about 3 m^3 of water by ton of cane). This supposes a saving of money but adds problems like bigger costs for the required electric power, the water and the additional manual labor. Some studies also confirm that during cane washing some percentage of sucrose is lost [5]. Moreover, this technique requires availability of a piece of land to install settling tanks of sand and mud to reuse the water.

A promising strategy to achieve the reduction of the impact of the fallen cane and the extraneous material takes root in achieving a better operative efficiency in harvesting the cane, fundamentally in sectors with fallen cane, previous to the entrance of the raw material to the factories.

At present, the quantization of the amount of fallen cane in productive squares is an expensive and inefficient process. It requires sending evaluators to the fields. They must explore big surfaces of land to obtain a representative sampling that permits an estimation of fallen cane incidence. The movement around the field produces damage, requires many man hours of work and generates estimations with a high degree of uncertainty.

Other strategy consists in carrying out flights over the cultivated lands with specialists who estimate the percentage of fallen cane according to what is observed from the air and their experience. This strategy is subjective and depends on the available specialists in each area of interest.

Up to now, a report of some methodology with measurable precision to determine the affected area with fallen cane has not been observed. Such methodology would allow to plan a better way of harvesting the affected area and calculate its associate costs.

With this goal in mind and in conjunction with the National Institute of Agricultural Technology of Argentina (INTA), the group has worked on the application of computational methods on high-resolution aerial images of the cultivated lands. Those methods were implemented for the purpose of adjusting image characteristics and extracting useful information to identify areas with fallen cane and areas with standing cane. Because of the characteristics of the task and the available data, it was proposed to address the problem with image processing techniques and machine learning methods.

2. Objective

The objective of the work is to develop specific techniques in order to identify areas with homogeneous characteristics in multispectral aerial images. The main interest is to identify areas with fallen cane and areas with standing cane to make harvesting plans of sugarcane productive lands.

3. Materials

A set of images was obtained from a Sky Arrow 650TCNS ERA airplane. The airplane was equipped with a Global Positioning System (GPS), a Geospatial MS4100 multispectral camera and a system of control and storage of data. The camera took images in three bands of the electromagnetic spectrum: green (530-580 nm), red (650-685 nm) and near infrared (770-830 nm) with a resolution of 1920 x 1075 pixels. The GPS provided the position, altitude and height in synchronism with the acquisition of each image.

A flight was carried out on 5th May 2008, moment of the sugarcane growing cycle where the presence of fallen cane is clearly evident. The flight was carried out in the solar midday to have homogeneous illumination in the surface. The flight plan was designed for an altitude of 1200 m, resulting a pixel size of 0.7 meters by side.

The Red and Near Infrared channels were used to calculate the Normalized Difference Vegetation Index (NDVI) [12] of each image. The NDVI is defined as follows: $NDVI = (NIR - Red)/(NIR + Red)$, where NIR and Red are the fractions of Near Infrared and Red (respectively) radiation reflected by the earth's surface and detected by the camera.

With the supervision of an agricultural engineer the images were labelled in the classes of interest, using the application Label Me [13]. The labelling classes were: standing cane (CP), fallen cane in the form of patches (CCP), fallen cane in big areas (CCA), fallen cane with phototropism (CCF) and areas of exposed ground (C). All the fallen cane types were grouped together in a fallen cane class (CC) because the main interest of the research is to distinguish between areas with fallen cane and areas with standing cane.

What resulted from this labelling was a set of labelled subimages, corresponding to the CP class or the CC class.

4. Methods

4.1. Spectral signature analysis

In remote sensing it is assumed that specific objects interact in a unique and characteristic way with the electromagnetic radiation given off by the sun and that reach the earth. This interaction describes the spectral response of the objects, which is also called spectral signature. In this work, the spectral response of vegetation is of interest.

The spectral response of an object located on earth surface depends on five factors: reflectivity, absorptivity, emissivity, transmissivity and dispersion. These concepts explain the interaction between the objects and the electromagnetic radiation.

The spectral response curve or spectral signature of vegetation shows low values at wavelengths corresponding to Red and Blue (visible light). On the other hand, it presents a lower peak at Green wavelengths. These peaks and valleys are produced by Red and Blue light absorbed by chlorophyll and other pigments.

The spectral response of sugarcane plants varies according to their state (see Figure 1). At visible wavelenghts of the electromagnetic spectrum, corresponding to Green and Red light, the reflectance of the canopy is lower than the bare soil; at Near Infrared wavelenghts the vegetation and the bare soil have different reflectances but are similar at Middle Infrared wavelengths. The Middle Infrared wavelength band is important for separating trash from others states of the crop.

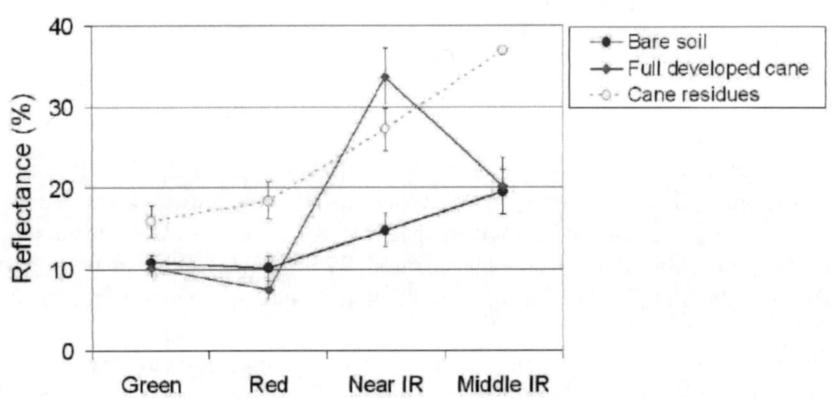

Figure 1: Mean and standard deviation of spectral response of sugarcane in different states and bare soil. [9]

From the photometric point of view, the Normalized Difference Vegetation Index (NDVI) shows healthy vegetation if it takes high values and values near zero shows bare soil and dry vegetation.

4.2. Preprocessing

In order to obtain standardized images with similar illumination level, the set of images was processed with an algorithm called Texture+Cartoon [4]. It decomposes a given image f in two components:

$$f = u + v \tag{1}$$

where u represents the geometric component, called Cartoon, and the component v represents the textural part with oscillating patterns.

For each pixel in the images it needs to be decided whether it belongs to the cartoon part or to the textural part. This decision is made by computing the Local Total Variation (LTV) of the image around the pixel and comparing it to the LTV of the same image around the pixel after applying a low pass filter.

The texture components of the images are considered a version with normalized brightness. They conserve the patterns given by the fallen and standing cane.

4.3. Feature extraction

Considering that the sugarcane cultivation has pronounced alterations in the spatial distibution, it was suggested to address the problem using texture characteristics. In [10] it was reported the use of the entropy, which is a common texture feature used in precision agriculture. Nevertheless, it provides insufficient information to distinguish the different states of the cane. For that reason, the present work proposes to incorporate and test other statistical descriptors divided in three groups: mean, standard deviation and entropy (Group 1); variance, smoothness, uniformity and entropy (Group 2); contrast, energy and homogeneity extracted from the co-occurrence matrix (Group 3). These descriptors are described in [2] and [7].

Using square windows of 11x11 pixels (the window size corresponds to an area of 7.7 meters per side in the sugarcane field), which were moved over the preprocessed images, the texture descriptors were calculated in the three channels of the images (Infrared, Red, Green) and in the NDVI. For each window, the different groups of statistical descriptors were calculated. Doing this in all the labelled images resulted in three data sets with vectors $\mathbf{x}$ corresponding to the different groups of texture characteristics and a label y corresponding to the state of cane associated with $\mathbf{x}$ (CC or CP).

The three groups of statistical descriptors were tested and it was evaluated their provision of information at the moment of distinguishing between the CC class and CP class. Eventually, one of the groups was selected to segment the aerial images.

4.4. Segmentation

To select the classification model, it was carried out a search in the specialized literature. The machine learning methods most commonly used to perform the classification task of mutispectral images are: Random Forests, Bagging, Boosting, Decision Trees, Artificial Neural Networks, Support Vector Machines and K Nearest Neighbors. However, when comparing the different methods it was found in several studies, described in [1], that the methods based on trees, in particular the Random Forests, have superior performance in classification tasks of multispectral images.

Using the labelled data set it was possible to train supervised learning methods based on trees. Primarily, it was used as classifier a simple model like a Decision Tree [14]. However, its high rate of misclassification motivated the use of a more complex model like Random Forest [3].

A Random Forest is a classifier consisting of a collection of tree-structured classifiers $\{h(\mathbf{x},\Theta_k), k = 1,...\}$ where Θ_k are independent identically distributed random vectors and each tree casts a unit vote for the most popular class at input $\mathbf{x}$.

Each new training set is drawn, with replacement, from the original training set. Then a tree is grown on the new training set using random feature selection. The trees grown are not pruned.

Bagging can be used to give ongoing estimates of the generalization error (PE^*) of the combined ensemble of trees. Given a specific training set T, form boostrap training sets T_k, construct classifiers $h(\mathbf{x}, T_k)$ and let these vote to form the bagged predictor. In each bootstrap training set, about one third of the instances are left out. For each $\{y, \mathbf{x}\}$ in the training set, aggregate the votes only over those classifiers for wich T_k does not contain $\{y, \mathbf{x}\}$. This is the out-of-bag classifier. Then the out-of-bag estimate for the generalization error is the error rate of the out-of-bag classifier on the training set.

Suming up, with the three data sets constructed with the different groups of texture characteristics $(\mathbf{x})$ and the labels y, three Random Forests with N trees each one were trained. The three models were compared and it was selected the one with better performance in the classification task. Afterwards, the selected model was used to segment the aerial images of sugarcane fields. For this, the texture characteristics were also calculated over windows of 11x11 pixels and the obtained vectors were put down the whole forest to get as a result if they belong to the class CC or CP. Doing this with all the pixels of the images resulted in a set of segmented images. Such images clearly showed regions with standing cane and regions with fallen cane in accordance with the precision of the model.

To measure their performance, the three trained models were cross-validated using 10-folds from the corresponding training set. An evaluation error was calculated with a set of reserved vectors which were not used for training. Finally, it was calculated the out-of-bag error during the training of each Random Forest model.

4.5. Post-processing

At this point, it would be useful to recall that the main objective of this work is to obtain information from the multispectral aerial images. That information may help to make decisions at the moment of harvesting the sugarcane fields. With the segmented images the agricultural engineers or producers should be able to make a harvesting plan. The goal of that plan is to harvest optimally the sugarcane, taking in mind the preharvest state of the plants.

The agricultural engineers could differentiate the harvesting of the standing sugarcane from the harvesting of the fallen cane if they consider that the size of the affected areas is considerable and worth it. They should also take into account other factors like how the harvesting machine works, the characteristics of the fields, operative costs, between others.

A post-processing stage was applied to the segmented images to improve the extraction of useful information. The resulting areas were made closed patchs with all the bounded pixels of the same class. Small sparse areas were eliminated. For that purpose, it was used an implementation of the Region Growing algorithm.

The algorithm receives as inputs an image that contains the region of interest, a point (x, y) inside the region and a threshold t. The region is iteratively grown by comparing all unallocated neighbouring pixels to the region, starting from the initial input point. The difference between a pixel's intensity value and the region's mean, is used as a measure of similarity. The pixel with the smallest difference measured this way is allocated to the respective region. This process stops when the intensity difference between region mean and new pixel becomes larger than t.

To get the desired results, it was proposed to apply this algorithm iteratively with different seed points. For seed points it was proposed to create a grid over the image with points equidistant both in the x coordinate and in the y coordinate. Then, for each point (x, y), it is was carried out the region growing algorithm. When detecting that a point (x, y) belonged already to a region identified in a previous iteration it was not reapplied the method.

The distance between the grid points in both coordinates (Δx and Δy) defined in some way the minimum size of the regions that could be identified. These distances depend on each case of study and this is the part of the process that requires assistance from the user. They were made somo tests with different values of Δx and Δy.

	Out-of-Bag Error	Cross-validation Error	Evaluation Error
Model 1	7,8%	7,3%	7,6%
Model 2	9,7%	9,4%	12,2%
Model 3	13,2%	12,8%	16,1%

Table 1: Error values of the Random Forests trained with the different groups of texture descriptors.

5. Results

The amount of trees used in each Random Forest was fixed to 50. For each group of texture descriptors, a set of 385 feature vectors was constructed. Each feature vector corresponded to one of the windows of 11x11 pixels obtained from the labelled images. Before training, from each set of vectors, a group of 77 (20%) vectors $(y, \mathbf{x})$ was reserved and not used for training. The remaining 308 vectors were used for training.

The resulting classification errors obtained with the trained models are listed in Table 1. Model 1 was trained with the descriptors: Mean, Standard Deviation and Entropy (Group 1); Model 2 was trained with the descriptors: Variance, Smoothness, Uniformity and Entropy (Group 2); and Model 3 was trained with the descriptors: Contrast, Energy and Homogeneity extracted from the co-occurrence matrix (Group 3).

As it can be seen in the error table, the model with better classification performance is the model trained with the descriptors Mean, Standard Deviation and Entropy (Model 1). For this model it was also calculated the Sensibility, resulting of 92%, and the Specificity, resulting of 94%.

Model 1 was selected and used to segment the aerial multispectral images in order to obtain information about the affected areas with fallen cane.

As an example, in figure 2a there is an multispectral aerial image taken from the available set of images. Figure 2b shows the segmented image and figures 2c, 2d, 2e and 2f show the resulted images after post-processing with differents values for Δx and Δy. The dark pixels correspond to the background of the picture. The blue pixels are fallen cane (CC class) and the green pixels are standing cane (CP class). It was also possible to calculate the real size of each detected area.

In Figures 3 is shown another example.

After applying the post-processing to different segmented images, about 10 to 16% of the pixels of each image changed of class with respect to the class assigned during the segmentation. This is the price that must be paid in order to obtain more valuable information from the segmentation.

6. Conclusions

In this paper, a pipeline to process multispectral aerial images of surgarcane productive squares and obtain data sets in order to estimate the state of the sugarcane (standing cane o fallen cane) was presented.

This pipeline includes 5 steps: image labelling, image filtering, feature extraction, image segmentation and postprocessing. The labelled datasets were built by means of a web tool to support the interaction with the experts of INTA. A Cartoon-Texture filter was used for the image filtering step. The filter was used to standardize the images respect to the brightness levels. The features extracted were groups of texture statistics from the original images and NDVI index, both used extensively for the agricultural engineer to characterize diferent cultures. Image segmentation process involved the use of machine learning methods with different complexity level. The Random Forest algorithm allowed to obtain a good performance for this application. The model trained with the characteristics Mean, Standard Deviation and Entropy had a cross-validation error of 7,3% and an evaluation error of 7,6%. The resulting segmented images presented several complex zones and small sparse areas in which the combine harvester can not work. For this problem, a post-processing by Region Growing was used.

Taking everything into account, the proposed approach has demostrated to provide valuable information for estimating the state of the sugarcane, apply different policy actions in relation to harvest methods, estimate associate costs, etc. In addition, its modular aspect guarantees the reusability of some processing steps to other images and cultures. As a result, the production of sugarcane and the manufacture of sugar could be more efficient and profitable.

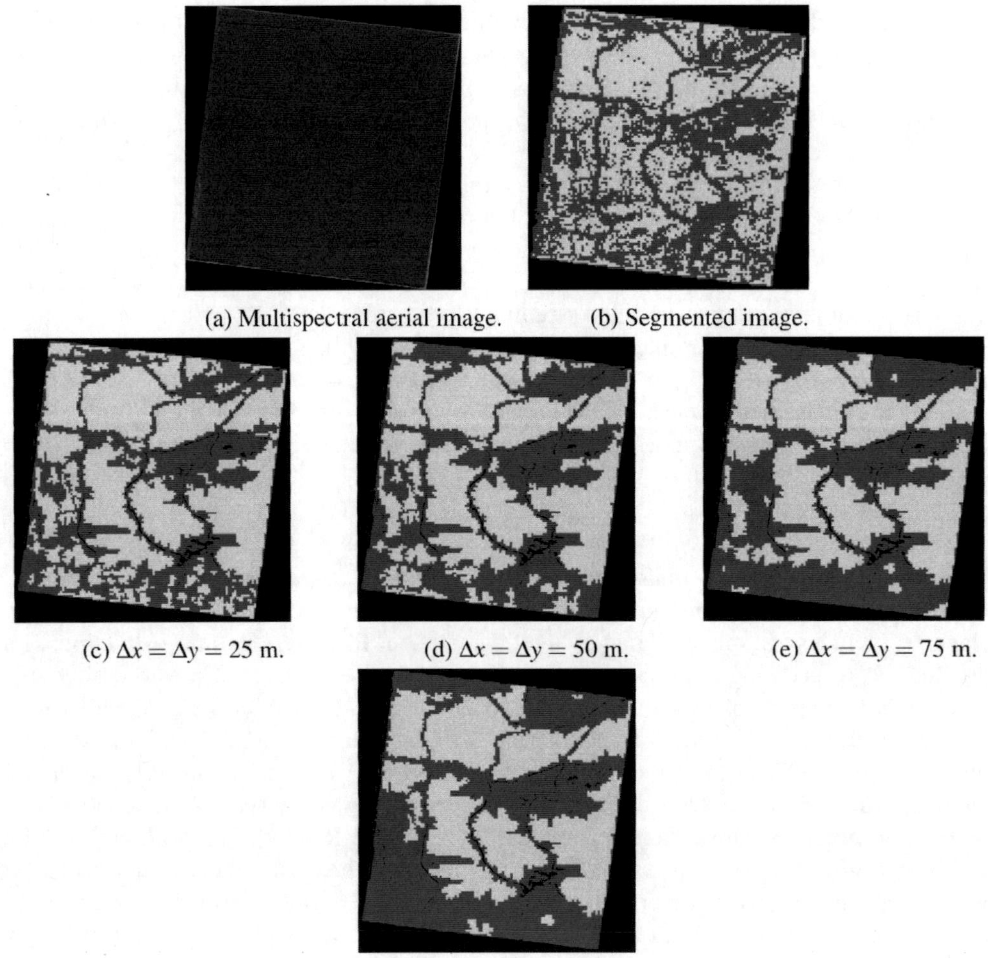

(a) Multispectral aerial image. (b) Segmented image.

(c) $\Delta x = \Delta y = 25$ m. (d) $\Delta x = \Delta y = 50$ m. (e) $\Delta x = \Delta y = 75$ m.

(f) $\Delta x = \Delta y = 100$ m.

Figure 2: Example images

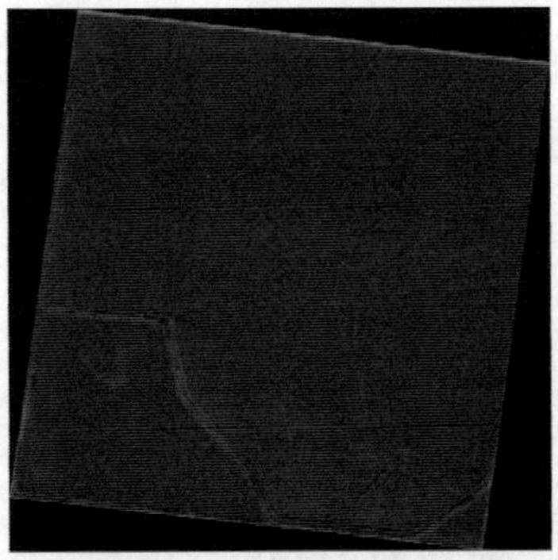

(a) Multispectral aerial image.

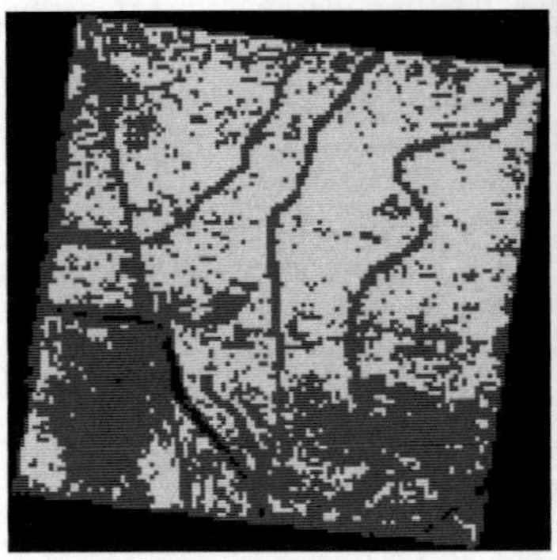

(b) Segmented image.

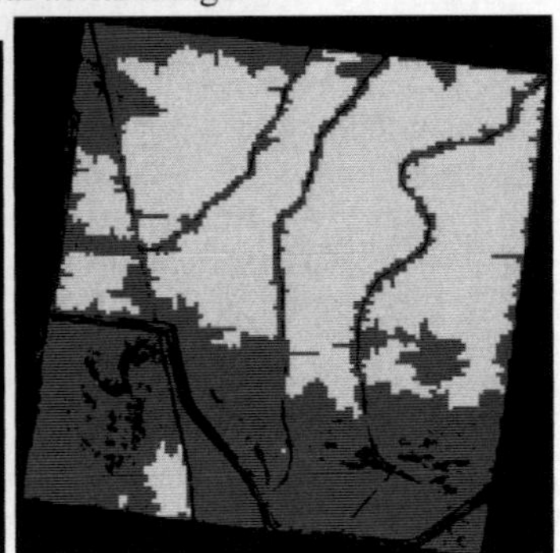

(c) Post-processed image ($\Delta x = \Delta y = 100$ m).

Figure 3: Example images

References

[1] Ozlem Akar and Ogur Gungor. Classification of multispectral images using random forest algorithm. *Journal of Geodesy and Geoinformation*, (2):105–112, 2012.

[2] Ángel Ruiz Alonso. Comportamiento y análisis de descriptores de texturas en imágenes MODIS, 2011.

[3] Leo Breiman. Random forests. *Machine Learning*, 45:5–32, 2001.

[4] Antoni Buades, Triet M. Le, Jean Michel Morel, and Luminita A. Vese. Fast cartoon + texture image filters. *IEEE Transactions on Image Processing*, 19(8), 2010.

[5] S J Clarke. Losses associated with cane yard operations and cane washing. *Proceedings of The South African Sugar Technologists' Association*, pages 139–144, 1991.

[6] Lênio Soares Galvao, Antônio Roberto Formaggio, and Daniela Arnold Tisot. Discrimination of Sugarcane Varieties in Southeastern Brazil with E0-1 Hyperion Data. *Remote Sensing of Environment*, (94):523–534, 2005.

[7] Rafael C. Gonzalez, Richard E. Woods, and Steven L. Eddins. *Digital Image Processing Using MATLAB*. Gatesmark Publishing, 2 edition, 2009.

[8] J E Larrahondo. Calidad en la caña de azúcar: el cultivo de la caña en la zona azucarera de colombia, cali. Technical report, Centro de Investigación de la Caña de Azúcar de Colombia (CENICAÑA), 1995.

[9] Valentine Lebourgeois, Agnes Begue, Pascal Degenne, and Eric Bappel. Improving harvest and planting monitoring for smallholders with geospatial technology: the reunion island experience. *International Sugar Journal*, 109(1298):109–119, 2010.

[10] D Perez, C Fandos, L Mazzone, F Soria, P Scandalaris, and J Scandalaris. Caña de azúcar en tucumán y argentina: evolución de algunos aspectos económicos y productivos de la campaña 2004. Reporte agroindustrial. Estadísticas y márgenes de cultivos tucumanos. Vol. 2, Boletín Num. 6, Instituto Nacional de Tecnología Agropecuaria, Estación Experimental Agroindustrial Obispo Colombres, Tucumán (Argentina), 2005.

[11] Noé Aguilar Rivera, Guadalupe Galindo Mendoza, and Javier Fortanelli Martnez. Evaluacin agroindustrial del cultivo de caa de azcar (saccharum officinarum l.) mediante imgenes spot 5 hrv en la huasteca mxico. *Revista de la Facultad de Agronoma*, 111(2), 2012.

[12] J W Rouse, R H Haas, J A Schell, and D W Deering. Monitoring vegetation systems in the great plains with erts. In *Third ERTS Symposium*, volume 1, pages 309–317. NASA, Washington D.C. (USA), 1973.

[13] B C Russel, A Torralba, K P Murphy, and W T Freeman. Labelme: a database and webbased tool for image annotation. *International Journal of Computer Vision*, 77(1–3):157–173, 2008.

[14] Agustín Solano, Gerardo Scheneider, Alejandra Kemerer, and Alejandro Hadad. Characterization of multispectral aerial images of sugarcane. *Journal of Physics: Conference Series*, 477, 2013.

[15] Gabriel Sustaita. *Modelo estratégico para la industria azucarera regional*. MBA, Universidad Católica del Norte Santo Tomás de Aquino, Fundación del Tucumán, Pontificia Universidad Católica de Valparaíso, 2005.

[16] Prasad S. Thenkabail, Ronald B. Smith, and Eddy De Pauw. Hyperspectral vegetation indices and their relationships with agricultural crop characteristics. *Remote Sensing of Environment*, 71(2):158–182, 2000.

[17] J Tonatto, E R Romero, M F Leggio Neme, J Scandalaris, J Alonso, P Digonzelli, L Alonso, and S Casen. Importancia de la calidad de la materia prima en la productividad de la agroindustria azucarera. Gacetilla Agroindustrial de la EEAOC, num. 67, Intituto Nacional de Tecnología Agropecuaria, Estación Experimental Agroindustrial Obispo Colombres, Tucumán (Argentina), 2005.

Segmentation Fusion Techniques with Application to Plenoptic Images: A Survey.

D. Evin[1,2], A. Hadad[1], A. Solano[1], B. Drozdowicz[1]
[1] Information System Laboratory, Engineering School, National University of Entre Ríos, Argentina
[2] Sensory Research Laboratory, INIGEM, UBA-CONICET

E-mail: {devin, ajhadad, asolano, bdrozdo}@bioingenieria.edu.ar

Abstract. The segmentation of anatomical and pathological structures plays a key role in the characterization of clinically relevant evidence from digital images. Recently, plenoptic imaging has emerged as a new promise to enrich the diagnostic potential of conventional photography. Since the plenoptic images comprises a set of slightly different versions of the target scene, we propose to make use of those images to improve the segmentation quality in relation to the scenario of a single image segmentation. The problem of finding a segmentation solution from multiple images of a single scene, is called segmentation fusion. This paper reviews the issue of segmentation fusion in order to find solutions that can be applied to plenoptic images, particularly images from the ophthalmological domain.

Introduction

The recent emergence of light field cameras in the consumer market [1, 2] has attracted the interest for developing applications which take advantage of the plenoptic images. Some of these applications include: face recognition [3], robotic navigation [4], 3-D microscopy [5], and computer graphics [6].

Conventional photographic cameras capture the intensity of light as it strikes their image sensing elements, while colour filters provide a second set of data, sorting the rays into different wavelengths. In addition, plenoptic cameras capture a third piece of information: angle. This allows cameras to go beyond, from focusing on a single plane to taking images at many different depths of a scene at once [7].

Combining multiple angle versions of a single scene to obtain information about the world is not a new idea. In fact nature has been doing this for thousands of years. The visual system of the mantis shrimp allows to these marine crustaceans to see 12 different colour channels, from ultraviolet to infrared, to distinguish linear and circular polarization, and to perceive depth using trinocular vision with each eye [8]. Another example is given by compound eyes of flying insects, which are composed by numerous simple-single aperture eyes. Although they have a lower spatial resolution compared to mammalian single-lens eyes, given the high flicker frequency fusion rate of compound eyes, their temporal resolving power is considerably higher than for a single aperture eye. Thus compound eyes are ideal for motion detection [9].

These combination of optics and computing processing to produce images that cannot be obtained with traditional cameras led to the appearance of new field: computational photography [10].

This paper outlines the state of the art of segmentation fusion techniques as a basis for a project aimed at integrating plenoptic images into ophthalmological diagnostic instruments, specifically using slit lamps. Its main purpose is to review the techniques already suggested in the literature to cope with the problem of using ***n*** images of a single scene to improve image segmentation in the context of ophthalmology. The rest of the paper is organized in the following order: the next section reviews the most referred algorithms and methods applied to solve the segmentation problem with a single image in ophthalmology. The third section describes the particularities of plenoptic imaging. The fourth section shows the strategies found to deal with the problem of segmentation fusion. Finally, the work presents the conclusions and future work.

Image Segmentation in Ophthalmology

Ophthalmological imaging is a subfield of medical imaging which makes use of specific methods and equipment to detail the state of different components of the human eye like the cornea, crystalline, iris, retina and optical nerve. **Figure 1** illustrates key functional and anatomical areas in human eye along with the most common diagnostic studies associated with their characterizations.

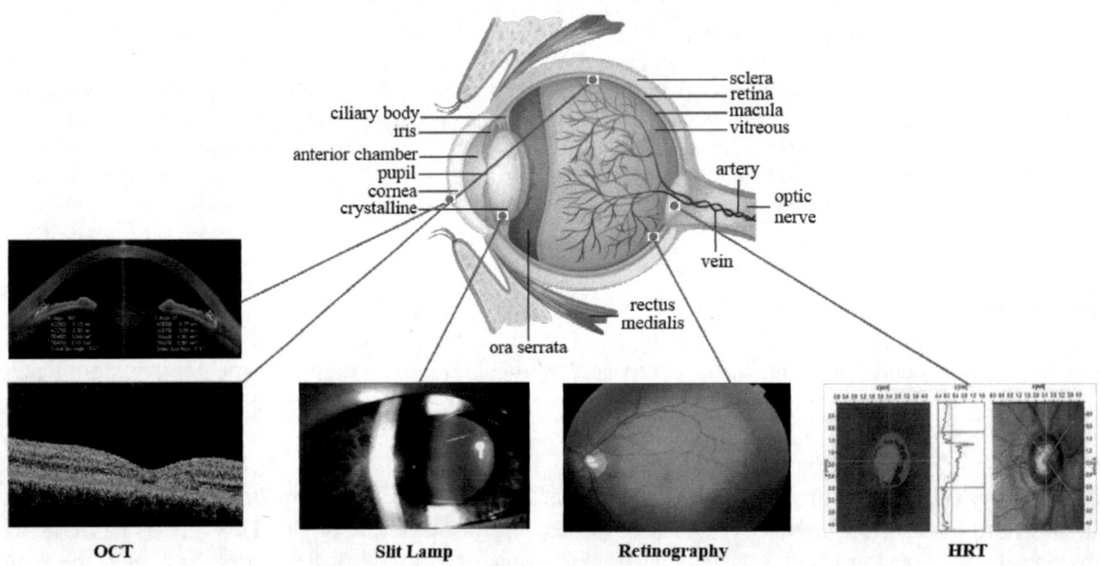

Figure 1. Human eye anatomy and main imaging methods

As can be seen in **Figure 1**, the human eyes have several structures and formations whose clinical characterization is fundamental to diagnose or to treat patients. The problem of segmenting those structures have its own body of research, and comprises the sub-tasks of pre-processing, feature extraction, and finally their localization and identification.

Hundreds of references related with the detection and segmentation of anatomic structures such as the optic disc, the fovea, and retinal blood vessels from digital colour fundus images can be found in the literature. But there are also other well studied cases related with the detection of abnormal elements such as exudates, microaneurysms, or haemorrhage.

In the following of this section, we present some work for each one of these tasks.

1.1. Optic Disc Segmentation

The optic disc or optic nerve head is the exit point for ganglion cell axons leaving the eye, which then form the optic nerve. **Figure 2** shows this landmark along with the macula, fovea and retinal blood tree. Authors of [11] propose a set of algorithms to recognise automatically main components of the fundus. The optic discs were located by identifying the area with the highest intensity variation of adjacent pixels, blood vessels were detected using a multilayer artificial neural network for which the inputs were derived from principal component analysis; and the fovea were identified using matching correlation and typical characteristics of this landmark such as the darkest area in the neighbourhood of the optic disc. They show that high sensibility and specificity can be achieved over a set of 112 cases.

In [12] authors use the active contour/snake model for the detection of optic disc. The high contrast of the optic disc is the key element in this approach, however this method may be sensitive to the initialization values for size and shapes. Later, these authors propose the detection of the optic disc using region growing and edge detection [13]. The results presented reveal high sensitivity and specificity for the delineation of fovea and the optic disc edges, but the method requires a precise selection of seed points for region growing.

In [14] a technique of unsupervised colour thresholding for detecting optic disc is used. Moreover, in this work authors try to find small yellowish structures, associated with the presence of exudates.

Several clustering strategies have also been proposed for the problem of optic disk detection. In [15] clustering techniques are applied for the initial localization, while the final definition of the optic disc is performed through circular Hugh transform and fuzzy approaches. The results in these tests are validated by ophthalmologists.

In [16] the optic disc detection is performed using k-nearest neighbour classifiers. This algorithm is a pixel level classification method, where intensity, edges and the outputs from Gaussian filters and Gabor wavelets are used as the set of attributes.

In [17] authors employs deformation techniques for the segmentation of the optic disc, which is subsequently used to distinguish between images from normal and glaucoma patients.

In [18] a technique for the segmentation of the optic disc based on histograms is presented. This technique includes the concept of fractal analysis, and authors found that their method is robust and computationally inexpensive.

In [19] a method for the automatic evaluation of the segmented optic disc is described.

The work of [20] integrates the local geometry of blood vessels with the image intensity for locating optic discs. A k-NN classifier is then applied to the optic disc segment. The results are relatively good for images with enough contrast but it is not so in case of low contrast

1.2. Fovea and Macula Segmentation

The macula is a pigmented oval area located in the central region of the retina, while the fovea is a small hole placed in the center of the macula, which contains the largest concentration of cone cells in the eye and is responsible for central, high-resolution vision. In [21] a method for macular segmentation using adaptive thresholds from optical coherence tomography is proposed. In this method various retinal layers are combined to detect the edges of the macula in images of normal and pathological retinas. The downside of this method is that it is not appropriate for low-quality images.

The work of [22] applies statistical methods for macular segmentation with the purpose of detecting macular structures with irregular sizes and shapes. These macular degenerations are typically associated with age.

1.3. Retinal Blood Vessels Segmentation

The work in [23] examines the relative importance of the red channel into the segmentation of blood vessels from retinal images. The segmentation of these vessels is obtained using red and green channels, and the results are compared against the output obtained using just the green channel. These results suggest that a differential treatment for colour channels is an approach worth considering for high precision segmentation.

The paper [24] proposes a vessel segmentation system based on the extraction of image ridges. These ridges, which coincide approximately with vessel center lines are used to compose primitives in the form of line elements. These elements are used to split the image into patches by assigning each pixel to the closest line element, and finally the feature vectors are computed making use of properties of the patches and the line elements. The feature vectors are classified using a k-NN classifier and sequential forward feature selection. Employing a set of 40 manually segmented images, authors find the performance of this system competitive with a couple of previously reported methods.

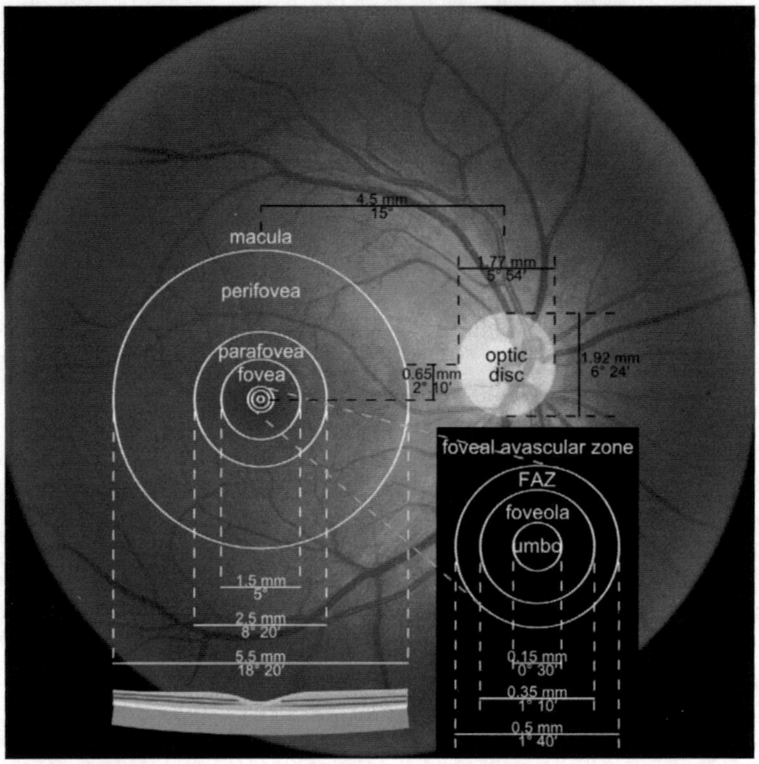

Figure 2. Macula, Optic Disc and Blood Tree in a Fundus Photography. Image by Photograph: Danny Hope from Brighton & Hove.

1.4. Microaneurysms, Exudates and Hemorrhage Segmentation

In [25] there is a proposal to detect retinal lesions from diabetic retinopathy patients. The pre-processing step eliminates background pixels and extracts the blood vessels and the optic disc from the original image. Then a filterbank extracts the set of candidate lesions. A feature set based on different descriptors, such as shape, intensity, and statistics, is formulated for each possible candidate region: this further helps in classifying that region. Finally an *m*-Mediods based modelling approach combined with a Gaussian Mixture Model is employed for the classification. It achieved a value of 0.981 as its area under the ROC curve, as against 0.977 for *m*-Mediods and 0.963 for GMM.

In [26] authors propose an exudate detection method which firstly perform normalization, denoising, and reflections/artifacts detection in the image; then selects between candidate segmentations based on mathematical morphology, and finally applies random forest to detect the exudates among the candidates. The method has been validated on three database, obtaining an AUC of 0.95.

In [27] the authors develop various fully automated systems for retinal haemorrhage measurement, but they ultimately found user interaction to be necessary to achieve satisfactory validity of segmentation.

In [28] authors compare the results of five different methods for microaneurysms detection, produced by five different teams of researchers on the same set of data. This set of data consisted of 50 training and 50 testing images, and beside the good results obtained for some types of aneurysms, the overall results shown that this is a challenging task for both the automatic methods as well as the human expert. The best performing system does not reach the performance of the human expert, indicating that there is room for improvement to solve this problem.

Plenoptic Imaging

Plenoptic imaging comprises a subset of computational photography approaches; specifically, those that aim at acquiring the dimensions of the plenoptic function with combined optical light modulation and computational reconstruction. Computational photography has grown tremendously in the last years in this interdisciplinary field, spanning optics, sensor technology and image processing. **Figure 3** shows a scheme of a plenoptic imaging system.

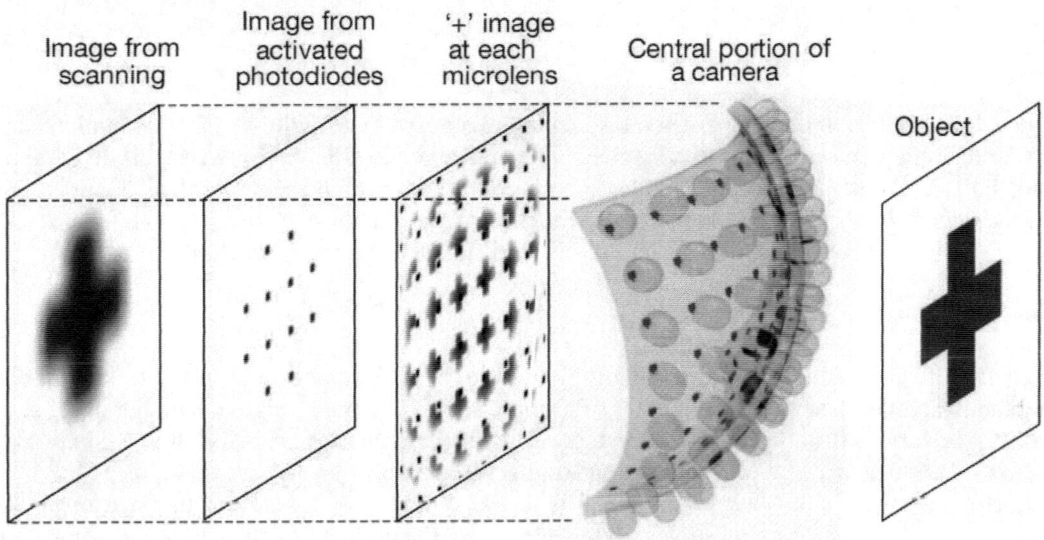

Figure 3. Conceptual View of Image formation in Light Field Cameras. Each microlens generates an image of the object ('+' pattern), with characteristics determined by the viewing angle. These images generates a proportional photo current at the corresponding photodiodes. The result is a sampled reproduction of the object. Image from [29]

The image formation involves two steps: measurement and processing. For the former, the camera has a sensor (CCD or CMOS) which provides with a directional sampling of the irradiance passing through each point in the sensor. Through the processing step, the captured light field can be used to recover a conventional image making focus on any depth of the original scene. The key rationale behind this processing comes from the projection-slice theorem, proposed by Bracewell in 1956 [30].

By the other hand, plenoptic photography also can be used to compose images from different perspectives. As under each microlens lays a set of pixels (as many as the microlens size or sensor resolution), they capture the rays coming from different directions of the same scene. Therefore, if the image is reconstructed employing the pixels from the same position under each microlens, it's possible to obtain the image in perspective, built upon the rays with the same angle from the scene. **Figure 4** depicts the same scene from two different perspectives.

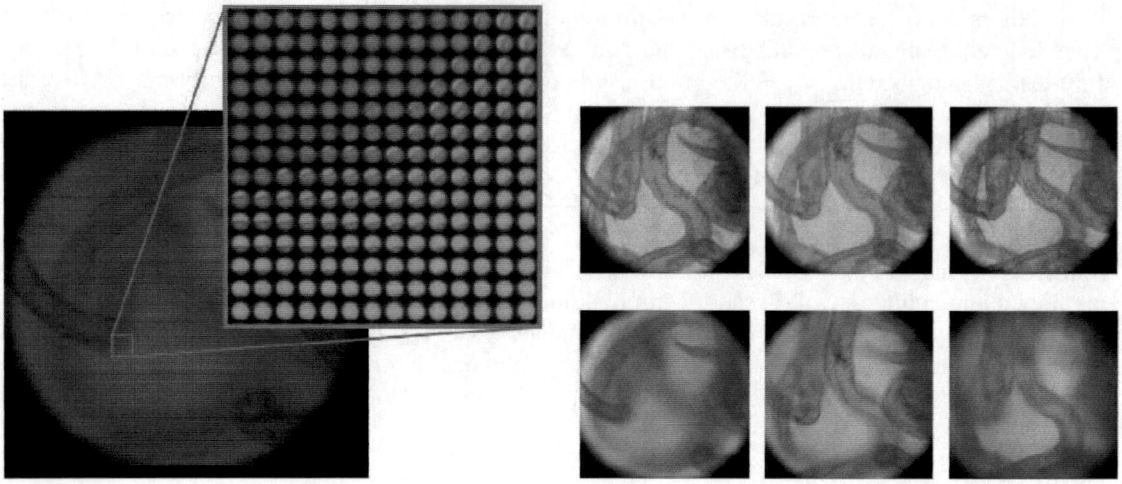

Figure 4. At left is a light field of several insect legs, captured in a single snapshot by a microscope into which a microlens array has been inserted. The objective was 25x/0.45NA. Magnifying a portion of this light field (inset), we see the circular subimages formed behind each microlens. Each is a view of the microscope's aperture. At right is a sequence of perspective views (top) and a focal stack (bottom), computed from this light field. (Extracted from [31])

Segmentation Fusion

Segmentation fusion is the common terminology used in the literature to refer to the task of finding a segmentation solution from multiple images of a single scene, or from a single image but using several algorithms, or various instances of the same algorithm. These scenarios are referred as *"n images, one algorithm"* and *"one image, n algorithms"* paradigms respectively.

Several strategies have been proposed to deal with the synthesis of a single solution from multiple segmentations. The most common strategy applied to solve the problem of segmentation fusion consist in obtaining *n* different segmentation hypotheses and applying a consensus strategy to gather and unify these *n* results into a single solution.

In [32] the authors propose an approach to the *"n images, one algorithm"* paradigm based on binary partition trees. This is an interesting work because instead of applying the segmentation fusion as a post-processing step, it attempts to handle the segmentation fusion applying the consensus decision during the construction of a hierarchical model representation which unifies the *n* images. The authors claim that solving the fusion decision during an early phase results in two benefits. First, it allows to define a hierarchical framework for the segmentation problem which is flexible and can be easily instantiated according to the field of application. Second, by operating the fusion on the internal data structures involved in the algorithmic construction of the hierarchy, instead of spatial regions of the segmentation maps, existing machine learning solutions for "non-spatial" data fusion can be used [33]. They use Binary Partition Trees (BPTs) [34], a technique widely applied in the field of remote sensing, for the hierarchical model. Authors illustrate and explain the benefits of their approach on two cases of remote sensing images: mono-date multi-imaging on urban areas, and multi-date imaging on agriculture areas. With the study cases they conclude that the experiments carried out on satellite multi-image datasets shown that the quality of the induced morphological hierarchies are sufficient to further perform improved segmentation, however, this preliminary work doesn't provide any quantitative comparisons with other segmentation schemes or with a gold standard.

In this sense, to count on apriori information as reference, several works [35-37] apply atlas-based segmentation methods. These methods have the property of segmenting the image with no well-defined relation between regions and pixels intensities. If the information about difference between these regions is incorporated in the spatial relationship between them, other regions, or within their morphometric characteristics, the atlas-based segmentation is expected to work well. In [35] the authors propose two approaches for combining multi-atlas segmentation and intensity modelling based on segmentation using expectation maximisation (EM) and optimisation via graph cuts. In [36] the authors present a framework to address the consequent problems of scale in multi-atlas segmentation. In this work a custom subset of atlases was selected for each query subject and provided more accurate subcortical segmentations than those given by non-selective combination of random atlas subsets. This approach was tested using a database of 275 atlases. An image based similarity criterion as well as a demographic criterion (age) in a leave-one-out cross-validation study were used. This work concludes that selecting atlases from large databases for atlas-based brain image segmentation improves the accuracy in the final segmentations. Additionally this work shows that the image similarity is a suitable selection criterion. Authors also give results based on selecting atlases by age which demonstrate the value of meta-information for such selection. Jia et al [37] present a multi-atlas-based framework for the segmentation of *n* target images. In this approach two strategies were described: a tree-based groupwise registration method for concurrent alignment of both the atlases and the target images, and an iterative groupwise segmentation method

Another typical approach for the segmentation task is the Markov Random Field (MRF) [38-40]. A MRF is similar to a Bayesian network in its representation of dependencies; the differences are that Bayesian networks are directed and acyclic, whereas Markov networks are undirected and may be cyclic. In [38] a multi-channel image segmentation method that uses a MRF region label model with adaptive neighbourhoods is described. Bayesian inference is applied to realize the combination of evidence from different knowledge sources. In such a way, the optimization of the shape of a neighbourhood set is achieved by following a criterion that makes use of hypothesis on the Markovian property by exploiting the local image content. The authors illustrates the results of this strategy through optical remote sensing data. Another segmentation algorithm based on MRF processing was proposed in [39]. The images are segmented initially by growing regions of similar colour values. Then, for refining initial clusters in a feature space they use a MRF processing. That processing works with the KNN classification rules among the neighbours of a pixel. It has been observed that, though using only chromatic information, good segmentation results are obtained. The luminance information improves the quality of segmentation in some cases. Results for different colour spaces such as the OHTA coordinate space, YIQ, CIELAB and UVW are also presented. On the average the performance in the OHTA space was better than the others. In [40] a simple MRF model for unsupervised image segmentation based on image features which has a new processing pipeline is proposed. The traditional two-component MRF model for that segmentation requires training data to estimate the necessary parameters for the model, and thus is unsuitable for unsupervised segmentation. The proposed pipeline solves this problem by introducing a function-based weighting parameter between the two components. Using this method, the simple MRF model is able to automatically estimate those model parameters, and to produce accurate unsupervised segmentation results. Experiments demonstrate that the proposed algorithm is able to segment various types of images (gray scale, color, texture) and it achieves an improvement over the traditional methods. In [41] authors propose a maximum-likelihood clustering for this task. The clustering registers all the images in a multisensor ensemble simultaneously. Experiments involving rigid-body and affine transformations show that the clustering method is more robust and accurate than competing pairwise registration methods. Moreover, the clustering results can be used to form a rudimentary segmentation of the image ensemble.

Another clustering approach was reported in [42]. The segmentation fusion procedure combine several segmentation maps associated to much simpler partition models in order to finally get a more reliable and accurate segmentation result. The label fields to be fused are obtained by K-means clustering on an input image, expressed in different colour spaces. The fusion strategy aims at combining these segmentation maps with a final clustering procedure using as input features the local histogram of the class labels, previously estimated and associated to each site and for all the initial partitions. This fusion framework has been successfully applied on the Berkeley image database. By the other hand [43] reports

its results using the same database, but the authors address the parameter selection problem by applying a general ensemble clustering methods in order to produce a consensus segmentation. The main contribution consists of applying and comparing a broad variety of representative and widely used ensemble clustering methods for the segmentation combination problem.
The first step consist in obtaining *m* segmentations using 3 state-of-art segmentation models to generate 3 ensembles: TBES ensemble (Texture and Boundary Encoding-based Segmentation), UCM ensembles (Ultrametric Contour Map) and TBES & UCM ensembles. Each ensemble is composed by 10 segmentations obtained by varying the parameter values of the segmentation algorithms used to generate the ensemble. The ensemble clustering methods used were: BOK (Best of K), BOEM (Best One Element Moves), EAC SL/AL (Evidence Accumulation), RW (Random Walker), Hypergraph based methods, Information theory based methods, Kernel based methods and Clustering based on semidefinite programming. The results obtained were compared with the human segmentations of each image.
The results suggest that good segmentation results may be obtained by using general ensemble clustering methods without knowing ground truth. In this context it must be emphasized that in many application scenarios, supervised learning is not applicable because the ground truth information is not available. Thus, ensemble clustering methods are preferred in scenarios where parameters of segmentation algorithms are unknown.
Finally, in [44] authors propose a method which exploits the semantic information of images. They propose a novel method for weakly supervised semantic segmentation. The training images are labelled only by the classes they contain, not by their location in the image. On test images instead, the method predicts a class label for every pixel. The main innovation of this proposal is a multi-image model (MIM) a graphical model for recovering the pixel labels of the training images. The model connects superpixels from all training images in a data-driven approach, based on their appearance similarity. For generalizing to new test images they integrate them into MIM using a learned multiple kernel metric, instead of learning conventional classifiers on the recovered pixel labels.

Conclusions and Future Work

In this work we inquire about the feasibility of applying segmentation fusion techniques to plenoptic ophthalmological images. In the first part we reviewed the problem of image segmentation in ophthalmology. We outline some of the strategies proposed to segment anatomic structures such as the optic disc, the fovea, and retinal blood vessels, and abnormal elements such as exudates, microaneurysms, or haemorrhage from digital colour fundus images. Then we detailed the characteristics of plenoptic images, and finally we described the problem of segmentation fusion along with some proposals and applications of this problem to different environments.

The segmentation fusion approaches found in the literature can be broadly classified into atlas-based methods, statistical characterizations approaches, and machine learning based techniques. For the purpose applying segmentation fusion into plenoptic images, we consider that both statistical and machine learning techniques would be appropriate, but only with unsupervised methods due to the lack of a gold standard for ophthalmic plenoptic image. We also identify the atlas-based approach as the more difficult to adapt to our target scenario because with the available data in this field there would be difficult to build reliable ophthalmic atlases. However, many heuristics from that approach would be exploited, for example the a priori information related with the shape, localization and geometric relations of the fovea, macula and blood tree.

References

[1] Lytro. The lytro camera. https://www.lytro.com
[2] Raytrix. 3D light field cameras. http://raytrix.de/
[3] Raghavendra R, Yang B, Raja KB and Busch C 2013 A new perspective-Face recognition with light-field camera *In International Conference on Biometrics (ICB), IEEE*, pp. 1-8

[4] Dong F, Ieng SH, Savatier X, Etienne-Cummings R and Benosman R 2013 Plenoptic cameras in real-time robotics. *The International Journal of Robotics Research,* **32**(2), pp. 206-217

[5] Levoy M, Ng R, Adams A, Footer M and Horowitz M 2006 Light field microscopy. *In ACM Transactions on Graphics (TOG)*, ACM **25**, pp. 924-934.

[6] Pearson J, Brookes M and Dragotti, P L 2013 Plenoptic layer-based modelling for image based rendering. *IEEE Transactions on Image Processing*, **22**(9), pp. 3405-3419.

[7] Savage N 2011 A new angle on imaging. *Spectrum, IEEE*, **48**(12), p. 15.

[8] Wetzstein G, Ihrke I, Lanman D and Heidrich W 2011 Computational plenoptic imaging. *In Computer Graphics Forum,* Blackwell Publishing. **30** (8), pp. 2397-2426.

[9] Sarkar M 2011 A biologically inspired CMOS image sensor. *Master's dissertation TU Delft, Delft University of Technology*.

[10] Zhou C and Nayar S K 2011 Computational Cameras: Convergence of Optics and Processing. *IEEE Transactions on Image Processing*, **20**(12), pp. 3322-3340.

[11] Sinthanayothin C, Boyce J F, Cook H L and Williamson T H 1999 Automated localisation of the optic disc, fovea, and retinal blood vessels from digital colour fundus images. *British Journal of Ophthalmology*, **83**(8), pp. 902-910.

[12] Agurto C, Murray V, Barriga E, Murillo S, Pattichis M, Davis H, Russel S, Abramoff M and Soliz P 2010 Multiscale AM-FM methods for diabetic retinopathy lesion detection', *IEEE Transactions on Medical Imaging*, **29**(2), pp. 502- 512.

[13] Kose C, Sevik U, Gencalioglu O, Ikibas C and Kayikicioglu T 2008 A statistical segmentation method for measuring age related macular degeneration in retinal fundus images, *Journal on Medical Systems*, **3**, pp. 1-13.

[14] Chrastek R, Wolf M, Donath K, Niemann H, Paulus D, Hothorn T, Lausen B, Lammer R, Mardin CY and Michelson G 2005 Automated Segmentation of the Optic Nerve Head for Diagnosis of Glaucoma, *Medical Image Analysis* **9**, pp. 297-314.

[15] Chutatape O 1988 Retinal Blood Vessel Detection and Tracking by Matched Gaussian and Kalman Filters, *Proceedings of the 20th Annual International Conference of the IEEE Engineering in Medicine and Biology Society*, **20**(6), pp. 3144- 3149.

[16] Heneghan C, Flynn J, Keefe MO and Cahill M 2002 Characterization of changes in blood vessel width and tortuosity in retinopathy of prematuriy using image analysis, *Medical Image Analysis*, **6**, pp. 407-429.

[17] Welfer D, Schacanski J, Cleyson MK, Melissa MDP, Laura WBL and Ruschel Marinho D 2010 Segmentation of the optic disc in color eye fundus images using an adaptive morphological approach, *Journal on Computers in Biology and Medicine*, **40**, pp. 124-137.

[18] Chakraborty D and Nikhil RP 2004 A neuro fuzzy scheme for simultaneous feature selection and fuzzy rule based classification *IEEE Transactions on Neural Networks*, **15**(1), pp. 110-123.

[19] Paulus D, Chastel S and Feldmann T 2005 Vessel segmentation in retinal images, *Proceedings of SPIE*, pp. 696-705.

[20] Massey E, Lowell J, Hunter A and Steele D 2009 Lesion boundary segmentation using level set methods, *In Advances in Computer Graphics and Computer Vision*. **1**, pp.1-8.

[21] Anzalone A, Bizzari F, Parodi M and Storace M 2008 A modular supervised algorithm for vessel segmentation in red-free retinal images, *Computers in Biology and Medicine*, **38**, pp. 913-922.

[22] Balanco M, Penedo MG, Barreira N, Penas M and Carreira MJ 2006 Localisation and extraction of the optic disc using the fuzzy circular Hough transform, *Springer Lecture Notes in Computer Science*, pp. 712-721.

[23] Reza AW, Eswaran C and Dimyati K 2010 Diagonosis of diabetic retinopathy: Automatic extraction of optic disc and exudates from retinal images using marker-controlled water shed transformation', *Journal of Medical Systems*. **35**(6), pp. 1491-1501

[24] Staal J, Abramoff M D, Niemeijer M, Viergever M and Van Ginneken B 2004 Ridge-based vessel segmentation in color images of the retina. IEEE *Transactions on Medical Imaging*, **23**(4), pp. 501-509.

[25] Akram MU, Khalid S, Tariq A, Khan SA and Azam F 2014 Detection and classification of retinal lesions for grading of diabetic retinopathy. *Computers in biology and medicine*, **45**, pp. 161-171.

[26] Zhang X, Thibault G, Decenciere E, Marcotegui B, Laÿ B, Danno R and Erginay A 2014 Exudate detection in color retinal images for mass screening of diabetic retinopathy. *Medical image analysis*, **18**(7), 1026-1043.

[27] Aslam T, Chua P, Richardson M, Patel P and Musadiq M 2009 A system for computerised retinal haemorrhage analysis. *BMC Research Notes*, **2**(1), p. 196.

[28] Niemeijer M, Van Ginneken B, Cree MJ, Mizutani A, Quellec G, Sánchez C and Abramoff M D. 2010 Retinopathy online challenge: automatic detection of microaneurysms in digital color fundus photographs. *IEEE Transactions on Medical Imaging*, **29**(1), pp. 185-195.

[29] Song YM, Xie Y, Malyarchuk V, Xiao J, Jung I, Choi KJ and Rogers JA 2013 Digital cameras with designs inspired by the arthropod eye. *Nature*, **497**(7447), pp. 95-99.

[30] Bracewell RN 1956 Strip Integration in Radio Astronomy. *Aust. J. Phys*. **9**(2), p. 198.

[31] Levoy M, Ng R, Adams A, Footer M and Horowitz M 2006 Light field microscopy. *ACM Transactions on Graphics*, **25**(3), pp. 924-93.

[32] Randrianasoa JF, Kurtz C, Desjardin E and Passat N 2015 Multi-image Segmentation: A Collaborative Approach Based on Binary Partition Trees. *In Mathematical Morphology and Its Applications to Signal and Image Processing*. Springer pp. 253-264.

[33] Topchy A, Jain AK and Punch W 2005 Clustering ensembles: Models of consensus and weak partitions. *IEEE Transactions on Pattern Analysis and Machine Intelligence*, **27**(12), pp. 1866-1881.

[34] Salembier P and Garrido L 2000 Binary partition tree as an efficient representation for image processing, segmentation, and information retrieval. *IEEE Transactions on Image Processing*, **9**(4), pp. 561-576.

[35] Lötjönen JM, Wolz R, Koikkalainen JR, Thurfjell L, Waldemar G and Soininen H 2010 Fast and robust multi-atlas segmentation of brain MRI. *Neuroimage*, **49**(3), pp. 2352-2365.

[36] Aljabar P, Heckemann RA, Hammers A, Hajnal JV and Rueckert D 2009 Multi-atlas based segmentation of brain images: atlas selection and its effect on accuracy. *Neuroimage*, **46**(3), pp. 726-738.

[37] Jia H, Yap PT and Shen D 2012 Iterative multi-atlas-based multi-image segmentation with tree-based registration. *Neuroimage*, **59**(1), pp. 422-430.

[38] Smits PC and Dellepiane SG 1997 An irregular MRF region label model for multi-channel image segmentation. *Pattern Recognition Letters*, **18**(11), pp. 1133-1142.

[39] Mukherjee J 2002 MRF clustering for segmentation of color images. *Pattern Recognition Letters*, **23**(8), pp. 917-929.

[40] Deng H and Clausi DA 2004 Unsupervised image segmentation using a simple MRF model with a new implementation scheme. *Pattern Recognition*, **37**(12), pp. 2323-2335.

[41] Orchard J and Mann R 2010 Registering a multisensor ensemble of images. *IEEE Transactions on Image Processing*, **19**(5), pp. 1236-1247.

[42] Mignotte M 2008 Segmentation by fusion of histogram-based-means clusters in different color spaces. *IEEE Transactions on Image Processing*, **17**(5), pp. 780-787.

[43] Franek L, Abdala DD, Vega-Pons S and Jiang X 2011 Image segmentation fusion using general ensemble clustering methods. *In International Conference on Computer Vision–ICCV 2010*. Springer, pp. 373-384.

[44] Vezhnevets A, Ferrari V and Buhmann JM 2011 Weakly supervised semantic segmentation with a multi-image model. *In International Conference on Computer Vision–ICCV 2011. Springer,* pp. 643-650.

Blood glucose regulation in diabetics. A flatness based nonlinear control simulation study.

Guillermo Cocha[1], Melina Podestá[2] Alejandro Mazzadi[3], Carlos Amorena[3] and Carlos D'Attellis[4]

[1] Departamento de Ingeniería Eléctrica , Universidad Tecnológica Nacional Facultad, Regional La Plata.

[2] Instituto de Ingeniería y Agronomía, Universidad Nacional Arturo Jauretche, Av. Calchaquí 6300, Florencio Varela,

[3] CESIMA, UNSAM, Buenos Aires, Argentina

[4] Grupo de Ing. Clínica U.T.N.; Dto. de Matemática, Univ. Favaloro; Centro de Matemática Aplicada, UNSAM, Buenos Aires, Argentina

E-mail: gcocha@frlp.utn.edu.ar

Abstract. Flat systems are a generalization of linear systems, but the techniques used for controlling flat systems are much different than many of the existing techniques for linear systems. In this paper we present the flatness-based control of blood glucose regulation in human system. A non-linear model, the Bergman Minimal Model, is used to represent the dynamics of blood regulation in humans, and because of the flatness property, the system variables can be expressed as functions of the flat output and their time derivatives and a control law is developed.

1. Introduction

In healthy people, blood glucose level is regulated by segregation of insulin in pancreas beta cells. Insulin is a hormone that activates glucose uptake in insulin sensitive cells. In Diabetes Mellitus Type I or advanced stages of Type II, the pancreas cannot segregate enough insulin to maintain blood glucose in normal level and this situation leads to hyperglycaemia, a condition given when blood glucose concentration overcomes 135 mg/dl. Diabetes Mellitus (DM) has been defined by World Health Organization (WHO) [1] as a chronic disease, which occurs when the pancreas does not produce enough insulin, or when the body cannot effectively use the insulin it produces. This leads to an increased concentration of glucose in the blood (hyperglycaemia).

Type 1 diabetes (T1DM), previously known as insulin-dependent or childhood-onset diabetes, is characterized by a lack of insulin production. Type 2 diabetes, formerly called non-insulin-dependent or adult-onset diabetes, is caused by the body's ineffective use of insulin. It often results from excess body weight and physical inactivity. Gestational diabetes is hyperglycaemia that is first recognized during pregnancy.

Traditional treatment consists in the administration of several doses of insulin day by day, in order to keep a normal blood glucose level. This therapeutic treatment, cannot avoid undetected peaks of blood

glucose. Also, the risk of hypoglycaemia, a situation where the patient can faint vanish or even go into a coma, is always present. Neither can this treatment prevent the progressive damage to the artery walls due to the effects of undetected peaks of blood glucose leading to ulterior complications such as coronary diseases, retinopathy, etc. The results of the Diabetes Control and Complications Trial (DCCT) [2] solidified the importance of continuous infusion of exogenous insulin based on mechanical pumps to the prevention of micro vascular complications of diabetes as the most effective tool to prevent complications in insulin-dependent Diabetes Mellitus [3].
Drawbacks of this approach are the skills required to operate the pump effectively and the rigid lifestyle that diabetic patients have to follow. These problems have shown the need of new developments, transparent to the user. A mechanical insulin pump, a glucose sensor and an algorithm of control conform a useful tool to sort the mentioned problems. In this work a model based control algorithm that can be integrated into an intelligent system that no requires patient doesn´t attention is presented.

2. Flatness of dynamical systems

A dynamical system is said to be flat if it is possible to find a set of variables, called the flat outputs, such that the system is (non-differentially) algebraic over the differential field generated by the set of flat outputs. Flatness was defined by Fliess et al. [4] and the word *flat* is used due to the fact that the linearizing outputs play an analogous roll as the plane coordinates in the context of the Frobenius theorem in differential geometry [5]. If the system has states $x \in \mathbb{R}^n$ and inputs $u \in \mathbb{R}^m$, then the system is flat if it is possible to find outputs $y \in \mathbb{R}^m$, $r = (r_1, \dots, r_m)^T \in \mathbb{N}^m$, of the form $y = h(x, u, \dot{u}, \dots, u^{(r)})$ such that

$$x = \varphi\big(y, \dot{y}, \dots, y^{(r)}\big) \tag{1}$$
$$u = \alpha\big(y, \dot{y}, \dots, y^{(r+1)}\big) \tag{2}$$

Flat systems are a generalization of linear systems in the sense that all linear, controllable systems are flat, but the techniques used for controlling flat systems are much different than many of the existing techniques for linear systems. Flatness is particularly well tuned for allowing one to solve the inverse dynamics problems and one builds off of that fundamental solution in using the structure of flatness to solve more general control problems. Consider a nonlinear system

$$\dot{x} = f(x, u) \tag{3}$$

where $x \in R^n$ is a vector of state variables, and $u \in R^m$ the input, $f(0,0) = 0$, and the rank of $\partial f / \partial u$ is equal to m. The system (1) is differentially (differentially independent, means that they are not related by any differential equation) flat if there exist a number m of real, analytic, scalar functions that depend on x, u and a finite number r of derivatives of u,

$$y_i = h\big(x, u, \dots u^{(r)}\big) \quad i = 1, \dots, m \tag{4}$$

so that the inverse of system (1), taking u as input and $y = (y_1, \dots, y_m)$ as output, does not have dynamics. This is, the state variables x and the input variables u can be expressed as A (n-dimensional) and B (m-dimensional) real and analytic functions of y

$$x = A\big(y, \dot{y}, \ddot{y}, \dots, y^{(r)}\big) \tag{5}$$
$$u = B\big(y, \dot{y}, \ddot{y}, \dots, y^{(r+1)}\big) \tag{6}$$

where the output *y* is called the flat or linearized output and *r* is an integer. The flat outputs have the following properties:

a. The components of *y* can be calculated from the state *x*, the inputs *u*, and a finite number of time derivatives of the later.
b. The components of *y* are differentially independent, i.e. they are not related by any differential equation.
c. The state variables *x*, the input variables *u*, and the time derivatives and functions of them can be calculated from the flat outputs *y* and a finite number of their time derivatives.

In [6] the concept of an affine flat input to a nonlinear system with a given output function is introduced. This approach can be seen as dual to the search for a flat output of a control system with a given input.
Definition.
1. The control system is said to be flat if there exists an output y = λ(x) such that the resulting SISO system

$$\begin{cases} \dot{x} = f(x) + g(x)u \\ y = \lambda(x) \end{cases} \quad (7)$$

has relative degree *n*. In that case, *y* is called a flat output of (3) defined by the output function *λ(x)*.
2. The observed system (3) is said to be flat if there exists an input vector field γ (x) such that the resulting SISO system

$$\begin{cases} \dot{x} = f(x) + \gamma(x)u \\ y = h(x) \end{cases} \quad (8)$$

has relative degree n. In that case u is called a flat input for (3) with input vector field $\gamma(x)$. Dynamics of the glucose/insulin system is represented by Bergman´s Minimal Model (BMM), a mathematical model [7] used to determinate the sensitivity of insulin action and metabolism of glucose in Intravenous Glucose Tolerance Test (IVGTT). The Bergman Minimal Model is given by the following equations

$$\dot{x_1} = -P_1x_1 - x_2(x_1 + G_b) + D(t), \quad (9)$$
$$\dot{x_2} = -P_2x_2 + P_3x_3, \quad (10)$$
$$\dot{x_3} = -n(x_3 + I_b) + \frac{u(t)}{V} \quad (11)$$

Details of the model are:
x_1 = blood glucose concentration over basal value.
x_2 = remote compartment.
x_3 = blood insulin concentration over basal value.
G_b = basal blood glucose concentration.
I_b = basal blood insulin concentration.
u(t) = infusion of external insulin.
n = transference rate.
t = time.
V = plasmatic volume.
P_1=1e^{-2} min^{-1}, P_2=3.33e^{-2} mU.L^{-1}min^{-2}, and P_3 =1.$^{33e-5}$min^{-1} are patient´s dependent parameters [13].
D(t) = Perturbation due to meals.

Journal of Physics: Conference Series **705** (2016) 012027 doi:10.1088/1742-6596/705/1/012027

. A flat output for the system is $y = x_1$. Substituting the flat output in (9), if *D(t)* is considered a unknown perturbation, results

$$\dot{y} = -P_1 y - x_2(y + G_b), \tag{12}$$

now, it is possible to obtain $x_2 = \frac{-P_1 y - \dot{y}}{y + G_b}$ and $\dot{x_2} = -P_2 \frac{-P_1 y - \dot{y}}{y + G_b} + P_3 x_3$. In the same way

$$x_3 = \frac{1}{P_3} \left\{ \frac{(P_1 \dot{y} - \ddot{y})(y + G_b) - (P_1 y - \dot{y})\dot{y}}{(y + G_b)^2} + P_2 \frac{-P_1 y - \dot{y}}{y + G_b} \right\}, \tag{13}$$

$$\dot{x}_3 = \frac{-n}{P_3} \frac{(P_1 \dot{y} - \ddot{y})}{(y + G_b)} + \frac{-n}{P_3} \frac{(P_1 y - \dot{y})\dot{y}}{(y + G_b)^2} - \frac{n P_2 (P_1 y - \dot{y})}{y + G_b} - n I_b + \frac{1}{V} u(t) \tag{14}$$

Isolating the control $u(t)$ from (14), the endogenous nonlinear control u_e (15) is obtained. A supervisor control u_s (16) in closed loop is added to u_e in order to improve robustness.

$$u_e = \frac{1}{V} \left(f(y, \dot{y}, \ddot{y}) \right) \tag{15}$$

$$u_s = \frac{abs(5 * G - G_b)}{5 * G - G_b} \tag{16}$$

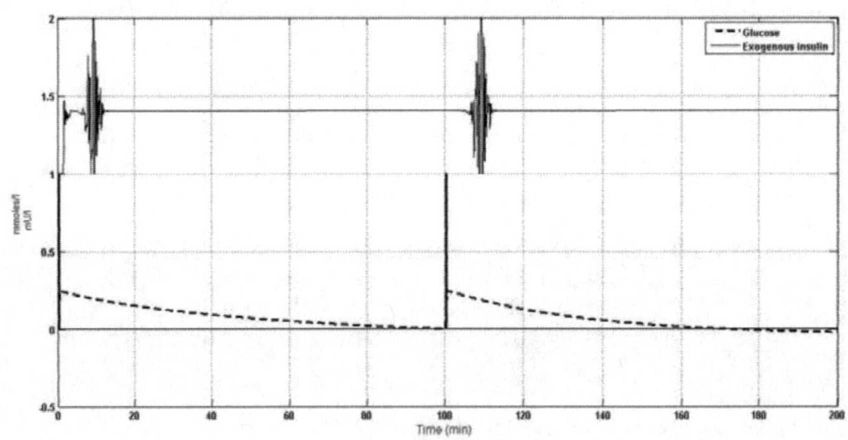

Figure 1: Blood glucose, infused insulin and perturbations at 0 and 100 minutes.

3. Results

Simulations were performed over a Matlab Simulink model. In Figure 1, a disturbance $D = 1.5 * exp(-0.05 * t)$, is added to (9) in t=0 min. and t= 100 min. to simulate a postprandial period of 200 minutes. In figure 2, similar experiences were performed over a period of 1400 minutes, approximately

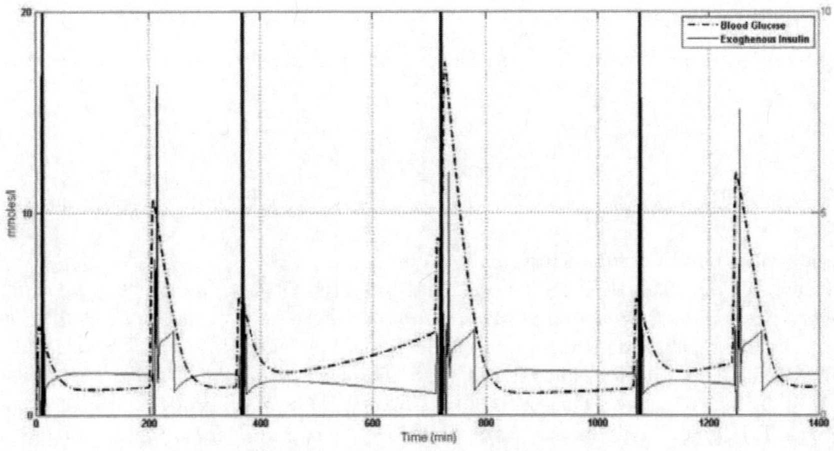

Figure 2: Blood glucose and infused insulin.

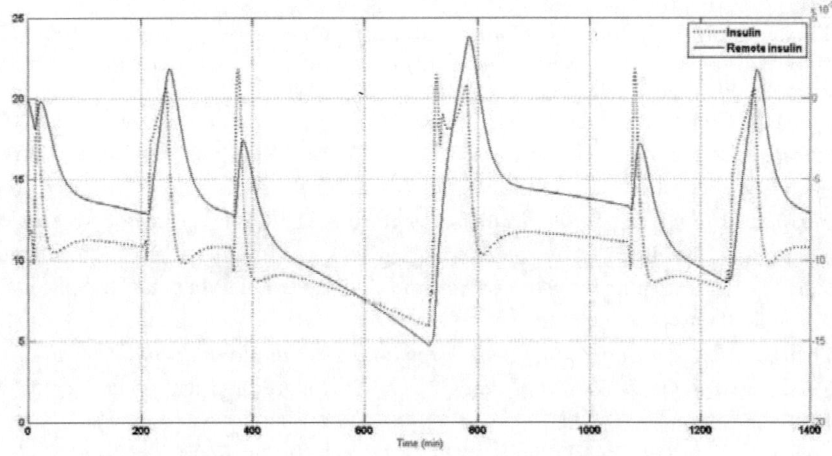

Figure 3: Blood insulin and instertitial insulin.

one day with several ingestions of carbohydrates. Two important results have been obtained:

a. Hypoglycemia, a mayor risk in glucose control, is not present. Figure 2 shows the carbohydrates ingestion, blood glucose along a day. Figure 3 show blood insulin and interstitial insulin concentration respectively. It is important to note that most of the time, interstitial insulin is below zero. This is a important indicator of lack of hypoglycemia. In others nonlinear control, for instance, based on feedback linearization (see [6],[7],[8],[9]) blood glucose is regulated but the risk of over insulin in blood is always present.

b. State observers are not need. In space state control techniques, it is assumed that all states are available. But blood glucose is the only measurable output and it is necessary to develop state observers to estimate the value of the rest of the state variables [10]. For this purpose, nonlinear observers must be developed. This can be avoided by the use of flatness control and taken blood glucose as the flat output.

As a future work, the same strategy must be adaptive and include a parameters estimation algorithm.

4. References

[1] http://www.who.int/topics/diabetes_mellitus/en/

[2] The Diabetes Control and Complications Trial Research Group, *The effect of intensive treatment of diabetes on the development and progression of long term complications in insulin-dependent diabetes mellitus*, New England Journal of Medicine, vol. 329,n 14, pp 977-986, 1993.

[3] D. M. Maahs, L. A. Horton, and H. P. Chase, *The Use of Insulin Pumps in Youth with Type 1 Diabetes*, Diabetes Technol. Ther. 2010 Jun; 12(Suppl 1), S-59–S-65.

[4] M. Fliess, J. Levine, P. Rouchon, *Flatness and Defect of Nonlinear Systems. Introductory theory and examples.* Journal of Control,Vol.61,No.6, (1995).

[5] M. Fliess, J. Levine, P. Martin and P. Rouchon, *On a Sur les Systémes Non Lineaires differcatiellement plats.* C,R.Acad Sel.París,Vol.315, (1992).

[6] S. Waldherr, M. Zeitz, *Conditions for the existence of a flat input.* Vol 81, Issue 3, (2008), pp 439-443.

[7] Bergman et. al, *Physiologic Evaluation of Factors Controlling Glucose Tolerance in Man: Measurement of Insulin Sensitivity and beta-cell Glucose Sensitivity from the Response to Intravenous*, J. Clinical Investigation.(1981), pp. 1456-1467 .

[8] G. Cocha, V. Constanza, C. D'attellis, "*Control No Lineal de la Diabetes Mellitus*". XIII Reunión de Trabajo en Procesamiento de la Información y Control, RPIC, (2009), Rosario, Argentina.

[9] G. Cocha, M. Podestá, C. D'attellis, *Regulación no lineal de la glucosa en pacientes diabéticos utilizando control no lineal globalizante*" .Congreso AADECA (2012), Buenos Aires, Argentina.

[10] G. Cocha, C. D'attellis, "*Control adaptativo de sistemas que admiten linealización exacta. Aplicación al sistema glucoregulatorio humano.*" Congreso de Matemática Aplicada ASAMACI 2011, Bahía Blanca, Argentina.

[11] G. Cocha, V. Constanza, C. D'attellis "*Sistema automático de infusión de insulina en pacientes diabéticos. Estrategia de control no lineal.*", XXII Congreso Argentino de Control Automático, (2010) Buenos Aires, Argentina.

[12] G. Cocha, V. Constanza, C. D'attellis "*Observadores No Lineales en el Control de la Diabetes Mellitus*", 1° Congreso de Bioingeniería, 5 y 6 de Noviembre de 2009, San José de Costa Rica, Costa Rica.

[13] R. Gillis, C. Palermo, H. Zisser, L Jovanovič, D. Seborg, and F. Doyle, III, *Glucose Estimation and Prediction through Meal Responses Using Ambulatory Subject Data for Advisory Mode Model Predictive Control*, Journal of Diabetes Science and Technology Volume 1, Issue 6, November 2007.

Journal of Physics: Conference Series **705** (2016) 012028 doi:10.1088/1742-6596/705/1/012028

Time Domain Estimation of Arterial Parameters using the Windkessel Model and the Monte Carlo Method

Vladimir Gostuski[1], Ignacio Pastore[1], Gaspar Rodriguez Palacios[1], Gustavo Vaca Diez[1], H. Marcela Moscoso-Vasquez[1,2] and Marcelo Risk[1,2]

[1] Department of Bioengineering, Instituto Tecnologico de Buenos Aires (ITBA), Av. Eduardo Madero 399, C1106ACD, Buenos Aires, Argentina.

[2] Consejo Nacional de Investigaciones Científicas y Técnicas (CONICET), Argentina.

E-mail: mrisk@itba.edu.ar

Abstract. Numerous parameter estimation techniques exist for characterizing the arterial system using electrical circuit analogs. However, they are often limited by their requirements and usually high computational burdain. Therefore, a new method for estimating arterial parameters based on Monte Carlo simulation is proposed. A three element Windkessel model was used to represent the arterial system. The approach was to reduce the error between the calculated and physiological aortic pressure by randomly generating arterial parameter values, while keeping constant the arterial resistance. This last value was obtained for each subject using the arterial flow, and was a necessary consideration in order to obtain a unique set of values for the arterial compliance and peripheral resistance. The estimation technique was applied to *in vivo* data containing steady beats in mongrel dogs, and it reliably estimated Windkessel arterial parameters. Further, this method appears to be computationally efficient for on-line time-domain estimation of these parameters.

1. Introduction

With the intent of practical use, this work aims to estimate the parameters of the arterial system, at a low calculation cost, prioritizing speed and simplicity. We chose the Windkessel model, a lumped model [1, 2], over tube models [3–5] and anatomically based distributed models [6–8], because it provides a quantitative point of view of the main parameters of a vascular system. Therefore it can be used for the systemic arterial system and the pulmonary arterial bed of all mammals.

The three-element Windkessel models the arterial system represented as analogous to an electric circuit, as shown in Figure 1. From this model we deduced the two parameters that we sought to estimate, the peripheral resistance R_p and the peripheral compliance C_p. There are numerous widely used and accepted methods for the estimation of these parameters [9]. The most common ones are: the decay time method [10], the stroke volume over pulse pressure method [11–13], the area method [14], the two-area method [15], the pulse pressure method [16, 17], the transient method [18].

The estimation of the parameters of the Windkessel offers a wide range of applications [9]. These include: calculating the ambulatory arterial stiffness index, calculating the load in artificial heart and valve studies [19–27], obtaining cardiac otuput from aortic pressure measurements [28] and theoretical analyses where a heart model and arterial system are coupled [29, 30].

2. Materials and Methods

Measurements of pressures from the left ventricle and the aorta arterial were taken to a group of mongrel dogs. Aortic (P_{ao}) and left ventricular (P_{lv}) pressures of the Windkessel model were registered with a sampling rate of 100 Hz during 4 seconds per measurement. Simulations were performed in *Scipy* [31] using a computer with an Intel Core i5 2.53 GHz processor, with 4 GB DDR3 RAM.

2.1. Windkessel Model

The three-element Windkessel model is a way to characterize part of the arterial system as an electrical circuit, the electrical diagram is shown in figure 1.

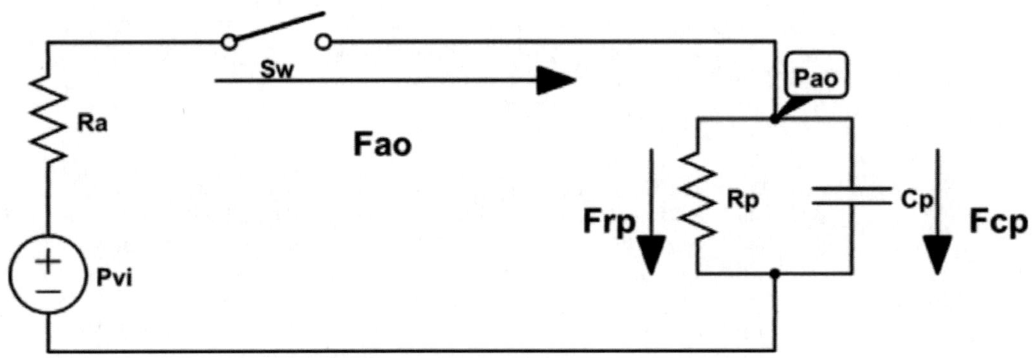

Figure 1. Three-element Windkessel model and its parameters, R_a, R_p, and C_p. S_w simulates de aortic valve by being closed during systole and open during diastole.

The three-element Windkessel employs equations derived from the analysis of the electrical circuit. Below are the final forms for the calculation of aortic pressure. These equations require three parameters and the left ventricular pressure as input data in order to simulate an aortic pressure curve. The equation for the simulation of aortic pressure P_{ao} during systole is:

$$P_{ao}[n+1]|_{sistole} = \alpha P_{lv}[n] + \beta P_{ao}[n] \tag{1}$$

where:

$$\alpha = \frac{\Delta t}{C_p R_a} \tag{2}$$

and

$$\beta = 1 - \frac{\Delta t (R_a + R_p)}{C_p R_p R_a} \tag{3}$$

And the equation for the simulation of aortic pressure P_{ao} during diastole is:

$$P_{ao}[n+1]|_{diastole} = (1-\gamma) P_{ao}[n] \tag{4}$$

where:

$$\gamma = \frac{\Delta t}{C_p R_p} \tag{5}$$

2.2. Monte Carlo Parameter Estimation and Curve fitting

The Windkessel model used in this case generates a P_{ao} with three inputs, which are the model's parameters. Generating a P_{ao} vector requires only the three input parameters, and considering that the flow is not taken into account in these calculations, there are infinite solutions for the three impedances (C_p, R_a and R_p) that would yield the same P_{ao}, each corresponding to a different flow.

In order to have a determined system, the aortic flow was taken into account calculating R_a as a function of it. To find this relation, R_a can be computed by means of equation (7), obtaining the difference between mean left ventricular pressure and mean aortic pressure during systole, divided by the measured aortic flow. This could also be approximated by using the Hagen–Poiseuille equation. This calculation, though hindering an optimal calculation of R_a, allows a much more precise approximation to the real values of the peripheral impedances, C_p and R_p. Moreover, another alternative allows R_p to remain constant, and computing R_p as the difference between mean arterial pressure and mean right atrial pressure, divided by the mean aortic flow. In this case, R_a could be calculated more precisely, if that is desired.

The aortic flow during systole results in:

$$F_{ao}[n]|_{sistole} = \frac{P_{vi}[n] - P_{ao}[n]}{R_a} \tag{6}$$

then R_a is:

$$R_a = \frac{\overline{P_{vi}}|_{sistole} - \overline{P_{ao}}|_{sistole}}{\overline{F_{ao}}|_{sistole}} \tag{7}$$

The Monte Carlo method was used to generate pseudo-random values for the peripheral parameters R_p and C_p. These pseudo-random values are generated within a fixed interval limited by the physiological boundaries for these parameters [15,32]. The initial intervals we set for R_p and for C_p were $[1, 20]$ and $[0.1, 5]$ respectively, selected in a way that the physiological values were included,and with some additional range in case of any anomaly.

An ammount of N sets were randomly generated and input into equations 1 and 4 along with experimental data for P_{lv}. This resulted in a set of N simulated aortic pressure curves as a function of time. Then, the curve which best fitted the experimental aortic pressure curve was selected employing the least squares method. Also, the *sixth* best curve was selected and compared with the best curve using an algorithm that shortened the interval that limits the pseudo-random variables generated.

During several iterations, the original parameter intervals were subsequently shortened to obtain a determined parameter value. To do so, the following algorithm was used:

```
if (6bv-bv)>0:
        if abs(bv-6bv)<=abs(bv-maxv):
                maxvalue=tbv
                if(bv-abs(bv-6bv))>minv:
                        minvalue=bv-abs(bv-6bv)
else:
        if abs(bv-6bv)<=abs(bv-minv):
                minvalue=6bv
```

```
if (bv+abs(bv−6bv))<maxv:
        maxvalue=bv+abs(bv−6bv)
```

The value of the parameters of the best fit curve *bv* were set as midpoints. Those corresponding to the *sixth* best fit curve *6bv* determined the distance from the midpoint to the upper limit and from the midpoint to the lower limit. If the new upper limit surpassed the upper physiological boundary, the former was set as the upper limit. The same was done for the lower limit.

The process of generating N curves and narrowing the parameter intervals was repeated until the top and bottom values of the intervals were within range of desired precision, we set it to ten decimal points. The estimated optimal values for the parameters thus correspond to the last iteration of the process.

3. Results

The estimated parameter values obtained with a precision of 10^{-4} for three different data sets are presented in Table 3. Here, the error is computed as the sum of the squares of the errors in each point and N represent the number of curves generated that allowed achieving the desired precision. Figures 2, 3, and 4 present the experimental data plotted against the best fit curve obtained, corresponding to the optimum values for R_p and C_p.

Table 1. Estimated peripheral resistance and capacitance with the proposed method.

Data Set	R_p (mmHg s/mL)	C_p (mL/mmHg)	Error	N
1	8.83463	0.2011	2.0880	1000
2	9.72960	0.1908	4.0660	1000
3	13.2225	0.1421	6.3319	1700

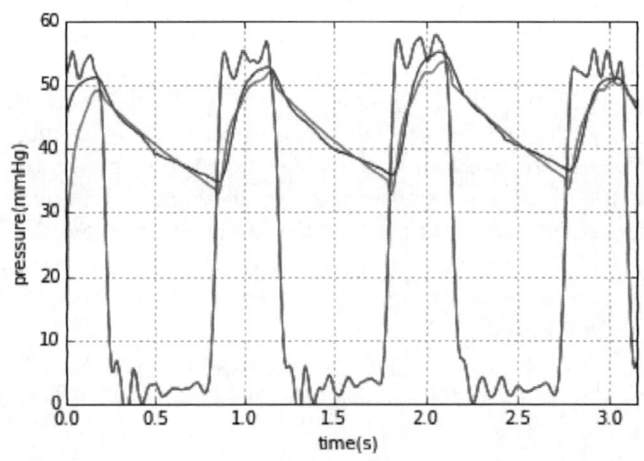

Figure 2. Estimated P_{ao} (red), experimental P_{ao} (blue) and P_{lv} (green) for Data Set 1.

Table 2 presents the computation time for all data sets varying the number of initial curves, N. Observing the calculation time for each data set, it was observed they had similar duration, with variations attributed to signal characteristics of each data set, specially the signal noise. It should be noted that, considering three heartbeats, the desired precision is achieved faster for

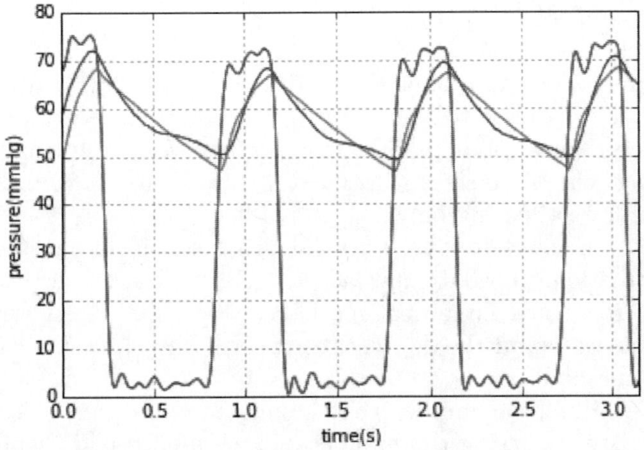

Figure 3. Estimated P_{ao} (red), experimental P_{ao} (blue) and P_{lv} (green) for Data Set 2.

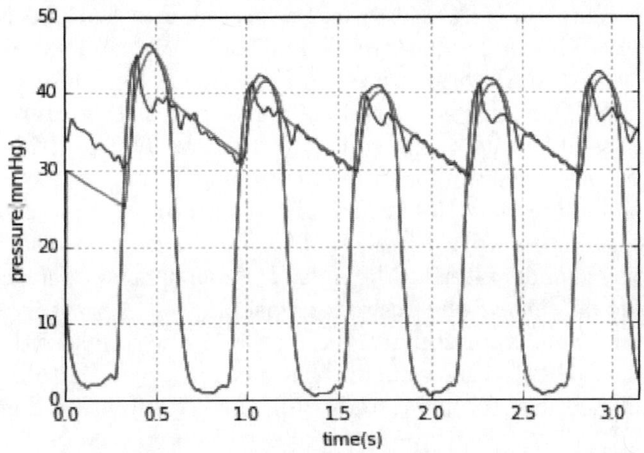

Figure 4. Estimated P_{ao} (red), experimental P_{ao} (blue) and P_{lv} (green) for Data Set 3.

the less-noisy signals (30 s for Data Sets 1 and 2, and 55 s for Data Set 3), thanks to a lower signal variability.

Table 2. Approximate calculation times for different number of Montecarlo runs (N).

N	time (s)		
	Data Set 1	Data Set 2	Data Set 3
100	5	4	5
200	9	10	8
500	19	20	18
1000	32	28	30
1700	65	80	55

4. Discussion

The aortic pressure curves were approximated with a relatively low error, implying that the parameters obtained should be close to the real values. However, we were not able to take any physical measurements of the arterial parameters to compare to our results. An alternative would involve estimating this parameters with other methods, but it is important to note that most of them are useful for estimating only C_p.

These physical measurements also have errors themselves which would be something to be compared with much attention. Our data also lacked the mean flow, which forced us to calculate the arterial resistance, R_a, using numerical methods. By creating an error function, which resulted being unimodal, we found the R_a which provided the minimum of the error function for several trials of our method.

Using the Monte Carlo method implies that results are subject to probability and could provide varying results. By increasing the number of generated pseudo-random values, N, this variation is reduced. Moreover, P_{ao} curves with noise and higher frequency components required a higher N value, and therefore, a higher computation time to reach a certain precision, as evidenced by the fact that the third data set (Data Set 3) required higher N values to reach the same precision than the first two sets (Data Sets 1 and 2).

It was also observed that when N is too large, precision is lost, because of an inherent complication of our interval limiting algorithm. When too many random aortic pressure curves are generated, the best and sixth best curve end up being too close and thus drastically shortening the interval limits for the next iteration. Sometimes this shortening is so drastic that it cuts the ideal curve out and the algorithm converges to parameters outside logical boundaries.

The desired precision was achieved for different number of starting sets depending on the quality of the signal. We choose $N = 1000$ to avoid any extreme variations of the Monte Carlo method and the issues that arise with large N. In order to improve the performance of our method when facing noisy signals, filtering the noise frequency component could be useful, and therefore a smaller N should achieve the desired precision.

For Data Sets 1 and 2, an approximate time of 7.5 s was required for estimating the parameters of a single heart-beat, which makes this method a good alternative for estimating these parameters on a clinical environment, and using these results as a diagnostic tool.

5. Author Contributions

Analyzed the data: VG IP GRP GVD. Contributed reagents/materials/analysis tools: VG IP GRP GVD HMV MR. Wrote the paper: VG IP GRP GVD HMMV MR.

References

[1] Frank O 1899 *Z. Biol-Munich* **37** 19
[2] Westerhof N, Elzinga G and Sipkema P 1971 *J. Appl. Physiol.* **31** 776–781
[3] Burattini R, Knowlen G G and Campbell K B 1991 *Circ. Res.* **68** 85–99
[4] O'Rourke M F 1967 *J. Appl. Physiol.* **23** 139–149
[5] Wetterer E and Kenner T 1968 *Grundlagen der Dynamik des Arterienpulses* (Springer)
[6] O'Rourke M F and Avolio A P 1980 *Circ. Res.* **46** 363–372
[7] Stergiopulos N, Young D F and Rogge T 1992 *J. Biomech.* **25** 1477–1488
[8] Westerhof N, Bosman F, De Vries C J and Noordergraaf A 1969 *J. Biomech.* **2** 121–143
[9] Westerhof N, Lankhaar J W and Westerhof B 2009 *Med. Biol. Eng. Comput.* **47** 131–141 ISSN 0140-0118
[10] Frank O 1899 *Z. Biol-Munich.* **37** 483–526
[11] Chemla D, Hébert J L, Coirault C, Zamani K, Suard I, Colin P and Lecarpentier Y 1998 *Am. J. Physiol-Heart. C.* **274** H500–H505
[12] Randall O S, Esler M D, Calfee R V, Bulloch G F, Maisel A S and Culp B 1976 *Aust. NZ. J. Med.* **6** 49–59
[13] Rosen I and White H 1926 *Am. J. Physiol.* **78** 168–184

[14] Liu Z, Brin K P and Yin F 1986 *Am. J. Physiol-Heart. C.* **251** H588–H600
[15] Segers P, Verdonck P, Deryck Y, Brimioulle S, Naeije R, Carlier S and Stergiopulos N 1999 *Ann. Biomed. Eng.* **27** 480–485
[16] Stergiopulos N, Meister J J and Westerhof N 1994 *Ann. Biomed. Eng.* **22** 392–397
[17] Stergiopulos N, Segers P and Westerhof N 1999 *Am. J. Physiol-Heart. C.* **276** H424–H428
[18] Toorop G P, Westerhof N and Elzinga G 1987 *Am. J. Physiol-Heart. C.* **252** H1275–H1283
[19] Aarnoudse W, van den Berg P, van de Vosse F, Geven M, Rutten M, van Turnhout M, Fearon W, De Bruyne B and Pijls N 2004 *Catheter. Cardio. Inte.* **62** 56–63
[20] Cochrane T 1991 *J. Biomed. Eng.* **13** 335–340
[21] Dumont K, Yperman J, Verbeken E, Segers P, Meuris B, Vandenberghe S, Flameng W and Verdonck P R 2002 *Artif. Organs.* **26** 710–714
[22] Geven M C, Bohté V N, Aarnoudse W H, van den Berg P M, Rutten M C, Pijls N H and van de Vosse F N 2004 *Physiol. Meas.* **25** 891
[23] Glower J, Cheng R, Giridharan G, Gillars K, Pantalos G, Litwak K, Ewert D and Koenig S 2004 *26th Annual International Conference of the IEEE Engineering in Medicine and Biology Society, 2004. IEMBS'04.* vol 2 (IEEE) pp 3773–3776
[24] Hildebrand D K, Wu Z J, Mayer Jr J E and Sacks M S 2004 *Ann. Biomed. Eng.* **32** 1039–1049
[25] Mol A, Rutten M C, Driessen N J, Bouten C V, Zünd G, Baaijens F P and Hoerstrup S P 2006 *Circulation* **114** I–152
[26] Rutten M, Wijlaars M, Strijkers G, Peters G, Bouten C and Baaijens F 2002
[27] Scharfschwerdt M, Misfeld M and Sievers H H 2004 *ASAIO J.* **50** 498–502
[28] Wesseling K, Jansen J, Settels J and Schreuder J 1993 *J. Appl. Physiol.* **74** 2566–2573
[29] Burkhoff D and Sagawa K 1986 *Am. J. Physiol-Reg. I.* **250** R1021–R1027
[30] Stergiopulos N, Meister J J and Westerhof N 1996 *Am. J. Physiol-Heart. C.* **270** H2050–H2059
[31] Jones E, Oliphant T, Peterson P *et al.* 2001– SciPy: Open source scientific tools for Python [Online; accessed 2015-06-09] URL `http://www.scipy.org/`
[32] Segers P, Brimioulle S, Stergiopulos N, Westerhof N, Naeije R, Maggiorini M and Verdonck P 1999 *Am. J. Physiol-Heart. C.* **277** H725–H731

Arterial pulse pressure amplification described by means of a nonlinear wave model: characterization of human aging

M Alfonso[1], L Cymberknop[1], R Armentano[1, 2,] F Pessana[1, 2,] S Wray[1], W Legnani[1]

[1] Buenos Aires Regional Faculty and from School of Advanced Studies in Engineering Sciences, National Technological University, Buenos Aires, Argentina
[2] Faculty of Engineering, Exact and Natural Sciences, Favaloro University, Buenos Aires, Argentina.

Abstract. The representation of blood pressure pulse as a combination of solitons captures many of the phenomena observed during its propagation along the systemic circulation. The aim of this work is to analyze the applicability of a compartmental model for propagation regarding the pressure pulse amplification associated with arterial aging. The model was applied to blood pressure waveforms that were synthesized using solitons, and then validated by waveforms obtained from individuals from differentiated age groups. Morphological changes were verified in the blood pressure waveform as a consequence of the aging process (i.e. due to the increase in arterial stiffness). These changes are the result of both a nonlinear interaction and the phenomena present in the propagation of nonlinear mechanic waves.

1. Introduction

Pulse Pressure Amplification (PPA) is conventionally understood in clinical practice as an increase in the entire amplitude of Pulse Pressure (PP) as it travels distally. By definition, it is better seen as a distortion than an amplification, translated into an alteration of the waveform morphology. Aging is the main non-modifiable factor associated with a decreased PPA. Additionally, PPA changes in subjects with traditional cardiovascular risk factors [1]. After reaching the age of 60, a substantial decrease in diastolic pressure and an increase in systolic pressure as a consequence of a progressive increase of arterial stiffness (AS) are observed [2]. This is the result of a greater load and demand of oxygen in the ventricle, together with an arterial degenerative process caused by the breaking of arterial wall elastin fibers [3]. This elasticity loss is responsible for an increment of the propagation velocity of Blood Pressure (BP) pulse waves along the arterial conduits. Furthermore, BP values measured in peripheral sites (PBP), i.e. brachial or femoral arteries, overestimate the value of the central blood pressure (CBP), measured in the aorta. The amplification of PBP and CBP, is greater at low propagation velocities and decreases with aging [1].

BP representation through nonlinear waves called "solitons" was introduced initially by Yomosa et al [4]. These waves propagate with no changes in their structure and interact with each other while maintaining their original properties. The nonlinear representation of the overlap of two or three solitons for BP waves is described in detail by Laleg et al [5]; mathematically, this behavior is described using the Korteweg and DeVries equation (KdV). This representation of pressure pulse waves as a combination of solitons, captures many of the phenomena observed in BP propagation,

---

[1] Manuel Alfonso. malfonso@frba.utn.edu.ar.

such as peaking (increase in amplitude), steepening (decrease in width) and the variation of wave propagation velocity.

The aim of this work is to analyze the applicability of a compartmental model for BP propagation previously used by our group [6], regarding PPA changes associated with arterial aging. Firstly, PPA phenomena will be described based on a synthesized soliton combination. Secondly, the effects of increasing vascular stiffness on the PPA will be tested on the model. Finally, the system will be validated with data from individuals categorized in age groups.

2. Materials and Methods

Our group has previously proposed the use of a compartmental model of the arterial tree [6] from which simulations of nonlinear wave propagation can be performed throughout a path of different arterial segments. Each of the segments is considered as a thin-walled elastic tube, where BP pulse wave (P) behavior is described by the partial differential equation obtained by Crepeau et al [7]:

$$P_Z + d_0 P_T + d_1 P\, P_T + d_2 P_{TTT} = 0, \tag{1}$$

where Z and T are the variables of space and time respectively, P_Z and P_T are the space and time partial derivatives, and:

$$d_0 = \frac{1}{c_0},\ d_1 = -\left(\alpha + \frac{1}{2}\right)\frac{1}{\rho {c_0}^3},\ d_2 = -\frac{\rho_w h_0 R_0}{2\rho {c_0}^3}, \tag{2}$$

$$c_0 = \sqrt{\frac{E h_0}{2\rho R_0}}, \tag{3}$$

where c_0 is the Moens-Korteweg propagation velocity in an elastic tube, ignoring nonlinear effects; E is the elastic modulus; h_0 is the initial thickness; R_0 is the initial radius and ρ_w is the density of the arterial wall, respectively. Finally, ρ is the blood density and α is a nonlinearity coefficient. The numerical evolution was performed by means of a pseudo-spectral high order approximation method developed by Cox et al [8], whose properties were previously evaluated by our group [9].

1.1. Simulation using compartmental modelling

The considered arterial path for BP propagation started at the aortic arch and continued with the subclavian, axillary and brachial artery segments, reaching the radial artery [6]. For the analysis of the model behavior, a synthesized BP wave was defined by combining two solitons, consistent with the normal BP values for an average adult (120/80 mmHg for systolic/diastolic pressure, and a heart period between 0.7 and 0.8 seconds). The resulting wave can be expressed as follows [7]:

$$P(Z,T) = 12\frac{d2}{d1}\frac{a_1^2 f_1 + a_2^2 f_2 + 2\left(\frac{a_1-a_2}{a_1+a_2}\right)^2 (a_2^2 f_2 f_1^2 + a_1^2 f_1 f_2^2)}{\left(1+f_1+f_2+\left(\frac{a_1-a_2}{a_1+a_2}\right)^2 f_1 f_2\right)^2} \tag{4}$$

$$f_j(Z,T) = \exp - \left(a_j\left(T - S_j - Z\left(do + a_j^2 d_2\right)\right)\right), \tag{5}$$

where a_j and s_j are the amplitude and position of each soliton. The AS values were adjusted in order to observe the variations in BP, in terms of peaking and steepening phenomena. As established in the previous section, PPA appears to be affected by vascular aging, showing a smaller increase in the systolic peak of the PBP with respect to the CBP systolic peak. In this sense, it is worth mentioning that the diastolic pressure has been considered as a constant value along the arterial system.

1.2. Validation

The model was validated using BP waveforms that belong to two well-differentiated age groups. Radial artery BP was measured, using the tonometry technique (Millar Inc., Houston, Texas, USA), in 8 subjects (ages 20 to 29 years [n=4] and ages 50 to 69 [n=4]) who had no prior cardiovascular risk factors. The waveforms were obtained in basal position and calibrated using sphygmomanometric measurements. CBP was determined by means of a transfer function using a previously validated algorithm (SphygmoCor, Atcor Medical, Illinois, USA).

For all individuals (young and adults), CBP wave was numerically propagated through the mentioned arterial path. The original AS values were iteratively adjusted, in order to obtain the best fit of the observed PPA, with a maximum error of 1 mmHg. This allowed analizing the ability of the model to reproduce the PPA variations generated in the BP pulse wave propagation as a result of aging. Furthermore, in order to compare the morphology between simulated and measured PBP waveforms, normal mean square error (NMSE) was calculated. Finally, data were expressed as mean ± standard deviation. An unpaired Student-t test was adopted, where a null hypothesis probability of $p<0.05$ was considered as statistically different.

3. Results

Figure 1 shows the changes in PPA for a synthetic simulated pressure wave in relation to the variations of AS. As it can be observed, the BP wave morphology (proximal to the heart) is affected by the peaking and steepening phenomena when arriving to the periphery.

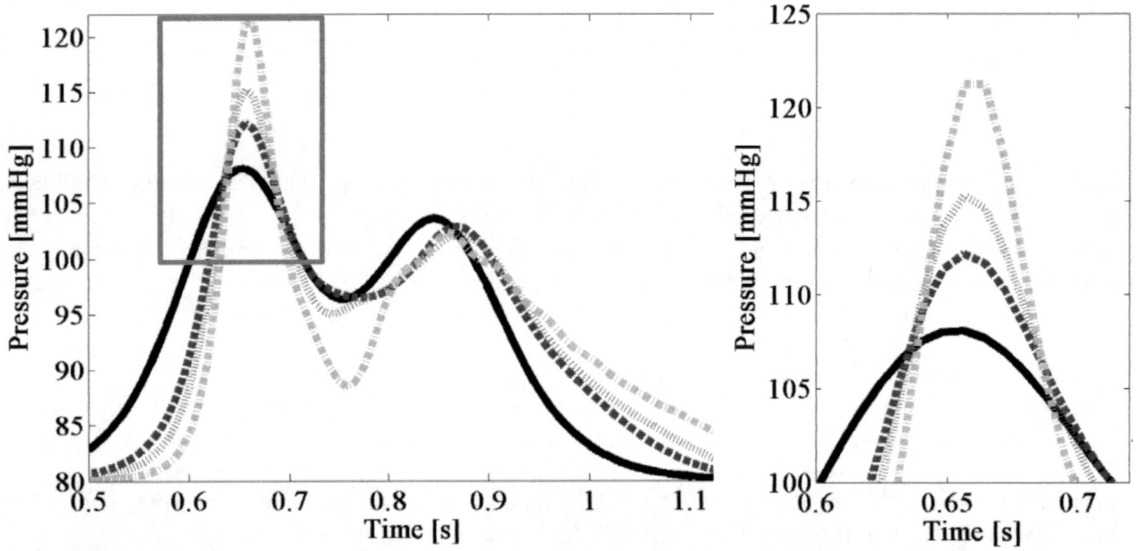

Figure 1. Changes in the synthesized blood pressure (BP) as a result of the application of the model and the increase in the stiffness modulus. Central BP (—), radial BP with initial stiffness (- • -), radial BP with stiffness increased by 50% (• • •), radial BP with stiffness increased by 100% (- - -).|

Measured values of systolic CBP and PBP and calculated PPA are shown in Table 1. AS_{it}/AS constitutes the relation between the original AS [10] and the adjusted AS obtained as a consequence of the iterations (AS_{it}). The calculated normalized mean square error (NMSE) between the measured and the obtained waveforms is shown in the last column.

Table 1. Calculated parameters: Systolic CBP (SCBP), Systolic PBP (SPBP), pulse pressure amplification (PPA), increment in [10] AS

Individual	SCBP	SPBP	PPA	ASit/AS	NMSE
Young_01	98.64	117.53	18.89	1.00	0.77
Young_02	105.62	126.7	21.08	1.625	0.80
Young_03	108.5	122.16	13.66	1.625	0.84
Young_04	86.64	109.42	22.78	1.063	0.66
Mean±SD	99.85±9.73	118.95±7.38	19.10±3.96	1.33±0.34	0.77±0.08
Adult_01	135.5	144.33	8.83	3.50	0.90
Adult_02	137.9	147.63	9.73	3.00	0.64
Adult_03	122.44	130.7	8.26	3.00	0.90
Adult_04	143.94	151.79	7.85	4.50	0.95
Mean±SD	134.95±9.06*	143.61±9.13*	8.67±0.81*	3.50±0.71*	0.84±0.14

* P<0.05 was considered as statistically significant

Figure 2 and Figure 3 show the measured CBP, PBP and the propagation of CPB with the corresponding adjusted AS_{inc} in two of the considered subjects. In Figure. 2, pulse propagation of CBP through the application of the model for a young individual is shown. It can be observed that the systolic peaks in comparison to the corresponding of PBP are equal. In Figure. 3, the CBP pulse of an adult individual was also propagated. For all cases, the corresponding PPA of the adult group has resulted smaller than in the young group, while the AS_{inc} became higher (Table 1).

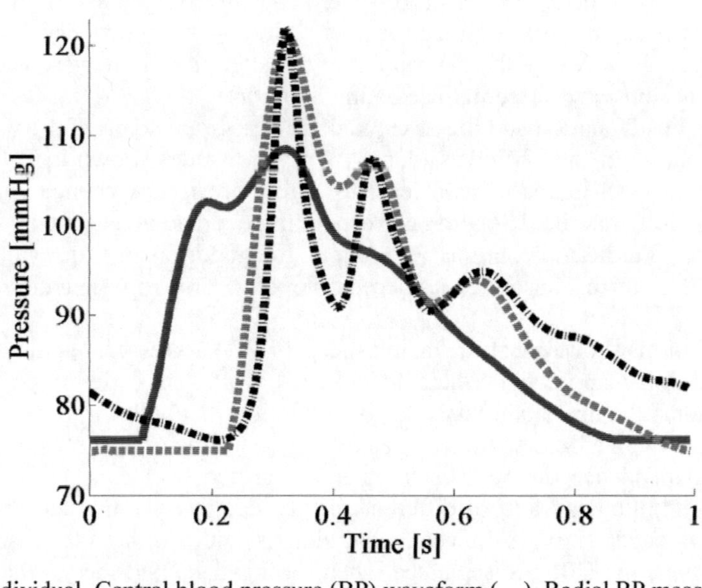

Figure 2. Young individual. Central blood pressure (BP) waveform (—). Radial BP measured for each subject (- - -). Result of the application of the model with adjusted AS_{inc}(- • -).

Journal of Physics: Conference Series **705** (2016) 012029 doi:10.1088/1742-6596/705/1/012029

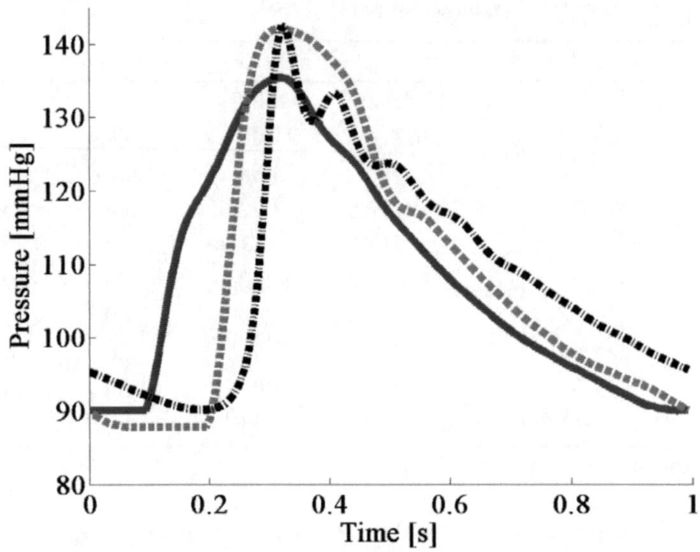

Figure 3. Adult individual. Central blood pressure (BP) waveform (—). Radial BP measured for each subject (- - -). Result of the application of the model with adjusted AS_{inc}(- • -).

4. Discussion

In this study, the compartmental model proposed in [6] –in which pressure waves were considered as a combination of solitary waves– was utilized in order to evaluate the effects of aging on BP waves, more precisely in PPA phenomena. BP waveforms were considered from its origin (CBP) as nonlinear waves, and then propagated through a defined vascular path. For our knowledge, a model based on the propagation of this type of waves for the description of the alteration of the vascular wall as a result of aging, has not been previously reported in the existing literature.

Arterial stiffening, radius dilatation and increased wall thickness are some of the well-known changes that characterize the aging of the cardiovascular system. The values shown in Table 1, in which the PPA becomes lower and AS increases as a result of aging are in accordance with previous results [11,12]. It can be observed that the AS values corresponding to an adult subject are doubled in relation with young individuals. Furthermore, in the last column, the goodness of fit provided by the NMSE is between 70-80 percent, which constitutes an acceptable waveform representation for the developed model.

The performed simulations showed that the main aspects of PPA have been emulated. It is noteworthy that, unlike the traditional approach (where the effects of PPA are described through the wave reflection phenomenon [2]), this study analyses the evolution of nonlinear waves traveling from the aortic arch, and whose interaction determines the morphology of the peripheral wave.

The morphological dependence on AS can be easily described by means of soliton theory, where waves with different amplitudes travel at different speeds, due to an amplitude-velocity relationship. In this sense, at low speeds (i.e. AS in young individuals with no vascular disease), the different solitary waves that shape the CBP are separated when reaching the periphery, where the peaking and steepening phenomena are evidenced. Increased AS, caused by aging, diminishes this separation and consequently a smaller PPA is accused.

5. Conclusion

In the present work, a nonlinear wave model was used to evaluate human aging. AS adjustments were performed to recreate morphological changes and obtained PPA variations are in accordance with previous studies. Additionally, soliton theory was used to explain changes experimented by CBP, from

a totally different approach. As a result, this model allows aging simulation and can be useful for explaining the clinical implication of PPA.
The analysis of BP morphological variations in terms of soliton wave propagation is very promising and deserves further studies.

6. References

[1] Avolio A P, Bortel L M V, Boutouyrie P, Cockcroft J R, McEniery C M, Protogerou A D, Roman M J, Safar M E, Segers P and Smulyan H (2009). Role of Pulse Pressure Amplification in Arterial Hypertension Experts' Opinion and Review of the Data *Hypertension* **54** 375–83

[2] Nichols W, O'Rourke M and Vlachopoulos C (2011). *McDonald's Blood Flow in Arteries, Sixth Edition: Theoretical, Experimental and Clinical Principles* (London: CRC Press)

[3] Armentano R, Megnien J L, Simon A, Bellenfant F, Barra J and Levenson J (1995). Effects of hypertension on viscoelasticity of carotid and femoral arteries in humans *Hypertension* **26** 48–54

[4] Yomosa S (1987). Solitary Waves in Large Blood Vessels *J. Phys. Soc. Jpn. - J PHYS SOC JPN* **56** 506–20

[5] Laleg T-M, Crépeau E and Sorine M (2007). Separation of arterial pressure into a nonlinear superposition of solitary waves and a windkessel flow *Biomed. Signal Process. Control* **2** 163–70

[6] Alfonso M R, Cymberknop L J, Legnani W, Pessana F and Armentano R L (2014). Conceptual model of arterial tree based on solitons by compartments *Conf. Proc. Annu. Int. Conf. IEEE Eng. Med. Biol. Soc. IEEE Eng. Med. Biol. Soc. Annu. Conf.* **2014** 3224–7

[7] Crepeau E and Sorine M (2005). Identifiability of a reduced model of pulsatile flow in an arterial compartment *44th IEEE Conference on Decision and Control, 2005 and 2005 European Control Conference. CDC-ECC '05* 44th IEEE Conference on Decision and Control, 2005 and 2005 European Control Conference. CDC-ECC '05 pp 891–6

[8] Cox S M and Matthews P C (2002). Exponential time differencing for stiff systems *J. Comput. Phys.* **176** 430–55

[9] Manuel R Alfonso W E L (2011). A Numerical Study For Improving Time Step Methods In Psudospectral Schemes Applied To The Korteweg And De Vries Equation *Mecánica Comput.* **XXX** 2763–75

[10] Avolio A P (1980). Multi-branched model of the human arterial system *Med. Biol. Eng. Comput.* **18** 709–18

[11] Black J and Hastings G (1998). *Handbook of Biomaterial Properties* (Boston, MA: Springer US)

[12] Langewouters G J, Wesseling K H and Goedhard W J (1985). The pressure dependent dynamic elasticity of 35 thoracic and 16 abdominal human aortas in vitro described by a five component model *J. Biomech.* **18** 613–20

A 2D model of axial symmetry for proximal tubule of an average human nephron: indicative results of diffusion, convection and absorption processes

J F Insfrán[1], S Ubal[2] y J Di Paolo[3]

[1,2,3]Grupo Biomecánica Computacional (GBC)
Facultad de Ingeniería – UNER
Ruta 11, km 10, 3100, Oro Verde, Entre Ríos, Argentina.
http://www.bioingenieria.edu.ar/grupos/biomecanica_computacional/

[1]jfinsfran@bioingenieria.edu.ar,
[2]subal@bioingenieria.edu.ar
[3]jdipaolo@bioingenieria.edu.ar

Abstract. A simplified model of a proximal convoluted tubule of an average human nephron is presented. The model considers the 2D axisymmetric flow of the luminal solution exchanging matter with the tubule walls and the peritubular fluid by means of 0D models for the epithelial cells. The tubule radius is considered to vary along the conduit due to the trans-epithelial pressure difference. The fate of more than ten typical solutes is tracked down by the model. The Navier-Stokes and Reaction-Diffusion-Advection equations (considering the electro-neutrality principle) are solved in the lumen, giving a detailed picture of the velocity, pressure and concentration fields, along with trans-membrane fluxes and tubule deformation, via coupling with the 0D model for the tubule wall. The calculations are carried out numerically by means of the finite element method. The results obtained show good agreement with those published by other authors using models that ignore the diffusive transport and disregard a detailed calculation of velocity, pressure and concentrations. This work should be seen as a first approach towards the development of a more comprehensive model of the filtration process taking place in the kidneys, which ultimately helps in devising a device that can mimic/complement the renal function.

1. Introduction

In the human organism almost all the cells are surrounded by extracellular fluid called internal environment. Survival of tissues depends on maintaining electrolyte composition, osmotic balance and volume of the liquid within very narrow ranges. In the excretory system, the kidneys are the main organs responsible in formation of the urine for its further excretion. In this process, the kidney filters the blood plasma forming a plasmatic ultrafiltrate, a solution without cellular components or proteins. Then, the kidneys reabsorb substances from the ultrafiltrate, some of which must remain in the internal environment, and secretes others that must be eliminated from it. Urine, is the result of the renal filtering, reabsorption and secretion processes and, once produced, it is excreted through a series of ducts that direct it towards the bladder where it will be stored until the next urination. Within the kidney, these processes are performed in basic functional units of micrometer scales called nephrons. There are about one million nephrons per kidney in a healthy adult.

A nephron is essentially composed by a filtering component, called glomerulus, and a tubule in whose walls reabsorption and secretion of substances occurs. The luminal fluid that continues its path inside the tubule is then excreted. The urine is therefore formed by mixing the tubular liquid from all nephrons that are performing their functions. The tubule of a nephron is usually divided into segments: proximal convoluted tubule, descending limb of loop of Henle, thick and thin segments of ascending limb of loop of Henle, a short segment that contains a special group of cells called the macula densa, the distal convoluted tubule, and then, there is a collecting duct. Among the main electrolytes exchanged into the tubular walls include the cations sodium, potassium, hidrogenion and the anions chloride and bicarbonate. Other substances without net charge are also exchanged like urea, creatinine and glucose. There are substances such as pesticides, pharmaceuticals and food additives, which after being ultrafiltered don't interact with tubular walls and simply circulate through the tubules to be excreted, [1].

There are studies that models and analyzes the behavior of the nephron and surrounding structures involving many physiological details through constitutive equations for the exchange surfaces and constitutive equations of the substances involved as in [2], [3] or [4]. However, in those works it's not take into account the bi and three-dimensional characteristic of diffusive phenomena in the tubular lumen, because it's not their objective. A detailed understanding of the diffusive mass transport phenomenon and its effect on the spatial distribution of concentrations is essential for physically explain the approach of dissolved substances to the tubular walls.

In this paper a simplified model of the proximal convoluted tubule of an average human nephron with emphasis on the diffusion, convection, absorption and excretion of primary urine processes in the tubular lumen for a set of substances typically present in plasma ultrafiltrate is presented. For simplicity, a two-dimensional axisymmetric model is used to model the intratubular flow in combination with two dimensionless models to represent the epithelium of the tubular wall covered by a peritubular fluid whose characteristics are known. The tubular radius is considered variable along the length of the tubule with respect to the transepithelial pressure difference. This model considers the diffusive phenomena involved for a total of 15 intercellular and luminal solutes, 16 peritubular solutes and 18 intracellular solutes typically present in such compartments. In the luminal domain Navier-Stokes and Nernst-Planck ecuations are resolved considering the principle of electroneutrality to calculate pressure, velocity and concentration profiles, tubular transmembrane flows and deformation. In the remaining domains transepithelial and paracellular flows are resolved. The structures before, after or outside the proximal convoluted tubule are not considered. Calculations are performed using the finite element method. Agreement of calculated results with those of other authors to classic models that do not consider the diffusive fluxes and details of the concentration profiles, pressure and velocity, are observed. This last allows having a first approach to a comprehensive enough comparison model to aid in the design of microdevices aimed to complement and mimic the complex functions of the kidneys relating to the homeostasis maintenance of the internal environment.

2. Metodology

For modeling the proximal convoluted tubule of the nephron the advection-diffusion system of figure 1 is used, wherein can be seen an axis of axial symmetry. This geometry actually represents a surface that revolutionized generates two concentric cylinders, one hollow and the other filled with the same axis of symmetry as shown in figure 2. There are three distinct areas: Tubular lumen tubular, tubular renal epithelium and peritubular solution.

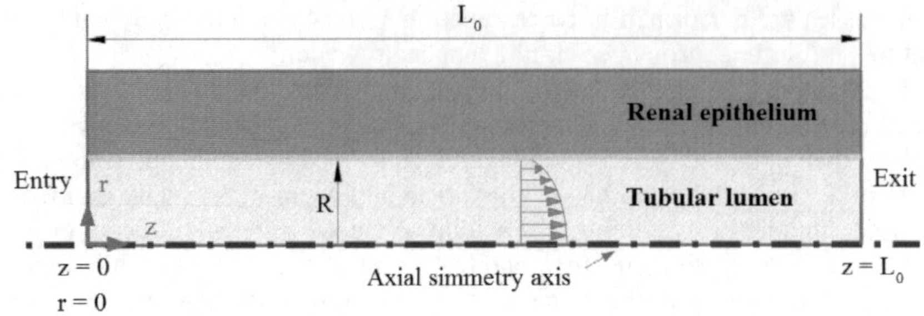

Figure 1. Simplified model of a proximal convoluted tubule.

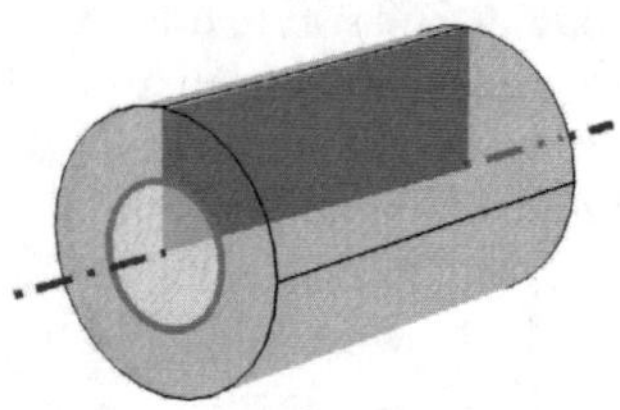

Figure 2. Section in the cilindrical model of the convoluted proximal tubule.

In the tubular lumen a two-dimensional model for a multicomponent fluid is used. The fluid is a dilute aqueous solution that, while flowing, exchanges different substances with the tubular walls in a steady state regime. Angular symmetry is assumed in all dependent distributed variables with respect to the revolution angle, thus only radial position ($r \in [0, R]$) and and longitudinal position ($z \in [0, L_0]$) , from the tubule entry, are considered independent variables. In the renal tubular epithelium region a subdivision is performed in two domains: epithelial cells and intercellular space, or interspace zone. Both are represented by two zero-dimensional models and the approximation of matter amounts it's done through constant average values in space. The peritubular solution is modeled as a solution of known characteristics. Below there is a list of compartments in which the model is divided with their identification symbols:

- Luminal compartment (L): 2-D model.
- Cell compartment (C): 0-D model.
- Intercellular compartment (I): 0-D model.
- Peritubular bath (P): Represented by a set of known features of interest.

The compartments indicated interact through intercompartmental contact surfaces. The interface between luminal compartment and cell compartment is the apical cell membrane (A_{LC}), between the luminal compartment and intercellular space compartment is the tight junction surface (A_{LI}), between cell compartment and intercellular compartment is the lateral cell membrane or paracellular membrane (A_{CI}), between intercellular compartment and peritubular solution is the basement membrane (A_{IP}) and between cell compartment and peritubular solution is the basal cell membrane (A_{CP}). The characteristics for these exchange surfaces are adapted from [2] and are shown in table 1.

Table 1. Exchange areas between compartments.

Exchange surface name (m.: membrane)	Symbol	Area in $cm^2\ cm^{-2}$ of epithelium	Requires luminal domain integration
Apical cell membrane	A_{LC}	36	Yes
Tight junction surface	A_{LI}	0.001	Yes
Paracellular membrane	A_{CI}	36	No
Basement membrane	A_{IP}	$0.02 \times [1.0 - 0.01\ mmHg^{-1}(p_I - p_C)]$	No
Basal cell membrane	A_{CP}	1	No

The last column of table 1 shows those areas which require integration schemes to properly interrelated amounts between compartments dimensionally different. Pressure in I compartment, p_{I}, is considered constant and equal to -23 mmHg. And pressure in C compartment, p_{C}, is considered equal to the intraluminal pressure, p.

Figure 3 shows a relation scheme between compartments with its exchange areas.

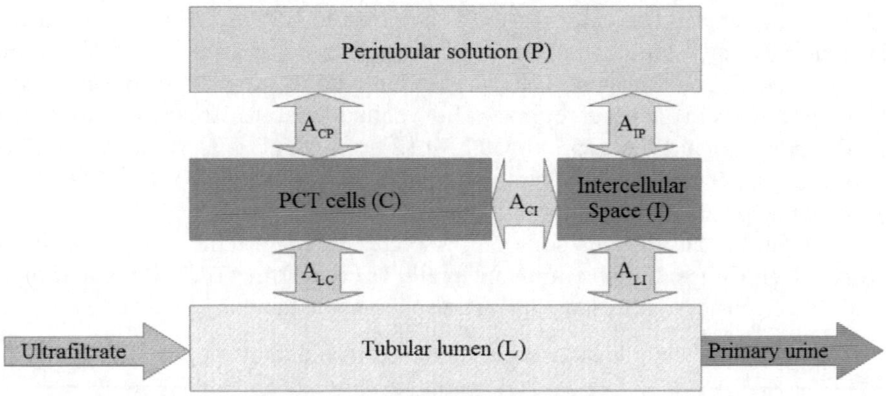

Figure 3. Compartmental scheme.

The luminal radius is $\mathrm{R} = 10.6[1 + 0.03\ \mathrm{mmHg}^{-1}(p_{\mathrm{L}} - p_{\mathrm{P}})]\ \mu\mathrm{m}$ and the tubular length $\mathrm{L}_0 = 12$ mm. The aspect ratio between length and radius within the tubular lumen is $\mathrm{L_0 R^{-1}} \approx 800$. The volumes are V_{C} for cellular compartment and V_{I} for intercellular compartment. The values for these volumes were taken from [2] and are shown in table 2.

Table 2. Volumes for epitelial compartments.

Compartment name	Volume symbol	Value in $\mathrm{cm^3\ cm^{-2}}$ of epithelium
Cell compartment (C)	V_{C}	1.00×10^{-3}
Intercellular compartment (I)	V_{I}	$0.7\times10^{-4}\times[1.0 - 0.1(p_{\mathrm{I}} - p_{\mathrm{C}})]$

Due to the micro dimensions of the model, is established as a validity criterion to the continuous material hypothesis that the Knudsen number must be less than 0.01, according to [5]. In this system the Knudsen number is calculated as the ratio of the molecular mean free path (λ) of molecules that compose the circulating fluid and the radius (R), i.e. $\mathrm{Kn} = \lambda \mathrm{R}^{-1}$. If the molecular mean free path estimated value is about[4] $\lambda \approx 0.7$ nm then the Knudsen number is $\mathrm{Kn} \approx 5\times10^{-5}$ and fulfill the criterion because it's about 200 times less than 0.01.

A homogeneous fluid occupy the tubular lumen, meaning that all functions are continuous in this domain. This fluid is considered to be circulating with a velocity profile governed by the Navier-Stokes equations (1) and continuity equation (2) for an incompressible fluid. The non-contact force effects are assumed to be null, [6].

$$\rho \frac{\mathrm{d}\mathbf{v}}{\mathrm{d}t} = -\boldsymbol{\nabla} p - \mu \Delta \mathbf{v} \tag{1}$$

$$\boldsymbol{\nabla} \cdot \mathbf{v} = 0 \tag{2}$$

[4] $\lambda = (2^{0.5}\pi n a^2)^{-1} = (2^{0.5}\pi P M_{water} N_{av} a^2)^{-1} = (2^{0.5} 3.1415\ 55140\ \mathrm{mol\ m^{-3}} 6\times10^{23}(10^{-10}\mathrm{m})^2)^{-1} \approx$ 0.7 nm. This is an approximate calculation for liquid water because it's the most abundant species as solution's solvent. "a" is a molecular radio estimate for water molecule.

Journal of Physics: Conference Series **705** (2016) 012030 doi:10.1088/1742-6596/705/1/012030

Figure 1 shows the approximately parabolic velocity profile that appears when these equations are solved being null the fluid absorption at the tubule luminal surface. Has to be consider here the radial symmetry of the geometry to correctly interpret the schema.

The used density is $\rho = 1000\ \mathrm{kg\ m^{-3}}$. The used viscosity for the circulating fluid is $\mu = 1.2\ \mathrm{cP} \equiv 1.2\ \mathrm{mPa\ s}$. This value approaches the blood plasma viscosity according to [7]. To solve these equations a parabolic velocity profile is considered at the entrance and generates a rate about $0.5\ \mathrm{nl\ s^{-1}}$ entering the tubule. The exit pressure is assumed to be 8 mmHg and the solvent flow rate was calculated using expressions in which solute concentrations aside and another membrane is coupled to the transmembrane transport. The calculated Reynolds number is less than 1 in the luminal domain indicating a flow in laminar regime. This regime is considered valid for the entire fluid domain since the entry length L_e is approximately $0.124\ \mu\mathrm{m}$ which is less than the total length L_0 of the tubule by a factor of 9.6×10^4 [6]. The assumptions assumed to solve the equations (1) and (2) permits them to be independent of the Nernst-Planck equations.

The circulating fluid is an aqueous solution of solutes listed in table 3 along with their relevant features for this model. Concentrations in the intercellular compartment I, cell compartment C and the peritubular bath P are considered constant and are also shown in table 3.[5]

Table 3. Chemical species considered in plasmatic ultrafiltrate.

Chemical specie	Index i	Chemical formula	PM in $\mathrm{g\ mol^{-1}}$	z_i [a]	D_i [b] in $\mathrm{m^2\ s^{-1}}$	Domain	c_C in $\mathrm{mmol\ l^{-1}}$	c_I in $\mathrm{mmol\ l^{-1}}$	c_P in $\mathrm{mmol\ l^{-1}}$
Hydron	1	H^+	1.01	+1	9.68×10^{-9}	L, C, I, P	4.69×10^{-5}	4.59×10^{-5}	4.95×10^{-5}
Sodium	2	Na^+	22.99	+1	1.39×10^{-9}	L, C, I, P	19.60	140.30	140.0
Potassium	3	K^+	39.10	+1	2.04×10^{-9}	L, C, I, P	138.1	4.66	4.90
Ammonium	4	NH_4^+	18.04	+1	2.04×10^{-9}	L, C, I, P	0.23	0.18	0.20
Chloride	5	Cl^-	35.45	−1	2.11×10^{-9}	L, C, I, P	16.3	112.0	113.2
Bicarbonate	6	HCO_3^-	61.02	−1	1.23×10^{-9}	L, C, I, P	25.0	25.6	24.0
Hydrogen phosphate	7	HPO_4^{2-}	95.98	−2	7.94×10^{-10}	L, C, I, P	8.50	2.98	2.97
Dihydrogen phosphate	8	$H_2PO_4^-$	96.99	−1	9.16×10^{-10}	L, C, I, P	2.52	0.86	0.93
Formate	9	HCO_2^-	45.02	−1	1.51×10^{-9}	L, C, I, P	0.52	0.77	1.00
Cellular anion	10	Imp^-	-	−1	-	C	27.7	-	-
Carbon dioxide	11	CO_2	44.01	0	2.00×10^{-9}	L, C, I, P	1.49	1.49	1.50
Urea	12	$CO(NH_2)_2$	60.06	0	1.67×10^{9}	L, C, I, P	4.96	4.91	5.00
Carbonic acid	13	H_2CO_3	62.02	0	1.40×10^{-9}	L, C, I, P	4.36×10^{-3}	4.36×10^{-3}	4.41×10^{-3}
Ammonium	14	NH_3	17.03	0	1.90×10^{-9}	L, C, I, P	3.48×10^{-3}	2.70×10^{-3}	2.82×10^{-3}
Formic acid	15	H_2CO_2	46.03	0	1.41×10^{-9}	L, C, I, P	0.91×10^{-5}	2.04×10^{-4}	2.85×10^{-5}
Glucose	16	$C_6H_{12}O_6$	180.2	0	9.40×10^{-10}	L, C, I, P	15.1	7.79	5.00
Cellular acid	17	ImpH	-	0	-	C	41.1	-	-
Protein	18	Prot	-	0	-	C, P	68.9	-	2.00

[a] Electric charge's number[6] z_i in e.
[b] Pseudobinary diffusivity in water.

Diffusivity data are estimated using classical expressions of Stokes-Einstein and Wilke-Chang equations and using ion mobility measurements given in [8] and [9]. Furthermore, are considered the chemical species called "Cellular acid (ImpH)", a corresponding "Cellular anion (Imp^-)" and "protein (Prot)". The two firsts represent confined solutes within the tubular cells that accumulate acid-base effects in cell that don't come from the acid-base effects of the others modeled buffers. In addition,

[5] The molecular weights given in this table have been estimated by http://es.webqc.org/mmcalc.php.
[6] $1\ \mathrm{e} \approx 1.6 \times 10^{-19}$ C and it's equivalent to the electric charge of an electron. Should not be confused z_i symbol for the number of electric charge of the solutes with the z symbol for longitudinal space coordinate the tubule.

Imp^- is assumed unreactive in this model so the amount of mass, considering the amounts involved in acid-base reactions, is constant within the cell space. The solute Prot allows modeling oncotic effects of protein confined at the C and P compartments. For these substances is not necessary to indicate diffusivities because they are inside dimensionless compartments.

The bidimensional model proposed for the luminal compartment has three edges and a symmetry axis as shown in figure 1. The left's contour (in green) represents a radial section of the surface through which fluid enters. The right's contour (in red) represents a radial cut surface where the fluid leaves the control volume (in light blue). The upper boundary (in orange) shows a longitudinal section of the inner tubular wall where it is assumed the possibility of a normal solvent flow, i.e. the solvent speeds are normal to this wall. The solvent's intercompartmental flux density N_{vOD} is calculated according to equation (3). In exchanges C↔I, C↔P and I↔P the calculation is straightforward because conditions in compartments C, I and in peritubular fluid P are constants. Moreover, in exchanges L↔C and L↔I it is required a successive approximation scheme to the equilibrium solution because the solute and solvent streams are coupled. In equation (3) the first term represents the effect of the hydrostatic pressure, the second represents the oncotic pressure effect of proteins and the third the effect of dissolved solutes associated with its corresponding reflection coefficients. All terms are affected by the water permeability of the considered exchange surface.

$$N_{vOD}(z) = L_{pOD}.A_{OD}.\left[(p_O - p_D) + (\pi_D - \pi_O) + R_g.T\sum_{j=1}^{NEQ}\sigma_{OD,i}\left(c_{D,i} - c_{O,i}\right)\right] \quad (3)$$

The tubular wall ability in absorption or secretion of solutes to and from the luminal compartment is considered normal thereto and its magnitude is governed by the transport equation (4), [2].

$$N_{OD,i}(z) = N_{vOD}.\left(1.0 - \sigma_{OD,i}\right).\bar{c}_{OD,i} + h_{OD,i}.\zeta_{OD}.A_{OD}.\frac{c_{O,i} - c_{D,i}.e^{-\zeta_{OD,i}}}{1 - e^{-\zeta_{OD,i}}} + \sum_{j=1}^{NEQ} L_{OD,i,j}.A_{OD}.(\bar{\mu}_{O,j} - \bar{\mu}_{D,j}) + N_{act,OD,i} \quad (4)$$

Where: O is the origin compartment, D is the destination compartment, $N_{OD,i}(z)$ is the molar flux from O to D of the *i*-th specie and is given as a longitudinal coordinate z function, N_{vOD} is the density of solvent volumetric flow from O to D, $\sigma_{OD,i}$ is the reflection coefficient for the *i*-th solute in the plasmalemma, $\bar{c}_{OD,i}$ it is the logarithmic average for concentration of solute *i* over the exchange surface, $h_{OD,i}$ is the membrane permeability for solute *i*, ζ_{OD} is the normalized electric potential difference, A_{OD} is the area that separate both compartments, $c_{O,i}$ and $c_{D,i}$ are solute *i* concentration in the O and the D side of the plasmalemma respectively, NEQ is the total number of chemical species, $L_{OD,i,j}$ is the coupled transport coefficient, $\bar{\mu}_{O,j}$ and $\bar{\mu}_{D,j}$ are the electrochemical potentials of the solute *j* in the O and D compartments respectively, $N_{act,OD,i}$ is the active transport of the *i* solute from the compartment O to compartment D.

In equation (3) the first term shows the effect of membrane reflection coefficient in determining the convective transport. The second term includes the Goldman ratio for determining transmembrane ion fluxes. The third term represents the coupled transport of species according to the model of linear nonequilibrium thermodynamic. The fourth term is an active transport of solutes. This equation and the above definitions also apply to the transports C↔I, I↔P and C↔P with the simplification that there are not dependent variables of *z* coordinate. In these cases the average values of the parameters in compartment C, compartment I and the fluid P are used to calculate intercompartmental matter flow's.

The hypothesis are: (a) the system is in steady state, (b) the electric potential gradient in the domain is zero, (c) the solution within the domain are considered dilute, (d) the temperature is constant, (f) the electroneutrality principle is met, (g) the chemical activity coefficient is one, (h) the density is constant, (i) can be assumed the hypothesis of material continuity (Knudsen number less than 0.01 according to [5]), (j) the dissociation of carbonic and dihydrogen phosphate acids at their respective resulting bases and hydron are instant. In addition, due to the hypothesis (c), it is also accepted that the contributions of speeds, shear stresses and other solutes forces are negligible compared to the corresponding quantities in the solvent (water), and thus is considered valid the the approximation of pseudobinary behavioral for dissolved chemical species.

The diluted solution hypothesis (c) is assumed because the mol percentage of all solutes in the solution is about 0.5%, while the mol percentage of water solvent in entire solution is 99.5%. With this concentration ratio it is reasonable to assume that molecules of each solute interacts most of the time with solvent molecules and, almost never with other solute molecules. Thus, for each solute can be considered a binary behavior between this one and the solvent. Because the solution is not strictly binary in the sense of having only two chemical species (one solute and one solvent), it is called solution with pseudobinary behavioral because for each solute can be thought a binary behavior between this and the solvent. This assumption greatly simplifies the equations used to describe the advective-diffusive behavior of substances.

To calculate the concentration profiles in the luminal domain is used the convection-diffusion equation associated with the fundamental principle of stoichiometry (5) assuming the electroneutrality hypothesis (6) in the domain. The total surface current density flow (7) is assumed to be null for all surface exchanges. Convective flow of substances is established in $z = L_0$ and concentrations are fixed on the surface of fluid entry to the tubular lumen, at $z = 0$. In the axis of symmetry it is assumed that the concentration gradients are zero in the radial direction r, see (8).

$$\nabla \cdot \mathbf{N}_i = \mathrm{R}_i \text{ with } \mathbf{N}_i = -D_i \nabla c_i + c_i \mathbf{v} \text{ and } \textstyle\sum_{i=1}^{\mathrm{NEQ}} \mathrm{R}_i \mathrm{PM}_i = 0 \quad (5)$$

$$\textstyle\sum_{i=1}^{\mathrm{NEQ}} \mathrm{z}_i c_i = 0 \text{ in luminal domain L} \quad (6)$$

$$\mathbf{j} = \textstyle\sum_{i=1}^{\mathrm{NEQ}} \mathrm{z}_i (-D_i \nabla c_i) \quad (7)$$

$$\left.\frac{\partial c_i}{\partial r}\right|_{r=0} = 0 \text{ for } i = 1..NEQ \quad (8)$$

The buffer reactions pairs carbonic acid-bicarbonate and hydrogen phosphate-dihydrogen phosphate are considered. It is assumed that the dissociation of acids is instantaneous and always in chemical balance in the whole domain in accordance with constants $pK_{HCO_3^-} = 6.1$ and $pK_{HPO_4^{-2}} = 6.8$. The reaction rate of hydration and dehydration of carbon dioxide is considered finite, and for that reason are provided the first-order reaction constants, k_h and k_d, that allow to calculate the reaction rates R_{CO_2} y $R_{H_2CO_3}$.

$$CO_2 + H_2O \underset{k_\mathrm{d}}{\overset{k_\mathrm{h}}{\rightleftarrows}} H_2CO_3 \rightleftarrows HCO_3^- + H^+ \quad (6)$$

Solving this diffusive-advective-reactive system means to find the concentration distributions for all chemical species dissolved in the luminal domain Ω_L and finding the amount of solute and solvent that flows through all intercompartmental membranes, including the surface of tight junctions linking apical cell membranes in the wall of the tubular lumen. The Ω_L domain with the domain Ω_C (cellular compartment) and Ω_I (intercellular compartment) constitute the domain Ω given by its union.

The remaining parameters used to obtain the results are in table 4.

Table 4. Additional parameters and modeling equations[7].

Inlet concentrations in mmol.l^{-1} and reflection coefficients $\sigma_{LI,i}$ in tight junction[a]								
Solute	$c_{\mathrm{inlet},i}$	$\sigma_{LI,i}$	Solute	$c_{\mathrm{inlet},i}$	$\sigma_{LI,i}$	Solute	$c_{\mathrm{inlet},i}$	$\sigma_{LI,i}$
H^+	4.94×10^{-5}	0.20	HPO_4^{2-}	2.970	0.90	H_2CO_3	1.51	0.90
Na^+	140	0.75	$H_2PO_4^-$	0.926	0.90	NH_3	2.86×10^{-3}	0.30
K^+	4.90	0.60	HCO_2^-	1.000	0.30	H_2CO_2	2.84×10^{-4}	0.70

[7] Los datos de la tabla 4 se tomaron de [2] a excepción de los coeficientes de reflexión unitarios en nota *a* y la constante $pK_{HCO_3^-}$ que en el citado documento vale 3.57 y no concuerda con los datos bibliográficos.

NH_4^+	0.20	0.60	Imp^-	0.000	-	$C_6H_{12}O_6$	5.00	1.00
Cl^-	113	0.30	CO_2	1.200	0.90	ImpH	0.00	-
HCO_3^-	24.2	0.90	$CO(NH_2)_2$	5.000	0.70	Prot	0.00	-

Equilibrium equations in buffers, with $pK_a = -\ln K_a$

Chemical equilibrium for pair $HCO_3^- - H_2CO_3$: $pK_{HCO_3^-} = 6.1$ $\quad [H^+] = K_{HCO_3^-}[H_2CO_3][HCO_3^-]^{-1}$

Chemical equilibrium for pair $HPO_4^{2-} - H_2PO_4^-$: $pK_{HPO_4^{-2}} = 6.8$ $\quad [H^+] = K_{HPO_4^{2-}}[H_2PO_4^-][HPO_4^{2-}]^{-1}$

Oncotic pressure (π_D) [b]: $\pi_D = RT[Prot]_D$, with $R = 8.314$ Pa m^3 mol^{-1} K^{-1}, $T = 310$ K

Electrochemical potential $(\bar{\mu}_{D,j})$ [b]: $\bar{\mu}_{D,j} = RT \ln c_{j,\mathrm{D}} + z_j F$, being z_j the charge number, F the Faraday constant and ψ the electric potential.

Equation for active Na^+ transport:

$$J_{CI-Na^+}^{act} = 10.8\left\{\frac{[Na^+]_{\mathrm{C}}}{[Na^+]_{\mathrm{C}} + K_{Na^+}}\right\}^3\left\{\frac{[K^+]_{\mathrm{I}} + [NH_4{}^+]_{\mathrm{I}}}{[K^+]_{\mathrm{I}} + [NH_4{}^+]_{\mathrm{I}} + K_{K^+}}\right\}^2$$

$$K_{Na^+} = 0.2 \times 10^{-3}\left\{1 + \frac{[K^+]_{\mathrm{C}}}{8.33 \times 10^{-3}}\right\}$$

$$K_{K^+} = 0.1 \times 10^{-3}\left\{1 + \frac{[Na^+]_{\mathrm{I}}}{18.5 \times 10^{-3}}\right\}$$

[a] The reflection coefficients $\sigma_{LC,i}, \sigma_{CI,i}, \sigma_{CP,i}$ and $\sigma_{IP,i}$ are taken as unit.
[b] The subscript D can take values C and P.

The above equations are solved using a numerical scheme based on the finite element method implemented in the commercial software COMSOL Multiphisics®. The results are shown below corresponding to a mesh of 1640 triangular elements with quadratic basis and form functions in all cases, and cubic functions to describe the velocity field.

3. Results

The inner tubular radio is shown in figure 4. The integral over the revolution surface that it generates is $A_{\mathrm{pti}} = 0.831\ \mathrm{mm}^2$, which is considered as internal tubular wall area. Taking an epithelium thickness of 8 µm microns, then the epithelial area is calculated from this radius R showed in the graph. Epithelial area is $A_{\mathrm{ep}} = 1.43\ \mathrm{mm}^2$.

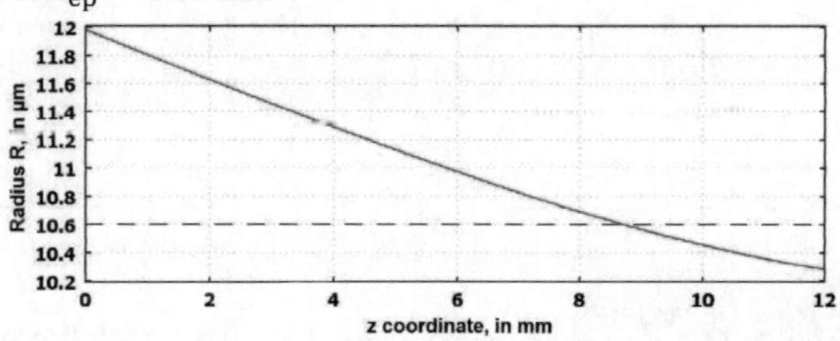

Figure 4. Deformed tubular radius R, in violet color, and undeformed tubular radius R, in dotted lines.

The tubular lumen pressure and the transepithelial density of volumetric flow is shown in figures 5 and 6. The resulting pressure drop in the proximal tubule is about $\Delta P = P_{\mathrm{in}} - P_{\mathrm{out}} = 5.5$ mmHg , being reasonable compared to a typical filtration pressure of 20 mmHg, [1]. The tubular wall, reabsorbs approximately 80% of the solvent flow at the inlet agreeing with [2].

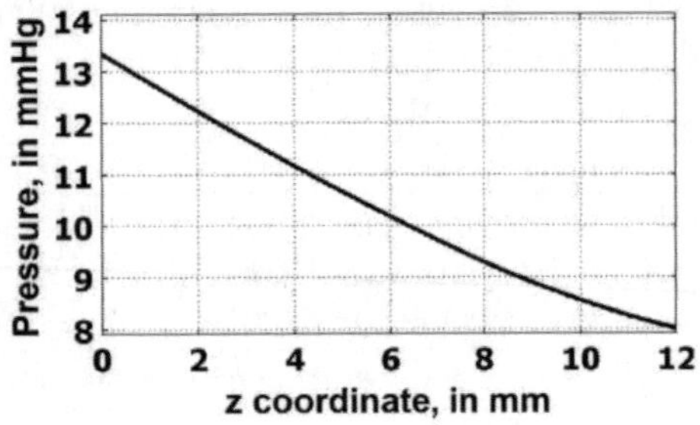

Figure 5. Intratubular pressure.

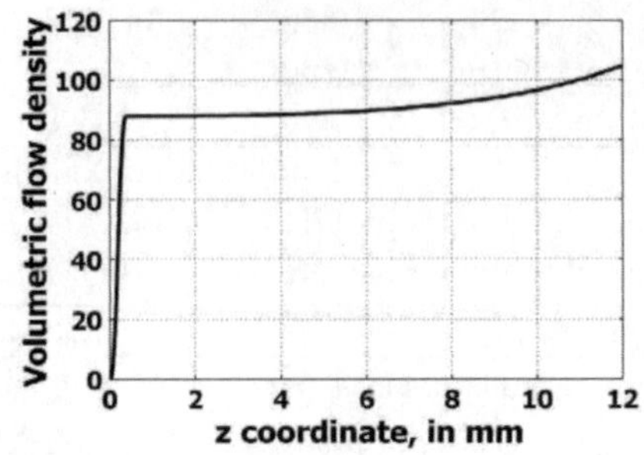

Figure 6. Transepitalial volumetric flow density in nl s^{-1} cm^{-2} of epithelium.

Membrane flows obtained can be seen in table 5. Intercompartmental compensation flows is observed. About 46 % of solvent water passes through the tight junctions. In [2] the water reabsorption in tight junctions is around 57 %.

Table 5. Total membrane's flows for solute and solvent.

	Solvent volume flow in nl s^{-1} cm^{-2} of epithelium				
	LI	*LC*	*CI*	*CP*	*IP*
$\frac{1}{A_{ep}}\int_{A_{pti}} N_{vOD}\,dA$	40.5	46.5	46.6	-0.112	87.1
	Solute flows in pmol s^{-1} cm^{-2} of epithelium				
Solute	*LI*	*LC*	Solute	*LI*	*LC*
H^+	6.72×10^{-4}	–0.154	Na^+	312.0	3523
K^+	134.0	0.000	NH_4^+	5.750	0.000
Cl^-	3094	0.000	HCO_3^-	663.0	0.000
HPO_4^{2-}	81.30	0.000	$H_2PO_4^-$	34.30	0.000
HCO_2^-	27.40	0.000	CO_2	28.50	0.000
$CO(NH_2)_2$	137.0	0.000	H_2CO_3	0.0283	0.000
NH_3	0.0785	0.000	H_2CO_2	1.700	0.000
$C_6H_{12}O_6$	0.000	136.0			

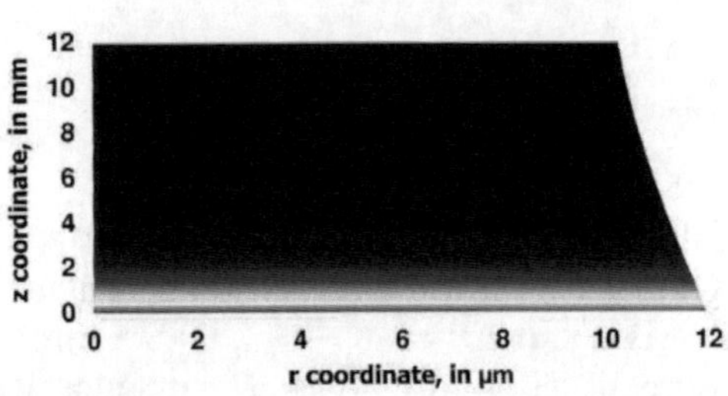

Figure 7. Glucose concentration in tubular domain in mmol/l. The concentration is 5 in red zones and is 0 in blue zones. Compare with blue curve in figure 8.

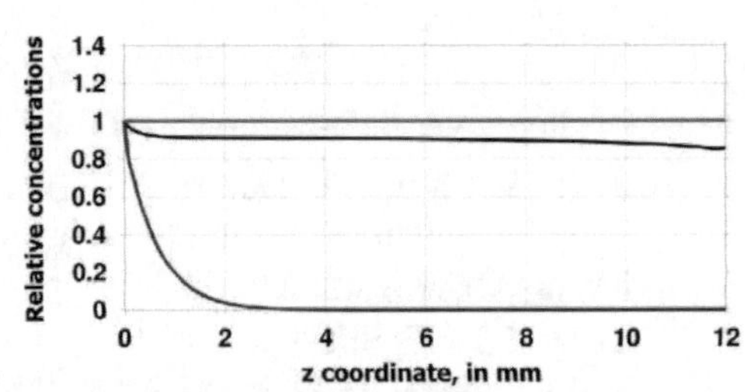

Figure 8. Concentrations normalized according to inlet concentration: in blue glucose, in black ammonium, and in red sodium, potassium and bicarbonate overlap each other.

Are shown the glucose concentration distribution in domain in figure 7 and the concentration variations on the inner tubular wall for some important solutes in figure 8, where the values are represented as fractions of the inlet concentrations. In figure 8, concentrations are taken on the tubular wall and for bicarbonate, potassium and ammonium concentrations there are differences regarding [2].

At entry, pH is 7.31 and at exit is about 7. The osmolarity remains constant. Essentially perfect compensation charges is observed throughout the luminal domain.

4. Discussion

The 2D model shows in greater detail the concentration, pressure and velocity distributions in the tubular lumen respect to one-dimensional classical models. Morphological differences are observed for concentration distributions regarding those given by other authors, who use radial average values. These morphological differences are attributed to: concentrations are constant in the compartments 0D, reflection coefficients are assumed to be identical in all exchange surfaces, except in tight junctions, all the instantaneous chemical equilibrium equations present in existing models are not included yet, the membrane potential that influences the transmembrane solute flows is not applied yet in this model.

On the other hand, the model retains the essential characteristics of pre-existing models, in relation to terms present in equations. At the same time incorporates additional spatial dimensions at the lumen compartment, in apical cell surface and in tight junctions, and incorporates the simulation of the diffusion effects. This makes more direct the comparison of concentration distributions with those that take place in artificial assistance devices.

5. Conclusions

In this paper a simplified model of a proximal convoluted tubule of an average human nephron, considering the radial coordinate and diffusive fluxes in the tubular domain, is presented. The Navier-Stokes and Nernst-Planck equations were resolved in a tubular domain of variable radius coupled to transepithelial pressure, and assuming the condition of electroneutrality. Accordance with data provided by other authors was found for integral values of intercompartmental flow profiles, solvent's volumetric flow, pressure, pH, luminal osmolarity and distribution of the most abundant solutes in the tubular lumen, namely, sodium, chloride and bicarbonate.

It is projected incorporating the remaining phenomena, mentioned in the discussion section, to the bidimensional model in order to obtain greater consistency in the integral values in relation to those provided by one-dimensional models currently available for proximal convoluted tubule.

6. Referencias

[1] Guyton A C and Hall J E 2006 *Tratado de fisiología médica* (Madrid: Elsevier)
[2] Weinstein A M, Weinbaum S, Duan Y, Du Z, Yan Q and Wang T 2007 Flow-dependent transport in a mathematical model of rat proximal tubule *American Journal of Physiology - Renal Physiology* **292** F1164–81
[3] Nieves-González A, Clausen C, Layton A T, Layton H E and Moore L C 2013 Transport efficiency and workload distribution in a mathematical model of the thick ascending limb *American Journal of Physiology - Renal Physiology* **304** F653–64
[4] Edwards A, Seguin N and Tournus M 2012 A finite-volume scheme for a kidney nephron model *ESAIM: Proceedings* **35** 287–92
[5] Tabeling P 2010 *Introduction to microfluidics* (Oxford; New York: Oxford University Press)
[6] Whitaker S 1981 *Introduction to fluid mechanics* (Malabar, Florida 32950: Robert E. Krieger)
[7] Fung Y C 1998 *Biomechanics: Motion, Flow, Stress, and Growth* (New York: Springer)
[8] Poling B E, Prausnitz J M and O'Connell J P 2000 *The Properties of Gases and Liquids* (New York, NY: McGraw-Hill Professional)
[9] Atkins P, Paula J de and Friedman R 2008 *Quanta, Matter and Change: A Molecular Approach to Physical Chemistry* (New York: W. H. Freeman)

Journal of Physics: Conference Series **705** (2016) 012031 doi:10.1088/1742-6596/705/1/012031

Left ventricular hypertrophy index based on a combination of frontal and transverse planes in the ECG and VCG: Diagnostic utility of cardiac vectors

Maria Paula Bonomini[14], Fernando Juan Ingallina[2], Valeria Barone[2], Ricardo Antonucci[2], Max Valentinuzzi[1], Pedro David Arini[1,3]

[1] Intituto de Ingeniería Biomédica (IIBM), Facultad de Ingeniería, UBA, Buenos Aires, Argentina,

[2] Instituto de Investigaciones Médicas "Dr. Alfredo Lanari", UBA, Buenos Aires, Argentina,

[3] Instituto Argentino de Matemáticas "Alberto P. Calderón", CONICET, Buenos Aires, Argentina

E-mail: paulabonomini@gmail.com

Abstract. The changes that left ventricular hypertrophy (LVH) induces in depolarization and repolarization vectors are well known. We analyzed the performance of the electrocardiographic and vectorcardiographic transverse planes (TP in the ECG and XZ in the VCG) and frontal planes (FP in the ECG and XY in the VCG) to discriminate LVH patients from control subjects. In an age-balanced set of 58 patients, the directions and amplitudes of QRS-complexes and T-wave vectors were studied. The repolarization vector significantly decreased in modulus from controls to LVH in the transverse plane (TP: 0.45±0.17mV vs. 0.24±0.13mV, $p<0.0005$; XZ: 0.43±0.16mV vs. 0.26±0.11mV, $p<0.005$) while the depolarization vector significantly changed in angle in the electrocardiographic frontal plane (Controls vs. LVH, FP: 48.24±33.66° vs. 46.84±35.44°, $p<0.005$, XY: 20.28±35.20° vs. 19.35±12.31°, NS). Several LVH indexes were proposed combining such information in both ECG and VCG spaces. A subset of all those indexes with AUC values greater than 0.7 was further studied. This subset comprised four indexes, with three of them belonging to the ECG space. Two out of the four indexes presented the best ROC curves (AUC values: 0.78 and 0.75, respectively). One index belonged to the ECG space and the other one to the VCG space. Both indexes showed a sensitivity of 86% and a specificity of 70%. In conclusion, the proposed indexes can favorably complement LVH diagnosis

1. Introduction

LVH indexes based only on the electrocardiogram are known to present low sensitivity and high specificity. Although it is well known that repolarization is also modified with LVH [1, 2, 3], attempts to include this information in electrocardiographic indexes [4] fell into disuse, focusing all efforts on different features of the depolarization phase [5, 6].

There has also been a longstanding disagreement as to whether the ECG or VCG spaces appears as more informative, with reports showing the VCG as better [7, 8], similar [6] or poorer [9] than the ECG for LVH diagnosis.

[4] Present address: Instituto de Ingeniería Biomédica, Facultad de Ingeniería, Universidad de Buenos Aires (UBA), Paseo Colón 850, Argentina.

Herein, we have combined depolarization and repolarization information in the transverse and frontal planes to construct a set of LVH indexes and compared their diagnosis performance in the ECG and VCG spaces. Therefore, we compared the performance of these indexes in two sets of equivalent planes: (1): the electrocardiographic frontal plane (FP) versus the vectorcardiographic frontal plane (XYP), (2): the electrocardiographic transverse plane (TP) versus the vectorcardiographic transverse plane (XZP).

2. Materials and Methods

2.1. Population

A total of 58 subjects without intraventricular conduction disturbances were retrospectively studied. Two groups were balanced for age characteristics: the LVH group (31 subjects, mean age 68.5±12.3 years old) and the control group (27 subjects, mean age 60.6±13.2 years old). The hypertrophy group included patients with left ventricular mass indexes greater than 125 g/m^2 in males and greater than 110 g/m^2 in females, as calculated by Devereux's formula from echocardiography on M-mode [10]. These patients lacked a coronary artery disease history. Healthy subjects, without clinical or echocardiographic evidence of cardiovascular disease, comprised the Control group. Besides echocardiographic data, 5-minute 12-lead ECG recordings were taken from all the subjects. Patients were recruited in the medical institution *"Instituto de Investigaciones Medicas, Dr. Alfredo Lanari" of the University of Buenos Aires* and in all cases informed consent was signed.

2.2. ECG Preprocessing

Signal preprocessing was applied to the 12 standard ECG leads, implementing QRS-detection and normal beat selection according to the method given in reference [11]. The QRS-complexes and T-waves were located and delineated using the wavelet transform based method described in [12] and baseline wandering attenuation was treated by cubic spline. Noisy beats were rejected when differences in mean isoelectric level with respect to adjacent beats were larger than 300 μV.

VCG was synthesized by means of the Kors Matrix [13]. The XZP was obtained from the VCG signals while the TP was derived from the standard ECG signals. More specifically, the precordial leads V_6 and $V_{1\text{-}2}$ were used, being the latter an average between leads V_1 and V_2.

2.3. Cardiac vectors

Cardiac vectors were measured in all planes following the same methodology as the one used here for the TP. Segmentation of the QRS-complex and T-wave was accomplished for each i_{th} beat using a single lead criterion, where the respective QRS and T onsets were taken at the earliest reliable QRS and T, either for V_6 or for $V_{1\text{-}2}$. The offsets, in a symmetric way, were accepted respectively as the latest reliable QRS-complex and T-wave offsets for the same leads, as in the previous case. On these segmented waves, the QRS-loop and the T-wave loop for each i_{th} beat were constructed and both cardiac vectors computed for the loop samples n $\in W_i^{\text{QRS}}$ and n $\in W_i^{\text{T}}$, representing the QRS-complex and T-wave windows respectively. The samples $n_{\text{max}}^{\text{QRS}}(i)$ and $n_{\text{max}}^{\text{T}}(i)$ at which the respective QRS-complex and T-wave cardiac vectors resulted maximum were also computed. Thereafter, the angle and modulus of the QRS-complex main cardiac vector for the i_{th} beat, $R_H^{\alpha}(i)$ and $R_H^{m}(i)$ respectively, were defined as,

$$R_H^{\alpha}(i) = \text{atan}\left(\frac{V_{1\text{-}2}(n_{\text{max}}^{\text{QRS}}(i))}{V_6(n_{\text{max}}^{\text{QRS}}(i))}\right) \tag{1}$$

$$R_H^{m}(i) = \sqrt{V_6(n_{\text{max}}^{\text{QRS}}(i))^2 + V_{1\text{-}2}(n_{\text{max}}^{\text{QRS}}(i))^2} \tag{2}$$

where,

$$n_{\max}^{\mathrm{QRS}}(i) = \arg\max_{n} \left[\sqrt{V_6(n)^2 + V_{1\text{-}2}(n)^2} \right] \tag{3}$$

where n $\in W_i^{\mathrm{QRS}}$

Notice that subscript H instead of T is used for denoting the transverse plane. This is so to avoid confusions with repolarization parameters.

Analogously, the angle and modulus of the T-wave maximum cardiac vector for the i_{th} beat, $T_H^{\alpha}(i)$ and $T_H^{m}(i)$ respectively, were defined as in (1)-(3) for n $\in W_i^{\mathrm{T}}$.

The angle and modulus of the main depolarization vectors for the i_{th} beat, $R_F^{\alpha}(i)$, $R_{XZ}^{\alpha}(i)$, $R_{XY}^{\alpha}(i)$ and $R_F^{m}(i)$, $R_{XZ}^{m}(i)$, $R_{XY}^{m}(i)$ respectively, as well as the angle and modulus of the T-wave maximum cardiac vectors for the i_{th} beat, $T_F^{\alpha}(i)$, $T_{XZ}^{\alpha}(i)$, $T_{XY}^{\alpha}(i)$ and $T_F^{m}(i)$, $T_{XZ}^{m}(i)$, $T_{XY}^{m}(i)$ were defined as in (1)-(3).

2.4. LVH indexes
LVH indexes consisted of relations between angle and modulus of the cardiac vectors combining phases (depolarization/repolarization) and planes (frontal/transverse in ECG; XYP/XZP in VCG). Afterwards, Receiver Operative Characteristic (ROC) curves were computed for every index, and a subset with all those indexes with the area under the curve (AUC) values greater than 0.70 was selected for further analysis.

2.5. Statistical Analyses
All data were expressed as Mean±SD. The D'Agostino-Pearson normality test was applied to quantify the discrepancy between the distribution of the indexes and an ideal Gaussian distribution. In order to determine the statistical power of each marker to discriminate health from hypertrophy, a non-parametric two-tailed Mann-Whitney test was applied between controls and LVH patients. When p- value was <0.05, differences were considered statistically significant.

3. Results
Figure 1 shows the dominant depolarization and repolarization vectors for the control and the hypertrophy groups for every plane in both ECG and VCG spaces. These vectors resulted from averaging the i_{th} beat-dominant vectors accross all patients within each group. The complete numerical values and statistical significance can be found in Table 1.

LVH indexes based on amplitude and angle relationships of the above calculated vectors and their corresponding ROC curves and areas were computed. Figure 2 shows the AUC values for every index under study in both the ECG and VCG space. Notice that four out of ten indexes produced AUC values greater than 0.70. The latter indexes were: T_H^{α}/T_H^{m}, R_H^{α}/T_H^{m}, R_F^{α}/T_H^{m} and R_H^{α}/T_F^{m}. Notice that three out of four indexes in this selected subset combined depolarization and repolarization information in both representations: the frontal and transverse planes and all the three of them belonged to the ECG space.

The ROC curves for the four pairs of LVH indexes with AUC values greater than 0.70 together with their ECG(VCG) counterparts are displayed in Figure 3. The optimal cut-off point in the ROC curves were computed as the point nearest to the top left-hand corner. This selection maximizes the sensitivity and specificity sum, when it is assumed that the 'cost' of a false negative result is the same as that of a false positive one [14]. Characterization of the cut-off points are also shown in the paired format (Sensitivity; Specificity).

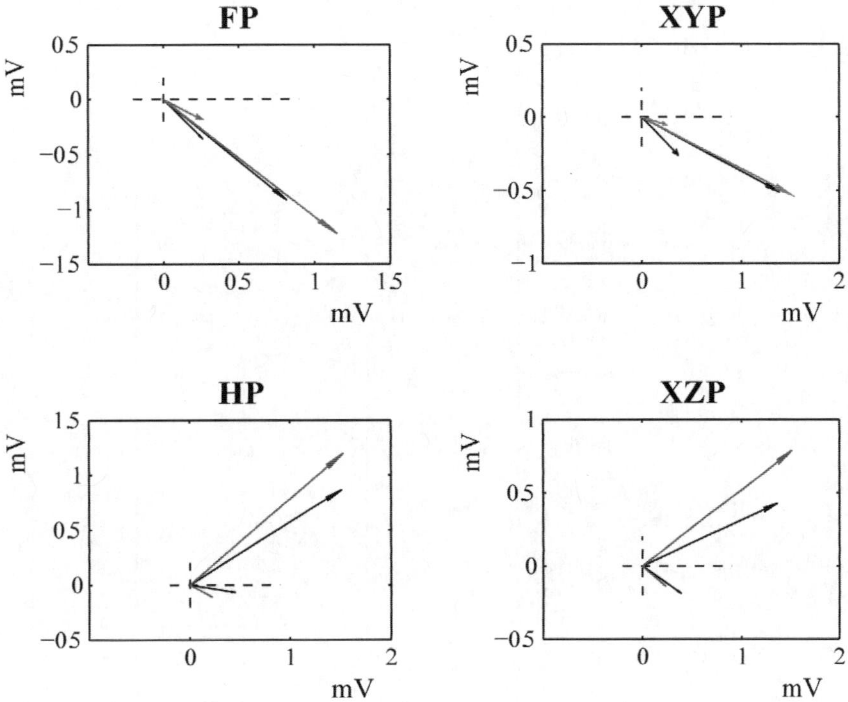

Figure 1. Mean depolarization and repolarization vectors for controls (solid lines) and LVH (grey line) groups.

Table 1. Mean±SD values for every parameter in control and LVH groups. * p<0.05, # p<0.005, & p<0.0005. 1: Frontal Plane (FP), 2: transverse plane (TP), 3: XY plane (XYP), 4: XZ plane (XZP)

	Controls				LVH			
	$R^{\alpha}[^{o}]$	R^{m}[mV]	$T^{\alpha}[^{o}]$	T^{m}[mV]	$R^{\alpha}[^{o}]$	R^{m}[mV]	$T^{\alpha}[^{o}]$	T^{m}[mV]
1	48.24±33.66#	1.22±0.44*	54.08±25.17*	0.44±0.20#	46.84±35.44	1.68±0.79	35.78±42.20	0.30±0.12
2	-29.83±28.00	1.74±0.41	8.45±33.80	0.45±0.17&	-38.12±24.58	1.93±1.09	26.09±50.22	0.24±0.13
3	20.28±35.20*	1.47±0.35	35.01±16.61	0.45±0.17&	19.35±12.31	1.59±0.74	12.20±62.00	0.25±0.10
4	-17.39±31.37	1.42±0.34	25.38±26.25*	0.43±0.16&	-27.49±35.18	1.70±0.82	30.85±57.56	0.26±0.11

4. Discussion

Four ECG-based LVH index combining features from the QRS-complex and the T-wave in both transverse and frontal planes were presented. Although it has been well described that LVH alters both depolarization and repolarization phases [15, 3], no utilization of the T-wave amplitude in LVH diagnosis has been reported in the clinical practice. ECG describes the electrophysiological remodelling induced by LVH, consisting of a ventricular conduction delay [16] and a prolongation of the action potential duration [17]. Thus, the final repolarization changes such as amplitude, axis and loop morphology result from the interplay with the depolarization modifications induced by LVH. These changes were classified into primary and secondary repolarization changes by Bacharova et al. [3] . This depolarization/repolarization interplay, makes it even harder to obtain clear ECG/VCG patterns when studied the cardiac

Journal of Physics: Conference Series **705** (2016) 012031 doi:10.1088/1742-6596/705/1/012031

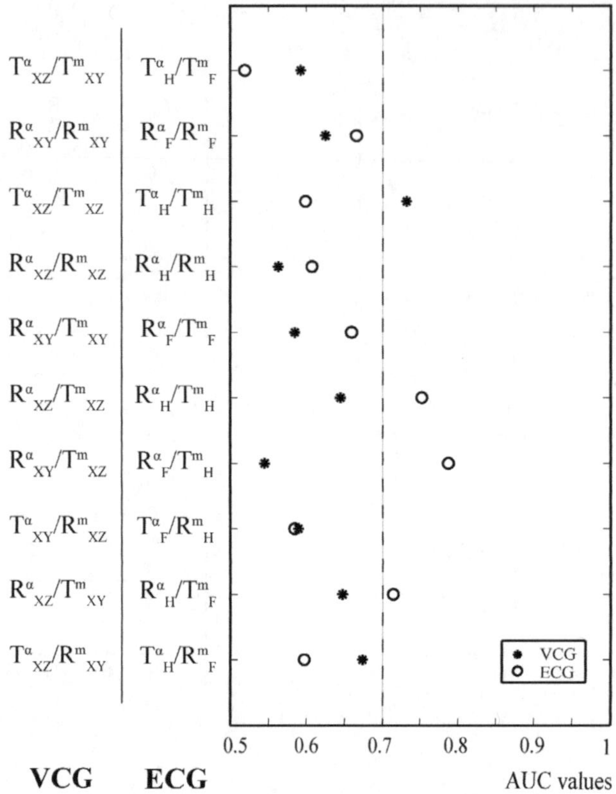

Figure 2. AUC values for all the LVH indexes proposed. AUC values for every index are plotted for the ECG (asterisks) and the VCG (circles) counterparts.

phases isolatedly. The main hypothesis applied herein was that cardiac vectors compensate for this particular fluctuations and show a more robust behavior to the LVH-induced changes when studied simultaneously.

On the other hand, the best performance of R_F^{α}/T_H^{m} is greatly supported by the literature. The Romhilt-Estes score includes the electrical axis in the frontal plane [18] and the amplitude of the T-wave has been a LVH marker in hypertrophic patients without ECG criteria for LVH [19]. Finally, even though angle and modulus of the vectors conforming the index were calculated from the depolarization and repolarization loops, which are mathematical constructions, the equivalent elements, namely, the left axis deviation in the frontal plane and the T-wave amplitude in the transverse plane (V_6 in this case) can be easily obtained in the clinical scenario, providing an efficient LVH index fully available with a 12-leads ECG record.

5. Conclusion

Most of the LVH indexes with a good performance to separate out LVH patients from controls belonged to the ECG space, vindicating the ECG for LVH diagnosis. The best electrocardiographic LVH index derived from this analysis, R_F^{α}/T_H^{m} combined depolarization and repolarization parameters suggesting that repolarization would greatly contribute to LVH diagnosis.

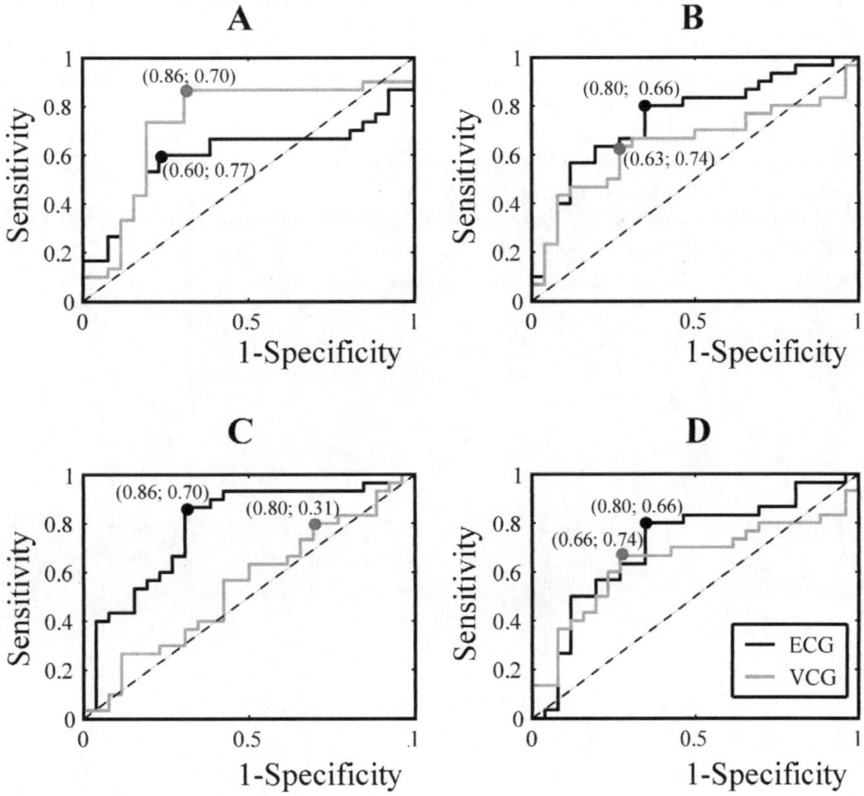

Figure 3. ROC curves for the subset of LVH indexes with AUC values greater than 0.7. A): $R^{\alpha}_{XZ}/T^{m}_{XZ}$, B): R^{α}_{H}/T^{m}_{H}, C): R^{α}_{F}/T^{m}_{H} and D): R^{α}_{H}/T^{m}_{F}. Every index is plotted with the ECG (black) and VCG (grey) counterparts.

Acknowledgements

This work was supported by the Consejo Nacional de Investigaciones Científicas y Técnicas, under Project PIP-538 and PICT 2008 #2108 Agencia Nacional de Promoción Científica y Tecnológica, Argentina.

6. References

[1] Bacharova L 2009 *Journal of Electrocardiology* **42** 388–91
[2] Bacharova L 2009 *Journal of Electrocardiology* **42** 593:96
[3] Bacharova L, Szathmary V and Mateasik A 2010 *J. of Electrocardiology* **43(6)** 624–33
[4] Casale P, Devereux R, Kligfield P, Eisenberg R, Miller D, Chaudhary B and Phillips M 1985 *J Am Coll Cardiol.* **6** 572–80
[5] Sokolow M and Lyon T 1949 *Am. Heart. J.* **37** 161–86
[6] Romhilt D, Greenfield J and Estes E 1968 *Circulation* **XXXVII** 15–19
[7] Bristow J, Porter G and H G 1961 *Am Heart J.* **62** 621
[8] Pérez Riera A, Uchida A, Filho C, Meneghini A, Ferreira C, Schapacknik E, Dubner S and Moffa P 2007 *Clin Cardiol.* **30(7)** 319–23
[9] Borun E, Chapman J and Massey F 1966 *Am. J. Cardiol.* **18** 656
[10] Devereux R B and Reichek N 1977 *Circulation* **55** 613–8
[11] Pan J and Tompkins W J 1985 *Biomedical Engineering, IEEE Transactions on* **BME-32** 230–236
[12] Mendieta J 2012 *Algoritmo para el delineado de senales electrocardiograficas en un modelo animal empleando*

tecnicas avanzadas de procesamiento de senales Master's thesis Facultad de Ingenieria, Univ. de Buenos Aires

[13] Kors J A, Van Herpen G, Sittig A C and Van Bemmel J H 1990 *European Heart Journal* **11** 1083–1092

[14] Altman D 1991 *Practical Statistics for Medical Research* (26 Boundary Row, London: Chapman Hall)

[15] Hart G 1994 *Cardiovascular Research* **28** 933–946

[16] Cooklin M, Wallis W R J, Sheridan D J and Fry C H 1997 *Circulation Research* **80** 765–771

[17] McIntosh M, Cobbe S, Kane K and Rankin A 1998 *Journal of Molecular and Cellular Cardiology* **30** 43 – 53

[18] Romhilt D and Estes E J 1968 *Am Heart J* **75** 752–58.

[19] Dilaveris P, Gialafos E, Poloniecki J, Hnatkova K, Richter D, Andrikopoulos G, Lazaki E, Gialafos J and Malik M 2000 *Clinical Cardiology* **23** 600–606

A new algorithm for epilepsy seizure onset detection and spread estimation from EEG signals

Antonio Quintero-Rincón[1], Marcelo Pereyra[2], Carlos D'Giano[3], Hadj Batatia[4], Marcelo Risk[1,5]

[1] Department of Bioengineering, Instituto Tecnologico de Buenos Aires (ITBA), Av. Eduardo Madero 399, C1106ACD, Buenos Aires, Argentina.
[2] Department of Mathematics, University of Bristol, University Walk, Clifton, Bristol BS8 1TW, UK.
[3] Fundación contra las Enfermedades Neurológicas Infantiles (FLENI), Argentina.
[4] University of Toulouse, IRIT - INP-ENSEEIHT, Toulouse, 2, rue Charles Camichel, B.P. 7122, 31071 Cedex 7, France.
[5] Consejo Nacional de Investigaciones Científicas y Técnicas (CONICET), Argentina.

E-mail: `aquinter@itba.edu.ar`

Abstract.
Appropriate diagnosis and treatment of epilepsy is a main public health issue. Patients suffering from this disease often exhibit different physical characterizations, which result from the synchronous and excessive discharge of a group of neurons in the cerebral cortex. Extracting this information using EEG signals is an important problem in biomedical signal processing. In this work we propose a new algorithm for seizure onset detection and spread estimation in epilepsy patients. The algorithm is based on a multilevel 1-D wavelet decomposition that captures the physiological brain frequency signals coupled with a generalized gaussian model. Preliminary experiments with signals from 30 epilepsy crisis and 11 subjects, suggest that the proposed methodology is a powerful tool for detecting the onset of epilepsy seizures with his spread across the brain.

1. Introduction

The term epilepsy derives from the Greek term *epilambanein* which means to seize, and it denotes the predisposition to have recurrent, unprovoked seizures. Seizures can be symptomatic; that is, result from specific precipitants such as fever, strokes, metabolic disturbances (e.g. hypoglycaemia, drug abuse/withdrawal), trauma, infections in the central nervous system, and acute head injury. In epilepsy, however, seizures are unprovoked and expected to be recurrent [46]. Appropriate diagnosis and treatment of epilepsy is a main public health issue. According the World Health Organization [1], there are more than 50 millions people worldwide that suffer from some form of epilepsy, nearly 80% of them are in developing regions, where it is believed that 3 out of 4 people with this conditions do not get appropriate diagnostic and treatment [1].

The International League Against Epilepsy (ILAE) [10] defines *"epileptic seizure as a transient occurrence of signs and/or symptoms due to abnormal excessive or synchronous neuronal activity in the brain"*. The elements defining an epileptic seizure include its mode of onset and termination, its clinical manifestations, and its abnormal enhanced synchrony [32]. Physical manifestations of epilepsy result mainly from the synchronous and excessive discharge of electricity by a group of neurons behaving abnormally in the cerebral cortex. Epileptic seizures usually have a sudden onset, they spread within seconds and, in most cases, are brief. The precise manifestation of a seizure depends on the location in the brain where it originates, and on how far and fast it spreads. The correct identification of this location and spread information is key to a proper treatment.

Electroencephalography (EEG) is a non-invasive and widely available biomedical modality that can be used to diagnose epilepsy and plan treatment. One particular task that has received attention during the last years is the detection of epileptic seizures from EEG signals measured on the brain scalp. It is widely acknowledged that detection can be improved by summarizing the EEG signals using relevant descriptors; that is, quantities computed from the EEG signals that capture the main relevant features of the seizure, and whose values help discriminating between normal and abnormal brain activity. The most relevant EEG features studied in the literature can be classified in three groups: a) frequency spectral properties such as band power and edge frequency; b) signal morphological properties such as wave amplitude, sharpness, line length, and zero crossings; and c) statistical descriptors such as statistical moments, correlation times and information entropies.

In this work we study a new algorithm for epilepsy seizure onset detection and spread estimation. The algorithms is based on a statistical generalised Gaussian model that represents the activity in each of the brain's main frequency bands, which are identified by using a multi-resolution temporal wavelet representation. The remainder of this document is structured as follows. Section 2 describes the state of the art in seizure onset detection (SOD). Section 3 defines the univariate generalised Gaussian statistical model. Section 4 describes the proposed the epilepsy seizure onset detection and spread estimation algorithm. In Section 5 the proposed methodology is demonstrated on real EEG signals from patients suffering from epileptic seizures. Discussions and conclusions are finally reported in Section 6.

2. Seizure Onset Detection

Seizure onset detection (SOD) methods stem from the observation that EEG descriptors allow discriminating between normal and abnormal brain activity; that is, between the expected rhythms of the brain and varying degrees of thalamocortical interdependence, and the activity resulting from seizures and burst suppressions. Epileptic attacks have two clinical manifestations that may reflect abnormal brain activity: *Ictal* and *Inter-Ictal* discharges, in which the impaired consciousness plays an important role (this is the inability to respond normally to exogenous stimuli by virtue of altered awareness and/or responsiveness) [24]. The Interictal discharges are a positive indicator of epilepsy and, depending on the duration of EEG recordings and the inclusion of different states of vigilance, they can be observed in up to 90% of cases [37].

The Ictal discharges are clinical signs used to detect the onset seizure from the epileptogenic zone in the brain cortex, SOD may enable physicians to better therapy with drug treatment, diagnostic and alert procedures, since is intended to recognize the start of a seizure, with the shortest possible delay, but not necessarily with the highest possible accuracy. The SOD when used with intracranial recordings, for seizure identification and retrospective analysis of seizures, often in the context of presurgical evaluations, it can be relatively straightforward to detect with reasonable sensitivity and specificity since it is dealing with events often lasting over a minute [22]. Another use of SOD with extracranial recordings to distinguish between primary and secondary irritative areas may be difficult, because a interictal discharge can spread via normal anatomical connections between cortical areas, through the commissural fibers or via subcortical structures, finally this may lead to widespread or bilateral occurrence of interictal discharges [45].

Seizure onset detection was first investigated in the seventies by Viglione et al. [50] and Liss et al.[30], and with later contributions by Ktonas et al. [27] and Gotman et al. [14]. Moreover, several works studied linear and nonlinear prediction techniques to separate transients from background activity. For example, filter techniques [3], power spectrum techniques [11], cross-correlation techniques [52], principal and independent component analysis techniques [33], [20] have been investigated. Other examples include techniques based on wavelet representations [15], state space reconstruction [35], correlation measures [43], signal dimension [17], density [29], integral [41], mutual prediction [44], Lyapunov exponents [19], synchronization [40], similarity measures [18], recurrence quantification measures [47], and nonlinear predictability [4]. We refer the reader to [49] for a comprehensive treatment of measurement, models, detection and prediction. Other important surveys of the literature in this topic can be found in [23], [15], [26], [7], [21], [53], [34], [28], [8], [48].

Modern SOD methods can be grouped into the following categories: 1) *Template matching*: These are techniques based on finding events tha match previously selected spikes; the detection is made whenever the cross correlation of the EEG with a template exceeds a threshold [31]; 2) *Parametric methods*: These techniques are based on traditional signal processing and consider that a seizure has occurred when the difference between the EEG signal and a predicted value (based on the assumption that the background is stationary) exceeds a threshold [2]; 3) *Mimetic Methods*: these techniques seek to copy the human expert (i.e. neurophysiologist) and operate by monitoring the value of parameters computed from each wave and applying thresholds [13], [6], [25]; 4) *Morphologic Analysis*: these techniques are based on the characterisation of the waveforms in terms of sharpness, amplitude, duration, convexity, frequency bands or time-frequency representations of spikes [13], [16]; 5) *Syntactic methods*: these techniques are based on the detection of the presence of structural features [51]; 6) *Neural Networks*: this approach adopts a machine learning perspective to learn transients related to epileptic seizures [42], [38]; 7) *Expert systems*: this approach detects seizures by mimicking an expert's knowledge and reasoning process[12]; 8) *Data mining techniques*: this approach also adopts a machine learning perspective to train a classifier [8], [9], [48]; 9) *Clustering Techniques*: in this case detections are based on hierarchical agglomerative processes and self-organising maps [3], [48]; 10) *Knowledge-based rules*: these techniques seek to incorporate knowledge from neurophysiologists who adopt spatial and temporal rules [5], [7], [5], [48].

3. Generalized Gaussian Distribution

The univariate generalised Gaussian distribution (GGD) is a flexible statistical model that has found numerous applications in science and engineering, and which recently was use to model epilepsy signals in [39]. This statistical model has 3 parameters, with a probability density function (PDF) given by:

$$f(x;\mu,\sigma,\mathcal{B}) = \frac{\mathcal{B}}{2\mathcal{A}(\sigma)\Gamma(\frac{1}{\mathcal{B}})} exp - \frac{|(x-\mu)|^{\mathcal{B}}}{2\sigma^2} \quad (1)$$

with

$$\mathcal{A}(\sigma) = \sigma\sqrt{\frac{\Gamma(\frac{1}{\mathcal{B}})}{\Gamma(\frac{3}{\mathcal{B}})}}, \quad \Gamma(z) = \int_0^{+\infty} e^t e^{-z-1} dt, \quad z > 0$$

and where $\mu \in \mathcal{R}$ is a location parameter, $\sigma \in \mathcal{R}^+$ is a scale parameter and $\mathcal{B} \in \mathcal{R}^+$ is a shape parameter that controls the shape of the density tail. The density (1) has many properties that are useful for modelling signals: i) the value of (1) approaches zero as x approaches positive and negative infinity, ii) the mean and mode of (1) coincide and are given by μ, iii) (1) is symmetric with respect to the mean value μ. Notice that the GGD parametric family comprises many popular distributions that are commonly used in biomedical signal processing. For example, setting $\mathcal{B} = 1$ leads to a Laplacian or double-exponential distribution, $\mathcal{B} = 2$ to a Gaussian or normal distribution, and $\mathcal{B} \to \infty$ to a uniform distribution. We refer the reader to [36] for a comprehensive treatment of the mathematical properties of the GGD statistical model.

4. Methodology

The first step of the proposed methodology is to represent the EEG signals in a time-frequency domain. This step allows decomposing each one of the EEG channels in the signal array into five frequency bands associated with the five *brain rhythms* that clinicians typically use to capture different types of neurological activity. We believe that this decomposition is very useful for SOD because there are rhythms or frequency patterns that appear often at the beginning of the epileptic crisis and that are difficult to quantify visually. We have used a time-frequency representation associated with the following bands: *Delta*: 0.5-4Hz, *Theta*: 4-8Hz, *Alpha*: 8-13Hz, *Beta*: 13-30Hz and *Gamma*: 30-45Hz. In order to achieve this we used a multilevel 1-D wavelet decomposition filter constructed using a Dauchebies wavelet decomposition with five resolution scales. From the wavelet decomposition it is then possible to determine how the energy in the signal is spread over the different *brain rhythms*, and how this distribution of energy evolves over time. Precisely, this information is captured by modelling the signals associated each each brain rhythm with a GGD statistical model, whose parameters $\mathcal{A}$ and $\mathcal{B}$ are estimated from the observed data, see equation (1). *The shape parameter* $\mathcal{A}$ *is closely related to the variability of the brain activity and is therefore a good descriptor for performing SOD detection.*

We evaluated the performance of the proposed seizure onset detector and spread estimation algorithms by using the Children's Hospital Boston database [45], which consists of 36 EEG recordings from paediatric subjects with intractable seizures. In this work we used 30 crisis from 11 subjects. A set including two to five bipolar EEG recordings sampled at 256Hz were available for each subject. Each recording contained a seizure event with a labeled onset that was detected by an experienced neurologist, who worked backward from the observed clinical onset to find the electrographic onset. Each signal was edited to have an epoch with the following characteristics: 2 minutes before crisis, crisis at minute 2 and 2 minutes after crisis.

For each epoch we know where the crisis begins and can calculate the onset delay by comparing the average amplitude with the background, similar to [16]. Once onset delay estimation has been performed for each brain rhythm and each channel (by performing SOD with the respective parameter $\mathcal{A}$), we collect this information in a table that allows identifying channels with low delay and thus understanding how the seizure originated and propagated. This is illustrated in Figures (1) to (6) and Table (1). The proposed algorithm can be summarised as follows:

(i) Separate each EEG signal in the array of signals into the different physiological brain frequency bands.

(ii) Estimate the GGD parameters $\mathcal{A}$ and $\mathcal{B}$, see eq. (1).

(iii) Use the parameter A to calculate the delays of each SOD for each channel and each frequency bands.

(iv) Create a table with the information and calculate each delay for each channel by frequency bands.

(v) Organize the table ascendently.

(vi) Calculate the SOD candidate channels for the different brain areas (i.e. Frontal, Parietal, Temporal and Occipital).

(vii) The candidates for the onset are the channels that are common to all other channels by frequency band in a determined brain area; and the other channels are the possible spread accompanying the crisis.

(viii) Repeat these steps for all the rhythms bands.

Table 1. Onset Delay (in seconds) by Frequency Bands, the symbol (-) means that there is no clear difference in the beginning of the onset. In this example, according to the data, the channels (13) FP2-F8, (14) F8-T8 and (15) T8-P8 are the best candidates by brain area for SOD because it have minor delays; in this case F8 is common to all channels in this brain area, therefore is the best option for be the onset and the other channels form the spread.

Channel Number	Channel Name	Delta	Theta	Alpha	Beta	Gamma
1	FP1-F7	-0.099	.001	.002	.13	.131
2	F7-T7	-	.069	.234	.131	.058
3	T7-P7	-	.037	.215	.369	.371
4	P7-O1	-	-	-	.224	.162
5	FP1-F3	-	.037	.011	.133	.133
6	F3-C3	-	-	-	.221	.132
7	C3-P3	-	-	-	.378	.14
8	P3-O1	-	-	-	.534	.552
9	FP2-F4	.083	.06	.037	.212	.098
10	F4-C4	-	.001	-	.125	.110
11	C4-P4	-	.018	.029	-	.103
12	P4-O2	-	.035	-	-	-
13	FP2-F8	.105	.012	.025	.021	.101
14	F8-T8	.073	.087	.134	.097	.1
15	T8-P8	.081	.018	.013	.099	.103
16	P8-O2	-	.036	.016	.253	.086
17	FZ-CZ	-	-	-	.034	.141
18	CZ-PZ	-	-	-	-	-
19	P7-T7	.086	.085	.04	2.36	.35
20	T7-FT9	-	.069	-	.372	.377
21	FT9-FT10	.005	.001	.016	.098	.117
22	FT10-T8	.075	.066	.01	.102	.101
23	T8-P8	.065	.032	.015	.098	.117

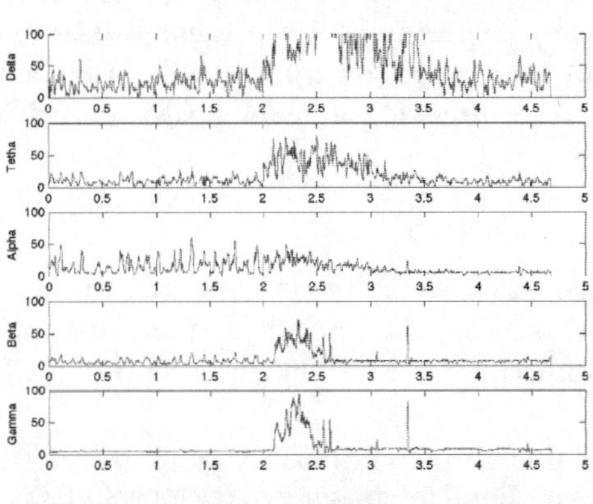

Figure 1. $\mathcal{A}$ for Channel 13.

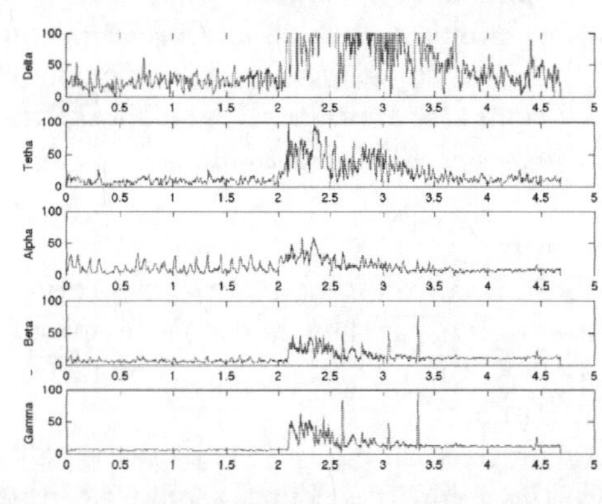

Figure 2. $\mathcal{A}$ for Channel 14.

No distinctions regarding the types of seizure onsets were considered; the data contains focal, lateral, and generalized seizure onsets. Furthermore, the recordings were made in a

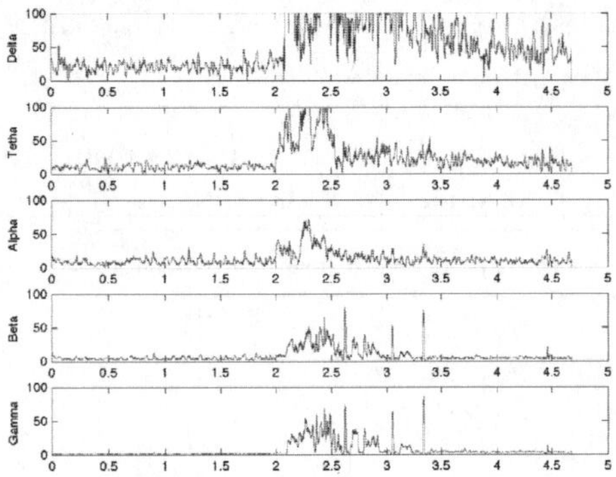

Figure 3. $\mathcal{A}$ for Channel 15.

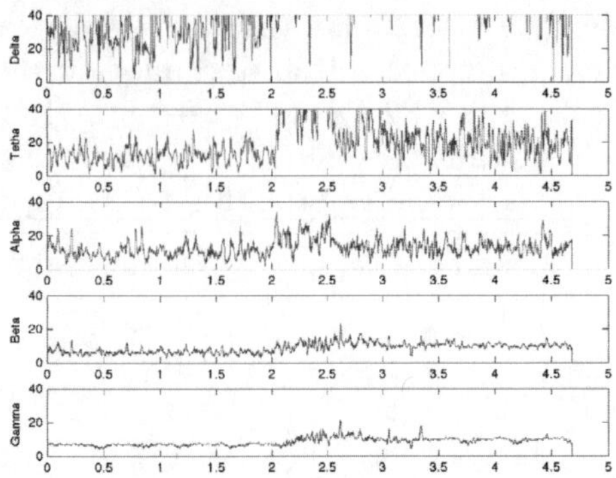

Figure 4. $\mathcal{A}$ for Channel 16.

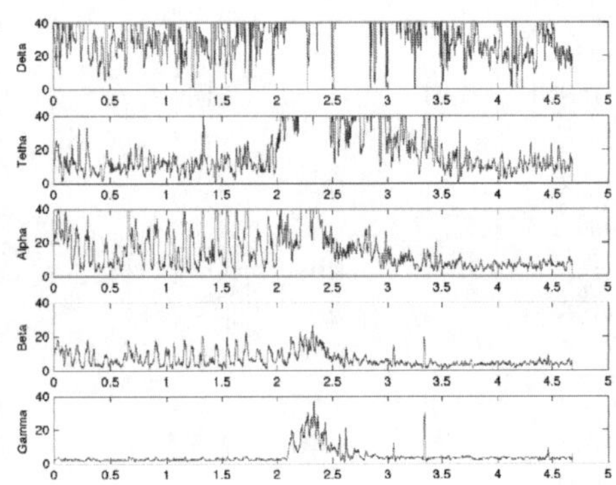

Figure 5. $\mathcal{A}$ for Channel 10.

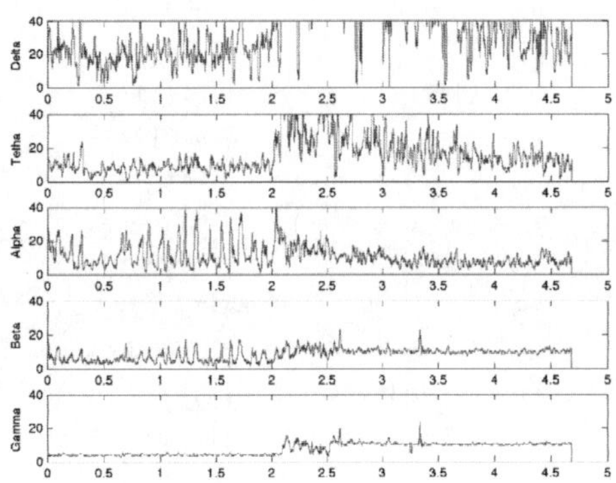

Figure 6. $\mathcal{A}$ for Channel 11.

routine clinical environment, so non-seizure activity and artifacts such as head/body movement, chewing, blinking, early stages of sleep, and electrode pops/movement are present in the data. The set of recordings lasted on average 35 minutes for 30 subjects in total; 2 hours for 4 subjects; and 12 hours for 2 other subjects. Taken together the recordings account for 60 hours of EEG recordings and 139 seizures [45].

5. Results

The performance of the algorithm proposed for seizure onset detector and spread estimation was assessed using the Children's Hospital Boston database described above, and by comparing results with those obtained by visual inspection by an experimented neurologist relying on EEG and MRI data.

For illustration figure (8) shows the EEG and MRI image for one patient; this information allows the reconstruction of a model of the brain and the determination of the location where the seizure was originated, as well as how is spread throughout the brain, which allows is the information that our algorithm seeks to estimate directly from the EEG.

In this SOD example, we can see the $\mathcal{A}$ parameter in the x-axis, se Figures (1) to (6), which correspond to right temporal brain area. Our algorithm shows that the seizure starts in channel F8 and then spreads across the channels F8-T8-P8, F4-C4-P4, as is shown the data in table (1) and Figure (7).

Channels with the minor delay in the right temporal brain area: (13) FP2-F8, (15) T8-P8, (22) FT10-T8, (14) F8-T8, (12) P4-O2.

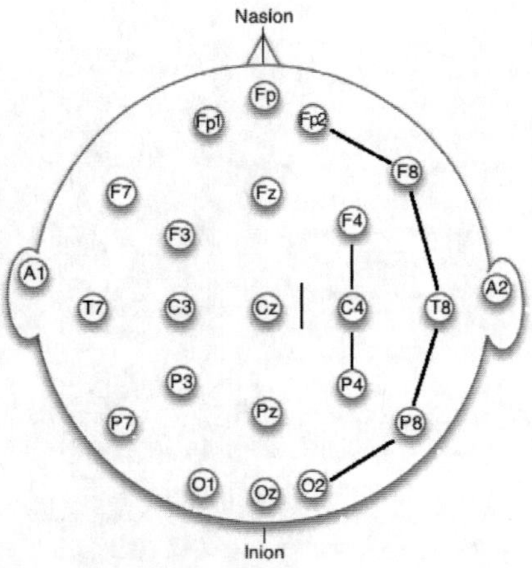

Figure 7. EEG electrodes Array, shows that the seizure starts in channel F8 and then spreads across the channels F8-T8-P8, F4-C4-P4

Spread by Delta Bands: (10) F4-C4, (21) FT9-FT10, (11) C4-P4, (23) T8-P8, (14) F8-T8, (22) FT10-T8, (15) T8-P8, (9) FP2-F4, (19) P7-T7, (1) FP1-F7, (13) FP2-F8.

Spread by Theta Bands: (1) FP1-F7, (21) FT9-FT10, (13) FP2-F8, (15) T8-P8, (11) C4-P4, (23) T8-P8, (16) P8-O2, (3) T7-P7, (5) FP1-F3, (9) FP2-F4, (22) FT10-T8, (2) F7-T7, (20) T7-FT9, (19) P7-T7, (14) F8-T8.

Spread by Alpha Bands: (1) FP1-F7, (22) FT10-T8, (5) FP1-F3, (15) T8-P8, (23) T8-P8, (16) P8-O2, (21) FT9-FT10, (13) FP2-F8, (12) P4-O2, (9) FP2-F4, (19) P7-T7, (14) F8-T8, (3) T7-P7, (2) F7-T7.

Spread by Beta Bands: (13) FP2-F8, (17) FZ-CZ, (14) F8-T8, (21) FT9-FT10, (23) T8-P8, (15) T8-P8, (22) FT10-T8, (10) F4-C4, (1) FP1-F7, (2) F7-T7, (5) FP1-F3, (9) FP2-F4, (6) F3-C3, (4) P7-O1, (16) P8-O2, (3) T7-P7, (20) T7-FT9, (7) C3-P3, (8) P3-O1, 19 P7-T7.

Spread by Gamma Bands: (2) F7-T7, (16) P8-O2, (9) FP2-F4, (14) F8-T8, (13) FP2-F8, (22) FT10-T8, (11) C4-P4, (15) T8-P8, (10) F4-C4, (21) FT9-FT10, (23) T8-P8, (1) FP1-F7, (6) F3-C3, (5) FP1-F3, (7) C3-P3, (17) FZ-CZ, (4) P7-O1, (19) P7-T7, (3) T7-P7, 20 T7-FT9, (8) P3-O1.

6. Discussion

The preliminary results reported in this work suggest that the proposed algorithm is potentially useful for onset detection and spread estimation, and that performing the analysis at the level of the brain activity rhythm bands can improve the identification of the area of the brain affected. This suggests that the algorithm is potentially interesting for devising automatic processing systems.

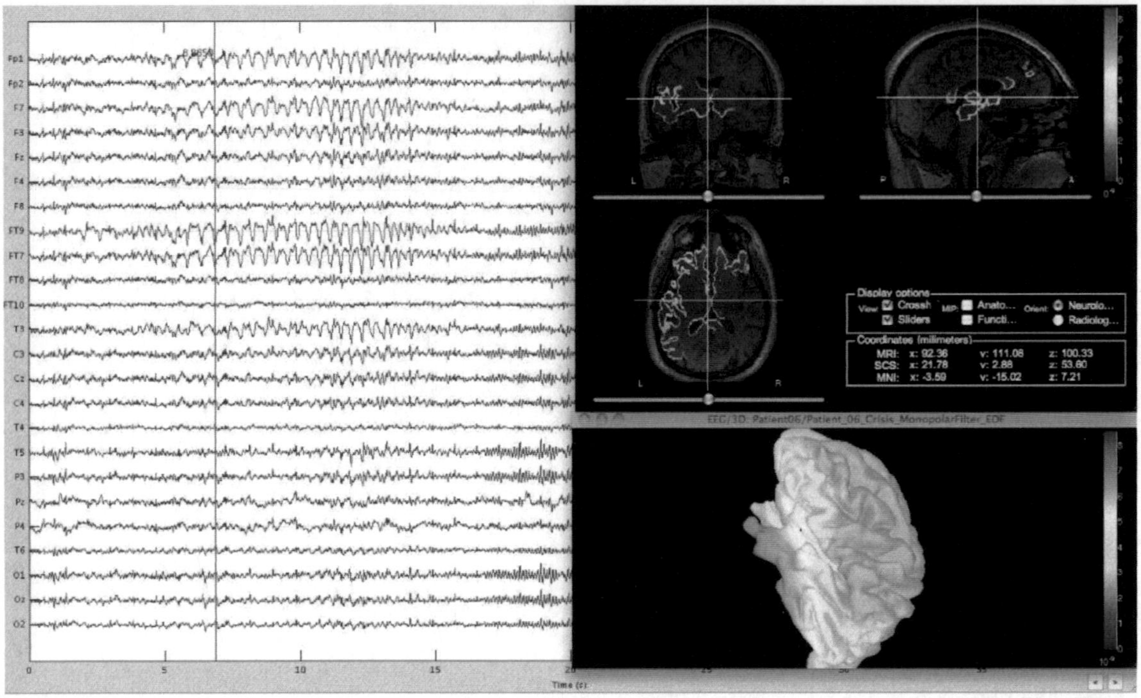

Figure 8. For illustration, the figure shows the EEG of a patient (left) with his corresponding MRI (right top). A reconstruction of the brain is shown with his corresponding brain spread (left bottom). On the right of each image is a color map which indicates the intensity of each sour activation.

Another interesting research questions to explore in more depths, are related with the analysis of the frequency bands detected before the onset seizure, and which channels are active despite not participating directly in the detection of the seizure onset.

Perspective for future work include an extensive evaluation of the proposed methodology, as well as performing comparisons with other detection methods from the state of the art, and the development of fusion techniques to combine detections from several algorithms to increase robustness to noise and to artefacts.

7. Acknowledgements

Part of this work was funded by the SuSTaIN program - EPSRC grant EP/D063485/1 - at the Department of Mathematics, University of Bristol.
Marcelo Pereyra holds a Marie Curie Intra-European Fellowship for Career Development.
The authors acknowledges to Gabriela Ugarnes, from Fundación contra las Enfermedades Neurológicas Infantiles (FLENI), Argentina.

8. References

[1] World health organization. mental health action plan 2013-2020. who document production services, geneva, switzerland.
[2] Kaoru Arakawa, Derek H. Fender, Hiroshi Harashima, Hiroshi Miyakawa, and Yoichi Saitoh. Separation of a nonstationary component from the eeg by a nonlinear digital filter. *Advances in Behavioral Biology*, 33(7):724–726, 1986.
[3] William P. Birkemeier, A. Burr Fontaine, Gastone G. Celesia, and Ken M. Ma. Pattern recognition techniques

for the detection of epileptic transients in eeg. *IEEE Transactions on Biomedical Engineering*, 25(3):213–217, 1978.

[4] K. J. Blinowska and M. Malinowski. Non-linear and linear forecasting of the eeg time series. *Biological Cybernetics*, 66(2):159–165, 1991.

[5] B. Davey, W.R. Fright, G.J. Caroll, and R.D. Jones. Expert system approach to detection of epileptiform activity in the eeg. *Medical and Biological Engineering and Computing*, 27:365–370, 1989.

[6] P. Guedes de Oliveira, C. Queiroz, and F.H. Lopes da Silva. Spike detection based on a pattern recognition approach using a microcomputer. *Electroencephalography and Clinical Neurophysiology*, 56(1):97–103, 1983.

[7] Alison A. Dingle, , Richard D. Jones, and Grant J. Carroll W. Richard Fright. A multisatge system to detect epileptiform activity in the eeg. *IEEE Transactions on Biomedical Engineering*, 40(12):1260–1268, 1993.

[8] Mahmoud El-Gohary, James McNames, and Siegward Elsas. User-guided interictal spike detection. *30th Annual International Conference of the IEEE in Engineering in Medicine and Biology Society*, pages 821–824, 2008.

[9] Themis P. Exarchos, Alexandros T. Tzallas, Dimitrios I. Fotiadis, Spiros Konitsiotis, and Sotirios Giannopoulos. Eeg transient event detection and classification using association rules. *IEEE Transactions on Information Technology in Biomedicine*, 10:451–457, 2006.

[10] Robert S. Fisher, Carlos Acevedo, Alexis Arzimanoglou, Alicia Bogacz, J. Helen Cross, Christian E. Elger, Jerome Engel Jr., Lars Forsgren, Jacqueline A. French, Mike Glynn, Dale C. Hesdorffer, B.I. Lee, Gary W. Mathern, Solomon L. Moshe, Emilio Perucca, Ingrid E. Scheffer, Torbjorn Tomson, Masako Watanabe, and Samuel Wiebe. Ilae official report: A practical clinical definition of epilepsy. *Epilepsia*, 55(4):475–482, 2014.

[11] R.S. Fisher, W.R. Webber, R.P. Lesser, S. Arroyo, and S.R.S. Uematsu. High-frequency eeg activity at the start of seizures. *Journal of Clinical Neurophysiology*, 9(3):441–448, 1992.

[12] John R. Glover, Narasimha Raghavan, Periklis Y. Ktonas, and James D. Frost Jr. Context-based automated detection of epileptogenic sharp transients in the eeg: Elimination of false positives. *IEEE Transactions on Biomedical Engineering*, 36:519–527, 2003.

[13] J. Gotman and P. Gloor. Automatic recognition and quantification of interictal epileptic activity in the human scalp eeg. *Electroencephalography and Clinical Neurophysiology*, 41(5):513–529, 1976.

[14] J. Gotman, J.R Ives, and P. Gloor. Automatic recognition of inter-ictal epileptic activity in prolonged eeg recordings. *Electroencephalography and Clinical Neurophysiology*, 46(5):510–520, 1979.

[15] Jean Gotman. Computer analysis of the eeg in epilepsy. *Clinical applications of computer analysis of EEG and other neurophysiological signals, Elsevier Amsterdam*, pages 171–204, 1986.

[16] Jean Gotman. Automatic seizure detection: improvements and evaluation. *Electroencephalography and Clinical Neurophysiology*, 76(4):317–324, 1990.

[17] Peter Grassberger and Itamar Procaccia. Characterization of strange attractors. *Physical Review Letters*, 50:346–349, 1983.

[18] L.M. Hively, P.C. Gailey, and V.A. Protopopescu. Detecting dynamical change in nonlinear time series. *Physical Letter A*, 258:103–114, 1999.

[19] Leonidas D. Iasemidis and J. Chris Sackellares. Chaos theory and epilepsy. *Neuroscientist*, 2:118–126, 1996.

[20] Anthony J.Bell and Terrence J. Sejnowski. An information-maximization approach to blind separation and blind deconvolution. *Neural Computation*, 7(6):1004–1034, 1995.

[21] K.K. Jerger, T.I. Netoff, J.T. Francis, T. Sauer, L. Pecora, S.L. Weinstein, and S.J.: Schiff. Early seizure detection. in: Journal of clinical neurophysiology. *Journal of Clinical Neurophysiology*, 18(3):259–268, 2001.

[22] Christophe C. Jouny, Piotr J. Franaszczuk, and Gregory K. Bergey. Improving early seizure detection. *Epilepsy and Behavior*, 22(1):S44–S48, 2011.

[23] J.D. Frost Jr. Automatic recognition and characterization of epileptiform discharges in the human eeg. *Journal of Clinical Neurophysiology*, 2(3):231–249, 1985.

[24] Jerome Engel Jr. *Epilepsy Board Quick Review Selected Tables and Figures from Seizures and Epilepsy*. Oxford University Press, 2014.

[25] Periklis Y. Ktonas. Automated analysis of abnormal electroencephalograms. *Critical Reviews in Biomedical Engineering*, 9(1):39–97, 1983.

[26] Periklis Y. Ktonas. Automated spike and sharp wave (ssw) detection. *Methods of analysis of brain electrical and magnetic signals. Elsevier Amsterdam*, pages 211–411, 1987.

[27] P.Y. Ktonas and J.R. Smith. Quantification of abnormal eeg spike characteristics. *Computers in Biology and Medicine*, 4(2):157–163, 1974.

[28] Klaus Lehnertza and Brian Litt. The first international collaborative workshop on seizure prediction summary and data description. *Journal of Clinical Neurophysiology*, 116:493–505, 2005.

[29] David E. Lerner. Monitoring changing dynamics with correlation integrals: case study of an epileptic seizure. *Physica D: Nonlinear Phenomena*, 97:563–576, 1996.

[30] S. Liss. Apparatus for monitoring and counteracting excess brain electrical energy to prevent epileptic seizures and the like. *US patent No. 3850161*, 1973.

[31] D.S.L. Lloyd, C.D. Binnie, and B.G. Batchelor. Pattern recognition in eeg. *Advances in Behavioral Biology*, 5:153–166, 1972.

[32] Hans O. Luders. *Textbook of Epilepsy Surgery.* CRC Press, 2008.

[33] J. Maier, G. Dagnelie, H. Spekreijse, and B.W. van Dijk. Principal components analysis for source localization of veps in man. *Vision Research*, 27(2):165–177, 1987.

[34] Patrick E. McSharry, Leonard A. Smith, and Lionel Tarassenko. Comparison of predictability of epileptic seizures by a linear and a nonlinear method. *IEEE Transactions on Biomedical Engineering*, 50(5):628–633, 2003.

[35] A.M. Murro, D.W. King, J.R. Smith, B.B. Gallagher, H.F. Flanigin, and K. Meador. Computerized seizure detection of complex partial seizures. *Electroencephalography and Clinical Neurophysiology*, 79(4):330–333, 1991.

[36] Saralees Nadarajah. A generalized normal distribution. *Journal of Applied Statistics*, 32(7):685–694, 2005.

[37] Ivan Osorio, Hitten P. Zaveri, Mark G. Frei, and Susan Arthurs. *EPILEPSY, The Intersection of Neurosciences, Biology, Mathematics, Engineering, and Physics.* CRC Press Taylor and Francis Group, 2011.

[38] Clement C. C. Pang, Adrian R.M. Upton, Glenn Shine, and Markad V. Kamath. A comparison of algorithms for detection of spikes in the electroencephalogram. *IEEE Transactions on Biomedical Engineering*, 50(4):521–525, 2003.

[39] Antonio Quintero-Rincón, Hadj Batatia, Marcelo Pereyra, and Marcelo Risk. Detection of onset in epilepsy signals using generalized gaussian distribution. *Fifth International Conference on Advances in New Technologies, Interactive Interfaces and Communicability*, 2014, 2014.

[40] R. Quian Quiroga, J. Arnhold, and P. Grassberger. Learning driver-response relationships from synchronization patterns. *Physical Review E*, 61(5):5142–5148, 2000.

[41] M. Le Van Quyen, J. Martinerie, M. Baulac, and F. Varela. Anticipating epileptic seizures in real time by a nonlinear analysis of similarity between eeg recordings. *NeuroReport*, 10(10):2149–2155, 1999.

[42] R.C.Eberhart, R.W. Dobbins, and W.R.S. Webber. Eeg wavefonn analysis using casenet. *Proceedings of the Annual International Conference of the IEEE Engineering in Engineering in Medicine and Biology Society. Images of the Twenty-First Century*, 6:2043–2047, 1989.

[43] Michael G. Rosenblum, Arkady S. Pikovsky, and Jurgen Kurths. Phase synchronization of chaotic oscillators. *Physical Review Letters*, 76:1804–1807, 1996.

[44] Steven J. Schiff, Paul So, Taeun Chang, Robert E. Burke, and Tim Sauer. Detecting dynamical interdependence and generalized synchrony through mutual prediction in a neural ensemble. *Physical Review Letters*, 54:6709–6724, 1996.

[45] Ali Shoeb, Herman Edwards, Jack Connolly, Blaise Bourgeois, S. Ted Treves, and John Guttagf. Patient-specific seizure onset detection. *Epilepsy and Behavior*, 5:483–498, 2004.

[46] W. Henry Smithson and Matthew C. Walker. *ABC of Epilepsy.* BMJ Books, 2012.

[47] Nitza Thomassona, Thomas J. Hoeppnerb, Charles L. Webber Jr., and Joseph P. Zbiluta. Recurrence quantification in epileptic eegs. *Physical Letter A*, 279:94–101, 2001.

[48] Alexandros T. Tzallas, Markos G. Tsipouras, Dimitrios G. Tsalikakis, Evaggelos C. Karvounis, Loukas Astrakas, Spiros Konitsiotis, and Margaret Tzaphlidou. *Automated Epileptic Seizure Detection Methods: A Review Study.* Epilepsy - Histological, Electroencephalographic and Psychological Aspects. Dr. Dejan Stevanovic Editions InTech, 2012.

[49] Andrea Varsavsky, Iven Mareels, and Mark Cook. *Epileptic seizures and the EEG Measurement, Models, Detection and Prediction.* CRC Press, Taylor and Francis Group, 2011.

[50] S.S. Viglione, V.A. Ordon, and F. Risch. A methodology for detecting ongoing changes in the eeg prior to clinical seizures. *21st Western Institute on Epilepsy*, 1973.

[51] Russell Walters, José C. Principe, and Seung-Hun Park. Spike detection using a syntactic pattern recognition approach. *Proceedings of the Annual International Conference of the IEEE Engineering in Engineering in Medicine and Biology Society*, 6:1810–1811, 1989.

[52] Norbert Wiener. *Cybernetics or control and communication in the animal and the machine.* The MIT Press, 1965.

[53] Scott B. Wilson and Ronald Emerson. Spike detection: A review and comparison of algorithms. *Journal of Clinical Neurophysiology*, 113(12):1873–1881, 2002.

Clinical values dataset processing through cluster analysis to find cardiovascular risk

C M Bucci[1,2], **W E Legnani**[1,2], **R L Armentano**[1,2,3]
[1] Biologic-Engineering Research and Development Group, Electronic Department, Buenos Aires Regional Faculty, National Technological University, Medrano 951 (C1179AAQ), Buenos Aires, Argentina.
[2] Non-Linear Systems and Modelling Group, Centre for Signal and Image Processing, National Technological University, Medrano 951 (C1179AAQ), Buenos Aires, Argentina.
[3] Electronic Department, Favaloro University, Solis 453 (C1078AAI), Buenos Aires, Argentina.

E-mail: cbucci@rec.utn.edu.ar
E-mail: wlegnani@rec.utn.edu.ar
E-mail: armen@ieee.org

Abstract. The scope of this work is to show another way to grouping population with clinical variables measured in health centres and to assign a cardiovascular risk indicator. To do this, two different datasets were used, one coming from France and another coming from Uruguay. The well proved Framingham index was used to validate the results. The preliminary results are very auspicious to encourage the research and get deeper knowledge of the cardiovascular risk indicators.

1. Introduction

The cardiovascular system plays a role in maintaining homeostasis - that is, a stable environment - inside the body. This system can perform, or indicate to other systems that carry out short-term rapid adjustments in response to the demands requested by the body due to the development of various activities and changing external conditions. For example, when the blood supply increases, the flow must be reduced to other organs, or heart rate should be increased. Throughout these adaptations, blood pressure must remain constant to maintain vital functions of all body tissues. To perform these adjustments, the cardiovascular system communicates with other organs through a complex network of mechanisms for monitoring and alarms. This sends signals about their condition and, in turn, receives messages that control its operation (Zaret et al., 1992).

Consequently, it is very interesting the development of models and / or signal processing tools for predicting cardiovascular risk, both in order to understand the possible mechanisms that increase it, to be able to early intervention through prevention campaigns or perform therapeutic treatments.

2. Cardiovascular System Description

2.1. State of art

The cardiovascular system is divided into the systemic circulation provided by the left ventricle and the pulmonary circulation provided by the right ventricle. Each movement can be conveniently divided into four parts: the heart pump, arteries, veins and microcirculation. While each part separately analyzes always remember that it is a closed system with each of the parties interacting in varying degrees with each of the other system. For example, the same amount of blood must flow through the left and right side of the heart. There may be temporary changes that lead to the redistribution of blood in the circulatory system, but these differences cannot be sustained for long and in that sense we must quickly restore homeostasis. The heart is a unique organ that is divided into the left and right sides. The arteries are the major blood vessels that carry blood from the heart to the microcirculation of the tissue which must be balanced and then the veins carry blood back to the heart. In mechanical terms the great arteries and veins are different from other vessels because of its size and inertial characteristics rather than viscous effects in blood flow between them (Formaggia et al., 2009).

2.2. Cardiovascular Risk

Cardiovascular diseases come up suddenly affecting the heart (heart attack) or brain (stroke) at a given time.

Everyday life coupled with poor eating habits and sedentary lifestyle leads these diseases. Its effects are manifested in the arteries of the heart, brain, kidneys and lower limbs.

The main factors that influence these diseases are cholesterol, hypertension, smoking, diabetes, stress, physical inactivity, obesity, heredity and heart rate. It should also be noted that the existence of several factors increases the risk of cardiovascular disease. Most of them are very serious and can leave sequels or even lead to death. Often it generates irreversible consequences.

Therefore early detection of patterns to determine a potential risk is vital to diagnose a preventive treatment.

2.3. Framingham Method

The *Framingham Heart Disease Epidemiology Study* is designed to measure certain selected constitutional factors and certain of the conditioning factors in a large number of "normal" persons selected at random and to record the time during which these selected factors act and interact before clinical cardiovascular disease result (Gordon and Kannel, 1968).

3. Cluster Analysis Modelling Based On Real Datasets

3.1. Cluster Analysis

The Cluster Analysis (CA) is related to the identification of groups of data sets. The main objective of these techniques is the division or partition of a set of observations into a number of subgroups or clusters which is called "cluster" so that all observations within a subgroup are similar to each other and differentiate into the fullest extent of the observations belonging to other clusters, while observations from different groups need not be similar to each other (Timm, 2002). Consequently, the objective of this methodology is the identification of hidden structures that are hidden in the data sets (Everitt, et al., 2011) and (Fielding, 2007). The non-hierarchical clustering procedures usually follow the following steps (Everitt, 1993), (Decker and Lenz, 2007) and (Abonyi and Feil, 2007).

1. Select the k (n-dimensional) centroids of the clusters or seeds.
2. Allocate each observation to the nearest centroid using some measure of distance, (Usually Euclidean).
3. Relocate each observation to one of the k clusters based on a preset criterion.
4. To close the process if you cannot perform any relocation of the observations in any cluster or reallocation satisfies the convergence criterion given, otherwise return to step 2.

The non-hierarchical clustering method begins by selecting k centroids or seeds. These k seeds can be the first k observations that may be taken at some level separation is otherwise identify them at random, other algorithms begin with random seed and then through some very well defined algorithm relocated (Timm, 2002). Once the seeds are selected, each of the observations is evaluated to assign it to one of the clusters defined by them.

In step 2 the seed may or may not be updated. In this way you can use two tests to determine whether to perform the replacement. An observation can replace one or a few seeds if the distance between seeds is less than the distance between an observation and the next nearest seed. The displaced seed becomes an observation for recalculation of the centroids. If an observation fails this test may pose a second instance of evaluation (Timm, 2002).

The observation replaces the nearest seed if Euclidean distance there from all next seed is greater than the shortest distance from the nearest seed remaining seeds.
In the first instance all observations are associated with k clusters. Then this process is repeated until the changes in the seeds of small clusters are based on some predetermined convergence criterion.

3.2. Description of Real Datasets

In this paper were used two distinct datasets coming from different countries, one of them from France (dataset A) and another from Uruguay (dataset B). Of course people included in dataset A has different lifestyle, feeding and genetic characteristics to people included in dataset B, the use of these different sources of information was used in order to compare results of classifying cardiovascular risk. The dataset A is compound by a matrix with 618 patients (rows) and 10 measured clinic variables (columns) and the dataset B consist of 123 patients (rows) and 11 measured clinic variables (columns).

3.3. Modelling Validation

One of the more extended methods for validating the number of clusters for aggregating the values which comes from the dataset during a non-hierarchical clusterization process is to compute the silhouette index or silhouette coefficient.
This index is used with a plot that means of assessing the quality of a cluster solution, enabling the investigator to identify "poorly" classified objects and so distinguishing clear-cut clusters from weak ones. Silhouette plots for cluster solutions obtained from different choices for the number of groups can be compared, and the number of groups chosen so that the quality of the cluster solution is maximized (Everitt, et al., 2011).

To find detailed information about the silhouette concept the reader can see (Fielding, 2007, Everitt, et al., 2011, Kogan, 2007 and Kaufman and Rousseeuw, 1990).

3.4. Framingham Index

The aim of the *Framingham logistic model*, deals with on common factors identification or characteristic that contribute to cardiovascular disease, through monitoring of development (evolution), over a period of time which a large group of participants has not yet developed symptoms of cardiovascular illness, or suffered heart attacks or strokes (Armentano 2011).

The variables involved are *Sex* (dichotomous variable), *Age* in years, *Serum Cholesterol LDL* (mg / dl), the *fraction of cholesterol linked to high density lipoprotein HDL*, *Systolic Pressure*, *Diastolic Pressure*, the presence of *Diabetes* (dichotomous variable: Yes, No) and *Smoking* (dichotomous variable: Yes, No) (Armentano 2011):

4. Design and Implementation

4.1. Patients Dataset A

The dataset A is coming from different patients taking at the Pole Cardiovasculaire Hopital Europeen Georges Pompidou, Paris, France, at which has measured the clinical variables described in the Table 1. The size matrix size was described previously.

Table 1. Original dataset matrix 618 rows x 10 columns.

Variable	**Age**[a]	**Sex**[b]	**SBP**[c]	**DBP**[d]	**LDL**[e]	**HDL**[f]	**Smoke**[g]	**Diab.**[h]	**Fram.**[i]	**SCa**[j]
Patient #1	40	1	116	70	93	46	1	0	0,029596604	0
Patient #2	49	1	156	94	155	25	0	0	0,145509165	0
Patient #3	44	1	132	88	200	51	1	0	0,156247042	0
.....	.....	.....	.....	.....	.....	.....	.....	.....	.....	.....
Patient #618	46	1	136	92	200	30	0	0	0,212639732	1

[a] Age in years, [b] Sex (dichotomous variable), [c] SBP Systolic Pressure, [d] DBP Diastolic Pressure, [e] LDL Serum Cholesterol, [f] HDL the fraction of cholesterol linked to high density lipoprotein, [g] Smoke (dichotomous variable: Yes, No), [h] Diab. the presence of Diabetes (dichotomous variable: Yes, No), [i] Fram. calculated value of Framingham Index, [j] SCa measured value of Score Calcic

The first step is the matrix preparation for the experiment in which case the binary data such as "yes" or "no" in case of measures variables of "smoke", "diabetic" and "sex" was eliminated.
In order to prepare the data matrix to the clusterization algorithm is to apart the column containing the Score Calcic index to be used as comparison variables of the results. With these considerations a 618 rows by 5columns data matrix was used during the clustering process as is showed in Table 2.

Table 2. Study dataset matrix 618 x 5

Variable	**Age**	**SBP**	**DBP**	**LDL**	**HDL**
Patient #1	40	116	70	93	46
Patient #2	49	156	94	155	25
Patient #3	44	132	88	200	51
.....	.....	.....	.....	.....	.....
Patient #618	46	136	92	200	30

Once obtained the final matrix it is proceeded to get the correct number of clusters that the system needs to make a correct separation of population. To carry out with this job the silhouette coefficient explained before was applied, and the results are showed in Table 3.

Table 3. Silhouette coefficient.

Cluster #	**Silh. Coeff.**
2 Clusters	0.4946
3 Clusters	0.4515
4 Clusters	0.4180
5 Clusters	0.3899

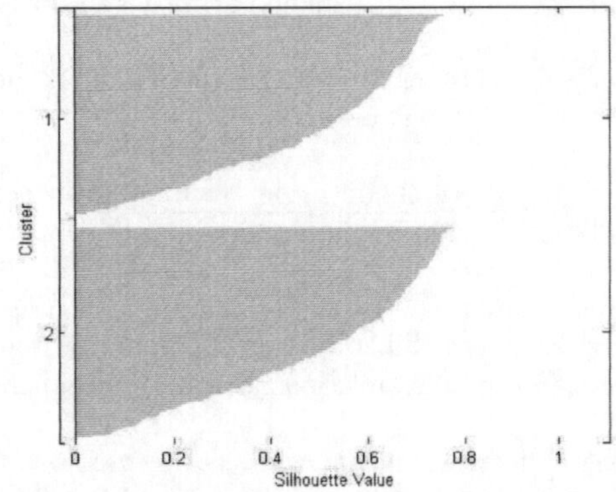

Figure 1. Silhouette coefficient for 2 clusters, for dataset A

As can see in Table 3, the correct numbers of clusters are 2 because it is the highest indicator of the four coefficients. As much as it sticks to 1 more reliable will be the result selected (J Kogan, 2007). It is possible to show a graphical representation (Everitt, et al., 2011) of this result as can see in Figure 1.

The subdivision of the dataset A into two clusters was performed by the technique named k-means cluster analysis.
After running the clustering process it is obtained 2 groups of patients grouped in cluster #1 and cluster #2. A quick verification of the correct grouping is showed making a scatter plot representation of the results (Figure 2).
The result shows that the best cut is represented by the LDL and SBP due to both groups of patients are well separated without mixing patients of each cluster. The rest of scatter plots with different variables do not generate two well separated clusters.

Table 4. Arithmetic mean for each cluster

Cluster #	**Age**	**SBP**	**DBP**	**LDL**	**HDL**
Cluster #1	48	141	90	159	48
Cluster #2	47	133	85	216	47

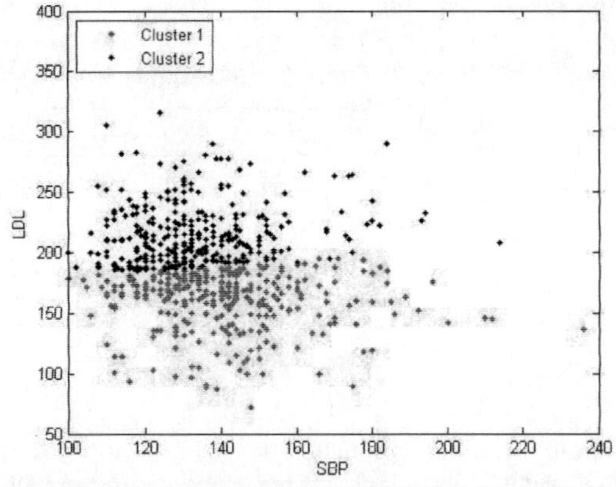

Figure 2. Matrix 618 x 5 (scatter plot LDL and SBP), for dataset A

Now it is easy to determinate graphically which patient in which clusters are by grouping patients in cluster #1 in red colour on the bottom area and cluster #2 in black colour on top area. Considering a boundary cutting line value of 190.50 on the LDL axis is possible to define that 73 % of patients in cluster #2 (246 patients) are located above this cutting and 27 % (66 patients) are below it. It is considered that 100 % (372 patients) of cluster #1 are also below of the cutting line.

The calculated "arithmetic mean" for each variable included in cluster #1 (306 patients) and #2 (312 patients) is showed in Table 4

It shows that the "arithmetic mean" factors are in general in the same level, the biggest difference between cluster #1 and #2 is in the variable LDL.

An additional validation will be done by implementing of Welch's T-Test. This procedure is used only when it can be assumed that the two population variances are different (the sample sizes may or may not be the same) and therefore must be estimated separately (Welch, 1947).

The output of the test implemented has two possible results: 1 or 0. If 1, the population A and B are both different means which imply that A and B are well separated, then it is possible for instance to use "Age" in order to separate the population in two groups, or could be interpreted as the age is a factor to classify cardiovascular risk. Otherwise, when the result is 0 two groups aren't well separated, and the means are equal at a certain statistical level of significance.

The Table 5 shows the result T-Test with 2 different quantity of population for comparing cluster #2 (312 patients) and cluster #1 (306 patients). In this case the population could be well separated implementing Age, SBP, DBP, LDL, Fram.

Table 5. T-Test result 618 x 5 (Adding Fram. comparison)

Variable	**Age**	**SBP**	**DBP**	**LDL**	**HDL**	**Fram.**
T-Test	1	1	1	1	0	1

Now it will be necessary to insert the measured SCa column and then repeating again the same procedure as follows:

Table 6. Study dataset matrix 618 x 6

Variable	**Age**	**SBP**	**DBP**	**LDL**	**HDL**	**SCa**
Patient #1	40	116	70	93	46	0
Patient #2	49	156	94	155	25	0
Patient #3	44	132	88	200	51	0
.....	.....	.....	.....	.....	.....	.....
Patient #618	46	136	92	200	30	1

The output of the silhouette coefficient is showed in Table 7 and Figure 3 where the highest indicator of silhouette coefficient is represented for the model of 2 clusters:

Table 7. Silhouette coefficient

Cluster #	**Silh. Coeff.**
2 Clusters	0.9283
3 Clusters	0.8302
4 Clusters	0.3700
5 Clusters	0.4252

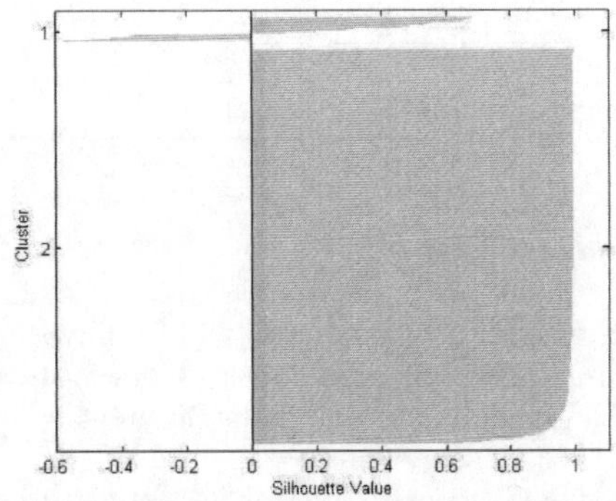

Figure 3. Silhouette coefficient for 2 clusters.

Two groups of patients will be determined for cluster #1 an cluster #2 will by using k-means. Scatter plot will show graphically the best cut for instance of variables LDL and SCa.

Assuming a boundary cutting line value of 328 on the SCa axis (LDL 195.075) is possible to define two groups of patients well separated in cluster #1 and #2 following Figure 4. A total of 35 patients grouped 100 % in cluster #1 are located above this cutting with the higher values of SCa. The rest of 583 patients are 100 % all grouped in cluster #2.

Table 8. Arithmetic mean for each cluster

Cluster #	**Age**	**SBP**	**DBP**	**LDL**	**HDL**	**SCa**
Cluster #1	56	146	89	193	48	606
Cluster #2	47	136	87	187	48	29

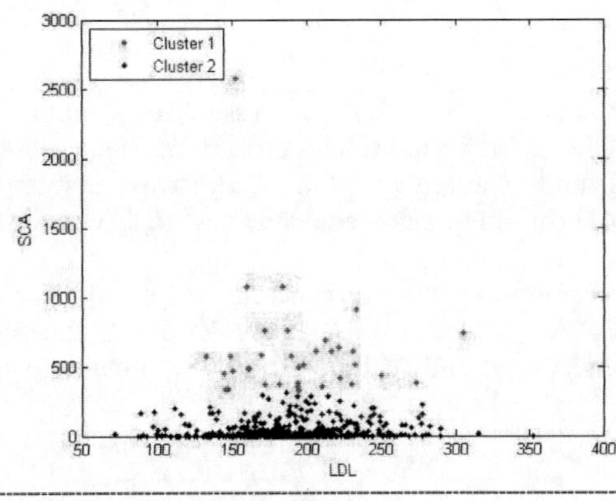

Figure 4. Matrix 618 x 6 (scatter plot LDL and SCa).

The calculated "Arithmetic mean" for each variable included in cluster #1 and #2 is showed in Table 8

It shows that the "arithmetic mean" factors are in general in the same level, the biggest difference between cluster #1 and #2 is in the variable SCa.

The Table 9 shows the result T-Test with 2 different quantity of population for comparing cluster #1 (35 patients) and cluster #2 (583 patients). In this case the population could be well separated implementing Age, SBP, SCa, Fram.

Table 9. T-Test result 618 x 6 (Adding Fram. comparison)

Variable	Age	SBP	DBP	LDL	HDL	SCa	Fram.
T-Test	1	1	0	0	0	1	1

4.2. Patients Dataset B

The experiment for dataset B will be the same than dataset A by means of repeating all the same processes. The only difference is that patients considered come from Uruguay and the clinical variables measured are a little bit different.

The dataset B consists of real measured variables matrix in different patients analyzed at Republic University, Uruguay. The matrix size is about 123 patients located at the rows and 11 columns representing the clinic variables according to Table 10.

Table 10. Original dataset matrix 123 rows x 11 columns.

Variable	Age[a]	SBPp[b]	SBPa[c]	DBP[d]	LDL[e]	HDL[f]	BMI[g]	PWV[h]	CIMTr[i]	CIMTl[j]	AAP[k]
Patient #1	52	129	121	79	259	47	26.3	16.03	0.957	0.976	13
Patient #2	53	111	106	74	253	47	28.7	9.54	0.600	0.593	10
Patient #3	42	162	156	104	252	36	23.3	8.63	0.572	0.553	32
.....	.....	.....	.....	.....	.....	.....	.....	.....	.....	.....	.....
Patient #123	60	140	128	72	54	70	23.1	11.97	0.864	0.811	15

[a] Age in years, [b] SBP Systolic Pressure (Peripheral), [c] SBP Systolic Pressure (Aortic), [d] DBP Diastolic Pressure, [e] LDL Serum Cholesterol, [f] HDL the fraction of cholesterol linked to high density lipoprotein, [g] BMI body mass index, [h] PWV pulse wave velocity, [i] CIMTr carotid intima-media thickness (right), [j] CIMTl carotid intima-media thickness (left), [k] AAP Aortic Augmented Pressure

In preparation for the experiment it will be also necessary to eliminate some measured variables such as SBP(A), BMI, PWV, CIMT(R), CIMT(L) and AAP in order to obtain the same 5 variables in the columns, then finally getting Table 11 as a matrix of 123 x 5 variables.

Table 11. Study dataset matrix 618 x 5

Variable	Age	SBP	DBP	LDL	HDL
Patient #1	52	129	79	259	47
Patient #2	53	111	74	253	47
Patient #3	42	162	104	252	36
.....	.....	.....	.....	.....	.....
Patient #123	60	140	72	54	70

The output of the Silhouette coefficient is showed in Table 12 and Figure 5 where the highest indicator of Silhouette coefficient is represented for the model of 2 clusters:

Table 12. Silhouette coefficient

luster #	**Silh. Coeff.**
2 Clusters	0.5789
3 Clusters	0.4985
4 Clusters	0.4206
5 Clusters	0.4226

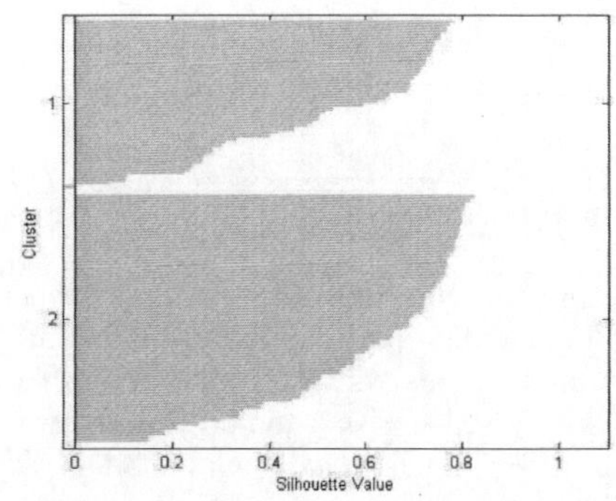

Figure 5. Silhouette coefficient for 2 clusters.

Two groups of patients will be determined for cluster #1 an cluster #2 by using k-means and with the best cut of scatter plot for instance of variables LDL and SBP.
Following the Figure 6, the boundary cutting line value will be 159 on the LDL axis which defines two groups of patients separated in cluster #1 and #2. In cluster #2 are included a total of 49 patients (98 %) located above this cutting and 1 patient (2 %) below it. The rest of 73 patients grouped 100 % in cluster #1 are located below cutting line.

Table 13. Arithmetic mean for each cluster

Cluster #	Age	SBP	DBP	LDL	HDL
Cluster #1	50	126	79	124	70
Cluster #2	51	122	78	192	53

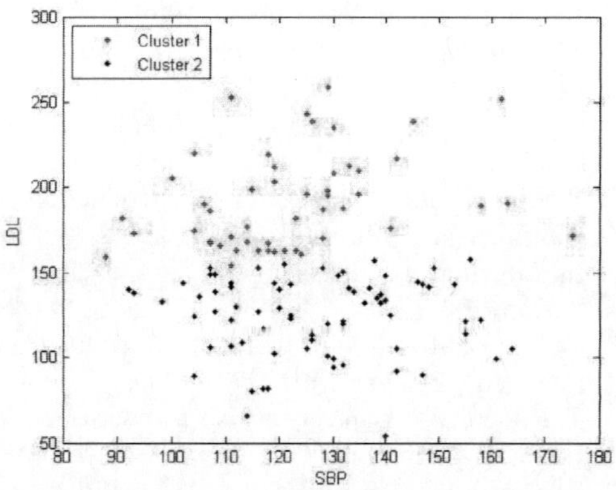

Figure 6. Matrix 123 x 5 (scatter plot LDL and SBP).

Table 13 shows that the "arithmetic mean" factors are in general in the same level, the biggest difference between cluster #1 and #2 is in the variable LDL.

The Table 14 shows the result T-Test with 2 different quantity of population for comparing cluster #2 (50 patients) and cluster #1 (73 patients). In this case the population could be well separated implementing LDL, HDL.

Table 14. T-Test result 123 x 5 (Adding PWV comparison)

Variable	**Age**	**SBP**	**DBP**	**LDL**	**HDL**	**PWV**
T-Test	0	0	0	1	1	0

There are no changes (less than 1%) in the results when is introduced PWV in order to implement a running clustering algorithm for new Dataset matrix of 123 patients x 6 variables.

5. Summary Report

In the present work regarding two independent datasets has been clearly exposed that the clustering algorithm can be used to classify standard clinical measurements in order to obtain a cardiovascular risk pattern. In this way the cluster with this proposal appears a cluster containing patients with the low cardiovascular risk another cluster characterized by high cardiovascular risk. To validate the results the Framingham index was applied. The number aggregate (clusters) was justified by means of the silhouette coefficient.

The SCa clinical value has shown a good discriminator between clusters as was seen in the cluster dataset A. In another way if SCa is not available as dataset B the values of LDL was used as proxy discriminator to differentiate the two clusters obtained for classification of clinical cases.

6. References

[1] The Framingham Study, An Epidemiological Investigation of Cardiovascular Disease – Section 1, T Gordon and W B Kannel, 1968
[2] Cluster Analysis, 5th Edition, King's College London, UK, Wiley, B S Everitt, S Landau, M Leese, D Stahl, 2011
[3] Cluster and Classification Techniques for the Biosciences, Cambridge University Press, A Fielding, 2007.
[4] Finding Groups in Data - An Introduction to Cluster Analysis, Vrije Universiteit Brussel, L Kaufman and P Rousseeuw, 1990.
[5] Introduction to Clustering Large and High-Dimensional Data, Cambridge University Press, J Kogan, 2007.
[6] The generalization of "Student's" problem when several different population variances are involved", B L Welch, 1947
[7] The unequal variance t-test is an underused alternative to Student's t-test and the Mann–Whitney U test, G D Ruxton, 2006
[8] Performance of five two-sample location tests for skewed distributions with unequal variances. M W Fagerland, L Sandvik, 2009
[9] A note on preliminary tests of equality of variances. British Journal of Mathematical and Statistical Psychology, D W Zimmerman, 2004.
[10] T-Tests, non-parametric tests, and large studies—a paradox of statistical practice. BioMed Central Medical Research Methodology, M W Fagerland, 2012.
[11] Integrated e-Health Approach Based on Vascular Ultrasound and Pulse Wave Analysis for Asymptomatic Atherosclerosis Detection and Cardiovascular Risk Stratification in the Community, D B Santana, Y A Z´ocalo and R L Armentano, 2011.
[12] Cardiovascular Mathematics, Modeling and simulation of the circulatory system, Springer-Verlag Italia, L Formaggia, A Quarteroni, A Veneziani, 2009
[13] Heart Book, University of School Medicine, Yale, B L Zaret, M Moser, L S Cohen 1992

Multiresolution analysis (discrete wavelet transform) through Daubechies family for emotion recognition in speech.

D Campo[1,2], O L Quintero[2], M Bastidas[2]

[1] Dipartimento di Ingegneria Navale, Elettrica, Elettronica e delle Telecomunicazioni (DITEN). Information and Signal Processing for Cognitive Telecommunications ISIP40 Italy Genova.

[2] Mathematical Modeling Research Group at Mathematical Sciences Department in School of Sciences at Universidad EAFIT, Medellín Colombia.

Emails: dcampoc@eafit.edu.co , oquinte1@eafit.edu.co, mbastida@eafit.edu.co

Abstract. We propose a study of the mathematical properties of voice as an audio signal. This work includes signals in which the channel conditions are not ideal for emotion recognition. Multiresolution analysis- discrete wavelet transform – was performed through the use of Daubechies Wavelet Family (Db1-Haar, Db 6, Db8, Db10) allowing the decomposition of the initial audio signal into sets of coefficients on which a set of features was extracted and analyzed statistically in order to differentiate emotional states. ANNs proved to be a system that allows an appropriate classification of such states. This study shows that the extracted features using wavelet decomposition are enough to analyze and extract emotional content in audio signals presenting a high accuracy rate in classification of emotional states without the need to use other kinds of classical frequency-time features. Accordingly, this paper seeks to characterize mathematically the six basic emotions in humans: boredom, disgust, happiness, anxiety, anger and sadness, also included the neutrality, for a total of seven states to identify.

1. Introduction

Emotions have been studied by different knowledge areas; among them are medicine, psychology, neuroscience and the social sciences. An emotional state can be evidenced in the human physiological alteration. In this sense, there are scientific studies that use different measures in order to identify emotional changes in a person; it is the case of brain potentials, facial micro expressions, and galvanic skin conductance together with breathing and blood pressure. In addition to the physiological signals mentioned above, it has been found that emotions can be analyzed by voice signals. In order to detect emotions in audio signals, some scientific researches use traditional features extracted from the primary signal, among them are the formant frequencies, fundamental frequency, signal energy in time, signal spectrum by Fast Fourier Transform (FFT) and the Mel Frequency Cepstral Coefficients (MFCCs). The researches that consider these types of features usually have some limitations when the gender of the speaker is considered [1-18]. Those results show that this kind of features are speaker dependent, this causes loss of generality when detecting emotions. In addition, predictive models based on traditional voice characteristics generally not exceed 70% accuracy on complete databases which consider several speakers and the six basic human emotions: boredom, disgust, happiness, anxiety, anger and sadness. There is a variety of researches that use traditional features as inputs of inference systems based on fuzzy logic, Artificial Neural Networks (ANN), Support Vector Machine (SMV), K-nearest neighbors (KNN), Hidden Markov Models (HMM) and Gaussian Mixture Models (GMM) [19-21]. In addition to the traditional classification features, there are some other scientific papers [22-24] that use a signal decomposition based on the wavelet transform in order to extract features that provide higher results in terms of speaker independence and overall accuracy percentage in the recognition of complete emotional databases. Those results demonstrate that the use of the wavelet decomposition technique has advantages over the spectral analysis that can be done using the Fourier transform. One of the most important advantages of wavelet decomposition is that it allows extracting information from the signal to be analyzed. In that way, wavelets act as filters that allow the extraction of high and low frequencies from an initial signal. Depending on the wavelet that is used; the filter acting over the audio signal will have certain properties. In this vein, the present article seeks to employ selected wavelets to distinguish certain features of an emotion. There have been identified some scientific studies in which wavelet decomposition is used to recognize emotions over speech signals. In the first [25] is considered a database of emotions from 5 native languages of Assam, they use an ANN classifier. Another research considers a database in Mandarin and they use a HMM classifier [27]. Additionally, there is a work [26] in which the wavelet transform is used in conjunction with some statistical techniques on the Berlin database.

The studies that involve wavelet analysis in speech emotion recognition (SER) have an accuracy of over 80%; and they use wavelet decomposition together with other techniques that allow differentiating emotional states in speech. In addition, these studies usually tend windowing the audio signals [24], [27], [28] in order to find features that allow detecting emotions in short speech expressions.

This paper presents an analysis of emotions in audio signals using wavelet transform without another kind of features. It is hypothesized that wavelet decomposition is sufficient for classifying emotions in audio signals. Also, the audio signals are analyzed without using windows over the initial signal, thus, it is possible to find global voice characteristics. Finally, this article also seeks to accomplish the construction of an inference system based on artificial intelligence that uses wavelet decomposition features as inputs. This work was performed using the Berlin Emotional Speech Database (Emo-DB), which has proven to be reliable in the emotional content which is reflected in its audio signals [29].

The paper is organized as follows. In Section 2 there is a review of the essentials issues of wavelet transform and it is also shown the principal properties of the wavelet family used in this work. In section 3 is shown the proposed methodology for emotion identification in audio signals. Section 4 shows the results obtained using the proposed method and finally some conclusions are exposed in section 5.

2. Multiresolution analysis

2.1. *Theoretical aspects of discrete wavelet transform*

Let $\psi(x)$ be a mother wavelet, the operations of translation and dilation are defined by the following expression:

$$\psi_{jk}(x) = 2^{j/2}\psi(2^{j}x - k) \tag{1}$$

Where $\psi_{jk}(x)$, for all j and k forms an orthonormal basis for $L^2(R)$ function. This space, for fixed j is defined by the following expression:

$$V_j = \left\{\sum_{k=-\infty}^{\infty} a_k\psi_{jk}(x)/\sum_k|a_k|^2 < \infty\right\} = n\left\{\psi_{jk}(x), k \in Z\right\} \tag{2}$$

A function or signal $f(x) \in L^2(R)$ can be expressed by the form

$$f(x) = \sum_{j=-\infty}^{\infty}\sum_{k=-\infty}^{\infty} \alpha_{jk}\psi_{jk}(x) \tag{3}$$

The α_{jk} are the wavelet transform coefficients of f(x). Every orthonormal mother has an auxiliary wavelet function $\phi(x) \in L^2(R)$ called father wavelet or scaling wavelet function and satisfies the following properties:

$$\int_{-\infty}^{\infty} \phi(x)dx = 1$$
$$\|\phi(x)\|^2 = \int_{-\infty}^{\infty} |\phi(x)|^2 dx = 1 \quad (4)$$
$$\langle \phi(x)\phi(x-n)\rangle = \int_{-\infty}^{\infty} \phi(x)\phi(x-n)dx = \delta(n)$$

V_0 is defined as the subspace of $L^2(R)$ generated by the set of moved scale functions $\phi_k(x)$, such that:

$$V_0 = \text{gen}\{\phi_k(x) = \phi(x-k), k \in Z\} \quad (5)$$

Then any function V_0 can be expressed as a linear combination of base functions, that is to say, a function $f(x) \in L^2(R)$ can be projected onto the subspace V_0 so that the development of this projection in terms of the moved scale functions is an approximation of f(x) such that $P_0 f = f_0(x) = \sum_{-\infty}^{\infty} a_k \phi_k(x)$,where $a_k = \langle f(x)\phi_k(x)\rangle$. In general, closed subspaces V_j are obtained:

$$V_j = \text{gen}\left\{\phi_{jk}(x) = 2^{\frac{j}{2}}\phi(2^j x - k), k, j \in Z\right\} \quad (6)$$

Thus, if $f(x) \in L^2(R)$ can be projected onto the subspaces V_j so that the development of such a projection in terms of moved scale functions constitutes an approximation of f(x), such that $P^j f = f_j(x) = \sum_{-\infty}^{\infty} a_k \phi_{jk}(x)$, where $a_k = \langle f(x)\phi_{jk}(x)\rangle$. V_j subspaces are associated with scale function $\phi(x)$, where the effect of the indices k and j on $\phi(x)$ is as follows: k moves the function along x axis and j compresses in 2^j the function along the x axis. Once projected a function or signal f(x) in the V_j subspaces, an approximation of f(x) is obtained and relevant information associated to frequency details of f(x) is missed [35].In this vein, W_j subspaces are defined as the orthogonal complement of V_j in V_{j+1}. In this way, f(x) can be rewritten in terms of an approximation subspace V_j in the j resolution and the remaining orthogonally terms containing the finer details associated with the subspace W_j associated with the mother wavelet $\psi(x)$. Apply the Wavelet Packet Decomposition (WPD) on a signal is equivalent to divide it into a group of approximation coefficients (V_j) and detail coefficients (W_j), where j defines the number of times this decomposition process is performed.

2.2. Daubechies Family of Wavelets

Wavelet packets were introduced by Coifman, Meyer and Wickerhauser [30] by generalizing the link between multiresolution approximations and wavelets [31]. Accordingly, a wavelet is a signal (or waveform) with limited duration whose mean value is zero. The wavelets tend to be irregular and asymmetric. The analysis by the Discrete Wavelet Packet Transform (DWPT) consists in decomposing a signal into scaled functions transferred in time from an original wavelet, called mother wavelet [32]. Signals with sharp changes are best analyzed by irregular wavelets [33]. Such wavelets are empowered to analyze local areas of large signals. For the calculation of the wavelet transform a mother wavelet must be chosen and it will serve as a prototype for all windows that are used in the process. There is an important collection of wavelet functions that have proved to be extremely useful, among which are: Haar, Daubechies, Biorthogonal, Coiflets, Symetrics, Morlet, Mexican Hat and Meyer. In order to carry out a WPD over a signal the wavelet collection should form an orthonormal base which relies on the concept of multi-resolution analysis [34]. In this sense, this paper uses 4 orthonormal wavelets in order to perform a WPD on audio signals from the Emo-DB.

This work involves an analysis over audio signals in which is possible to take advantage of the properties of those signals in order to perform a methodology which can deal with the characterization of the emotional content into them. As is well known, human voice has a wave nature, more specifically, the speech is basically generated when sound pressure waves are channeled or restricted in various ways by manual control of the anatomical components of the human vocal system [32]. Base on that, in this paper is considered a wavelet analysis over the audio signals exploiting the properties they have to express periodic signals into decomposed new signals. Wavelet analysis has proven to be a better approach than Fourier transforms at the time to characterize particular emotions in speech signals [25], [26] and [27].

3. Statistical features extraction and neural networks classifiers

3.1. Statistical features extraction

As is well known, statistics and probability are used in Digital Signal Processing to characterize signals and processes that generate them [30]. In the treatment of signals is very common to take into consideration high order statistics to extract information about the signal. The most conventional techniques involve second order statistics and the signal spectrum [31].

This work uses some statistical measurements over the decomposed signals in order to extract relevant features that allow recognizing the speaker emotion through the use of the Daubechies wavelet family. The principal statistical measurements extracted are: Maximum values, mean, median, interquartile range, standard deviation, zero frequency crossings, kurtosis, sknewness. The statistical features were extracted for different levels of wavelet decomposition. The statistical signal classification is an important component for the categorization of input data into identifiable digital pattern classes using extraction significant features [33], those statistical methods have been used in many areas of signal processing, including speech recognition [34]. In the present proposal is used a statistical hypothesis test for mean difference in order to separate two classes, each emotion represents a class and it is considered all the possible pairs among the 7 emotional states to perform the test.

3.2. Artificial Neural Network Classifier

The use of ANNs as classifiers in speech recognition problems has been considered by researchers due to the accuracy that this method can present when it is implemented correctly and the capacity that it has to generalize the information responding appropriately to inputs that differ from those with which the training step of the network was done [35]. The use of ANN has been proved to have better performance than other methods in the classification of emotions over speech signals considering both frameworks: speaker dependent and independent focuses [36]. It represents a motivation to choose an ANN as an emotion classifier. It was identified a work [37] that proposes a set of Bayesian classifiers considering every pair of emotions over 5 principal states (anger, happiness, neutral, sadness and surprise) in Japanese speech. A pairwise classification of emotions is performed in that article.

In the present work a pairwise selection of emotions is performed; this focus allows solving a multi-class problem using a binary approach in which the final features are joined in a final neural network structure whose outputs are the membership value to each emotion of interest.

4. Methodology

First, all the audios were decomposed in ten levels using the wavelets Db1, Db6, Db8 and Db10. A big set of features was extracted in each level and a statistical hypothesis testing was performed for each pair of emotions in order to select the most relevant features that allow to differentiate the emotions. Specifically the test performed was the Student's t-test, which is used to determine if two set of data are different from each other. In some sense, this process was executed in order to do a projection of the big set of features into a littler one.

Figure 1. Methodology for the decomposition and feature selection

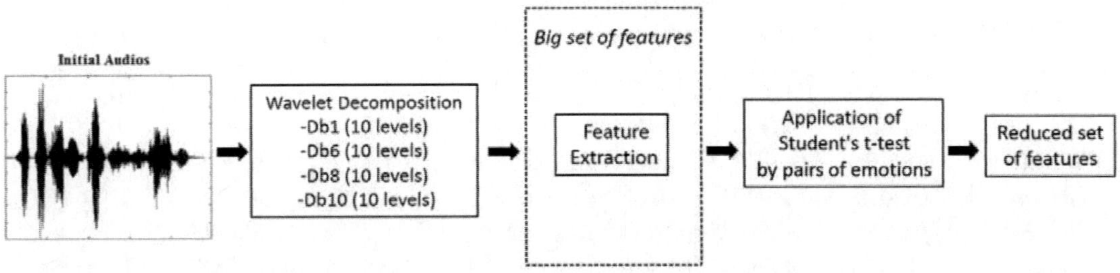

A simple ANN was trained based on the new set of features selected by the statistical hypothesis method. The final ANN structure used in this work is as follows: the output layer has 7 neurons due to the total numbers of emotional states. The number of inputs is 42 due to the process that was performed using the statistical hypothesis testing, in which it was assumed that two features for each pair of emotions are enough to classify the seven emotions of the database. Finally, it was considered only one hidden layer with 50 neurons whose transfer function is the hyperbolic tangent sigmoid.

Figure 2. Final architecture of the used ANN

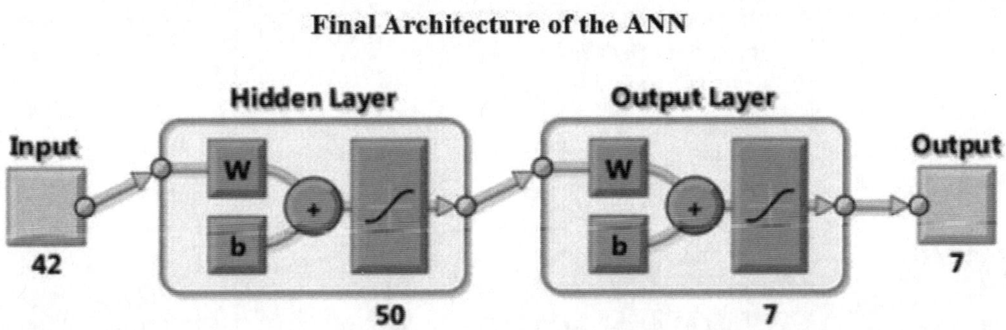

The assumption that is taken into account in this work is based on the fact that is possible to identify a specific emotion inside a group if we are able to classify correctly each possible pair of emotion into the group.

5. Results

The confusion matrix for the first proposed ANN is shown in the table 1. From the results, it can be seen that the most difficult emotional state to detect among the seven proposed is happiness with 85.71%, while the easiest one to detect is sadness with 93.33%.

Thus, the Table 2 shows the error rates per couple of emotions was obtained. From the table, it can be seen that the higher rate of error occurs when trying to differentiate audio signals associated to boredom and neutrality. This was expected due to the selection of the features that was done in the beginning, where it was observed that features associated with the detection of the pair Boredom-Neutral had a lower percentage of accuracy. It can be seen in Table 2 there are five pairs of emotions are classified with 100% accuracy, i.e. with 0% error, these pairs of emotional states are Boredom-Happiness, Disgust-Sadness, Happiness-Sadness, Neutral-Anger and Anger-Sadness.

If the results of the confusion matrices there is that neutrality is detected with the same precision in both cases. Boredom, happiness and anxiety are better classified by the first proposal Consequently the total accuracy for this proposal is 90.96%.

Table 1. Confusion matrix for the first ANN proposal

Emotional states	**Boredom**	**Disgust**	**Happiness**	**Anxiety**	**Neutral**	**Anger**	**Sadness**
Boredom	89.74%	00.00%	00.00%	01.28%	06.41%	01.28%	01.28%
Disgust	02.22%	91.11%	04.44%	00.00%	02.22%	00.00%	00.00%
Happiness	00.00%	02.86%	85.71%	02.86%	01.43%	07.14%	00.00%
Anxiety	01.45%	01.45%	01.45%	92.75%	02.9%	00.00%	00.00%
Neutral	06.25%	00.00%	00.00%	02.5%	91.25%	00.00%	00.00%
Anger	00.00%	00.79%	04.76%	01.59%	00.00%	92.86%	00.00%
Sadness	01.67%	00.00%	00.00%	01.67%	03.33%	00.00%	93.33%

Table 2. Error percentage by pair of emotional states using the first ANN proposal.

Pair of emotional states	Error percentage
Boredom - Disgust	1.11%
Boredom - Happiness	0.00%
Boredom - Anxiety	1.36%
Boredom - Neutral	6.33%
Boredom - Anger	0.64%
Boredom - Sadness	1.47%
Disgust - Happiness	3.65%
Disgust - Anxiety	0.72%
Disgust - Neutral	1.11%
Disgust - Anger	0.39%
Disgust - Sadness	0.00%
Happiness - Anxiety	2.15%
Happiness - Neutral	0.71%
Happiness - Anger	5.95%
Happiness - Sadness	0.00%
Anxiety - Neutral	2.75%
Anxiety - Anger	0.79%
Anxiety - Sadness	0.83%
Neutral - Anger	0.00%
Neutral - Sadness	1.66%
Anger - Sadness	0.00%

6. Conclusions

Feature's extraction from voice as audio signals in no ideal conditions is a challenge for emotion recognition purposes. Wavelet approximation through multi resolution analysis, results to be an accurate methodology for decomposition of the signal in several levels so called approximation and details. The use of statistical measurements over this signal decomposition for features selection, allows the recognition of several emotional states through the use of an Artificial Neural Network as classifier.

Future work will present this approach over audio signals for real applications.

References

[1] J. E. LeDoux 2000 *Annual Review of Neuroscience*, 155-184.
[2] V. Bostanov and B. I. Kotchoubey 2004 *Psychophysiology*, 259-268.
[3] G. Bush , P. Luu and M. I. Posner 2000 *Trends in Cognitive Sciences,* 215-222.
[4] J. Kissler, C. Herbert, I. Winkler and M. Junghofer 2009 *Biological Psychology*, 75-83.
[5] M. Esslen, R. M. Pascual, D. Hell, K. Kochi and D. Lehmann 2004 *Neuroimage,* 1189-1203.
[6] P. Ekman, 1970 *California Mental Health Research Digest*, 151-158.
[7] L. Zhang, M. Jiang, D. Farid and M. A. Hossain 2013 *Expert Systems with Applications* 5160 - 5168.
[8] A. Majumder, L. Behera and V. K. Subramanian 2014 *Pattern Recognition*, 1282–1293.
[9] G. K. Verma and U. S. Tiwary 2014 *NeuroImage*, 1-11.
[10] L. Farwell, 2013 *Encyclopedia of Forensic Sciences (Second Edition)*, 144-149.
[11] E. Leon, G. Clarke, V. Callaghan and F. Sepulveda 2004 *Computers & Graphics*, 635–642
[12] Y. Zhao, L. Zhao, C. Zou and Y. Yu, 2008 *Journal of Electronics (China),* 840-844.
[13] D. D. Joshi and M. B. Zalte 2013 *Journal of Electronics and Communication Engineering*, 34-37.
[14] V. Sethu, E. Ambikairajah and J. Epps 2013 *EURASIP Journal on Audio, Speech, and Music Processing.*
[15] M. Malkawi and O. Murad, 2013 *Human-centric Computing and Information Sciences.*
[16] M. Grimm, K. Kroschel, E. Mower and S. Narayanan 2007 *Speech communication* 787-800.
[17] R. Fernandez and R. W. Picard 2005 *INTERSPEECH* 1-4.
[18] J. Rong, G. Li and Y.-P. C. Phoebe, 2009 *Information Processing and Management,* 315-328.
[19] D. Bitouk, R. Verma and A. Nenkova 2012 *Speech Communication*, 613-625.
[20] H.-C. Koong Lin, . M.-C. Hsieh, . L.-C. Loh and C.-H. Wang, 2011 *Journal of Ambient Intelligence and Humanized Computing.*
[21] K. Sreenivasa Rao, S. G. Koolagudi and R. R. Vempada, 2012 *International Journal of Speech Technology,* 143-160.
[22] T. L. Nwe, S. W. Foo and L. C. De Silva, 2003 *Speech Communication*, 603-623.
[23] L. Chen, X. Mao , Y. Xue and L. L. Cheng, 2012 *Digital Signal Processing*, 1154-1160.
[24] M. El Ayadi, M. S. Kamel and F. Karray, 2011 *Pattern Recognition* 572-587.
[25] A. B. Kandali, A. Routray and T. K. Basu, 2009 *International Journal of Speech Technology* 1-13.
[26] V. N. Degaonkar and S. D. Apte, 2013 *International Journal of Speech Technology*. 1-5.
[27] Z. Han and j. Wang, 2013 *25th Chinese Control and Decision Conference*, 3156-3159.
[28] S. Ntalampiras and N. D. Fakotakis, 2012 *IEEE Transactions on Affective Computing*. 116-125.
[29] F. Burkhardt, A. Paeschke, M. Rolfes, W. Sendlmeier and B. Weiss, *A Database of German Emotional Speech.*
[30] S. W. Smith, 1997 *California: California Technical Publishing.*
[31] T. Chonavel, 2002 *Statistical Signal Processing*, (London: Springer).
[32] P. Sinha, 2010 *Speech Processing Embedded Systems*, (New York : Springer).
[33] G. J. Miao and M. A. Clements, 2002 *Digital Signal Processing and statistical Cassification*, (Norwood: Artech House).
[34] L. Rabiner and B. H. Juang, 1993 *Fundamentals of Speech Recognition*, (New Jersey : Prentice-Hall International).
[35] I. Sethi and A. Jain, 1991 *Artificial Neural Networks and Statistical Pattern Recognition*, (ELSEVIER SCIENCE PUBLISHERS).
[36] T. Iliou and C. N. Anagnostopoulos, 2009 "Comparison of different Classifiers for Emotion Recognition," in *Panhellenic Conference on Informatics*, Corfu.
[37] J. Cho and S. Kato, 2011 "Detecting Emotion from Voice using Selective Bayesian Pairwise Classifiers," in *IEEE Symposium on Computers & Informatics* , Kuala Lumpur.

Double Fourier analysis for Emotion Identification in Voiced Speech

D. Sierra-Sosa[1], M. Bastidas, D. Ortiz P., and O.L. Quintero

Mathematical Modeling Research Group, GRIMMAT, School of Sciences, Universidad EAFIT, Carrera 49 N° 7 Sur-50, Medellín – Colombia.

E-mail: dsierras@ eafit.edu.co

Abstract. We propose a novel analysis alternative, based on two Fourier Transforms for emotion recognition from speech. Fourier analysis allows for display and synthesizes different signals, in terms of power spectral density distributions. A spectrogram of the voice signal is obtained performing a short time Fourier Transform with Gaussian windows, this spectrogram portraits frequency related features, such as vocal tract resonances and quasi-periodic excitations during voiced sounds. Emotions induce such characteristics in speech, which become apparent in spectrogram time-frequency distributions. Later, the signal time-frequency representation from spectrogram is considered an image, and processed through a 2-dimensional Fourier Transform in order to perform the spatial Fourier analysis from it. Finally features related with emotions in voiced speech are extracted and presented.

1. Introduction

Spectrograms have being widely used over the years due to the broad band of applications in several areas of signal processing [1-5]. In voice speech signals, spectrograms allows for identify subtle cues related with utterance acoustics, providing the means to identify the speaker, separate speech from background noises, recognize and transcript spoken word to text, among others [6-7]. A particular characteristic when synthesizing speech recordings by using spectrograms, is formant transitions occurrence, this structures often related to consonants in spoken word, arrange in well-defined patterns and groups portraying speech features, which can be analyzed to extract information from signal [1, 6, 8].

Image processing techniques can be employed to analyze spectrograms, given the particular structures from power spectrum densities distributions [9-14]. By considering spectrograms as images, several features from signals can be extracted and recognized. For instance, the principal components can be classified in terms of Fourier descriptors with a high recognition rate [7]. Nonetheless, Fourier analysis provide either good temporal accuracy or good frequency accuracy, leading to other approaches as multi-scaled filtering and correlation methods to synthesize variations in color, texture and shape [15]. Also, mathematics morphology operators like erosion and dilation have being used to reduce spectrogram noise, which combined with two-dimensional non-linear filtering, allows for automatic speech recognition and enhancement [10, 11].

Signal processing derived from the time-frequency representation of spectrogram images, by following image processing methods, allows for identify and recognize features induced by emotions in voiced speech [16, 17]. We propose an analysis technique to identify emotion related information on spectrograms, based on the spatial Fourier analysis from binary threshold spectrogram images preserving the formant transition structures. Double Fourier analysis results, related with the particular characteristic from each synthetic emotion in Berlin Database of Emotional Speech are presented. This database includes ten different phrases, spoken by ten different speakers, portraying six synthetic emotions: Anxiety, Disgust, Happiness, Boredom, Anger and Sadness.

2. Proposed Method

From recorded voiced speech signals, spectrograms are obtained by using short time Fourier Transform (STFT), with different Gaussian windows, depending on signal sampling. These windows allows for obtain sufficient accuracy in both time and frequency, which resolution limits are described by Heisenberg's uncertainty principle [18]. Also, the windows are overlapped a 50% in order to adequately describe signal changes.

The spectrogram power spectral density information is converted to a gray-level image, where each matrix position is considered a pixel with values ranging from 0 to 255 [19]. The latter is then converted to a 0 and 1 valued matrix by using a binary threshold [20]. This procedure allows for preserve the spatial distribution and shape information from formants. Despite the evident information loss in the binary threshold process, the formant shape and spatial distribution is sufficient information to relate speech acoustic cues with emotion induce features in voiced speech.

A spatial 2D Fourier Transform is obtained from the binary image processed from each voiced speech record. The power spectral densities from these spatial transforms portray particular features, related with each synthetic emotion available in Berlin Database of Emotional Speech. These power spectral functions are considered as new gray-level images, containing the spatial frequency distributions from each spectrogram image. An average image is obtained for each emotion. In these average images the emotion-related feature become apparent.

Figure 1 depicts the proposed signal processing method. From voiced speech audio recording a STFT is obtained, the signal is considered as an image in following steps and synthesized accordingly. The power spectral density gray level image is binarized using a threshold, to extract formant shape and distribution information. A spatial 2D Fourier transform is obtained by means of a Fast Fourier Transform algorithm, and an average image from power spectra function is obtained for each synthetic emotion. A vertical line profile, crossing the average image origin, is evaluated. A qualitative evaluation from line profiles allows for infer that emotion-induced features in voiced speech may be extracted when synthesized by the proposed method; as each emotion in Berlin Database of Emotional speech contain recordings from both different speakers and phrases.

Figure 1. Block Diagram for voiced speech synthesis and analysis using double Fourier method

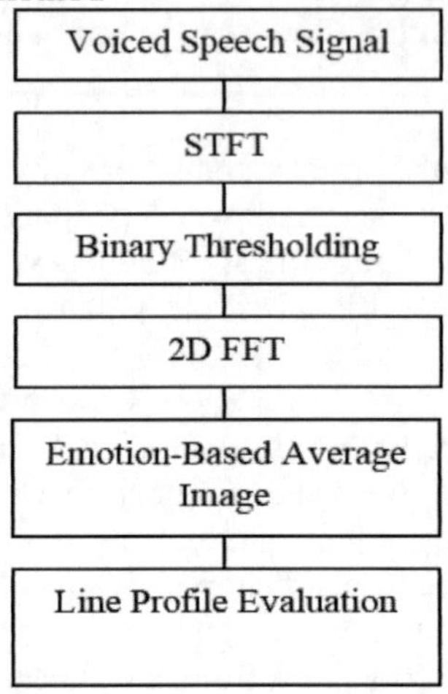

3. Results

The results obtained by the Double Fourier analysis are depicted in Figure 2, where the rows in descending order display the spatial power spectrum densities corresponding with the synthetic emotions anxiety, disgust, happiness, boredom, anger and sadness respectively; the columns are related with audio sampling and STFT windows parameters, in the first column the results obtained with voiced speech recordings sampled at 8 KHz with a 256 samples sized Gaussian window are presented, in the second and third column portray the results for recordings sampled at 16 KHz, with 512 and 1024 samples sized Gaussian windows respectively. All Gaussian window where overlapped a 50%.

Albeit slight changes in the spatial spectral density distributions, induced by voiced speech sampling and STFT window parameters. Each synthetic emotion portrays a recognizable particular structure.

When the recordings are processed by using the proposed Double Fourier Analysis, the emotion particular structures become apparent. These shapes, when conducting a qualitative analysis, are independent from audio sampling and STFT Gaussian window parameters. The latter, broaden the spatial frequencies distribution, but each particular structure is preserved.

The spatial spectral density distribution qualitative analysis was conducted by synthesizing the vertical line profiles from averaged images. In these line profiles, particular structures become apparent for each synthetic emotion, which cannot be related with speakers or phrases, because both different speakers and phrases were included in the Double Fourier Analysis process. Figure 3 present the vertical line profiles for each emotion, the left column depict the results related with emotional voiced speech sampled at 8 KHz, with 256 samples sized Gaussian window in the STFT; the right column present the results when using recordings sampled at 16 KHz, with 512 samples sized Gaussian window. Both windows were overlapped a 50%.

From figure 3, it should be noted that, despite slightly changes in the line profiles corresponding to each sampling rate, both results portray characteristics that can be tracked over particular emotional speech recordings, such as local maxima and local maxima positions, center peak width and shape. With this information a particular line profile shape can be related with each emotion in Berlin Database of Emotional Speech.

An important fact is that those particular profiles associated to each emotion (extracted through our double Fourier approach) can be easily introduced to a classifier in order to perform the task of emotion recognition.

We hypothesize that this methodology can be used as emotion recognizer independent from language because of the physical features extracted from the signal in both time frequency domain and spatial frequencies.

Figure 2. Spatial power spectral densities obtained by the Double Fourier Analysis of emotional voiced speech

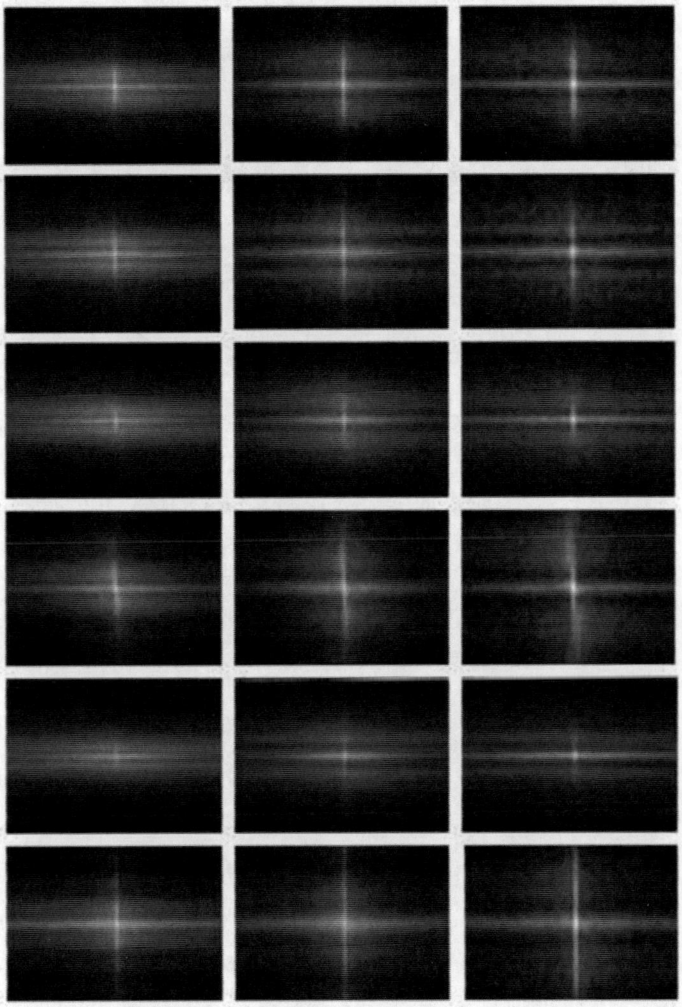

Figure 3. Spatial power spectral distributions and the corresponding vertical line profile for each emotional voiced speech, sampled at 8 KHz on the left, and 16 KHz on the right column

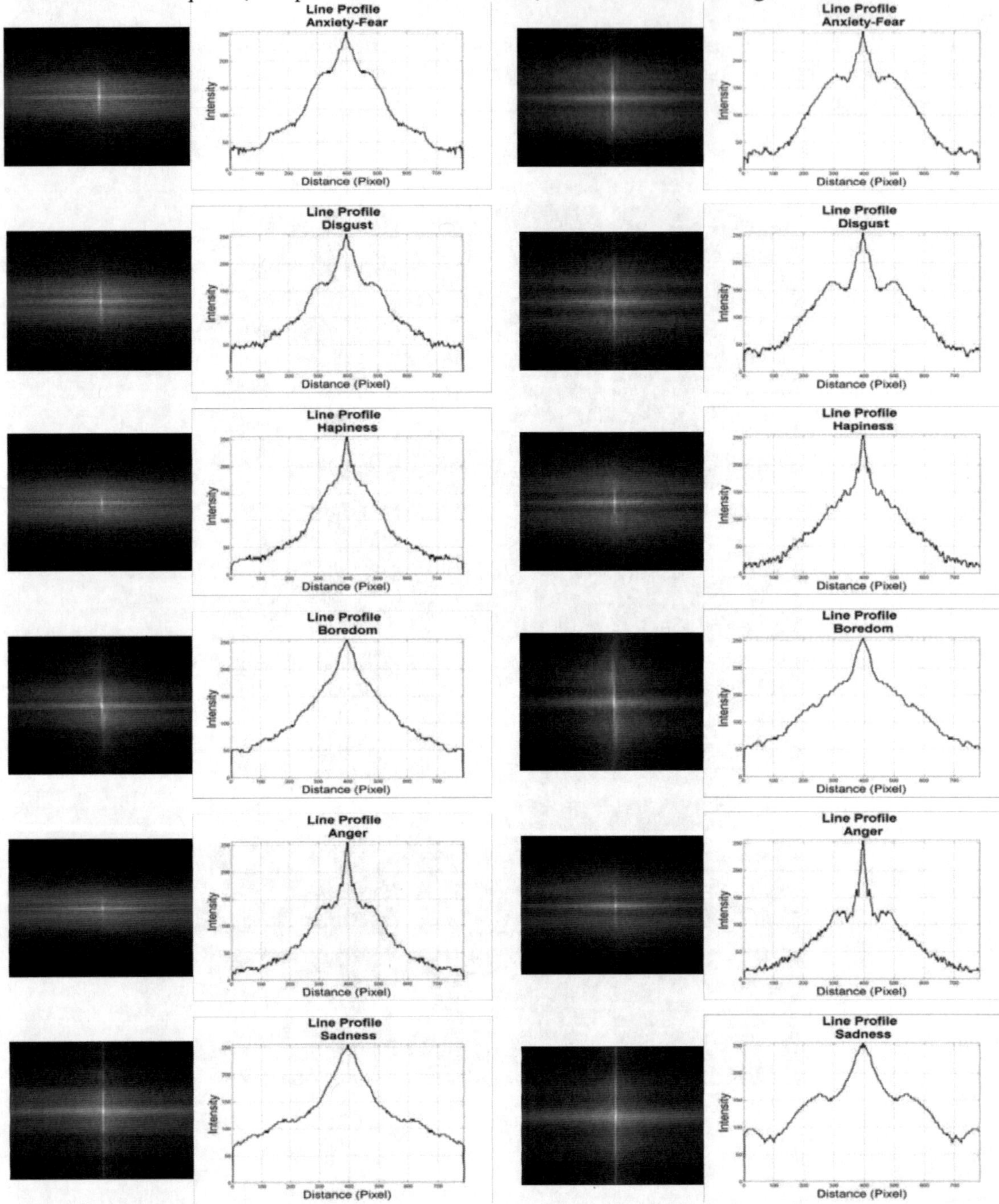

4. Conclusion

Evidences that emotions can be recognized when conducting a Double Fourier Analysis were presented. Spectrograms from voiced-speech recordings were obtained by using the short time Fourier transform, each of these time-frequency representations were considered as images, then from 2-

dimensional Fourier Transforms, obtained for each particular case, a spatial frequency distributions analysis were conducted. Particular structures related to emotion-induced features in voiced speech, become apparent in spatial power spectral density distributions, when synthesized by proposed method. Those structures are preserved even when signals are sampled at different frequencies, and different Gaussian windows are employed in the STFT to obtain spectrograms.

We are currently working on emotion classification when using Double Fourier Analysis, looking forward to extract and recognize emotional content in voiced speech, taking into advantage the particular features portrayed in spatial power spectra density functions.

Acknowledgement

Authors acknowledges Universidad EAFIT (Colombia) and Mathematical Modeling Research Group, where this research was performed.

References

[1] Borst, J.M. 1956. *Journal of the Audio Engineering Society* 14-23.
[2] Bolt, R.H., F.S. Cooper, David, E.E., Denes, P.B., Pickett, J.M., Stevens, K.N 1969 *Science*, **166** 338-342.
[3] Bolt, R. H., Cooper, F. S., David Jr, E. E., Denes, P. B., Pickett, J. M., & Stevens, K. N 1970 *The Journal of the Acoustical Society of America,* **47(2B)** 597-612.
[4] Mergu, R. R., & Dixit, S. K. 2011 *International Journal of Computer Applications*, **15(4)** 28-32.
[5] Oppenheim, A. V. 1970 *IEEE*, **7(8)** 57-62.
[6] Benesty, J., Sondhi, M. M., & Huang, Y. 2008 *Springer Science & Business Media*
[7] Pinkowski, B. "Principal component analysis of speech spectrogram images". Pattern recognition, 30(5), 1997, pp. 777-787.
[8] Welling, L., & Ney, H. 1998 *Speech and Audio Processing, IEEE Transactions on,* **6(1)** 36-48.
[9] Al-Darkazali, M., Young, R., Chatwin, C., & Birch, P. 2013 *In SPIE Defense, Security, and Sensing. International Society for Optics and Photonics.* 87480G-87480G
[10] Cadore, J., Gallardo-Antolín, A., & Peláez-Moreno, C. 2011 *In Advances in Nonlinear Speech Processing. Springer Berlin Heidelberg* 224-231
[11] Cadore, J., Valverde-Albacete, F. J., Gallardo-Antolín, A., & Peláez-Moreno, C. *Cognitive Computation* **5(4)** 426-44.
[12] Jin, X. C., & Wang, Z. F.2006 *18th International Conference on Pattern Recognition* **4** 278-281.
[13] Dennis, J., Tran, H. D., & Li, H 2011 *Signal Processing Letters, IEEE,* **18(2)** 130-133.
[14] Nilufar, S., Ray, N., Molla, M. K. I., & Hirose, K. 2012 *IEEE International Conference on Speech and Signal Processing (ICASSP),* 501-504
[15] Caelli, T., & Reye, D.1993 *Pattern recognition*, **26(4)** 461-470.
[16] Wang, K. C. The Feature Extraction Based on Texture Image Information for Emotion Sensing in Speech. Sensors, 14(9), 2014, pp. 16692-16714.
[17] Přibil, J., & Přibilová, A. 2010 *Measurement Science Review*, **10(3)** 72-77.
[18] Prandoni, P., & Vetterli, M. 2008. *Signal processing for communications. CRC Press.*
[19] Otsu, N. 1975 *Automatica* 23-27.
[20] Dennis, J., Tran, H. D., & Li, H. 2011 *Signal Processing Letters, IEEE*, **18(2)** 130-133.

Journal of Physics: Conference Series **705** (2016) 012036 doi:10.1088/1742-6596/705/1/012036

Discrete Kalman Filter based Sensor Fusion for Robust Accessibility Interfaces

I Ghersi[1,2,3], M Mariño[2], M T Miralles[1,2]

[1] Pontificia Universidad Católica Argentina (LaBIS-Facultad de Ciencias Fisicomatemáticas e Ingeniería-UCA), Av. A. Moreau de Justo 1500, CABA, Argentina.
[2] Universidad de Buenos Aires (CIDI-Facultad de Arquitectura Diseño y Urbanismo-UBA), Intendente Guiraldes 2160, Pabellón III, Cuidad Universitaria, CABA, Argentina.

e-mail: igghersi@gmail.com

Abstract. Human-machine interfaces have evolved, benefiting from the growing access to devices with superior, embedded signal-processing capabilities, as well as through new sensors that allow the estimation of movements and gestures, resulting in increasingly intuitive interfaces. In this context, sensor fusion for the estimation of the spatial orientation of body segments allows to achieve more robust solutions, overcoming specific disadvantages derived from the use of isolated sensors, such as the sensitivity of magnetic-field sensors to external influences, when used in uncontrolled environments. In this work, a method for the combination of image-processing data and angular-velocity registers from a 3D MEMS gyroscope, through a Discrete-time Kalman Filter, is proposed and deployed as an alternate user interface for mobile devices, in which an on-screen pointer is controlled with head movements. Results concerning general performance of the method are presented, as well as a comparative analysis, under a dedicated test application, with results from a previous version of this system, in which the relative-orientation information was acquired directly from MEMS sensors (3D magnetometer-accelerometer). These results show an improved response for this new version of the pointer, both in terms of precision and response time, while keeping many of the benefits that were highlighted for its predecessor, giving place to a complementary method for signal acquisition that can be used as an alternative-input device, as well as for accessibility solutions.

1. Introduction

Recent technological advances have converged towards a wave of massively-available consumer devices displaying highly intuitive user interfaces, based on alternate control means, such as face detection, motion and spatial-orientation measurements through 3D MEMS (Micro ElectroMechanical System) sensors [1][2], etc. Within this context, these devices have the potential to be adapted for the provision of advanced accessibility functions, forming customizable environmental control units, as well as to provide new input-means for the control of dedicated medical devices [3], through the integration of these advanced, minimally invasive mechanisms, in the detection of user commands. Aside from their potential as isolated inputs for human-machine interfaces, the group of sensors that are of interest for the present work (MEMS accelerometers, magnetometers and gyroscopes, and cameras) reach their highest impact when combined in real time. Sensor fusion, achieved through algorithms of variable

[3] Corrsponding author

complexity, allows to combine a number of estimations over a variable from isolated, supplementary sensors, in such a way that a joint estimation, more reliable than those obtained by the individual sensors, is obtained. As an example, the detection of the attitude or spatial orientation of body segments requires the combination of information from a 3D-accelerometer with a 3D-magnetometer, since the first is not capable of detecting changes in orientation around the vertical axis. Meanwhile, magnetometers are susceptible to interferences and deviations in their measurements, depending on the environment in which they are implemented [4], resulting in the need to take cautionary measures for their isolated deployment in highly-sensitive systems.

The development of a pointer-type interface, intended for the control of mobile-device applications, and based on face-detection and MEMS orientation sensors, has been presented in [5]. This solution allows to control an on-screen pointer with head movements. However, the potential susceptibility for the response of this system under variable or uncontrolled environments, as well as the specific interest in implementing it as a proposal for accessibility solutions, has fueled the search for alternate means to supplement the functioning of this system, when the reliability of one of its components is significantly degraded.

The present work shows results of the development of an alternate version of this pointer-type interface, taking advantage of and combining signals from different sensors (camera and MEMS gyroscope), with a Discrete Kalman Filter, in order to obtain a new solution, with a low relative computational cost, that displays an improved performance, as well as a higher independence to outer interferences, while keeping many advantages found in the preliminary version of this interface. This new solution can be implemented as a supplement to the first version of this system, and forms a new alternative-input and accessibility controller, valuable on its own for the command of devices and applications.

2. Materials y Methods

2.1. Materials

The aim of the proposed systems is to sum advantageous features from mobile devices (part of an expanding, increasingly available market), with minimally-invasive external peripherals, in order to enhance their joint interfacing capabilities between user and device. Prototypes and test stages presented in this work were developed on a 7.0 inch tablet, with a dual-core 1 GHz processor, Android 4.1.1 OS, and a 1024x600 pixels screen. Figure 1 shows the architecture of the system: external sensors (3D magnetometer-accelerometer, for the initial version of the system, and 3D gyroscope for the new version presented in this work), were interfaced through a dedicated circuit board, with a 16-bit microcontroller (dsPIC33F, from Microchip) in its core. Information from this board was transmitted to the tablet device through a wireless TCP link, through a PC, for initial test stages of both systems, while a dedicated USB link has also been achieved between the tablet and the external sensors. Both versions of the system take advantage of face-detection functionalities, through the frontal camera of the tablet device, and the native Android library for such purposes [6]. Face-detection is performed over low-resolution frames (160x120 pixels).

2.2. Initial Version: Mouse/pointer interface controlled by head turns

The first pointer interface solution, deployed on this platform, featured the integration between registers from a 3D MEMS orientation sensor (magnetometer-accelerometer), and information provided by the open-source face-detection library. While this MEMS module is subjected to limitations as specified in section 1, its development made it possible to explore the required strategies for the integration of all data sources (camera, orientation sensors in the tablet and external orientation sensor), while serving as a reference for future controlled tests with other versions of the system. Also detailed in figure 1, the proposed strategy for the combination of measurements and estimations between these input-devices, is described as follows:

- The face detector acquires information concerning relative position between the screen and the user's face. The position of the face in the image that is obtained from the frontal camera is indicative of its relative position in two dimensions (x: horizontal, y: vertical), while the size of the face in the image is indicative of the distance between user and screen (z axis).
- The external orientation sensor, fixed to detect head movements (i.e. integrated with a headphone), is useful in the estimation of relative orientation between user and screen, around three rotation axes (referred to as Heading or Yaw, for horizontal head turns, θx, Pitch for vertical head turns, θy, and Roll around the remaining axis). In order to acquire this information, the system obtains the difference between the external sensor (user) and a similar module, fixed to the interface screen (mobile devices usually feature these useful orientation sensors).

In this way, the proposed system is capable of obtaining relative position and orientation information between user and device, through an approximation of the perspective from which the user interacts with it. It has been stated, in [5], that an interface that responds to this information is capable of giving place to a more intuitive and robust interactivity, resulting in an on-screen pointer that remains effective while the user explores the environment surrounding the device, always responding to its movements: by turning its head, the user points to an absolute position on the screen. The coordinates of the pointer in the screen is a result of the two-axis projection of the vector that represents the spatial orientation of the user's head, while taking into consideration relative distance and positions between user and device (see figure 1 and [5] for a detailed description of the algorithm).

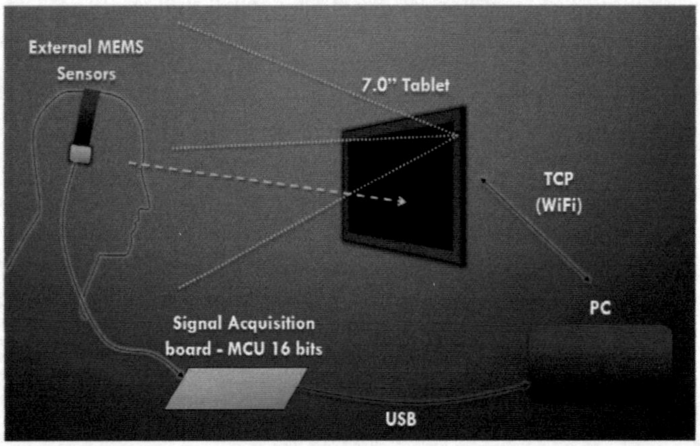

Figure 1. System components and pointing-interface strategy [5].

2.3. Updated Version: Camera-Gyroscope Sensor Fusion based on Discrete Kalman Filter
The face detector used for the initial pointer interface acquires information about the *relative position* between user and screen. An alternative interpretation of this information, however, would allow *detecting head turns from an initial position*, since, when turning the head in any direction, its center is displaced over the image acquired by the camera (figure 2). If there are not significant displacements during this interaction, and movements are limited to free head turns, which is considered acceptable for the target population of this work (accessibility interfaces), the face detector can be alternatively used as an input for orientation and distance estimations, without the need to use algorithms with a higher computational cost for such purpose.

These new relative-orientation estimations, acquired from the output of the face detector in both axes, can be realized in many ways. As an example, a basic approximation supposes that the sine of the angle between user and screen (see figure 1), is related to the center of the face in the image, with respect to its origin coordinates (PX-X0, for the x-axis), and the estimated distance between the user and the screen, after adequate projections and unit-conversions between pixels and centimeters. Equations (1) and (2) express these estimations, while it should be stated that, for the expected range of motion required for the control of this device in both axes, which falls below ± 15° (7.0-inch screen, at 30 cm to 50 cm distances), other variants, including linear approximations between displacements and angles could be used:

$$\theta x_{FD} = \sin^{-1}((\mathrm{PX_{cm}} - \mathrm{X0_{cm}})/(\mathrm{Dist_{cm}} + 20)) \quad (1)$$

$$\theta y_{FD} = \sin^{-1}((\mathrm{PY_{cm}} - \mathrm{Y0_{cm}})/(\mathrm{Dist_{cm}} + 20)) \quad (2)$$

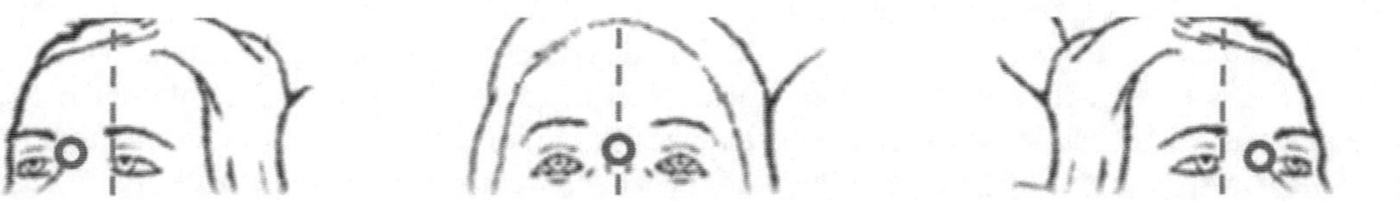

Figure 2. Angle-estimation strategy for horizontal head turns (θx) based on face detection. From an initial position (center, X0), heading angles are approximated from the relative displacements to this point. Image adapted from [7].

Based on the calibration of device-specific parameters (screen resolution, relative position and camera field of view), the interpupillary distance, in pixels, estimated by the face detector, is used to estimate the distance between user and device, in cm ($\mathrm{Dist_{cm}}$) [8], through a linear estimation [5]. The constant that is added to this variable in equations (1) and (2) (20 cm), represents the distance between the center of the face and its rotation axis. A similar anthropometric variable (distance of the eyes to the back of the head) varies between 14 cm (5th female percentile) and 19 cm (95th male percentile), for adults [9], so the selected value represents a potential maximum among users.

Besides dealing with the issue of being derived from low-resolution images (160x120 pixels), and suffering potential noise, resulting from the image-processing algorithm, this proposal for angle estimations suffers non-linearity, associated to relative displacements between the center of the camera and the screen. However, it has been proposed as a simple estimation, because it is stable, while noisy, and supposes a minimum added computational cost, and its limited response is expected to show improvements when integrated with other sensors.

The sensor-fusion proposal involves the 3D digital gyroscope L3G4200D, with 16-bit outputs, integrated programmable high-pass filter, and used in the range of ±250 °/s, this new MEMS sensor taking the place of the orientation sensor in the previous version (section 2.2). Under this strategy, the camera is capable of providing head-turn estimations, (θx_{FD}, θy_{FD}), while these are limited and do not contemplate variable perspectives. In order to do this, the starting point of the pointing device is calibrated during the initialization of the system. Coming from low-resolution images, face-detection results can be noisy, but represent a stable reference on the user's relative orientation. On the other hand, the gyroscope gathers angular-velocity measurements from these head turns ($\omega = d\theta/dt = \dot{\theta}$), supplementing the noisy information from the camera. However, the direct estimation of angles from this sensor would result in slow drifts in the estimated signals, as a result of noise or zero-level inaccuracies during signal processing stages. As a result, these signals appear to be supplementary, and their fusion is expected to result in a positive joint response.

Sensor fusion between these information sources has been implemented through a Discrete Kalman Filter [10] on each axis, an algorithm that allows estimating variables or states, based on measurements from multiple sources, when these are combined through a known model for the behavior of the system and its estimated uncertainties. A large number of viable models for MEMS sensors have been implemented for research purposes, in the search for more precise position and orientation estimations in the field of navigation systems [11]. In this case, the proposed system model is a simple relationship between angular velocity and heading or pitch angle, depending on the axis, in discrete time. Equations (3) and (4) show the Discrete Kalman Filter model:

$$\theta_{k+1} = \theta_k + \dot{\theta}_k \Delta t + w_k \quad (3)$$

$$z_k = \theta_k + v_k \quad (4)$$

considering: θ_k: Angle, estimated process state at instant k
$\dot{\theta}_k$: Angular Velocity, as measured by the gyroscope
Δt: Time interval between measurements, in seconds
z_k: Angle measurement, as obtained from the face detector (θx_{FD}, θy_{FD})
v_k: Measurement noise w_k: Process noise

Based on this model, the recursive algorithm of the filter [10], performed over the corresponding signals in each axis, allows updating and projecting head-angle estimations. These estimated angles in each axis (θ_x, θ_y) are later converted to on-screen coordinates in pixels, through an inverse operation from equations (1),(2). During this step, it is possible to amplify or attenuate the sensitivity to these head turns, which, when allowed by the resolution of the system, could be found valuable for accessibility implementations, over users that display a lesser degree of mobility in the neck. According to the specifications of the system, it is possible to tune certain filter parameters (the noise estimations for the model and measurements, Q and R, which affect w_k and v_k), in order to prioritize the gyroscope or the camera, when the other's reliability is compromised.

For this case, static and non-linearity errors in the estimations of angle from the face-detector have been accepted, because this source complies with the need to rely on a pointer that is stable in the long term, which the gyroscope alone cannot achieve. In this point, it should be noted that, for a user interface such as the one described, the noise that could affect the position of the on-screen pointer is far more detrimental to the experience in its use than the actual accuracy of the orientation estimations, which is why the contribution of the gyroscope is critical in this sense. In case of counting with sufficient motor control, the user could detect and compensate for small inaccuracies in the pointer coordinates, as long as these are not significant, and the pointer can be relied with enough stability in response to its movements. However, there would be no possibility for the user to compensate for a highly changing position, even if its mean value was actually an accurate measurement of the relative angle. Table 1 shows a comparison between properties and algorithms involved in both system versions:

Table 1. Comparative analysis between system versions – Features and algorithms

Version / ***Parameter***	***Initial*** *(section 2.2)*	***Kalman*** *(section 2.3)*
Sensors	3D Mangetometer-Acelerometer Camera (160x120 pixels)	2D Gyroscope Camera (160x120 pixels)
Algorithms	FaceDetect (native) 3D-orientation estimation [4] Discrete Signal Filtering User-screen distance estimation [5] Pointer position estimation [5]	FaceDetect (native) User-screen distance estimation [5] User-screen angle estimation Sensor Fusion (Kalman)
Update Rate	FaceDetect Maximum >10 Hz, Average 7.46 Hz MEMS Sensors $\geq$ 100 Hz On-screen Pointer Position ~ 16 Hz	
Perspective Detection	Distance and relative-positon estimation, between user and screen	Relative distance estimation, between user and screen

2.4. Test application and analysis strategy

In order to assess the properties and characteristics of both versions of the system, a test application for Android OS was developed with Eclipse IDE. This application shows five options on screen, representing five buttons, and a central, yellow number, which changes randomly, indicating which option the user should point to with the on-screen pointer (crosshair). This application gathers information concerning the requested option, pointer coordinates (x_p: horizontal, y_p: vertical), time stamps, raw sensor data, and an indicator of the detection or absence of a face, as detected by the face-detector. Figure 3 shows a screenshot of this test application (right), with an xy chart showing results of its use (left), with the corresponding trajectory of the on-screen pointer, concentrated over each option.

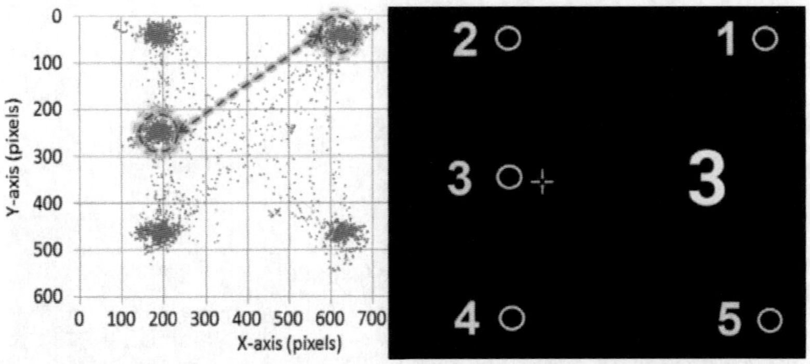

Figure 3. Test application screenshot (right), with x_p, y_p (on-screen pointer coordinates) plot (left). Pointer positions are concentrated around each option.

From this application, information about pointer speed and reliability can be obtained, which in turn allows comparing both system versions, estimating their respective performance and potential. The upper side of figure 4 is a time chart displaying the same xy pointer coordinates, through the use of the interface. The bottom of the figure shows another time chart, with the progression of requested options during its use. Based on these signals, pointer stability (range and standard deviation), of special interest for this work, is obtained after changes in the requested option, once the transition begins to stabilize (yellow rectangle, compatible with the yellow circle in figure 3), represented by the definitive entrance of the pointer into a broad proximity of the option. The duration of this stable period is, on the other hand, indicative of the possibility of implementing a selection strategy for the interface, based on holding the pointer over an option for a certain period of time.

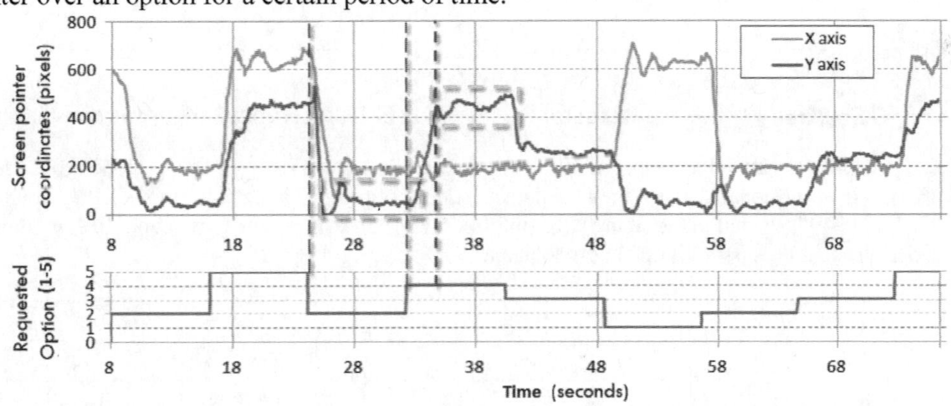

Figure 4. Test application data from initial version. Top: screen pointer position, in pixels, over time (x_p: horizontal, y_p: vertical). Bottom: requested option (1-5). The region of interest for stability and reliability assessments is highlighted (yellow).

3. Results

3.1. 1D response: Horizontal Head Turns, sensor fusion behavior and general performance

Figures 5, 6 and 7 display three cases of the estimation of horizontal head turn angles, based on the gyroscope, the face detector, and through the Discrete Kalman Filter, as well as through measurements provided by the orientation sensor of the initial version of the system, as a reference.

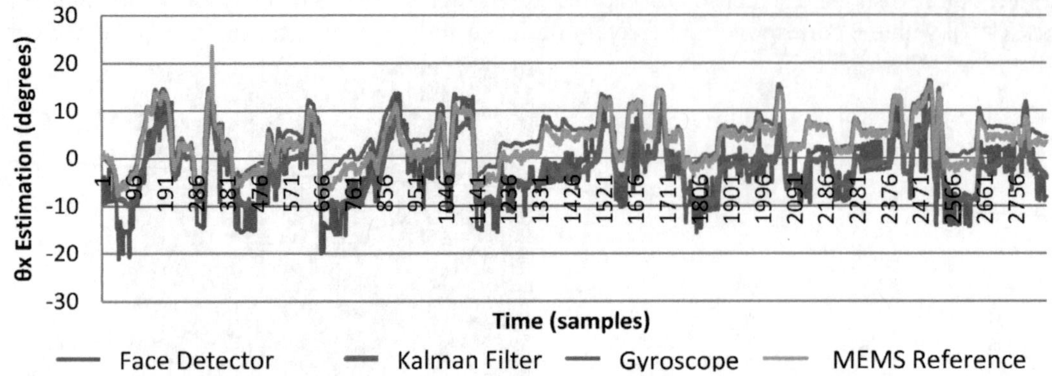

Figure 5. Horizontal head-turn angle estimations (θx), based on face detection (purple), angular velocity from the gyroscope (red), and Kalman Filter (blue), together with a reference, obtained from the 3D MEMS magnetometer-accelerometer (green): the sensor fusion signal results in an improved version of the face-detection estimation.

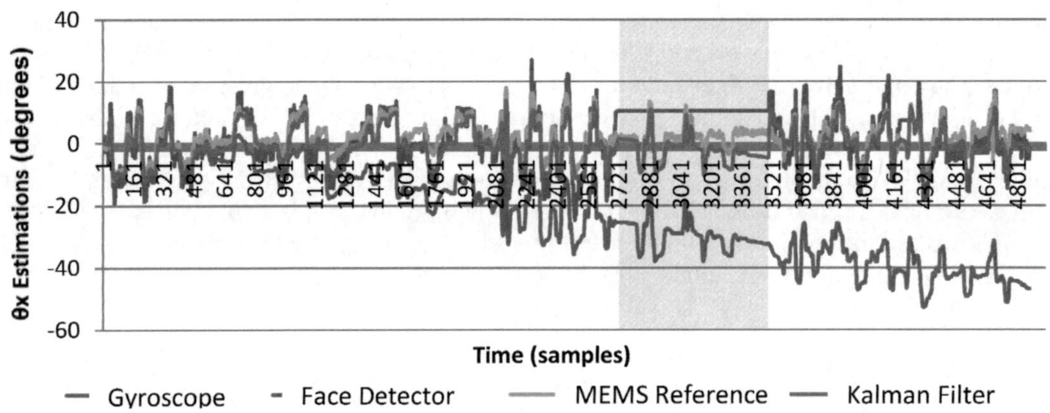

Figure 6. Discrete Kalman Filter response: θx angle estimations from the gyroscope alone would result in variable drifts (red). Additionally, when the face detector does not provide an update (forced condition between samples 2761 and 3497, shaded in the figure), the output of the filter follows the gyroscope. In sum, the overall solution is more robust than its individual components.

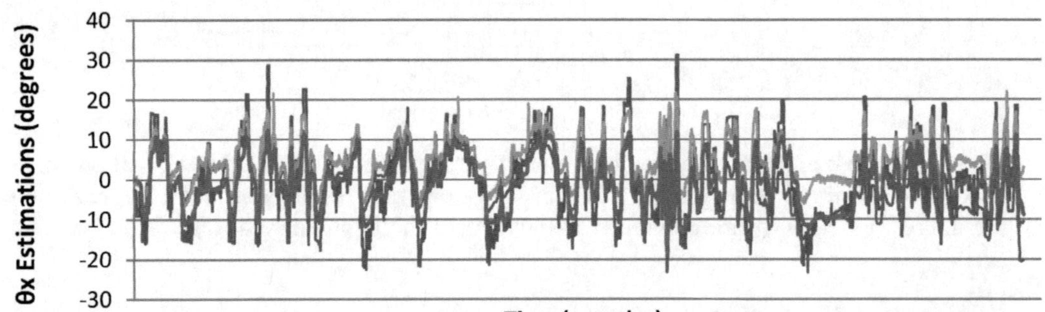

Figure 7. Discrete Kalman Filter – Horizontal head-turn estimation.

As a result of these implementations, the following considerations are highlighted:

- The signal that is derived from the Kalman filter, on the horizontal axis, is less noisy than the one provided by the face-detection algorithm (figure 5), as well as the original reference.

- This new integrated estimation compensates for the drift derived from the gyroscope (figure 6)
- In terms of noise and drift, the on-screen angular response, derived from the Kalman filter, proves satisfactory.
- It is necessary to consider that, unlike the angular velocity signal, which is periodically updated, the output of the face detector can provide no updates on its information over an interval, given possible changes in the image acquired by the camera during the control of the pointer. The nature of the parameters from the Kalmn filter, however, made it possible to propose a mechanism to mitigate this possible lack of information, through the real-time change of filter parameters Q and R. Figure 6 exemplifies how, when no face is detected, the filter adapts to prioritize information from the gyroscope (this is achieved by changer filter parameter R). This condition is visible in the shaded section of figure 6.

While it does not mitigate every imperfection from its component signals, the implemented Kalman filter results in an improved version, with a fast response, of the signal generated from the face detector (figure 7), based on information from a completely different sensor than the ones implemented on the initial version of the system (gyroscope).

3.2. 2D-response: Test Application and Comparative Performance
Figure 8 shows results of the implementation of the discrete Kalman Filter, in estimating the on-screen pointer position in two axes.

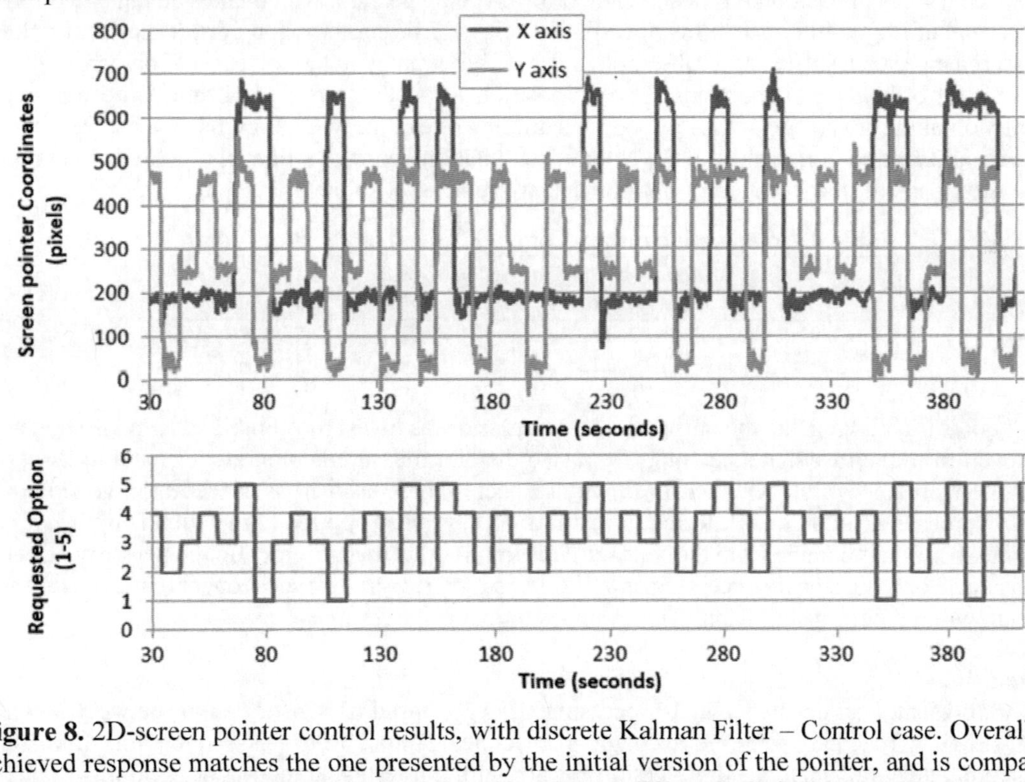

Figure 8. 2D-screen pointer control results, with discrete Kalman Filter – Control case. Overall, the achieved response matches the one presented by the initial version of the pointer, and is compatible with its expected use, with the benefit of stemming from a different sensing strategy.

For this implementation, no additional pre-processing strategies were applied on the face-detection and angular-velocity signals, only recurring to the programmable high-pass filter, available for the L3G4200D gyroscope, prior to acquiring its outputs. In order to compare the relative performance between the two versions of the pointer, responses acquired from their use with the test application, for a control case (user with training in the use of both systems), were assessed. This test scenario allowed mitigating the impact of adaptation and learning periods that new users of any interface go through, and is proposed as a preliminary strategy for evaluating relative virtues and flaws between these versions. Results achieved by the control case, in use of the pointer, and under analogous conditions to those shown in [5], show a stable pointer, which can be positioned in any desired spot, and that responds with no perceptible delays to gestures and head movements, in both directions. One of the disadvantages found for the initial version of the pointer consisted in the compromise between the precision of the pointer and its response-time.

Table 2 shows a comparison between characteristic parameters, derived of the analysis of responses from the initial and updated (Kalman) versions of the pointer, for the control scenario. These results confirm the presence of a significant improvement in the performance of the new pointer in the horizontal axis: both standard deviations (±11 pixels) and ranges of the pointer (46 pixels), once stabilized in the vicinity of the desired option, showed significantly lower values than those achieved by the previous version of the system. Moreover, a similar performance in the vertical axis was observed between both versions. Additionally, it should be noted that the updated version achieved similar performance levels in both axes, unlike the response from the initial version.

In terms of pointer speed and control characteristics, the user is able to hold the pointer in each position, with the detailed precision, for a mean period of stability of 6.15 seconds (minimum 3.73 s, maximum 7.09 s), before a change in the requested option occurred. This result is comparable with the average 5.98 s, achieved in use of the initial version (4.28 s minimum, 7.05 s maximum), while a faster response to changes in position has been verified for this new version, which takes advantage of sensor fusion instead of nested filters, in terms of response delays. This combination of precision and stability, is considered more than sufficient for the application of this strategy in the selection of on-screen options through means of holding the pointer over each position for a certain period of time. Furthermore, the possibility of taking advantage of the high-speed sampling rate of the MEMS gyroscope for the detection of gestures and motion-commands, in the external acquisition board, is still available in this version, as an alternate means for the selection of options that are selected with the pointer.

Table 2. Comparative results from pointer test application, control

Version	*Pointer Deviation (x, y) (pixels)*	*Pointer Range (x, y) (pixels)*	*Stable Period (s)*
Initial	(±22, ±12)	(91, 55)	5.98 ± 0.61
Kalman	(±11, ±12)	(46, 55)	6.15 ± 0.73

Lastly, it is highlighted that the purpose of this scenario was to test the combined response that could be achieved from the raw signals coming from two disjoint measurement strategies, and to assess the exclusive performance of the Kalman filter, with respect to these, as well as to the initial version of the pointer. Therefore, no additional filtering or restrictions have been applied to the outputs of the Kalman filters or to the signals coming from the sensors. This implies that the aforementioned performance could take advantage of basic signal-processing and filtering strategies, prior or subsequent to the sensor fusion algorithm, without posing significant compromises to the response time.

4. Conclusions

This work presented an updated signal-processing strategy, based on sensor fusion (between a camera and a MEMS gyroscope) achieved through a Discrete Kalman Filter. Based on this new signal processing strategy, a supplementary version for a pointing interface aimed at accessibility solutions, developed in previous work, has been achieved, presenting relative advantages under a similar working principle. Advantages found for this updated version, with respect to the original pointer proposal, include the following:

- Given its composition, its performance is less susceptible to external interferences and to its environment. This version only requires visibility of the face of the subject, and the correct placement of the sensor for its use, without needing further adjustments.

- The precision of the system has been greatly improved on the horizontal axis, and a comparable response has been achieved in the vertical axis, for the fixation of the pointer over an option, even when this version takes one less reliable source for the estimation of the angle, derived from the face detector.
- The delay in the response of the pointer has been greatly reduced, while holding this stability and precision, making the user experience of the product a more intuitive one.
- This version holds the robustness of the original version, through the combination of supplementary information sources, as well as the possibility to incorporate gesture detection (made possible by the high-speed update rates of the gyroscope signals) for the activation of options selected with the pointer.

In sum, a robust strategy for the real-time control of an on-screen pointer, directed to accessibility interfaces over small screens has been achieved, with an increased independence in its performance to its environment of deployment. The proposed combination between image-processing and external sensors, based on simple estimation models, has significantly increased the potential for the systems that are currently in development. These diverse sources of information supplement each other, and make for a more robust solution overall, bringing forward the possibility to add embedded gesture detection based on these sensors for the comprehensive control of the pointer, for future versions of the system.

5. References

[1] Vittorio S A 2001 Micro-Electromechanical Systems, *Cambridge Scientific Abstracts*.

[2] Sim N Gavriel C Abbott W and Faisal A 2013 The Head Mouse-Head Gaze Estimation 'in-the-wild' with Low-Cost Inertial Sensors for bmi use *6th International IEEE/EMBS Conference on Neural Engineering*.

[3] World Health Organization 2010 *Medical Devices: Managing the Mismatch-an Outcome of the Priority Medical Devices Project*, WHO Press, Geneva.

[4] ST Microelectronics 2010 Application Note 3192, Using LSM303DLH for a tilt compensated electronic compass, Rev. 1.

[5] Ghersi I Miralles M T Mariño M 2014 Augmented-Reality and Alternative-Input System based on Simultaneous Head Position and Pose Estimation, *IFMBE Proceedings,* Vol 49, 2015, pp 222-225, DOI 10.1007/978-3-319-13117-7_58.

[6] Apache License Agreement at http://www.apache.org/licenses/

[7] Bachman T 2006 Patient Positioning: More than Just "Turn Every 2h", *Barrow Quarterly,* Vol 22, N° 3, Barrow Neurological Institute.

[8] Dodgson N A 2004 Variation and extrema of human interpupillary distance. *SPIE Proc.* vol. 5291 Stereoscopic displays and Virtual Reality Systems XI, California, USA, pp 36-46.

[9] Peebles L Norris B 1998 *Adultdata: The Handbook of Adult Anthropometric and Strength Measurements*, Department of Trade and Industry, Great Britain.

[10] Grover Brown R Hwang P Y C 1997 *Introduction to Random Signals and Applied Kalman Filtering*, 3rd Edition, Wiley, United States.

[11] Sabatini A 2011 Kalman-Filter-Based Orientation Determination Using Inertial/Magnetic Sensors: Observability Analysis and Performance Evaluation, *Sensors,* Vol. 11, 9182:9206.

Acknowledgements

This work was carried out under –and supported in part by- UBACyT 20620120100011BA 2013-2016 project, Facultad de Arquitectura, Diseño y Urbanismo, Universidad de Buenos Aires (FADU-UBA), and by Facultad de Ciencias Fisicomatemáticas e Ingeniería, Pontificia Universidad Católica Argentina (UCA).

A simple but efficient voice activity detection algorithm through Hilbert transform and dynamic threshold for speech pathologies

D. Ortiz P.[1], Luisa F. Villa, Carlos Salazar, and O.L. Quintero
Mathematical Modeling Research Group, GRIMMAT, School of Sciences, Universidad EAFIT, Carrera 49 NO 7 Sur-50, Medellin — Colombia.

E-mail:dpuerta1@eafit.edu.co

Abstract. A simple but efficient voice activity detector based on the Hilbert transform and a dynamic threshold is presented to be used on the pre-processing of audio signals. The algorithm to define the dynamic threshold is a modification of a convex combination found in literature. This scheme allows the detection of prosodic and silence segments on a speech in presence of non-ideal conditions like a spectral overlapped noise. The present work shows preliminary results over a database built with some political speech. The tests were performed adding artificial noise to natural noises over the audio signals, and some algorithms are compared. Results will be extrapolated to the field of adaptive filtering on monophonic signals and the analysis of speech pathologies on futures works.

1. Introduction

Many authors label the section of the speech as *voiced*, where the vocal chords vibrate and produce sound, *unvoiced*, where the vocal chords are not vibrating, and *silenced* [1] [2]. The union of these three sections is important within the tools for audio analysis because they delimit the recognition of the speech and the specific characteristics of the speaker [2]. This process of identifications of *voiced/unvoiced* and *silenced* is known as voice activity detection [3]. As this work advance, the sections of *voiced/unvoiced* will be named as speech and *silenced* sections as silences.

A silence can be defined as the absence of audible sound or as a sound with a very low intensity [4]. These silences allow identify and separate the main components inside of communication channels marking the boundaries of the prosodic units and exposes the rate at which the speaker delivers his speech. It is possible to specify the silent pauses inside of a speech as the lack of the physical perturbation of the sound wave in a medium of propagation, indicated in the audio signal as the lack of amplitude. However, the low amplitude of the silence do not imply a totally absence of sound inside of the audio signal.

It is important to provide a methodology that accomplishes to discriminate properly the silence speech sections, considering the previously mentioned about the presence of sound with low amplitude in the silent pauses. These sounds of low amplitude are known as noises, which can be described as disturbances that interfere in the signal obtained by altering their real values. From this, the following hypotheses are proposed: Is it possible to differentiate speech sections with the silent pauses in a noisy signal? Is it possible to design an adaptive system to different types of noise that they can be present in different audios and discriminate speech and silence?

For the voice activity detection, it is common to apply different techniques that depend of the information of the obtained signal. Some features like energy, the zero-crossing rate and the coefficients of linear prediction, can be combined in such a way that the distance between them would indicate if the analysed segment is speech or silent pauses [1] or used with a threshold, fixed or dynamic, to detect the speech [5]. Other methods used probability distributions of the noise present in the silences [2] [6].

This work uses signal own features like the zero-crossing rate and the signal energy in a particular window, in order to determinate a dynamic threshold. The zero-crossing rate indicates the number of times that the signal passes, in a time gap, by the value of zero, giving a simple measure of frequency content of the signal and the signal energy represents the amplitude variations. Once this information is obtained, it is used a modification of the methodology proposed in [7] to obtain a dynamic threshold that consist of the convex combination of the maximum and minimum of each of the property calculated. Finally, a second convex combination of the two thresholds is performed. Once the threshold is obtained, it is compared with the signal coverage obtained from the Hilbert transform and determines what speech (*voiced/unvoiced*) is and what *silence* is.

This work was developed under two objectives, adaptive filtering over monophonic for the pre-processing of noisy audio signal with no reference of noise and spectral overlapping as it is shown in [8] and the analysis of speech pathologies. The second objective is planned as future work to detect speech pathologies as stuttering [9].

2. Methodology

As mentioned previously, for the detection of the speech and silences sections we propose the combination of three features from the signal: the zero-crossing rate, the signal energy and the signal coverage from the Hilbert transform.

2.1. Zero-crossing rate

The zero-crossing rate is a simple measure of the frequencies in a certain signal. In speech sections, frequencies are of high amplitude and low band; therefore, the rate will be small, different to the silence [10].

$$Z_j = \sum_{i=(j-1)\cdot N+1}^{j\cdot N} |sgn[x(i)] - sgn[x(i-1)]| \tag{1}$$

N is the size of the window to measure.

2.2. Mean square error of the energy

To compute the energy is was used the mean square error of the same signal, because this gives in detail the peaks on speech and the valleys that points silences. The energy in a time window is define like

$$E_j = \left[\frac{1}{N} \sum_{i=(j-1)\cdot N+1}^{j\cdot N} x^2(i)\right]^{1/2} \tag{2}$$

N is the size of the window to measure.

2.3. Signal covering

For the signal covering it was used the modulus of the analytic signal defined as

$$|\psi(t)| = (g(t)^2 + \hat{g}(t)^2)^{1/2} \tag{3}$$

Where $g(t)$ is the original signal and $\hat{g}(t)$ is the Hilbert transform of $g(t)$. The Hilbert transform is defined as

$$\hat{g}(t) = \mathcal{H}[g(t)] = g(t) * \frac{1}{\pi t} = \frac{1}{\pi}\int_{-\infty}^{\infty} \frac{g(t-\tau)}{\tau} d\tau \quad (4)$$

In figure 1 it can be seen an example of the coverage of the original signal using de modulus of the analytic signal.

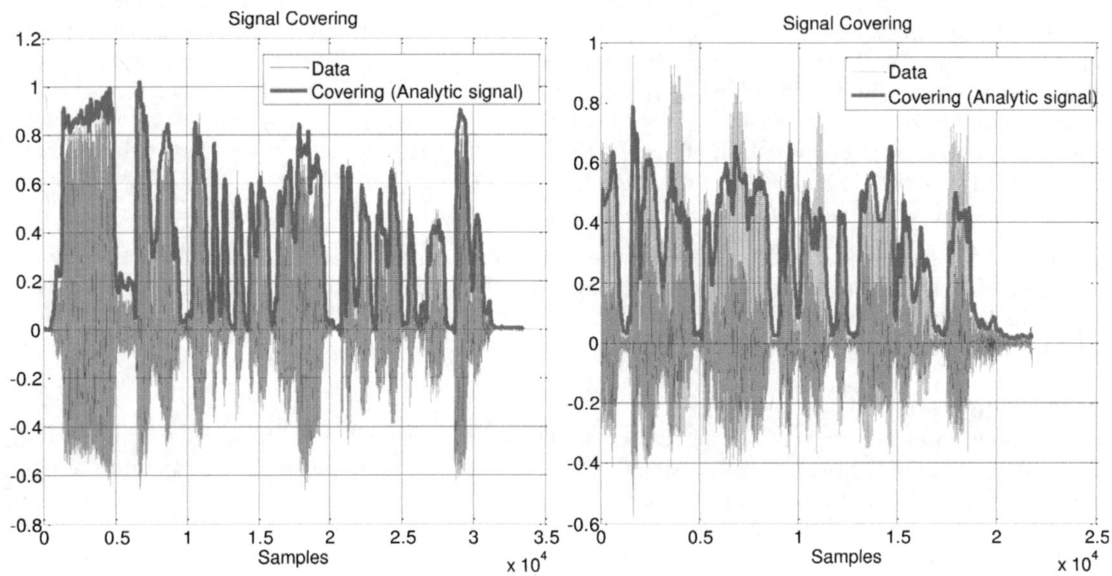

Figure 1. Signal covering for 2 different audio samples.

2.4. Dynamic threshold

For the calculation and implementation of the dynamic threshold, zero crossings and energy as dynamic features of the signal are used. First, both of them are extracted using overlapped time windows so non-stationary changes can be measured correctly, then, these data vectors are normalized, so the maximum value will be 1 and they can be compared with the signal information.

Once the data is normalized, a modification of the method proposed in [7] is used. This method consists in a convex combination of the maximum and minimum levels of the characteristic in each window. The zero-crossing rate and the energy threshold is defined by

$$\begin{aligned} E_{th}(j) &= (1-\lambda_E)\cdot E_{max} + \lambda_E \cdot E_{min} \\ Z_{th}(j) &= (1-\lambda_Z)\cdot Z_{max} + \lambda_Z \cdot Z_{min} \end{aligned} \quad (5)$$

Where λ is a scaling factor that control the process of estimation and j indicates the window. For diferents types of signals this value may vary depending of its characteristics [7], then, a scaling factor that depend directly of the signal

$$\lambda_E = \frac{E_{max} - E_{min}}{E_{max}} \qquad \lambda_Z = \frac{Z_{max} - Z_{min}}{Z_{max}} \quad (6)$$

It's possible that the minimum values of these two features can change until to find a value almost zero. In this case, the thresholds don't adapt properly to the signal changes, i.e., if it finds a value close to zero (that is the minimum in all the information of the signal), the threshold for the energy and the zero-crossing rate, will be kept constant and low, which will give incorrect information in the case that there are silent pauses with noise of amplitude and high frequencies. To avoid this, the minimum value of (6) is increased slightly and is defined by

$$\begin{aligned} E_{min}(j) &= E_{min}(j-1) \cdot \Delta_E(j) \\ Z_{min}(j) &= Z_{min}(j-1) \cdot \Delta_Z(j) \end{aligned} \tag{7}$$

The parameter Δ is define as

$$\Delta(j) = \Delta(j-1) \cdot \alpha \tag{8}$$

where α is a growth factor. Once this threshold is obtained for the energy and the zero-crossing rate, it is defined the global threshold for discriminate the silent pauses in a speech like a convex combination of the two previous thresholds

$$TH(j) = (1-p) \cdot E_{th}(j) + p \cdot Z_{th}(j) \tag{9}$$

where p is the scaling factor of the convex combination. Once the dynamic threshold is obtained of the signal it is possible to compare the coverage of the same signal obtained from (3). If the coverage is below of the limit, the audio section is considered a silence, if it is above, is considered a speech section.

3. Silence detection procedure

Once calculated the features of the signal (Zero Crossing rate, energy and signal covering) and obtained the dynamic threshold, the procedure for detecting the silences sections by comparing the threshold and coverage was established.

1. First, signal data is normalized, followed by a pre-filtering band-pass with cut frequency 100-3200 Hz.
2. For the dynamic threshold, first the maximum and minimum variables for the energy and the zero-crossing rate are determined. For the energy, the maximum will be the average of the data and the minimum variable will be the minimum value. For the zero-crossing rate, if the first value is equal to zero, it will be taken the average as maximum value, if is different to zero, will be the first value of the data. For the minimum variable, will be the minimum zero-crossing rate of the data, if these is equal to zero, this variable is taken as an epsilon $\epsilon > 0$ (a low number closed to 0).
3. Once the maximum and minimum are determined, it follows to determine the threshold for the energy and the zero-crossing rate for each overlapped window. In this case, the overlapping was set in 90% of the size of the window. Later, the total threshold of the window is calculated using (9).
4. The complete signal coverage is determined from of the analytic signal by using the Hilbert transform, then, a decimation over the analytic signal is made to smooth the covering.
5. Finally, the dynamic threshold is compared with the coverage obtained in step 4. If the threshold is above of the coverage this audio section is taken as silent pauses; if the threshold is below, is taken as speech.

4. Results and analysis

4.1. Test

For the test, it was made a database with different political speeches published on the internet. These speeches were recorded in noisy environments that can disturb the voice activity detection and the noise has spectral overlapping with the real signal of the speech. The data base has a sample rate of 8 KHz, and to analyse the correct voice activity detection, the speech and silence section where identified manually by an expert operator as shown in figure 2. To test the robustness of the algorithm it was added artificial Gaussian white noise with SNR of 5 dB, 15dB and 20dB measure using the energy of the noise and the signal, and the result was compared with a benchmark algorithm found in [11]. An error measure was used to calculate the performance of the algorithm. This measure was made by comparing the samples in the signal identified by the expert as speech or silence and results of the algorithm. Another measure used was the number of silences identified by the algorithm.

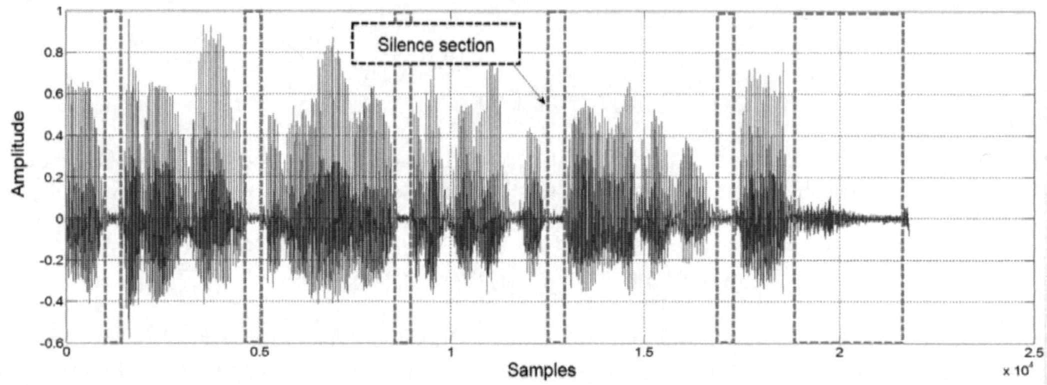

Figure 2. Silence section identified by an expert for the test.

To obtain the analytic signal it was established a decimation factor of 10 from different tests, watching that this describe in a good way the signal coverage. For the extraction of the features, which define the dynamic threshold, were used small windows of 12.5 ms or 100 samples over the sample rate (8 KHz, 8000 samples per second) with an overlapping of the 90%. Were used small windows with the objective of that abrupt changes do not alter the measure and the overlapping allows following precisely the characteristics behavior. The growth factor α for the minimum energy was settled in 1.0001, so the minimum energy grows up in a low rate, and the scaling factor p in 0.1 to prioritize the measure of the energy threshold.

4.2. Results and analysis

For the first test, signals where used without adding synthetic noise, as it was mentioned before, audio signals has their natural noise. Results can be seen on table 1 where the percentage of error and the number of silences identified by both algorithms is presented.

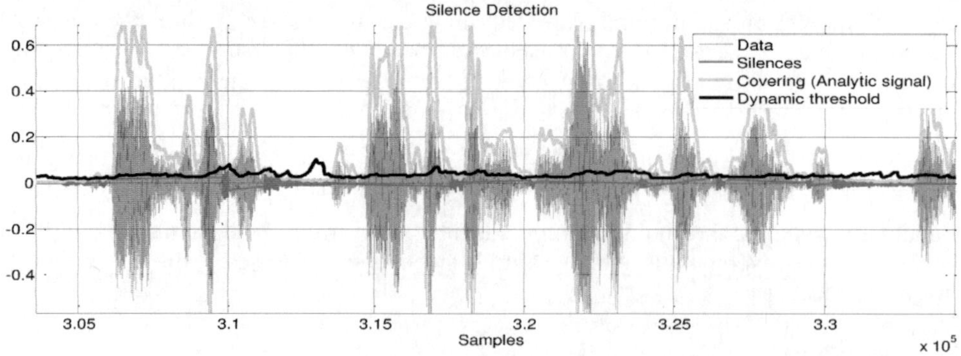

Figure 3. Behaviour of the proposed algorithm. On green the coverage of the signal is shown. On black, the dynamic threshold. The speech section can be find on blue, and silences on red.

Table 1. This table shows the results for the VAD of 10 audio signals of the data base. Method 1 is the algorithm presented in this work. Method 2 is the benchmark method found on literature [11]. Here the percentage of accuracy is presented and the number of silences identified

	Natural noise				
	Method 1		Method 2		
Audios	%	N. Sil.	%	N. Sil.	N. Silence Real
Audio 1	16.30	3	16.30	3	4
Audio 2	8.30	6	23.30	3	7
Audio 3	13.50	6	31.00	2	8
Audio 4	6.34	9	27.34	3	10
Audio 5	31.25	5	42.91	1	12
Audio 6	16.37	6	38.24	1	8
Audio 7	3.52	7	28.52	2	7
Audio 8	5.325	11	22.82	5	12
Audio 9	9.45	7	21.12	3	6
Audio 10	1.71	10	1.92	5	10

As it is shown, the performance of the proposed method is much better than the algorithm proposed on [11]. The percentage of accuracy has an average of 11.21% and the number of silences identified are closed to the values to those founded by the expert. In figure 3 can be seen the behavior of the proposed algorithm, showing the combination of the dynamic threshold and the covering of the signal to identify the silence section.

One of the objectives of using a dynamic threshold is that it can adapt to the spectral characteristics over the signal. As it can be seen on figure 3 dynamic threshold can change over time and by different kind of spectral overlapping with natural noise in this case.

The algorithm was also tested contaminating the audio signals with Gaussian white noise with SNR of 5 dB, 15 dB and 20 dB. Results can be observed on table 2 to 4.

Table 2. This table shows the results for the VAD of 10 audio signals contaminated with white Gaussian noise with SNR of 20 dB. Method 1 refers the proposed algorithm. Method 2 is the benchmark method found on literature [11].

	Gaussian Noise SNR 20 dB				
	Method 1		Method 2		
Audios	%	N. Sil.	%	N. Sil.	N. Silence Real
Audio 1	18.70	5	18.70	3	4
Audio 2	12.35	9	22.35	3	7
Audio 3	36.25	14	36.25	2	8
Audio 4	36.25	18	32.75	3	10
Audio 5	5.12	12	37.21	1	12
Audio 6	8.11	9	34.36	1	8
Audio 7	3.83	7	28.83	2	7
Audio 8	11.55	15	23.22	5	12
Audio 9	18.96	8	24.79	3	6
Audio 10	15.58	13	22.58	5	10

Table 3. This table shows the results for the VAD of 10 audio signals contaminated with white Gaussian noise with SNR of 15 dB. Method 1 refers the proposed algorithm. Method 2 is the benchmark method found on literature [11].

	Gaussian Noise SNR 15 dB				
	Method 1		**Method 2**		
Audios	%	N. Sil.	%	N. Sil.	N. Silence Real
Audio 1	55.49	8	29.24	3	4
Audio 2	07.27	8	22.27	3	7
Audio 3	35.35	13	39.73	2	8
Audio 4	49.37	21	35.37	3	10
Audio 5	7.23	13	36.40	1	12
Audio 6	2.26	8	32.89	1	8
Audio 7	16.79	9	31.79	2	7
Audio 8	26.24	18	23.32	7	12
Audio 9	20.86	8	26.70	3	6
Audio 10	17.51	13	28.01	4	10

Table 4. This table shows the results for the VAD of 10 audio signals contaminated with white Gaussian noise with SNR of 5 dB. Method 1 refers the proposed algorithm. Method 2 is the benchmark method found on literature [11].

	Gaussian Noise SNR 5 dB				
	Method 1		**Method 2**		
Audios	%	N. Sil.	%	N. Sil.	N. Silence Real
Audio 1	98.24	11	45.74	3	4
Audio 2	13.77	9	23.77	3	7
Audio 3	46.65	15	42.27	2	8
Audio 4	56.43	22	38.93	3	10
Audio 5	17.04	15	40.37	1	12
Audio 6	3.44	8	34.07	1	8
Audio 7	40.62	12	40.62	2	7
Audio 8	35.69	20	26.94	7	12
Audio 9	50.44	12	32.94	3	6
Audio 10	22.83	8	33.33	5	10

Table 5. Average Error comparison of the two methods

	Average error %			
	Natural Noise	**SNR 20 dB**	**SNR 15 dB**	**SNR 5 dB**
Method 1	11.21	16.67	23.84	38.51
Method 2	27.08	28.10	30.57	35.90

Table 6. Standard deviation Error comparison of the two methods

	Standard deviation error %			
	Natural Noise	**SNR 20 dB**	**SNR 15 dB**	**SNR 5 dB**
Method 1	8.68	11.52	17.97	27.20
Method 2	11.45	6.65	5.70	6.95

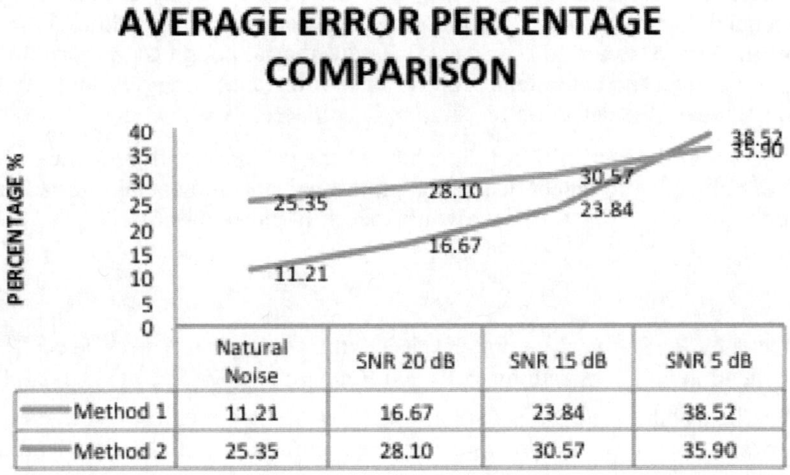

Figure 4. Average percentage error comparison between the two methods.

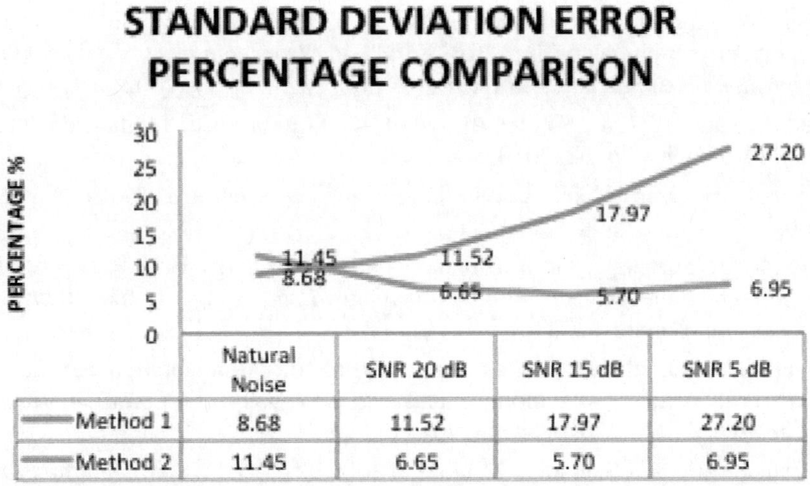

Figure 5. Standard deviation percentage error comparison between the two methods.

By the results is clear that the proposed algorithm is robust under the different test performed. On tables 2 and 3, results shows that the algorithm is consistent with the first test where no noise was added. The low percentage of error shows that the algorithm is robust against the noise with low and middle energy. Also the number of silence section detected keeps closed to the real.

Although in test with SNR of 20 dB and 15 dB shows good result, is important to note that as energy of the signal increases, the percentage of error increases too. As shown in table 5, the performance of the proposed algorithm gets worse as the energy of the noise increase, but it remains at a low percentage.

It can be observe in Figure 5, the value of the standard deviation of error increases as the noise level in the audio, which means that the method is prone to failure in the presence of noisy signals unlike the other method in which the standard deviation of error remains constant.

5. Conclusions

Considering that noise is a natural phenomenon when getting the information, is important to build tools that can adapt to this noises without inconvenient. Comparing this test with real life, different kind of noise can be found when getting the information to analyze such other voices, short circuits and others.

Voice activity detection take an important place in issues such as emotion detection in patients with diseases or emotional disorders, in remote monitoring of these patients, in pathologies of the vocal tract, and others. From the analysis carried out, it can be said that although the algorithm proposed has a simple structure, it is robust and consistent against noise of different energies so it can be implemented in different applications for the detection of pathologies related to speech.

These results could be used for stablish relationships of the presence and frequency of these segments in a speech with the objective to detect deception, emotional states in social interaction, shortcomings of affective disorder or pathologies associated with speech like the stuttering.

6. References

[1] B. S. Atal and L. R. Rabiner, "A Pattern Recognition Approach to Voiced-Unvoiced-Silence Classification with Aplications to speech Recognition," *IEEE,* pp. 201-212, 1976.

[2] G. Saha, S. Chakroborty and S. Senapati, "A New Silence Removal and Endpoint Detection Algorithm for Speech and Speaker Recognition Applications," Indian Institute of Technology, Khragpur.

[3] F. G. Germain, D. L. Sun and G. J. Mysore, "Speaker and Noise Independent Voice Activity Detection," Proceedings of Interspeech, Lyon, 2013.

[4] R. V. Prasad, A. Sangwan, H. S. Jamadagni, C. M. C., R. Sah and V. G. , "Comparison of Voice Activity Detection Algorithms for VoIP," in *Proceedings of the Seventh International Symposium on Computers and Communications (ISCC'02)*, 2002.

[5] E. Verteletskaya and B. Simak, "Voice Activity Detection for Speech Enhancement Applications," Acta Polytechnica, Praha, 2010.

[6] S. G. Tanyer and H. Özer, "Voice Activity Detection in Nonstationary Noise," *IEEE,* pp. 478-482, 2000.

[7] K. Skahnov, E. Verteletskaya and B. Simak, "Dynamical Energy-Based Speech/Silence Detector for Speech Enhancement Applications," in *Proceedings of the World Congress on Engineering*, Londres, 2009.

[8] D. Ortiz and O. L. Quintero, "Una aproximación al filtrado adaptativo para la cancelación de ruidos en señales de voz monofónicas," in *XVI Congreso Latinoamericano de Control Automático, CLCA 2014*, Cancún, 2014.

[9] P. Pichot, Diagnostic and Statistical Manual of Mental Disorders, Washington, D.C.: American Psychiatric Association, 1994.

[10] R. G. Bachu, S. Kopparthi, B. Adapa and B. D. Barkana, "Separation of Voiced and Unvoiced using Zero crossing rate and Energy of the Speech Signal".

[11] Z. H. Tan and B. Lindberg, "Low-Complexity Variable Frame Rate Analysis for Speech Recognition and Voice Activity Detection," IEEE Journal of Selected Topics in Signal Processing, vol. 4, p. 5, January 2010.

Evaluation of agreement between temporal series obtained from electrocardiogram and pulse wave.

GM Leikan[1]; E Rossi[1]; MCuadra Sanz[2]; D Delisle Rodríguez[3]; MC Mántaras[1]; J Nicolet[1]; D Zapata[1]; I Lapyckyj[1]; L Nicola Siri[1]&MS Perrone[1]
[1]Facultad De Ingeniería-Bioingeniería-Universidad Nacional de Entre Ríos-Argentina.
[2]Centro de Investigación y DesarrolloTecnológico de la Industria de la Electrónica y la Informática-Colombia.
[3]Centro de BiofísicaMédica-Universidad de Oriente-Cuba.

E-mail: msperrone@bioingenieria.edu.ar

Abstract. Heart rate variability allows to study the cardiovascular autonomic nervous system modulation. Usually, this signal is obtained from the electrocardiogram (ECG). A simpler method for recording the pulse wave (PW) is by means of finger photoplethysmography (PPG), which also provides information about the duration of the cardiac cycle. In this study, the correlation and agreement between the time series of the intervals between heartbeats obtained from the ECG with those obtained from the PPG, were studied. Signals analyzed were obtained from young, healthy and resting subjects. For statistical analysis, the Pearson correlation coefficient and the Bland and Altman limits of agreement were used. Results show that the time series constructed from the PW would not replace the ones obtained from ECG.

1. Introduction

It has been shown a significant relationship between the autonomic nervous system (ANS) and the operation of the cardiovascular system [1]. To quantify this relationship, the development of quantitative indicators of autonomic activity has been necessary [2].

A common way to estimate certain aspects of cardiovascular function is through a surface electrocardiogram (ECG) at rest. It is used for diagnosis of various cardiovascular diseases by analyzing, among other things, the morphology of the signal. Extract the average heart rate, defined as the number of beats per minute, is also possible, and the average period defined as the average length of RR intervals per minute, although this indicator is rarely used in clinical practice.

Since RR intervals are not constant and these fluctuations are influenced by the respiratory cycle and ANS activity, variations of the consecutive RR intervals can be studied using time series constructed from the sequences of such intervals. From these time series is possible to obtain indicators that describes what is known as heart rate variability (HRV).

The quality of measurements of biological variables is crucial in clinical practice. The inherent variability of the variable itself, the associated physiological process, the measuring system and the operator are determining the value of conclusions that emerge from measurements obtained. This reinforces the needingof ensure that new measurement techniques be reliable and,eventually, interchangeable [3], [4], [5],[6].

While ECG signal is a reliable primary source for obtaining the intervals associated with heartbeat, there are other physiological signals which also contains information of heart rate, as pulse wave (PW) obtained by Photoplethysmography (PPG). To obtain the PW, normally used to get information of cardiovascular physiology, requires a much simpler patient's instrumentation than that used in ECG recordings, as it consumes less time and its implementation is less complex, besides being more comfortable for the patient [7].

However, it is necessary know how reliable is the information obtained from PW signal to generate equivalent time series to those obtained from ECG. There are several publications where PW is used as a primary source of information for HRV study. Some of those publications considered that both methods are equivalent for this purpose [8], [9], [10], [11], [12].

The aim of this study is to determine if time series obtained from PPG and those obtained from ECG are interchangeable. Then, the series were analyzed using Pearson's correlation coefficient and Bland-Altman'sagreement method[3].

2. Materials and methods

Studies were conducted in young, healthy subjects, in supine position. 13 recordings were performed, each between 5 and 8 minutes long.Sample consists of 7 women and 6 men, aged between 19 and 27 years (mean 23.7 ± 2.3 years). The applied protocol establishes a rest period of 5 minutes before data acquisition start. The ANGIODIN V2-Plus was used as acquisition device, previously validated [13]. Each recording consists of 3 signals, ECG (Lead I) and PPG (left and right) acquired simultaneously with 10-bit resolution and 1 [kHz] sampling frequency per channel. This sampling frequency is higher than minimum recommended by the Task Force[14].

The equipment used performs an analog filtering of signals before digitalization. A digital processing which includes filtering and detection of fiduciary points is then performed. In case of ECG's time series, the fiduciary point used is QRS complex's R wave's maximum. In PPG signal, three possible fiduciary points were studied: the systolic point (P), the maximum positive derivative point (D) and the diastolic point (F). P and D points are commonly used [15], but diastolic point's detection is very uncertain when the signal is affected by motion artifacts, so it was discarded. The detection's algorithms and the construction of time series were implemented in Matlab®, where PPG and ECG records were processed with the same techniques of signal analysis.

Fiduciary points that were studied are indicated in **Figure 1**: R points obtained from ECG, D and P points obtained from PW.

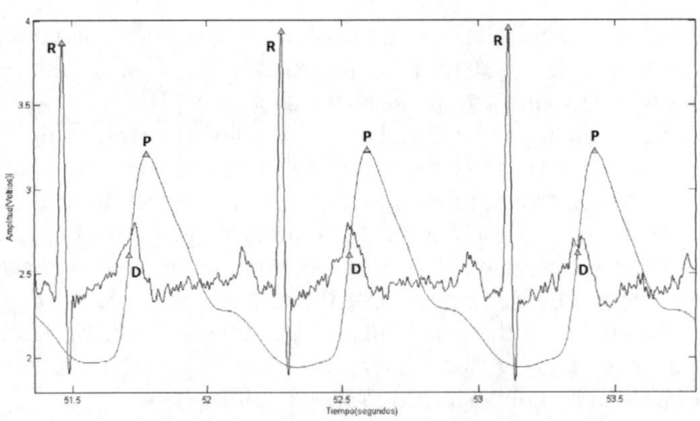

Figure1: Fiduciary points on ECG and PW signals.
R: wave**R**from ECG; **D**: maximum positive derivative point y **P:** systolic point from PPG.

Time series corresponding to each recording are constructed after event detection. According to recording's duration and individual's heart rate, each series has between 250 and 400 samples. For this study, time series of all subjects were concatenated, to verify correlation in the detection of intervals;

Journal of Physics: Conference Series **705** (2016) 012038 doi:10.1088/1742-6596/705/1/012038

thus a single time series with 4500 intervals for each fiducial point was obtained. The time series obtained from ECG is called RR, whereas in case of PPG signals, are called PP (PPG1), PP (PPG2), DD (PPG1) and DD (PPG2), being PPG1 and PPG2 the signals obtained from left hand and right hand, respectively.

Were calculated Pearson's correlation coefficients and Bland-Altman's limits of agreement [2] of these series, with RR as reference.

Bland-Altman's limits of agreement assesses if the proposed measurement method can replace the reference method by analyzing the differences between one or another method measurements. In turn, this method can identify trends and behavior of the measuring system. The decision about the interchangeability of methods depends on the application field; in our case, we assessed if differences between them were significant.

3. Results and discussion

The results of correlation and agreement analysis are presented in numerical and graphical form to facilitate understanding. As shown in **Table 1**, all correlation coefficients are greater than 0.9988, which indicates excellent correlation.

Compared series	r
RR vs DD (PPG1)	0.9988
RR vs PP (PPG1)	0.9990
RR vs DD (PPG2)	0.9989
RR vs PP (PPG2)	0.9991

Table 1: Pearson's correlation coefficient between temporal RR series (obtained from ECG) and temporal PP and DD series (from PPG1 -left hand- and PPG2 -right hand-).

High correlation does not imply agreement, so it was decided to implement BlandAltman agreement's graphics, a method which represents average of measurements in abscissa and difference between measurements in ordinate. In **Figure 2** the Bland Altman analysis between time series of RR intervals (ECG) and DD intervals (PPG1) can be seen.The average of measurements is close to zero, 8.4 [μs], and the standard deviation (SD) is around 5.5 [ms], as is shown in the figure.

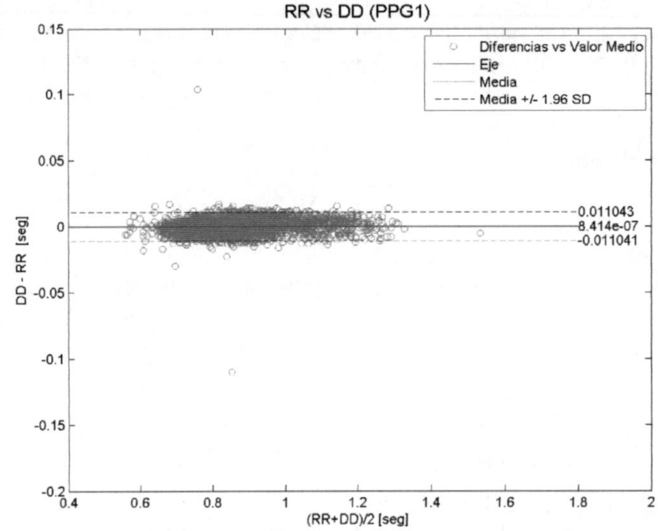

Figure2: Agreement analysis between time series of RR intervals (ECG) and DD intervals (PPG1).

The same analysis performed between time series of RR intervals (ECG) and PP intervals (PPG1) is shown in **Figure 3**. The average difference between measurements is close to zero, -9.6 [μs], and SD is around 4.9 [ms].

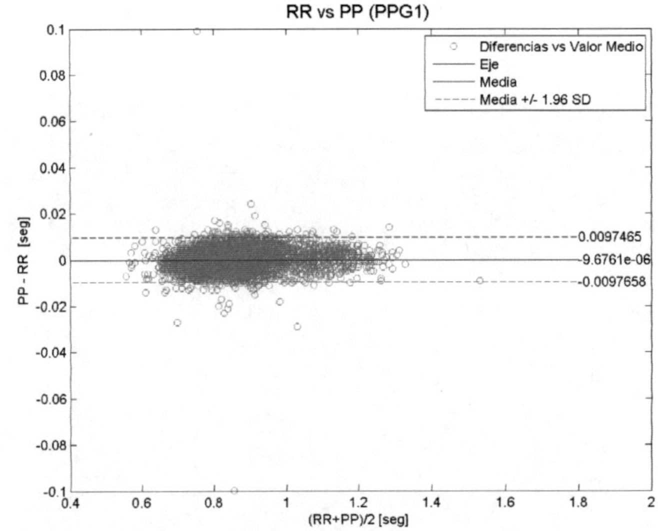

Figure3: Agreement analysis between time series of RR intervals (ECG) and PP intervals (PPG1).

The agreement analysis between time series of RR intervals (ECG) and DD intervals (PPG2) show an average difference between measurements of -1 [μs] and SD of 5 [ms] (**Figure 4**).

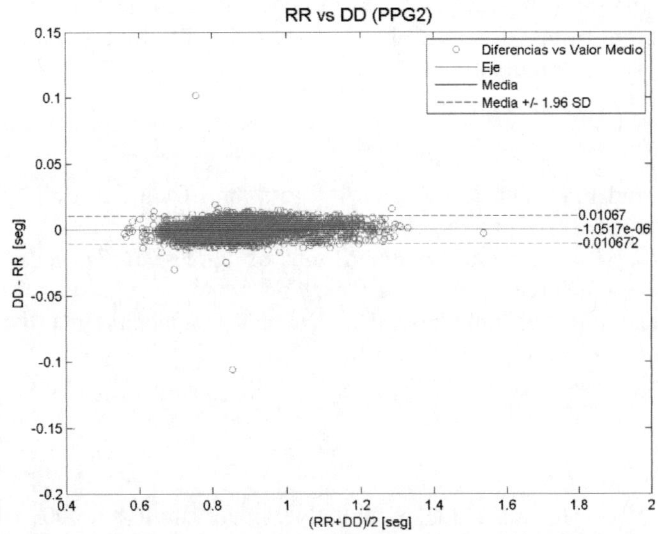

Figure 4: Agreement analysis between time series of RR intervals (ECG) and DD intervals (PPG2).

The results shown in **Figure 5** belongs to agreement analysis between time series of RR intervals (ECG) and PP intervals (PPG2). The average difference is around -7.57 [μs] and SD of 4.8 [ms].

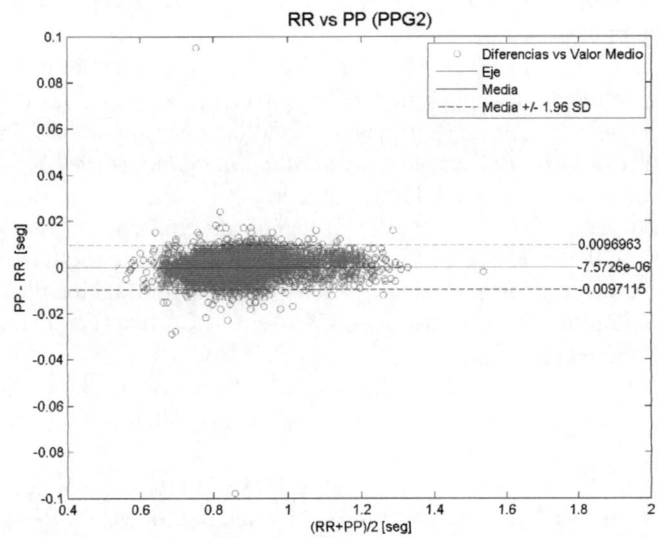

Figure 5: Agreement analysis between time series of RR intervals (ECG) and PP intervals (PPG2).

Discrepancies between either method's measurements have a SD between 2.2 [ms] and 5.5 [ms]. Limits of agreement (average ± 1.96 SD) exceeds the maximum sampling period suggested by reference bibliography (4 [ms], corresponding to 250 [Hz]) [14].

It is known that autonomic modulation over PW is more complex than heart rate regulation and is affected by other regulation systems, so it is necessary to elucidate if discrepancies between measurements are inherent to the physiological process or to the measurement methodology.

If it is considered that the cause of differences can be attributed to ANS vasomotor regulation, but also to the neurohumoral systemic regulation (catecholamines, angiotensin, vasopressin, etc.) and neurohumoral local regulation (prostaglandins, histamine, nitric oxide, etc.) [16], it is possible that time series obtained both from ECG and PW can reflect different dynamics and mechanisms associated to peripheric vascular regulation.

4. Conclusions

Considering that ECG and PPG signals contains information of cardiac cycle duration, it could be infer that results of analysis both signals, in this aspect, would be equivalent. Although Pearson's correlation coefficients shows high correlation, limits of agreement exceeds maximum sampling period recommended by Task Force in HRV analysis. This work's results suggests that PW's obtained time series can't replace those arising from ECG to HRV analysis, but they can be useful to investigation of vasomotor regulation.

References

[1] Estévez M. (a). Visiónactualizadadel Sistema NerviosoAutónomo. 2007 [fecha de consulta: 6 de abril de 2015]. Disponible en: <http://infomed20.sld.cu/wiki/doku.php?id=librosabiertos:vision_actualizada_del_sistema_nervioso_autonomo&rev=11927976>

[2] McKinley, P. S., Shapiro, P. A., Bagiella, E., Myers, M. M., De Meersman, R. E., Grant, I., & Sloan, R. P. (2003). Deriving heart period variability from blood pressure waveforms. *Journal of Applied Physiology (Bethesda, Md.: 1985)*, *95*(4), 1431–1438. doi:10.1152/japplphysiol.01110.2002

[3] Bland, J. M., & Altman, D. G. (1986). Statistical methods for assessing agreement between two methods of clinical measurement.*Lancet* (Vol. 1).

[4] BuendíaLozada, E.R.P. (2011). Reproductibilidad del instrumento HC / The HC instrumentreproducibility.*RevistaInternacional de Medicina y Ciencias de la ActividadFísica y el Deporte vol. 10 (41) pp. 1-13.* Disponible en: <http://cdeporte.rediris.es/revista/revista41/artreproductibilidad203.htm>

[5] Pita Fernández S, PértegasDíaz S, La fiabilidad de las medicionesclínicas: el análisis de concordancia para variables numéricas. Unidad de EpidemiologíaClínica y Bioestadística. ComplexoHospitalario-Universitario Juan Canalejo. A Coruña (España) Actualización 12/01/2004. www.fisterra.com.

[6] Jaime Latour[1], Víctor Abraira[2], Juan B. Cabello[3] y Javier López Sánchez[2]. Las medicionesclínicasencardiología: validez y errores de medición.*RevEspCardiol 1997; 50: 117-128.*

[7] Bolanos, M., Nazeran, H., &Haltiwanger, E. (2006). Comparison of heart rate variability signal features derived from electrocardiography and photoplethysmography in healthy individuals. *In Annual International Conference of the IEEE Engineering in Medicine and Biology - Proceedings (pp. 4289–4294).*doi:10.1109/IEMBS.2006.260607

[8] Posada-Quintero, H. F., Delisle-Rodríguez, D., Cuadra-Sanz, M. B., &Fernández de la Vara-Prieto, R. R. (2013). Evaluation of pulse rate variability obtained by the pulse onsets of the photoplethysmographic signal.*Physiological Measurement, 34(2), 179–87.*doi:10.1088/0967-3334/34/2/179

[9] Lu, G., & Yang, F. (2009). Limitations of oximetry to measure heart rate variability measures.*Cardiovascular Engineering*, *9*(3), 119–125. doi:10.1007/s10558-009-9082-3

[10] Lu, G., Yang, F., Taylor J.A., Stein J.F. (2009). A comparison of photoplethysmography and ECG recording to analyse heart rate variability in healthy subjects. *J Med Eng Technol. 2009;33(8):634-41.*doi: 10.3109/03091900903150998.

[11] Gil, E., Orini, M., Bailón, R., Vergara, J. M., Mainardi, L., & Laguna, P. (2010). Photoplethysmography pulse rate variability as a surrogate measurement of heart rate variability during non-stationary conditions.*Physiological measurement (Vol. 31).* doi:10.1088/0967-3334/31/9/015

[12] Johnston, W., & Mendelson, Y. (2005). Extracting heart rate variability from a wearable reflectance pulse oximeter. *Proceedings of the IEEE 31st Annual Northeast Bioengineering Conference, 2005.*doi:10.1109/NEBC.2005.1431971

[13] CuadraSanz, M.; Rossi, E.; Delisle Rodriguez, D.; Mántaras, M.C.; Perrone, M.S.; Fainstein, D.; Zapata, D.; Vázquez SeisDedos, C.; Nicola Siri, L., Sistema para el registrosimultáneo de la onda de pulso y el electrocardiogramaorienta al estudio de la regulaciónautonómica. *Proceedins FIE'08, N°66, pag 1-4, ISBN 978-84-00-08680-0, Universidad de Oriente, Santiago de Cuba, Cuba.*

[14] Electrophysiology, T. F. of the E. S. of C. the N. A. S. of P. (1996). Heart Rate Variability: Standards of Measurement, Physiological Interpretation, and Clinical Use. *Circulation, 93 (5), 1043–1065.* doi:10.1161/01.CIR.93.5.1043

[15] Schäfer, A., &Vagedes, J. (2013). How accurate is pulse rate variability as an estimate of heart rate variability?*International Journal of Cardiology, 166, 15–29.* doi:10.1016/j.ijcard.2012.03.119

[16] Guyton, A.C. and Hall, J.E. "Medical Physiology" 11th. Edition 2006 , Cap.10 (123-124).

Journal of Physics: Conference Series **705** (2016) 012039 doi:10.1088/1742-6596/705/1/012039

Development of an apnea detection algorithm based on temporal analysis of thoracic respiratory effort signal

C R Dell'Aquila, G E Cañadas, L S Correa, E Laciar

Gabinete de Tecnología Médica, Universidad Nacional de San Juan, San Juan, Argentina

E-mail: carlos.dellaquila@unsj.edu.ar, gcanadas@unsj.edu.ar, lcorrea@gateme.unsj.edu.ar, laciar@gateme.unsj.edu.ar

Abstract. This work describes the design of an algorithm for detecting apnea episodes, based on analysis of thorax respiratory effort signal. Inspiration and expiration time, and range amplitude of respiratory cycle were evaluated. For range analysis the standard deviation statistical tool was used over respiratory signal temporal windows. The validity of its performance was carried out in 8 records of Apnea-ECG database that has annotations of apnea episodes. The results are: sensitivity (Se) 73%, specificity (Sp) 83%. These values can be improving eliminating artifact of signal records.

1. Introduction

Breathing can be defined as the function which aims is to provide necessary amount of oxygen (O2) to the body's cells and remove carbon dioxide (CO2) resulting from the cell's combustion [1]. Pathologies as the asthma, chronic obstructive pulmonary disease (COPD) and sleep apnea/hypopnea syndrome (SAHS) are examples of deficiencies in breathing process. In particular, SAHS consists in cessation of the respiratory activity by ten seconds (10s) or more which is produced five times or more by hour during the sleep [2]. The apneas can be the results of an obstruction in airways (Obstructive Apnea), absence of airflow and inspiration effort caused by failure in central nervous conduction (Central Apnea) and a combination of both (Complex Apnea).

The prevalence of this pathology in general population is 3.1 to 7.5% in men and 1.2 a 4.5% pre-menopausal women, while in children is 1 to 3% [2]. This statistics only shows SAHS in patients diagnosed but it is estimated that there is a lot of children and adults with this disease, and are not diagnosed.

SAHS patients have an excessive daytime sleepiness, and an increased probability to suffer job or traffic accidents and a higher prevalence of cardiovascular diseases, like hypertension, myocardial infarction and stroke [3].

Polysomnography (PSG) is considered to be the "Gold standard" study for diagnose SAHS. It consists in acquired biomedical signals during one or more nights. Usually a PSG study include signals registers of EEG, EOG, ECG, EMG, oronasal airflow, oxygen saturation (SaO2) and abdominal and thoracic respiratory effort signal [4]. In particular, the last is an indirect technique for continuous and dynamic measurement of respiratory volume and it is based on inductive plethysmography method [5].

Even though PSG is an effective method and allows diagnose others pathologies too, it is an expensive study. In addition, It is uncomfortable for the patient due to lot of signals that can be acquired and it can't get a normal sleep, which decreases the diagnostic effectiveness of this technique. Because of these drawbacks arises the need to reduce the number of signals and for it we must develop algorithms to detect apnea events from them.

In this work the development of an apnea detection algorithm based on temporal analysis of thoracic respiratory effort signal is proposed. The algorithm must have a low computational load to enable its use in very low power consumption microcontrollers and allow, in the future, their use in portable medical equipment. In these, the processing speed and memory resource are limited as well as the energy consumption because these devices are powered with batteries.

Even though there are others works about apnea detection, they use other methods with high computational load and require more type of signals. Zamarrón et al. developed an algorithm for detect apneas using spectral analysis of heart frequency obtained from pulse oximetry signal (PPG) [6]. Ravelo et al. proposed two algorithms based on support vector machine (SVM) and mixed Gaussian models for detect apneas from RR intervals from ECG signal [7]. In addition, there are works which use only the PPG signal to detect apneas, that use different types of thresholds for detection [8] and logistic regression classifier [9].

2. Materials

2.1. Database

For the validation of the designed algorithm were used eight records of the Apnea-ECG Database which is available free in Physionet's web site [10]. This database consists of 70 records of patients (age: 45 ±18 years, 57M and 13F) with breathing diseases including SAHS. The duration of each record is among 401 and 578 minutes and have annotations done by experts showing the QRS complex of ECG and apnea episodes. In this last case, there are marks that indicate if there are current apnea (A) or normal breathing (N).

The eight records used for the validation (a01 to a04, b01, and c01 to c03) are the only ones that have electrocardiography signal, thoracic and abdominal respiratory effort signal (SR and Resp A) obtained by inductive plethysmography method, oxygen saturation (SpO2) and oronasal airflow measured by nasal thermistor. The other records weren't used, because they only have ECG signals [11].

It is important to mention that the plethysmography signals are not calibrated and therefore the algorithm designed uses only relatives values, no absolute.

3. Methods

3.1. Signal pre-processing

A digital filter is applied to SR signal first, to obtain the spectral range that defines the respiratory process, which comes up to 0.5Hz [12]. The 4to order Butterwoth low-pass filter has been selected with cut-off frequency in 0.5 Hz. The figure 1 shows the effect of the filter on the signal.

Then 80 seconds windows has been analysed to obtain locals maximum and minimum of the filtered signals and calculate the inspiration and expiration time.

First derivative method is used in order to determine local extrema, i.e. finding the zero crossings of the derivative signal and according to the crossing, it is made from positive to negative if it is a maximum and if it goes from negative to positive, it is a minimum.

Although this is a robust method for detect local maximums and minimums, it can detect local extremes produced by noise or interferences in the signal as be show in figure 2. To solve this problem the instant range of the signal is calculated. It consists in getting the difference of amplitude between a consecutives minimum and maximum and then it is applied a median filter (MF) of 250ms to the resulting temporal series. Finally the filter output is compared with each range and if it is less than

30% of the filter output then associated extremes are considered incorrect. The figure 3 shows the effect of this stage.

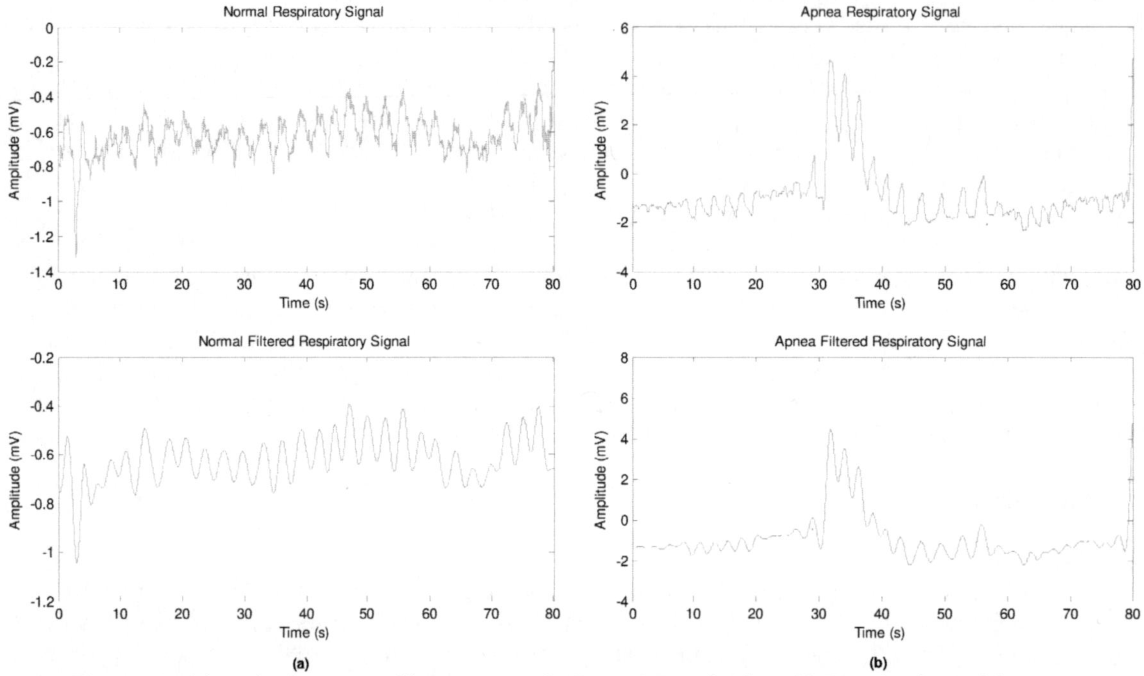

Figure 1. 4th order low-pass filter output. a) Normal Respiration. b) Respiration with apnea

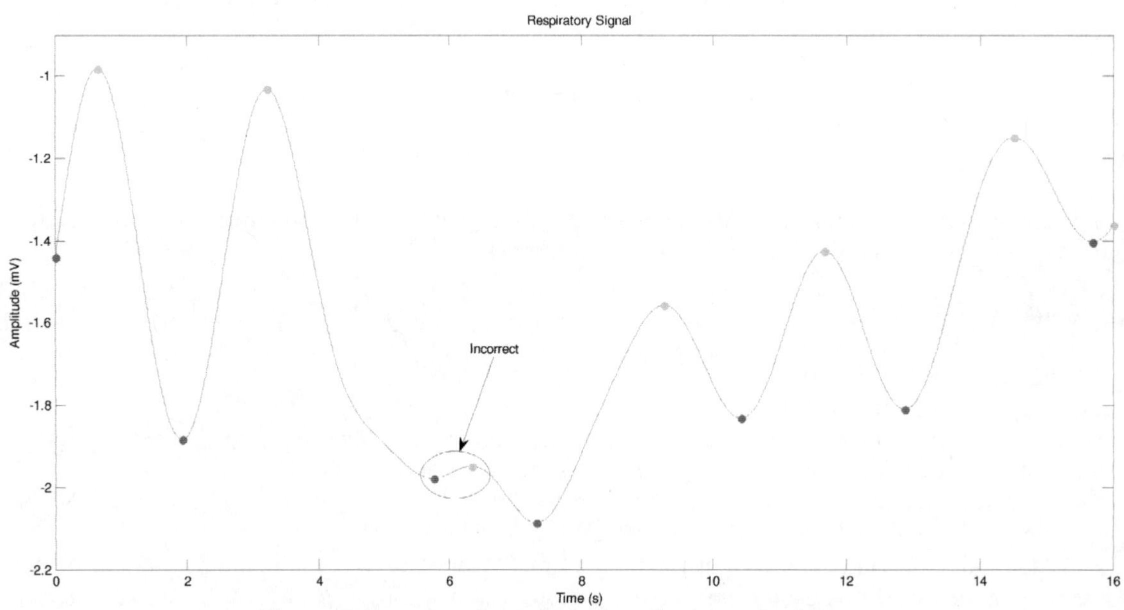

Figure 2. Incorrect detection of maximum and minimum

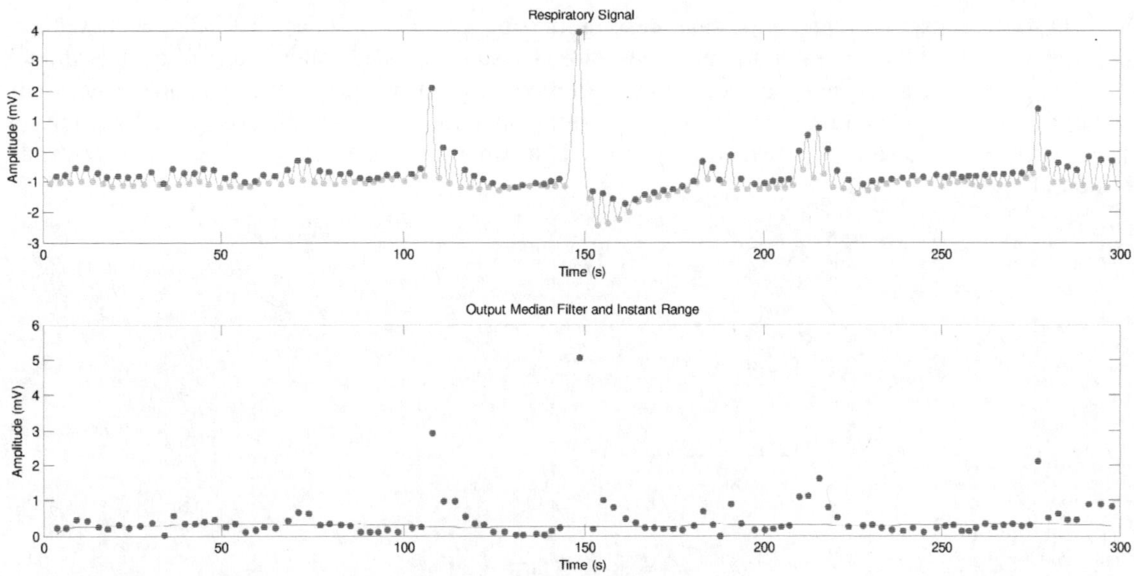

Figure 3. Median filter output

3.2. Respiratory Period

The respiratory cycle is formed by inspiration time (IT) and expiration time (ET). The IT can be estimated as the time between a minimum and the maximum. ET is estimated as the time between this maximum and the next minimum. The figure 4 shows a complete respiratory cycle.

Respiratory rate in adults varies between 12-20 respirations per minute [1]. The minimum period resulting is 3 seconds but it is no equally distributed between IT an ET, the ratio in normal respiration is 2:1 respective. Therefore, the minimum time between a maximum and minimum is one second (1s). The algorithm eliminates all pair of maximum and minimum with difference less than one second.

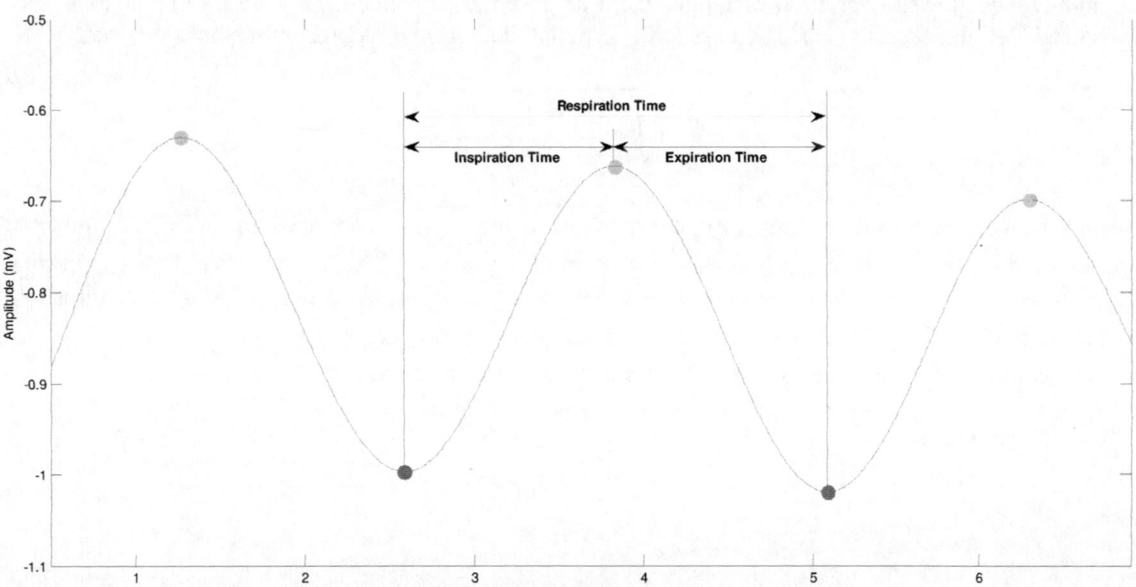

Figure 4. Inspiration and Expiration time.

3.3. Decision Rule

Two rules were used to classify normal respiration and apneas episodes. The first is based on analysing the IT and ET values. In normal respiration, ET shouldn't be greater than 3.4 seconds and IT to 1.67 seconds. The algorithm analyses a portion of record (40 seconds, by example) and considers all normal respiration cycles if the periods are not greater than 4 seconds. This rule arise from the analysis of normal respiration records in which those periods of time times are between 3 and 4 seconds, as shows in figure 5.

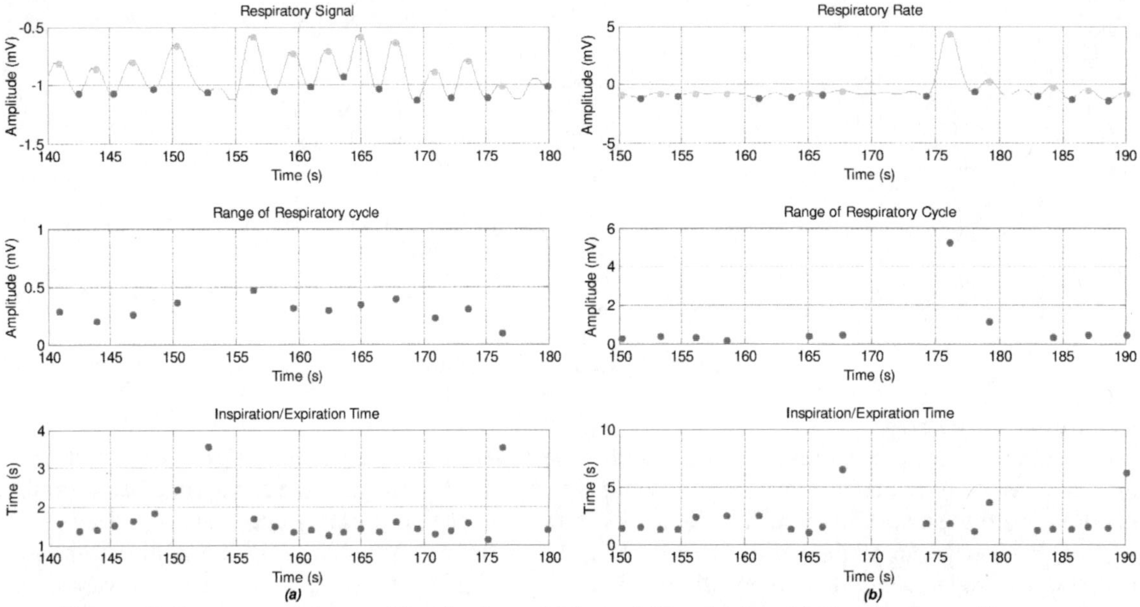

Figure 5. Ranges and times of inspiration. a) Normal Respiration. b) Respiration with apnea.

For second rule the variation of ranges were analysed and Standard Deviation (S) statistical tool was used, defined equation (1). Then based in normal and abnormal records a threshold was set.

$$S = \sqrt{\frac{1}{N-1}\sum_{i=1}^{N}\left|x_i - \overline{x}\right|^2} \tag{1}$$

Where x_i is each range obtained by algorithm, $\overline{x}$ is the average of ranges and N is the number of these. In this work S is computed over 40 second's windows. The figure 5(a) shows a normal respiration with $S_a = 97\mu V$ and figure 5(b) shows an apnea episode with $S_b = 1.95mV$. A threshold value is established from the average of those values, that is $S_{th} = 792\mu V$.

The complete decision rule is showed in figure 6, as a flow diagram.

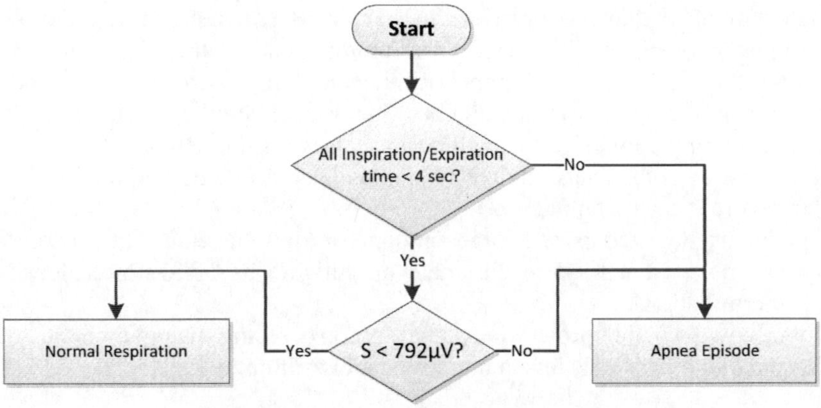

Figure 6. Flowchart of Decision Rule

4. Results
The algorithm was tested with the thresholds set in previous section in 8 records of Apnea-ECG Database and the windows analysis were centred in the episodes marks available in the records. The windows analysis is 40 seconds.

In order to compare the performance of the propose detection method, the following statistical parameters were calculated: Specificity (Sp), Sensitivity (Se), Positive Predictive Value (Vpp) and Negative Predictive Value (Npp) which were computed as follows [14]:

$$Sp\% = 100 \cdot \frac{TN}{(\mathrm{TN+FP})} \tag{2}$$

$$Se\% = 100 \cdot \frac{TP}{(\mathrm{TP+FP})} \tag{3}$$

$$PPV\% = 100 \cdot \frac{TP}{(\mathrm{TP+FP})} \tag{4}$$

$$NPV\% = 100 \cdot \frac{TN}{(\mathrm{TN+FN})} \tag{5}$$

Where True Positives (TP) is the number of apnea episodes detected correctly and True Negatives (TN) is the number of normal respirations detected correctly. Then the False Positives (FP) is the number of normal respiration detected as apnea episodes and False Negatives (FN) the episodes no detected. Table 1 summarises the statistics results of apnea detection method.

Table 1. Performance of the Proposed Algorithm

Sp%	Sen%	VPP%	VPN%
84	73	80	78

5. Discussion and Conclusion
In this preliminary work we have designed an algorithm capable to detect apnea episodes through a simple method and non-invasive as inductive plethysmography. In addition the algorithm doesn't need a calibrated instrument. The detection is performed in the time domain, finding the period of respiratory signal and the standard deviation of the respiratory cycle's instant range over 40 seconds temporal window.

The performance of algorithm has been evaluated with 8 real records of Apnea-ECG database which has been validated by experts. The results are summarized in Table 1 and reach a Specificity of 84% and Sensitivity of 73%, those are coherent with the simplicity of the algorithm and they only use a signal. It is noteworthy that other works use more signals with similar results. In [8] the Sp is 75% and Se is 73% but the computational load is high because uses neural networks.

It is important to mention that pulse oximetry signals has variations that are not exclusive of the apnea episodes, therefore it causes an increment of FP number [9].

One of the problems detected is the large amount of noise artifacts in the records analysed. Although in this paper we have tried to eliminate or mitigate its effects, the algorithm has some sensitivity to these perturbations.

In a future work, we will improve the signal pre-processing using thoracic and abdominal respiratory effort signal in order to achieve a more robust algorithm.

References

[1] Guyton C G and Hall J E 2006 *Tratado de Fisiología Médica*. 11ª Edición. (Elsevier).

[2] Nogueira F, Nigro C, Cambursano H, Borsini E, Silio J and Ávila J 2013 Guías prácticas de diagnóstico y tratamiento del síndrome de apneas e Hipopneas obstructivas del sueño *Medicina (B. Aires)* **73** 349-362.

[3] Polanco P M and Pinsky M R 2008 Cardiovascular issues in respiratory care: clinical applications of heart-lung interactions *In: Yearbook of Respiratory Care Clinics and Applied Technology*. World Federation of Respiratory Care and Applied Murcia (Molina de Segura) 396-401 ISBN 9788461259014.

[4] Iber C, Ancoli-Israel S, Chesson A and Quan S F (Eds.) 2007 The AASM manual for the scoring of sleep and associated events: Rules, terminology and technical specifications (1st ed.) Westchester, IL: American Academy of Sleep Medicine

[5] Castro L, Echenique A and Laciar E 2010 Desarrollo de un plestimógrafo inductivo para la monitorización no invasiva de la actividad respiratoria *Rev. Arg. Bioing.* **16** 27

[6] Zamarrón C, Romero P V, Gude F, Amaro A and Rodriguez J R 2001 Screening of obstructive sleep apnoea: heart rate spectral analysis of nocturnal pulse oximetric recording *Respir Med* **95** 759-765.

[7] Ravelo A, Travieso C, Lorenzo F, Navarro J, Martin S, Alonso J and Ferrer M 2006 Application of support vector machines and gaussian mixture models for the detection of obstructive sleep apnoea based on the RR series *WSEAS Trans. Comput.* **5** 121–124.

[8] Gil E, Vergara J M and Laguna P 2008 Detection of decreases in the amplitude fluctuation of pulse photoplethysmography signal as indication of obstructive sleep apnea syndrome in chilren. *Biomed. Sig. Process. Control*. **3**. 267-277.

[9] Alvarez D, Hornero R, Marcos J V and del Campo F 2010 Multivariate analysis of blood oxygen saturation recordings in obstructive sleep apnea diagnosis. *IEEE Trans. Biomed. Eng.* **57** 2816–2824.

[10] Goldberger A L, Amaral L A N, Glass L, Hausdorff J M, Ivanov P C, Mark R G, Mietus J E, Moody G B, Peng C-K and Stanley H E 2000 PhysioBank, PhysioToolkit, and PhysioNet: Components of a new research resource for complex physiologic signals *Circulation* **101** 215–220.

[11] Penzel T, Moody G B, Mark R G, Goldberger A L and Peter J H 2000 The Apnea-ECG Database. *Comput. Cardiol.* **27** 255–258.

[12] Redmond S J and Heneghan C 2006 Cardiorespiratory-Based Sleep Staging in Subjects With Obstructive Sleep Apnea. *IEEE Trans. Biomed. Eng.* **53** 485-496.

[13] Klintworth A, Ajtay Z, Paljunite A, Szabado S and Hejjel L 2012 Heart rate asymmetry follows the inspiration/expiration ratio in healthy volunteers *Physiol. Meas* **33** 1717.

[14] Altman D G 1993 Some common problems in medical research *Practical statistics for medical research*, (London: Chapman & Hall) chapter 14 pp 396-439.

Application of Data Mining techniques to relate Cardiovascular Risk and Coronary Calcium

F N Lujan, L J Cymberknop, M Alfonso, W Legnani, R Armentano Feijoo

National Technological University – Buenos Aires Regional Faculty,
Medrano 951 (C1179AAQ) Capital Federal, Argentina

E-mail: lujan.facundo@hotmail.com

Abstract. Introduction: Knowledge Discovery in Databases (KDD) constitutes a process that allows data sets to be modeled and analyzed in an automated and exploratory manner. In this sense, data mining can be considered the main core of this procedure. Objective: In this study, a classification of clinical subjects (cluster) based on the comparison of parameters associated to cardiovascular risk factors was performed by means of KDD-based algorithms. Materials and Methods: the K-means algorithm, Hierarchical Agglomerative Clustering and Kohonen's Self-organizing Maps were applied to the database in order to obtain relationships based on the dissimilarity of its constitutive fields. Results: Four different clusters were obtained, represented by a group of well-defined clustering rules. Conclusion: KDD can be used to extract relevant data from clinical databases, which are strongly correlated with well-known cardiovascular risk markers.

1. Introduction

Knowledge Discovery in Databases (KDD) consists in the automated and exploratory analysis and modeling of big data repositories. KDD is an organized process aimed at identifying useful and novel patterns from complex datasets. In this sense, Data Mining (DM) constitutes the core of this process since it includes inference algorithms used to explore data and to understand, analyze and make predictions about the phenomena involved [1]. A subset of these algorithms may be defined as Clustering Algorithms. They have various applications ranging from data compression and vector quantization [2] to pattern discovery and recognition [3], among others. Their implementation yields a differentiated set of clusters (resulting from the application of specific rules). The notion of what constitutes a proper cluster depends on its application and there are various methods to obtain them. Such methods, in turn, must meet different criteria, both ad-hoc and systematic [4].

In biomedical terms, coronary diseases are the main cause of death worldwide. According to the world health statistics 2014 published by the World Health Organization (WHO), coronary heart disease (ischemic) is the leading cause of premature death worldwide [5]. These diseases affect the vascular conduits that irrigate the cardiac muscle (myocardium), where the ischemic event is produced by atherosclerotic obstructions of the arterial walls. This event prevents blood supply to the cardiac muscle, thus causing cell death (in ongoing situations) [6].

In the cardiovascular health, the need to obtain information for making decisions has become critical [7] together with the increasing importance of having accurate markers from a Digital Clinical History System [8]. In this regard, DM algorithms have been successfully applied in the prediction of clinical events in patients with chronic diseases [9], in the evaluation of the effectiveness of specific treatments for certain types of cancer [10], and in increasing the level of accuracy of medical evaluations (making differential diagnoses) [11], among other applications.

The main aim of this work was to obtain a series of clusters from a database (generated for the prediction of atherosclerotic events) with a specific set of clinical parameters and risk factors, by applying DM techniques. This implementation and correlation with sophisticated parameters of cardiovascular evaluation was accompanied by an analysis based on inclusion rules, which was not influenced by subjective evaluations from healthcare professionals.

2. Materials and Methods

2.1. Clustering Algorithms

The term clustering refers to the set of techniques and tools used for fractioning or partitioning data in a database. In order to do this, each cluster is grouped with those elements which are most alike and, at the same time, with those which are most different from the elements of the other clusters. In turn, the term centroid refers to the point equidistant (in Euclidean terms) to all the objects belonging to such cluster [12]. In this work, this type of algorithms has been used to establish a taxonomy of subjects from clinical parameters obtained in a non-invasive way.

2.2. k-means algorithm

Let us consider a dataset $X = \{x_1, x_2, \ldots, x_n\}$, $x_n \in \Re^d$, where the objective is to partition the dataset in M disjunct clusters $C_1, C_2, \ldots, C_M$. The k-means algorithm determines the local minimum solution to the clustering error, defined as the sum of the Euclidean distance between each point of data x_n and the center m_k of the cluster X_n to which it belongs. Analytically, the clustering error is defined as:

$$E(\boldsymbol{m_1}, \ldots, \boldsymbol{m_M}) = \sum_{i=1}^{N} \sum_{k=1}^{M} I(\boldsymbol{x_i} \in C_k) \, \|\boldsymbol{x_i} - \boldsymbol{m_k}\|^2 \quad (1)$$

where $I(Y) = 1$ if Y is true and 0 in another case.

2.3. Hierarchical Agglomerative Clustering

The idea behind Agglomerative Hierarchical Clustering (HAC) is to start with each object in a cluster of its own and then repeatedly merge the closest pair of clusters until we end up with just one cluster containing everything [13]. The basic algorithm is given in Table 1.

Table 1. Hierarchical Agglomerative Clustering Basic Algorithm

1. Assign each object to its own single-object cluster. Calculate the distance between each pair of clusters.
2. Choose the closest pair of clusters and merge them into a single cluster (so reducing the total number of clusters by one).
3. Calculate the distance between the new cluster and each of the old clusters.
4. Repeat steps 2 and 3 until all the objects are in a single cluster.

If there are N objects there will be N −1 mergers of two objects needed at Step 2 to produce a single cluster. However the method does not only produce a single large cluster, it gives a hierarchy of clusters as we shall see. In this case, the linkage criterion used was the decrease in variance for the cluster being merged[14].

2.4. Kohonen's Self-organizing Maps

Kohonen's self-organizing maps (SOM) are important neural network models for dimension reduction and data clustering. "Self-Organizing" is because no supervision is required. SOMs learn through unsupervised competitive learning on their own. "Maps" is because they attempt to map their weights to conform to the given input data. The nodes in a SOM network attempt to become like the inputs presented to them. In this sense, this is how they learn. SOM can learn from complex, multidimensional data and transform them into a topological map of much fewer dimensions typically one or two dimensions.

2.5. Determining the number of clusters

As is known, determining the number of clusters is one of the major problems in the use of clustering algorithms. In this context, this problem has been addressed in two directions. On the one hand, we have used the "elbow method" to determine k, where k is the number of clusters to seek the k-means algorithm.

The elbow method is based on the observation that increasing the number of clusters can help to reduce the sum of within-cluster variance of each cluster. This is because having more clusters allows one to capture finer groups of data objects that are more similar to each other. However, the marginal effect of reducing the sum of within-cluster variances may drop if too many clusters are formed, because splitting a cohesive cluster into two gives only a small reduction. Consequently, a heuristic for selecting the right number of clusters is to use the turning point in the curve of the sum of within-cluster variances with respect to the number of clusters. Technically, given a number, k > 0, we can form k clusters on the data set in question using a clustering algorithm like k-means, and calculate the sum of within-cluster variances, var(k). We can then plot the curve of var with respect to k. The first (or most significant) turning point of the curve suggests the "right" number [15].

As seen in Figure 1, the first (and most important) peak corresponds to four clusters. Thus, it is used k = 4 for implementing the k-means algorithm.

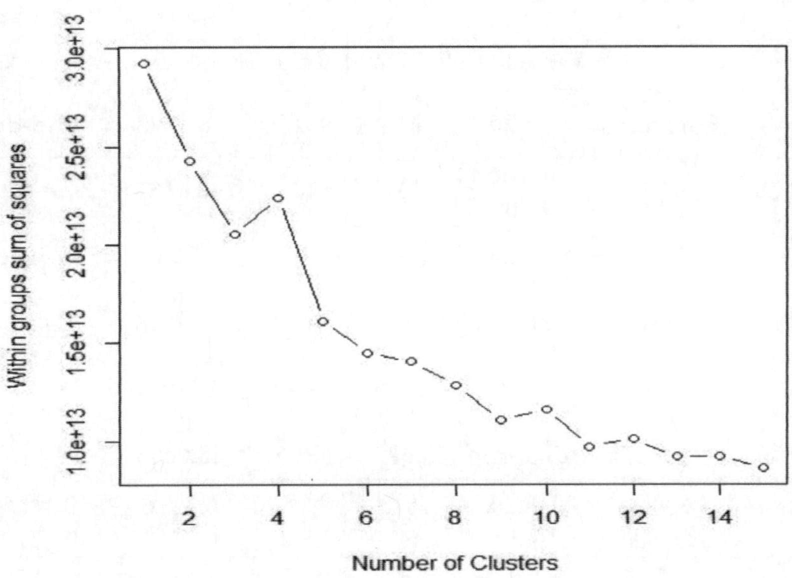

Figure 1. Number of clusters - Within groups sum of squares

On the other hand, the method for determining the number of clusters defined by the HAC algorithm has been implemented as part of the algorithm itself, taking into consideration the biggest jump in the dendrogram.

2.6. C4.5 algorithm

Once every cluster had been defined, the C4.5 algorithm was applied in order to obtain inclusion rules. This algorithm, developed by Ross Quinlan [16], is an extension of the ID3 algorithm previously developed by him. C4.5 is used to generate a *decision tree*, which takes an object or situation described by a set of attributes as input and then provides a "true/false" decision. Therefore, this set of rules defines the *natural* behavior of such input. In this study, the classification field obtained from the application of the *k-means* algorithm was defined as the input and the remaining attributes were defined as "descriptors".

2.7. TANAGRA Software

Within the currently available options, the Tanagra plat-form (Entrepôts, Représentation et Ingénierie des Connaissances, Lyon, France) was chosen because it is a free soft-ware tool used for academic and DM specialized research purposes. This platform allows multiple supervised learning algorithms to be implemented. The main objective of the Tanagra project is to provide students and researchers with an entirely accessible DM algorithmic structure, which makes it easy to overcome the intrinsic programming problems in this domain.

2.8. Database

We have a database belonging to the French PCV-METRA cholesterol follow-up program (Prévention Cardio-Vasculaire en Médicine du Travail), a comprehensive evaluation of cardiovascular risk factors and non-invasive detection of infra-clinical atherosclerosis during a day of hospitalization. The subjects involved in the study were contacted between 1995 and 1996. None of them had a history (or symptoms) of cardiovascular disease. The database used was composed of 618 records and the clinical parameters evaluated by the algorithm are detailed in Table 2. In addition, the existing data include a quantification of cardiovascular risk based on the Framingham model (CVR) and the determination of coronary arterial calcium (CAC) [17]. Under such terms, the mean values for both parameters were obtained for the group of subjects belonging to each cluster.

Table 2. Definition of attributes

Attribute	Abbreviation	Type	Min	Max	Average	Std-dev	Std-dev /avg
Age	AGE	Numeric Ordinal	30	70	48,0162	7,6317	0,1589
Systolic Blood Pressure	SBP	Continue	100	236	137,2540	18,7956	0,1369
Diastolic Blood Pressure	DBP	Continue	50	146	87,8528	12,0556	0,1372
Low density lipoproteins	LDL	Continue	72,375	352,575	188,0516	38,2125	0,2032
High density lipoproteins	HDL	Continue	20,625	127,5	48,0803	12,2573	0,2549
Smoking habits	SMO	Binary	0	1	0,3916	0,4885	1,2475
Diabetes	DIAB	Binary	0	1	0,0518	0,2218	4,2828

All attributes described above have been converted to numeric-continuous power in order to be properly processed by the data mining algorithms. Furthermore, in order to standardize the attributes are applied to the function:

$$f(x) = \frac{x - average(x)}{\sigma} \tag{1}$$

3. Results

After the k-means algorithm, detailed in previous sections was applied (McQueen method [18]), 4 clusters composed of 187, 153, 32 and 246 data records respectively, were obtained. They are characterized by the calculated centroids, whose values are shown in Table 3. The algorithm was executed in 5 Trials and 40 as the maximum number of iterations. The 5 most significant inclusion rules for each cluster may be observed in Table 4. Thus, the mean values of CVR and CAC corresponding to each cluster are described in Table 5.

As it can be observed, the 4 clusters are clearly defined, represented by their inclusion rules, which are met by more than 80% of the cases studied.

Table 3. Clusters represented by Centroids after the application of an algorithm for data mining

Attribute	C1	C2	C3	C4
AGE	45.711	52.588	53.156	46.256
SBP	128.829	158.673	149.063	128.801
DBP	82.551	100.386	94.375	83.240
LDL	191.282	175.449	173.688	195.302
HDL	45.293	53.213	42.949	47.674
SMO	1.000	0.307	0.250	0.000
DIAB	0.000	0.000	1.000	0.000

Table 4. Inclusion Rules

Cluster No.	Rule	Effectiveness
1: Smokers	Non-diabetic Smoker Normal to moderate blood pressure	186 cases
2: Hypertensives	Non-diabetic High blood pressure	109 cases
3: Diabetics	Diabetic	32 cases
4: Normals	Non-diabetic Non-smoker Normal blood pressure	233 cases

Table 5. Mean values corresponding to Cardiovascular Risk (CVR) and Coronary Arterial Calcium (CAC)

Attribute	C1	C2	C3	C4
CVR	0.169	0.179	0.284	0.109
CAC	46.989	104.007	100.625	43.439

On the other hand, the HAC algorithm has been executed as another entry point in the data analysis process. The resulting output was four clusters with centroid that can be seen in Table 6. Furthermore, in order to compare the results of K-means and HAC, a comparative table of the classification obtained by each method has been prepared (see Table 7).

Table 6. HAC Clusters Represented By Centroids After The Application Of An Algorithm For Data Mining

Attribute	C1	C2	C3	C4
AGE	53.156	53.500	45.560	45.513
SBP	149.063	156.642	128.696	129.154
DBP	94.375	99.216	82.435	83.467
LDL	173.688	177.290	191.418	194.650
HDL	42.949	53.900	45.053	47.157
SMO	0.250	0.309	1.000	0.000
DIAB	1.000	0.000	0.000	0.000

Table 7. HAC & K-Means Clusters Cross Table

		K-MEANS				
		C1	*C2*	*C3*	*C4*	***Sum***
HAC	*C1*	0	0	32	0	32
	C2	4	144	0	14	162
	C3	183	1	0	0	184
	C4	0	8	0	232	240
	Sum	187	153	32	246	618

As a result of and by means of the cluster obtained, a hierarchy of attributes was generated, which determines inclusion of each data record in its corresponding cluster (Table 8).

Table 8. Attribute hierarchy

Attribute
Diabetes
Smoking habits
Systolic blood pressure
Diastolic blood pressure

Finally, SOM algorithm was ran with a 2x2 MAP structure, an starting learn rate of 0.2

Table 9. SOM Clusters Represented By Centroids After The Application Of An Algorithm For Data Mining

Attribute	C1	C2	C3	C4
AGE	45,663	46,366	52,471	53,156
SBP	129,144	128,829	158,242	149,063
DBP	82,711	83,142	100,346	94,375
LDL	192,992	195,246	173,450	173,688
HDL	45,203	47,690	53,299	42,949
SMO	1,000	0,000	0,307	0,250
DIAB	0,000	0,000	0,000	1,000

Table 10. SOM & K-Means Clusters Cross Table

		K-MEANS				
		C1	*C2*	*C3*	*C4*	*Sum*
SOM	*C1*	184	3	0	0	187
	C2	0	2	0	244	246
	C3	3	148	0	2	153
	C4	0	0	32	0	32
	Sum	187	153	32	246	618

4. Discussion

CVR estimation is mainly used for raising public aware-ness of the occurrence of diseases with high morbidity and mortality rates as is the case of cardiovascular diseases, over a 5 to 10-year timespan. In the present clinical practice, CVR is estimated by means of stratified tables where cardiovascular risk factors are evaluated through a scoring system for men and women separately. Equations based on the Framingham model, the SCORE algorithm and the PROCAM model, among others, constitute some of the standardized predictors for the evaluation of CVR [19]. Coronary and cerebrovascular events, as a result of atherosclerosis development, occur in a sudden and asymptomatic manner. The reduction of the occurrence of coronary and cerebrovascular events depends on early modifications of factors such as tobacco, diet and sedentary lifestyle, together with periodic controls of cholesterol, glycemia and blood pressure [20].

In addition, in this work, DM techniques were applied with the aim to cluster subjects from a clinic database (patients who attended a cardiovascular check-up) without the direct intervention of healthcare professionals. Thus, we have implemented three different clustering algorithms in order to assess the existence of a classification / hierarchy in the data. Then these results were compared to identify any relationship between them.

Thereby, observing the tables centroids K-means algorithm, HAC and SOM (Table 3, Table 6 and Table 9, respectively), and then the comparison chart between these algorithms (Table 7 and Table 10), one can observe a strong relationship between clusters.

In clinical terms, it may be inferred that there are 4 clearly differentiated clusters. Cluster 4 corresponds to healthy subjects (non-smokers, non-diabetics, with normal blood pressure). Cluster 1 is composed of smokers mainly. Cluster 3 is identified with diabetic subjects and, finally, Cluster 2 is made up of hypertensive subjects.

These results are in line with the clinical diagnoses that any general practitioner would provide [21]. However, this study shows that cluster 4, characterized by low risk subjects (see Table 5, CVR 10.9%) are also subjects with low artery calcium levels, which is a specific marker for the presence of atherosclerosis. The hypertensive group (cluster 2) mostly composed of non-smokers (70%) has a higher CVR (17.9%) as a result of their high blood pressure but with a CAC 2.5 times higher than that of normal subjects. Cluster 1, composed of 100% smokers has a CVR similar to that of hypertensives. Nonetheless, they show a slight increase in CAC, which would imply a low predisposition for coronary disease. Finally, cluster 3, composed of 100% diabetics, has the highest CVR (28.4%) as well as a high marker for atherosclerosis (CAC 100).

It was observed that, when using 3 clusters or more, there is always a group exclusively composed of diabetics. In this sense, the analysis of the results shows that diabetic subjects have physiopathological characteristics significantly different from those of a non-diabetic subject. Finally, the methodology selected has the advantage that it is not biased by the opinion of the healthcare professional, and, as a result, it is free from subjectivities that could be present in the analysis and examination performed by the physician [22]. As a disadvantage, it must be acknowledged that this analysis requires that the number of groups to be identified by the algorithm be defined beforehand; and therefore, the ideal number of clusters is obtained through successive runs.

5. Conclusion

An original way to obtain relevant information from clinical databases is showed as feasible in this work. Such information is strongly correlated to frequently used cardio-vascular risk markers, as a comparison technique. The basic clustering analysis done here is enough to aim futures lines of research. However, further studies are required in order to optimize the determination of the number of groups as well as a better correlation between the attributes (cardiovascular risk factors) and the evaluation of such risk. Among other improvements, exists clustering schemes that could contribute in a best way to classify the clinical cases based upon sophisticated clustering algorithms.

References

[1] Bramer M 2013 *Principles of data mining* (Berlin: Springer)
[2] Gersho A and Gray R M 1992 *Vector quantization and signal compression* (Berlin: Springer)
[3] Duda R O and Hart P E 1973 *Pattern classification and scene analysis* vol 3 (New York: Wiley)
[4] Mikut R 2008 *Data Mining in der Medizin und Medizintechnik* vol 22 (Karlsruhe: KIT Scientific Publishing)
[5] World Health Organization 2014 *Estadísticas Sanitarias mundiales 2014 – una mina de información sobre salud pública mundial* Available: http://apps.who.int, last visit: 07/06/15
[6] Cabrera Fischer E 2013 *Bases de la fisiología para ingeniería* (Buenos Aires: CEIT).
[7] Bellazzi R and Zupan B 2008 *Predictive data mining in clinical medicine: current issues and guidelines (*International journal of medical informatics 77(2) 81-97)
[8] Luna D Soriano E González Bernaldo de Quirós F 2007 *Historia Clínica Electrónica* vol 27 N° 2 (Buenos Aires: Revista del Hospital Italiano de Buenos Aires)
[9] González Bernaldo de Quirós F Luna D Baum A Plazzotta F Otero C Benítez S 2012 *Incorporación de tecnologías de la información y de las comunicaciones en el Hospital Italiano de Buenos Aires* vol 27 N° 2 (Buenos Aires: Revista del Hospital Italiano de Buenos Aires)
[10] Reparaz D Merlino H Rancan C Rodríguez D Britos P V and García Martínez R 2008 *Determinación de la eficacia de la braquiterapia en tratamiento de cáncer basada en minería de datos* (Buenos Aires: X Workshop de Investigadores en Ciencias de la Computación)
[11] Kuo W J Chang R F Chen D R and Lee C C 2001 *Data mining with decision trees for diagnosis of breast tumor in medical ultrasonic images*. (Neatherlands: Kluwer Academic Publishers - Breast cancer research and treatment, 66(1), 51-57)
[12] Forgy E 1965 Cluster analysis of multivariate data: Efficiency vs. interpretability of classifications (United Kingdom: Biometrics 21, 768)
[13] Kaufman L Rousseeuw P J. 1990 *Finding Groups in Data: An Introduction to Cluster Analysis* (1 ed.) (New York: John Wiley. ISBN 0-471-87876-6)
[14] Ward J H Jr 1963 *Hierarchical Grouping to Optimize an Objective Function* (Massachusetts: Journal of the American Statistical Association 58 236–244)
[15] Bholowalia P and Kumar A 2014 *EBK-Means: A Clustering Technique based on Elbow Method and K-Means in WSN* (Pennsylvania: International Journal of Computer Applications, 105(9))
[16] Quinlan J R 1993 *C4. 5: programs for machine learning* vol 1 (Massachusetts: Morgan kaufmann)
[17] Chironi G Simon A Megnien J-L Sirieix M-E Mousseaux E Pessana F Armentano R L 2011 *Impact of coronary artery calcium on cardiovascular risk categorization and lipid-lowering drug eligibility in asymptomatic hypercholesterolemic men* (Netherlands: International Journal of Cardiology 151: 2. 200-204)
[18] MacQueen, J. 1967 *Some methods for classification and analysis of multivariate observations*. (California: Proceedings of the fifth Berkeley symposium on mathematical statistics and

probability (Vol. 1, No. 14, pp. 281-297))
[19] Lloyd-Jones D M 2010 *Cardiovascular Risk Prediction Basic Concepts, Current Status, and Future Directions* (Philadelphia: Circulation 121, 1768–1777)
[20] World Health Organization 2007 *Prevention of Cardiovascular Disease. Guidelines for assessment and management of cardiovascular risk* Available: http://apps.who.int, last visit: 07/06/15
[21] Srinivas K Rani B K and Govrdhan A 2010 *Applications of data mining techniques in healthcare and prediction of heart attacks* (Indian: International Journal on Computer Science and Engineering (IJCSE), 2(02), 250-255)
[22] Soni J Ansari U Sharma D and Soni S 2011 *Predictive data mining for medical diagnosis: An overview of heart disease prediction* (Indian: International Journal of Computer Applications, 17(8), 43-48)

High efficiency processing for reduced amplitude zones detection in the HRECG signal

N. Dugarte[1,2], A. Álvarez[2], J. Balacco[2], G. Mercado[2], A. Gonzalez[2], E. Dugarte[1,2], A. Olivares[1]

E-mail: ndj0227@gmail.com, irb@frm.utn.edu.ar

Summary – This article presents part of a more detailed research proposed in the medium to long term, with the intention of establishing a new philosophy of electrocardiogram surface analysis. This research aims to find indicators of cardiovascular disease in its early stage that may go unnoticed with conventional electrocardiography. This paper reports the development of a software processing which collect some existing techniques and incorporates novel methods for detection of reduced amplitude zones (RAZ) in high resolution electrocardiographic signal (HRECG).The algorithm consists of three stages, an efficient processing for QRS detection, averaging filter using correlation techniques and a step for RAZ detecting. Preliminary results show the efficiency of system and point to incorporation of techniques new using signal analysis with involving 12 leads.

1. Introduction

At the present, the cardiovascular diseases are a leading cause of death [1]. One of the most used for the study of cardiac pathologies methods lies in the electrocardiographic assessment [2]. The cardiac signals measurement is a noninvasive technique that is realized capturing the signal with electrodes placed on the skin surface and the information obtained is plotted in a record called electrocardiogram or ECG [3]. Modern studies show the relationship between cardiovascular disease and time intervals in ECG signal elements [4].Relations between the ST and ischemia, QT changes in drugs response, QRS alterations depending on the cardiopathy and other indices, suggest treatment or patient intervention based on the specialist medical evaluation [5] [6].

But recent studies have also shown that ECG conventional has limitations. An example is the case of patients with left ventricular wall abnormalities motion and reduced ejection fraction. This situation increases cardiac risk and frequently only is diagnosed when the advanced cardiovascular damage is done. Typical case linked to Chagas disease [7].

The conventional ECG is acquired to display signals with a bandwidth from 0.05 to 120 Hz. For which the digitized signal from 300 to 500 samples per second with 8 bits per sample resolution. This resolution allows an ECG graph with sufficient definition for human viewing, but it is shown that many significant details cannot be captured. A typical signal that goes unnoticed is the late potential at the QRS complex end content [8], see Fig. 1.

[1] Grupo de Ingeniería Biomédica (GIBULA) de la Universidad de Los Andes (ULA), Mérida, Venezuela.
[2] Instituto Regional de Bioingeniería, Universidad Tecnológica Nacional – Facultad Regional Mendoza.

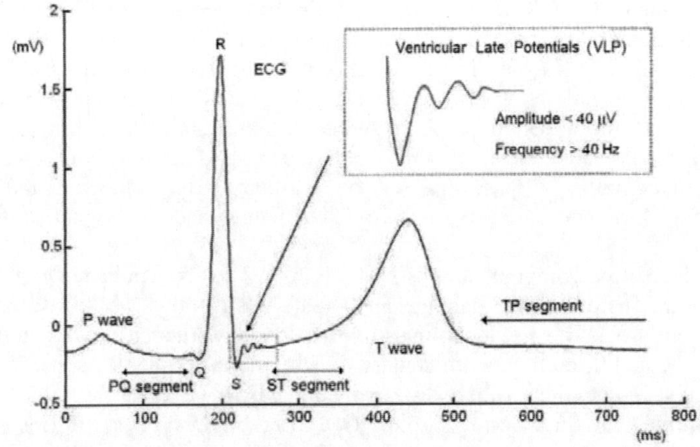

Fig. 1, Ventricular late potentials detection [8].

Some time, with the aim to increase the details of trace, high resolution electrocardiography (HRECG) is used [9]. The HRECG is used to perform the signal acquisition with higher definition and a greater bandwidth when required in conventional electrocardiography. One application of ECGAR allows the detection of myocardial zones affected by some pathology kind. This technique is also known as reduced amplitude zones (RAZ) detection [10] [11]. It involves in high frequency components of QRS complex contained (HF-QRS) analysis, looking for abnormal frequency components decrease in the range of 150 to 250 Hz. The Fig. 2 shows an example of HF-QRS signal averaged and filtered of lead I, at healthy subject.

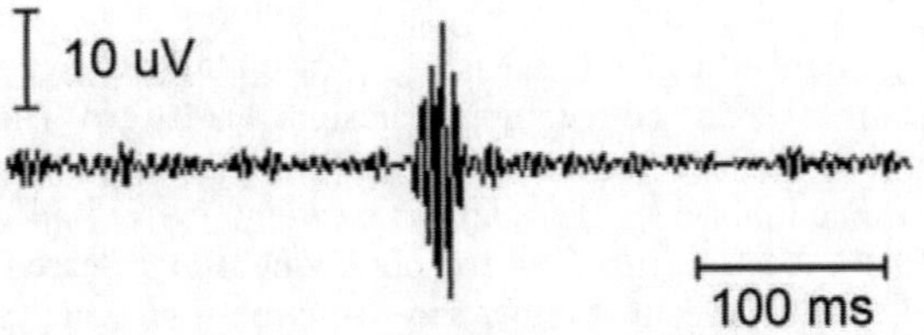

Fig. 2, HFQRS signal from a healthy patient.

A RAZ occurs when are present two local maxima of the upper envelope or two local minima of the lower envelope at least, in the signal analyzed (originally defined by Abboud [10]), see Fig. 3.

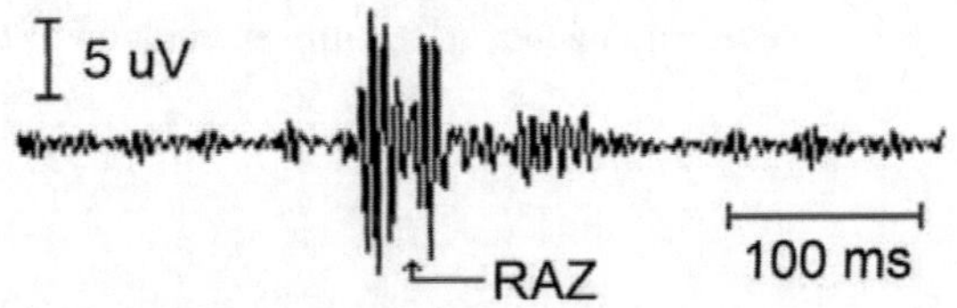

Fig. 3, HFQRS with a RAZ zone of a heart disease patient.

The RAZ testing is a relatively new, non-invasive, economic and feasible technique in real time. However it has not been massively implemented given the complexity and time involved on processing with this technology. This technique is more accurate than ECG conventional to detect heart disease such as affections of coronary arteries and myocardial ischemia.

Another important feature arises from the 12 leads HRECG simultaneous analysis [12]. Previous data Maehara et al. [13] and Delgado et al. [14] suggest that 12 leads HRECG can be highly sensitive and specific for detecting a cardiomyopathy presence and early manifestations of cardiac damage. Schlegel [15] and other researchers have shown that 12 leads electrocardiography increases notoriously the sensitivity in the detection of diseases caused by coronary artery damage and left ventricular hypertrophy. Schlegel has increased positive predictive value in detecting left ventricular systolic dysfunction.

The high resolution electrocardiogram analysis involving 12 leads would be particularly useful to detect early manifestations of cardiac damage in patients whom otherwise could not be diagnosed. There are few studies associating HD techniques of analysis in a practical system and that in turn can be massively implemented in public health centers. Furthermore, frequently some kind of algorithms and procedures have such complex that they hinder their use by themselves.

Since 2009, the Biomedical Engineering Group (GIBULA) in University of Los Andes (ULA) at Bolivarian Republic of Venezuela together with the Regional Institute for Bioengineering (IRB) of National Technological University (UTN) at Republic Argentina, have generated a series of products with technological innovation in 12 lead HRECG acquisition hardware and software signal analysis.

These research and scientific development aims to achieve a more efficient and easy implementation of systems with the objective of wide spreading their use on massive populations around the world.

This article reports one of the algorithms developed in the framework of a larger project. This algorithm implements techniques for RAZ detection, in efficient software and with friendly features for the operator of the instrument.

2. Methodology

The associated work between GIBULA and IRB has developed an electrocardiograph that allows the 12 leads HRECG simultaneous acquisition signal. The project developed under the name DIGICARDIAC [16] [17], is an instrument consisting of a multichannel acquisition hardware with high noise rejection in common-mode and a computer designed software. The system works under an environment designed like an electronic health record (EHR) applied to cardiology [18].

The DIGICARDIAC instrument, besides presenting excellent performance as a tool for clinical diagnosis, allows manipulating the acquisitions for deployment at research field. Additionally, the EHR software is designed with features that allow activating the signal analysis module. This module can incorporate any type of signal analysis application. This is done in order to provide the specialist a more detailed assessment of the clinical examination.

The algorithm aims to bring together some of the criteria set by researchers, in an efficient software that analyzes the signal HRECG looking for RAZ. The Fig. 4 diagram, show the steps of system operation, which they are: heartbeat detection, reducing of baseline bleed by averaging filtering using correlation techniques and the step for RAZ detecting. The developed processing is executed by analyzing a branch at a time, but the research points to the implementation of the 12 leads.

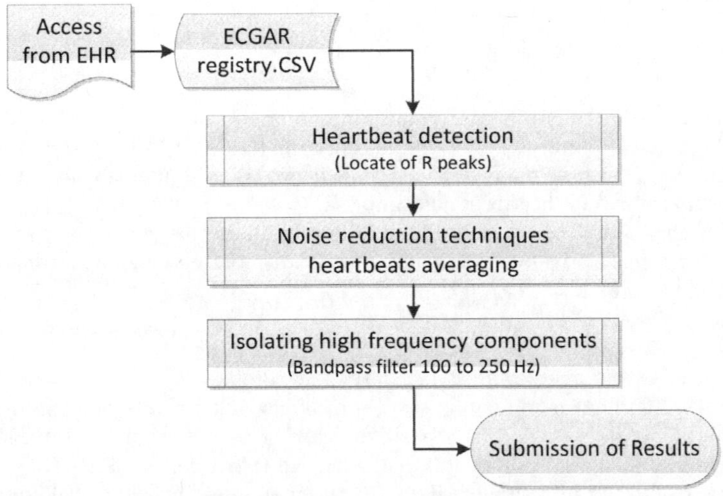

Fig. 4, Work sequence at developed algorithm.

2.1. Heartbeat detection

Heartbeat detection is designed for detecting the QRS occurrence through electrocardiogram. The processing of Pan and Tompkins [19], involves converting the QRS complex in well-defined peaks while are discarded the other ECG components. Then the system applies adaptive thresholds of time and amplitude to discard false QRSs. Dolinsky and Stoyanov [20] used two criteria in QRS detection, one based on the evaluation of peaks slope and the second by the biphasic waveforms occurrence to identify ectopic beats.

Other techniques have been highlighted as Ivaylo Christov I. [21], Mohamed Ben Messaoud [22], Chouhan and Mehta [23], Dugarte et al. [24], Lena Wahab [25], and other researchers who have achieved a good approximation at QRS location.

For obtained the QRS is developed an algorithm by combining some used criteria by Pan and Tompkins with some used criteria by Dotsinsky and Stoyanov. The obtained software is relatively simple to implement, efficient, and involves low computational requirement. The procedure involves R peaks detecting on a signal with altered morphology, where the QRS amplitude is highlighted at positive single pulse while the rest of the components of each pulse is attenuated.

The process starts with a recursive high-pass filter [26], with cutoff frequency of 0.65 Hz. It is defined by equation (1), where X forms the input signal, Y output data, C1 and C2 are filter coefficients.

$$Y(i) = C1[X(i) - X(i-1)] + C2[Y(i-1)] \quad (1)$$

The (2) and (3) equations allow calculation the C1 and C2 coefficients, were Fc is cutoff frequency and T is sampling period.

$$C1 = \frac{1}{1+tg(Fc \cdot \pi \cdot T)} \quad (2)$$

$$C2 = \frac{1-tg(Fc \cdot \pi \cdot T)}{1+tg(Fc \cdot \pi \cdot T)} \quad (3)$$

The next step is to process signal with a low-pass filter at cutoff frequency of 35 Hz. The filter implemented is defined by equation (4), where X forms the data input signal, Y data output to 1 /'N represents the filter coefficients. Fc = 35 Hz is achieved with N = 26.

$$Y(i) = \frac{1}{N}[X(i) + X(i-1) + \cdots X(i-(N-1))] \tag{4}$$

The application in cascade the filters described in equations (1) and (4) is performed to work as a band-pass filter with cutoff frequency between 0.65 and 35 Hz. With this filtering a signal with low noise and good correction of baseline is obtained, but cannot be used in diagnosis because it has been altered, however it is optimal for heartbeat detecting.

Next, the filtered signal is derived to magnify the QRS amplitude changes. By applying equation (5), an approximation of signal derivative is obtained, where X represents the signal input and Y is the derived signal.

$$Y(i) = 2\,X(i) + X(i-1) - X(i-3) - 2\,X(i-4) \tag{5}$$

The next step in the algorithm is to apply a rectifier filter, which consists of squaring the derivative signal. This is done in order to obtain only positive peaks. Then, the signal is subjected to low-pass filter, identical to described by equation (4) but with cutoff frequency of 15 Hz. This allows the integration of peaks obtained in a single pulse of high amplitude, but without altering the length of QRS complex occupied space by. The delay caused by the filter is removed by introducing the appropriate time correction factor.

The Fig. 5 shows, three seconds of signal original at the top, in middle at result of deriving and the final at result obtained with equation 5. In the image, it stands the peaks generated as response of QRS complex, while the remaining signal components almost have disappeared. Note that no matter the original signal morphology, even if the QRS complexes are negative or deformed, the resultant signal always generates a positive pulse for each QRS interval found.

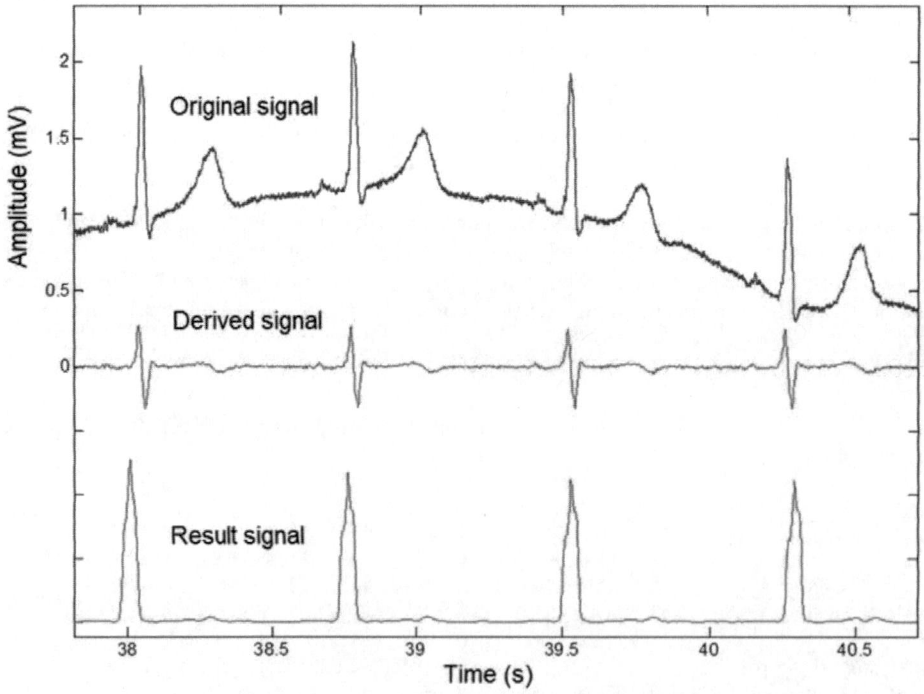

Fig. 5, Results obtained in the firsts stages the signal processing.

The next stage is to find and evaluate the pulses generated with the above processing to verify if it is a QRS, a generated artifact by noise or an ectopic beat. It is to locate the maximum amplitude in each of the signal pulses and apply some criteria defined by adaptive thresholds. The first criterion is established by amplitude comparisons. This adaptive threshold is obtained by averaging the amplitude of the last 6 peaks R accepted as true. Amplitude discrimination is done considering that an R peak is true if its amplitude is contained within a range of ± 20% of the adaptive threshold. The second evaluation criterion is established function of time. It is based on the RR interval time cannot change by more than 30% between one heartbeat and the next. The adaptive threshold time is obtained at average the last 6 RR accepted as true.

In the next step, the markers are relocated in each QRS of original signal. The Fig. 6 shows a fragment of 3 seconds of original signal, with marker lines indicating the R peaks location in each QRS.

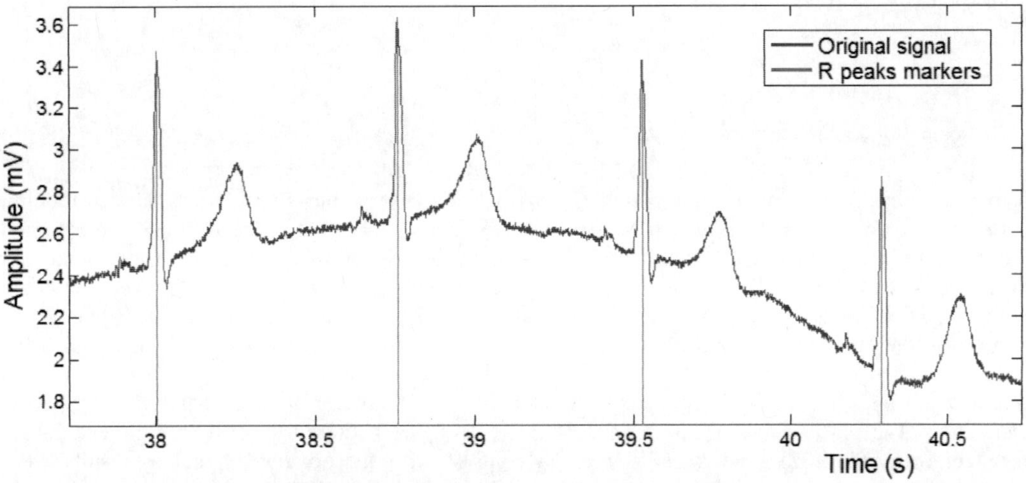

Fig. 6, Original signal with R peaks markers.

2.2. Noise reduction by averaging heartbeat

For analysis and measurement of potential intra-QRS abnormal [10] [11] [12] [13] is requires that the ECG signal is properly filter. The noise reduction technique by averaging [11] is to perform an averaging of the heartbeat at deemed with deformation minimum. It is assumed that the signal noise is uncorrelated between a beat and the following; therefore, considered random noise, it is progressively removed by complex averaging more.

The first processing step is to perform the automatic selection of a fragment that contains a complete record heartbeat, to be used as comparison template. The template interval starts 0.2*RR seconds before the R peak and reaches 0.4*RR seconds after the R peak, the selected beat.

The automatic template selection process initially takes the registry beat third. If the percentage of accepted beats less than 60% in register at end of iterations, this template is discarded, the next beat is taken as a possible template and the process is repeated. If the system cannot find an optimal heartbeat after 5 discarded templates, it is considered that the registry has too many alterations to apply the method.

The precise location of intervals for averaging is performed by comparing selected template with each heartbeat found. This comparison is performed by coefficient correlation Pearson calculating [8] [11] [27]. Pearson's coefficient can be calculated using the relationship given by equation (6).

$$r_{xy}(i) = \frac{cov(X,Y)}{\sqrt{\sigma_x^2 * \sigma_y^2}} \; donde: \; -1 \leq r_{xy}(i) \leq +1 \qquad (6)$$

Equation (6) specifies $r_{xy}(i)$ as the Pearson correlation coefficient. This coefficient is the comparison between X and Y signals. Where X is template vector and Y represents the beat vector bounded by template alignment.

Pearson coefficient can generate values between +1 and -1, where 1 is overall positive correlation, 0 indicates no correlation and -1 is totally negative correlation [27]. The i range is defined by the times number that template is compared with the beat. In this procedure is $-50 \geq i \geq 50$, where i = 0 indicates that the template and the beat are matching at the peak R.

The covariance ratio cov(X,Y) can be calculated using equation (7). The vectors variance σ^2_z can be found by equation (8), where Z may values take the X o Y vectors, N identifies the samples number of Z, and n are the vectors samples.

$$cov(X,Y) = \frac{1}{N}\sum(X(n) - \bar{X})(Y(n) - \bar{Y}) \qquad (7)$$

$$\sigma_z^2 = \frac{1}{N}\sum(Z(n) - \bar{Z})^2 \qquad (8)$$

To preserve the medical interest components in SAECG signal, the beats with a less 0.98 correlation value are discards [11]. If the Pearson coefficient value is above 0.98 is proceed to add the obtained interval with the accumulation vector and the variable that is used as accepted beats counter is increases. When finished a beat cycle analysis, is proceed to correlation calculate of next beat with the template. This process is repeated until all beats contained in the register are analyzed. Upon completion the processing is proceed to divide the accumulation vector between the accepted beats number.

The obtained signal retains all frequency components of ECGAR recording but with high noise attenuation. The Fig. 7 shows 0.55 seconds of ECGAR signal acquired de of a sick patient. In the image of original record a heartbeat is seen in contrast with the obtained signal as result. Were analyzed 171 beats, 28 beats automatically discarded and the resulting signal it is obtains averaging 143 beats.

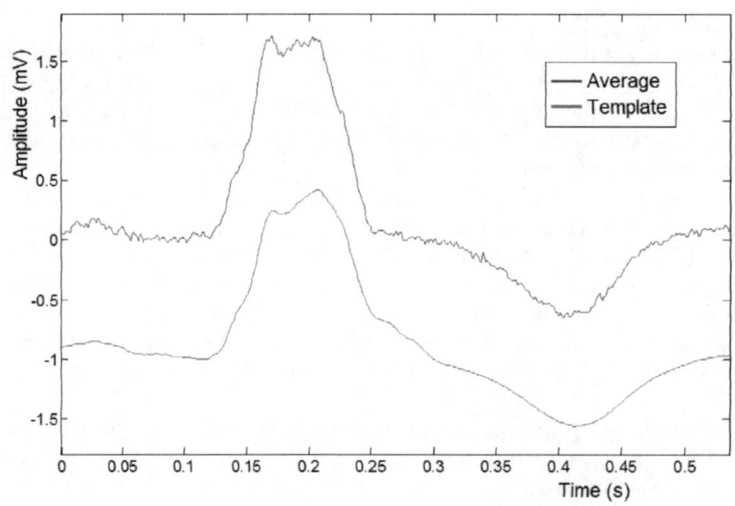

Fig. 7, Original signal and obtained signal from 143 average beats.

2.3. RAZ detecting step

It consists of applying a band-pass filter at obtained signal in 2.2 section. An FIR type filter as described by Equation (9) is used. The filter coefficients are calculated for cutoff frequencies between 150 and 250 Hz.

$$Y(n) = \sum_{k=0}^{N-1} b_k * X(n-k) \quad (9)$$

The next stage of processing, detects the maximums and minimum in sign contour and draw lines envelope highlighting as correspond to the QRS interval. The Fig. 8 shows a fragment of obtained result after band-pass filter applying to averaged signal shown in Fig. 7. The plot corresponds to the QRS interval. The RAZ marks are manually positioned by the specialist in order to turning points highlight identified.

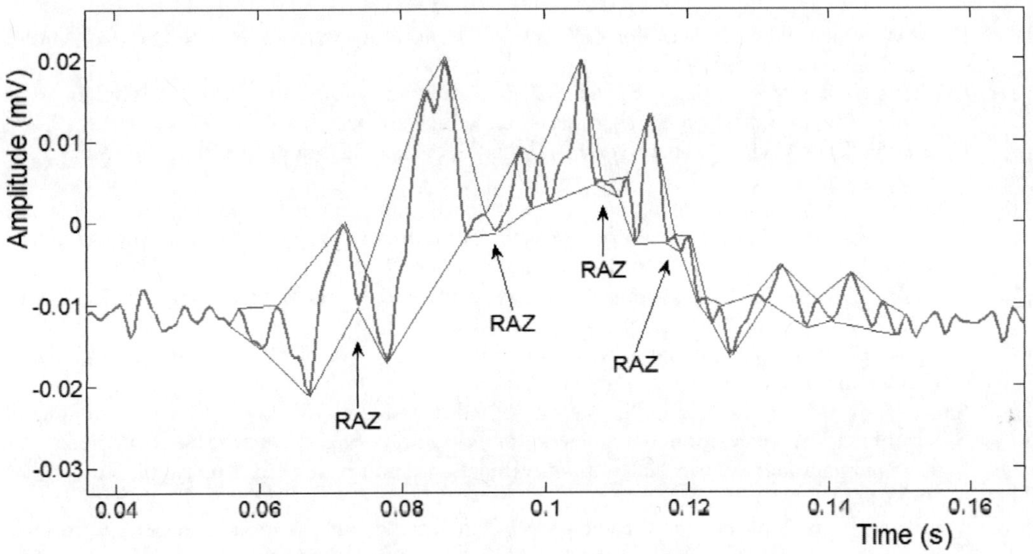

Fig. 8, Zoom of HFQRS with RAZ zones marks.

3. Results

The system evaluation was performed by acquired signals analyzing with DIGICARDIAC instrument in two test groups. In the first group, the software was tested with acquisitions made in the laboratory under controlled conditions, a group of 12 volunteers offered as control patients. In these tests, not found heartbeat detection errors. Of cardiologist evaluation no indicative results of cardiac pathologies. The automatic analysis does not present RAZ indicators.

The second tests group involves analyzing at 30 patients acquisitions in a medical center. The errors obtained in QRS detection are below 0.1%, which represents a performance comparable with most processing algorithms reviewed in the literature [19] [20] [21] [22] [23], but with the premise to get the results in a relatively shorter time. Evaluation of high frequency components, by the system, presented evidence that 12 of the 30 patients had RAZ markers. The cardiologists evaluating showed that 10 of the 12 patients had proven pathologies and the remaining two were reported on inconclusive studies, but with signs of sick possible.

4. Conclusions

The importance of the elimination of noise in the signal is very useful in the application of advanced signal analysis. Evaluations of high frequency components in QRS complex are indicators of diseases that can be detected before the damage is too severe and therefore constitute an application that can help save lives.
The used technique in processing is relatively simple, however it is possible to use more complex algorithms to improve the estimate. Preliminary results consulted in literature point out that the 12 leads technique is highly sensitive and predictive.
Given the technical capability of simultaneous acquisition of 12-lead DIGICARDIAC instrument, only is required to develop appropriate algorithms for detecting indicators of cardiovascular disease in its early stage.

5. References

[1] Pan American Health Organization (PAHO) 2007 Mortality rate, available at: www.paho.org, Last login: September 2010.
[2] Cardiovascular Information Center 2007 Anatomy of the heart, Available at: http://www.texasheartinstitute.org/, Last login: June 2009.
[3] Ganong W 1999 Manual Moderno de Fisiología Médica, 17ª Edition, McGraw-Hill, México, D. F.
[4] Schlegel T 2010 Accuracy of advanced versus strictly conventional 12-lead ECG for detection and screening of coronary artery disease, left ventricular hypertrophy and left ventricular systolic dysfunction, BMC Cardiovascular Disorders, Vol. 10 (20), pp. 1471-2261.
[5] Clifford G, Azuaje F and McSharry P 2006 Advanced Methods and Tools for ECG Data Analysis, Artech house inc. Londres.
[6] John S 2004 Biosignal and Biomedical Image Processing, Matlab Based Applications, Marcel Dekker Inc. USA.
[7] Salles G, Cardoso C, Xavier S, Sousa A and Hasslocher M 2003 Electrocardiographic ventricular repolarization parameters in chronic Chagas' disease as predictors of asymptomatic left ventricular systolic dysfunction, Pacing Clin Electrophysiol, Vol. 26, pp. 1326-1335.
[8] Vargas M 1996 Contribución al estudio de los micropotenciales cardíacos, Doctoral thesis of Electronic Engineering Department, Polytechnic University of Catalonia, Spain.
[9] Gutiérrez O 2001 El electrocardiograma de alta resolución, Costa Rican cardiology journal, Vol. 3, no. 2, pp. 48-53.
[10] Abboud S, Belhassen B, Miller H, Sadeh D and Laniado S 1986 High frequency electrocardiography using an advanced method of signal averaging for non-invasive detection of coronary artery disease in patients with normal conventional electrocardiogram, Journal Electrocardiology, Vol. 19, pp. 371-80.
[11] Schlegel T, Arenare B, Starc V, Greco E, Poulin G, Moser Dand Delgado R 2005 The best identifiers of cardiomyopathy in short duration ECG recordings: high frequency qrs reduced amplitude zone score, QT interval variability, low frequency RR interval power and heart rate turbulence slope, journal cardiology, Vol. 12, Suppl. D: 1-4.
[12] Schlegel T, Kulecz W, DePalma J, Feiveson A, Wilson J, Rahman M and Bungo M 2004 Real-Time 12-Lead High-Frequency QRS Electrocardiography for Enhanced Detection of Myocardial Ischemia and Coronary Artery Disease, Mayo Clin. Proc. Vol. 79, pp. 339 – 350.
[13] Maehara K, Kokubun T, Awano N, Taira K, Ono M, Furukawa T, Shimizu Y, and Maruyama Y 1999 Detection of abnormal high-frequency components in the QRS complex by the wavelet transform in patients with idiopathic dilated cardiomyopathy, Japanese Circulation Journal, Vol. 63, pp. 25-32, 1999.

[14] Delgado R, Poulin G, Vrtovec B, Eastwood C, Radovancevic B, Franklin W, Kar B and Schlegel T 2004 The utility of high frequency QRS electrocardiogram in the diagnosis of cardiomyopathy, Journal Am. College of Cardiology, Vol. 43 (5A), pp. 208-209.
[15] Schlegel T 2010 Accuracy of advanced versus strictly conventional 12-lead ECG for detection and screening of coronary artery disease, left ventricular hypertrophy and left ventricular systolic dysfunction, BMC Cardiovascular Disorders, Vol. 10 (20), pp. 1471-2261.
[16] Dugarte N, Medina R and Rojas R 2011 Adquisición y Procesamiento de la Señal Electrocardiográfica de Alta Resolución para la Identificación de Enfermedades Cardiovasculares como el Mal de Chagas, Electronic Journal of Free Knowledge and Licensing "EL CLIC", Vol. 1 (4), pp. 55-66.
[17] Dugarte N, Medina R, Rojas R and Álvarez A 2010 Certificación del Sistema ECGAR para su Aplicación en Centros de Salud, IV Ibero-American Congress of Electrical Engineering Students (CIBELEC), Art IB-01, Mérida, Venezuela.
[18] Dugarte N, Medina R and Rojas R 2015 Open Source Cardiology Electronic Health Record Development for DIGICARDIAC Implementation, 11th International Symposium on Medical Information Processing and Analysis (SIPAIM), Cuenca - Ecuador.
[19] Pan J and Topkins W 1985 A Real Time QRS Detection Algorithm, IEEE Transactions on Biomedical Engineering, Vol. 32 (3), pp. 230-236.
[20] Dotsinsky I and Stoyanov T 2004 Ventricular beat detection in single channel electrocardiograms, Biomedical Engineering on Line, doi: 10.1186/1475-925X-3-3.
[21] Christov I 2004 Real time electrocardiogram QRS detection using combined adaptive threshold, Biomedical Engineering on Line, doi: 10.1186/1475-925X-3-28.
[22] Messaoud M 2007 On the Algorithm for QRS Complexes Localisation in Electrocardiogram, Journal of Computer Science and Network Security, Vol. 7 (5), pp. 28-34.
[23] Chouhan V and Mehta S 2008 Detection of QRS Complexes in 12-lead ECG using Adaptive Quantized Threshold, Journal of Computer Science and Network Security, Vol. 8 (1), pp. 155-163.
[24] Dugarte N, Cuadros J, Rubén M, Rubén R, Jugo D and Nuñes T 2012 Sistema de adquisición multicanal y análisis de la señal electrocardiográfica de alta resolución aplicado a pacientes chagasicos, XI International Congress of Numerical Methods in Engineering and Applied Sciences (CIMENICS), pp. 405-408, Venezuela.
[25] Wahab L 2006 Detection of QRS complexes in ECG signals based on Poisson Transform and Root Moments, 5th International IEEE EMBS Special Topic Conference on Information Technology in Biomedicine(ITAB), pp. 1350-1356, Ioannina - Epirus, Greece.
[26] Cuadros J 2012 Procesamiento avanzado de la señal electrocardiográfica de pacientes chagásicos, Master Thesis of Group of Biomedical Engineering, Faculty of Electrical Engineering, University of Los Andes, Mérida, Venezuela.
[27] Nieves A, Domínguez F 2010 Probabilidad y Estadística para Ingeniería, McGraw Hill, Spain.

Noise Pollution Control System in the Hospital Environment

LM Figueroa Gallo[1], JM Olivera[2]
Gabinete de Tecnologías Médicas, Departamento de Bioingeniería, Facultad de Ciencias Exactas y Tecnología, Universidad Nacional de Tucumán

E-mail: lfigueroagallo@herrera.unt.edu.ar

Abstract. Problems related to environmental noise are not a new subject, but they became a major issue to solve because of the increasing, in complexity and intensity, of human activities due technological advances. Numerous international studies had dealt with the exposure of critical patients to noisy environment such as the Neonatal Intensive Care Units; their results show that there are difficulties in the organization in the developing brain, it can damage the delicate auditory structures and can cause biorhythm disorders, specially in preterm infants. The objective of this paper is to present the development and implementation of a control system that includes technical-management-training aspects to regulate the levels of specific noise sources in the neonatal hospitalization environment. For this purpose, there were applied different tools like: observations, surveys, procedures, an electronic control device and a training program for a Neonatal Service Unit. As a result, all noise sources were identified -some of them are eliminable-; all the service stable staff categories participated voluntarily; environmental noise measurements yielded values between 62.5 and 64.6 dBA and maximum were between 86.1 and 89.7 dBA; it was designed and installed a noise control device and the staff is being trained in noise reduction best practices.

1. Introduction

Noise pollution is a major issue to face and resolve as far as that human activity has increased, due to technological advances, both in complexity and in intensity. This transformation is also reflected on the development of the health care system which has changed from strictly palliative to preventive care, being the last one a fundamental pillar in the patient quality of life with its consequent increase in life expectancy. The personal notes of the British nurse Florence Nightingale -pioneer of modern nursing- marked the beginning on investigations of noise control in the hospital environment with her famous phrase *"Unnecessary noise is the most cruel abuse of care which can be inflicted on either the sick or the well"* [1]. This phrase becomes, once and again over time, a guide for those who want to deepen the knowledge to improve patient care and emphasize on prevention to avoid new diseases.

In the last 40 years, several papers had reflected the ongoing work of research groups around the world associating the continuous exposure to excessive noise levels –in critical patients as in health workers- with deficiencies in the disease recovery process, diminished quality of life and hindering the normal execution of tasks, affecting attention, communication and learning. These studies had allowed to identify at-risk populations that are more susceptible to suffer hazardous consequences caused by this air polluting agent, such as those who are hospitalized in General Intensive Care Units, Coronary Units or catheterism, newborns with low birth weight and/or pathological preterm infants most of which have not fully developed their hearing. Other papers had analyzed different methods of measuring and registering -both in the environment and inpatient unit-, analysis of national and international regulations, among other aspects [2, 3, 4, 5, 6]. However, up today there is no an unified criteria to lead such investigations [7] and it is possible to state that it is not enough to have a high level technology and a trained staff, it is necessary the presence of an integral healthcare humanized vision, involving direct and indirect responsible actors of patient care, for taking effective control and reduction noise actions.

[1] To whom any correspondence should be addressed.

Since 2007, in San Miguel de Tucumán-Argentina, the Gabinete de Tecnologías Médicas (GaTeMe) of the Departamento de Bioingeniería of the Facultad de Ciencias Exactas y Tecnología of the Universidad Nacional de Tucumán (DBI-FACET-UNT), the Electromedicine Department of the Sistema Provincial de Salud (SIPROSA) and the Department of Human Resources of the Ministry of Public Health work jointly solving the emerging problems.

The Instituto de Maternidad (IMNSM) is a public institution, in the northwest of Argentina, considered as Reference in its specialty; is one of the ten high complexity hospitals in Tucumán and is classified "Level 3" according to the Organization and Functioning of Neonatology and Neonatal Intensive Care Units (NICU) services guidelines of the National Ministry of Health –R.M. 306/02- and is the last step before to derive, the most complex cases, to Hospital Juan Garrahan in Buenos Aires, Argentina; it has approximately 1250 permanent employees including medical, technical and administrative staff. In 2012, the new Neonatology Service (NS) was inaugurated, it provides healthcare to all newborns (NB) without social insurance or those whose health condition cannot be solved in other provincial health institutions, whether public or private. It has 60 Neonatal Internation Units –incubators and infant warmers- distributed as follows: 25 in Intensive Care, 23 at Intermedia Care, 12 at Growth and Nutritional Recovery, and 3 at Pre-Medical Discharge. This institution attends more than the half of births of the province and this number varies between 11,000 and 13,000 per year, and has one of the highest birth rates in Argentina, with a daily NICU census among 55- 70 patients [8].

The GaTeMe has made several noise levels measurements and their results were published few time ago, they were of 24 h duration and the average value obtained was 63 dBA, consistent with results found in the international literature for different healthcare institutions [9, 10, 11].

In comparison to other controls of pollutants present in the environment, the environmental noise control is not treated as a priority; perhaps because most of its effects become evident in the long term and are not often noticed by those who suffer them until they have turned severe and irreversible. Currently, the information about dose-response relationship in human beings is not homogeneous and it can be found endless criteria around de globe, however there are no doubts about the harmful effects of noise. Therefore, some actions are needed to limit and monitor the noise exposure based on an appropriate scientific assessment, particularly the dose response relationship because it is the basis in an evaluation process and risk management [12].

Finally it should be noted that a critical patient, such as premature or pathological newborn (NB), exposed to a hostile acoustic environment such as NICU, hinders the organization of the developing brain. In case of being under excessive, continuous and high-pitched noise levels, it could show up some injury in the sensitive hearing structures with a high risk of hearing loss and biorhythms disorders.

The purpose of this paper is to present the development and implementation of a system, which addresses technological, management, and training aspects in order to monitor and control the noise pollution levels focused to neonatal hospitalization.

2. Materials y Methods

The different stages for the developing of the noise control system are detailed below. They could be applied in any health institution, whether public or private, when a noise reduction policy is desired.

2.1. Adaptation period

In this stage, the working group that is external to NS, should become familiar with the daily tasks of the service staff, working in partnership to generate a collaborative working environment. It must be avoided any change in habits or behavior of the staff, as well as the characteristics and dynamics of its functioning. The IMNSM is a Maternity with Family Centered Policy (MFC) and this is the reason why it has the following features: unlimited access for parents during the 24 hours, prearranged visit schedule for grandparents and siblings, residence for mothers and a multidisciplinary team formed by 4 professionals under "Newborn Individualized Developmental Care and Assessment Program" training (NIDCAP) -in order to apply the Family-centred developmentally supportive care (FDSC)-. NIDCAP is a FDSC reference method and nowadays this team is the only group on Argentinian Northwest enrolled in this

training that focuses into the observation, characterization of the physical space and adequate care of patients, trying to improve the interaction between family -that takes care of the NB- and staff [13].
Once the IMNSM Research and Teaching Committee approved the project and previous to any action in the NICU, a series of interviews were performed, to key actors of the project -Medical Director of NS, NIDCAP team, Technical Director-. This allowed to detect needs in order to develop tools to enable a prompt recognition, properly based, to decide the actions to be taken. There were developed and implemented both, a survey to the NS staff -performed by applying a set of techniques designed to gather data about selected issues in a systematically way- and in vivo-in situ observations to register any event that disturb the NICU environment [14].
It was designed a descriptive cross-sectional study; the population was composed by the SN staff located at the IMNSM´s 3rd floor that, fully or partially, develops their tasks at these facilities, inasmuch as their activities create the micro and macro environment of the service.
The survey design was leaded by GaTeMe, was reviewed, approved and authorized by the NIDCAP team, the Head of SN and the IMNSM Research and Teaching Committee. The participation was voluntary, including “closed-ended questions” -the subject must choose one or more answers from a pre-set list- and “open-ended questions” -subject can respond freely using their own words-.
The application methodology was self-administered -no interviewer is required -, saving resources and allowing the simultaneous filling of several questionnaires at each turn. This also reduced the bias that would produce the interviewer presence, because the instrument was filled at his absence -on a separated physical space- providing privacy, anonymity and giving time to consider his answer to the interviewee. The survey was divided into 5 sections -Employee information, tasks performed, noise sources, noise level knowledge, suggestions and personal opinion- [15].
The second tool -“observation”- was the systematic, valid and reliable record of behavior or overt conduct [16]. Two ways of implementation can be chosen:

- Participatory observation: where the observer interacts with the subjects observed
- Non-participatory observation: where the observer does not interact with the subject observed.

Non-participatory observations were chosen at two different moments: initial reconnaissance observations and during environmental noise measurements. The most significant issues noted were the work rate, staff attitudes, tasks performed, handling ability of Medical Technology (MT), number of people present at NICU, among others, in order to correlate or identify particular variables or noise sources that could not be detected or stated by staff in the survey.

2.2. Inspection and Environmental Noise (EN) Measurements
The GaTeMe developed noise measurements procedures and reports for Neonatal Services according to ISO 17025 / IRAM 301 guidelines. These documents state the steps to perform EN measurements at NS, recommending its use in NICU services because they attend the most complex and unstable patients. The procedures are based on the standards IRAM 4113-1 & 2 / ISO-1996 1 & 2, IRAM 4074-1 & 2 / IEC 61672-1 & 2, IRAM 4220-1 &2 / IEC 60601-1 & 2-19 and the American Academy of Pediatrics Recommendations (AAP) [17]. There were performed a series of A frequency weighted sound pressure level (SPL) measurements, sampled every second for 24 hours, on days detailed below: March 26 (Day 1: Start at 11:00 h), March 27 (Day 2: Start at 21:00) and March 30 (Day 3 : Start at 13:00 h) of 2014. Measurement periods were selected in order to record during the weekend, visiting schedule -grandparents and siblings- and Medical interconsultation -regular activity among internal medicine specialists-.
The instruments were placed without disrupting the staff work, the microphone was located at 1.20 m above the floor based on the patient location and evaluating the most critically inpatient unit -as closely as possible to the incubator to get data of the nearby environment under working conditions-.
The AWIC tool -web application for the clinical engineering area- provided the information about the NS equipment [18]. The available MT manuals were evaluated to study their operating principles, modes of using and configuration of basic parameters, in order to find repetitive errors or problems in their operation that turn them into noise sources, for the developing of both, control and noise reduction

protocols.
The infrastructure facilities inspection found that the NS covers an area of 820 m2 and is made up of external brick walls and double-glazed glass panels on the north side –bordering Mate de Luna avenue- which is basically an air sealed camera enclosed between two layers of glass of 3 + 3 mm each one. This build, sets and guarantees a volume of dry air between the layers, and this air chamber reduce heat transfer between indoor and outdoor while reducing the entry of solar radiant energy and noise transfer. The new structure, built during remodeling inside the floor, was executed by using masonry partitions with iron structures, plasterboard ceilings without any insulation or sound absorption components on the upper dome formed by such plates and the original ceiling. Internal divisions were solved applying dry construction method using shock and impact resistant, waterproof, and fireproof materials.

2.3. Environmental Noise Warning Device

Among the solutions that can be proposed to mitigate this problem, some are administrative and others lead to design and develop specific supporting technologies. Worldwide, there are control-monitoring level devices for noise pollution for those environments where there is a need to warn or indicate when a preset safety noise limit has been reached –values defined by legislation or international recommendations-. Today, there are no devices -made in Argentina- with this features, therefore, health care institutions must contact sales providers or import these devices. Therefore, health institutions face high international prices and difficulties to import, which discourages incorporating this technology. In search of a solution, the academic field have designed and developed a device called "Electronic Ear", whose purpose is to monitor the noise level and activate a warning visible light signal. The values referred below as safe and dangerous are configurable, allowing a gradual adjustment, always aiming to get to the lowest levels possible.

For this application the chosen limits are:

1. If the measured value <Safe value, the visible light signal is GREEN
2. If the safe value < Measured value < Dangerous value, the visible light signal is YELLOW
3. If the measured value > Dangerous value, the visible light signal is RED

This device can be used in critical services as well as in any other environment in which, after making an EN assessment, monitoring results indicate that there is one of the following situations:

1. The use of Personal Protective Equipment (PPE) its required
2. No PPE needed, but there are high noise levels during working hours, or pregnant women during their gestational period exposed to high noise levels.
3. Health institutions with patients on treatment and / or recovery where low noise levels are needed to guarantee their evolution and improving the care quality.

It should be noted that this device is a facilitator for the implementation of a "culture of silence", with all the benefits that this implies. Figure 1 shows a simplified block diagram of the "Electronic Ear Vincent 1.0".

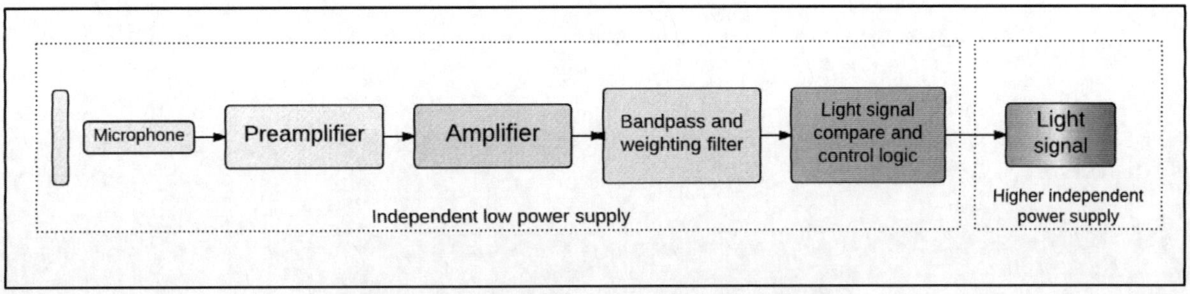

Figure 1: "Electronic Ear" simplified block diagram

The 1st block captures the sound signal via an omnidirectional electret microphone, with flat frequency response in the range of audible frequencies and a sensitivity of 18 mV output for a signal of 1 Pa at 1 kHz. The acoustic signal, transduced into an electrical signal, is amplified by a ultra-low noise operational amplifier and conditioned by filter cascades that consist of a band pass filter to remove low frequency components above 16 KHz (because they are not of interest for the actual purposes), a notch filter to eliminate 50 Hz noise, and finally an "A-weighted" response active filter. The frequency weightings are defined in the standards to which a noise measurement instrument is designed, in this case the filter transfer function is defined in the standard IEC 61672-1 by the following equation [19]. This standard also specifies the performance and tolerances for the frequency weighting curves to be used.

$$A(f) = 20.log_{10}\left[\frac{f_4^{\,2}.f^4}{(f^2+f_1^{\,2})(f^2+f_2^{\,2})^{1/2}(f^2+f_3^{\,2})^{1/2}(f^2+f_4^{\,2})}\right] - A_{1000}$$

In this equation the constants f1 and f4 are calculated from the solution of two biquadratic equations (See section 5.4.9 of the Standard) while f2 and f3 constants represent poles that are obtained from linear equations (See section 5.4.10 of the Standard). The A_{1000} normalized constant, expressed in decibels, represents the power gain needed to provide a 0 dB frequency weightings at 1 kHz and its value defined in the IEC is -2.000 dB, while $f_1 = 20{,}60\ [Hz], f_2 = 107{,}7[Hz], f_3 = 737{,}9\ [Hz]$ and $f_4 = 12194[Hz]$. With this values, the active signal processing network filter -to adapt the system response to the human ear- was synthesized applying the Butterworth approximations and was implemented by using ultra-low noise operational components.

The weighted signal is amplified to obtain easily manageable electrical values and is squared to obtain its effective value. In this stage, the weighted noise value could be read but the instant noise variations are so fast and therefore the operator will face serious difficulties to register it. In order to measure sound pressure levels of fluctuating sounds, it can be selected one of these two methods:

- Peak values: Is the fastest analysis method that takes as reference the peak levels of the audio signal; it is not the same as the Maximum Sound Level -referred to as the Lpeak or sometimes Lpk-. There is no time-constant applied and the signal has not passed through an RMS circuit or calculator. This is the true Peak of the sound pressure wave.
- Root Mean Square (RMS) Values: Takes as reference the average value of the audio signal -shows the range where most of the energy is concentrated-, not only the peaks at particular instants.

As the physiological response of the human ear can be equated to measure RMS value of fluctuating sounds -for example, if a person hears a thunder of few seconds and in that time heard a thunder higher in intensity than the first, the subject perception will be such that the increase in volume is similar to the RMS value of both sounds and not as a maximum or peak value (PEAK) like in electrical analysis- it was used a logarithmic output True RMS-to-DC Converter chip.

The A weighted output signal and two adjustable DC voltage values -representing the thresholds defined as safe and dangerous- are driven to the PIC16F628A microcontroller Inputs in a comparator mode that in addition it was programmed to change between 3 externally configurable comparison times -1, 2 or 3 seconds- according to the staff desire. This module controls the visible warning signal made of high intensity RGB LED strips. All the parts and components were mounted in a molded cabinet of 2mm thick steel sheets welded to avoid the entry of dust and/or contaminants. It consists of a rear cover; a frame and a divider that serves as a support for the frontal acrylic on which has been recorded an image -ear shape- in order to facilitate the warning interpretation (see Figure 2).

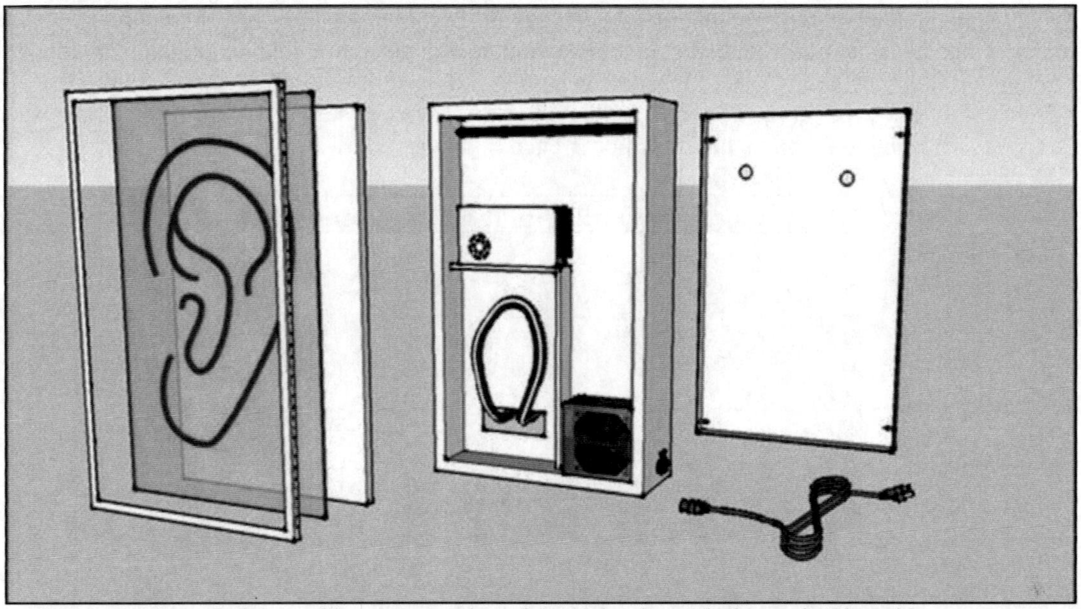

Figure 2: 3D graphic representation of the Noise Control Electronic Dispositive

2.4. Capacitation

This is an effective mechanism to motivate the employees, particularly for those workers that believe that they have not reach their maximum capacity yet. PAHO has a pedagogical line, work-study methodology centered, designed to give a "in-service training" to health care staff because they consider that is not enough to offer training programs to institutions -taught courses and conferences at work environments often lack commitment to effective work processes-.

Based on these guidelines, the training process began forming a workgroup and performing meetings with the participation of the IMNSM Directors Board and NIDCAP team as a space to discuss-analyze-propose-authorize the applicable possible solutions, which are currently in development, for the implementation and support of the IMNSM NIDCAP program.

The designed training program based on the NFI guidelines and the previously performed team tasks includes themes like “Introduction and theoretical concepts of FDSC”, “Professional role in the FDSC”, “Synactive Theory of Infant Development”, “Benefits of FDSC”, “Macroenvironment-Microenvironment and family”, “Light as an environment pollutant: Harmful effects, measurement units, typical NICU values”, “Noise as an environmental pollutant: Harmful effects, measurement units, typical NICU values”, “Interventions to improve the care quality”, “NICU noise and light intensity: situational diagnosis”, “Recommendations and strategies to minimize noise: Parents and staff”, “Family role as a natural caregiver”, “The importance of teamwork to implement the FDSC”.

3. Results

The period of adaptation allowed a deep knowledge of the actual operation of the NS, allowing to identify critical points and to detect the main high noise pollution levels generators. The NS mainly perform a permanently overloaded healthcare work, where the days of highest demand correspond to the beginning of each week, especially after of a 72 h long weekend, and along the year the workload rate shows a pronounced increase due seasonal diseases. In this context, MT is an important noise source either in normal or wrong operation or due to insufficient knowledge, wrong configuration of alarms and control parameters.

Modern noise measurement instruments acquire, store and analyze the data (data logging) simplifying the field tasks, becoming in other instrument of the NS equipment set. However, in this case, it was necessary the presence of an operator to register the noise events that, applying his expert judgment, were representative during the identification of noise sources process. It was extremely important the interchange and contact with staff previously to any installation, operation and record of the noise measurement equipment to facilitate the understanding of the objective and to generate a cooperative environment avoiding the alterations in the staff behavior and customs –as an example, stating that it does not register voice or conversations-. Therefore the non-participatory observations allowed to categorize the noise sources according to their frequency of occurrence as shown below:

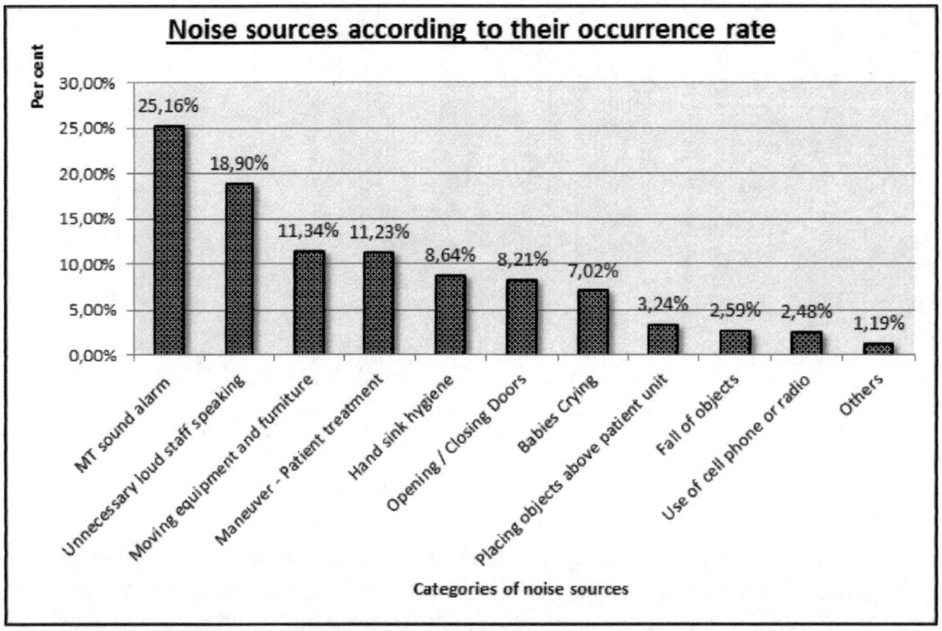

Figure 3: Causes and Noise sources identified at the observations according to their frequency of occurrence

All the data acquired were analyzed searching for time zones with an increase or decrease trend and correlate them with the operation of the NS. The results of this analysis are in table 1; it shows the quantity of time in which the 70 dBA and 80 dBA were exceeded -extremely high EN values for NS-. It can be noted that the greatest density values that exceeded 80 dBA are located on the 3rd Day which corresponds to Sunday-Monday measuring period, and the mayor noise pollution time slot was Monday from 7:00 to 12:00 am. This is correlated with the visits of specialists and external interconsulting physicians, diagnosis and NB monitoring procedures, exceeding the values registered during visiting of siblings and grandparents.

Table 1: Average equivalent SPL for 24 h, total time were the 70 (dBA) and 80 (dBA) in NICU were exceeded

	SPL (dBA)	T>70 (dBA)	T>80 (dBA)
Día 1	62.94	20'15''	1'14''
Día 2	62.53	23'35''	0'09''
Día 3	64.60	72'01''	8'45''

For the patient macroenvironment analysis, were performed both the data acquiring and the NIDCAP observations techniques. It can be noted that during some regular maneuvers carried out on newborn, there is a tendency to a heart rate increase, significant changes in oxygen saturation and respiratory rate; as can be seen in the following graphs.
For the survey, there was the voluntary participation of 121 of the 195 employees of the service (62.05%) and all occupations were represented. The analysis of their answers determined that the respondents perceive a change in the noise level during day hours (60.48%) highlighting morning shifts and particularly on Mondays during the performing of different medical procedures. This allowed to change the preconception that visits are the cause of the largest noise pollution compared to other sources, and this was an important tip for the capacitation design to staff and subsequent implementation of the MFC policy.

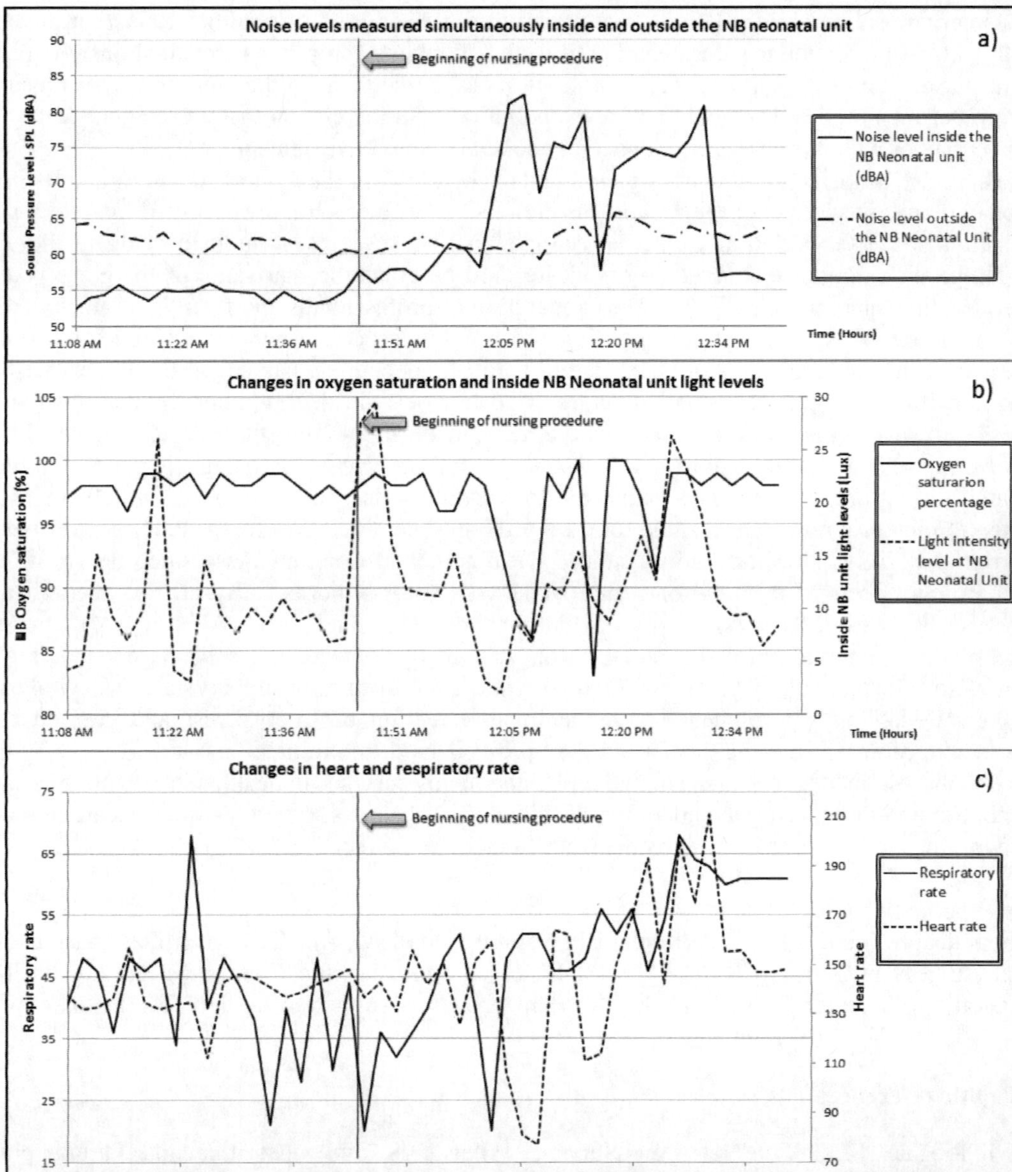

Figure 4: Results obtained at the NIDCAP observation. a) Noise levels inside and outside the Infant Incubator; b) Changes in oxygen saturation percentage and light levels inside the Infant Incubator; c) Changes in heart and respiratory rate before and during a nursing procedure.

At present, the training has been completed by 80% of staff and at first it was leaded to medical specialties, nursing and kinesiology chiefs. Currently is being elaborated a differential training, leaded to the non-health care staff to whom this problematic and solutions must be presented from a different perspective.
For the installation of the "Electronic Ear" it has being developed a three stages measurement procedure: prior to installation, during installation and tracking operation over time at control intervals –today in application-. The last two stages are sources of further comments and improvement proposals as a continuous feedback process for the multidisciplinary teamwork.

4. Conclusion

The data analysis shows that the equivalent SPL values are higher than those recommended by the AAP and other international recommendations, but similar to those reported by health care institutions all over de world; it also shows similar values compared to those obtained in previous evaluations –since 2009-. This work performs an evaluation about the role of social component in the environmental noise levels, relationship between the levels of light and noise pollution, and the effects that those agents can generate on human behavior with its subsequent impact on neonatal macro environment.
The noise sources identification process was systematized, allowing their standardization and its potential replication in any service or health care institutions. The methodology adopted allowed a quick familiarization of the service to any external actor or in case of external audit to evaluate the acoustic situation. Regarding the noise sources observed, it could be conclude that some of them are potentially minimized or eliminable, principally with the cooperation of professionals and family –as an example, less loudness of alarms-.This is the reason to bring the training process first to all chiefs of service and specialties. Each hierarchical actor of the different medical specialties, has begun the in-service training for workers with the objective "small changes in behavior will deeply improve the hospitalization conditions". However, there are other sources that depend on the architectural design of the NICU and may not be possible to easily modify -e.g. sound of falling water for hands hygiene-, limiting the implementation of some actions. The main way to mitigate noise pollution is raising awareness via a continuous training programs that include topics like Family-centered developmentally supportive care – also referred as FDSC- , Newborn Individualized Developmental Care and Assessment Program -known as NIDCAP- and effects of the Environmental Noise (EN) on newborns (NB), all this accompanied by Best Practices Policies Design with axes on training both staff and family. Through this project the mitigation of environmental noise was tackled from data analysis instead of subjective analysis plus the designing of specific technology to suit the requirements of the institution support staff. It is vital to create a collaborative working environment, strengthening the feeling of safety, trust and confidence from parents to health care staff, turning them as an example of behavior to imitate.
The set of tools described can be adapted and replicated in any services in health care institutions, with the hope of contribute in improving the operation of critical care services, providing quality care and the best possible working environment for health care staff, family and patients.

Acknowledgements
The authors deeply thanks to Luis A. Rocha EE, Sofía Casella LIC, Ana Torrego MD, Sebastián Álvarez KIN., Sebastián Díaz Carrasco BME, IMNSM Research and Teaching Committee and all the IMNSM staff especially to Director Rossana Chala MD. This project is supported by the UNT Research Council (CIUNT) No. 26 / E547-2 Project.

Bibliographic references

[1] Nightingale, F. (1859) Notes on Nursing: What it is, and what it is not, Philadelphia: J.B Lippincott Company.
[2] Philbin, K. et al (1999) 'Recommended Permissible Noise Criteria for Occupied, Newly constructed or Renovated Hospital Nurseries', Journal of Perinatology, vol. 19, no. 8 parte 1, pp. 559 -563.

[3] Topf, M. (2000) 'Hospital noise pollution: an environmental stress model to guide research and Clinical interventions', Journal of Advanced Nursing, vol. 31, no. 3, pp. pp. 520-528
[4] Darcy, E. et al. (2008) 'A Descriptive Study of Noise in the Neonatal Intensive Care Unit',
[5] Advances in Neonatal Care, vol. Vol. 8, no. 3, June, pp. pp. 165-175, Available: doi: 10.1097/01.ANC.0000324341.24841.6e
[6] Hsu, T. et al (2011) 'Noise Pollution in Hospitals: Impact on Patients', Journal of Clinical Outcomes Management, vol. 19, no. 7, July, pp. 301-309.
[7] Ryherd, E. et al (2012) 'Noise Pollution in Hospitals: Impacts on Staff, Journal of Clinical Outcomes Management, vol. 19, no. 11, Noviembre, pp. 491-500.
[8] Hasek Nogueira, M.F. et al (2011) 'Noise Measurement in NICUs and Incubators with Newborns: A systematic Literature Review', Rev. Lat. de Enf., vol. 1, no. 19, January-February, pp. 212-221, Available: ISSN: 0104-1169
[9] Ministerio de Salud Pública de la Provincia de Tucumán (2012) Memoria Anual 2012, San Miguel de Tucumán: Gobierno de la Provincia de Tucumán.
[10] Olivera, J.M. et al (2009) 'Contaminación Acústica en salas de neonatología', Anales del XVII Congreso Argentino de Bioingeniería y VI Jornadas de Ingeniería Clínica, Rosario, 1-4.
[11] Olivera, J.M. et al (2011) 'Acoustic pollution in hospital environments', Journal of Physics: Conference Series, ISSN 1742-6596, Universidad Nacional de Mar del Plata, Setiembre de 2011.
[12] Olivera, J.M. et al (2013) 'New approach to evaluate Acoustic Pollution in hospital environments', Journal of Physics: Conference Series. IOP Science. Vol.477 n°1. p1 - 10. EISSN 1742-6596. Memorias del XIX Congreso Argentino de Bioingeniería
[13] World Health Organization WHO (1999) 'Guidelines for Community Noise', Genova, Suiza.
[14] Hospital Materno Infantil Ramón Sardá y UNICEF (2011) Maternidad Segura y Centrada en la Familia (MSCF) Conceptualización e implementación del modelo, 1st edition, Buenos Aires: UNICEF.
[15] Zapata, O. A. (2005) Herramientas para elaborar tesis e investigaciones socioeducativas, 1st edition, México D.F.: Editorial Pax México.
[16] Figueroa Gallo L.M, Olivera J.M., Rocha L. A. "Noise Pollution Control in Neonatal Service Units: A multidisciplinary approach". IFMBE PROCEEDINGS del VI Congreso Latinoamericano de Ingeniería Biomédica, CLAIB 2014, Paraná, Entre Ríos, Argentina, October 2014, pp1-4, Ed. SPRINGER, ISSN: 1680-0737
[17] Hernández Sampieri, C.R et al (1991) Metodologias de Investigación, Primera Edition, México D.F.: Mc. Graw Hill.
[18] American Academy of Pediatrics (AAP) (1997) 'Noise: A Hazard for the Fetus and Newborn', vol. 100, no. 4, Available: Impreso: ISSN: 0031-4005 / Online: ISSN: 1098- 4275
[19] Alarcón, M.V. et al (2011) 'Aplicación Web para el Área de Ingeniería Clínica', Proceedings del XVIII Congreso Argentino de Bioingeniería SABI 2011, VII Jornadas de Ingeniería Clínica, Mar del Plata.
[20] IEC, International Electrotechnical Commission (2002) Norma IEC 61672:2002, Ginebra.
[21] Davini, M. C. (1989) Bases Metodológicas para la educación permanente del personal de Salud, neoBuenos Aires: Organización Panamericana de la Salud (OPS).

Formative Assessment and Professional Training: Reflections from a Mathematics course in Bioengineering

C Carrere[1,2], S Milesi[1], I Lapyckyj[1], E Ravera[1], L Escher[1], A Miyara[1,3], G Pita[1,2], M Añino[1,2]

[1] Department of Mathematics, School of Engineering, National University of Entre Ríos, Oro Verde, Entre Ríos, Argentina.

[2] Engineering Education Research and Development Group (GIDEI), School of Engineering, National University of Entre Ríos, Oro Verde, Entre Ríos, Argentina.

[3] School of Exact Sciences, Engineering and Surveying, National University of Rosario, Rosario, Santa Fe, Argentina.

E-mail: carrerecarolina@bioingenieria.edu.ar; emilianoravera@bioingenieria.edu.ar

Abstract. Bioengineering is currently considered an interdisciplinary professional field which provides solutions to different problems arising in the area of health care. Its strategic importance is widely acknowledged since its developments and proposals could help diminish the level of technological dependence in the sector. The fast pace of innovation in the area of biomedical technology gives rise to permanent reflection on the learning goals and teaching strategies proposed by educators in the different training stages of a bioengineer. In this context, learning assessment appears as a controversial issue which needs to be debated and rethought. This paper describes the reflections of teachers of a Mathematics course within a Bioengineering program around the question, What approach to assessment favors the student's participation, autonomy and training as a future bioengineer? The investigation was carried out in the framework of a Participatory Research Action project and helped us to redesign assessment activities from a different perspective.

1. Introduction

While the universe of scientific and technological advances related to a bioengineer's job changes rapidly and the dynamics of this system is perceived as dramatic and challenging, the academic world in which such a professional is trained seems to cling to old traditions. This is more conspicuously perceived in the assessment process, especially in the first years of the program which are dominated by basic disciplines, in which tests (be they midterm or final) in their different formats are firmly entrenched and seem to mark the pace of the courses, especially those of Mathematics. This is at least our perception as teachers of the Department of Mathematics at the Engineering School of the National University of Entre Ríos. A first disruption of the assessment system was observed when computer technology arrived to the University and the Engineering Schools incorporated Computer Science Laboratories; little by little, assignments completed with specific software gained an increasing space in the planning of Mathematics courses, although they did not displace midterm tests or match their importance.

While it seems natural to question the currency and training value of the contents we teach, reflecting on the assessment methodology we implement in our courses and its relationship with the training of the future bioengineer leads us to uncharted territory we are apprehensive to step into. When we take courage and ask ourselves, "What to assess?," a series of questions arise that little by little unveil the complexity of assessment: How?, When?, For what?, For whom?, Whom does it benefit?, What's the goal?, In what context?

Students, on their part, make queries from the beginning about the assessment system and understandably ask: "How many midterm tests?", "What are they like?", "What topics do they cover?", "How much time are we given to solve them?", "What support can we use?". They also devote much time to collecting tests from past courses and in many cases this material replaces textbooks, teacher-prepared slides and guides, and their own in-class output. In this scheme they would seem to develop skills for academic life, focusing their efforts on getting the necessary grade to attain the status of regular or promoted students, pushing learning into the background.

We ask ourselves then, how does our assessment style influence our students' learning? Is it consistent with the course's aims? Is it aligned with teaching strategies? Does it contribute to the development of critical thought, or does it favor learning by heart? Does it foster the necessary abilities to complete undergraduate training in Bioengineering? Does it strengthen the skills necessary to enter professional practice? The question list seems endless and even then we are neglecting a fundamental question: What do we understand by assessment? That could be the starting point for unraveling a thread from which we can get the answers sought.

In this work we describe a portion of the exploratory path we have trodden in our activity as Mathematics teachers from the courses of Vector Calculus and Differential Equations at the Bioengineering program. We wish to highlight that our team is interdisciplinary as it is made up of an Electrical Engineer (the Professor in charge of the courses), a Professor of Mathematics (the Associate Professor of the courses), a Chemical Engineer (one of the Practical Applications Teachers), three young Bioengineers who beside working as Practical Applications Teachers at the Department of Mathematics are completing their postgraduate studies in different specializations of Bioengineering, two advanced Bioengineering students, a teacher from the Department of Humanities and the Pedagogical Advisor of the School of Engineering, who holds a BA in Education Science. This investigation is framed in the methodological principles of Action-Research. In the following sections we will describe this methodology and how study and research seminars helped us construct a theoretical framework of reference that allowed us to find new meanings and approaches to assessment which favor student participation, their autonomy and their training as future bioengineers.

2. –The methodology: Action-Research

Various works published in the fields of Psychology, Social Psychology, Education in general and Health-Care Education in particular state that the term, "Action-Research," was first defined by Kurt Lewin, who carried out studies on rural medicine, biology and philosophy in Germany. At the onset of religious persecution in that country he decided to move, in 1933, to the United States. He worked at the Massachussets Institute of Technology, where he established the Center for Group Dynamics. The literature recognizes that the first Action-Research process took place in the 1940s when Lewin took on an urgent challenge: that of introducing changes into the feeding habits of a population affected by the scarcity of certain products in the context of World War Two. After different experiences, Lewin concluded that better results are obtained when the community or group of people affected by the problem are incorporated into the quest for solutions and their implementation.

In the educational field, these ideas were revisited and reassessed in the 1970s by UK pedagogists Lawrence Stenhouse and John Elliot, who picked up Lewin's model. To these educators, Action-

Research (AR) relates to practical, everyday problems experienced by teachers. Elliot posits that AR solves the problem of the relationship between theory and practice since it enables the teacher to problematize, analize and propose improvements from and for their own day to day practice. That is to say that the teachers undertake a researcher's role as s/he systematically plans, acts, observes, reflects and inquiries into her/his teaching practice to produce knowledge and changes aiming to improve it [1].

With regard to the research process, this is structured in cycles, each of which includes four phases or moments: (1) phase of diagnosis and INITIAL REFLECTION, which helps identify "problematic" situations and establish priorities; (2) phase of reviewing the conceptual framework and PLANNING an improvement or innovation; (3) phase of implementation of the planned ACTION; and (4) phase of observation and appraisal of results. At this point, a new instance of reflection starts a second research cycle. A review of the theoretical framework and of the knowledge acquired in practice will allow actions to be redesigned so as to move forward with the process [2].

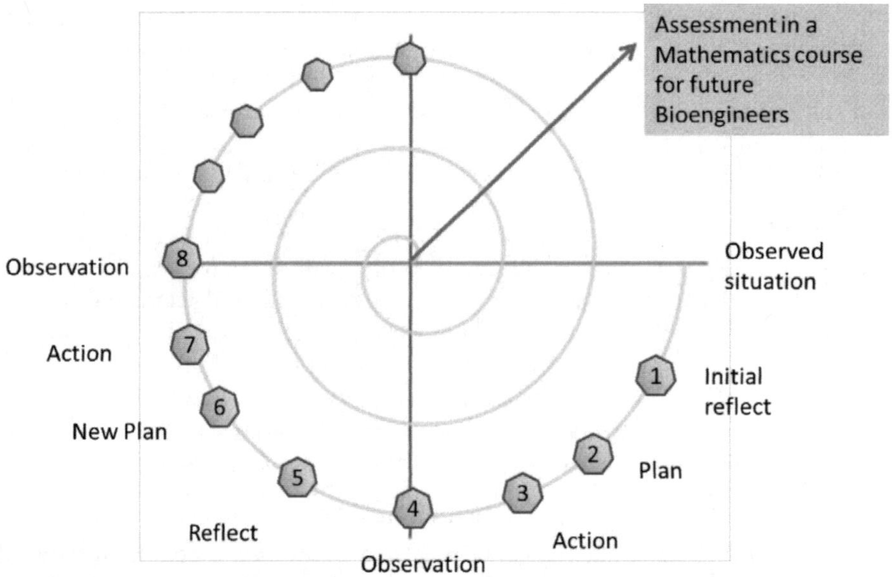

Figure 1. Illustrates the phases of the Action-Research process. During it and little by little, cycle after cycle the participating teachers-researchers uncover different aspects and strive to understand the outcome of the assessment methodology applied, detecting unsatisfactory aspects and preparing new action plans that allow assessment to be oriented towards a formative approach at the service of learning.

This methodology combines three processes, those of knowing, acting and teaching, all of which involve the whole academic community that participates. It aims to construct knowledge that will bring about a change in those aspects of educational practice that emerge during the research as possible hindrances to learning.

In our case we have started a spiral process (Figure 1) with the aim to investigate assessment in the mentioned courses to understand it deeply and detect aspects that might be negatively influencing the learning process so as to improve them.

Our *Collaborative Work Seminar* is an integral part of the process. Through teamwork, which fosters and encourages each participant's reflection, we started to analyze learning assessment in the courses we teach. In our project the *Seminar* constitutes an environment of study, investigation and shared decision making affecting the different research stages. Through it each participating teacher gets a deeper grasp of the theoretical and practical skills that allow them to research into the problem in the context of their own teaching practice to better understand it and propose mechanisms for improvement and transformation.

In the next section, we briefly describe the theoretical framework of reference collaboratively constructed by the participating teachers.

3. The construction of the theoretical framework of reference

The questions posed in the introduction oriented the literature review. Reading and discussing different materials allowed us to discover various approaches and perspectives that have enriched our conception of assessment. After the extensive analysis done, we prepared a synopsis of the situations described by different authors and selected certain conceptual guidelines.

The meaning of assessment in the educational field has changed over time, shaped by social, economic and political situations, as well as by scientific and technological advances [3].

In higher-level institutions where professionals are trained, such as Engineering schools, the assessment process has both a formative and a certifying function [4], [5]. From a formative viewpoint, assessment can be understood as the window through which teachers and students get information on the progress of the teaching and learning processes. Based on what they observe, teachers introduce changes or new activities while students modify their studying strategies so as to bridge the gap between the reality observed and the objectives of the course or program.

From a certification viewpoint, assessment is a process oriented towards ensuring that students have acquired the basic skills and knowledge to move forward in the program, so that upon ending his or her studies the new graduate has the abilities required to enter professional life. From a conceptual viewpoint and in teaching tradition, this kind of assessment is called Summative or Accumulative Assessment, and its results are used to decide grades or student promotion at course end.

In both the international and national contexts, we find a wealth of research and documents describing unbalances or pathologies in the assessment process in higher learning. They highlight that the certifying function is often privileged over the formative one, and assessment is not usually interpreted in a wider, more open sense, but as a synonym of grading or other terms such as marks, exam, test, control among others [6],[7]. Assessment is done only at the end of each period and simply to test the level of "learning" achieved. Thus, assessment only affects the student and has a belated and scarcely relevant influence on the learning process.

In the last few years and in an international context, a move to extend the meaning of assessment in higher learning can be observed, aiming to enhance its function as a promoter of learning. In education science papers different terms have arisen which are closely related to this function of assessment: Formative Assessment, Alternative Assessment, True Assessment, Assessment for Learning, Forming Assessment, Learning-Oriented Assessment.

The theoretical principles of Formative Assessment are found in the works of Sadler, Black and Wiliam [8], [9]. These authors defined formative assessment as that comprising "all those activities undertaken by teachers, and/or by their students, which provide information to be used as feedback to modify the teaching and learning activities in which they are engaged" (1998: 7). This definition helps to identify the three essential aspects of Formative Assessment indicated in Figure 2:

- The precise communication of the objective to be met, that is to say, clear indications to the student about the goal or expected learning and about the criteria that will be used to assess their work.
- Information on the student's current situation with regard to the proposed goal, be it through teacher feedback or through self-assessment, so that the students can know their starting point before undertaking a new effort and continuing their learning process.
- Improvement strategies and student orientation as needed so that they will be able to advance, overcome difficulties and get closer to the goal.
- Following these conceptual guidelines, a new assessment plan for the Vector Calculus and Differential Equation courses was prepared. This is described in the following sections.

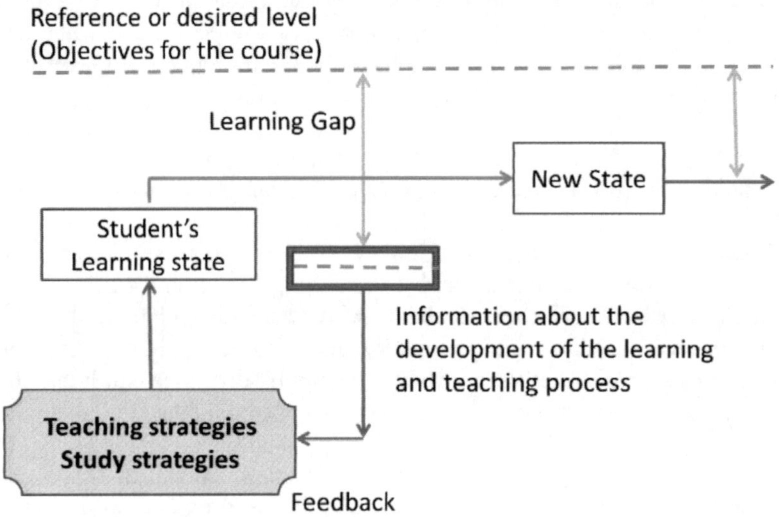

Figure 2: Illustrates the different stages of Formative Assessment; their results provide information to improve the system.

4. Assessment Process Planning

Analysis of the course objectives

From the conceptual reference framework considered, the design of a new assessment plan begins with an explicit statement of the course objectives. A document analysis was carried out to orient the updating and redesign of objectives so that they would become consistent and aligned with the Professional Profile, the current job market and the future prospects of the program. To achieve this, we reviewed the guidelines set by Resolution N° 1603 (2004) of the Ministry of Education, Science and Technology, as well as the Bioengineer Professional Profile established in the in the 2008 Curriculum, the Informative Bulletins issued by our School's Extension Department and the document published by the Federal Council of Engineering Deans made public as the "Valparaíso Agreement" about Generic Graduation Competencies for an Argentinian Engineer.

From the Professional Profile, two central activities of a bioengineer's job can be identified: solving problems in the health-care area and designing equipment or systems to be applied in medicine using Basic Science and Technology skills.

At this point, Mathematics provides a language in which other disciplines such as Physics and technologies express themselves, and is present in scientific papers in the field. It also provides a logical framework for the Mathematical Models that often arise in the first stages of a design process and in problem solving. The use of appropriate software to do calculations and make both symbolic and graphical representations is part of a bioengineer's job, so we consider it important to benefit from its advantages to teach and learn Mathematics.

From these considerations the following goals are set for the third- and fourth-semester Mathematics courses:

- Applying the language, theoretical principles, concepts and fundamental methods of Vector Calculus and Differential Equations to problem resolution.
- Applying Vector Calculus and Differential Equations to devise Mathematical Models for biological, physical or other Bioengineering-related phenomena, with a degree of difficulty appropriate to the third and fourth semesters of the course, so that the student can understand their scope and constraints and interpret the results obtained in the context considered.Using mathematical software for problem resolution.
- Increasing the students' confidence in their own reasoning and strengthening their abilities for reflection and critical analysis.

At the UNER Engineering School, an electronic bulletin is weekly published which includes a section titled "Opportunities for Students, Graduates and Teachers." In order to gather information about the abilities and attitudes that the job market demands from Bioengineering graduates, offers published by the bulletin were analyzed from April 2014 to April 2015. We broke down requirements into two categories; the first one included the skills required to work in specific areas while the second comprised other personal abilities mentioned in the bulletins as "personal competencies," "personal characteristics," "soft skills," "transversal skills." A frequency analysis was performed on this category which is shown in Figure 3. This study in part supported the inclusion of the following goals:

- Reinforcing oral and written communication strategies.
- Working effectively in teams.

On the other hand, the cited CONFEDI document indicates that the fast pace of technological change and scientific advance in Bioengineering-related fields demands undergraduate students to develop skills and abilities to learn continuously and autonomously, assess their own learning, find the resources to improve it as needed, do a literature search on different media, select the relevant material and carry out a comprehensive and critical reading thereof.

Planning formative instances

Once the goals have been clarified, formative assessment instances are planned within the conceptual framework considered. Such instances are understood as "observation windows" through which information about the learning process is collected, which will help develop the improvements needed to meet the course goals. The following formative spaces are thus introduced:

Formative assessment in the context of the practical class: In the traditional design of practical classes the teacher does an introductory review of theoretical concepts, then solves a few examples on the blackboard and proposes other exercises or problems for the students to work on. This scheme has been replaced by a new one in which the student takes center stage. The class consists of a debate with student presentations on the blackboard. The discussion is centered on a piece of written output we have come to call the *Weekly Report*, which is prepared beforehand by the students working in small teams. The report consists of completing three or four exercises, but it has certain features that distinguish it from a regular assignment: In the first place it is not assessed with a grade, errors are not punished, and it is not required that it be presented fully completed and with the correct result. The wording of the problems encourages an analysis of the data provided, and the students are oriented through questions so that they will go through the different stages of each problem resolution, justifying each step with theoretical concepts or properties. During the resolution process, the students become aware of their own difficulties, which must be incorporated into the report to be shared during the discussion in the practical class. The written report is individual and consists of solutions or solution attempts, as well as the doubts and queries that have arisen, which are shared in the classroom debate. The group doing a presentation receives feedback from their classmates (*peer assessment*) and also from the teacher, who strives to create an appropriate environment for errors and misunderstandings to be expressed without fear. Furthermore, students perform *self assessment* by confronting their own developments and results with the debate conclusions, taking note in their worksheets of errors, correct answers and clarifications. At the end of the class, the teacher collects the reports to do a general review, prepare a list of difficulties, make suggestions and mark aspects not observed during self-assessment. Then the report is handed back to the students with the written feedback. When the assignment returns to the student it has undergone three instances of formative assessment.

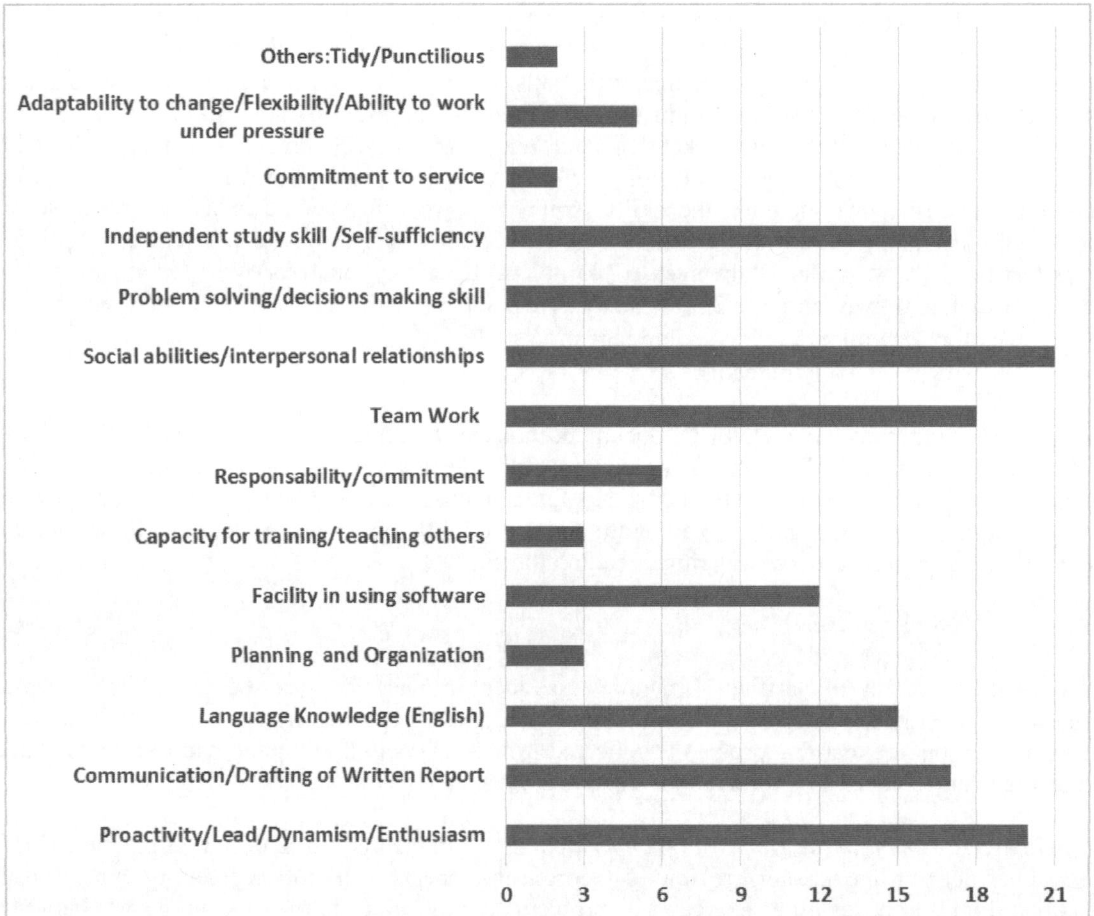

Figure 3: Frequency analysis of Transversal Skills required by the job market through the Informative Bulletins of the UNER Engineering School from April 2014 through April 2015.

Formative assessment in the context of the practical class: In the traditional design of practical classes the teacher does an introductory review of theoretical concepts, then solves a few examples on the blackboard and proposes other exercises or problems for the students to work on. This scheme has been replaced by a new one in which the student takes center stage. The class consists of a debate with student presentations on the blackboard. The discussion is centered on a piece of written output we have come to call the *Weekly Report*, which is prepared beforehand by the students working in small teams. The report consists of completing three or four exercises, but it has certain features that distinguish it from a regular assignment: In the first place it is not assessed with a grade, errors are not punished, and it is not required that it be presented fully completed and with the correct result. The wording of the problems encourages an analysis of the data provided, and the students are oriented through questions so that they will go through the different stages of each problem resolution, justifying each step with theoretical concepts or properties. During the resolution process, the students become aware of their own difficulties, which must be incorporated into the report to be shared during the discussion in the practical class. The written report is individual and consists of solutions or solution attempts, as well as the doubts and queries that have arisen, which are shared in the classroom debate. The group doing a presentation receives feedback from their classmates (*peer assessment*) and also from the teacher, who strives to create an appropriate environment for errors and misunderstandings to be expressed without fear. Furthermore, students perform *self assessment* by confronting their own developments and results with the debate conclusions, taking note in their worksheets of errors, correct answers and clarifications. At the end of the class, the teacher collects the reports to do a general review, prepare a list of difficulties, make suggestions and mark aspects not observed during self-assessment. Then the report is handed back to the students with the written feedback. When the assignment returns to the student it has undergone three instances of formative assessment.

Formative assessment and computer lab assignments: The second homework assignment consists of the Computer Lab Assignments, in which problems are proposed that require the selection of a mathematical model and computer simulation thereof. These techniques are usually applied by bioengineers in the different stages of a design process, be it of an algorithm or a device. Computer simulation allows parameters to be changed and behaviors to be observed in order to make decisions conducive to an optimal design. The learning of these techniques is fostered in the Lab assignments through problems of appropriate complexity.

Initially, they were handed in at a scheduled date after a few instances of office-hour consultations. Currently, they are submitted through the virtual campus and draft submissions are allowed. Thus, the teacher observes the process and intervenes, not for correcting or grading at this stage, but to suggest revisions, improvements and deeper analyses. The virtual campus favors formative assessment, as it adapts to the student's schedule and allows digital drafts to be uploaded which contain graphs and code the teacher can provide feedback on as needed, thus fostering the student's reflection on what they did so as to improve on it.

5. Actions implemented

The Formative Assessment instances planned were implemented in coordination with the Summative Assessment activities. The latter include two Midterm Tests and the submission of the final version of the Computer Lab Assignments. Figure 4 shows the distribution of these activities along the fourteen weeks allocated to each course.

Week	Formative Assessment Tasks	Evidence about the Learning Process	Summative Assessment Tasks	Formative Assessment and Computer Lab Assignments
1	• Drafting of the weekly written report • Class discussion • Self-assessment	Information for planning the Revisión Week		Computer Lab Assignment Nº 1 Team work with draft submissions through the virtual campus and teacher's feedback (Platform Moodle)
2				
3				
4				
5				
6			Computer Lab Assignment Nº 1	
7	**Revision Week**		**Midterm Test Nº 1**	**Analysis of frequent errors and return.**
8	• Drafting of the weekly written report • Class discussion • Self-assessment	Information for planning the Revision Week		Computer Lab Assignment Nº 2 Team work with draft submissions through the virtual campus and teacher's feedback (Platform Moodle)
9				
10				
11				
12			Computer Lab Assignment Nº 2	
13	**Revision Week**		**Midterm Test Nº 2**	**Analysis of frequent errors and return.**
14	**Revision Week**			**Make-up Test**

Figure 4: Illustrates the articulation between Formative Assessment and the activities of the original Assessment Plan which only included Summative Assessment.

After the midterm tests, the teaching team works on the errors observed; an Error Categorization and an analysis of the most frequent ones have been prepared. We have worked with the following categories: Errors associated to the analysis of the problem wording, Reasoning errors (in the logical sequence of the solution procedure), Conceptual and/or methodological errors related to topics from the course, Errors associated to lack of verification, Errors associated to mathematical language and its link to other sciences, Basic calculation errors. Activities in the review week are centered on this information, which is useful for those students who need to take a makeup test, and also for teachers as it shows the difficulties that persist. Thus, the planned activities have incorporated teacher feedback, student self-assessment and work with peers into a fabric structured by the contents, the learning goals stemming from the bioengineer's professional responsibilities and the time available (14 weeks).

6. Results

The original assessment plan included Summative Assessment only and students were promoted if they obtained a minimum 80 points out of 100 in Midterm Tests and Laboratory Assignments. By inserting the formative dimension we incorporated a third requirement: Attending and participating in 80% of the Debates and submitting 80% of the Weekly Reports. At the end of the course, the student attains one of the following statuses: Regular, Promoted or External. This approach had a positive impact on the dynamics of the classes, and also on academic performance. The average of students who attained the Regular status jumped from a historical 56% to 62% in the Vector Calculus course taught in the first semester of 2014 and 74% in the course of 2015, while in the Differential Equations course taught in the second semester of 2014, the percentage of students who achieved the Regular status was 87%. Qualitative changes were also observed in student participation when: completing an assignment (written reports) not rewarded with a grade, explaining the resolution procedure for an exercise (rather than just trying to get to the answer), working with academic honesty in written reports and drafts sharing errors with teachers and classmates.

7. Conclusions

The Action-Research process described has brought about a change in our way of understanding assessment. It has not been easy to modify traditions that permeate assessment in engineering programs; overcoming such resistances has been possible only through long-term collaborative work. The project has included spaces to reflect on learning accomplishments, the teaching practices that made them possible, and the assessment tools used, and those reflections have provided both feedback and proposals for improvement. Much remains to be done, however; the teaching team is currently reshaping midterm tests and final exams, developing new assessment instruments, and seeking ways to improve the level of student participation in assessment, which will expectedly contribute to training critical Bioengineers, able to test their own knowledge and undertake further training along their life.

Acknowledgments
Thanks are due to our Pedagogical Advisor, Marisol Perassi, for her valuable and permanent collaboration, and to Diana Waigandt of the Humanities Department at UNER Engineering School for her unfaltering support.

References

[1] Elliott, J 2007 Assessing the quality of action research. *Research Papers in Education* **22**(2) 229-246.

[2] McKernan, J. 2013 *Curriculum action research: A handbook of methods and resources for the reflective practitioner* (London and New York : Routledge)

[3] Palou de Maté, M C 1998 *La evaluación de los aprendizajes en el debate didáctico contemporáneo* (Buenos Aires : Paidós)

[4] Guardia, P 2013 Sentidos y controversias en torno a la evaluación de los aprendizajes en educación superior. *Argonautas Revista Digital de Educación superior* **3** 17 – 30.

[5] Fernández March, A., Maiques March, J. M., and Ábalos Garcerán, A. 2012 Las buenas prácticas docentes de los profesores universitarios: estudio de casos *Revista de Docencia Universitaria* **10** (1) 43-66.

[6] Zabalza, M. A., & Beraza, M. Á. Z. 2002 *La enseñanza universitaria: el escenario y sus protagonistas* (Madrid:Narcea Ediciones)

[7] Guerra, M. Á. S. 2012. *Paradojas de la evaluación del alumnado en la universidad. Promover el cambio pedagógico en la universidad* 243-270 (Madrid:Pirámide.)

[8] Black, P., & Wiliam, D. 1998 Assessment and classroom learning. *Assessment in education* **5**(1) 7-74.

[9] Wiliam, D. (2011). What is assessment for learning? *Studies in Educational Evaluation* **37**(1), 3-14

Inclusive EducativeTechnologies, for people with disabilities

AM Echenique[1], JP Graffigna[1], E Pérez[1] , N López[1], D Piccinini[2], H Fernández[1], A Garcés[1].

E-mail: amechenique@gateme.unsj.edu.ar

Abstract: The conventional educational environment imposes barriers to education for people with disabilities, limiting their rights, which is a non-discriminative education. In turn, hampers their access to other rights and creates huge obstacles to realize their potential and participate effectively in their communities. In this sense Assistive Technology provides alternative solutions, in order to compensate for a lost or diminished ability. Thus the necessary assistance is provided to perform tasks, including those related to education, improving the inclusion. In this paper some researches had been made in the *Gabinete de TecnologiaMedica*, in the *Facultad de Ingenieria* of the *Universidad Nacional de San Juan* in order to solve this problem. The researchers are classified by type of disability; sensory (visual and auditory) or motor. They have been designed, developed and experienced through various prototypes that have given satisfactory results. It had been published in national and international congresses of high relevance.

Introduction

Theresearch interest in disability problems emergesfrom the statistic data of the World Health Organization (WHO) [1] which reveals that from 93 to 150 millionchildren (from 0 to 14 years) live with some disability. The global population affected by some kind of disability has many difficulties in the normal development of personal, social, educative and/or labor activities. Historically, many children and adults with disabilities have been excluded from the conventional educational opportunities. In the majority of the countries, the first efforts to provide education and professional training were undertaken, in general, through special schools, that habitually were designed for specific impairments, like the schools for the blind. These institutions assist a small proportion of those people who needed help and they were not efficient in relation with the costs: they were generally located in urban zones and they used to isolate people from their families and communities. The situation started to change just when the inclusion of the disabled into the educational system was required by law. All the countries should have as priority that people with some disability receive a good quality education in an inclusive environment. In the context of theUnited Nations Convention on the Rights of Persons with Disabilities (CRPD), the right of being included in general educational systems and receiving individual support whenneeded was recognized. Systemic changes are required to eliminate barriers and to give supporting services and reasonable adjustments in order to guarantee that disabled people are not excluded from

1 Gabinete de Tecnología Médica, Facultad de Ingeniería, Universidad Nacional de San Juan.
2 Grupo de Análisis, Desarrollos e Investigaciones Biomédicas, Fac.Regional San Nicolás, Universidad Tecnológica Nacional, Argentina.

conventional educational opportunities. It is recommended to increase access to adequate, maintainable, affordable and accessible technical assistance [1].

The general acceptance of this reality has let, in Argentina, to the creation of national and provincial laws that considers the issue of social reintegration of people with disabilities and older adults. For instance, the section 75, subsection 23 of the National Constitution, states: "the Congress is responsible for: legislate and promote positive action measures which will ensure real equality of opportunities and treatment, and full enjoyment and exercise of the rights recognized by this Constitution and existing international treaties on human rights, particularlyas regards children, women, elderly and persons with disabilities…".

A particular way of responding to this situation is through activities on research, development and technological innovation (I+D+i)promoting the search of innovative solutions for the rehabilitation of these individuals. However, the public and private institutional resources and efforts of Argentina, to confront this issue and provide real conditions of equality, are scarce.

Particularly, in the province of San Juan, there are different centers for the care of people with disabilities and the elderly. Through the collection of data of different institutions, various limitations and difficulties were revealed. The available technology is habitually unattainable because of the lack of information or inadequate economicresources that these institutions and people themselves have. These restrictions are more sharpened if it is considered that these people always need technological tools for their functional performance. That is to say, this need should be satisfied within the institutions that help to the rehabilitation and during the time remaining out of them.

In an attempt to give specific solutions to this part of the population, the *Gabinete de TecnologíaMédica*of the *Universidad de San Juan* has been working from the year 2003, in the area of *Ingeniería*and*Tecnología de Rehabilitación* (Engineering and Rehabilitation Technology). There, many human resources trained andgraduate and undergraduate about this topic, that shares its knowledge and experiences with the objective of, on the one hand, design, develop and implement innovative scientific-technologic tools, and on the other hand, adapt existent devices, aimed at satisfying specific needs of people affected by different kinds of disabilities and of the elderly. In this way, the goal of this is to provide alternative tools for the people to interact with their environment, increasing their autonomy. In this sense, the assistive technologies, properly applied, can contribute to overcome constraints in the activityimposed by the disability.

The aim of the present work is to present different assistive technologies developed in the *Gabinete de TecnologiaMédica* (GATEME), with the objective of giving alternatives of help to people who have sensory or motor disabilities, in the access to the inclusive education.

1. AssistiveTechnology

The fundamental objective of a rehabilitation device is to improve the independence of the person with a disability; that is why the design and development of these devices is based not only on the user skills (that is, it should focus on the other functions of the affectedpeople and not on the lost functions), but also on the activity to what is designated, and the context or environment where it will be used[2].Paying attention to these concepts, a system of assistive technology is defined as that integral system resulting from the interaction between the assistive device, the human operator with some disability and context where the functional activity will be done (Figure 1).

The man-machine interface is the component of major importance in the assistive technological systems. An interface may be defined [3] as the boundary shared by the components that interact in a system through the communication or informationexchange on both sides through the boundary. Consequently, there are endless interfaces, in all the application areas.

The man-machine interfaces may then be defined as the boundary between the person and any device through which the information exchange is produced. This exchange is bidirectional and includes the interface from the person to the device to control it and the interface that gives information with reference to the environment or to the operation if the device. The information exchange that exists in an input way for being able to operate the device is done through a control interface between the user and the device. This control interface varies depending on the needs of the user, the task to be performed and the device to be used. Furthermore, the information exchange from the device to the person is done through visual, audible and tactile signs, and these play an important role in the feedback to the user.

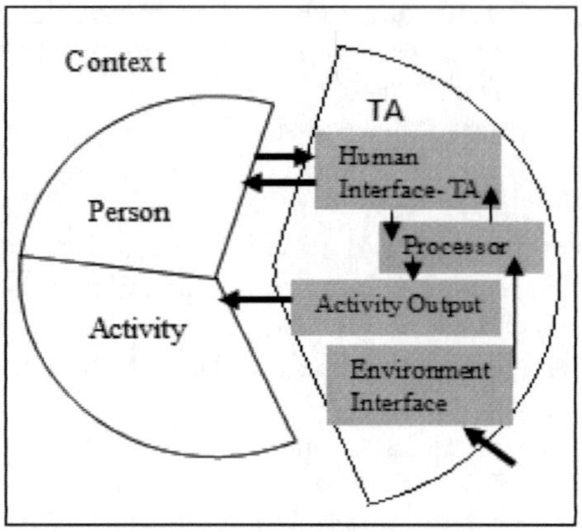

Figure 1. Analysis of the role of Assistive Technologies in the HAAT(Human Activity Assistive Technology)model taken from [2].

2. Developed Interfaces

This section presents the development of different interfaces applied in systems of assistive technologies, like alternative tools for people with neuromotor and sensorial disabilities. These interfaces have as objective giving different solutions for educational purposes. Interfaces based on biological signals and artificial vision are developed for neuromotor disabilities, and for people with sensorial disabilities, the developed interfaces are based on tactile stimulation and augmentative systems. These systems may be defined as speech therapy and educational instruments for people with various communication and/or language disorders, and its goal is the teaching of a structured set of non-vowel codes that allow functions of representation, using specific procedures of instruction.

2.1. Neuromotor Disabilities

2.1.1. Interfaces based on Biological Signals

The biological signals often used in these applications are electroencephalogram (EEG), electromyogram (EMG) and electrooculography (EOG), because they have already demonstrated to greater or lesser degrees, a skill of classification and discernment of the volunteer's intentionality.

The electrooculography (EOG) is employed in the case of people with severe motor disabilities, where the damage is at brain or spinal level and they are not able of doing voluntary movements with their extremities.To recognize the signs sent by visual fluctuations, a system of electrodes is employed, one placed near the muscles of the eyes. In this way, the visual fluctuations produce variations in the field current, that is reflected in a more positive potential when the eye turns to the electrode, and less positive, when it turns in the opposite direction [4]. These interfaces end in the production of control commands for peripheral devices, computers, among others, with the aim of improving the communication of the user with the surroundings and allowing the access to various educational, recreational and visual environment computer systems[5], [6], [7].

The signals EOG were employed in the commandof the mouse through an Accessibility Interface for **Control of the Mouse Pointer** [8], of a personal computer and potentially of other devices.

The communication between the circuit of acquisition and the PC is implemented wirelessly. A circuit of acquisition and conditioning of electrooculographic and electromyographic signals has been designed accompanied by a processing software of the signals with which the user may move the mouse pointer and click in any personal computer.

These signals are acquired through superficial electrodes in the face of the patient, Figure 2, and then expanded and filtered in a conditioning circuit.

The acquisition circuit consists of three channels, one for the acquisition of signals from the eye vertical movements, another for the acquisition of horizontal movements, and the third for the acquisition of electromiographic signals. In the second stage, the signals from the three channels are digitalized and wirelessly sent to the PC, to provide the actions to the mouse pointer (horizontal, vertical and diagonal pointer movements, and click).

Figure 2. Accessibility Interface for the Mouse Pointer Control.

2.1.2. Vision-BasedInterface

The Vision-*B*ased Interfaces(VBIs) are perceptive interfaces that use the vision as a channel of communication from the user to the computer. The Vision-based Interfaces havecameras as constituent parts that are non-invasive entry tools, one standard computer to process the images, control and obtain drive commands. The module of image processor consists of the monitoring or recognition of the region or portion of the object of study, to bring the parameters that enter to the control module, where the events are generated. These are the entries to the action command module that generate the actions for some specific task.

The Vision-Based Interface: an alternative tool for Children with Cerebral Palsy [9]; develop, detect and follow hand or foot movements of the user. This interface consists of a webcam that captures the image of the zone to be sensed and applying image processing algorithms it detects the movements of these regions. The movements detected in Figure 3, are translated into cursor movements in the computer screen through which different PC utilities are commanded. This accessibility tool uses the residual abilities and it has been successfully tested by people with severe motor disabilities.

Figure 3. Vision-Based Interface

2.2. Sensorial Disability

2.2.1. Visual Disability

One of the major problems that people with visual disabilities confront regarding educational activities is the access to reading material. In this sense, the Braille system for literacy is widely used by blind and deaf-blind people; and it is characterized by its universality and relative simplicity.

In the case of blind people, where the augmentative devices are inadequate, the use of other sense is required, like the touch or audition sense [10]. The touch presents series of advantages as regards auditory stimulation that at the same timeliberates the channel for communication. In this aspect, the "Touch Display" [11] can be mentioned. It is the equivalent for a visual "display" where the information is perceived by the contact with the skin. These devices use various means of achievingthat stimulus, for example: temperature, electricity, compressed air or mechanical variation (of pressure or vibration). The heat as stimulation medium raises the problem of thermal inertia. The mechanical devices have difficulties related to the deterioration, fatigue, heat, size and production, etc. The electrical stimulation needs to supress the sense of electrical shock and an adequate electrode-skin interface. It presents less manufacturing and maintenance problems than mechanical devices;it costs are lower, and it gives more robustness and useful life to the system [12].

Based in these considerations, an **Electrodermal Braille System**[13] was developed. This allows that any digital text (books, documents, Internet information, etc.) could be loaded in a PC and subsequently read in the belt through the Braille code. It is a platform of electrodes that simulates 32 characters or Braille engramsby electrostimulation.

The feelings to receive these characters are generated by the electrostimulation of the skin receptors of the fingertips and replace the sheet of paper, generating feelings of pressure similar to raised dots of traditional texts in Braille. Blind people have experienced it with satisfactory results (Figure 4).

Following with the assistive technologies to the blind, a Mobility device to help blind skaters [14] has been developed. The device facilitates the mobility and orientation of people with severe visual disabilities in leisure and entertainment activities. Particularly it consists of a communication system to gives information to a person who goes figure skating through a wireless connection and a touch display based on vibratingmotors. There are devices that use actuators from different parts of the body, but there is not a specific application to the presented problem.

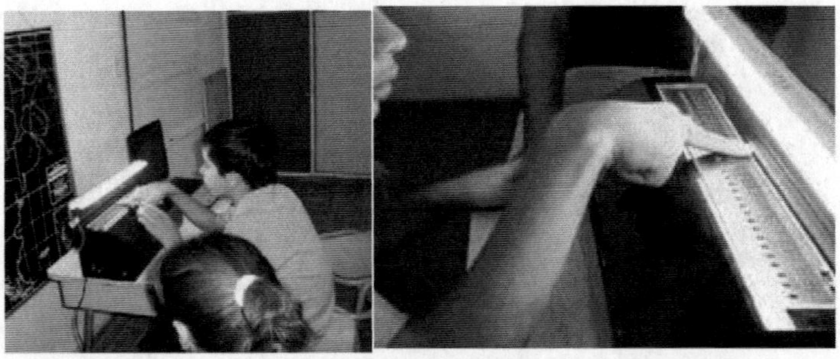

Figure 4. Electrodermal Braille Belt Experimentation

As the activity requires high mobility and attention from the skater, vibrating mechanical stimulation has been used. The vibrating motors are powerful actuators that can give clear and defined stimulus while the person is performing the choreography. Through the touch display is expected transmit the necessary information to orientate in the rink, preventing collisions and allowing the skating in a team. The device is composed of three parts: transmitter, receiver and touch display (see Figure 5). The transmitter receives the events indicated by the coach and sends the information to the receiver wirelessly. The receptor receives the information and activates the touch display to notify the skater. The coach gives information through four channels that correspond to four vibrating motors placed on a vest of elastic bands that are put on the body of the skater and allow the modification of its path through a command set establishedpreviously. The purpose of the elastic band is that it fits and shapes the body of the user and finally achieves a more direct contact. The transmitter is made up of four keys, indicator lights and the antenna.

Figure 5. Vest and Receptor

The device has been tested in the training track and in the competition track too with highly satisfactory results. In Figure 6, both situations can be observed.

The developed device meets the requirements proposed by the coach and the skater. Both the transmitter used by the coach and the receiver used by the skater are compact. The touch display is comfortable and aesthetically covered by the clothing. The functioning is perfect because it achieves the objective, enabling an adequate performance and a continuous use during the training and public exposures. The implemented tool gives to the skater the possibility of listening to the music played in the choreography and receivingthe coach feedback for the mobility and orientation through touch. Moreover, it offers the opportunity of participate in professional competences and do group choreographies. The coach and the trainer have said that the development has significantly improved in the performance and security during the skating.

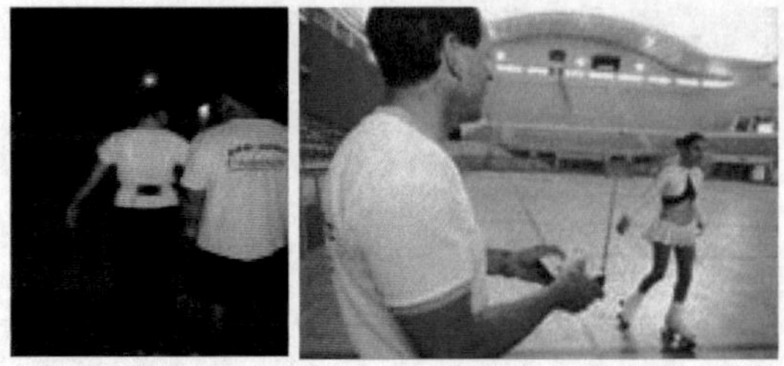

Figure 6. Experimentation in the training and competition track.

2.2.2. Auditory disability

In this subline, it has been designed and developed a ***cheap wireless communication device for the education of the hearing-impaired***[15]. The communication system consists of a transmitter and one or more receivers for hearing-impaired users. The signal is transmitted from the sender directly to the ear of the user by transmission of wireless signals moduled in FM, in the range of frequiencies that goes from 88 MHz to 108MHz.

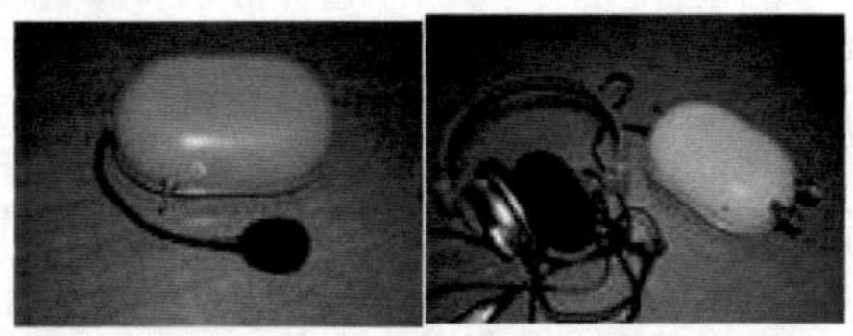

Figure 7. Wireless transmitter and receiver.

The main difference between this device and a conventional wireless hearing aid is that this minimizes the negative effects of the background noise and distance, and is very convenient in the educational context. The developed hardware (Figure 7) is being used in the "Escuela de Educación Especial paraSordos José A. Terry", with children with pure hearing loss and in cases related to other disabilities, with successful outcomes.

In order to optimize the developed prototype, a new prototype design has been started that could adapt the audio signals with the specific needs of the user. That is, according to the audiometric response of each user, it will adjust the receiver in its response in frequency. To do that, there is a stage of programmable amplification and filtering in the receiverwith individual audio output connected to each speaker of the headphones. The receiver also has a potentiometer to modify the volume levels, fine adjustment of frequency and a power indicator led (see Figures 8 and 9).

The device is portable, small and light, which allows the teacher and students mobility, offering more "natural" school situations for the hearing impaired learners.

Figure 8. Headphones for the Receiver. **Figure 9**. User with the Receiver.

2.3. Mild Cognitive Impairment in Infants

Autism Spectrum Disorders (ASD) or Pervasive Development Disorders (PDD) are a group of diseases that have the following characteristics in common: social interaction affection, little imaginative activity, verbal and non-verbal communication disrupted, and stereotyped and repetitive behaviors. This deterioration generates problems in the adaptation to the family, school and community life. According to Autism and Developmental Disorders Monitoring (ADDM), in the year 2008, the ASD have a prevalence of 11.3 in a thousand, what means 88 childrenborn, one has this disorder. Due to technological advances it is possible to help children with autism to communicate and interact in their environment.

2.3.1. SicaaKids System

A platform with a specific design for children with ASDis presented, and it is evaluated. The software was developed and designed according to the needs stated by professionals of the service of pediatric speech therapy in the Hospital Guillermo Rawson, hospital of the province of San Juan, Argentina [16]. These professionals state that software is required with a window showing different useful activities to work with the child to stimulate the communication. The objective of this work is to evaluate the designed graphic interface, to know if its design is simple and easy to access for these patients.

To evaluate this software it was used a calculating performance TP which is based in the Fitt's Law, and can be used to compare the user efficiency in specific tasks.Then, to obtain the opinion of the users they were surveyed following the model of the SUS (ScaleUsabilitySystem) questionnaire.

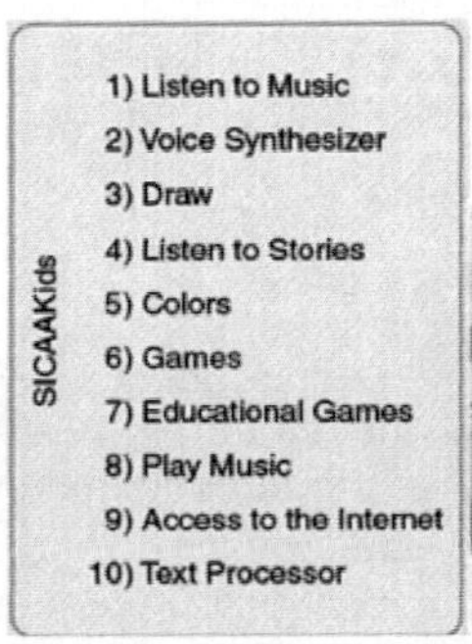

Figure 10. Software functions.

Figure 11.The main panel of the SICAAKidssystem.

The design is simple and attractive, with representative pictures that relate the icons with the underlying functions.The scenario chosen is a child's room, with the aim of relating the pictures. All the embedded programs (music, educational games, and others) are public license. The system covers the entire screen, not to distract the user with other images. Each numbered object on the screen (figure 11) of the main panel of the SicaaKids can be associated with the list in figure 10.

These functions are implemented according to the needs and activities developed by the phonoaudiological service of the Hospital.

The techniques for the quantitative evaluation of the software use that were employed were the TP and SUS scale, for more details see [4]. The evaluation has been done with children that do phonoaudiological therapies in the abovementioned hospital. The software evaluation involves the work with underage youth users, and that's why it was needed that the parents or tutors were engaged in the evaluation and signed the informed consent. The sample consisted of 8 patients, divided into two groups, group 1: subjects A, B, C, and D that use the computer frequently. They are 10, 8 and 7 years old respectively and the ASD diagnostic is mild. Group 2: subjects E, F, G and H who do not use the computer, of 5, 7, 8 and 7 years respectively and ASD moderate diagnostic.

The experiment has been done in two phases, one phase in aWindows® desktop environment and the other in the SicaaKids environment. The TP has been calculated for each selected icon in the Sicaa environment and in the Windows®, too. To do this the time that the user lasts in arriving to the indicated icon has been calculated. The TP values have been higher in the SiccaKids environment than in Windows, owing to the better relation that is obtained between the movement precision and the speed to do that. The survey shows that the software has had a great acceptability of the subjects and the majority of the volunteers think that the accessibility of the activities was easily with SiccaKids.

2.4. Uses of existent technology

Another important aspect to emphasize is the assistive technology, the capacity of using existent devices to potentiate institutional capacities. The research team has been working with that aim in the implementation of procedures that allow having the technological resources out of the University to potentiate the accessibility of the students to higher education. In that sense, the following aspects have been examined:

- Braille impression of school notes: through a university volunteer project, the University gave a high-speed Braille printer to the AsociaciónSanjuaninaparaCiegos.
- Magnetic rings: from some years ago the team has been created human resources to the development of magnetic rings, from an initiative in some classrooms with the requirement of academic units.
- Basic accessibility and specific technology: the research team advises on the technology acquisition to improve academic environments of students with disabilities (adapted chairs, specific software, etc.)

3. Conclusions

In this work, is has been address the issue of inclusive education for people with sensor, motor and mild cognitive disabilities, taking into account the more relevant problems and necessities of the affected people to bring technological solutions. Considering the strengths of the investigation group technical helps has been proposed, applying Assistive Technologies with the aim of restoring or substituting partially or totally the auditory, visual or motor functions, improving the realization of normal or habitual actions of people with special abilities.

To resolve the problems identified, it has been designed and develops devices and systems to care for specific disabilities considering utilities, costs and needs. Some of these technologies have been transferred to different institutions that are involved in the attention of people with special abilities. It is worth stressing that these institutions have been helping actively in the evaluation of the developed technologies.

4. References

[1] Informe Mundial sobre Discapacidad. OMS 2011

[2] AM Cook, SM.Hussey "Assistive Technologies: Principles and Practice", 2° edition. United States of America, Mosby Inc. Editions, 2003.

[3] RW. Bailey. "Human Performance Engineering", 2° ed., New York, Englewood Cliffs, Prentice Hall, 1989.

[4] C. R Cassemiro. y C.G Arce, "Comunicación visual por computador en la esclerosis lateral amiotrófica", Archivos Brasileños de Oftalmología, 2004, Vol. 67, No. 2, pp. 295.300.

[5] M. Johnson "Recent trends in robot-assisted therapy environments to improve real-life functional performance after stroke", 2006.Journal of NeuroEngineering and Rehabilitation.

[6] H. Krebs, N. Hogan "Therapeutic Robotics: A Technology Push", 2006. Proceedings of the IEEE.Vol. 94, No. 9.

[7] M. Matarić, J. Eriksson, D. Feil-Seifer, C. Winstein "Socially assistive robotics for post-stroke rehabilitation", 2007. Journal of NeuroEngineering and Rehabilitation, 4:5.

[8] Sergio Bajinay, Roberto Zanetti, Eugenio Orosco, Natalia M. López. "Equipo inalámbrico de adquisición de señales biológicas para el comando del mouse". XVIII Congreso Argentino de Bioingeniería, VI Jornadas de Ingeniería Clínica. 28 al 30 de septiembre del 2011. Mar del Plata, Argentina.

[9] M. González, D. Mulet, E. Perez, C. Soria & V. Mut. "Vision Based Interface: an Alternative Tool for Children with Cerebral Palsy" 32nd Annual International Conference of the IEEE EMBS Buenos Aires, Argentina, August 31 - September 4, 2010. ISBN: 978-1-4244-4124-21995.

[10] C. Lenay, O. Gapenne y S. Hanneton: Sensory Substitution: Limits And Perspectives. Université de Technologie de Compiègne. 2003.

[11] F. Vidal-Verdú, y M. Hafes: Graphical Tactile Displays for Visually-Impaired People. IEEE Trans Neural S and Rehab Eng, V15, N1, 2007..

[12] H. Kajimoto, K. Naoki, M. Taro, T. Susumu "Tactile Feeling Display using Functional Electrical Stimulation", In Proc. of The 9th Int. Conf. on Artificial reality and Telexistence, pp.107-114, 1999.

[13] A. M Echenique, J. P Graffigna, and V. Mut "Electrocutaneous Stimulation System for Braille Reading". 32nd Annual International Conference of the IEEE Engineering in Medicine and Biology Society, (EMBC 2010), 31/8 al 4/09/2010.4 pp 5827-5830 ISBN 978-1-4244-4124-2: ISSN: 1557-170X

[14] Acerbi A, Graffigna JP, Polimeni G, Fernández HH: "Dispositivo de ayuda a la movilidad de patinadores ciegos". XVI Congreso Argentino de Bioingeniería & V Jornadas de Ingeniería Clínica, 2007 ISBN: 978-950-605-505-9.

[15] A. Garcés, H. Polimeni, H. Fernández (2007). "Educación de Hipoacúsicos, Equipo de bajo Coste de comunicación inalámbrica", Mundo Electrónico, nº 389, Barcelona. ISSN: 0300-3787 Pág.: 52-59., Apr. 1965.
[16] E. Perez, D. Piccinini, N. López, S. Ponce. Evaluación de plataforma de Accesibilidad para Niños con Trastornos del Espectro Autista. VI Congreso Latinoamericano de Ingeniería Biomédica. CLAIB 2014. Paraná, Entre Ríos, Argentina. ISBN 978-950-698-343-7

Journal of Physics: Conference Series **705** (2016) 012045 doi:10.1088/1742-6596/705/1/012045

Design of a prototype device for remote patient care with mild cognitive impairment

M Sanchez-Ocampo[1], B Segura-Giraldo[2], R Floréz-Hurtado[3], C Cortés-Aguirre[4]

[1] Biomedical Engineer, Universidad Autónoma de Manizales
[2] Electrical Engineer, M Sc. In Physics, PhD. Engineering
[3] Electrical Engineer, M Sc., In Physics Instrumentation
[4] Electrical Engineer, M Sc., In Physics Instrumentation

Email: maria.sanchezo@autonoma.edu.co
bsegura@autonoma.edu.co
rubenfh@autonoma.edu.co
carcort@autonoma.edu.co

Abstract

This paper describes the design of a prototype telecare system, which allows to provide home care to patients with mild cognitive impairment and thus ensures their permanence in their usual environment. Telecare is oriented towards people who require constant attention due to conditions of advanced age, illness, physical risk or limited capabilities. Telecare offers these people a greater degree of independence.
QFD methodology is used to develop electronic devices intended to monitor the environment and physiological state of the user continuously, providing communication between the telecare system and a monitoring center in order to take the most appropriate actions in any abnormal event.

1. Introduction

The term "Assistive technology", was defined in 2004 by WHO (World Health Organization) as "*An umbrella term for any device or system that allows individuals to perform tasks they would otherwise be unable to do or increases the ease and safety with which tasks can be performed*" [1]. Therefore, it is important to differentiate the categories of technology used for assistance. Telemedicine is defined as "*The delivery of health care services, where distance is a critical factor, by all healthcare professionals using information and communication technologies for the exchange of valid information for diagnosis, treatment and prevention of disease and injuries, research and evaluation, and for the continuing education of health care providers, all in the interests of advancing the health of individuals and their communities*" [2]. Telecare which is defined as "*a continuous source of home care that allows users to stay in their usual environment, thereby uprooting is avoided while protection is provided in crisis situations personal, social and health* " [3], therefore telecare supports independent life and the welfare of the elderly or persons with limited capabilities. The most significant concerns of older adults cognitively and physically restrained by caregivers are medical treatments in the event of an accident, which is why technology interventions have the potential to reduce adverse consequences, avoiding the transfer of patients to health centers of high complexity and allowing to prolong the period of life at home. Considering the needs of each individual whom are specific and therefore the appropriate technological interventions must be made, by redesigning the environmental conditions of life at home and providing the independence necessary for these individuals. Based on work by Barlow [4], the content of telecare monitoring includes staff, technology assistance and electronic interaction between information and communication, this creates a challenge at the time of designing a system that meets the characteristics of flexibility and modularity that can adapt to the routines and habits of adults.

According to the World Health Organization "it is expected that in developing countries the number of older adults who have lost their autonomy will have quadrupled by 2050. There are many elderly people who cannot live alone because they have difficulty moving or have poor physical or mental health" [5]. Globally about 25% of people over the age 60 are diagnosed with some form of dementia [6]. In Colombia, Pradilla, G., Vega, B., Phidias, E. (2003), Neuro Epidemiological study that included the cities of Bogota, Cali, Medellin, Barranquilla and Bucaramanga, found the following prevalence per 1,000 population: dementias 13.1% and 4.7% of Parkinson disease. Among these dementias is a mild cognitive impairment, defined as greater cognitive loss (attention, memory, language and executive functions), expected for its age and educational level, which does not significantly interfere with daily life activities [8]. It is an entity with a high prevalence in the population, which requires attention and care, involving patients, families and many groups of professionals, where diagnosis and treatment can be complex especially in the early stages. Therefore, the integration of new technologies in the area of health, where homes should be equipped with technology to monitor the environment and the individual's condition in real time and improve the safety of home and the quality of life of the user.

The objective of this paper is to present the design of a telecare device to monitor people over 60 years with mild cognitive impairment, using the QFD (*Quality Function Development*) methodology. The system consists of sensors and transducers located in strategic places of the individual's home, which allow continuous monitoring of the environment and the user, providing adequate assistance should an unusual situation occur, and allows a study of habits at the same time. These alerts, the moment when issued will be sent to a central module and a module hospital, where information is stored and appropriate measures be taken to provide assistance, in addition to sending an alert to the person in charge of patient care.

The Telecare system should be treated as an integration of electronic devices and network architecture of internal and external communication, which allows the management of information in a comprehensive way avoiding data loss and providing continuous monitoring. In Figure 1 you can see the interconnection between devices, the user and the central patient device.

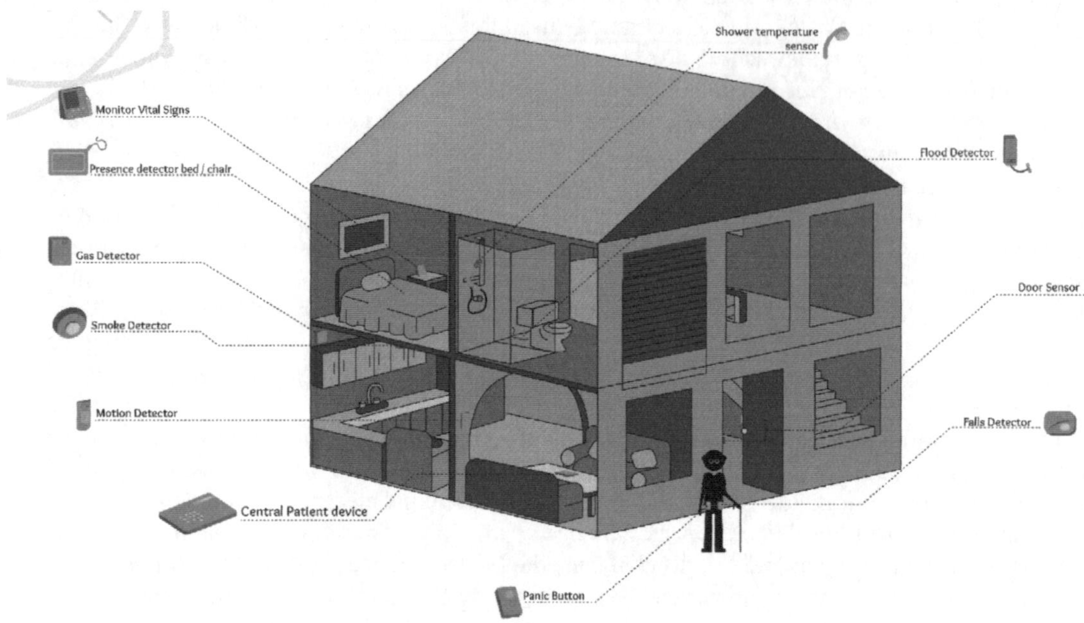

Figure 1. Devices in patient's home

The central patient device manages information generated by peripheral devices and sensors that are in the environment, which are:

- **Gas sensor:** Detects dangerous levels of gases, should the person be distracted or with memory loss a person can leave an open a gas tap causing an accident.
- **Door Sensor:** It is designed to monitor the safety of people with cognitive problems that often leave home at inappropriate times. The device is programmed to expect a sequence of events before generating an alarm.
- **Shower temperature sensor:** Detects water temperature, generating a visible alarm to the person in case the shower temperature is excessively high. Usually people with cognitive losses, as a consequence, develop a type of sensory deprivation syndrome [9] , where they lose the perception of heat or cold to the skin.
- **Smoke / Fire Detector:** Detects potentially harmful smoke particles.
- **Flood Detector:** Detects water levels caused by leaving the water open, broken pipes or electrical appliances.
- **Presence detector bed / chair:** Detects when a person stands or sits on the bed or chair, knowing their habits, generating an alarms in case the person does not get up or lie down for an unscheduled hour.
- **Motion detector:** Used to monitor the activity and inactivity of the user in the home, combined with other sensors such as the fall detector and presence detector in bed / chair can minimize unwanted effect of an event such as a fall, thus reducing the time of arrival of aid and treatment.

Similarly, in the system are monitoring devices for patients, among them are:

- **Falls Detector:** Older adults who have cognitive impairment have an increased risk of falls associated with motor impairment, visual impairment, low bone mineral density, fragility among others, so it is important to monitor users and when the system detects a fall, sends an alert to the monitoring center and the caregiver.
- **Panic Button:** Generates a warning message when the user requires or is in an emergency situation
- **Monitor Vital Signs:** It has special sensors to monitor physiological variables, such as body temperature, heart rate, oxygen saturation and blood pressure. This monitoring is done depending on the requirements of each patient and sends an alert when any of these parameters change.

Some of the mentioned devices are being designed and implemented in full as will be described later, while others are being adapted to the system considering that with commercial devices, the manufacturing cost decreases significantly.

For the connection between the devices that make up the remote assistance system, the different communication protocols should be take into account. The first is between peripheral devices and the Central patient device which will be made by radio frequency, based on the modulation technique Gaussian Frequency Shift (GFSK) [10], which is a modulation method for digital communication coding data as a series of changes in a carrier signal frequency. To code a signal by means of this technique, the Enhanced ShockBurst, a baseband protocol is implemented, which allows automatic packet data bidirectionally to transfer between transceivers, the main configured as a receiver (PRX) and others are configured as transmitters (PTX). The communication starts when the PTX sends a data packet and ends when it receives an acknowledgment packet (ACK) from the PRX. If PTX does not receive the acknowledgment packet the original data packet is retransmitted after a set time interval and the PTX is put in receive mode waiting for ACK, preventing data loss when an event alarm is detected.

The second communication protocol, is the SMPP (Short Message Peer to Peer), which is a standard protocol designed to provide a flexible interface for data communication between a central message and a central messaging server [11]. This protocol supports GSM technology (Global System for Mobile).

2. QFD methodology

The QFD methodology (deployment of the quality function) was used to design the prototype patient telecare system module and consists of the systematization of product requirements and defines the technical and operational characteristics to satisfy those needs. The following describes each of the stages.

Stage I: Clarify and define the task
Device specs:

- Maintain two-way communication between peripheral devices and the Central patient device, and between the latter and the registered external modules, central module and module hospital.
- Send a text message when an alarm event occurs.
- user-friendly interface for the user.
- Continuous feeding (24 hours a day / 7 days a week).

Common questions about device performance were:

- How often the device will be used?

- Daily (24 hours a day / 7 days a week)
- Would it be mobile device?
- Environmental peripheral devices and the Central patient device would be fixed in certain places of abode and peripherals; patient would be mobile.
- What kind of power supply does it have?
- The Central patient device will be supplied with 110V power supply with battery backup and peripheral devices will be battery-powered.
- How will the information obtained from peripheral devices be shipped?
- Radiofrequency.
- How will the information of an alarm event be sent?
- GSM network, via text messages.

The device must be versatile, easy to use for users, reliable and easy to configure for the different needs of each individual, in addition to real communication at present time and an alarm event. As a device often used, it must have a system that allows the user to know the state of the batteries of peripheral devices to prevent data loss.

The characteristics of "WHAT´S" featured devices are described as.

Those “WHAT’S":

- Establish two types of two-way communication in real time.
- The user peripheral devices must be small and light.
- Protect the user, and orient it for management.

Stage II: Determine the functions and structures

At this stage the possible solutions or mechanisms to help meet requirements are discussed:

- Allow easy connection between peripheral devices and the Central patient device.
- Optimally manufactured components to reduce size and weight.
- Transmission of data in real time between Central patient device and the core modules and hospital.
- Pleasing design.
- Marks on the connection points to guide the proper use of the device.

Taking the above information to obtain the "HOW’S" or characteristics this device should have in the end:

- Peripheral Devices lightweight and portable patient.
- Communication between peripheral devices and the Central patient device is made by the RF transceiver module.
- Communication between Central patient device and the central module and hospital will be made through a GSM Shield, which allows the sending and receiving text messages.
- constant measurement of the variables is done by specific sensors for every need.

Stage III: Looking for the principles of solution and its variants
At this stage the solution to the telecare system characteristics are shown.

Table 1. Solution module features

No.	Characteristics	How to do it
1	Light	Reduction of components and protective polymer surface
2	RF Communication	Module NRF24L01
3	Sending alarm messages	Shield GSM / GPRS M95
4	Constant medical measuring of physical variables and the detection of changes in physical phenomenon	Arduino Mini, specific sensor or detector for each event

Stage IV: Split into realizable modules

Depending on the area the appropriate solution from the point of view of the user considering the following specified:

- Operation (A)
- Robustness (B)
- Cost (C)
- Manufacturing (D)
- Maintenance (E)
- Assembly (F)

This will take into account the product life cycle. The following table shows only the importance of each product feature, with each of the above parameters, and thus it determines what should be focused on when developing the device. The importance is given a rating on a scale of 1 to 10, with 1 being the least and 10 being the most important.

Table 2. Relevance of features

No.	Feature	A	B	C	D	E	F
1	Light	10	8	9	8	5	10
2	RF Communication	10	10	8	4	3	6
3	Sending messages	10	5	8	4	3	6
4	Detection and measurement variables	10	10	8	7	9	10
		10,00	8,25	8,25	5,75	5,00	8,00

As shown in Table 2, the most important features are operating, robustness, cost and assembly. Here these are the most influential for the development of the device.

3. Results and discussion.

In Figure 2, the block diagram of the overall system Telecare, where the RF link between peripheral devices and the Central patient device, which in turn sends a text message by GSM communication to the central module and hospital when an alarm event is generated.

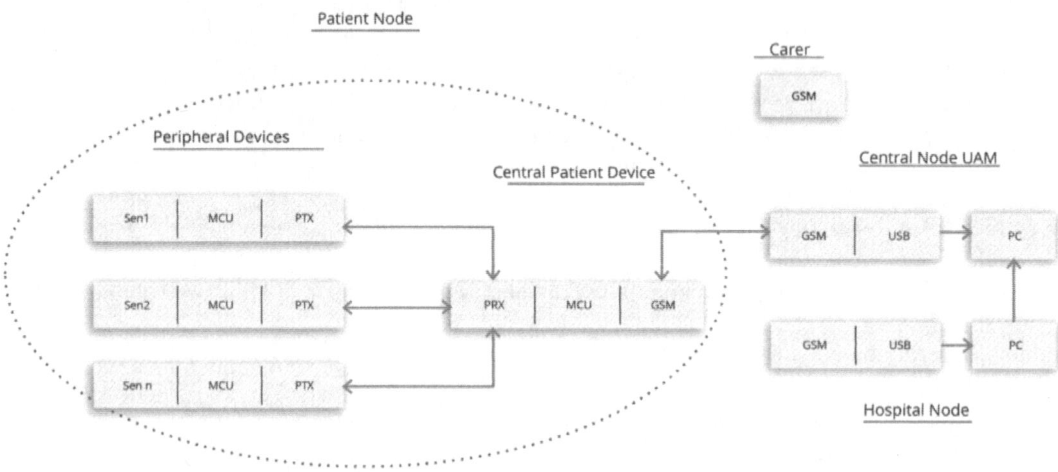

Figure 2. Telecare system – General diagram

The central patient device has both hardware and software components, which must meet the above design features.

Hardware:

- **Central patient device:** its main function is to receive the information from the sensors by nRF24L01 transceiver configured as PRX, so that it can check the communication between peripheral devices and avoid collisions or loss of data. Upon the arrival of a Patient Alert module it sends a text message via the Shield GSM / GPRS to external modules M95.
- **Communication external modules (Central node / Hospital node):** receiving alarm messages generated by the Central patient device is made by using the Shield GSM / GPRS M95, where it is stored in a database to keep track of events.
- **Peripheral Devices:** as mentioned, some of these peripheral devices are completely designed and the architecture design is very similar for all of them, ie each device must have a radio frequency module for communication, microcontroller and specific sensor or detector for each device. Among these are:
- **Falls Detector:** for the design and implementation of one IMU (Inertial Measurement Unit) of nine degrees of freedom, where an algorithm is implemented to detect falls and differentiate each of the activities of daily life in which are used: standing, siting, lying down, picking up an object and walking. This device will be placed on the patient's waist and have a button that will serve as a panic button.

- **Presence detector bed / chair:** This detector requires, the RF module and the microcontroller, and a device capable of detecting user presence in bed or chair, using a piezoresistive materials, which has the characteristic of changing its resistance when applying a force.

- **Shower temperature sensor:** 10k NTC thermistor is used to monitor the temperature provided by the shower, where temperature ranges are established for which notice is given to the user visually, should the increase in water temperature rise. In figure 3, one can observe the temperature ranges set.

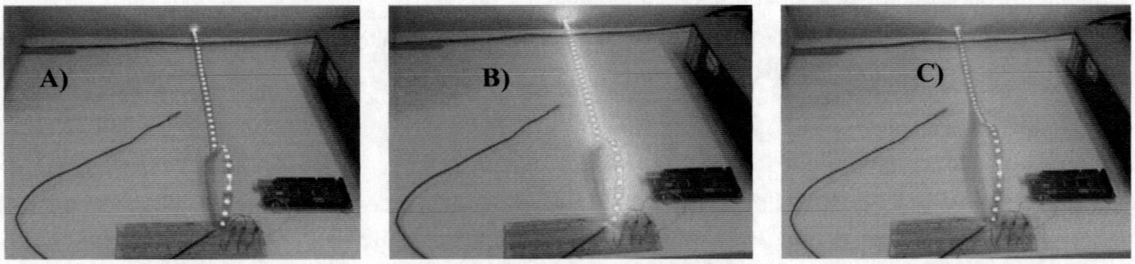

Figure 3. Temperature range

A) Temperature: ≤ 17°C. **B) 17°C < Temperature: ≤ 40°C** **C) Temperature: > 40°C**

- **Vital Signs Monitor:** for this device a pulse sensor was used, where the variables of heart rate and oxygen saturation are obtained. Taking the values of the graph by mathematical calculations the value of pulse pressure P. Equation (1) is obtained. Where Pd is the diastolic pressure and Ps is the systolic pressure.

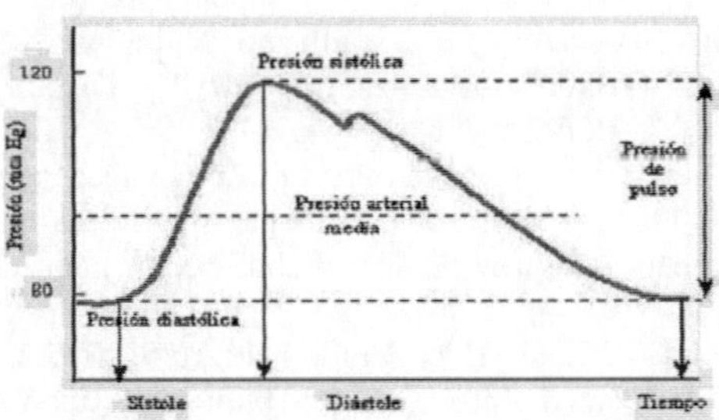

$$P = \int Pd + \frac{1}{3}(Ps - Pd) \qquad (1)$$

Figure 4. identification of parameters in the wave of heart rate. [12]

Besides this, the vital signs monitor also has an implemented body temperature sensor integrated with LM35, this generates the emergency message when finding the value of the temperature outside established ranges. (Hypothermia: Less than 36 ° C, average: 36.1 ° C - 37.2 ° C, hyperthermia: Higher than 37.2 ° C)

Software

To program the various peripheral devices including tele-assistance system, it was decided to use the software and development boards Arduino Mini Arduino Nano, with ATMega328P microcontroller, which have their own programming language, because of the benefits it offers, to be able to develop a modular, simpler, clearer and with fewer lines of program code.

Central patient device architecture of the code is more complex so the used programming language is C language, using a microcontroller Freescale FRDM - KL25Z, since the features of this algorithm are essential, because values are shown the moment that an alarm is generated, and sends an SMS alert to the hospital and central module and also allows the voice channel to activate if necessary as well as storing history records.

4. Conclusions

The implementation of a prototype module patient with peripheral devices, allowing the status of the patient and their environment, to be notified in any alarming situation to the central module located at the Universidad Autónoma de Manizales and the nearest hospital to the patient's home. The implementation of these devices are so they can perform continuous monitoring to individuals who have been diagnosed with cognitive impairment, such that they are not apart from their home, in many cases reducing travel costs for both the patient and medical staff.

This prototype module is versatile because each person is an individual case and the module must be adjusted to the needs of each individual.

5 Acknowledgements

The authors express their thanks to: Colciencias support, Young Investigator through call 645 of 2014; Furthermore, the design of this monitoring device is derived from the project “Implementación del programa para diagnóstico y control de enfermedades crónicas no transmisibles y cáncer de cérvix y mama, con el apoyo de TIC en el departamento de Caldas” financed by the department of Caldas (Sistema General de Regalías - SGR) and executed by the Universidad Autónoma de Manizales (UAM) and the Universidad de Caldas.

6. References

[1] A glossary of terms for community health care and services for older persons. Recuperado: Abril 30, 2015. Disponible en: http://www.who.int/kobe_centre/ageing/ahp_vol5_glossary.pdf
[2] WHO (World Health Organization). Telemedicine. Opportunities and developments in member states. Report on the second global survey on eHealth. Global Observatory for eHealth series. Volume 2. ISBN 978 92 4 156414 4 ISSN 2220-5462© World Health Organization 2010. Disponible en: http://www.who.int/goe/publications/goe_telemedicine_2010.pdf
[3] Miriam Atienza Mañas, C. C. (2014). In Paraninfo, Teleasistencia (Primera ed., p. 2). España: Paraninfo.
[4] Barlow, J., Curry, R., Wardle, D., Bayer, S., & Trejo Tinoco, M. (2004). Implementing telecare: Strategic analysis and guidelines for policy makers, commissioners and providers. In Public sector national report . London: Audit Commission.
[5] Organización Mundial de la Salud. (n.d.). Día Mundial de la Salud. Recuperado Febrero 6, 2015. Disponible en: Día Mundial de la Salud 2012 - ¿Está usted preparado?: http://www.who.int/world-healthday/2012/toolkit/background/es/index2.html
[6] Carbonell, C. (2000). La depresión en el anciano. Actas Españolas de Psiquiatría.
[7] Pradilla A. Gustavo, Vega A. Boris E., León-Sarmiento Fidias E. (2003). Estudio neuroepidemiológico nacional (EPINEURO) colombiano. Rev Panam Salud Pública
[8] Gauthier, S., Reisberg, B., Zaudig, M., Petersen, R. C., Ritchie, K., Broich, K., & Winblad, B. (2006). Mild cognitive impairment. The Lancet, 367(9518), 1262-1270.
[9] Guía practica clínica. Recuperado Mayo 7 de 2015. Disponible en: http://www.cenetec.salud.gob.mx/descargas/gpc/CatalogoMaestro/144_GPC_DEMENCIA_AM/Imss_144_08_grr_demencia_am.pdf
[10] Implementation of Digital Signal Processing: Some Background on GFSK Modulation. Recuperado Junio 6. Disponible en: http://wwwhome.ewi.utwente.nl/~gerezsh/sendfile/sendfile.php/gfsk-intro.pdf?sendfile=gfsk-intro.pdf
[11] Short Message Peer to Peer Protocol Specification v3.4, Document Version: 12 de Octubre de 1999, Issue 1.2.
[12] Imagen recuperada el 27 de Febrero de 2015. Disponible en: http://ocw.unican.es/ciencias-de-la-salud/fisiologia-humana-2011-g367/material-de-clase/bloque-tematico-1.-fisiologia-del-aparato/tema-6.-circulacion-arterial.-presion-arterial/tema-6.-circulacion-arterial.-presion-arterial

Evaluation of a Framework to Implement Electronic Health Record Systems Based on the openEHR Standard

Diego A. Orellana[1], Alberto A. Salas[1], Pablo F. Solarz[1], Luis Medina Ruiz[1] and Viviana I. Rotger[2]

[1] Instituto de Bioelectrónica, Fac. de Medicina, Universidad Nacional de Tucumán, Argentina
[2] Departamento de Bioingeniería, Fac. de Cs Exactas y Tecnología, Universidad Nacional de Tucumán, Argentina

E-mail: pablosolarz@gmail.com

Abstract. The production of clinical information about each patient is constantly increasing, and it is noteworthy that the information is created in different formats and at diverse points of care, resulting in fragmented, incomplete, inaccurate and isolated, health information. The use of health information technology has been promoted as having a decisive impact to improve the efficiency, cost-effectiveness, quality and safety of medical care delivery. However in developing countries the utilization of health information technology is insufficient and lacking of standards among other situations. In the present work we evaluate the framework EHRGen, based on the openEHR standard, as mean to reach generation and availability of patient centred information. The framework has been evaluated through the provided tools for final users, that is, without intervention of computer experts. It makes easier to adopt the openEHR ideas and provides an open source basis with a set of services, although some limitations in its current state conspire against interoperability and usability. However, despite the described limitations respect to usability and semantic interoperability, EHRGen is, at least regionally, a considerable step toward EHR adoption and interoperability, so that it should be supported from academic and administrative institutions.

1. Introduction

The production of clinical information about each patient, coming from complementary studies, reports, derivations, admissions, diagnostics, symptoms, etc. is constantly increasing, and it is predictable the advance of such situation because science and technology contributions on healthcare. It is noteworthy that the information is created in different formats and at diverse points of care.

The use of health information technology (HIT) has been promoted as having a decisive impact to improve the efficiency, cost-effectiveness, quality and safety of medical care delivery, by making wider and faster access to information, enhancing decision support systems, making easier access to follow care with specialists, more reliable prescribing, etc.

Particularly, researchers have examined the benefits of EHRs (Electronic Health Record Systems) by considering clinical, organizational, and societal outcomes which include improvements in quality of care, reduction in medical errors, and other improvements in patient-level measures that describe the appropriateness of care [1]. However in developing countries the utilization of HIT is insufficient and lacking of standards, among other situations, having as consequences that health information is fragmented, incomplete, inaccurate, and isolated. This leads to information silos, so that the information contained inside them cannot be used for patient care or data analysis [2]

There are some sparse efforts in Latin America (LA) to promote and to adopt standards improving conditions for an increasing adoption of e-health solutions and for interoperability as in Brazil and Uruguay from public health authorities, Chile and Argentina through private and public actors [3]. The efforts are scarcer when solutions are patient centered. Solutions based on standards such as RIM HL7 [4] and openEHR [5] are in that way, but the construction of service layers are complex and expensive, and it is necessary highly trained human resources. For such service layer we are looking for open source solutions based on standards. In this sense, the framework EHRGen [6] it is perhaps a considerable advance and promised to go in the right direction.

In the present work we evaluate the framework EHRGen, based on the openEHR standard, and having in mind its use in the public health sector, as mean to reach generation and availability of patient centered information. The evaluation refers to the services provided to develop EHR systems.

The remainder of this paper is organized as follows: In section 2, we describe the essential of the openEHR standard. In section 3 we describe the EHRGen framework. In section 4 we present the clinical domains selected to evaluate EHRGen and the evaluation results, and finally, in section 5, we present the conclusions and some proposals.

2. The openEHR standard

Two of the most important problems related to healthcare information systems are evolution and interoperability. Software engineering has addressed the first with modern agile lifecycles, multilayer architectures and abstract designs. However, in highly dynamic context such as healthcare, that approaches are not enough, probably because, the changes in medicine in three aspects: 1) breadth, due to new information is being discovered, 2) depth, finer grained details are discovered and became relevant, 3) complexity, because new relationships are added among concepts [7]. The second problem, interoperability, has been addressed by standardized communication protocols, thesaurus and more recently, reference models of information [4][5][8]. However, communication protocols lack of sufficient semantic richness to handle interrelated information that has to be transmitted as a set, thesaurus only address specific terms and the reference models became very complex when implementing domain concepts, as evidenced by their low use by applications [9].

An alternative to address those problems, particularly evolution and interoperability, is to separate information and knowledge into two levels of model. The former is most familiar to developers - the level of software object models and database schemas, denoted here by the term reference model (RM) - and is used to build information systems. Concepts such as role, address, party, observation, evaluation, etc. are in the RM. It must be small in size, in order to be comprehensible, and contain only non-volatile concepts in order to be maintainable. The second level is the knowledge level (KM), requiring its own formalism(s) and structure, and it is where the numerous, volatile concepts of most domains are expressed [10]. Concepts such as Patient, Medical doctor, Blood Pressure, Lipid studies, obstetric summary, etc. are in the KM.

The standard openEHR implements that two level approach. KM is built on domain concepts such as: patient, hospital, pressure, body weight, etc. all expressed using constraints on instance structures of an underlying RM. Such constraints are expressed in a model called archetype model.

Persistence and instance data are in the RM, and changes in KM (from now Archetype Model – AM-) does not affect the RM. Changes in AM mean create, update or delete concepts made as archetypes. Figure 1 shows RM and AM as layers of the software architecture, which are specified by the standard. It is added a service model layer (SM), which represents the software layer of processing, presentation, configuration tools, etc.

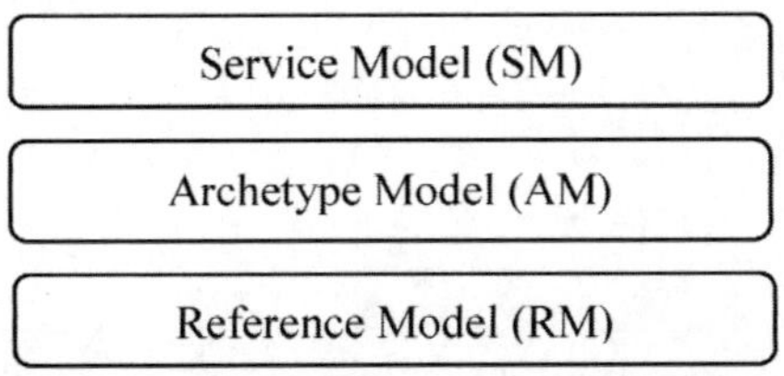

Figure 1. RM and AM are specified by openEHR standard. SM represents the services added by the application.

The figure 2 (took from [11]) shows the relationship among models and user roles. On the left, an expert on information technology creates an information layer by instantiating the RM in a particular implementation platform. On the right, a domain expert creates archetypes and templates. Archetypes are made with references to RM classes. Templates are particular configurations on archetypes, they permits to reuse archetypes for application in different use cases, specialties, etc., by grouping, predefining values, etc. Medical terminologies are added to use controlled vocabularies. Archetype Definition Language (ADL) is the specific language to create archetypes.

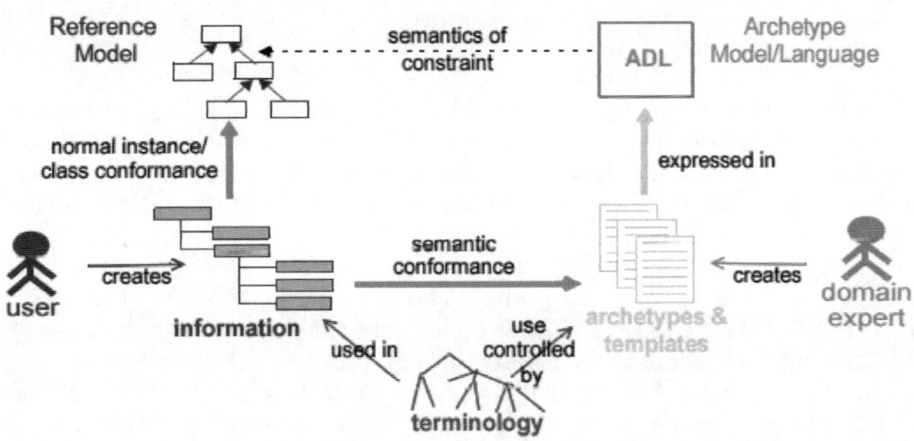

Figure 2. Architecture elements in relation to user roles.

From the functional point of view, the most important parts of the RM, are the demographic RM and the EHR RM. As it is shown in left side of Figure 3 (took from [12]) classes of demographic model are: Roles, Agent, Person, Organization, Address, etc. They are the base for demographic archetypes as Patient, Healthcare Provider, Healthcare consumer, Payer, etc.

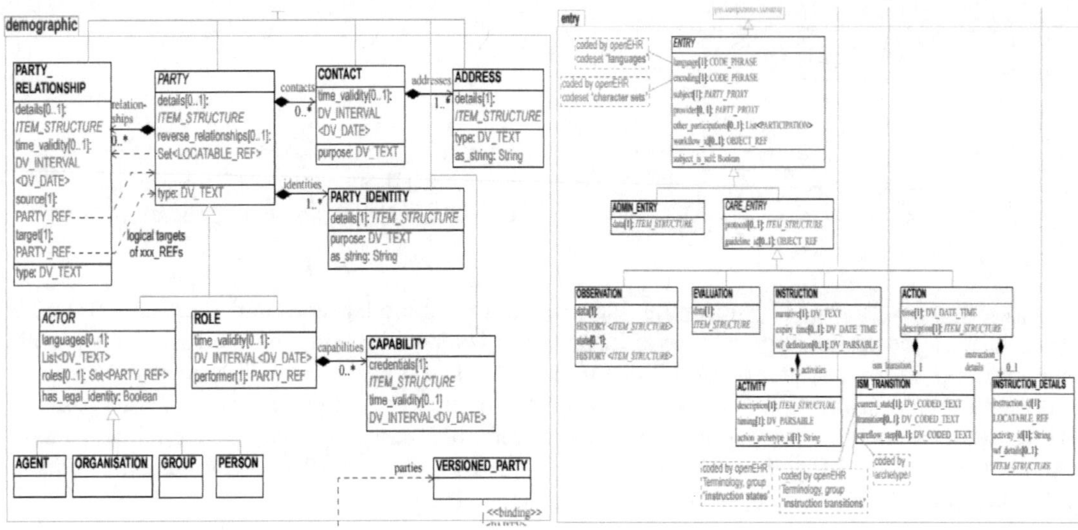

Figure 3. Left: Class diagram of the demographic part of RM. Right: Class diagram of the EHR part of RM.

Figure 3 on right side (took from [13]) shows, in UML (Unified Modeling Language), part of the EHR RM, from the entry package, with the classes Observation, Evaluation, Instruction and Action which are subtypes of an Entry of a clinical record, and corresponds to stages of a hypothetical clinical process. They are the base for the Entry archetypes as Blood Pressure, Lipid studies, obstetric summary, etc.

The concepts in RM: Roles, Agent, Person, Organization, Address, Entry, Observation, Evaluation, Instruction, Action, are made up with abstract attributes: DV_TEXT, DV_DATE, DV_CODED_TEXT, ITEM_STRUCTURE, etc. which are independent of languages and platforms. It is expected they are stable although domain concepts evolve.

Domain concepts, represented as archetypes, are more unstable and follow the dynamic of medical concept as described in section 1. In fig 4 we show, in ADL language, part of the definition of the archetype Temperature, which is an Observation. Some of the attributes of the composing temperature are, as can be seen, ELEMENT and C_DV_QUANTITY. The definition is usually managed from a tool in the service layer with a more amenable interface.

```
ELEMENT[at0004] matches {-- Temperature
        value matches {
          C_DV_QUANTITY <
              property = <[openehr::127]>
               list = <
                      ["1"] = <
                              units = <"°C">
                              >
                      ["2"] = <
                              units = <"°F">
                      precision = <|1|>
```

Figure 4. Part of the definition of an archetype

Interoperability in openEHR must be at RM and AM layers to achieve interoperability both at semantic level (corresponding to AM) and at syntactic level (corresponding to RM). Applications which will exchange information should use a set of agreed archetypes made on a RM basis. To this end there is a repository of archetypes called Clinical Knowledge Manager (CKM) at [14], which is an international, online clinical knowledge resource with a set of reusable archetypes. Templates however do not need to be agreed because they are related to each particular application and they are not the clinical concepts to be exchanged.

3. EHRGen

EHRGen[6] is an open code framework that allows the creation of EHRs oriented by clinical knowledge management which implements the openEHR standard. As such, EHRGen produces an information model, which implements the RM. The KM is implemented with archetypes which can be downloaded from the international CKM, or developed with framework such as founded in [15]. Both archetypes, took from CKM and created *ad hoc,* are made by clinical domain experts and define one part of the Knowledge Base (KB) of the EHRs. The archetypes created may be reused in different contexts. In addition, KB has terminologies which constrain and specifies the clinical language to reach understandability and to contribute to semantic interoperability. Templates are component of the KB too and can be implemented by EHRGen. Templates allow, on one hand to aggregate archetypes according to a particular need, and on the other hand to filter part of archetypes.

The architecture of EHRGen [16] is based on the implementation of the information repository under the RM of openEHR standard and the implementation of the repository for the KB. The components of the EHRGen architecture are shown in the fig 5.

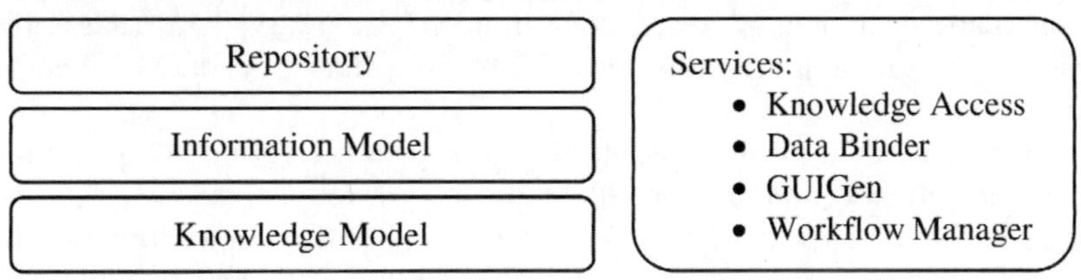

Figure 5. EHRGen architecture.

Knowledge Access (KA) provides access to the KB, Information Model (IM) implements the class model of the openEHR standard which is related to the repository and the documents by means of object-relational mappings, while Data Binder creates IM structures and validates data coming from the user interface. Workflow Manager assists the user in the process to create, update and view patient information. GuiGen allows to automate the interface component creation. There are other components of the EHRGen architecture not considered here because they do not participate in our evaluation.

The software technologies with which EHRGen was made up are: JDK6, Grails 1.37. For the repository the options are MySQL or PostgreSQL.

The methodology to create an electronic health record system, proposed by EHRGen developers [16], consists of the following processes:

1. Base modeling: to create (or to select) the archetypes which will represent the clinical concepts.
2. Templates modeling: to create (or to select) the templates by selecting and composing the archetypes that made up the clinical records according to each particular clinical domain.
3. Creation of workflows: to define user access interfaces and stages according to the different healthcarer profiles.
4. Creation of domains: a domain allows us to manage (and to reuse) a group of templates according to a clinical domain.
5. User interface generation: EHRGen support automatic generation in HTML format, according to templates and the GuiGen components configuration.

Journal of Physics: Conference Series **705** (2016) 012046 doi:10.1088/1742-6596/705/1/012046

6. The production or use of the software system: that is, creation, updating and visualization of clinical records.

EHRGen does not provide tools for the first process. For creating and editing archetypes we can use the Archetype Editor de Ocean Informatics [15]. The second, third, fourth y sixth processes are carried out with the EHRGen framework and are described and evaluated in the next section. The fifth process is fulfilled practically without user intervention because the framework creates the HTML interfaces with preconfigured components and the templates definitions.

4. Evaluation of EHRGen

In this section we show an evaluation of the framework EHRGen described previously. The evaluation refers to the process of creation of EHR Systems. To that end we have considered two clinical domains as perspectives: 1) A form to track diabetic patients elaborated by the public healthcare authority of Argentina, 2) An emergency medical history from a public hospital from the province of Tucumán, Argentina.

The first domain contains data such as: Physical Examination (Weight, Height, Head Circumference, Abdominal girth, feet exam, cardiac frequency, respiratory frequency), Laboratory test results (blood glucose, cholesterol, HDL, LDL, creatinine, etc.), Education to patient (states of the education such as: taught, assimilated, accomplished, etc.), Current Medication, Smoking habits, Explorations (ECG, dental exam, eyes exam, etc.), Prescribed medication, etc.

The second domain has the following data: chief compliant, personal and family history, physical exam, patient values (cardiac frequency, respiratory frequency, axillary temperature, blood pressure, laboratory test results, oxygen saturation, Glasgow scale), fulfilled procedures, consultation request, prescribed medication ,etc.

Both domains share some data but have some different data too. Thus, we pretend to create two perspectives on a unique patient centered database.

The elicitation tasks were relative to the objective of carry the paper forms to computer user interface, thus there were considered only issues about term interpretations, content of fields, etc., but not respect to changes in format or content. Both clinical forms were considered for respective template designs. Previously we have analyzed the archetypes that should be to composite the templates.

By following the processes indicated in the previous section, we begin specifying archetypes. For space reasons we synthesize the process with few cases which represents all found situations.

Many of the data components of both forms, diabetes and emergency, there exist in archetypes in the CKM, for example: Physical exam (called openEHR-EHR-OBSERVATION.exam.v1), Laboratory test results (in various archetypes called: openEHR-EHR-OBSERVATION.lab_test-blood_glucose.v1, openEHR-EHR-OBSERVATION.lab_test-blood_gases.v1, openEHR-EHR-OBSERVATION. lab_test-esr.v1, etc.), weight (openEHR-EHR-OBSERVATION.body_weight.v1), temperature (openEHR-EHR-OBSERVATION.body_temperature.v1), Blood pressure (openEHR-EHR-OBSERVATION.blood_pressure.v1), family history (openEHR-EHR-EVALUATION. family_history.v1), Glasgow scale (openEHR-EHR-OBSERVATION.glasgow_coma_scale.v1), procedures (openEHR-EHR-ACTION.procedure.v0), education (openEHR-EHR-ACTION.health_education.v1), etc.

Data not included in CKM archetypes, must be included in new archetypes according to clinical concepts specified by domain experts. There are only few archetypes in such a sense. Two of them:

- Patient history (or patient antecedents): according to domain experts contains only a free text field.
- Patient destination: Describes the immediately death, derivation, inpatient or outpatient condition.

The created archetypes are added to the local KB.

The next step is to design the templates. A template is generated with the interface shown in figure 6. It has an identification, a name and the archetypes that the template must contain.

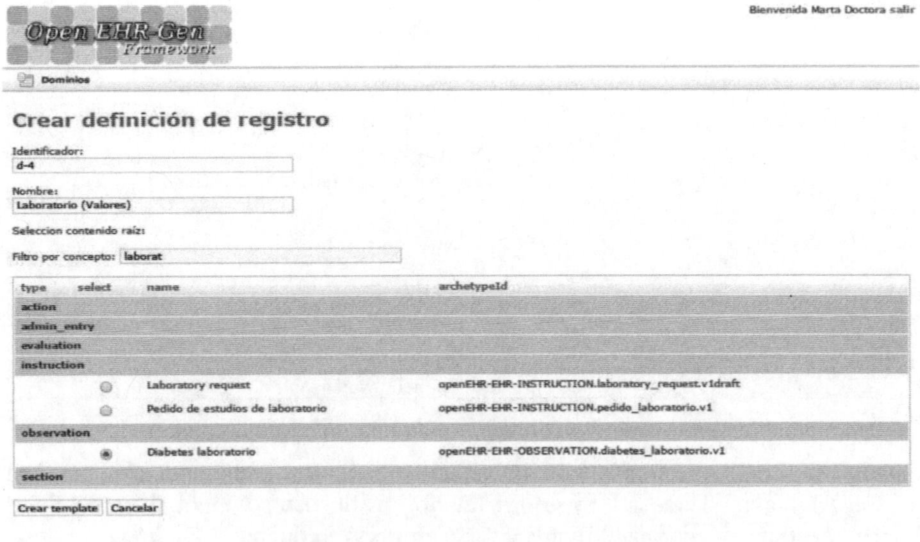

Figure 6. Template creation interface.

As templates are directly related with the user interfaces, their design must into account, besides archetypes, the custom grouping of data in medical history. Some selected groups are: Patient history, Physical exam, Laboratory. The first one corresponds directly with the previously *ad hoc* designed archetype patient_history. Physical exam has an archetype in the CKM, described as "Use to record details about findings on physical examination of the subject of care. This may include a narrative description of the findings, a framework in which to nest detailed CLUSTER examination archetypes, and a clinical interpretation of the findings". Thus, the cluster may contain other archetypes as: blood pressure, weight, etc., then we may to generate a template with such archetype. But it is not possible to use archetypes with cluster as data type, at least until version OpenEHRGen v0.8 beta1

An alternative way is to generate a template with the archetypes: blood pressure, weight, etc., but that is not possible either, because the template creator tool of the framework allows incorporating only one archetype. That is, there not be possible with the framework tools to create templates with more than one archetype.

A third alternative design is to create one template per archetype. In that manner we could to use the CKM archetypes and to develop an interoperable system both at syntactic level because RM, and at semantic level because AM. But in this last case we found two problems related with the usability of the system. To create one template for each archetype means that final user of system will have one screen per archetype, i.e. one screen for blood pressure, one for height, one for weight, one for glucose, etc. transforming the current use of a unique preprinted form in tens of screens. The second problem is related to the fact that definitions of archetypes are the result of consensus from various point of views, thus the use of all parameters in each particular application would be an unnecessary overcharge. In the table 1 it is showed the elements of the blood pressure archetype. They are 14 parameters, some of this are only used in rare applications. In majority of application, especially in emergency, it is only necessary diastolic and systolic values, and other values are unavailable or unnecessary.

Table 1: data components of the CKM blood pressure

Blood Pressure	
Data	Systolic Diastolic Mean Arterial Pressure Pulse Pressure Comment
State	Position Confounding factors Exertion Sleep status Tilt
Protocol	Cuff size Location
Events	Any event 24 hour average

An alternative solution, which was our choice, is to create an archetype with the data elements required for the template. That archetype has the following data: weight, height, blood pressure, abdominal girth, feet exam, cardiac frequency, respiratory frequency. Then we have created the template shown in fig 7, on that archetype basis.

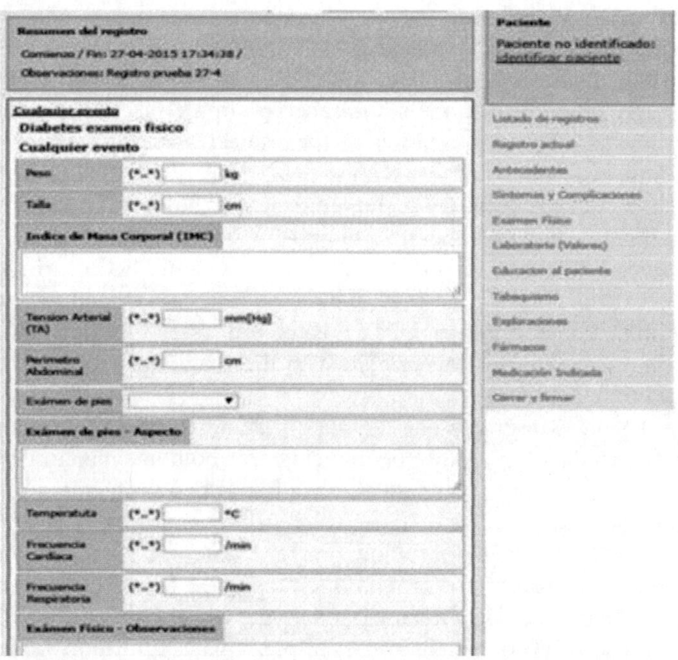

Figure 7. Template for physical exam in diabetes domain.

The methodology, consisting in developing *ad hoc* archetypes, because the limitation of one archetype per template of the templates creation tool, was generalized for both diabetes and emergency domains.

In fig 8, on left, we show the main screen of the diabetes domain, and on right, the emergency domain. In both, the list below "REGISTRO CLÍNICO" (Clinical registry) corresponds to the templates (and archetypes) designed.

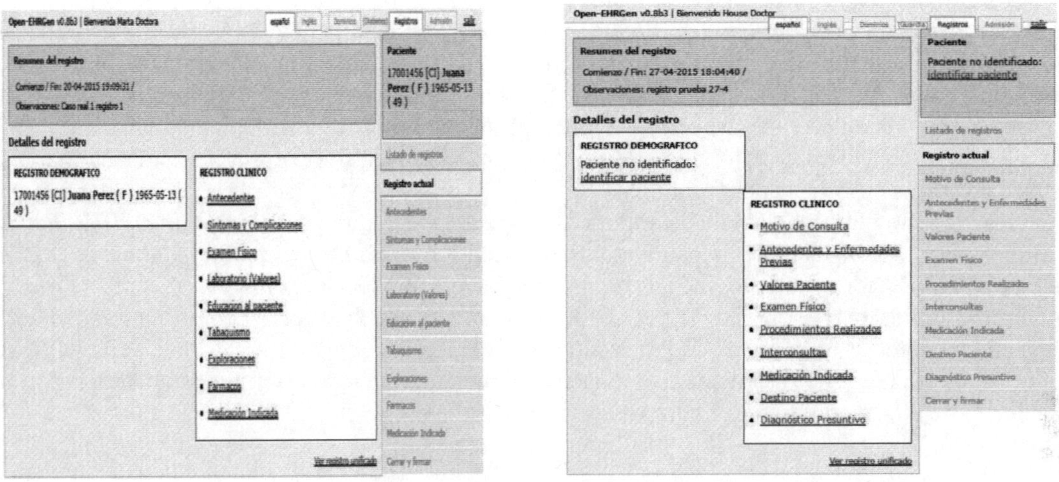

Figure 8. Main screens of domains diabetes (left) and emergency (right).

The diabetes and emergency domains, share the RM, because all archetypes are designed based on its elements. However, although there is much information shared by those domains, and particularly by the designed archetypes and templates, the use of *ad hoc* archetypes entails interoperability problems at knowledge level, because they are not shareable with systems operating with CKM archetypes. Moreover, the archetypes physical_exam_diabetes and physical_exam_emergency are slightly different, for example, the *feet exam* element there exist only in the first and the *Glasgow scale* element there exist only in the last, thus, the data components of shared data are interoperable, but semantic of data are not. For example, there is not defined semantic to interpret which both archetypes contains temperature, because it is not a concept in a knowledge common model; however the values, comments and measurement units are shared in RM.

A different situation occurs with the archetypes Patient and Prescribed Medication, they are the same in both domains, thus although the templates would be different, concepts in both domains can be shared.

5. Conclusions

We have evaluated a framework to generate EHRs. The EHRs generation process includes: software creation artifacts, data repositories and user interfaces. The framework also allows to create the necessary workflows for the development of record editing tasks and it puts in the stage for speaking a tool for electronic medical records based on a standard. The framework was evaluated through the provided tools for final users, that is, without intervention of computer programmers or experimented users about documents with configurable metadata. To evaluate the framework we have used two clinical domains as perspectives: 1) a form to track diabetic patients, 2) an emergency medical history. They have both some common and some different data elements, and they have some concepts involved in both domains which are implemented in CKM, while some are not.

The framework allows solving EHRs generation based on requirements such as: knowledge about the fundamental concepts of the openEHR standard, CKM and local knowledge databases access, technical skill respect to the framework use.

The standard openEHR proclaim the reuse of concepts already implemented. Thus, concepts such as temperature, weight, etc., should be incorporated to the designed grouping of domain concepts, called templates, and should not be built from scratch. But the current state of the framework imposes some limitations over that idea. 1) Archetypes with CLUSTER as datatype are not allowed, 2) templates cannot incorporate more than one archetype and 3) archetypes must be incorporated at all without being possible to filter unnecessary data. Those limitations have made necessary the creation of ad hoc archetypes, even when many components are already in CKM, breaking down semantic interoperability even between our test domains: diabetes and emergency. Moreover, many designed archetypes are no sense as clinical concepts because they are a group of very different concepts sharing particulars use cases, as is the case of laboratory values. An option would have been a lot of user interfaces. However, the reference model remains unmodified, allowing data interchange, although with a considerable effort. In addition, we consider that the current state of automatic user interface creation conspire against usability.

It is clear that despite the described limitations respect to semantic interoperability, EHRGen is, at least regionally, a considerable step toward EHR adoption and interoperability. It makes easier to adopt the openEHR ideas and provide an open source basis with a set of services.

We consider that EHRGen should be turn into a framework which receives contributions from other groups and the interest from public health institutions, having in mind that it is a considerable progress aimed at addressing the problems of dispersion, incompleteness, lack of standardization that characterize health information in Latin America.

References

[1] Nir Menachemi and Taleah H Collum. *Benefits and drawbacks of electronic health record systems.* Risk Manag Healthc Policy. 2011; 4: 47–55. Doi: http://dx.doi.org/10.2147/RMHP.S12985

[2] Daniel Luna, Alfredo Almerares, John Charles Mayan, Fernán González Bernaldo de Quirós, Carlos Otero. *Health Informatics in Developing Countries: Going beyond Pilot Practices to Sustainable Implementations: A Review of the Current Challenges.* Healthc Inform Res. 2014 Jan;20(1):3-10.

[3] Beatriz de Faria Leão. Electronic Health Record Initiatives in South America. 11th International HL7 Interoperability Conference - IHIC2010 May 14 - 15, 2010. RJ, Brasil

[4] Reference Information Model – Health Level 7. RIM HL7. http://www.hl7.org/implement/standards/rim.cfm

[5] openEHR. http://openehr.org/

[6] https://code.google.com/p/open-ehr-gen-framework/

[7] Rector, A. L. Clinical Terminology: Why Is It So Hard? Yearbook of MedicalInformatics 2001.

[8] Unified Medical Language System UMLS. http://www.nlm.nih.gov/research/umls/

[9] Barry Smith and Werner Ceusters. *HL7 RIM: An Incoherent Standard.* Studies in Health Technology and Informatics, vol. 124 (2006), 133–138

[10] Beale T. *Archetypes: Constraint-based Domain Models for Future-proof Information Systems.* Eleventh OOPSLA Workshop on Behavioral Semantics: Serving the Customer (Seattle, Washington, USA, November 4, 2002).

[11] openEHR architecture overview. Release 1.0.1. http://openehr.org/releases/trunk/architecture.

[12] openEHR architecture demographic reference model. Release 1.0.1.http://openehr.org/releases/trunk/architecture/rm/demographic_im.pdf

[13] openEHR architecture electronic health record reference model. Release 1.0.1. http://openehr.org/releases/trunk/architecture/rm/ehr_im.pdf

[14] Clinical Knowledge Manager CKM. http://openehr.org/ckm/#

[15] Ocean Archetype Editor. http://ocean-archetype-editor.software.informer.com/

[16] Pazos P. EHRGen: Generador de Sistemas Normalizados de Historia Clínica Electrónica Basados en openEHR. 41 JAIIO - CAIS 2012 - ISSN: 1853-1881. Pp. 212-228.

Medical Image Processing Server applied to Quality Control of Nuclear Medicine.

C. Vergara[1], J.P. Graffigna[1], E. Marino[2], S. Omati[2], P. Holleywell[1]
[1] Gabinete de Tecnología Médica - Facultad de Ingeniería - Univercidad Nacional de San Juan, Argentina
[2] Fundación Escuela de Medicina Nuclear – Mendoza - Argentina
E-mail: carlosvergara1786@gmail.com, jgraffig@gateme.unsj.edu.ar, eam.marino@gmail.com, somati@fuesmen.edu.ar

Abstract. This paper is framed within the area of medical image processing and aims to present the process of installation, configuration and implementation of a processing server of medical images (MIPS) in the Fundación Escuela de Medicina Nuclear located in Mendoza, Argentina (FUESMEN). It has been developed in the Gabinete de Tecnología Médica(GA.TE.ME), Facultad de Ingeniería-Universidad Nacional de San Juan. MIPS is a software that using the DICOM standard, can receive medical imaging studies of different modalities or viewing stations, then it executes algorithms and finally returns the results to other devices. To achieve the objectives previously mentioned, preliminary tests were conducted in the laboratory. More over, tools were remotely installed in clinical enviroment. The appropiate protocols for setting up and using them in different services were established once defined those suitable algorithms. Finally, it´s important to focus on the implementation and training that is provided in FUESMEN, using nuclear medicine quality control processes. Results on implementation are exposed in this work.

1. Introduction.

This work is part of the area of processing and management of medical images. The objective is to present the implementation in a clinical environment of a Medical Image Processing Server (MIPS) developed by the investigation group of GATEME [1], that can receive medical imaging studies from different computers, perform processing algorithms developed and return the image result to other devices.

Medical image processing and analysis involves improving segmentation, quantification, registration, visualization and management. To do this, different algorithms are applied, which produce diagnostic information manually, semiautomatic or automatic. Although algorithms are important for quantifying measures, full analysis of the images is associated with a suitable visualization system and physician experience. The growth of the graphics hardware and specific visualization software have allowed from the virtual navigation to manipulation and calculation of parameters.

The management of medical images is important for the application of clinical algorithms. There is a strong demand for this area that involves communication, storage and image compression techniques. This allows storage and sends medical imaging at the time and place required. Hundreds of Mb data volumes involving these studies require appropriate management within clinical environments. For this Picture Archiving and Communication Systems (PACS) provide connectivity between different resources of a health center [2]. These systems operate on the international standard DICOM (Digital Image and Communications in Medicine) [3].

The PACs main components are:

PACS Central Server: the main system comprises hardware and it is responsible for coordinating tasks associated with storage systems, studies communication between different components, the connection with other information systems (HIS-Hospital Information System and RIS - Radiological Information System), among others.

Viewer Workstation: visualization and analysis tools applied to digital images. These stations can be used for diagnosis or review. Hardware features are realted with these tasks.

Storage Server: this server is the essential element of PACS and its function is the storage, query and retrieval of all studies, generated by different modalities. This device has great capacity of storage, a high speed network interface and specific software (databases, etc.). It must provide all the DICOM communication with different imaging modalities (CT, MRI, PET, etc.), other PACS servers and DICOM workstations. The tasks are associated with on line storage, backup and archive.

Printing System: The printing systems allow viewing image on paper or film. It is important in diagnostic center that provide services to other health institutions, however there is a tendency to record in digital media (DVDs, etc.). Printing systems consist of film printers or paper printers with high resolution and big size. They are connected to the network via DICOM applications which allow different configurations of the image.

MIPS is a system for medical image processing that meets specifications of DICOM standard. It has following functions: receiving studies, processing algorithms and sending results. Operation details are described below.

This work links the development of processing algorithms created in GATEME research group and its clinical application in a real environment. This makes possible to transfer these tools, allowing improve quality of diagnosis and/or treatment. Specifically, we search to evaluate performance of MIPSMIPS, validation of the developed algorithms and synergies between research and development areas and clinical environments with high demands.

This publication describes installation process of MIPSMIPS using algorithms on FUESMEN PACS. While there have already evaluated all the tools separately, this work examines the operation of the MIPSMIPS on a high performance hardware, adapting to clinical needs and real-time setting. In the following sections materials used, methods and tests for the implementation and the results obtained are described.

2. Materials.

Institutional PACS with all component and interconnections are seen in Figure 1. Each of these blocks, including the MIPS, are made up of hardware and software (Table 1), which are described in this section. Some elements were already operating in FUESMEN and others have been installed specifically for this project.

2.1. Storage Server/Cluster.

The Cluster is a group of computers linked together by a high-speed network that behaves as if it were a single computer. FUESMEN's Cluster is composed of 3 machines in which two PACS Storage Servers are virtualized. One is the primary server, called Server 1, which has the following characteristics: 4 core 2GHz, 4GB of RAM and 50TB NAS for storage. A NAS (Network Attached Storage) is a server whose function is to store files. This Server recives all images of the different modalities and it is where all workstations make queries.

The second virtualized server (Server 2) has 2 cores 2GHz, 2GB RAM and 512GB of storage. This server receives studies multislice CT and MRI of Server 1 and redirects these studies to different workstations depending on DICOM characteristics.

Moreover, there is a third Storage Server (Server 3) that is not virtualized within the Cluster and has the following features: Intel Core i3 3.60GHz 64 bit x 4, 8 GB of RAM and a RAID 5 storage unit of 28TB (built by software). This computer performs the Backup primary server (Server 1). RAID (Redundant Array of Inexpensive Disks) is an array of independent disks that are connected to form a single logical storage unit.

The operating system used by all servers is Debian 7 Wheezy. The servers are connected in a LAN at speeds of 1Gbps.

Table 1. FUESMEN's PACS.

FUESMEN PACS		
HADWARE	**COMPONENTS**	**DESCRIPTION**
CLUSTER	SERVER 1	DCM4CHEE, MySQL
	SERVER 2	DCM4CHEE, MySQL
	MIPS	XAMPP, PhP, APACHE, PhPMyAdmin, MySQL,MIPS software
SERVER	SERVER 3	DCM4CHEE, MySQL
WORKSTATION	VW	K-PACS, eFilm, Osirix and Other
MODALITIES	MR,CT,US,MN	Manufactures Software

2.1.1. DCM4CHEE

This software is an image manager within the PACS that meet IHE and DICOM standard [4]. Its main function is to provide a storage service (STORE) and retrieving medical images stored in DICOM format.

This tool has been developed as an open source project distributed under GPL and can store all kinds of medical studies.

DCM4CHEE also provides search and retrieval of stored studies (Query / Retrieve) and it allows applying schemes and access to medical images using multiple AE (Application Entities).

For the control and operation, MySQL, Apache and PHP are used. These tools are described below.

2.1.2. APACHE, PHP ,PHPMYADMIN

This set of applications allows to manage the database DCM4CHEE and MIPS. Apache is an open source web server for development of web pages and services [5]. This is a program designed for hypertext transfer data such as web pages with all elements (text, widgets, banners, etc). These servers use the HTTP protocol and are installed on a computer that has Internet access. The same is combined with PHP [6] (HypertextPreprocessor) which is a programming language commonly used embedded in HTML and is available under GPL licence. With this language it is possible to generate dynamic web pages with access to other resources, such as databases.

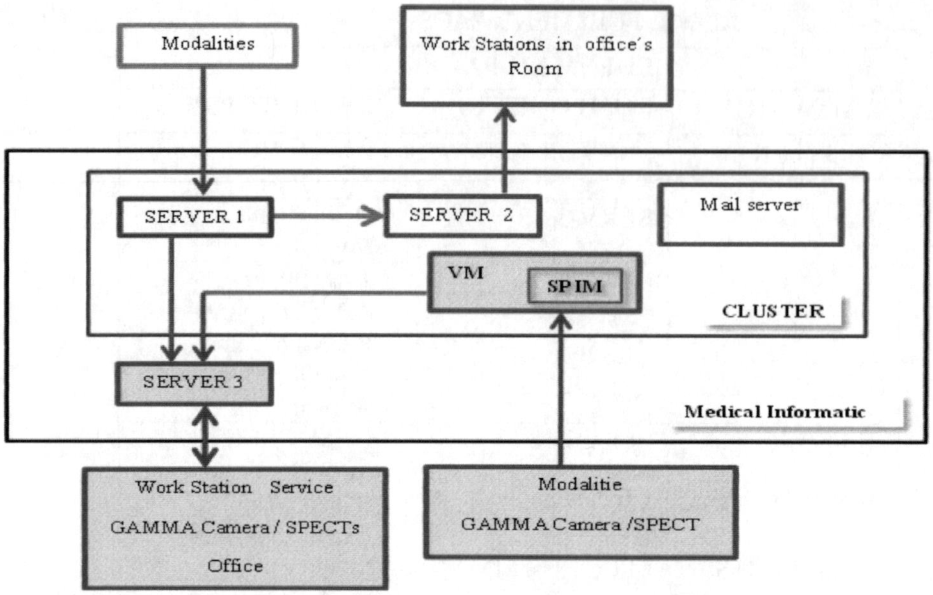

Figure 1. General Scheme of implementation developed.

For configuration data, a popular tool, called phpMyAdmin, is used formanagement of MySQL database; this is used by MIPS and DCM4CHEE.

MySQL [7] is a relational database management system (RDB). A RDB stores data in separated tables according to a model resulting of analysis and design system. The tables are linked to related data combined in a consultation process. It uses queries sublanguage SQL, using a GPL(GNU General Public License) license and operates as a server in a network environment.

2.2. Viewing Workstation(VW)

These devices are composed by powerful workstation with specific software. These viewing applications operate with centralized facilities for the reception and transmission of medical images. They are DICOM compliant and interact with the rest of PACS platforms. Most important requirements are: interfaces for high-speed video (using GPU), high volume RAM, high-speed microprocessor and multiple cores.

The VW can view and manipulate images, supporting a variety of formats (DICOM, .png, .jpg, etc.) of different image systems. They have lots of post processing operations: filtering, view manipulation, 3D reconstruction, window settings, etc. They can create local databases with the received studies and export the images in different formats. Within the FUESMEN, the VW use the following applications: EFilm, KPACS, Osirix and other proprietary applications provided by manufacturers of medical equipment.

2.3. Medical Image Processing Server (MIPS)

This server is based on connectivity specifications of DICOM standard and has the ability to manage and execute various image processing applications. These process are executing autonomously using a make-decision system. The MIPS uses algorithms developed in orther to attend specific clinical or operative problems.

After acquisition of a study by any modality, it is possible to send images to MIPS for processing before any radiologist can see images in viewing workstation. Thus it is possible use idle time to perform tasks of any complexity.

The MIPS has the ability to serve multiple users simultaneously, therefore, it uses threads of execution to perform each operation. Thus, any job (request for association, receiving, shipping or processing) uses a separate thread. Threads promote overall stability of the system because if a process fails, this causes disruption of ythis job but not the rest of the system. Descrption of its operation can be seen in Figure 2, obtained from DICOM Conformance Statement of MIPS, indicating compatibility with standard.

The MIPS can meet different kinds of processing orders, that because the system is configurable by user from a MySQL database [7], which it has registered all processes and their configurations. Thus MIPS defines a TCP listening port for each processing operations declared and remains waiting for new orders. When a user requires a given process, simply he sends a study, initiating DICOM association to server. Order is send to IP address, using a port corresponding to chosen process.

The MIPS is installed in a virtual machine on Cluster FUESMEN and it has following features: 2 processors KVM Kommon of 2.4GHz, 8GB RAM, 64-bit OS, a disc of 199Gb. The virtual machine is operated remotely from the GATEME (San Juan) and FUESMEN (Mendoza).

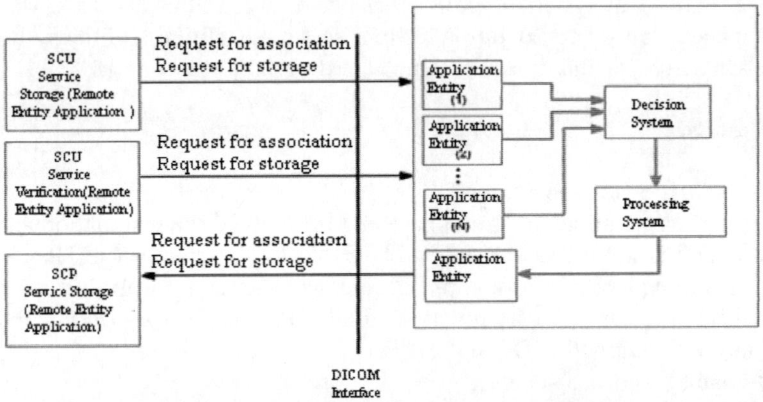

Figure 2: Diagram of Operation MIPS.

2.4. Processes

GATEME together with FUESMEN has developed eight automatic processing algorithms. The purpose of algorithms is to provide solutions to some specific issues of FUESMEN with its corresponding implementation through MIPS. As such, these algorithms require as input one or more DICOM studies and its output is processed study.

MIPS formally established as a condition that the execution of each process can be performed through a command line using the following syntax:

Executablefile <Operation><OutputFolder><FolderStudy1><FolderStudy2> ....<FolderStudyN>

The implementation of the executable depends on programming language, toolboxes used and platform that supports the system.

3. Methodology.

The work involved installation of MIPS in a dedicated virtual machine on Cluster (Figure 1), which it is localizated in the Department of Medical Informatics (FUESMEN). Thus MIPS is part of the computer resources available to FUESMEN PACS.

Once MIPS is installed and configured, the application remains full active waiting images that arraive to the corresponding port. All processes were analysed and five algorithms related to Quality Assurance in Nuclear Medicine were implemented and assestment. The generated images in the service of gamma / SPECT camera were sent through the high-speed network to the MIPS for the corresponding port. After processing, resulting images are sent to the storage server (Server 3). From there, stored images can be retrieved by any VW FUESMEN to be viewed.

The methodology used for the installation and working of the server in real clinical enviroment required several phases. First a theoretical phase consisted of search reference and analysis of theoretical and practical aspects of the problem to solve. Particularly it focused on the study of the DICOM protocol, developed algorithms, operation of MIPS and MySQL engine.

In a second phase, preliminary tests were performed and software tools in the laboratory were evaluated prior to installation in FUESMEN. For this purpose were installed, on different workstations, visualization applications, storage serversand MIPS with associated tools (Apache, PHP, MySQL). In addition eight algorithms oriented to three different problems to implement in the MIPS were evaluated.

The third phase involved the installation and implementation of MIPS in FUESMEN and different activities were carried out to launch system in 24x7 mode. This involved design of the protocols, configuration of existing equipment, testing each protocol on services and staff training.

In the last phase, analysis of system operation was performed. In this regard, it was evaluated initial impact and it was made a data systematization for to assess the usefulness of the system.

In the next section activities and results obtained are described.

4. Testing and Results.

4.1. Laboratory tests: MIPS-Storage Server.

This test is part of the preliminary testing phase and it involved the installation and interconnection of MIPS and DCM4CHEE storage server on PCs of GATEME's Laboratory. Then, it was carried out necessary adjustments to networking. Both applications were setted up using network information (IP, port, application entity title, etc.). Finally it was send DICOM image from DCM4CHEE to MIPS, and then, preccesed study was returned to DCM4CHEE.

The following Figure 3 shows a scheme.

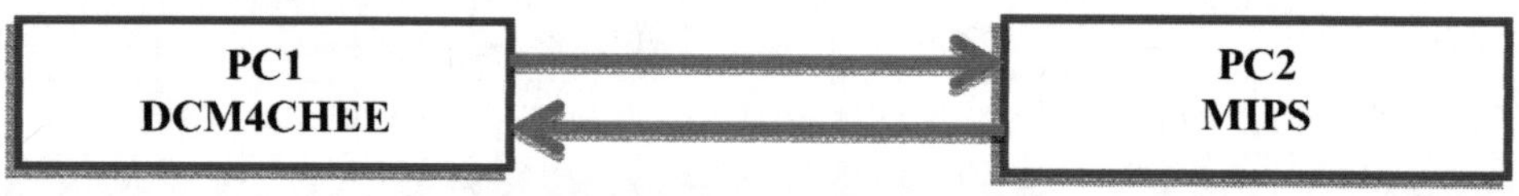

Figure 3: MIPS-DCM4CHEE test.

4.2. Laboratory tests: Adaptation of algorithms developed for MIPS.

The adequacy of processing algorithms was carried out generating an executable file with multiple input arguments. These arguments correspond to input and output folders of images. Depending on the languages and libraries used, this can be more or less complex. In this paper it was proposed to adapt 8 automatic algorithms about 3 different areas.

First we evaluated the corresponding executable to the lesions classification of multiple sclerosis [9]. This algorithm allows receiving MRI volumes; making Brain segmentation and finally classifing those corresponding to injuries (Figure 4a). This segmentation process uses SPM library [8], which present several drawbacks in the compilation stage. While they have tried different ways to solve these problems (versions, consultation forums) has not been able to conclude this task.

Five automatic algorithms for the quality control of nuclear medicine [10] are then used. They were adapted to receive DICOM images and deliver the results. These tasks could be run on MIPS. A resulting image can be seen in Figure 4b.

It was implemented two algorithms that perform automatic control of CT volumes used for radiotherapy [11]. They had been evaluated on MIPS (in the laboratory) in the development stage and it did not present difficulties to implement on high-performance server. The resulting images can be seen in Figure 4c.

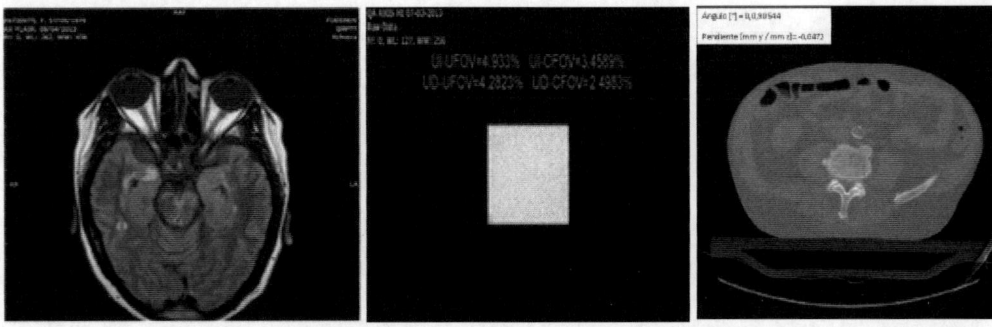

(a) (b) (c)

Figure 4: Output image's algorithms.

In order to probe these algorithms in in clinical settings, it has been implemented protocols to use in nuclear medicine. They aim to make quality control of nuclear medicine equipment routinely. Its implementation allows evaluating a large number of SPECT or Gamma cameras located in different diagnostic centers of FUESMEN.

The implemented controls are required to evaluate features and performance from the imaging test. These tests are scheduled routinely (weekly, monthly, quarterly or annually). If the tests show negative results should apply corrective action.

Below each of the five algorithms used are described [10].

4.1.1. Gamma camera's algorithms

Spatial Resolution algorithm: It is a feature of the scintillation camera which describes the ability to solve two separate points or two lines sources of radiation as separate entities.

Planar Uniformity algorithm: Describe the degree of uniformity of the density of accounts in the image when the detector is flooded with a uniform flow of incident radiation.

4.1.2. SPECT's algorithm:

Uniformity Tomographic algorithm: Quantifies the uniformity of a reconstructed uniform section activity.

Spatial Resolution algorithm: is defined in terms of the full width at half maximum (FWHM) of the reconstruction by the analytical method of filtered back projection (FBP) of a point source in a transversal section. It determines the system's response against a circular tomographic image point.

Center of Rotation (COR) algorithm measures imbalance of the COR, this is a parameter of SPECT system. While it is not important that the center of rotation is exactly in the center of the image, this shift should be taken into account in the image reconstruction.

4.2. Laboratory tests: Viewing Workstation - MIPS - Storage Server.

In this test, phantom images were used to test all processes of quality control of nuclear medicine implemented with MIPS. Each process requires a particular configuration and the images are different in each case. The test consisted of sending the image corresponding to each process from a VW to the MIPS, and that it return the resulting image DICOM to another VW or Storage Server (See Figure 5). All components are connected in the same LAN.

First it was made the check operation of each protocol, and then performance tests were conducted with many images sent simultaneously and / or sequentially. This allowed analyzing the server overload response. In all tests the server was stable.

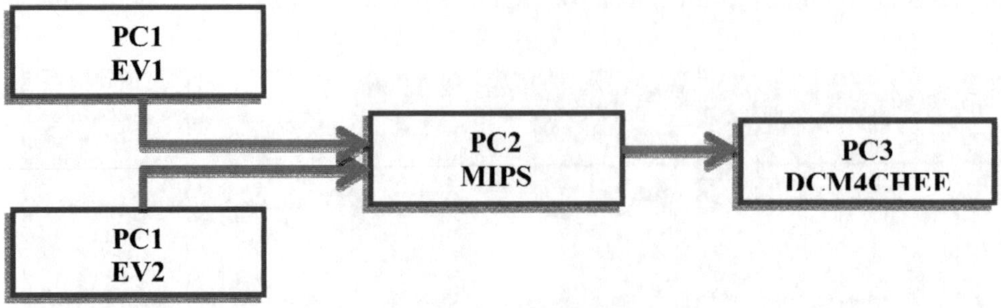

Figure 5: Test configurationVW-MIPS-Storage Server.

4.3. MIPS test in FUESMEN's PACS.

This test was performed in implementation phase where the MIPS was running over FUESMEN's PACS. The images generated from the gamma camera / SPECT modalities were sent to MIPS for processing by nuclear medicine quality control algorithms..

The resulting images were sent to Server 3 by MIPS, then visualized from workstations. This helped to check system performance. Simultaneously and using remote connection software was connected to the virtual machine and it accessed command window MIPS. This enabled real-time monitoring and their behavior during testing via interactive debugging mode. Figure 6 shows the implementation.

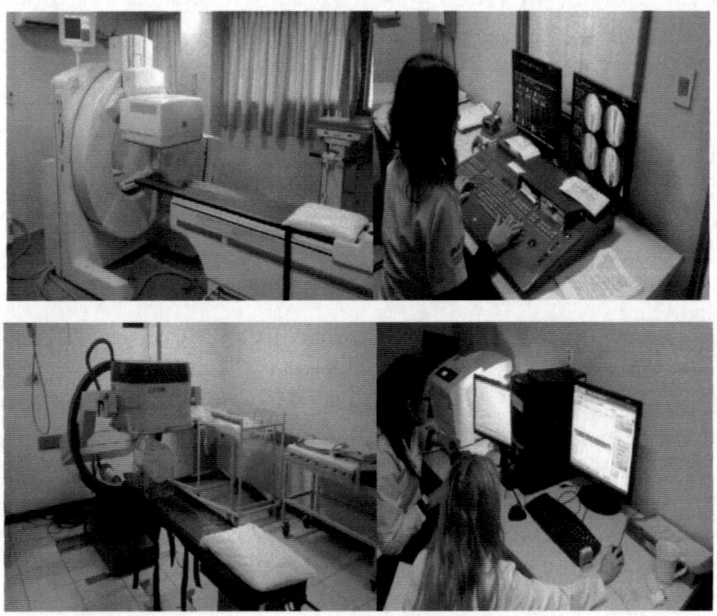

Figure 6: Quality control processes in Nuclear Medicine.

4.4. Training.

The training was divided into three stages. On the one hand training staff of Medical Informatics for the incorporation of new algorithms, then training those responsible for setting the viewing stations and finally the training of users algorithms. The first two stages have been fully implemented, adjusting the protocol for the incorporation of new algorithms.

The first protocol is aimed at managing the virtual machine, the maintenance of MIPS, high and low process as well as any problems that might arise. The second protocol is aimed at technicians who set up WV. Finally is performing the training phase of end users. In the case of the algorithms implemented on the server, staff are physics, medics and technicians of gamma/SPECT camera, that they are users of these tools.

5. Conclusion.

The main objective of this work was to implement a tool for processing studies in clinical enviroment using the DICOM standard. This server, called MIPS, can receive studies from different clients, initiate processing algorithms on them and return results of the operation to a common access server.

For the server installation was coordinated with the Medical Informatics Service to determine the hardware requirements and software necessary for the operation of MIPS Computer. Within the cluster of FUESMEN a stable high-performance virtual machine with remote access from GATEME was prepared.

During the preliminary stage in GATEME, all tools are installed at the laboratory: MIPS, viewing software configured as workstation (eFilm, KPacs, etc.) and DCM4CHEE as storage server. All tools were installed in the GATEME's lab in order to evaluate system behavior experimentally. A series of tests of transmission between DCM4CHEE, MIPS and WV were carried out with positive results. Then we proceeded to adapt the algorithms to be implemented on the clinical environment server. Only one of 8 algorithms evaluated is pending adaptation.

Finally MIPS was installed on the virtual machine with all applications. Then, staff was trained in accordance to the programed tasks This stage is being finalized with the training to end users.

This project enabled the integration of numerous studies conducted in recent years. This has facilitated the link between different medical institutional facilities in FUESMEN with GATEME research area.

A future is expected to generate new automatic algorithms for fast implementation in clinical environment.

References:

[1] Barberís, L.S.; Graffigna, J.P.; Lucena, F. Módulo de comunicación DICOM para librerías de procesamiento de imágenes médicas (Bioingenieria). Gabinete de Tecnologia Medica-Facultad de Ingenieria.Univercidad Nacional de San Juan.2008.

[2] Yongmin Kim.Steven Horii: Handbook Of Medical Imaging. 3 Edicion. Volume III: Display A PACS. Editors SPIE PRESS. 2004

[3] American College of Radiologists and National Electrical Manufacturers Asociation: Digital Imaging and Communications In Medicine (DICOM) Standard. Draft Versión 2013.

[4] DCM4CHEE. http://www.dcm4che.org/confluence/display/WEA/Documentation

[5] APACHE. http://www.apache.org/dist/httpd/docs/httpd-docs-2.4.10.

[6] PHP.http://php.net/manual/php4.php

[7] MySQL.MySQL 5.6 Reference Manual.http://dev.mysql.com/doc/refman/5.6/

[8] SPM. Statical Parametring Mapping. http://www.fil.ion.ucl.ac.uk/spm/.

[9] Filipuzzi, M.; Rodrigo, F; Graffigna, J.P.; Isoardi, R; Noceti, M.. Normalización de Estudios de RMN aplicada a Clasificación de lesiones de Esclerosis Múltiple.. Chile. Valparaiso. 2012. Libro. Artículo Breve. Jornada. 3 Jornadas Chilenas de Ingeniería Biomédica JCIB2012. Universidad de Valparaiso.

[10] Marino, E.A.; Arenas, G; Graffigna, J.P; Mosconi, S; Barberis, S. Realización de test de aceptación y desarrollo de software de analisis automático de calidad de imagen para cámara gamma/SPECT de simple y multiple cabezal. Gabinete de Tecnologia Medica-Facultad de Ingenieria.Univercidad Nacional de San Juan.2011.

[11] Velasques, M.F.; Graffigna, J.P.; Aguero, H.; Holeywell, P.M. . Evaluation of Stretcher Alignment in Radiotherapy using Computed Tomography. IEEE LATIN AMERICA TRANSACTIONS.: IEEE-INST ELECTRICAL ELECTRONICS ENGINEERS INC. 2013 vol.11 n°1. p201 - 206. issn 1548-0992

Formal Logic and Flowchart for Diagnosis Validity Verification and Inclusion in Clinical Decision Support Systems

M Sosa[1], L Grundel[1] and F Simini[1]

[1] *Núcleo de Ingeniería Biomédica, Facultades de Medicina e Ingeniería, Universidad de la República,* 11600 Montevideo, Uruguay

E-mail: msosavitabar@fing.edu.uy

Abstract. Logical reasoning is part of medical practice since its origins. Modern Medicine has included information-intensive tools to refine diagnostics and treatment protocols. We are introducing formal logic teaching in Medical School prior to Clinical Internship, to foster medical practice. Two simple examples (Acute Myocardial Infarction and Diabetes Mellitus) are given in terms of formal logic expression and truth tables. Flowcharts of both diagnostic processes help understand the procedures and to validate them logically. The particularity of medical information is that it is often accompanied by "missing data" which suggests to adapt formal logic to a "three state" logic in the future. Medical Education must include formal logic to understand complex protocols and best practices, prone to mutual interactions.

1. Introduction

Medical education in the twenty-first century faces the challenge of maintaining a high level of quality without losing the empathy of the doctor-patient relationship. The amount of acquired and research generated knowledge increases exponentially, making it difficult to update people and institutions.

Concomitantly, information technologies provide a possibility to go beyond the difficulties of information management. Medical reasoning has always included logical thinking. In particular, the intellectual processes resulting in diagnostic and therapeutic decisions may be seen as following a tree-type structure in its many forms.

The wide span of procedures to choose from, all with many pieces of information to consider calls for the inclusion of formal logic in the study and practice of medicine. We suggest here how two examples related to common diagnostics may be prepared for inclusion into clinical decision support systems (CDSS). This formal modality is also better adapted for electronic health records (EHR) and other documentation support media.

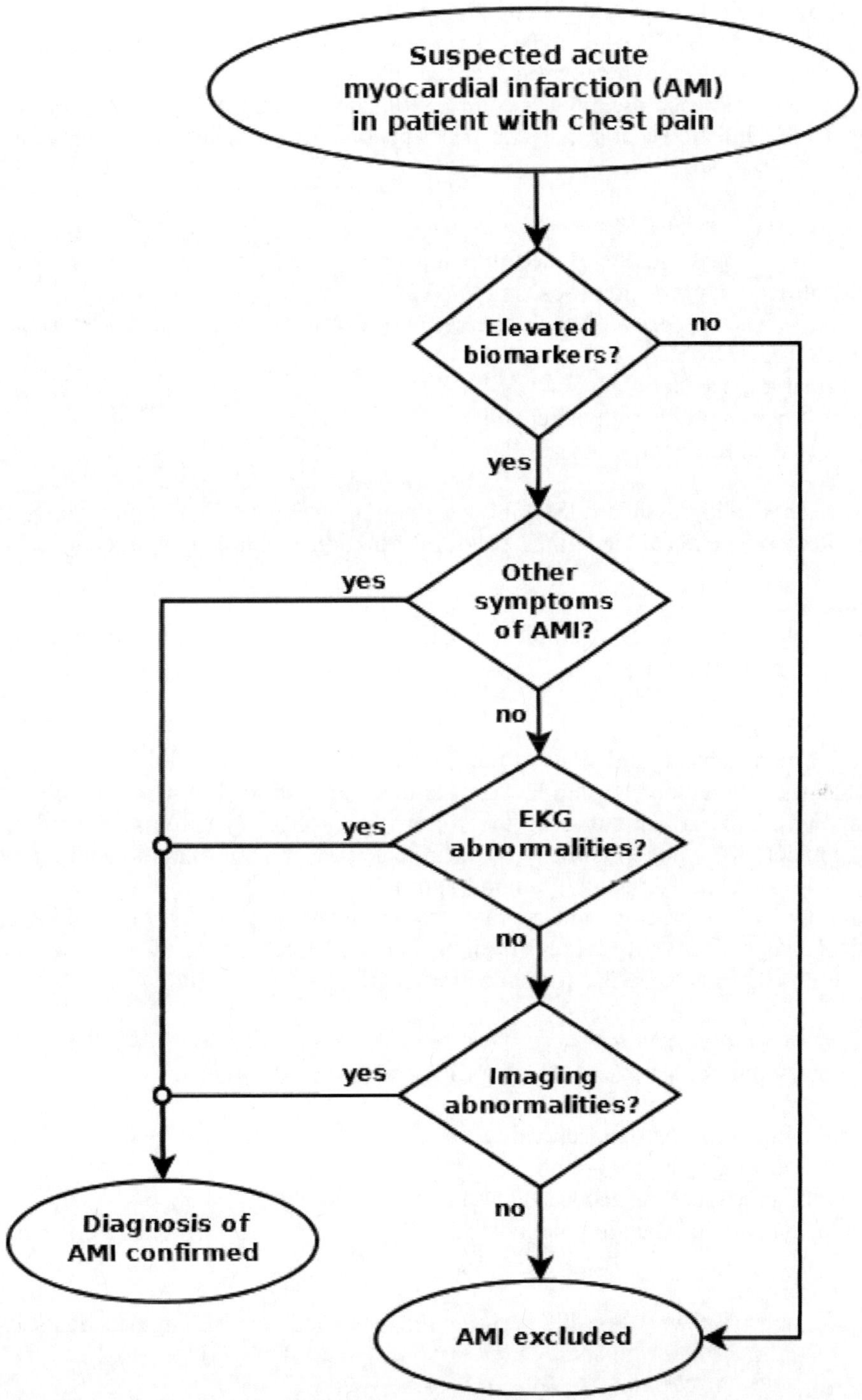

Figure 1. Flowchart for Acute Myocardial Infarction diagnosis. (OGTT: Oral glucose tolerance test).

2. Objectives

The objective of this study is to provide elements for discussion on the usefulness of teaching formal logic in medical training. Medical students can improve and systematize their medical reasoning, specifically during diagnostic and therapeutic decision-making.

As a corollary, this approach assimilates complex diagnostic and therapeutic processes to logical formulae, truth tables and flowcharts that can be easily incorporated into clinical decision support systems.

3. General method

The method we have adopted consists of the following steps:

I. identification of the premises contained in a statement or protocol;
II. study of the link between each premise and identification of logical connective operators (AND, OR, IF, etc.);
III. formal logical expression;
IV. construction of the truth table, and
V. analysis of the results.

In addition, flowcharts follow standard ISO 5807: 1985 nomenclature [1], to allow better and quicker understanding of complex clinical guidelines, using Information Technology applications.

4. Implementation examples

Consider two simple and frequent examples of diagnostic processes: acute myocardial infarction (AMI) and diabetes mellitus (DM).

4.1. *Acute myocardial infarction*

According to AMI management guidelines of the European Society of Cardiology (ESC) published in 2013 [2], a diagnosis is based on elevated levels of at least one cardiac biomarker above percentile 99 (especially troponin), plus at least one of the following criteria: symptoms of cardiac ischemia, electrocardiographic (EKG) abnormalities (ST segment elevation or pathological Q waves) and imaging abnormalities (echocardiogram or angiography).

Standard medical practice incorporates these recommendations in the doctor's mental process intuitively appealing to its logical training and experience, that overlaps the pure logical reasoning. The following procedure is a formalization of the diagnostic activity.

4.1.1. Definition of variables

The premises contained in the statement are identified and assigned letters to each:

- p : increase of cardiac biomarkers (troponin)
- q : symptoms of myocardial ischemia
- r : electrocardiographic abnormalities
- s : alterations observed in echocardiogram or angiography
- t : acute myocardial infarction diagnosis

4.1.2. Logical formulation

The formula resulting from the relationship between the identified variables in the statement is:

$$[p \wedge (q \vee r \vee s)] \rightarrow t \qquad (1)$$

where $\wedge$ represents the logical AND, $\vee$ the logical OR and $\rightarrow$ implication or conditional.

Table 1. Truth table for Acute Myocardial Infarction diagnosis

p	q	r	s	t	q\/r	(q\/r\/s)	(p/\(q\/r\/s))	(p/\(q\/r\/s))=>t
1	1	1	1	1	1	1	1	1
1	1	1	1	0	1	1	1	0
1	1	1	0	1	1	1	1	1
1	1	1	0	0	1	1	1	0
1	1	0	1	1	1	1	1	1
1	1	0	1	0	1	1	1	0
1	1	0	0	1	1	1	1	1
1	1	0	0	0	1	1	1	0
1	0	1	1	1	1	1	1	1
1	0	1	1	0	1	1	1	0
1	0	1	0	1	1	1	1	1
1	0	1	0	0	1	1	1	0
1	0	0	1	1	0	1	1	1
1	0	0	1	0	0	1	1	0
1	0	0	0	1	0	0	0	1
1	0	0	0	0	0	0	0	1
0	1	1	1	1	1	1	0	1
0	1	1	1	0	1	1	0	1
0	1	1	0	1	1	1	0	1
0	1	1	0	0	1	1	0	1
0	1	0	1	1	1	1	0	1
0	1	0	1	0	1	1	0	1
0	1	0	0	1	1	1	0	1
0	1	0	0	0	1	1	0	1
0	0	1	1	1	1	1	0	1
0	0	1	1	0	1	1	0	1
0	0	1	0	1	1	1	0	1
0	0	1	0	0	1	1	0	1
0	0	0	1	1	0	1	0	1
0	0	0	1	0	0	1	0	1
0	0	0	0	1	0	0	0	1
0	0	0	0	0	0	0	0	1

Note 1: Column "t" refers to the diagnosis of acute myocardial infarction (AMI) and the last column represents the validity of the logical proposition (diagnosis) based on different combinations of values of the variables involved.

Note 2: Diagnosis is based on the result of cardiac biomarkers, presence of cardiac ischemia symptoms, EKG alterations and imaging abnormalities.

4.1.3. Truth table

The truth table resulting from the formula (1) is represented in Table 1 including all possible combinations of values of the variables. The last column indicates whether diagnosis "t" made in each case was correct (1) or incorrect (0).

4.1.4. Particular case management and logic verification

Consider the following particular case: "You are the Physician on call at the emergency room, and you see a patient complaining of a strong precordial pain. When performing an EKG you discover an elevated ST segment in some of the derivations. When the laboratory results arrive you confirm that troponin levels are very high. You therefore diagnose acute myocardial infarction". In this case, the logical formula is:

$$\{\{[p \wedge (q \vee r \vee s)] \rightarrow t\} \wedge (q \wedge r \wedge p)\} \rightarrow t \tag{2}$$

resulting in the truth table shown in Table 2. The result indicated in the last column is a tautology, therefore, we conclude that the reasoning is valid.

4.1.5. Flowchart

Figure 1 represents the flowchart of the diagnostic criteria for AMI example.

Table 2. Truth table for Acute Myocardial Infarction diagnosis in a particular case.

p	q	r	s	t	q/\r	(q/\r/\p)	q\/r	(q\/r\/s)	(p/\(q\/r\/s))	((p/\(q\/r\/s))=>t)	(((p/\(q\/r\/s))=>t)/\(q/\r/\p))	(((p/\(q\/r\/s))=>t)/\(q/\r/\p))=>t
1	1	1	1	1	1	1	1	1	1	1	1	1
1	1	1	1	0	1	1	1	1	1	0	0	1
1	1	1	0	1	1	1	1	1	1	1	1	1
1	1	1	0	0	1	1	1	1	1	0	0	1
1	1	0	1	1	0	0	1	1	1	1	0	1
1	1	0	1	0	0	0	1	1	1	0	0	1
1	1	0	0	1	0	0	1	1	1	1	0	1
1	1	0	0	0	0	0	1	1	1	0	0	1
1	0	1	1	1	0	0	1	1	1	1	0	1
1	0	1	1	0	0	0	1	1	1	0	0	1
1	0	1	0	1	0	0	1	1	1	1	0	1
1	0	1	0	0	0	0	1	1	1	0	0	1
1	0	0	1	1	0	0	0	1	1	1	0	1
1	0	0	1	0	0	0	0	1	1	0	0	1
1	0	0	0	1	0	0	0	0	0	1	0	1
1	0	0	0	0	0	0	0	0	0	1	0	1
0	1	1	1	1	1	0	1	1	0	1	0	1
0	1	1	1	0	1	0	1	1	0	1	0	1
0	1	1	0	1	1	0	1	1	0	1	0	1
0	1	1	0	0	1	0	1	1	0	1	0	1
0	1	0	1	1	0	0	1	1	0	1	0	1
0	1	0	1	0	0	0	1	1	0	1	0	1
0	1	0	0	1	0	0	1	1	0	1	0	1
0	1	0	0	0	0	0	1	1	0	1	0	1
0	0	1	1	1	0	0	1	1	0	1	0	1
0	0	1	1	0	0	0	1	1	0	1	0	1
0	0	1	0	1	0	0	1	1	0	1	0	1
0	0	1	0	0	0	0	1	1	0	1	0	1
0	0	0	1	1	0	0	0	1	0	1	0	1
0	0	0	1	0	0	0	0	1	0	1	0	1
0	0	0	0	1	0	0	0	0	0	1	0	1
0	0	0	0	0	0	0	0	0	0	1	0	1

4.2. *Diabetes Mellitus*

The Latin American Association Diabetes (ALAD) guidelines published in 2013 [3] comprise diagnostic criteria for diabetes mellitus (DM). In a patient with no suggestive symptoms, the disease is diagnosed with a fasting blood glucose level (in venous blood, after 8 hours of fasting) with a value greater than or equal to 126 mg / dL in two consecutive (both positive) test; with a value in oral glucose tolerance test (venous blood) greater than or equal to 200 mg / dL, repeated in two consecutive tests; or a value of glycosylated hemoglobin (HbA1c) greater than or equal to 6.5%. In a patient presenting symptoms suggestive of diabetes, diagnosis can be done with a glucose value greater than or equal to 200 mg / dL (venous blood at any time).

4.2.1. *Definition of variables*

Let the following premises be identified by letters for formal representation:

- p : fasting blood glucose ≥ 126 mg / dL on two occasions
- q : OGTT glucose ≥200 mg / dL on two occasions
- r : glycosylated hemoglobin ≥ 6.5%
- s : random blood glucose levels ≥ 200 mg / dL in patients with symptoms of diabetes
- t : diagnosis of diabetes mellitus

4.2.2. *Logical formulation*

Considering each of the premises and the relationships that link them according to the statement, we proceed to write the formula that represents it, choosing the appropriate connective with the following results and using the symbol ∨ as the OR operator:

$$(p \vee q \vee r \vee s) \rightarrow t \qquad (3)$$

4.2.3. *Truth table*

The truth table resulting from formula (3) is similar to given in Table 1. As in the AMI example, we have also represented here all possible combinations of values taken by the variables, allowing to know in what cases the diagnosis was right or wrong.

4.2.4. *Particular case management and logic verification*

"As a physician, you are seeing an asymptomatic patient with a fasting blood glucose of 127 mg / dl, and after 20 days the test is repeated yielding a value of 132 mg / dL, and therefore you make diagnosis of Diabetes Mellitus". In this case, the logical formula is:

$$\{[(p \vee q \vee r \vee s) \rightarrow t] \wedge q\} \rightarrow t \qquad (4)$$

The truth table for this case is similar to given in Table 2, the result of which is also a tautology. Therefore, we can say that the reasoning shown in the statement is valid, verifying the formal validity of the diagnostic criteria.

4.2.5. *Flowchart*

The flowchart for the diabetes mellitus diagnosis is included in Figure 2.

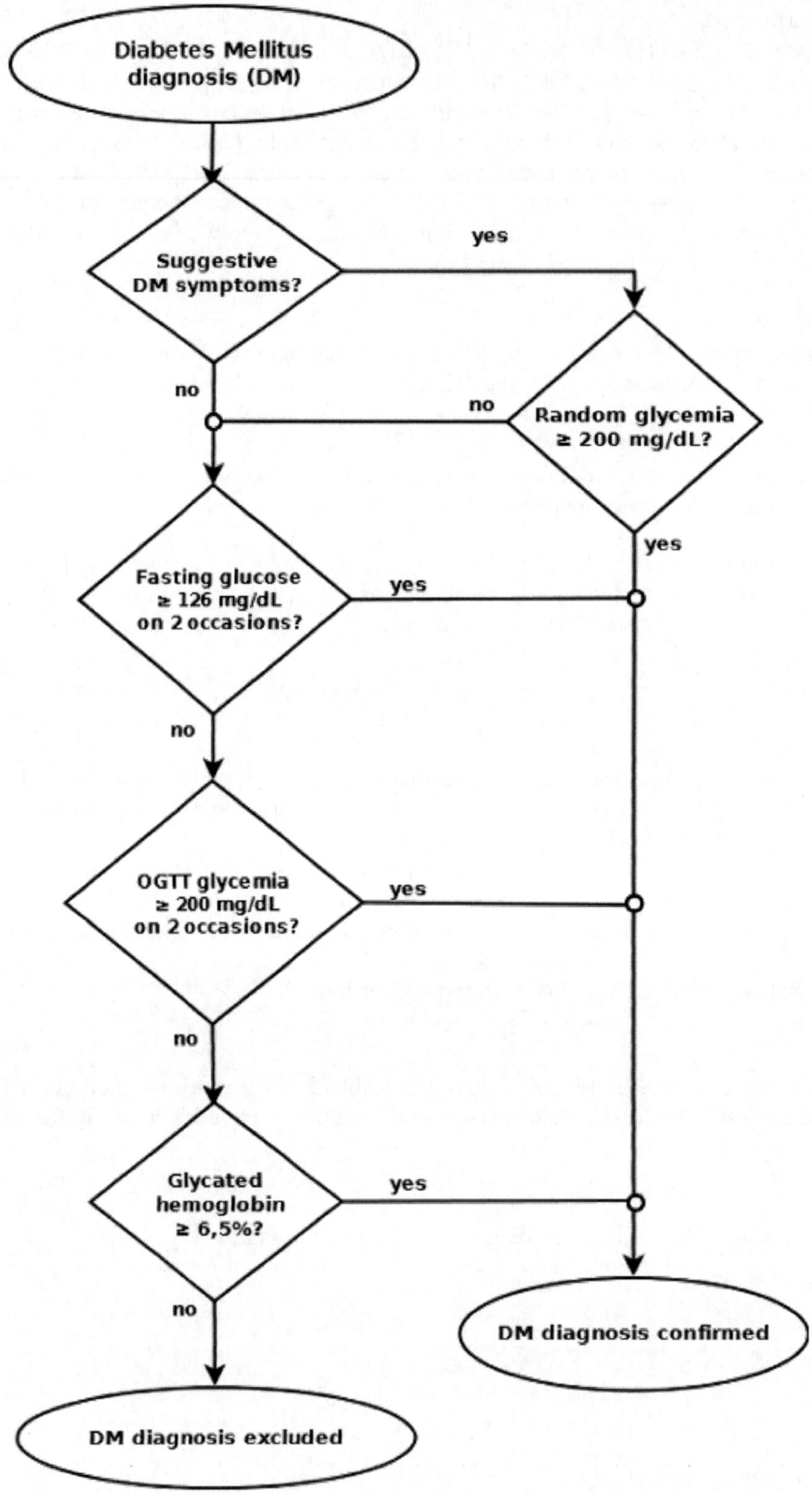

Figure 2 . Flowchart for Diabetes Mellitus (DM) diagnosis.

Journal of Physics: Conference Series **705** (2016) 012048 doi:10.1088/1742-6596/705/1/012048

5. Problem of "missing data" in medicine

In medicine, formal logic can help unravel confusing situations. Unlike other branches, in this discipline complete information is rarely available and, for various reasons, there are missing key variables, which would otherwise help decision making. The lack of data or either unknown or uncertain information are elements to consider. The usual practice of omitting an information element (ignoring it when negative for instance), is unfortunately common in medicine. If on one hand it allows to save time, on the other hand it may be a source of error. The assimilation of the lack of data to negative data is common and formal logic tools can help prevent it.

Furthermore the uncritical adoption of logic in the medical reasoning with formulations based on two values (true or false) results in errors, false optimistic statistics and inappropriate public health decisions. It is sufficient to think about a mortality rate with missing data of death in dead individuals, resulting in rates lower than real, simply because unknown data is assimilated to subjects that are alive.

6. Medical teaching

Since 2014 we have developed a course called "Medical Informatics" intended for advanced medical students, with the aim to introduce them to the understanding and use of theoretical and technological informatics tools linked to their profession and to the interconnection of biomedical equipment with electronic medical records. Two of the first topics that are taught in this course are formal logic and algorithms (and flowcharts) as we believe it is one of the bases that Medical School Students should inevitably know before getting to understand the operation of any computer system that handles reasoning and logical inductions. As a result of teaching this course on three occasions we have learned how important it is to introduce the group of future physicians into the use of powerful tools for diagnosis and inference. Complementing important issues in medicine, medical informatics course supports teaching research and scientific method, so as to practice it daily with patients, to the benefit of less errors.

7. Error reduction

The availability of diagnostic protocols and standards in formal logic schemes allows to reduce mistakes, leaving fewer items to memory and relying on the ability of synthesis and deduction on formal grounds. Although the medical profession includes the function and ability to solve diagnostic and therapeutic strategies driving, the formal logic structures is a possible support with the resources of the information technology available. These resources, although they do not differ in their essence from traditional medical reasoning, allow the physician to work with a large number of variables, in a reproducible way.

8. Discussion and conclusion

Medical practice in the presence of abundant data and even more numerous standards, research and interrelationships, can hardly be exercised without the help of logical assistance tools that operate as mental "orthoses", strengthening the response capacity and reducing errors. This first approach involves the teaching of formal logic in the Medicine study plan. The first three groups of about a total of 60 students have resulted, in addition to their own training, in the conformation of a set of concepts to transmit to future generations of students. The observation of the profit that claim to have received, along with their increased operational capacity in programs and applications dedicated to clinical work, confirm that the road started should be perfected. It will be necessary however quantitatively assess the impact of this training in formal logic comparing somehow the performance of students so formed with those who didn´t came to the training described here. The experience of other medical schools such as the Autonomous University of Mexico (UNAM) since 2010 points in the same direction [5] .

The rise of the expert systems (ES) developed in past decades [6] used extensively translating facts and logical chains based on "a priori" assumptions of medical diagnosis, logic that is included and encapsulated inside the ES. In this paper we propose instead the use of formal logic in medical training, as part of the conceptual tools for the use and development of diagnostic methods or deduction and setting up protocols on firm foundations. Either using computer systems or not attended practice of medicine. Because medicine can not be conceived in this historical moment away from the possibility of being benefited from the assistance of formal logic in proposals of increasing complexities, impossible to be solved by appealing to the simple memory of the professional.

In particular the complexity of certain diagnostic protocols with alternative associated treatments, arborescent with multiple variants, associated for example, to neonatal hypothermia therapy of hypoxic-ischemic encephalopathy of the newborn [4], are issues that will be addressed in this approach to reduce complexity and increase the rigor of its implementation. This can be achieved by the application of the methods of formal logic. We also see that by means of the rigorous application of protocols and documented observation of their results, they may deduct possible improvements and proposals for variants of treatment.

We call on readers to contribute ideas regarding the application of these methods to logical formalism, in view of its clinical adoption.

References

[1] International Organization for Standardization 1985 ISO 5807:1985 Information processing — Documentation symbols and conventions for data, program and system flowcharts, program network charts and system resources charts *Int. Organ. Stand.*

[2] Sociedad Europea de Cardiología 2014 Guía de práctica clínica de la ESC para el manejo del infarto agudo de miocardio en pacientes con elevación del segmento ST Grupo de Trabajo para el manejo del infarto agudo de miocardio con elevación del segmento ST *Rev. Española Cardiol.* **66** 1–46

[3] Asociación Latinoamericana de Diabetes 2013 Guías ALAD sobre diagnóstico, control y tratamiento de la Diabetes Mellitus Tipo 2 con Medicina Basada en Evidencia *Rev. la Asoc. Latinoam. Diabetes*

[4] Blanco D, García-Alix a., Valverde E, Tenorio V, Vento M and Cabañas F 2011 Neuroprotección con hipotermia en el recién nacido con encefalopatía hipóxico-isquémica. Guía de estándares para su aplicación clínica *An. Pediatr.* **75**

[5] Sanchez Mendiola M 2015, “Biomedical Informatics Education: is the glass half full or half empty?” MEDINFO 2015, 15th World Congress on Health and Biomedical Informatics” Sao Paulo, Brazil, **Keynote Lecture 04.**

[6] Simini F, Maillard F Breart G 1990 “Cesarean section odds ratios. European Journal of Obstetrics & Gynecology Reprodutive Biology, **34** p1-13

Multiplexed coding in the human basal ganglia

D.S. Andres[1,2,3,*], D. Cerquetti[3], M. Merello[3]

[1] Institute of Neuroinformatics, ETH and University Zurich, Zurich, Switzerland
[2] National University of San Martin, Buenos Aires, Argentina
[3] Institute for Neurological Research Raul Carrea, Fleni Institute, Buenos Aires, Argentina

Abstract— **A classic controversy in neuroscience is whether information carried by spike trains is encoded by a time averaged measure (e.g. a rate code), or by complex time patterns (i.e. a time code). Here we apply a tool to quantitatively analyze the neural code. We make use of an algorithm based on the calculation of the temporal structure function, which permits to distinguish what scales of a signal are dominated by a complex temporal organization or a randomly generated process. In terms of the neural code, this kind of analysis makes it possible to detect temporal scales at which a time patterns coding scheme or alternatively a rate code are present. Additionally, finding the temporal scale at which the correlation between interspike intervals fades, the length of the basic information unit of the code can be established, and hence the word length of the code can be found. We apply this algorithm to neuronal recordings obtained from the Globus Pallidus pars interna from a human patient with Parkinson's disease, and show that a time pattern coding and a rate coding scheme co-exist at different temporal scales, offering a new example of multiplexed neuronal coding.**

Keywords— **Neural code, time patterns, structure function, Parkinson's disease, basal ganglia.**

I. Introduction

The concept of neural code is a central paradigm in neuroscience [1-2]. However, a precise definition of what is meant by neural code, or neuronal encoding, can be elusive. A general definition of code can be applied to the particular study of the neural code, defining code as "a set of symbols and the rules for their combination, which can be used to represent a pattern of information in a manner different from that of the original signal" [3]. This definition can also be modified to "the smallest set of symbols" [4]. According to this, two main elements need to be identified in a spike train to be able to characterize its neural code: the symbols (words), and the rules governing their combination (grammar). In this sense, an analysis of the neural code is not necessarily bound to lead to an understanding of the meaning of such code in a given context (semantics), making a theoretical approach possible, independently from specific experimental settings [5].

A novel approach to the problem of finding the basic symbols of the neural code has been recently proposed by our group [6]. Based on the calculation of the temporal structure function, an algorithm was developed that can be used for finding the average word length of the neuronal code in a given spike train. (Our algorithm is publicly available at www.neurostruct.org.) Structure functions were originally applied by Kolmogorov to the analysis of fluids undergoing turbulence [7], and can be calculated in the spatial or the temporal domain. Structure functions have been shown to be robust to noisy and short time series, which makes them suitable for working with neuronal data, often consisting of short, highly variable spike trains [8]. More importantly, mathematical properties of the structure function allow determining if a signal is generated by an underlying monofractal, multifractal, or random process [9]. Of particular interest for the analysis of the neural code is the random case, where the slope of the structure function = 0. It has been argued for a long time that, if it can be assured that a given spike train is generated by a random process, then any kind of coding other than a time averaged measure (most probably a rate code) must be precluded [10]. This is precisely the kind of analysis that can be performed by the calculation of the temporal structure function. In the zero-slope region of this function a stochastic process is bound to generate the neuronal signal analyzed. Therefore, at this temporal scale it can be affirmed that any information transmitted must be necessarily coded in the rate of discharge, or a similar time averaged measure (for a more detailed discussion, see ref. 11). Conversely, regions of the function where a non-zero slope is found correspond to a temporal scale where complex geometrical properties are present (either monofractal, or multifractal). These scales are suitable for the transmission of complex time patterns (and hence, a time code). Furthermore, independent events cannot be transmitted at the temporal scales where a complex behavior is observed, given the high temporal correlation between spikes present at those. This is why any events transmitted within this temporal window are necessarily encoded in the same information unit, ultimately offering a measure of the word length of the code.

In this work, we propose to apply this kind of analysis to neuronal recordings obtained from the Globus Pallidus pars interna (GPi) from a human patient with Parkinson's disease. The GPi represents a main output center of the basal ganglia, the primary circuit that is altered by parkinsonism. To achieve a better understanding of the neuronal encoding by these neurons might help developing better pathophysiological models of Parkinson's disease, and hence a better treatment of this condition.

II. MATERIALS AND METHODS

A. *Algorithm*

Firstly, the temporal structure function is calculated based on interspike intervals (ISIS) time series obtained from single cell recordings. We define $I(t)$ as the t th interspike interval and $\Delta I(\tau) = I(t+\tau) - I(t)$, where τ is the number of ISIS between $I(t)$ and $I(t+\tau)$, also known as time lag increment. The structure function $S_q(\tau)$ is then calculated as:

$$(1)\ S_q(\tau) = \langle |\Delta I(\tau)| \rangle,$$

where $\langle \cdot \rangle$ accounts for the statistical average and τ varies between 1 and 1000. Here we only consider the first order structure function (q=1). A smoothing technique is then applied, averaging over a running window of 30 adjacent points. This step of the algorithm has the purpose of softening the effects of the high frequency variability typically present in neuronal signals. Finally, a breakpoint in the temporal structure function is looked for, where the slope abruptly changes from either positive or negative to zero on average. To determine this point, the first derivative of the smoothed structure function is calculated, S'. Then, the first point of S' is eliminated, to account for some cases where a high correlation is only found at a scale of $\tau = 1$, which we do not consider significant. Finally, the breakpoint τ_1 is defined as the smallest τ for which $S' < 0$ for three consecutive τ for cases where the initial part of the structure function is ascending, or the smallest τ for which $S' > 0$ for three consecutive τ, if the initial portion of the structure function is descending. At least 4 consecutive points are required to have either a positive or negative slope to consider an ascending / descending region of the function, respectively, to account for the effects of the high frequency variability. (Computer code is available by request. An online version of this algorithm can be tried at www.neurostruct.org.).

B. Data analysis

a) Neuronal recordings

A patient with Parkinson's disease, Hoehn&Yahr IV, and fulfilling Core Assessment Program for Surgical Interventional Therapies in Parkinson's Disease (CAPSIT) criteria, was routinely selected to undergo surgical implantation of DBS electrodes at our clinic. (Patient's age, gender and nationality are excluded to protect privacy.) The patient signed informed consent prior to the intervention, which was previously approved by the institutional ethics committee. The patient was not under the effect of any neurological medication at the time of the surgery. Briefly as described elsewhere, microrecording, stimulation, and neurosurgical procedures were performed while the subject was awake and under local anesthesia [12]. The patient underwent surgical target planning employing magnetic resonance imaging (MRI) and using a Leksell stereotactic system (Series G, Elekta, Sweden). Initial GPi targets were planned according to anatomical coordinates and corrected using MRI image fusion techniques. The patient showed a favorable outcome after the surgery.

Microelectrode recordings of the neuronal activity were obtained during the surgery. Microrecording was performed using platinum/iridium (Pt/Ir 80/20%) microelectrodes with nominal impedance of 0.8–1.2 megohms (mTSPBN-LX1, FHC Inc). The signal was amplified, conditioned and digitized with a dedicated acquisition system (1401plus, CED). The sampling rate used was 50 kHz. A preamplifier (remote probe) mounted onto a motorized microdrive (FHC 65-00-1 Stepper Drive and ST-M0-00 TMS Controller) located near the electrode tip to minimize pickup of electrical noise was connected to a differential amplifier with a built-in impedance meter (FHC IS-AM-00-01 Iso-Xcell 3 Amplifier) and an isolated stimulus generator (FHC IS-PL-06 Isolated Bipolar Pulsar Stimulator; FHC, Bowdoinham, ME). Acquired signals were submitted to a single unit's classification algorithm and separation into unit classes was carried out by means of wavelet analysis and a clustering algorithm (developed by Quian-Quiroga,[13]). Interspike intervals were extracted and used to construct time series of single units.

b) Surrogate data

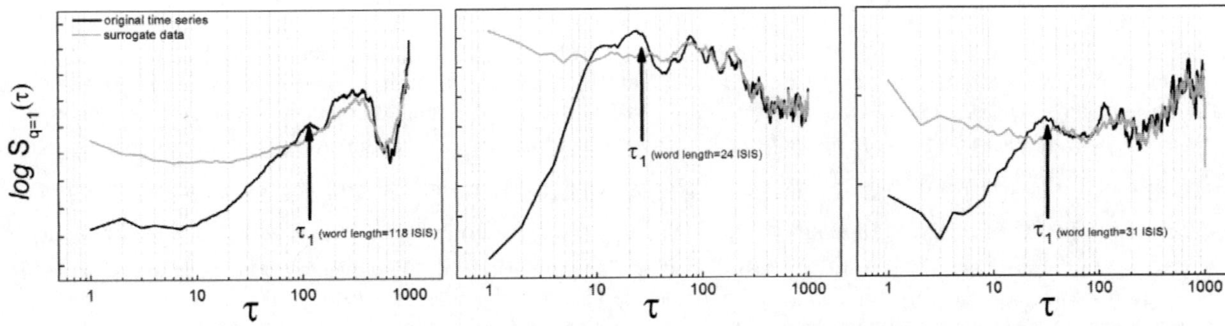

Fig. 1 Temporal structure function calculated for sample GPi neurons and the corresponding surrogate data. The word length (τ_1) can be observed, defined as the first breakpoint were the slope changes from positive to zero on average. The positive slope (related to fractal properties of the time series) is lost in the case of the surrogate data, indicating the dynamic origin of the complex behavior observed.

Surrogate data were generated applying a shuffling technique to the raw ISIS time series, in order to test the hypothesis of a non-random organization in the spike trains. Because our hypothesis states that the neuronal discharge only presents a temporal organization at the scales where the temporal structure function shows a non-zero slope, we shuffled the ISIS specifically at these scales, by changing each ISI for any other, randomly chosen at a distance smaller than τ_1 from the first one. This kind of shuffling preserves the probability distribution function (and all statistical properties) of the time series, while altering any dynamic properties (i.e., properties dependent on the specific temporal order of ISIS). If the temporal structure observed for a scale below τ_1 does depend on dynamic properties, the non-zero slope region should disappear in the analysis of the surrogate data.

III. Results

Nine single cell recordings were obtained from GPi cells. Time series length was 5424 ± 1485 (mean ± SD). Sample temporal structure functions of GPi neurons and the corresponding surrogate data are shown in figure 1. In 7 cases (78%) an initial region of the temporal structure with positive slope was observed, indicating the presence of complex (fractal) properties at small time scales. The rest of the cases presented a zero-slope structure function at small scales, meaning that no complex temporal organization was present at small time scales. Sample temporal structure functions of GPi neurons and the corresponding surrogate data are shown in figure 1.The value of τ_1 was 48 ± 37 ISIS (mean ± SD). The positive slope region disappeared in the surrogate data, proving the dynamic origin of the complex behavior observed.

IV. Discussion

A new paradigm is arising in the field of neuronal encoding, which has been called multiplexed coding. This refers to the simultaneous occurrence of different coding schemes in the same neuronal population, sometimes at different scales, with either complementary or redundant purposes [14-15]. An example of a multiplexed neuronal code at different time scales occurs in the basal ganglia. A rate code is known to play a role in the communication between the basal ganglia and the motor thalamus and cortex. This is evidenced by the fact that a higher frequency of activity of the output nuclei of the basal ganglia is correlated with a higher level of motor output [16-17]. However, the relevance of a time patterns coding scheme in the basal ganglia - thalamo - cortical circuit has also been proposed by many authors [18-22]. Recently, we reported that a rate and a time code co-exist in output structures of the basal ganglia in healthy and parkinsonian rats [6]. Both coding types were shown to play a role at different temporal scales, and these scales are altered in the parkinsonian animals. Here, we show a similar co-existence of different coding properties at small and large time scales in a human patient with Parkinson's disease. At small scales, a positive slope is observed in the temporal structure function in 78% of the neurons analyzed, which evidences the presence of complex temporal properties with fractal characteristics. This complex temporal organization of the neuronal discharge disappears in the surrogate (shuffled) data, demonstrating its dynamic origin. Regarding the transmission of information, complex time patterns can be transmitted at this scale, although independent information cannot. This obeys to the fact that ISIS are not independent from each other for $\tau < \tau_1$, but strongly correlated. This suggests the presence of a time patterns coding scheme, and allows defining τ_1 as the length of the smallest information unit of the code, or word length. Conversely, at large time scales (i.e., for $\tau > \tau_1$) a zero-average slope is observed, which can only be generated by random processes. In terms of neuronal coding this demonstrates an independence between successive ISIS, precluding any coding scheme different than a time-averaged code, as for example a rate code.

A note is in place regarding the limitations of our study, which are related to the fact that only one patient and few neurons were analyzed. The reliability of our results is enhanced by the fact that similar properties were previously observed in animal and simulated neuronal recordings [11-12]. Nevertheless, a future study including more recordings and patients is mandatory in order to confirm the results presented here.

V. Conclusions

In a majority of human (GPi) neurons with Parkinson's disease, our algorithm was able to detect two main temporal scales with different coding properties. At short time scales time patterns coding takes place, while at longer time scales a rate code is present. This shows the presence of a multiplexed neuronal code in the human basal ganglia. Neurostruct appears as a promising tools for the analysis of neuronal activity, in particular in Parkinson's disease.

Acknowledgments

The work of Daniela Andres is supported by Society in Science, The Branco-Weiss Fellowship, administered by ETH, Zurich, Switzerland. We acknowledge the work of the medical staff and technical personnel at Fleni Institute, Buenos Aires, Argentina, who helped making this work possible.

Conflict of Interest

The authors declare that they have no conflict of interest.

References

1. Attneave F (1954) Some informational aspects of visual perception. Psychol rev 61(3):183-193.
2. Mountcastle VE (1967) The problem of sensing and the neural coding of sensory events. In: Quarton GC, Melchenuk T, Schmitt FO (editors) The neurosciences, a study program. New York: The Rockefeller University Press, pp 393-408
3. Uttal WR (1969) Emerging principles of sensory coding. Perspect biology med 12(3):344-368
4. Panzeri S, Brunel N, Logothetis NK, et al (2009) Sensory neural codes using multiplexed temporal scales. Cell 33(3):111-120.
5. Andres DS (2015) The language of neurons: theory and applications of a quantitative analysis of the neural code. In: Horizons in Neuroscience Research, Vol. 16. New York: Nova Publishers, *in press*
6. Andres DS, Cerquetti DF, Merello M (2015) Neural code alterations and abnormal time patterns in Parkinson's disease. J Neural Eng 12:026004 (9pp)
7. Frisch U (1996) Turbulence, the legacy of A.N. Kolmogorov. Cambridge: University Press
8. Schulz-DuBois EO, Rehberg I (1981) Structure function in lieu of correlation function. Appl Phys 24(4):323-329

9. Vainshtein SI, Sreenivasan KR, Pierrehumbert RT, et al (1994) Scaling exponents for turbulence and other random processes and their relationships with multifractal structure. Phys Rev E 50:1823-1835.
10. Ferster D, Spruston N (1995) Cracking the neural code. Science 270:756-757
11. Andres DS, Gomez F, Ferrari FAS, et al (2014) A multiple time-scale framework for the understanding of the progression of Parkinson's disease. Phys Rev E 90:062709
12. Andres DS, Cerquetti D, Merello M (2010) Finite dimensional structure of the GPi discharge in patients with Parkinson's disease. Int J Neural Syst 21:1-12
13. Quian-Quiroga R, Nadasdy Z, Ben-Shaul Y (2004) Unsupervised spike detection and sorting with wavelets and superparamagnetic clustering. Neural Comput 16:1661-1687.
14. Lee J, Groh JM (2014) Different stimuli, different spatial codes: a visual map and an auditory rate code for oculomotor space in the primate superior colliculus. PLOS ONE 9(1):e85017
15. Zuo Y, Safaai H, Notaro G, et al (2015) Complementary contributions of spike timing and spike rate to perceptual decisions in rat S1 and S2 cortex. Curr Biol 25:357-363
16. Utter AA, Basso MA (2005) The basal ganglia: An overview of circuits and function. Neurosci Biobehav R 32:333-342
17. Albin RL, Young A, Penny JB (1989) The functions anatomy of basal ganglia disorders. Trends Neurosci 12:366–375
18. Montgomery Jr. EB (2007) Basal ganglia physiology and pathophysiology: a reappraisal. Parkinsonism Relat D 13:455-465
19. Darbin O, Soares J, Wichmann T (2006) Nonlinear analysis of discharge patterns in monkey basal ganglia. Brain Res 1118:84-93
20. Darbin O, Dees D, Martino A, et al (2013) An entropy-based model for basal ganglia dysfunctions in movement disorders. Biomed Res Int DOI 10.1155/2013/742671.
21. Dorval AD, Russo GS, Hashimoto T, et al (2008) Deep brain stimulation reduces neuronal entropy in the MPTP primate model of Parkinson's disease. J Neurophysiol 100:2807–2818
22. Lafreniere-Roula M, Darbin O, Hutchison WD, et al (2010) Apomorphine reduces subthalamic neuronal entropy in parkinsonian patients. Exp Neurol 225:455–458

Journal of Physics: Conference Series **705** (2016) 012050 doi:10.1088/1742-6596/705/1/012050

STN area detection using K-NN classifiers for MER recordings in Parkinson patients during neurostimulator implant surgery

L Schiaffino[1], A Rosado Muñoz[2] , J Guerrero Martínez[2] , J Francés Villora[2], A Gutiérrez[3], I Martínez Torres[3] y D R Kohan[4]

[1] Rehabilitation Engineering and Neuromuscular and Sensorial Research Laboratory (LIRINS) of the National University of Entre Ríos, Argentine.
[2] Digital Signal Processing Group, ETSE, Department of Electronic Engineering, University of Valencia, Valencia, Spain.
[3] Functional Neurosurgery Unit. La Fe Hospital, Valencia, Spain.
[4] Chair of Probability and Statistics of the Faculty of Engineering of the National University of Entre Ríos, Argentine.

E-mail: lschiaffino@bioingenieria.edu.ar

Abstract. Deep Brain Stimulation (DBS) applies electric pulses into the subthalamic nucleus (STN) improving tremor and other symptoms associated to Parkinson's disease. Accurate STN detection for proper location and implant of the stimulating electrodes is a complex task and surgeons are not always certain about final location. Signals from the STN acquired during DBS surgery are obtained with microelectrodes, having specific characteristics differing from other brain areas. Using supervised learning, a trained model based on previous microelectrode recordings (MER) can be obtained, being able to successfully classify the STN area for new MER signals. The K Nearest Neighbours (K-NN) algorithm has been successfully applied to STN detection. However, the use of the fuzzy form of the K-NN algorithm (KNN-F) has not been reported. This work compares the STN detection algorithm of K-NN and KNN-F. Real MER recordings from eight patients where previously classified by neurophysiologists, defining 15 features. Sensitivity and specificity for the classifiers are obtained, Wilcoxon signed rank non-parametric test is used as statistical hypothesis validation. We conclude that the performance of KNN-F classifier is higher than K-NN with $p<0.01$ in STN specificity.

1. Introduction

Deep Brain Stimulation (DBS) is an effective neurosurgery to overcome motor system alterations such as tremors, rigidity and dyskinesia appearing in advanced stages of the Parkinson's disease [1][2]. Stimulating DBS electrodes can be implanted in some brain basal nucleus, being the sub-thalamic nucleus (STN) which has proven a better clinical response in implanted patients [1]. An X-ray picture of a patient with implanted DBS electrodes is shown in figure 1. The STN is a small lens-shaped nucleus with an approximated size of 5.9[mm]×3.7[mm]×5[mm] located in the transition between the diencephalon and mesencephalon surrounded by other structures functionally called substantia nigra, red nucleus and zona incerta area [3].

An implanted DBS system acts as a brain pacemaker, generating electrical impulses specifically tuned for each patient for a better motor system response. Typically, the stimulation parameters are [1]: 2[V] biphasic rectangular pulse with a pulse width of 60 [μs] at a frequency of 130 [Hz].

The detection of the target area to fix the stimulation electrodes in the STN is a complex task performed by the neurosurgery team. Depending on the available technology at an institution, the surgeons can be assisted by different tools: previous magnetic resonance to surgery, generic atlas overlapped to obtained images, tomography images during surgery, 3D reconstruction for the fusion of resonance and tomography images (coregister), visual and acoustic analysis of MER acquired signals [2][4][5]. Microelectrodes are 100μm in diameter entering the brain by means of a mechanical system, crossing different brain functional structures as seen on figure 1 (left): anterior thalamus (TH), zona incerta area (ZI), STN and substantia nigra (SN). Each of the functional areas are characterized by different signal patterns [2]. Signals acquired from MER are non-stationary and they are formed by trigger sequences coming from: i) neural activity in form of spikes, ii) background neural activity and iii) artifacts [2].

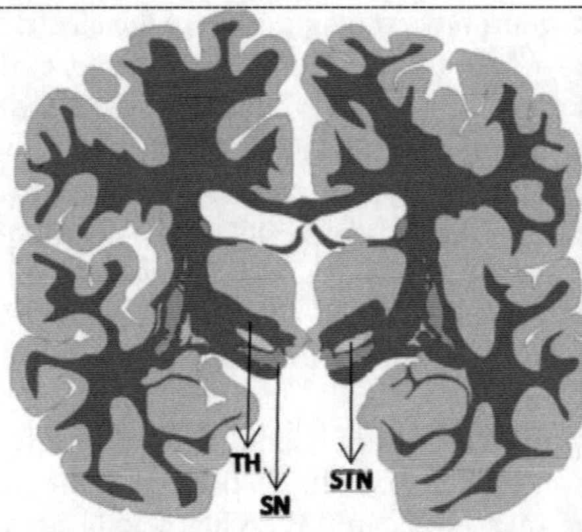

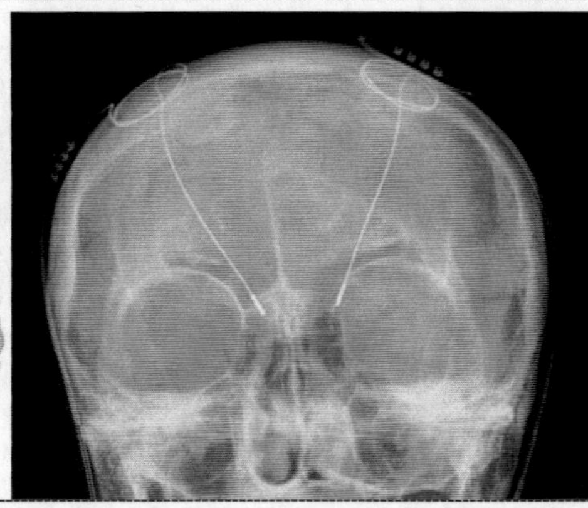

Figure 1. Left: Basal nuclei in the brain: Thalamus-TH, Subthalamus-STN and Substantia nigra-SN. Right: X-ray picture of a patient with a bilateral DBS implant.

Data mining tools are a useful to generate a model. Based on supervised learning and using previously acquired recordings, the model can successfully classify the STN area in new MER recordings. Guillen P. et al [6] were able to discriminate four sub-cortical zones (classes) in MER using statistical indexes further classified by 4.5 decision trees. However, it was robust but did not achieved a good precision in classification. Rodriguez – Sanchez A. et al [7] used non-linear indexes (correlation dimension, Hurst exponent and Lyapunov exponent) obtained from MER. The authors concluded that similar classification ratios could be achieved from statistical indexes than those obtained by non-linear indexes. Ciecierski K. et al [3] used decision trees to evaluate the Euclidean distance and the minimum variance algorithm to classify the mean square value into three groups: background noise amplitude, lowest and highest frequency in the background noise (calculated by means of the discrete Wavelet transform), and correlation between the pair of electrodes existing in the recording. The authors conclude that using these indicators is more stable and less statistically dependent than using spike detection. In addition, they show low classification ratios for some brain areas. Art Chaovalitwongse W. et al [2] propose three classifiers: Naive Bayes, K-nearest neighbours (K-NN) and decision tree using 13 statistical features and one frequency indicator obtained from the MER signals. They show that results obtained using the frequency indicator are poor and statistical features show better performance. From the three proposed classifiers, those obtaining better ratios were the decision tree and K-NN, achieving a sensitivity and specificity around 80% for the STN brain area. Up to our knowledge, there are no results for STN classification using Fuzzy K-NN (KNN-F) which improves the results of K-NN. This work describes STN detection using MER signals and comparing results between K-NN and KNN-F classifiers. Section 2 describes the methodology of the analysis and used data, followed by section 3 where the classification results are described. Finally, section 4 provides the discussion of the obtained results and the conclusions.

2. Methodology

2.1. Data recordings

La Fe Hospital in Valencia (Spain) leads a DBS implant program for Parkinson patients not responding to drugs. This study analyses MER neural activity recordings acquired from 8 patients during bilateral DBS surgery to implant the DBS stimulator in the STN area. Previously to surgery, the surgeons' team define the MER insertion trajectory supported by magnetic resonance images. This image also allows to define the target area where the stimulation electrode must be implanted. The surgery technique used at La Fe Hospital acquires tomography images at the time of surgery, which are merged with the initial resonance image (coregister) so that the position of the stimulation electrode can be accurately determined and finally positioned in the STN. Moreover, neurophysiologists analyse the electrical signal from MER during all the trajectory to evaluate the area where the electrodes are located in the brain: TH, STN or SN. This signal analysis allows to settle the start and the end of the STN area and they can even define the best position inside the STN according to the electrical signal patterns. In this work, the nerula activity recording was made in both brain hemispheres with 2 parallel MER electrodes per hemisphere (2 [mm] separation) using a Medtronic recorder model 3389 with platinum-iridium electrodes and an impedance between 2 and 6 [MOhm] at 1000 [Hz]. Recordings were obtained at 12 [kHz] sampling frequency using a 16-bit A/D converter at a total gain of 10000x with a 50 [Hz] mains filter. All recording were acquired with the MicroGuide Pro™ system from Alpha Omega which also determines the MER distance from the target area.

Data recording was done from 7 [mm] before to 4 [mm] after the target area. The MER advance was fixed in 0.2 [mm] steps with 1 second acquisition time in each step. Thus, 220 1-second registers was obtained from each patient. All recordings were later marked as in TH, STN or SN areas by the neurophysiologists and also validated by the acquired coregister images.

2.2. MER features obtained from recordings.

Signal processing was done in 1-second windows to calculate 15 features using Matlab®. Those registers showing artifacts and/or very low signal amplitude were discarded. For the remaining valid registers, the obtained parameters were, according to [2][4]:

1. Spike-independent features:

- basal: Basal amplitude value assessing the envolvent using the Hilbert transform.
- kurt: Signal kurtosis.
- CL: Sum of absolute value for the signal derivative.
- TH: Threshold in the signal amplitude.
- PK: Peaks number.
- RA: RMS value of amplitude.
- NE Non-linear mean energy.
- ZC: Zero crossings.

2. Spike-dependent features:

- SBI: Ratio between the number of interspike time intervals lower than 10 [ms] and intervals higher than 10 [ms].
- SP (pause ratio): Ratio between the number of interspike time intervals higher than 50 [ms] and intervals lower than 50 [ms].
- SPR (pause index ratio): Ratio between the accumulated interspike time in a time higher and lower than 50 [ms].

- SC: Spikes per second.
- SMAD: Mean spike differential amplitude as the mean value for the amplitude in two consecutive spikes.
- SSD: Standard deviation of interspike intervals.
- SF: Mean spike trigger frequency.

Thus, a signal database with all 15 parameters for each insertion step, MER electrode and patient was created, including the zone annotation class (TH, STN or SN) given by neurophysiologists. In case of zero or undetermined value of all parameters in a certain insertion step value, this step was removed from the database. As classifiers must have balanced register values in each class, 50 registers for each zone and patient were taken. Finally, the dataset was scaled between [0,1].

2.3. K nearest neighbours classifier (K-NN)

Let $R(z) \subset \Re^N$ a hypersphere with V volume centred in *z*. N_k is the number of samples for a training dataset T_k for the K-NN classifier and w_k the assigned class. The probability of having exactly *n* samples inside R(z) has a binomial distribution [8] according to equation (1).

$$E[n] = N_k \int_{y \in R(z)} p(y \mid w_k) dy \approx N_k V_{p(y|w_k)} \tag{1}$$

If a certain radius is chosen around z, able to generate a volume exactly containing K samples, then, the radius and its volume depends on the z position inside the measurement space [8]. Thus, we can describe V(z) instead of V, being the estimation of the density [9] that in equation (2).

$$\hat{p}(z|w_k) = \frac{K}{N_k V(z)} \tag{2}$$

In K-NN, the value of K is fixed to estimate the model and the minimum volume V(z) containing those K samples [9]. Expression in equation (2) shows that, in regions where the density estimation is high, a low volume is expected. This is similar to a reduced interpolation zone. On the other hand, when the density estimation is low, the sphere has to grow in size to include all samples [9].

Parameter K controls the balance between the observational error and the variance according to equation (3) [8].

$$\begin{aligned} &k \to \infty \quad \text{and} \quad N_k \to \infty \text{ to obtain a low variance} \\ &k / N_k \to 0 \quad \text{and} \quad N_k \to \infty \text{ to obtain a low observational error} \end{aligned} \tag{3}$$

The K-NN technique has a practical interest as it works with the data sample required to estimate the model without calculating the probability density. Being K_k the number of neighbour samples from class w_k, then, a conditional density estimator can be described in equation (4).

$$\hat{p}(z|w_k) \approx \frac{K_k}{N_k V(z)} \tag{4}$$

Combining (4) with the Naive Bayes classifier with uniform cost function [8], the estimated classification is obtained in equation (5).

$$\begin{aligned} &\hat{w}(z) = w_k \\ &k = \underset{i=1,\dots,K}{\arg\max} \left\{ \hat{p}(z|w_i) \hat{P}(w_i) \right\} = \underset{i=1,\dots,K}{\arg\max} \left\{ \frac{K_i}{N_i V(z)} \frac{N_i}{N_s} \right\} = \underset{i=1,\dots,K}{\arg\max} \{K_i\} \end{aligned} \tag{5}$$

Expression in equation (5) concludes that the assigned class to vector z is that with the highest number of neighbour samples from class w_k. closest to z. The K-NN method can be implemented with a relatively low computational cost even in case of high data samples [9]. The method greatly depends on the number of data samples but using the distance (typically Euclidean distance) as the parameter defining the class, a sub-optimal results might be obtained [8]. It is also a classifier which is very sensitive to noise in training data [9]. This work used PRTool under Matlab®; cross correlation (n=10) of training data gave a K value ranging from 1 to 15. Finally, K=3 was adopted as it provided the best performance in the classification for the database. This K value was already used in [2] for similar data as those used in this work.

2.4. Fuzzy K-NN classifier (KNN-F).

This variant of K-NN includes fuzzy logic and this fact allows a sample to obtain a certain degree of pertinence to different classes. Different application of fuzzy K-NN has been reported [10-12]. This method assigns distance functions from the k nearest neighbours to the sample. This method uses the same principle used in K-NN in which the chosen class for a sample is assigned by observing the neighbours with a higher degree of pertinence. Then, the final decision is more precise than K-NN where only distances were compared, especially in overlapped classes [12].

Being $W=\{x_1,\ldots,x_n\} \in \Re^N$ the set of n labelled data samples pertaining to c classes $W=\{Z_1,\ldots,Z_c\}$ where $\mu_i(x)$ (equation 6) is the membership function to be calculated for the input vector, and μ_{ij} is the membership function of the i-th class to vector j of labelled samples [11].

$$\mu_i(x) = \frac{\sum_{j=1}^{k} \mu_{ij}\left(1/\left|x - x_j\right|^{2/(m-1)}\right)}{\sum_{j=1}^{k} 1/\left|x - x_j\right|^{2/(m-1)}} \quad \text{where } i = 1,\ldots,c \tag{6}$$

The degree of pertinence of a sample to one of the classes is influenced by the inverse of the distances to the neighbours. This inverse of the distances can be used as a weight. In equation (6), m is the exponential weight representing the importance given to the centre due to the fact of being closer to the sample. Using cross correlation (n=10) of training data, the optimal m value ranging between 1 and 10 was obtained. Finally, m =2 was fixed as the best classification results were obtained. The nearest neighbour x_k related to the centre Z_i, is more influent when the membership degree is reflected. Assessing the membership degree is a certainty factor to assign a simple to a class. In this work, the KNN-F classifiers were obtained using Matlab®.

2.5. Performance of classifiers

Eight subsets (training and validation) were generated from the total features obtained from the MER recordings (section 2.2). As an example, subset 1 was created with all data from patient 1 (150 registers) for validation and all the other patients from 2 to 8 were used for training (1050 registers). This methodology was repeated for every patient as proposed in [2]. Using the 8 training subsets, 8 K-NN and 8 KNN-F classifiers were obtained, further obtaining the classifier performance for the validation data using a 3x3 confusion matrix. From all the 8 confusion matrices obtained for each classifier, 2 performance indexes were calculated: sensitivity and specificity. Sensitivity represents the ratio of registers detected in the right zone according to the annotated class. These values are shown in the diagonal values of the confusion matrix. Specificity shows those registers not belonging to a zone and they were not classified in that zone. The statistical analysis of the performance values was done using the Wilcoxon signed Rank non-parametric test for paired samples [8]. The tools used to obtain these values were PRTool and Matlab® statistical toolbox.

3. Results

Table 1 shows the results of sensitivity and specificity for the TH zone obtained with the two proposed classifiers (K-NN and KNN-F) using 15 features. The Wilcoxon signed rank test show $p<0.01$ for sensitivity and $p<0.05$ for specificity in the TH zone. In Figure 2 the box plot for both classifiers can be seen in the HT zone.

Table 1. Sensitivity and specificity for the TH zone using the K-NN and KNN-F classifiers trained with 15 features.

	Sensitivity K-NN	Sensitivity KNN-F	Specificity K-NN	Specificity KNN-F
Validation 1	0.73	0.85	0.79	0.91
Validation 2	0.76	0.82	0.80	0.89
Validation 3	0.73	0.88	0.69	0.91
Validation 4	0.73	0.86	0.83	0.93
Validation 5	0.66	0.86	0.71	0.90
Validation 6	0.71	0.91	0.77	0.94
Validation 7	0.72	0.86	0.84	0.84
Validation 8	0.79	0.84	0.71	0.84
Average	**0.73±0.038**	**0.86±0.027**	**0.77±0.058**	**0.90±0.037**

Table 2 shows the results of sensitivity and specificity for the STN zone obtained with the classifiers K-NN and KNN-F using 15 features. The Wilcoxon signed rank test shows a non-significant difference for sensitivity and $p<0.01$ for specificity in the STN zone. In Figure 3 the box plot for both classifiers can be seen in the STN zone.

Table 3 shows the results of sensitivity and specificity for the SN zone obtained with the proposed classifiers using 15 characteristics. The Wilcoxon signed rank test shows $p<0.05$ for sensitivity and non-significant difference in specificity in SN zone. In figure 4 the box plot for both classifiers can be seen in the SN zone.

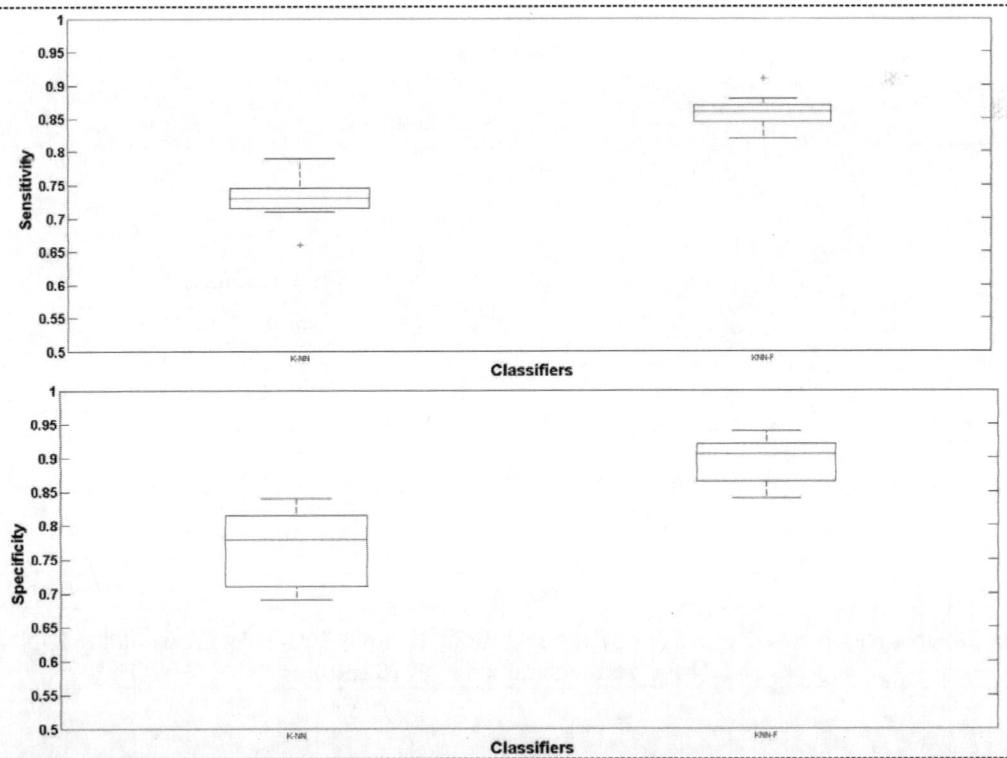

Figure 2. Sensitivity boxes diagrams (top) and specificity (below) for TH zone using the K-NN and KNN-F classifiers trained with 15 features.

Table 2. Sensitivity and specificity for the STN zone using the K-NN and KNN-F classifiers trained with 15 features.

	Sensitivity K-NN	Sensitivity KNN-F	Specificity K-NN	Specificity KNN-F
Validation 1	0.62	0.72	0.75	0.82
Validation 2	0.80	0.84	0.37	0.82
Validation 3	0.68	0.77	0.62	0.73
Validation 4	0.59	0.69	0.73	0.91
Validation 5	0.70	0.70	0.68	0.86
Validation 6	0.69	0.71	0.74	0.78
Validation 7	0.75	0.64	0.70	0.75
Validation 8	0.84	0.67	0.65	0.89
Average	**0.71±0.085**	**0.72±0.062**	**0.66±0.124**	**0.82±0.065**

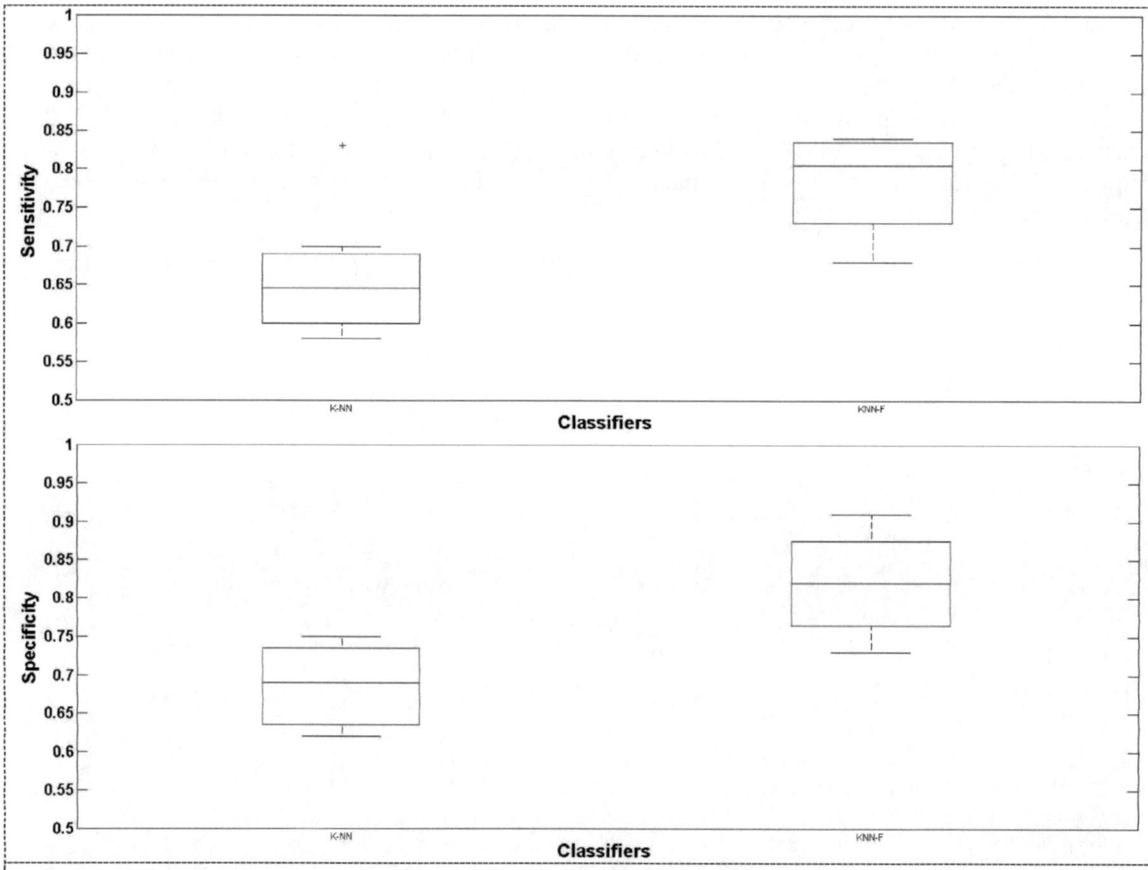

Figure 3. Sensitivity boxes diagrams (top) and specificity (below) for STN zone using the K-NN and KNN-F classifiers trained with 15 features.

Table 3. Sensitivity and specificity for the SN zone using the K-NN and KNN-F classifiers trained with 15 features.

	Sensitivity K-NN	Sensitivity KNN-F	Specificity K-NN	Specificity KNN-F
Validation 1	0.68	0.81	0.56	0.73
Validation 2	0.58	0.68	0.71	0.75
Validation 3	0.83	0.83	0.64	0.80
Validation 4	0.65	0.80	0.85	0.73
Validation 5	0.70	0.84	0.67	0.78
Validation 6	0.64	0.84	0.65	0.82
Validation 7	0.62	0.77	0.73	0.66
Validation 8	0.41	0.69	0.69	0.76
Average	**0.64±0.118**	**0.78±0.065**	**0.68±0.084**	**0.75±0.05**

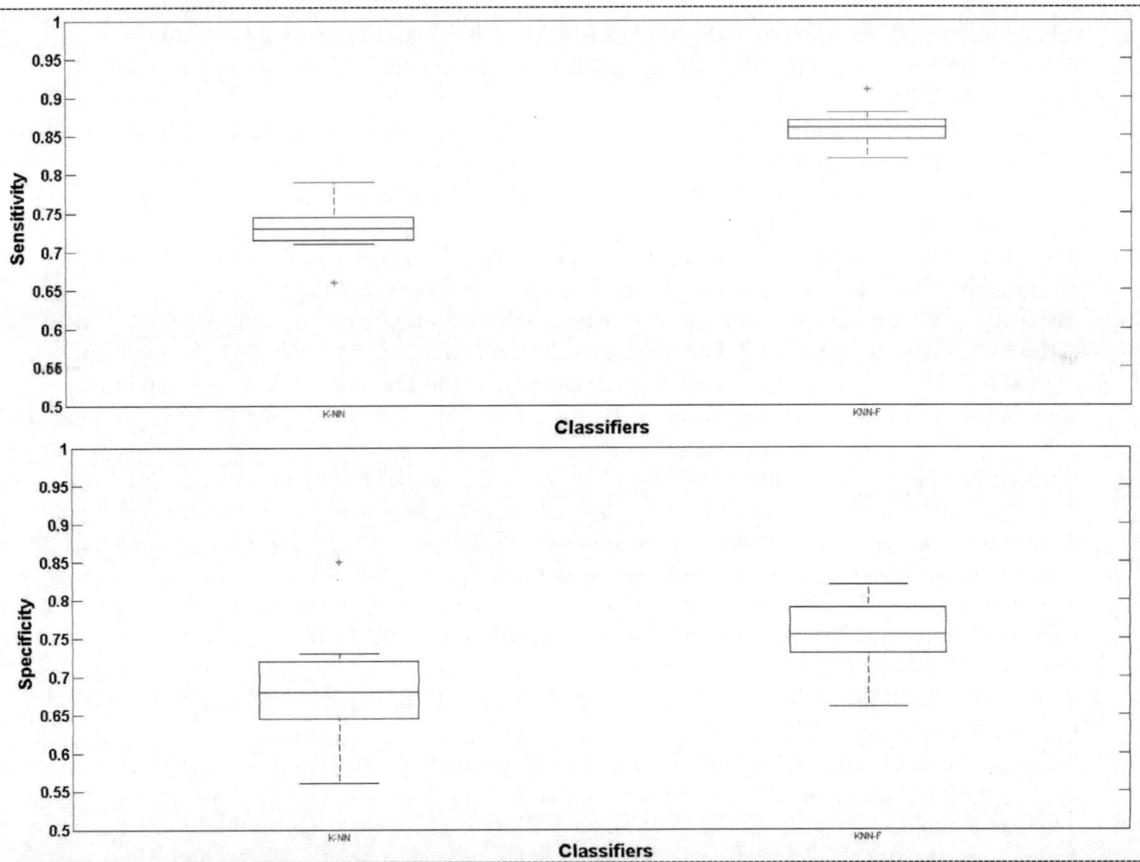

Figure 4. Sensitivity boxes diagrams (top) and specificity (below) for SN zone using the K-NN and KNN-F classifiers trained with 15 features.

4. Discussion and conclusions

The analysis of the results shows that the classifier KNN-F has better performance in all zones of classification. Specifically it has a very significant difference in sensitivity in the TH zone and significant difference in sensitivity of SN zone. For the zone of interest, that is the STN zone, the average sensitivity results were 0.72 ± 0.062 and average specificity results were 0.82 ± 0.065. Although we can not ensure a significant difference for sensitivity between the two methods, the statistical study shows a very significant difference in specificity. Probably it can be obtained higher sensitivities and specificities in the zone of interest and significant difference in sensitivity in the KNN zone if the number of patients for training classifiers it is increased. It is concluded that the performance of KNN-F classifier is acceptable and superior than KNN for the experimental conditions of this study and a group of 8 patients. Also it is possible, as a continuation of this work, combine KNN-F obtained by each training set using rules of maximum, minimum and average [10] which would provide a single more robust classifier with higher sensitivity and specificity, as it is proposed by [9] to other classifiers.

5. References

[1] Martínez Torres I Doctoral Thesis. 2014. Optimization and quantification of stimulation parameters in subthalamic nucleus deep brain stimulation for Parkinson's disease. Universidad de Valencia.

[2] Chaovalitwongse W A, Jeong Y-S, Jeong M-K, Danish S F and Wong S 2011 Pattern Recognition Approaches for Identifying Subcortical Targets during Deep Brain Stimulation Surgery *IEEE Intell. Syst.* **26** 54–63

[3] Ciecierski K, Raś Z W and Przybyszewski A W 2012 Foundations of Recommender System for STN Localization during DBS Surgery in Parkinson's Patients *Foundations of Intelligent Systems* Lecture Notes in Computer Science ed L Chen, A Felfernig, J Liu and Z W Raś (Springer Berlin Heidelberg) pp 234–43

[4] Guerrero-Martínez J. R-M A 2015 Characterization of Microelectrode Records in Deep Brain Stimulation Applied to Parkinson's Disease Patients *IFMBE* **49** 647–50

[5] Bronstein J M, Tagliati M, Alterman R L, Lozano A M, Volkmann J, Stefani A, Horak F B, Okun M S, Foote K D, Krack P, Pahwa R, Henderson J M, Hariz M I, Bakay R A, Rezai A, Marks W J, Moro E, Vitek J L, Weaver F M, Gross R E and DeLong M R 2011 Deep brain stimulation for Parkinson disease: an expert consensus and review of key issues *Arch. Neurol.* **68** 165

[6] Guillen P, Argaez M, Martínez-de-Pisón F, Velazquez L and Martinez-de-Pison F 2012 Data mining in the process of localization and classification of subcorticals structures *2012 6th Euro American Conference on Telematics and Information Systems (EATIS)* 2012 6th Euro American Conference on Telematics and Information Systems (EATIS) pp 1–5

[7] Rodriguez-Sanchez A, Delgado-Trejos E, Orozco-Gutierrez A, Castellanos-Dominguez G and Guijarro-Estelles E 2008 Nonlinear Dynamics Techniques for the Detection of the Brain Areas Using MER Signals *International Conference on BioMedical Engineering and Informatics, 2008. BMEI 2008* International Conference on BioMedical Engineering and Informatics, 2008. BMEI 2008 vol 2 pp 198–202

[8] Miner G, Nisbet R and IV J E 2009 *Handbook of Statistical Analysis and Data Mining Applications* (Amsterdam ; Boston: Academic Press)

[9] Heijden F van der, Duin R, Ridder D de and Tax D M J. 2005. Book *Classification, Parameter Estimation and State Estimation: An Engineering Approach Using MATLAB* (John Wiley & Sons)

[10] Piorno Campo J Master Thesis. 2009. Diseño de un nuevo clasificador supervisado para minería de datos. Universidad Complutense de Madrid.

[11] Keller J M, Gray M R and Givens J A 1985 A fuzzy K-nearest neighbor algorithm *IEEE Trans. Syst. Man Cybern.* **SMC-15** 580–5

[12] Kim S-Y, Sim J and Lee J 2006 Fuzzy k-Nearest Neighbor Method for Protein Secondary Structure Prediction and Its Parallel Implementation *Computational Intelligence and Bioinformatics* Lecture Notes in Computer Science ed D-S Huang, K Li and G W Irwin (Springer Berlin Heidelberg) pp 444–53

Acknowledgement
The authors wish to thank the Carolina Foundation, the University of Valencia, the Faculty of Engineering of the National University of Entre Ríos and the Autonomous University of Entre Rios for the aid received for the doctoral scholarship of Ms. Bioing. Luciano Schiaffino. Also thank Functional Neurosurgery Unit of La Fe Hospital (Valencia - Spain) by the advice received.

An approach to emotion recognition in single-channel EEG signals: a mother child interaction.

A Gómez[1], L Quintero[2], N López[3] and J Castro[4]*
[1] Mathematical Modeling Research Group, GRIMMAT, School of Sciences, Universidad EAFIT
[2] Mathematical Modeling Research Group, GRIMMAT, School of Sciences, Universidad EAFIT
[3] Medical Technology Laboratory, Faculty of Engineering, Universidad Nacional de San Juan
[4] Psychology, Education and Culture Research Group Faculty of Social Science Politécnico Grancolombiano University Institution

E-mail: [1] agomez13@eafit.edu.co, [2] oquinte1@eafit.edu.co, [3] nlopez@gateme.unsj.edu.ar and [4] jcastrom@poligran.edu.co

Abstract.
In this work, we perform a first approach to emotion recognition from EEG single channel signals extracted in four (4) mother-child dyads experiment in developmental psychology. Single channel EEG signals are analyzed and processed using several window sizes by performing a statistical analysis over features in the time and frequency domains. Finally, a neural network obtained an average accuracy rate of 99% of classification in two emotional states such as happiness and sadness.

1. Introduction

To provide an exact definition of emotions is difficult to establish [1],[2], but for convenience, we consider emotions as a neuronal response to external stimuli. The study of these stimuli is very important, because it could bring help to build a psychological profile of a person. This profile will can be usefull for the diagnosis of possible psychopathology of the individual, or will make suitable a prevention strategies with proper treatment, evaluating their behavior into a social or specific situations in the enviroment.

Different approaches have been reported for the determination of an emotional state, such as facial expression recognition, speech analysis, and interpretation of corporal gestures [3][4][5][6]. A different approach to the detection of emotions is through the analysis of electroencephalographic signals, making them the extent of the interaction between neurons to perform any action or work, in our case, "creating emotions"; the importance of emotion detection using these signals is that they are directly related to the brain activity and could be mapped to several areas of brain cortex. This will allows to determine or validate phisiologically

*This work was supported by Politecnico Grancolombiano and Universidad EAFIT.

the emotional state. Also, it is not necessary to perform a phonetic activity or a gestural expression, which is quite useful for the determination of lies or detecting emotions in infants.

This study seeks to determine when the emotional states of happiness, and sadness appears in a mother-child dyad with the analysis of some descriptive characteristics in time and frequency domain, after conducting an experiment to evoke emotions, using a single channel from a EEG signal, located in the x position of the standard 10-20 configuration.

The purpose of this work is to validate the psychological results adressed in [7][8][9] by a triangulation procedure of experts. Also, extrapolate some methodological results previosly obtained in emotion recognition from audio signals and finally to establish a procedure for lack of information in EEG signals in order to find the relevant features of the signals to proceed with a multi channel procedure.

It is necessary to give a definition of what is a dyad, Kenny define the dyad as "the fundamental unit of interpersonal interaction and interpersonal relations"[10]. That is, a group of two persons in this case mother and child. The importance of this dyad relays in the high impact of the mother's behavior in the social and emotional child development [11].

The importance of the study of these dyadic interactions focuses on the fact that can help to create a complete scheme on mental development of infants. In this case the mother-child dyad was analyzed for the evaluation of the interaction mother-child into a developmental psychological study. The study evaluated the capabilty of the emotional induction over a child from the mother without verbal communication. Also, detecting emotions evoked in the mother and its consequence on the evocation of emotions in her child[7][8][9]. Therefore, this research seeks for a classification system to estimate the emotional state of sadness and state of happiness in this dyad and validate the previous results, also develop an analysis method for EEG single channel signals.

The raw EEG signal usually has an amplitude of the order of μV and contains frequency components of up to 300Hz[12], in several studies report the optimal ranges for the detection of emotions in EEG signals in different frequency bands of the EEG signal. This bands are classified as: Delta band (0.5 to 4 Hz), Theta band (4 to 8 Hz), Alpha band (8 to 13 Hz), Beta band (13 to 30 Hz), and Gamma band (30 to 70 Hz)[12], [13]. Detecting emotions from EEG signals has been a study field in recent years [14][15][16][17][18][19][20][21][22].

Pentrantonakis used the Delta, Theta, Alpha and Beta band with a HOC analysis for determinated six emotions: Happiness, Surprise, Anger, Fear, Disgust, and Sadness with a minimal rate of classification for Fear with 75% and a maximum rate for Happiness with 99.373% [14]; Lee used Theta, Alpha, Beta and Gamma band for demostrated different patterns for differents emotional state, with different combination of this bands, performing a complete analysis of correlation, coherence, and phase synchronization index of each band in different emotional states, [15]; Varun executes a more conservative analysis as to the frequency range, taking all the frequency bands previously exposed in a range of 0.5 Hz to 100 Hz, implementing a classification between happiness, neutral, sadness and fear based on multiwavelet transform, with a mean acuracy of 80%[16]; Murugappan proposes the implementation of a Surface Laplacian filter to remove noises and artifacts and used an extraction method based on multi-resolution analysis of Wavelet Transform, to determinated between Disgust, Happiness, Surprise, Fear and Neutral with different classifier as LDA and KNN, comparing the number of channels of EEG from 62 to 8, with a mean acuracy of 85% [17]; and Li proposed the classification of two emotions, sadness and happiness using Common Spatial Patterns on the Gamma Band with a mean acuracy of 93.5%.[18]

This paper relates a study carried out on a small cohort of mother child "dyads" in order to detect a binary emotion of either happiness or sadness. The work revolves around using some standard statistical measures applied to a single channel EEG signal obtained from the subjects adult and child alike. Different window sizes were used to perform the statistical analyses and these were then used a features in a MLP neural network to classify outcomes into either happiness or sadness. The classifier gave a 99% success rate.

In this work we propose an approach to a methodology for emotion recognition in EEG signals from one channel information. Also, presents an easily and understandable solution with preprocessing of the signal, windowing and time-frequency classical features extration and finally a neural network classifier [23][24].

No specific EEG frequency bands were taken into account and for simplicity, statistical measurements were performed to extract relevant information from features. Signals are not only sampled from adults but also from children and emotional states were confirmed by human experts.

This work is organized as follows: Section 2 presents the methodology, preprocessing, features extraction and the classifier proposed. Section 3 presents the results obtained and finally the conclusions and future work will be adressed.

2. Methodology

The proposed methodology, looks for the extraction of simple features from the single channel signal. The first approach to the analysis of the information relays on the classical features extracted from EEG signals as descriptors of some physical phenomenon such as energy and frequency components. More sophisticated features will be considered in future work. The relevance of the proposed methodology is based on the use of time and frequency variables for feature extraction (typically used on voice features [25]; instead of the traditional features based on frequency bands used on EEG treatment. Also, the choice of these features related to the physical phenomena allows to perform a classification task with a high rate of accuracy. The classifiers were trained separating the complete set of data in three subsets for training (60%), test (20%) and a final set for validation (20%). Several MLP architectures were tested used at least the features provided by different sets obtained from three windows size on sampling stage as detailed in section II.A.This section presents the basic scheme for filtering and addresses some descriptive features in a simple way.

2.1. Pre-processing

The signal filtering process was implemented with three IIR filters: a high pass filter of 10th order with response Butterworth at 0.5Hz, a low pass filter of 10th order with response Butterworth at 70Hz, and a notch filter of 10th order with response Butterworth at 60Hz.

To perform the analysis of the signal, a windowing process is implemented with a Hamming window type, which satisfies the following equation:

$$w(n) = 0.54 - 0.46\cos\left(2\pi\frac{n}{L-1}\right) \tag{1}$$

Where L is the length of the window. An overlap of 50% was done, and three window sizes: 100ms, 40ms and 20ms; each windows sizes were essayed , in order to extract features as detalied below.

2.2. Descriptive Characteristics

For training a classifier system, it was decided to find some descriptive characteristics of the EEG signal in a temporal and frequency approach. Given the random behavior of the signal [12], some type of statistical analysis must be performed[26]. The employed features were:

2.2.1. IAV The integral average value of the signal:

$$IAV = \frac{1}{N}\sum_{i=1}^{N}|x_i| \tag{2}$$

Where N is the length of the samples, in this case the length of the window, and x_i the input signal.

2.2.2. RMS The root mean square of the signal:

$$RMS = \sqrt{\frac{1}{N}\sum_{i=1}^{N}(x_i)^2} \tag{3}$$

2.2.3. Mean The mean of the signal:

$$m = \frac{1}{N}\sum_{i=1}^{N}x_i \tag{4}$$

2.2.4. ZC The number of zero crossing of the signal:

$$zc = \sum_{i}^{N-1}(Y(i+1) - Y(i))^2 \tag{5}$$

Where Y is:

$$Y(i) = \begin{matrix} 1 & \text{if} & x(i) \geq 0 \\ 0 & \text{if} & x(i) < 0 \end{matrix} \tag{6}$$

2.2.5. MDF The median frequency of the signal:

$$MDF = \frac{1}{2}\sum_{i=1}^{N}P(i) \tag{7}$$

Where P is the Power Spectrum.

2.2.6. MNF The mean frequency of the signal:

$$MNF = \frac{\sum_{i=1}^{N} f(i)P(i)}{\sum_{i=1}^{N} P(i)} \tag{8}$$

Where P is the Power Spectrum, and f is the frequency.

2.2.7. Máx PS The maximun value of the Power Spectrum of the signal:

$$PM = MAX(P) \tag{9}$$

2.2.8. VAR The variance of the signal:

$$VAR = \sum_{i}^{N}(x(i) - m)^2 \tag{10}$$

2.2.9. SD The standard deviation of the signal:

$$VAR = \frac{1}{\sqrt{N}} \sum_{i}^{N} (x(i) - m) \tag{11}$$

2.2.10. WL The wavelength of the signal:

$$WL = \frac{1}{N} \sum_{i}^{N-1} |\delta x| = \frac{1}{N} \sum_{i}^{N-1} |x(i+1) - x(i)| \tag{12}$$

2.2.11. WL2 The wavelength of the standardized signal:

$$WL2 = \sum_{i}^{N-1} \frac{|x(i+1) - x(i)|}{(x(i) - m)} \tag{13}$$

2.3. Classifier
Several training options such us the training method the activation functions and neural networks architecture were evaluated. The final classifier is a well known multi layer perceptron (MLP) with an architecture of 11 inputs, 3 hidden layers, each one with 11, 11 and 10 neurons and a single output. Activation function were tansigmoidal and purelinear in output layer. Since the noisy behavior of the MLP's output, it is low pass filtered and a binarized. The low-pass filter is a moving average filter, given by the following equation:

$$x_f(i) = \sum_{4}^{i=1} x(i) \tag{14}$$

Where $x_f(i)$ is the filtered signal $x(i)$, as this filtered signal can take different values of 0 and 1, and our classifier has only two possible states: sadness and happiness, we use the binarizared function that is show in following expression:

$$x_b(i) = \begin{matrix} 1 & \text{if} & x_b(i) \geq 0.5 \\ 0 & \text{if} & x_b(i) < 0.5 \end{matrix} \tag{15}$$

2.4. Experimental Protocol
The study was conducted with 8 subjects, 4 women (mothers), and 4 children (3 males and 1 female) with a mean age of 22 months, each mother complete a form of Informed consent, and each mother was informed about the purpose of this experiment. The experiment was approved by the local Ethical Commitee.

The stimulus of happiness for the mother consisted into hear the happiest story of her own life and the stimulus of sadness for the mother consisted into hear the saddest story of her own life, each story was told in a previously session with a professional in charge of the experiment. The stimulus for the child in every emotional state was evoked by the presence of his mother face to face.

To perform the experiment the following protocol was followed: First, each mother was asked to make a recording in a room, the happiest moment of her life for the stimulus of happiness, and the saddest moment of her life to the stimulus sadness, then every dyad was placed face to face; the mother had headphones and listen in each case (happy or sad) the story that her previously recorded, causing the feeling of happiness or sadness, as the case, creating an evocation of emotions on her child who was staring at her. Each of these moments was videotaped, the

Table 1. Average of descriptive characteristics, being H happiness and S sadness, for mother and child.

		Mother		Child	
		H	S	H	S
100 ms	RMS	0.1085	0.1252	0.1284	0.1847
	IAV	0.1002	0.1152	0.1327	0.1704
	M	0.5319	0.5184	0.4591	0.4736
	ZC	0.2875	0.3181	0.3044	0.3609
	PM	0.0232	0.0264	0,0420	0.0563
	MNF	0.2492	0.2654	0.2490	0.3063
	MDF	0.0263	0.0328	0.0370	0.0655
	VAR	0.0419	0.0500	0.0389	0.0696
	SD	0.1273	0.1390	0.1314	0.1907
	WL	0.1408	0.1619	0.1801	0.2330
	WL2	0.3312	0.3582	0.3532	0.3666
40 ms	IAV	0.1043	0.1111	0.1232	0.1659
	RMS	0.0993	0.1061	0.1187	0.1578
	M	0.5171	0.5345	0.4651	0.4777
	ZC	0.2174	0.2730	0.2462	0.2801
	PM	0.0220	0.0223	0.0314	0.0441
	MDF	0.2150	0.2478	0.2302	0.2607
	MNF	0.0233	0.0250	0.0296	0.0500
	VAR	0.0292	0.0321	0.0350	0.0412
	SD	0.1182	0.1213	0.1356	0.1554
	WL	0.1280	0.1258	0.1443	0.1820
	WL2	0.2676	0.2877	0.2719	0.2671
20 ms	IAV	0.1002	0.1026	0.1087	0.1511
	RMS	0.0977	0.1000	0.1062	0.1502
	M	0.5102	0.5276	0.4448	0.4656
	ZC	0.1724	0.2013	0.1799	0.2113
	PM	0.0209	0.0208	0.0223	0.0342
	MDF	0.1807	0.1934	0.1855	0.1997
	MNF	0.0215	0.0221	0.0231	0.0353
	VAR	0.0234	0.0254	0.0280	0.0390
	SD	0.1051	0.1093	0.1184	0.1580
	WL	0.1089	0.1108	0.1218	0.1609
	WL2	0.2461	0.2677	0.2627	0.2735

measures of the EEG signals were taken from single channel to minimize the level of invasiveness of the experiment and it was taken at a sampling rate of 1000 Hz and It was counted with a team of psychologists who verified the validity of this protocol that meets the ethical principles of Helsinksi [7][8][9].

3. Results and Discussion

The emotion detection of sadness and happiness was performed with 3 different sizes of windows. In the Table 1 we present the averages of each of the extracted features, for happiness and sadness, with difference between the mother or child for differents length of windows.

It may be noted that each characteristic is affected by the window size, increasing its magnitude proportionally with the size of the window. It can be seen that each characteristic is

Table 2. Classification results for different descriptive characteristics individually and together, with different window sizes.

	100	40	20
RMS	56.7499	56.8474	57.0478
IAV	57.4737	56.6974	56.7725
M	56.6052	58.2705	57.9574
ZC	64.9796	58.8363	53.9109
PM	56.1709	56.4423	56.6932
MNF	57.2325	57.3768	56.9203
MDF	54.7877	53.8856	57.7474
VAR	59.1411	54.9250	55.8458
SD	58.3369	53.2726	56.1854
WL	59.0768	55.3408	55.7022
WL2	55.8224	56.3459	58.1234
All	**99.3566**	**99.9700**	**99.9486**

affected by the window size, but these changes are acceptably small, where it can been said after averaging all results, that could be used bigger window sizes for purposes of computational costs. With the matrix resulting from the extraction of descriptive characteristics we proceed to train a multilayer perceptron for each characteristics individually and one for the set of characteristics, which gives us a total of 12 MLP, the same procedure was performed with 3 different sizes of window. After having each MLP trained, was calculated the percentage of correct classification, the results are in Table 2.

It may be noted that each feature individually does not provide enough information for recognition of Happiness state or Sadness state, in contrast can be seen that the percentage of classification of all these characteristics together on average has a percentage of 99.7% correct classification. Which allows infer, that the size of the window may slightly affect the success rate of the classifier, but reduce computational costs, window sizes could be around 100ms to be affected by 0.4%, which can be disregarded in this case. This could be remedied by performing an analysis of emotional states in the previous and next segment, seeking to avoid irregularities and misclassification.

Future works related to single channel emotion detection will be focused on multiresolution analysis with Dabeuchies wavelet family to features extraction to improve accuracy and provide more phenomenon related characteristics.

4. Conclusions

The detection between the happiness and sadness emotional state, is possible through an analysis of a single channel of EEG. The analysis of the signal using temporal and spectral features, allows the classification of the emotional states with a high rate with a nonlinear classsifer as the MLP.

Moreover, the effect of window size on the results was also evaluated. Given the high classification rates, it can be concluded that the size of the analysis window is important but stable. Allowing the selection of larger window sizes to reduce computational costs, and trying to reduce the small effect of misclassification of the signal, assesing the segments before and after the emotional states obtained by the classifier.

This analysis can be used to study the mental mappings during the mother-child dyads interactions. As future work, we will study the detection of a greater number of emotions, and different number of channels.

Acknowledgments

The authors wish to acknowledge the volunteers and their children, which allowed this analysis possible.

References

[1] M. Cabanac and M. Cabanac, "What is emotion?" *Behavioural processes*, vol. 60, pp. 69–83, 2002.
[2] K. R. Scherer, "What are emotions? And how can they be measured?" pp. 695–729, 2005.
[3] Y. Sun, G. Wen, and J. Wang, "Weighted spectral features based on local hu moments for speech emotion recognition," *Biomedical Signal Processing and Control*, vol. 18, no. 0, pp. 80 – 90, 2015. [Online]. Available: http://www.sciencedirect.com/science/article/pii/S174680941400158X
[4] A. Milton and S. T. Selvi, "Class-specific multiple classifiers scheme to recognize emotions from speech signals," vol. 28, no. 3, pp. 727 – 742, 2014. [Online]. Available: http://www.sciencedirect.com/science/article/pii/S0885230813000594
[5] X. Feng and J. Watada, "Building a recognition system of speech emotion and emotional states," *2013 Second International Conference on Robot, Vision and Signal Processing*, vol. 0, pp. 253–258, 2013.
[6] T. Krishna, A. Rai, S. Bansal, S. Khandelwal, S. Gupta, and D. Goyal, "Emotion recognition using facial and audio features," in *Proceedings of the 15th ACM on International Conference on Multimodal Interaction*, ser. ICMI '13. New York, NY, USA: ACM, 2013, pp. 557–564. [Online]. Available: http://doi.acm.org/10.1145/2522848.2531746
[7] J. A. Castro Martinez, "Neurodinámica y autoorganización en la interacción socioemocional madre-hijo: aproximación de los sistemas dinámicos a los principios del desarrollo emocional infantil," vol. 2, p. 17.
[8] ——, "Sistemas dinámicos en la interacción emocional madre-hijo: primera fase," vol. 9, pp. 129–138.
[9] J. A. Castro Martinez and E. Londono Mora, "istemas dinmicos y regulacin emocional," vol. 6-11, pp. 131–150.
[10] D. Kenny, D. Kashy, and W. Cook, *Dyadic Data Analysis*, ser. Methodology in the social sciences. Guilford Press, 2006.
[11] L. A. Killeen, "Understanding parenting as a process: Frontal eeg alpha asymmetry as a measure of "online" maternal responsiveness to infant cues."
[12] S. Sanei and J. a. Chambers, *EEG Signal Processing*, 2007, vol. 1.
[13] M. Teplan, "Fundamentals of EEG measurement," *Measurement Science Review*, vol. 2, pp. 1–11, 2002.
[14] P. C. Petrantonakis and L. J. Hadjileontiadis, "Emotion recognition from EEG using higher order crossings." *IEEE transactions on information technology in biomedicine : a publication of the IEEE Engineering in Medicine and Biology Society*, vol. 14, pp. 186–197, 2010.
[15] Y. Y. Lee and S. Hsieh, "Classifying different emotional states by means of eegbased functional connectivity patterns," *PLoS ONE*, vol. 9, 2014.
[16] R. B. P. Varun Bajaj, "Classification of human emotions based on multiwavelet transform of EEG signals ," 2012.
[17] M. Murugappan, "Classification of human emotion from EEG using discrete wavelet transform," pp. 390–396, 2010.
[18] M. Li and B. L. Lu, "Emotion classification based on gamma-band EEG," pp. 1323–1326, 2009.
[19] Y. Liu, O. Sourina, and M. K. Nguyen, "Real-time EEG-based emotion recognition and its applications," in *Lecture Notes in Computer Science (including subseries Lecture Notes in Artificial Intelligence and Lecture Notes in Bioinformatics)*, vol. 6670 LNCS, 2011, pp. 256–277.
[20] K. Takahashi, "Remarks on Emotion Recognition from BioPotential Signals," in *Proceedings of the 2nd International Conference on Autonomous Robots and Agents*, 2004, pp. 186–191.
[21] X. Wang, D. Nie, and B. Lu, "EEG-based emotion recognition using frequency domain features and support vector machines," *Neural Information Processing*, pp. 734–743, 2011.
[22] E. O. Min-Ki Kim, Miyoung Kim and S.-P. Kim, "A Review on the Computational Methods for Emotional State Estimation from the Human EEG," *Computational and Mathematical Methods in Medicine*, vol. 2013, 2013.
[23] N. M. L. Celani, "Procesamiento de señales electromiográficas superficiales para el control de dispositivos robóticos," Ph.D. dissertation, Universidad Nacional de San Juan, March 2010.
[24] N. M. Lopez, A. Garces, F. di Sciascio, T. Freire Bastos, and M. E. Valentinuzzi, "Compensación de la fatiga muscular para aplicaciones en robótica de asistencia."
[25] P. Bustamante, N. Lopez, M. Perez, and O. Quintero, "Recognition and regionalization of emotions in the arousal-valence plane," *2015 IEEE EMBC*, 2015.
[26] D. Manolakis and V. Ingle, *Applied Digital Signal Processing: Theory and Practice.* Cambridge University Press, 2011.

Repeatability Evaluation of a Contrast Sensitivity System for Transfer to the Eye Clinic

N G Alcalde[1,2], L R Castillo[2,3], C Paz Filgueira[2,3], E M Colombo[2,3]
[1] Department of Bioengineering, Sciences and Technology School – National University of Tucumán(FACET – UNT). Avenida Independencia 1800 – CPA (T4002BLF) – Tucumán, Argentina.
[2] Department of Lighting, Light and Vision (DLLyV), FACET – UNT, Avenida Independencia 1800 CPA (T4002BLR) – Tucumán, Argentina;
[3] Institute of Investigation in Light, Environment and Vision (ILAV – CONICET - UNT) – Avenida Independencia 1800 1800 CPA (T4002BLR) – Tucumán, Argentina.

E-mail: noeg.alcalde69@gmail.com

Abstract: The Contrast Sensitivity Function (CSF) is a valuable tool which can be used to characterize functional vision and also for the diagnosis and management of patients with different eye diseases. In spite of its usefulness, the CSF is currently hardly ever used in clinical practice. The aim of this study was to validate the use of the system called FVC-100 (Tecnovinc-UNT-CONICET, Argentina), which calculates the CSF, in order to transfer this important tool to ophthalmological clinics. The validation was carried out through the design of a repeatability test and the subsequent analysis of the results. Furthermore, we evaluated the impact of different factors influencing the repeatability of the measurements such as age and previous training. The tests were based on the discrimination of sinusoidal gratings for different spatial frequencies (1, 4 and 12 c/°) in both eyes of 12 people, aged between 20 and 70. The results show that the calculated values of SC of each subject have a high repeatability and are not dependent on age or training. These results allow us to conclude positively regarding the effectiveness of the FVC-100, and to validate its use in clinics for the calculation of the FSC as a standard measure of functional vision quality.

1. Introduction

The concept of functional vision is associated with the capability of people to perform everyday tasks [1]. It is known that functional vision is not only related to visual acuity (VA), i.e. a measurement of the minimum detail within high contrast that a person can see, but also to contrast sensitivity (SC), that is the reverse of the minimum contrast that a person needs to distinguish a background stimulus, at different sizes and luminance levels. Measuring is carried out by using sine gratings of different spatial frequencies that covers the human range of vision, obtaining the Contrast Sensitivity Function (CSF). Bibliography has established that this function offers a valuable tool for the diagnosis and tracing of patients who have different diseases such as cataracts, amblyopia and glaucoma, among others [2 – 4]. However, and considering that nowadays there are commercial measurement systems, its use is reduced in clinics. In the past few years, interest regarding CSF has been renewed [5, 6], and there have been different tests evaluated for clinics [7, 8].

In several scientific and technological investigative works in carried out by our institution, DLLyV – ILAV (Department of Lighting, Light and Vision – Institute of Investigation in Light, Environment and Vision, FACET – UNT), that have led to several master`s thesis, final engineering projects and several scientific papers [5, 6, 9 – 12] the specifications that a system needs to be implemented successfully in Ophthalmological Clinics were analyzed and determined. As a result of this, a computerized system was developed, which we call FVC-100, to determine the CSF. This system has proved to be a valuable tool to clinical environment [11], and is continuously being improved on [13].

In this work, an experiment with the FVC-100 was designed, with the purpose of evaluating the repeatability degree in order to use it in a successful way in the clinic. In addition, this evaluation includes the effects of age and training on the repeatability degree. The hypothesis of this work is that a single measure would be accurate enough and carried out in an adequate time, parameters which are important, not only for patients but also for doctors and optometrists, were set in order to implement CSF as a clinic standard.

2. Materials and Methods

2.1. The Generator of Visual Stimulus

The system to be controlled is a computerized apparatus which ~~to~~ determines the contrast sensitivity function (FVC-100, Tecnovinc – UNT – CONICET, Argentina) It was developed entirely by the faculty where the present work was carried out. This system permits the generation of the visual stimulus and shows them on the screen of a CRT monitor LG brand, model FLATRON EZ T730SH with a widescreen of 17''. Using a video adapter/attenuator and, providing a luminance resolution of 0.01 cd/m², approximately 14 bits of resolution are obtained, which in turn permits reaching 3500 gray levels, needed to generate stimulus that can be presented with Michelson contrasts between 0.002 and 1, within errors of between 1.5% and 10% [5].

Figure 1. Sistema FVC-100 – Scheme and picture. Extracted image from C. Paz Filgueira and J. Santillán, 2014.

The viewing distance is 1.5 m from the monitor to minimize the demand for accommodation / convergence. The visual stimulus consists of sine gratings that are presented in a circle subtending 6.7 degrees which is located in the center of the CRT. The grating spatial frequency is modified by specific software and it lasts 500 ms. Furthermore, such gratings were randomly inclined 7 degrees to the left or to the right.

Figure 2. Typical sine grating shown in CRT monitor. Extracted image from C. Paz Filgueira and J. Santillán, 2014.

Since the task of the observer/patient is to discriminate the orientation of the grating, the system has a joystick control with two buttons, so that if the grating is visualized with a tilt to the left, the left button is pressed, otherwise, the right button will be pressed.

At a specific spatial frequency, each stimulus is generated by software that use an algorithm called QUEST (Quick Estimation by Sequential Testing) [12], which is an adaptive psychometric method. As the answer method is an alternative forced choice (2AFC), the observer has a 50% chance of guessing correctly and a 50% chance of guessing incorrectly. With each correct answer, the algorithm calculates and shows a lower contrast stimulus on the screen and when there is an incorrect response, the stimulus that is shown has a higher contrast than the preceding one. As mentioned before, this method is adaptive because each grating presented on the CRT is defined according to the answers given by the previous stimuli, so that the next stimulus is based on the current and all previous ones.

The measurement of the contrast threshold for each spatial frequency was performed with a total of 25 repetitions or stimuli presentation (see Figure 3), except in the case where the observer answers the first 17 reps correctly, then the test was increased to 42 repetitions (see Figure 4). The inverse of the final value obtained is the contrast sensitivity value for the spatial frequency that was measured. It should be clarified that each repetition involves the calculation of a new contrast value from the response of the observer/patient, and the generation of a new stimulus by the apparatus.

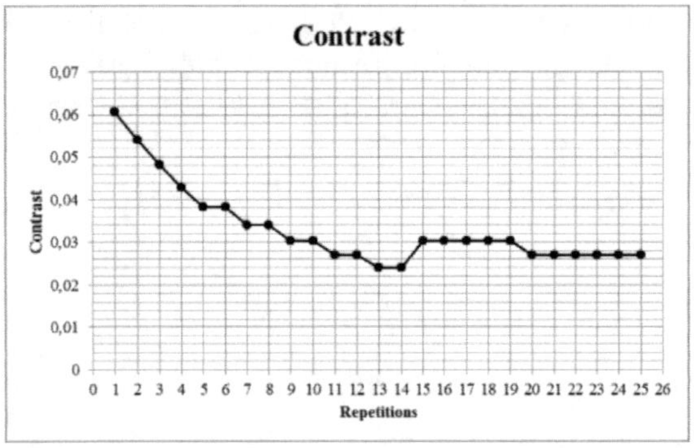

Figure 3. Contrast variation according to responses throughout a measurement for a given spatial frequency, with 25 steps.

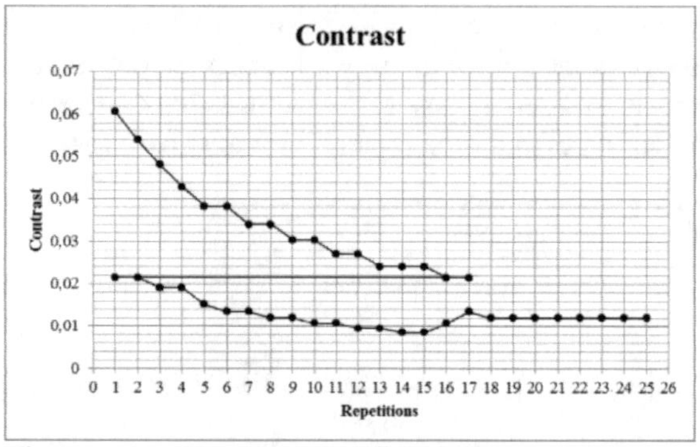

Figure 4. Contrast vs Repetitions when the orientation of the first 17 networks presented Have been answered correctly.

2.2. Participant Selection Criteria

Measurements were taken from 12 people aged between 20 and 70, for which we distinguish two wide age classes: from 20 to 49 and from 50 to 70 years old, considering the results from different studies describing the influence of the age of the subjects in determining contrast sensitivity [6, 13-15].

Within the inclusion criteria, it was considered that participants in these tests should be emmetropes or could have any refractive pathology such as myopia, hyperopia and/or astigmatism, but that they must be properly corrected at the time of the measurement. In this way, it was considered that they should have a Visual acuity (AV) greater than or equal to 0.8 per eye and, in order to control this requirement, all participants were measured with a Snellen chart from 3B Optic Instruments (see Figure 5, left). Furthermore, a OQAS System from Visiometrics, Terrassa, Spain (see Figure 5, right), was used, based on the double step technique to evaluate the retinal image quality. As a result of this measuring, the Object Scatter Index (OSI), was obtained and it was considered that participants should have an OSI lower than or equal to 1.5 [16].

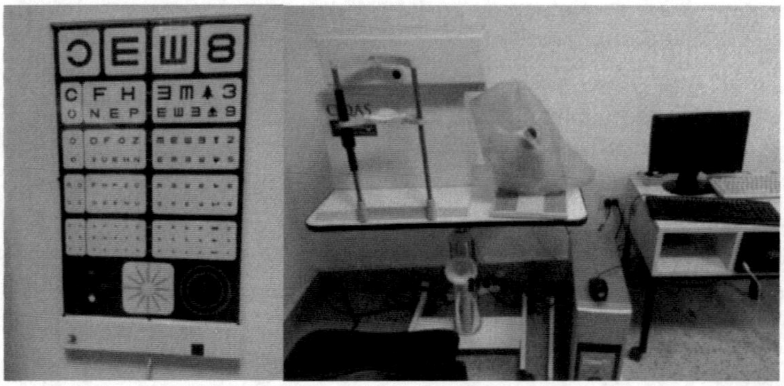

Figure 5. On the left, Snellen chart for VA measure. On the right OQAS System to measure the OSI (Object Scatter Index).

2.3. Experimental design to evaluate the repeatability, the age and training effects in contrast sensitivity measurements using FVC-100

The test was done in a room without ambient lighting. The illuminance between the subject's eyes (2.15 lux) and the average luminance of the stimulus and its immediate surroundings on the monitor (70 cd/m²) were measured performing the experiment under photopic adaptation [9]. The measurement of the illuminance was done with luxometer Minolta T1M brand and a luminance meter LMT 1009 was used for the luminance. CS Measurements were performed monocularly, occluding the fellow eye during the test. As the objective of this work was to evaluate FVC-100's repeatability, five measurements per eye were considered. Even though FVC-100 allows for a selection of up to six spatial frequencies (1, 2, 4, 8, 12 and 24 c/°), it was decided to perform the test with only three: 1c/°, 4c/° and 12c/°, that is, a low, a medium and a high spatial frequency. The experimental design consisted of each participant performing measurements with an only one frequency per day, and the order with which the measurements were performed was random. This approach was taken because it was necessary to ensure that, during successive measurements under the same stimuli pattern, the amount of time taken would be as short as possible for the subject, in order to avoid his/her fatigue and/or distraction during the measuring.

To assess whether there is any influence in terms of the results that denote if the measurements of participants with prior training are better or more repetitive than measurements of subjects that had never done this test before, participants were divided into two groups: those who had knowledge of the test and those who did not. Therefore, we established four groups of participants with an equal number of people in each group:

a) 20-49 years old with training
b) 20-49 years old without training
c) 50-70 years old with training
d) 50-70 years old without training

2.4. Computer Tools for Statistical Analysis

Firstly, the statistical tools provided by the Microsoft Office software package, particularly Microsoft Excel, were used to carry out the basic descriptive data analysis and graphics. To perform ANOVAs (Analysis of Variance) and the verification of the assumptions of the proposed model, we worked with free software, called R [17], which allowed us to obtain the results and reach the conclusions ~~to be~~ displayed below.

3. Results

Measurements were carried out on twelve subjects (24 eyes) of both age ranges, and it was observed, as the bibliography indicates, that there are meaningful differences between CS measurements for the two age groups and for considered spatial frequencies [6, 13 – 15].

In the following pictures (6, 7 and 8) we can observe the obtained results of CS per frequency (1, 4 and 12 c/°) and per eye for both groups of age, versus the repetitions. The two first rows correspond to young subjects (aged 20-49) and the last two rows are for older groups (aged 50-70). Furthermore, the groups of trained and untrained subjects can be seen separately in each row.

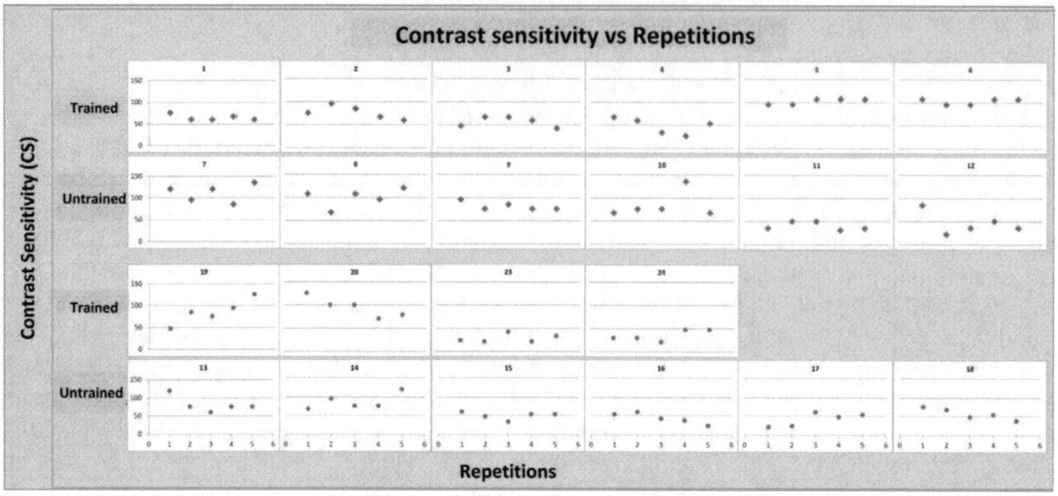

Figure 6. CS Results for 1c/°

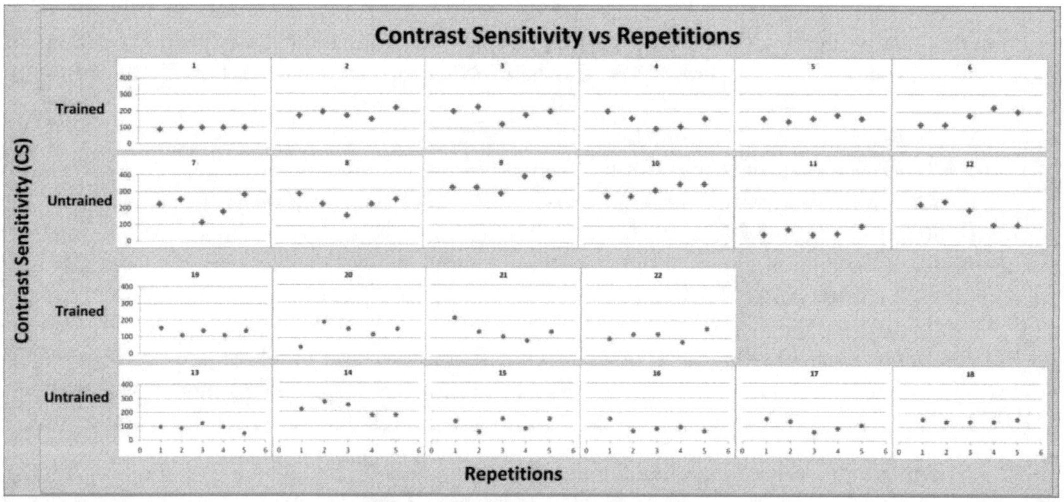

Figure 7. CS Results for 4c/°

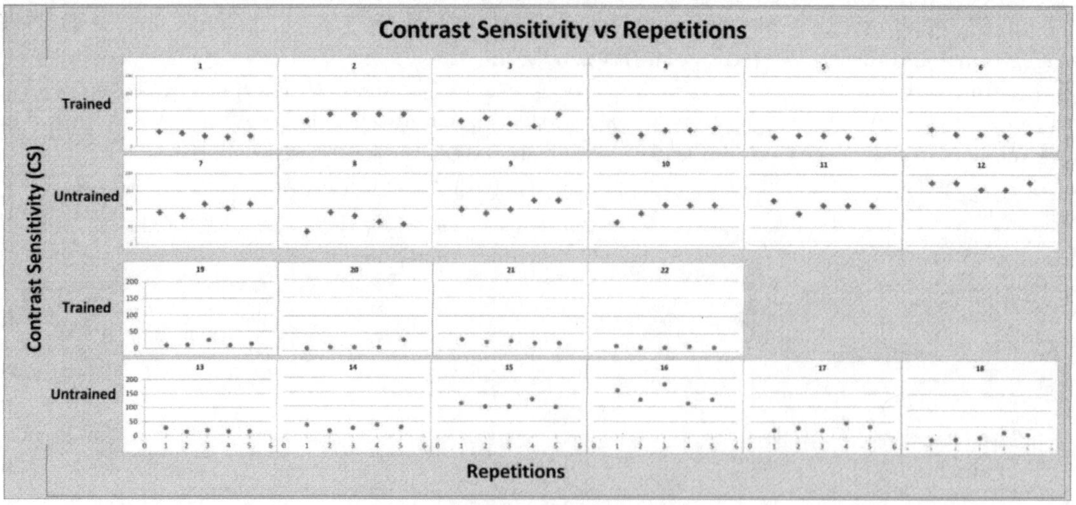

Figure 8. CS Results for 12c/°

From these graphics and a descriptive analysis, we can observe that measurements are approximately constants for 1c/º and 12c/º frequencies and that there is a higher variability of the measurements for 4c/º in both age groups, training independently.

After this an ANOVA was performed on each spatial frequency to assess the influence of Age, Training and Eye of each subject considering the average of the observations, with the result that age is significant for the three frequencies, which is a well known result. However, as the main objective of this work was to assess the repeatability of measurements, a second ANOVA was carried out that considers each observation (not the average). Considering then observations that are not independent because they come from the same subject, a mixed random effects model with a random intercept was adjusted, where a time/repetition factor captures said situation. After that, the model assumptions were verified.

Model specifications:

$$Y_{it} = \beta_{0i} + \beta_{1i}t_i + \beta_{2i}o_i + \beta_{3i}e_i + \beta_{4i}a_i + \varepsilon_{it} \qquad (1)$$

Where:

Y_{it} is the observation

β_{0i} is the random intercept: basic case of a Young subject, untrained and the task done through his left eye.

β_{1i} is the fixed coefficient of the variable time/repetition.

t_i is the time variable, which takes values from 1 to 5.

β_{2i} is the fixed coefficient of the variable "eye".

o_i is the "eye" variable, which takes values of 1 and 0 depending on whether the treatment is applied to the right or the left eye.

β_{3i} is the fixed coefficient for the variable training.

e_i is "training" variable, which takes values of 1 and 0 depending on whether the subject is trained or untrained.

β_{4i} is the fixed coefficient of "age" variable.

a_i is "age" variable, which takes values of 0 and 1 depending on whether the subject is Young or adult.

And finally ε_{it} is the error term.

The results obtained were as follows:

- For 1c/º:

Under this new model, only the effects of the age on the contrast sensitivity ($p<0.01$) are still meaningful. The "time" factor (number of repetitions) is not significant ($p=0.947$). The estimated confidence intervals for the coefficients referred to time, includes zero, so the goal of repeatability is reached.

The graphics corresponding to the validity of the model are shown below:

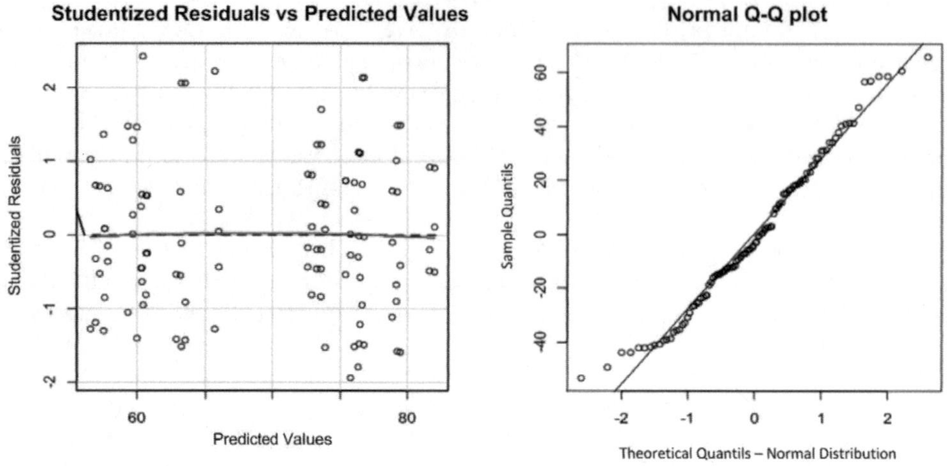

Figure 9. Model verification results for 1c/º.

Observing the chart on the left of Figure 9, it is evident that the model residuals have zero mean and constant variance. In the figure on the right of the same graph, the model residuals behavior is compared with the expected values under normality, as they concentrate around the oblique line representing the normal case. This assumption can also be accepted.

Conducting formal tests to validate the assumptions give as a result that they are acceptable, both in terms of normality of residuals (the Shapiro-Wilk test results in a p-value=0.068 and the tests to prove that the distribution is not symmetrical and mesokurtic are rejected) and its homoscedasticity, which implies that the stochastic variance of regression errors is the same for each observation.

- For 4c/º:

The ANOVA of repeated measures is significant again for the "age" factor on contrast sensitivity ($p<0.01$), but not for the time factor ($p=0.64244$). The estimated confidence intervals for coefficients referring to time include zero for the first four measurements, but not for the fifth although the effect is not meaningful. On this basis, we can affirm that the objective about the repeatability is reached.

In the graphical verification of the assumptions, those relative to homoscedasticity and normality are doubtful, as seen below:

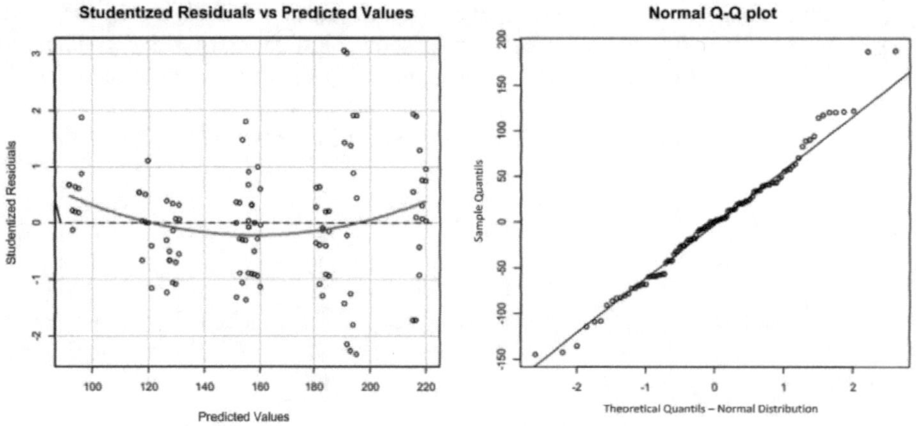

Figure 10. Model verification results for 4c/º.

In the graphic on the left of Figure 10, a "funnel" shape can be seen, with larger dispersion for higher values of observation. On the right thereof, the presence of values that deviate from normal is observed at the beginning and at the end. The formal tests suggest normality can be accepted (The Shapiro-Wilk test results with a p-value=0.2029) but not homoscedasticity (the p value from ncv.test is lower than 0.001). This aspect is improved by carrying out a data transformation by the logarithm, (having at the end a p-value=0.3364 with the ncv.test) and the conclusions arrived at regarding the relativity of the effects of time are the same.

- For 12c/º:

The ANOVA of repeated measures is significant again for age and training ($p<0.001$) and it is still not relevant for the observations of the effect of repetitions/time ($p=0.9627$). As the estimated intervals (at 90% of confidence) include zero, we cannot reject the hypothesis about the average effect of repetitions on the measurement being null. In this case repeatability is also supported by the functioning of FVC-100.

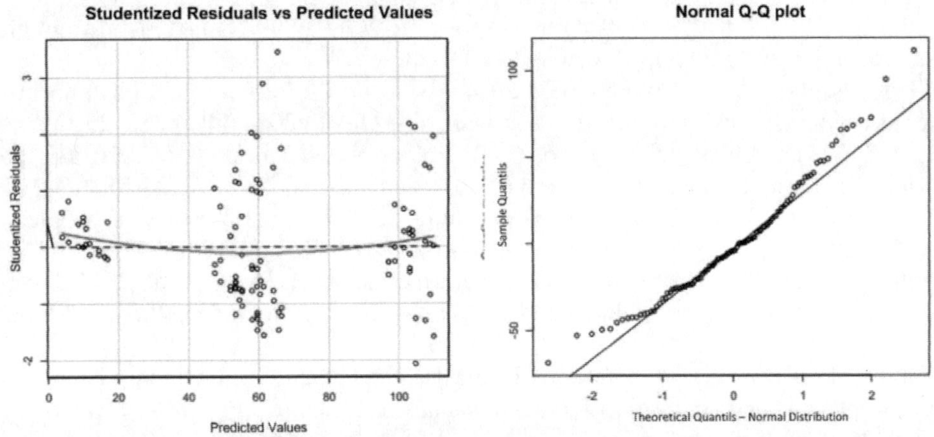

Figure 11. Model verification results for 12c/º.

Upon verification of the model assumptions and observing Figure 11, we can assume the validity of homoscedasticity and zero mean assumptions for residuals, but not that of the normal distribution (the p-value from the Shapiro-Wilk test is 0.005586). We can analyze that this absence of normality is due to the asymmetry observed in the data, reflected on points that are far enough away from the bisector on both ends. After a data logarithmic transformation, the situation improves (W de Shapiro-Wilk with a p-value=0.03416) without losing the validity of other assumptions nor the conclusions about the significance of the factors.

4. Discussions and Conclusions

The results obtained from this work show that for the three groups of spatial frequencies studied, the measurements are repeatable with time or repetition coefficients that are not statistically meaningful, and this condition does not depend on training or on the age of the participants/patients. This means that the changes on the observed value are not relevant between one measurement and another, making the FVC-100 a useful tool for the clinic, because one measurement alone would be enough to characterize a person's vision. However, as there is a slightly higher variability for the medium spatial frequency, since at 4c/º most people present their highest contrast sensitivity value and the contrast values are minimum, we recommend making another measurement at this frequency range to obtain a better estimation.

To conclude, we can say that the overall results are highly promising in terms of the objective of this study, as is the evidence of the robustness of measuring contrast sensitivity to be transferred to the eye clinic.

5. Bibliography

[1] Ginsburg AP 2003 Contrast Sensitivity and Functional Vision : International Ophthalmology Clinics [Internet] LWW Available from: http://journals.lww.com/internat-ophthalmology/Fulltext/2003/43020/Contrast_Sensitivity_and_Functional_Vision.4.aspx
[2] Ginsburg AP 2006 Contrast sensitivity: determining the visual quality and function of cataract, intraocular lenses and refractive surgery *Current Opinion in Ophthalmology* **17**(1):19–26.
[3] Pomerance GN and Evans DW 1994 Test-retest reliability of the CSV-1000 contrast test and its relationship to glaucoma therapy *Invest Ophthalmol Vis Sci*. **35**(9):3357–61.
[4] Tochel CM, Morton JS, Jay JL and Morrison JD 2005 Relationship between visual field loss and contrast threshold elevation in glaucoma *BMC Ophthalmology* **5**(1):22.
[5] Colombo EM, Issolio LA, Santillán J and Aguirre R 2009 What characteristics a clinical CSF system has to have? *Optica Applicata* **39**(2):415–28.
[6] Santillán J, Issolio LA and Colombo EM 2014 A statistical criterion to establish normal ranges for age in a contrast sensitivity function test *Optica Applicata* 2014 **44**(2):213–25.
[7] Bozkurt E, Yazıcı AT, Pekel G, Pekel E, Bayraktar Ş and Yılmaz ÖF 2010 The effects of posterior continuous curvilinear capsulorhexis on contrast sensitivity *Eye* **24**(5):805–9.
[8] Itoh N, Sagawa K, Okamoto A, Mitani S, Yoshida T 2007 Visual property of people with low vision: Contrast sensitivity function *Proceeding CIE 26th Session* Beijing D1-107-110.
[9] Colombo EM, Santillán J, Arellano W, Berman M, Sanchez R, Paz C, et al. 2011 Comparación de dos test psicofísicos de sensibilidad al contraste en un entorno clínico. *Psicol USP* **22**(1):45–66.
[10] Colombo EM, O'Donell BM, Santillán JE and Issolio LA 2013 Functional vision barriers: A new concept analyzed in terms of human visual performance *Psychology & Neuroscience* **6**(2):219–26.
[11] Paz Filgueira C and Santillán JE 2014 Sistema para la medición de funciones visuales: acondicionamiento y optimización para su transferencia a la clínica *Revista CET* **37**:1–9.
[12] Watson AB and Pelli DG 1983 QUEST: a Bayesian adaptive psychometric method *Percept Psychophycs* **33**(2):113–20.
[13] Higgins KE, Jaffe MJ, Caruso RC and de Monasterio FM 1988 Spatial contrast sensitivity: effects of age, test-retest, and psychophysical method *J Opt Soc Am A* **5**(12):2173–80.
[14] Johnson MA and Choy D 1987 On the definition of age-related norms for visual function testing *Applied Optics* **26**(8):1449.
[15] Ross JE, Clarke DD and Bron AJ 1985 Effect of age on contrast sensitivity function: uniocular and binocular findings *Br J Ophthalmol* **69**(1):51–6.
[16] Artal P, Benito A, Pérez GM, Alcón E, De Casas Á, Pujol J, et al. 2011 An Objective Scatter Index Based on Double-Pass Retinal Images of a Point Source to Classify Cataracts *PLoS ONE* **6**(2):e16823.
[17] R Core Team 2015 R: A language and environment for statistical computing R Foundation or Statistical Computing, Vienna, Austria. URL: http://www.R-project.org/.

Electroretinography: A biopotential to assess the function / dysfunction of the retina.

Quinteros Quintana[1,2,3], M.L; Benedetto[2], M.M; Maldonado[1,4], A.C; Vera de Payer, E.[1]; Contin, M.A[2]

[1]Facultad de Ciencias Exactas, Físicas y Naturales, UNC
[2]CIQUIBIC-CONICET, Facultad de Ciencias Químicas, UNC
[3]Becaria Doctoral Secyt, UNC
[4]CIEM-CONICET

mluzqq@efn.uncor.edu

Abstract. The Electroretinography (ERG) is a noninvasive technique that allows the assessment of functional integrity of the retina. The ERG recordings are biopotencials acquired in the corneal surface as a response of retinal tissue against controlled light stimuli. In clinical ophthalmology ERG is not commonly used but nowadays, because of the high incidence of degenerative diseases of the retina (RD), its use should be increased. Like other biopotentials as electrocardiography (ECG), electroencephalogram (EEG) and electromyography (EMG), ERG is a low amplitude signal, in this case a few hundred of microvolts (μV), which must be fitted and processed. The ERG signals are affected in morphology in the presence of pathologies that affects the integrity of the different retinal cell groups, for example due to some RD. In advanced cases of RD recordings can be abolished in the time domain; and yet in them it is believed that there is relevant clinical information making the ERG a great potential diagnostic tool.

1. Introduction

Biopotentials are electrical signals that can be measured between two points as electrical potential differences at the cellular level, tissue or an organism. Living tissue, when is adequately stimulated, responds with variations in its ionic composition causing local changes in current that varies with time. In some cases these local changes generate a depolarization that exceeds a threshold and can spread generating what is known as action potential. These variations, local and propagated can be registered with appropriate instrumentation and get analogical signals commonly called biomedical signals. In the field of Biomedical Engineering the importance of these kind of signals for diagnosis or research of different diseases is well known. ECG, EEG, EMG, among others are the most known and used biomedical signals, and their names depend on the cell group or tissue whose electrical activity is recorded.

Along human history there have been attempts to record all kinds of biopotentials and often those intentions were thwarted by the limitations of the electronics and instrumentation of the time. Gotch Francis (1903) and then Einthoven and Jolly (1908) were the first to determine the morphology of the electrical signal from the retina in response to flash light stimulus. It was then, in 1933, that Ragnar Granit identified different components in the cat's ERG, which later earned him the 1967 Nobel Prize in Physiology and Medicine. While some conceptions and descriptions were changed over the years, its foundations persist.

Even today some limitations due to electronics are remained. But if the minimum requirements that should be requested to the instrumentation are well known, reliable and reproducible recordings can be achieved. These limitations may be the reason why in clinical practice ERG technique is rarely used, even with the great need to find a diagnosis tool to know the existence of a degenerative disease of the retina (RD) and a clinical monitoring to know its evolution. In pathological conditions, the ERG recordings are diminished in amplitude and its latency time is increased, thus in advanced stages of the disease abolished recordings in the time domain can be found [1]. Our hypothesis is that the ERG could be a useful tool for confirming the diagnosis of RD, as it is known they are altered from very early stages of the disorder.

We believe that using basic techniques of signal processing such as Fourier Transform, the frequency spectra and hidden information in them can be analyzed.

1.1. The retina

The retina is a photosensitive tissue that belongs to the Central Nervous System (CNS) that lines the back of the eye. It consists of the neural retina and the retinal pigment epithelium (see Table 1). The thickness of the neural retina is between 100 and 200μm depending on the species. In the outermost layer photoreceptor cells (rods and cones) can be found. These types of cells perform the process of phototransduction, where light (photons) is transformed into a nervous signal that is sent to the brain through the optic nerve for integration and interpretation. The embryonic origin of the retina is the neural tube, the entire retina stems from its outgrowth. It consists of two layers that are called optical cup. The outer layer of the optical cup originates the pigment epithelium, while the inner layer gives rise to the rest of the retina [2].

Table 1. Cells and layers of the neural retina

Layer	Cells
Outer	Photoreceptors: rods and cones
Medial	Amacrines, bipolars and horizontals
Inner	Ganglions

1.2. Electroretinography

It is known that sensory receptor cells generate membrane potentials when they receive an excitatory stimulation. The extracellular potentials can be recorded noninvasively. This is the result of local conductance changes in the membrane of the activated cells originating input and output currents. These currents also flow through the extracellular space and create an extracellular potential. This current flowing through the conductive fluid surrounding a cell, whose activation has caused a local stream, is mainly directed towards the less activated cell parts; thus, when neurons are arranged in that way all of them are activated synchronously due to extracellular currents flowing in the same direction. The change of the extracellular potential resulting, called potential field, can be large enough so that it can be registered in a distal surface, for example the cornea [2]. Therefore the ERG is an electric potential generated by the retinal response to a flash of light and it is a measure of the average of biopotentials that reflects the medium and outer retinal layers functionality [1, 3-5], in addition it can be recorded in a distant tissue. The most studied parameters of this biopotential are the amplitude (microvolts μV) and the latency time (ms) [1, 3] of the main waves (a- and b-wave). Under dark-adapted conditions, in response to a stimulus such a bright flash, the a- and b-waves are generated. They are due to the activity of photoreceptor cells [6] and depolarization of the bipolar cells respectively [7]. It is considered that the ERG recordings are a powerful diagnostic tool to determine the progressive loss of retinal cells function in the clinical practice and in the investigation field, caused by different types of RD.

It is important to know the contribution of each type of retinal cells in the ERG recordings in order to determine if the ERG is an efficient tool for the assessment of the normal and pathological retinal activity. Radially oriented neurons (photoreceptors and bipolar cells) contribute more to the ERG than the cells that are oriented irregularly or laterally on the retina (horizontal and amacrine cells) [2].
The signal recording consists of different stages: a) acquisition, b) fitting and amplification and c) the signal processing. The acquisition should be performed following the standards of the International Society for Clinical Electrophysiology of Vision (ISCEV) for the clinical full field Electroretinography [8].

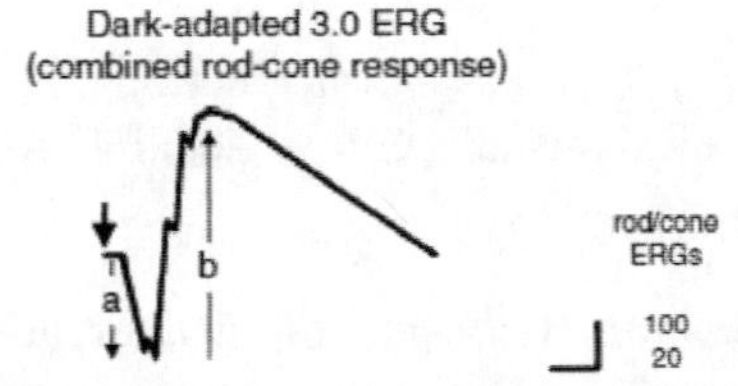

Figure 1: Response in dark-adapted conditions. ISCEV Standard (2008) [5]

1.3. Retinal degeneration.
RD is defined as the pathological process that causes death of retinal cells inducing the loss of tissue structure and possible blindness. Some of the diseases that are related to light stimulation:

1.3.1. Retinitis pigmentosa (RP): In most cases, the rods are most affected causing visual impairment ranging from mild retinal dysfunction (night blindness, reduced visual field side) to a possible total blindness [9-12].

1.3.2. Age related macular degeneration (AMD): It is one of the most common causes of irreversible blindness that affects about 50 million people worldwide. In humans is defined as an ocular disorder that destroys the macula, the part of the retina with the highest concentration of cones, responsible for visual acuity and color vision [10].

*1.3.3. Light pollution: A*rtificial lighting has drastic consequences for human such as retinal damage and disrupt of biological rhythms [13, 14]. Light pollution by day could accelerate the processes of death in RP and increases the incidence of AMD.

In early stages of the RD the main changes are the reduction of the amplitude of the main scotopic response waves the a- and b-wave (upon dark adaptation), with preservation of the photopic response (upon adaptation to light). In more advanced stages, both responses are affected in different ways, showing an important decrease in the a-wave. In stages near blindness, minimum ERG responses are registered barely differentiable from noise ($\approx 5\mu V$), which can be considered abolished recordings characteristic of the disease [15, 16]. This abolished registers could be hiding relevant information in time domain that could be useful for an early diagnosis. There are different types of ERG recordings that allow this analysis [17], however there is no clear definition of the changes they suffered due to the evolution of the RD, since no studies have been conducted on which and in what exact time of the disease the ERG recordings are altered.

1.3.4. RD Models: They are widely used to study the mechanisms of cell death in RP and AMD. Different methods of exposure (white light, monochrome, high or low intensity) in different animals (fish, amphibians, mammals (rat, mouse, etc.)) are used as models for different types of studies. In our laboratory we have demonstrated RD in an animal model causing photoreceptor cells (rods) death by constant exposure to 200 lux LED light. This could be due to constant activation of the phototransduction cascade by exposure to low intensity light without interruption.

1.4. Diagnosis of RD.
Nowadays there is not a technique for early detection of RD. When symptoms of diseases associated with RD appear may be a temporary affection of the visual system with the possibility of conducting a treatment to stop the progression or the permanent loss. Because of this in the clinical and experimental research, different early diagnostic techniques are being sought. That would allow the ophthalmologist to perform an accurate diagnosis in early stages of the disease.
The Association of affected people with RP from Euskadi ("AARPE"), established as a clinical examination protocol to confirm the diagnosis or monitor the evolution of the disease, the following list of tests:

1.4.1. Ophthalmological examination: It consists of inspection methods and evaluation of the integrity of the anatomical structures of the eye. Visual acuity, visual field, intraocular pressure (IOP), dark adaptation, color vision are measured.

1.4.2. Electrophysiological examination: It consists of an integrated assessment of the visual system functionality by measuring biopotentials in different anatomical structures that conducts the visual information. The techniques used are the ERG and visual evoked potentials (VEP).

The AARPE differentiate these techniques with those used for early diagnosis, and presumably the ERG "would allow early detection of new cases and the possibility of studying families or even isolated individuals from the initial stages when night blindness is the only symptom or even before symptoms appearance. The ERG can detect anomalies in phototransduction (electrochemical) before making changes in visual function (night blindness)". Furthermore it suggests that the ERG would be a useful tool to evaluate effects of treatments before detecting photoreceptors dysfunction. The intention is to capitalize the time earned with this tool to prevent or delay the arrival of early symptoms and further histological retinal damage [19].

2. Materials and Methods

2.1. Animals and induced retinal degeneration:
3 months aged male albino Wistar rats were bred in a bioterio under controlled temperature, humidity and illumination (80 lux) conditions. The animals were exposed to a 12:12 hs light-dark cycle (LD) with food and water *ad libitum*. Under these conditions LD control recordings were taken. Dark Control (DD): Animals were kept in total darkness for 4 days (DD4).
Experimental groups: Animals exposed to constant low intensity white-LED light (200lux) during 1 to 8 days: LL1, LL2, LL3, LL4, LL6, LL8 respectively.

2.2. ERG recording
ERG signals were recorded from control animals (n=40). The animals were anesthetized with ketamine hydrochloride and xylazine hydrochloride administrated intraperitoneally (xylazine, 200μl/100g) ¾ ketamine, ¼ xylazine (one hour of total anesthesia is achieved maintaining pupillary reflex and flicker). Pupils were fully dilated with a drop of tropicamide (Alcon Laboratories 0.5%), 10 minutes before the placement of the electrodes and topical anesthetic proparacaine (Alcon Laboratories 0.5%) was applied to ensure good electrode-corneal interface and avoid flicker. The acquisition was made in both eyes simultaneously.
Corneal (monopolar) electrodes were adapted to the size of the rats' eyes. The electrodes were hand made with malleable silver wires loop-shaped placed and held in the bulbar conjunctiva to avoid motion artifacts. The reference and ground electrodes used were needle monopolar electrodes placed subcutaneously in the ear and tail respectively. After positioning the electrodes to animals left 10 minutes in dark adaptation. The electrode placement was performed under red dim illumination. This maneuver does not significantly affects the dark adaptation. Scotopic ERGs were recorded from both

eyes simultaneously and 10 responses to flashes of unattenuated white light (5 ms, 0.1 Hz) from a full-field photostimulator LED (light-emitting diodes) set at maximum brightness (3 cd s/m2 without filter) were amplified and filtered (1.5Hz low-pass filter, 300Hz high-pass filter, notch filter activated) and averaged (Akonic BIOPC, Buenos Aires, Argentina). All the acquisitions were always performed at the same time of the day, so there is no interference with the circadian rhythm of the rat.

2.3. Temporal Analysis
The a-wave amplitude was measured from the pre-stimulus baseline, down to the negative peak signal, while the b-wave amplitude was measured from the baseline pre-stimulus to the peak positive deflection due to depolarization of bipolar cells followed by the a-wave. Latency times were measured from the moment of stimulation to the point where the maximum amplitude values of the a- and b-waves are taken, respectively called "a-wave latency" (TLA) and "b-wave latency "(TLB). Each record has length N = 510.

2.4. Frequency Analysis
Registers in time domain were saved with the .txt file extension. Then with the software MATLAB 2012b (Math-works, Natick, MA, USA) were imported and read. Fast Fourier Transform (FFT) was applied to the signals individually, as follows:

$$X(k) = \sum x(t)e^{-i(2\pi/N)tk}, \text{ k=0,1,...,N-1} \quad (1)$$

X(k) represents the FFT coefficients, x(t) is the ERG signal in the time domain, N is the number of samples. Each FFT coefficient contains the weight of the frequency component of the signal. Then the absolute value of X (k) can be plotted to show the frequency spectrum. N = 512 are taken with the addition of two equal endpoints at the position 510. The frequency analysis may be a useful tool to indicate events that in the time domain are not easily noticeable. Some main frequency components were identify and associated to the maximum power components [20-23], labeled 1, 2 and 3, and compared control signals against experimental groups.

2.5. Statistical Analysis
Mean values and standard deviations for the two main components of the ERG in the time and frequency domains were computed. To make comparisons a one way-ANOVA was performed and then a mean comparison tests were reported in each of the results. The statistical tool INFOSTAT (2013 version InfoStat Group, FCA, National University of Cordoba, Argentina) was used.

3. Results
The analysis in time and frequency domain are shown below. Also we show a comparison of the control recordings in the time domain against the RD ones.

3.1. Analysis in the time domain
As shown in Table 2 the analysis in the time domain allows the identification of the major components of the ERG recordings (a- and b-waves) by their amplitudes and latencies in normal registers, and in Table 3 the same analysis with the RD data is shown. In control recordings, the a-wave is at 42.18 ± 7.22 ms (mean ± standard deviation (SD) with n = 40) with an amplitude of -22.2 ± 6,7μV and b-wave 141.63 ± 27 , 61ms with an amplitude of 85.4 ± 29,82μV. Comparisons with the recordings of the experimental groups were performed. Figure 2 shows the ERG registers of the control and experimental groups, where the progression of the RD is shown. Throughout the days of constant light exposure, the amplitudes tend to decrease, and the latency time increase, reaching abolished records (in time domain), so that the main components are not easily identifiable; this is the reason why no values for the experimental groups LL8 and LL6 are presented. The results of statistical analyzes are presented in Figures 3 and 4. In Figure 3 the quantitative data obtained as mean ± SD of "n" values taken in each

experimental group are expressed. A one-way ANOVA was conducted to determine homogeneity of variance of the data and then applied Tukey post-hoc test as a test for means comparison between groups with p-value <0.05. Apparently, the RD would be altering mainly the latency of the two principal waves as shown in Figure 4. As expected, in the a-wave of LL4 a decrease in its amplitude is observed, simultaneously with an increase in the latency time, both statistically significant versus the control group, LL1, LL2 and even LL3. This would indicate a critical event of photoreceptor cells dysfunction.

3.1. Analysis in the frequency domain.
The power spectra of control ERG signals were compared against the experimental model ones. In all cases three principal low-frequency power components are found (from now on called Components 1, 2 and 3). The results of the statistical analysis are shown in Figure 5. Quantitative data are expressed as mean ± SD obtained from the "n" values taken from each experimental group. A one-way ANOVA was conducted to determine homogeneity of variance of the data and then applied Tukey post-hoc test for comparison of means between groups with p-value <0.05. No significant differences in the component 1 from all groups were reported. Therefore just the analysis of the components 2 and 3 are reported. In both frequency components it is shown that in LL3 there would be a significant difference in comparison with the other experimental groups and finally also in LL8. The coefficient of variation (CV) from the analysis performed on the component 2 shows that the results would be more reliable than the ones from the component 3.

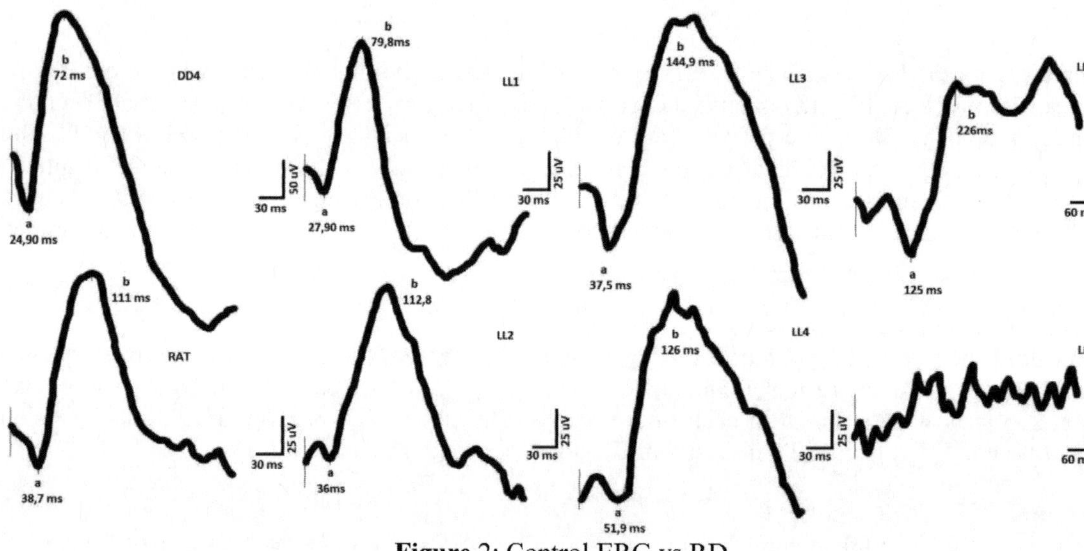

Figure 2: Control ERG vs RD.

Table 2: Analysis in time domain from the control group

Component	Latency time (ms)	Frequency (Hz)	Amplitude (μV)
a-wave	42.18±7.22	n/a	-24.1±9.27
b-wave	144.26±27.24	n/a	85.55 ± 29.82

Table 3: Results of the analysis in time domain from the experimental groups

	Component	Latency time (ms)	Amplitude (μV)
DD4	a-wave	56±14.18	-27.2±8.1
n=7	b-wave	135.43±20.16	83.77 ± 29.6
LL1	a-wave	36	-19.65
n=1	b-wave	133	93.1
LL2	a-wave	60	-20
n=3	b-wave	190	45
LL3	a-wave	60±0.94	-26±0.2
n=7	b-wave	214.67±2.31	59±0.9
LL4	a-wave	88.2±1.1	-10.77±2.62
n=7	b-wave	247.33±9.8	62.39±6.63

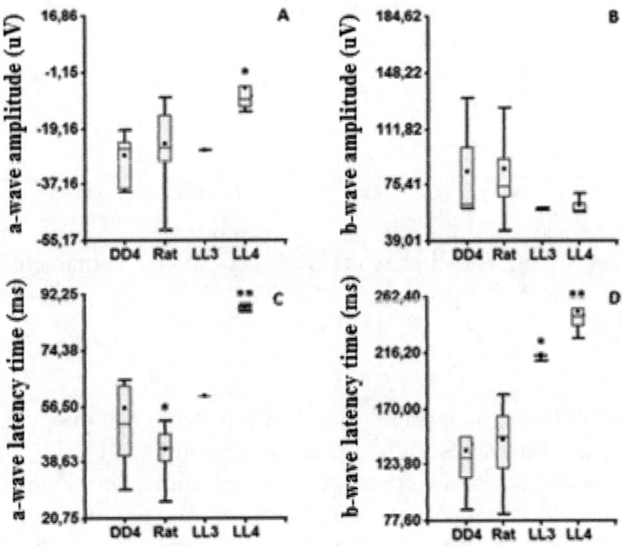

Figure 3: ERG characteristics comparison by experimental group. One-way ANOVA conducted and post-hoc test TUKEY *, **: p-value<0.05. **(A)** a-wave amplitude, **(B)** b-wave amplitude without significant differences, **(C)** a-wave latency time VC=14.9, **(D)** b-wave latency time VC=14.5

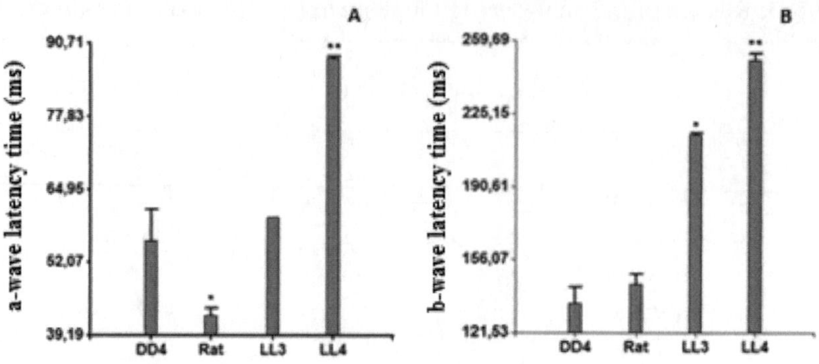

Figure 4: a- (A) and b-wave (B) latency time mean comparison.

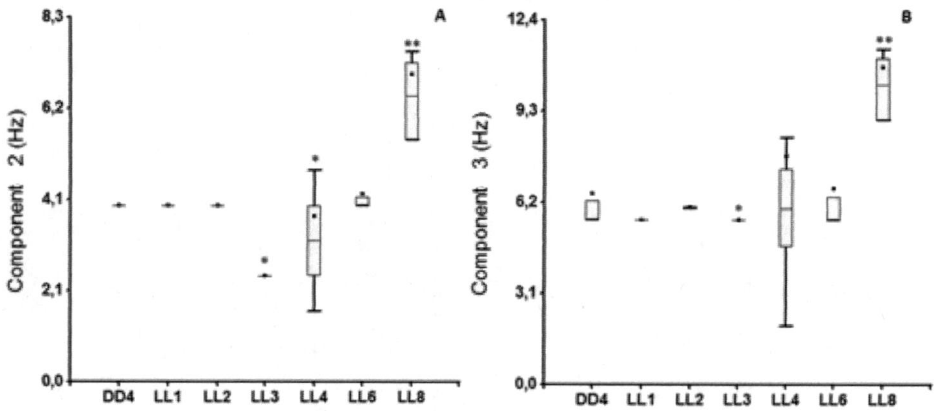

Figure 5: Comparison of frequency behavior. One-way ANOVA conducted and post-hoc test TUKEY: (A) Component 2 (*, **: p-value <0.05) VC=15.58 (B) Component 3 (*, **: p-value <0.05) VC= 25.07.

4. Conclusions

The RD is a major source of visual impairment caused by a lack of anatomical or functional integrity of the retina. The retina as a part of the CNS and is a sensible structure with little feasibility of regeneration in case of damage. The causes can be inherited or acquired, either by environmental factors or age. According to the World Health Organization (WHO) retinal diseases are the leading cause of visual disability in countries with medium-high and high socioeconomic development. The WHO suggests "make regular eye exams that allow early diagnosis of the disease and early treatment to prevent or delay the loss of visual function."

In the search of methods and noninvasive techniques for prevention and early RD detection in this paper we show that the ERG could be an appropriate tool to measure the function / dysfunction of the retina in its different cell layers.

From the results shown above we can conclude that both, in temporal and frequency analysis, the alterations caused by the RD become noticeable. According to Contin, et. al. (2013), there is a death of photoreceptors statistically significant by the 5th day (LL5) of constant light exposure. According to the results presented here it shows that LL4 could be a point of inflexion, where the cells of the retina, mainly photoreceptors, show a statistically significant dysfunction in both, time domain and frequency. Finding that point of inflection in the experimental model, the same analysis could be performed in patients with a diagnosed RD and may apply some existing treatments. Finally we conclude that there is relevant information in the ERG signals that could be useful and as in the time domain is not evident, with the using of more powerful signal processing tools, a prolonged monitoring of the RD would be possible.

5. References

[1] Rosaralis Paneca Santiesteban, Jorge E. Amaro Nodarse, Carlos E. Mendoza Santiesteban y Rosaralis Santiesteban Freixas. (2005). *Electrorretinograma. Valores normales con diferentes protocolos de estudio*. Instituto de Neurología y Neurocirugía. Ciudad de la Havana. Cuba.

[2] Ryan SJ, Hinton RD, Schachat AP, Wilkinson CP. *Retina*. New York USA; 2009.

[3] Rosaralis Paneca Santiesteban, Jorge E. Amaro Nodarse, Carlos E. Mendoza Santiesteban y Rosaralis Santiesteban Freixas. (2005). *Electrorretinograma estandarizado con respuesta de conos S y onda C en diabéticos sin retinopatía*. Instituto de Neurología y Neurocirugía. Ciudad de la Havana. Cuba.

[4] Marmor MF, Fulton A, Holder GE, Miyake Y (Chair), Brigell M, Bach M. *Estandar. (2008). ISCEV para Electrorretinografía Clínica.*

[5] Marmor MF, Holder GE, Seeliger NW, Yamamoto S. (2003). *Estándar para Electrorretinografía Clínica*. Doc Oftalmol 2004; 108: 107-114

[6] Penn RD & Hagins WA (1969). *Signal trnsmission along retinal rods and the origin of the electroretinographic a-wave*. Nature 223, 201-204

[7] Stockton RA & Slaughter MM (1989). *B-wave of the electroretinogram. A reflection of On bipolar cell activity*. J Gen Physiol 93, 101-122.

[8] Marmor MF, Holder GE, Seeliger NW, Yamamoto S. (2003). *Estándar para Electrorretinografía Clínica*. Doc Oftalmol 2004; 108: 107-114.

[9] Gutiérrez-Torre S.M. (1995). *Retinosis pigmentaria. Clasificación y tratamiento.* Servicio de publicaciones de la Universidad de Oviedo.

[10] Thomas E. Ogden, J. M. Jimenez Sierra. (1989). *Enfermedades retinales hereditarias: una guía de diagnóstico*. St. Louis: Mosby

[11] Miguel Ángel Marchena Fernández (2008). *El ratón pcd/pcd como modelo de degeneración de fotorreceptores. Estudio estructural y funcional*. Tesis Doctoral. Facultad de Biología. Departamento de Biología Celular Patológica. Universidad de Salamanca. España.

[12] Milam A.H., Li Z.Y. y Fariss R.N. (1998). *Histopathology of the human retina in retinitis pigmentosa.* Prog. Retinal Eye Res. 17: 175-205.

[13] Ukai H, Kobayashi TJ, Nagano M, Masumoto KH, Sujino M, Kondo T, et al. *Melanopsin-dependent photo-perturbation reveals desynchronization underlying the singularity of mammalian circadian clocks*. Nat Cell Biol 2007; 9:1327–34.

[14] Jewett ME, Kronauer RE, Czeisler CA. *Light-induced suppression of endogenous circadian amplitude in humans*. Nature 1991;350:59–62.

[15] Mora, J.C, Aregall, M.C. (2005). *Pruebas electrofisiológicas. Manual de Instrucciones*. Annals d'Oftalmologia 2005; 13 (1): 8-29 y Annals d'Oftalmologia 2005; 13 (2): 76-90.

[16] Rispoli E, Iannaccone A, Vingolo EM. *Low-noise electroretinogram recording techniques in retinitis pigmentosa*. Doc Ophthalmol. 1994; 88(1):27-37.

[17] G. Silva, D. Pepperberg. (2004). *Step response of mouse Rod Photoreceptors Modeled in Terms of Elemental Photic Signals*. IEEE. Transactions on Biomedical Engineering. Vol 51 N°1.

[18] Contin et. al (2013). *Photoreceptor damage induced by low-intensity light: model of retinal degeneration in mammals*. Molecular Vision 2013; 19:1614-1625

[19] A.A.R.P.E. Asociación de Afectados de Retinosis Pigmentaria de Euskadi. (2014). http://www.retinosispigmentaria.org/es/

[20] Mathieu Gauvin, Jean-Marc Lina, and Pierre Lachapelle, *Advance in ERG Analysis: From Peak Time and Amplitude to Frequency, Power, and Energy.* BioMed Research International, vol. 2014, Article ID 246096, 11 pages, 2014. doi:10.1155/2014/246096

Journal of Physics: Conference Series **705** (2016) 012053 doi:10.1088/1742-6596/705/1/012053

[21] Alberto Jesús Barrientos Castaño, Araceli Lantigua Cruz, Maritza Herrera Mora, Vivian Sistachs Vega, Milagros Sierra Hernández. (2012). *Electroretinography phenotype in retinitis pigmentosa patients.* Rev Cubana Genet Comunit, 2012; 6(1): 20-25.

[22] Hassan-Karimi, H., Jafarzadehpur, E., Blouri, B., Hashemi, H., Sadeghi, A. Z., & Mirzajani, A. (2012). *Frequency Domain Electroretinography in Retinitis Pigmentosa versus Normal Eyes.* Journal of Ophthalmic & Vision Research,7(1), 34–38.

[23] Gur M, Zeevi Y. *Frequency-domain analysis of the human electroretinogram*. J Opt Soc Am. 1980 Jan; 70(1):53-9.

From Modern Push-Button Hospital-beds to 20th Century Mechatronic Beds: A Review

I Ghersi [1,2,3] M Mariño [1] M T Miralles[1,2]

[1] Universidad de Buenos Aires (CIDI-Facultad de Arquitectura Diseño y Urbanismo-UBA), Intendente Guiraldes 2160, Pabellón III, Cuidad Universitaria, CABA, Argentina.
[2]Pontificia Universidad Católica Argentina (LaBIS-Facultad de Ciencias Fisicomatemáticas e Ingeniería-UCA), Av. A. Moreau de Justo 1500, CABA, Argentina.

e-mail: igghersi@gmail.com

Abstract. The aim of this work is to present the different aspects of modern high complexity electric beds of the period 1940 until 2000 exclusively. The chronology of the product has been strictly divided into three big stages: electric and semi-electric beds (until the 90's), mechatronic beds (90's until 2000) and, mechatronic intelligent beds of the last 15 years. The latter are not considered in this work due to the extension for its analysis. The justification for classifying the product is presented under the concepts of medical, assistive and mobility devices. Relevant aspects of common immobility problems of the different types of patients for which the beds are mainly addressed are shown in detail. The basic functioning of the patient's movement generator and the implementation of actuators, together with IT programs, specific accessories and connectivity means and network-communication shown in this work, were those that gave origin to current mechatronic beds. We present the historical evolution of high complexity electric beds by illustrating cases extracted from a meticulous time line, based on patents, inventions and publications in newspapers and magazines of the world. The criteria adopted to evaluate the innovation were: characteristics of controls; accessories (mattresses, lighting, siderails, etc.), aesthetic and morphologic properties and outstanding functionalities.

1. Introduction

The design of high complexity beds constitutes a specific field of research in the area of medical products, presently in expansion and development. Despite the historical existence of numerous manifestations of support surfaces for unwell and bedridden patients, a more defined point of origin for these products could be assigned to beds with adjustable siderails, which appeared in England between 1815 and 1825 [1]. Andrew Wuest and Son, registered, in 1874, a patent for a type of mattress frame with a hinged head that could be elevated [2]. Later, a series of efforts converged, originating the so-called push-button hospital beds in the 40s, decade from which the present work commences. At present, there is a whole chronology of products that start with the modern electric beds, in a strict sense, further moves to semi-electric models (that combine adjustments assisted by actuators with manual adjustment sections, still existing as alternatives of minor complexity) to finally arrive to the so called mechatronic beds, which incorporate software into specific motion programs, and display more advanced integrations between mechanics, electronics and computer science in its

[3] Corresponding Author

core. Turning into the 21st century, the intelligent mechatronic bed appears, which will be studied in a separate work. This bed includes a higher level of services and functions, including multiple actuators for surface movement, specific automation, monitoring and status data registering functions, as well as additional communication means between the patient and personnel of care.

This work presents: a) A putting into context of the product and its classification as such, b) A brief review of the immobility problem, common to most pathologies and patients (pressure ulcers) that give origin to the existence of these products; c) A basic scheme for the understanding of the required movements of the bed, as well as a model that provides specific functions of mobilization and control in addition to the basic ones (illustrated in a 1992 patent of the University of Buenos Aires); and, d) A diachronic study, starting from the early electric beds up to the mechatronic beds of late 20th century. The cases shown in this work were extracted from a time line that include the date of registry of patents, inventions and publications in journals and magazines of the world.

The adopted criteria to evaluate the innovative aspects of each product were: a) Control characteristics; b) Accessories (active or passive mattresses, lighting, siderail support, etc.); c) Aesthetic and morphologic properties; and, d) Outstanding functionalities.

2. Development of the study

2.1. High complexity electric beds: medical, assistive and mobility devices.

According to the definition of the US Federal Food, Drugs and Cosmetic Act a medical device is:

> *"an instrument, apparatus, implement, machine, contrivance, implant or other similar, including a component part, or accessory which is (...) intended for use in the diagnosis of disease or other conditions, or in the cure, mitigation, treatment, or prevention of disease (...) or intended to affect the structure or any function of the body (...) and which is not dependent upon being metabolized for the achievement of any of its primary intended purposes." [3]*

The World Health Organization (WHO) extends this criterion explicitly, incorporating this group into the diagnosis, treatment and healing of wound solutions, vital support systems, disinfection equipment and software developed for these purposes, among other definitions [4]. Therefore, high complexity electric beds are a specific type of medical device included in one of the 10,000 [4] categories covered by the mentioned definition. In recent years, the devices of high resolution image generators, tomography, magnetic resonance, and more recently, the definitive incorporation of the robotics in all of these, are examples of innovation in medical complex products. This trend has deepened the merging of medical devices with isolated IT and network systems [4], opening the path to a new age of products integrating electronics and communications. Thus, the evolution of the electric beds into intelligent mechatronics is paradigmatic.

In this context, it's possible to go in depth into medical devices up to the so called *assistive devices*. These are outlined in search of supporting or improving functions and to minimize the operational difficulties of its users. A definition for this type of devices is:

> *Any product (including devices, equipment, instruments, technology and software), produced specially or available in general, used by or for persons with disabilities, for their participation, protection, support, training, measuring or substitution of functions, structures or activities, or to avoid limitations or restrictions for their participation in activities" [4]*

Thus, high complexity electric beds constitute devices of excellent assistance, since they are particularly conceived to mitigate and to anticipate the secondary health conditions associated to immobility, both for patients transitorily immobilized as well as for patients suffering chronic conditions of variable severity.

An important subgroup among the *assistive devices* to which high complexity electrical beds belong to is the subgroup of *mobility devices*. These are products or technologies that allow disabled persons to displace, mobilize, support and change their corporal positions, constituting "*a prerequisite to reach equivalent opportunities, exercise their rights (...) and live in dignity*" [5]. These products can improve health of users reducing the risk of falls, injuries and other limitations that can result in complications and premature mortality. Due to the sensibility of the subject, the WHO proposed and assigned a number of valuable design principles for their provision and success, such as *acceptability, accessibility, adaptability, availability, low cost* and *quality [5]*.

2.2. From the immobility of the patient to the mobility of electric beds

High complexity electric beds respond with movements, to the lack of mobility of patients. But these movements are the product of a specific biomechanics, which is a particular field of knowledge in itself, tied to the evolution and improvement of the design of these products. The potential complications resulting from immobility are multiple: besides the changes in the mental condition of patients (depression) and pressure or of decubitus ulcers, muscular atrophy, bone demineralization, pneumonia, gas exchange dysfunction, thrombosis in deep veins, and edemas, are found among other serious conditions.

The duty of positioning a patient by regular intervals of time, achieving a better distribution of pressure, follows well-known protocols based on the practice and experience of the personnel of infirmary. This is an intensive task that generally falls upon the auxiliary personnel, assistants or patient's relatives. This additional dedication, necessary to the treatment and prevention of pressure ulcers, has a drastic impact on the associated costs of the medical system at multiple levels [6], being a serious concern for budget allocations of health organizations, at both national and international level.

2.2.1. Pressure ulcers

Pressure ulcers are a typical example of complications associated to lack of mobility and sensibility. Since 1932, it is considered that a 32 mm Hg tissue pressure can originate this kind of ulcers. Nevertheless, a set of additional risk conditions can produce ulcers in periods of less than two hours and at lower pressures [6]. Friction and tears, which can happen during mobilization techniques on patients, increase the risk of generating these ulcers, particularly in fragile skins [7]. Oppositely, controlled and slow movements provide minor predisposition for their appearance or worsening. When a person leans on its back, pressure ulcers generally appear in the rear part of the head, in the shoulders, elbows, coccyx, heels and toes. When leaning on a side, the ulcers could appear on ears, shoulders, pelvis, hip, knees and ankle bones [7]. Other external or environmental factors can boost the appearance in other zones (humidity, non-uniformity of the support surface, etc.). More than 100 risk factors, both intrinsic and extrinsic, have been associated to these complications on the basis of the evidence gathered by personnel of assistance and infirmary [6] that are recorded in scales made to evaluate the risk of occurrence. In general, the risk of suffering pressure ulcers appears when there is no possibility of moving independently, or when movements are limited, as well as when circulatory or sensibility problems appear, and under certain conditions of nutrition and skin moisture due to perspiration, or bad conditions of care and incontinence [7]. The risk can also appear in persons older than 70 years of age, or under dry skin conditions or confusion, among other difficulties [6].

Pressure ulcers occur both in particular housings and in long-term care environments, therefore high complexity electric beds will no longer be strictly related to hospital areas but to private houses of users, with the corresponding aesthetic changes.

2.2.2 Target population

Some distinguished subgroups of high tendency to suffer secondary conditions that high complexity electric beds are capable of responding, described in point 2.2.1 are: Patients with spinal cord injuries, both traumatic and non-traumatic, cerebrovascular diseases and cerebral palsy, patients with spinal damage [8], bariatric patients and older adults. The latter constitute a group of special risk presenting a higher predisposition to suffer ulcerations and secondary complications associated to immobility, and a major risk of falls and injuries of long recovery and treatment time.

Immobility is considered to be one of the *"four giants"* of geriatrics. One of four reasons of more prominent health and disability problems in this age group [9] together with depression, memory loss, and falls (also linked with recoveries, pain, and increased time of healing). In addition to the mentioned pathologies there are diseases that affect the cognitive skills of those individuals (dementia or Parkinson's disease, for example), or those patients transitorily submitted to restrictive conditions of activity (in postoperative and healing conditions, patients with pain, etc.).

2.2.3 Electric beds: Basic structure

The electric bed movements originate in the materialization of the experience achieved by nurses throughout the years of treating with diverse types of patients. It started by the study of the biomechanics of the own personnel of care, which was translated to technological principles such as the biomechanical principle of foil movement, inspired by the movements of the sheets when nurses change the position of the patients.

As an example, Figure 1 shows a view in perspective of the support structure of an electric bed, with its typical functional and formal characteristics [10]. The surface of support of the patient is divided, giving place to three sections that can be driven by the patient or by the caregivers: Back, thighs and calves. As usual in this type of product, a fourth section is fixed to an end of the back and another one to the corresponding support of thighs, fixing the support of the patient to its base level. This configuration, with four sections of support, avoids significant deformations in the mattress that is in contact with the patient, even when all the mobile surfaces are in limit position, and allows a uniform distribution of pressures along waist and hip. For operation, two actuators (or a dual actuator designed for such effects) allow to drive individually the back section and the leg section (adequately articulating thighs and calves). The movement of the leg section allows slight elevations, as well as descents -already in recent models- achieving chair positions and support for the removal of the equipment [11, 12].

The elevation of the support level of patients constitutes a problem that acquires special importance when they risk of suffering falls, both when trying to enter or exit the bed, as well as due to possible confusion or ignorance of the environment that can afflict them. This allowed the appearance of a segment of "low beds" in the market, created to prevent those problems, accompanied by other means of monitoring and control, among other equipment. Likewise, it turns suitable for the medical and support personnel to be able to rely on means that set the elevation of the patient, according to the different instances of treatment or care, as well as on ergonomic aspects of managing the patient and the product, consequently facilitating these tasks. Beside of having retractable siderails, electric beds can have active means to control the elevation of the base level of the patient, through actuators that mobilize “scissor” type mechanisms or through one or more extensible columns of controllable elevation (see Figure 1). Additionally, in case of having two (or possibly more) independent elevation mechanisms over the surface of support, the selective control of these allows to achieve specific positions as cardiac chair or Trendelenburg, which depend on changes in the inclination of the whole surface of support of the patient.

The distinctive parameters of these structures reached typical and accepted values consolidated by procedures that arose along their development and expansion, in the form of requirements on properties entailed for these products, in terms of dimensions, functioning, performance and safety [13]. In general, the frames of these beds measure 1.00 m of width and 2.00 m of length, and admit elevations between 0.40 m and 0.80 m for the base support level of the patient (independent to the thickness of the mattresses). The sections of the back support admit angles of up to 70 ° for its positioning. In particular, the mechanical conditions of the siderails are regular, to minimize risks of trapping and injuring during equipment operations, as well as of accidents such as those detected throughout the years [14].

At the end of the timeframe of interest of this study, and as functions were consolidated in different products, advanced versions of medical beds pointed at the incorporation of additional mechanisms of mobilization on this basic structure, valuable for improving patient care.

Figure 2 shows a patent scheme that presents the biomechanical principle of corporal rotation, directed to the mechatronic bed developed at the University of Buenos Aires [15, 16]. This product shows advanced virtues for the prevention of health conditions derived from immobility, as those mentioned in section 2.2.1. The figure details the lateral surfaces that can be driven to adopt multiple positions, by additionally supporting the regular control functions available for the support surface of the patient.

Figure 1. View in perspective of a high complexity electrical bed [10], with 4 sections a support surface and elevation control.

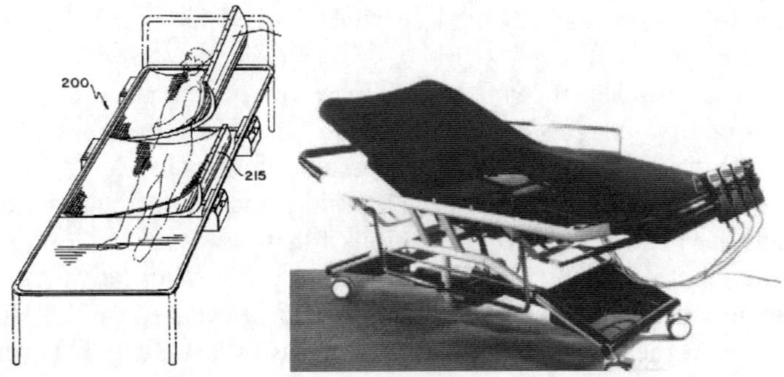

Figure 2. Left: View of the support structure of the patented biomechanical principle of corporal rotation [16]. Right: High complexity mechatronic bed model AC2000 with corporal rotation, developed by CIDI-FADU-UBA (*Centro de Investigación en Diseño Industrial de Productos Complejos)* [15, 17].

Electric beds, together with the complex technical difficulties for movement implementation and integration, software and communication of the patient with the environment, must respond to a set of socio-economic, political and aesthetic problems. The latter were also considered in the survey carried out in the proposed timeframe for the product shown hereafter.

2.3. Evolution and innovation

Hereby, we present the development of electric beds for hospital and household attention, following a determined time line (based on the date of submission of patents and inventions, and publication of related articles in magazines and newspapers). Only those products that were examples and marked innovation trends considered relevant along the 20th century were selected for this survey. The chronology was subdivided in three stages: early electric and semi-electric beds of the 40's to mid-70's; mechatronic beds in general; and from 2000, intelligent mechatronic beds. As any other subdivision, the present proposal is arbitrary and does not impose neither strict limits nor temporary rigid barriers. The most advanced interactivity (environment-patient) in use, the incorporation of software into the electromechanical functioning of mechatronic beds, and for the intelligent mechatronic beds, the incorporation of all sensors, interfaces and communication devices in a wide sense, was included as criteria, considering that they are fundamental for gathering information of the patient in real time.

The first half of the 20th century offers examples of electric beds with diverse functions. Figure 3a) shows an example of a hospital bed, designed to be controlled by means of a push-buttons panel. In addition to the control of the basic functions of the bed (as elevation of the support and legs section) there are many surprising and actual resources considered in this survey: Retractable sink, warm and cold water flow, control of lights and blinds, as well as the automation of the patient's residues disposal [18]. The electric beds with basic functions, formally closer to those pointed out in section 2.2.3, will progress globally [19], and advance into more options with additional functions such as Trendelenburg control and the total elevation of the surface of the patient (a tempting issue since late 40's [20].

Between 1946 and 1950 patents and inventions of equipment with additional functions and alternative actuators (height, Trendelenburg and oscillatory movement controls) were submitted. En 1946, Howard Hughes, after a plane accident and unsatisfied with the available equipment, created a hospital bed adapted to his needs, with 6 sections and 30 electric engines, and cold and hot water flow. In 1949, patents of interface of actuators, with control and basic functions, appear. In the 50's, the Hill-Rom Company, presently one of the most important enterprises, developed its first bed with electric engine (1952), and commercialized its first bed with full electrical functioning (1956) [21] (see Figure 3b)).

The decades of 1960 and 1970 were characterized by an increasing demand of medical attention, stimulating progress towards health services external to hospital care (community and residential health services) [22]. Other remarkable characteristics were the emerging incorporation of computers for information storage and processing, including the debate in regard to its potential and relevance, and the use of telecommunications for its transport to long distances [23] and centralized monitoring (incipient second stage). During these years, alternative models for high complexity care conditions appeared, such as the Circ-O'lectric Bed developed by Dr. Homer Stryker in 1958 [24], of general use between the late 50's and late 70's (see Figure 3c)). This remarkable design allowed the patient and person in care to control corporal rotations according to specific needs by means of electric actuators. Figure 4a) shows the first electric bed built and commercialized in Japan in 1962, illustrating the Far East market expansion [19].

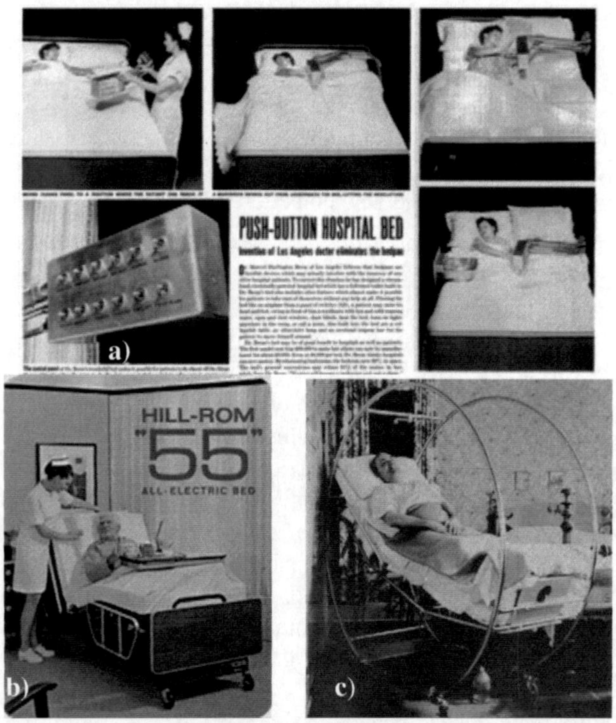

Figure 3: a). Push button hospital bed by General Electric, 1945 [18]; **b**) Hill-Rom's first bed with electric engine, 1952 [21]; **c**) Circ-O'lectric Bed, Stryker Company, 1958 [24, 25].

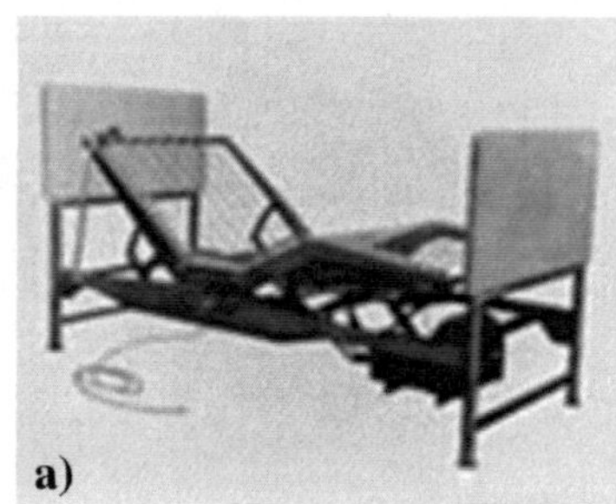

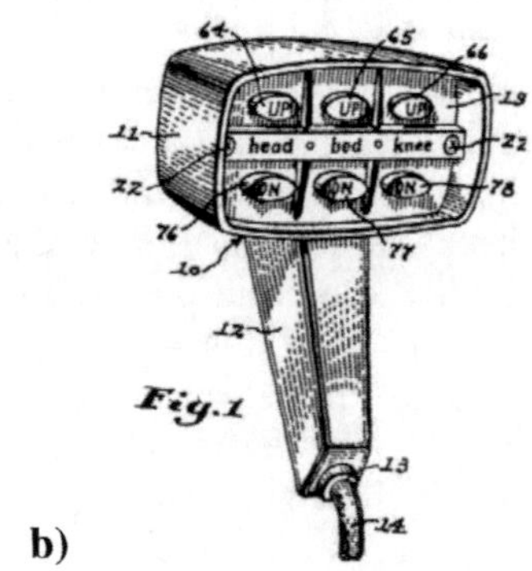

Figure 4 a). Bed built and commercialized in Japan, 1962 [19]; **b**) Patent of 1961, showing basic bed controller, with three functions [26].

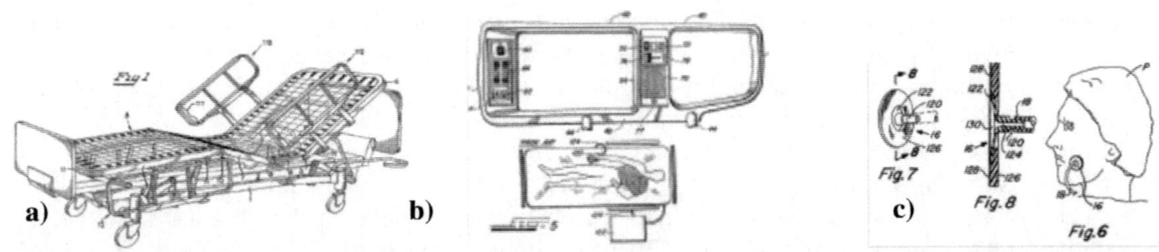

Figure 5 a) Image of a 1974 patent showing control panels on siderails [27]; **b)** Image of 1978 patent incorporating siderail controls for remote electrical devices [28]; **c)** 1980 patent that includes alternative means for personnel call [29].

From 1976, the evolution of beds is affected by the first medical devices regulation system (Medical Device Regulation Act). In 1977, the International Electrotechnical Commission published the IEC 60601 standard for medical electrical devices [4].

Between 1970 and 1979, beds in all hospitals incorporated siderails with side control panels, and some versions incorporated external equipment control (see Figure 5a) and 5b)). The incorporation of siderail control panels with increasing and variable functions, as expressed today, commenced in the decade of 1970 [21, 27] (see Figure 5b). Even when the debate on possible risks associated to these protections still remains [30]. Figure 5 shows a patent with the description of communication and detection means of a bed of those times. The first mattresses for preventing pressure ulcers appear. Attention outside hospitals is stimulated by promoting hospital beds for home use (see Figure 6a). In the 80's, there was an evolution and proliferation of patents of active and passive therapeutic mattresses, including models with built-in position detectors for the disabled, as well as mechanisms for weighing the patient while leaning in bed. Until then, remote controls were wired. Nowadays, this kind of control is offered as alternative. Patient-exit monitoring and control of external devices from the bed also appears as a concept in development in patents of this decade [28, 31]. The decades of 1980 and 1990 witnessed an exponential growth of medical care devices, including some models that became market icons (see Figures 7a and 7b). Permanent cardiovascular monitoring became common features in hospitals, and treatment was affected by technological progress [4].

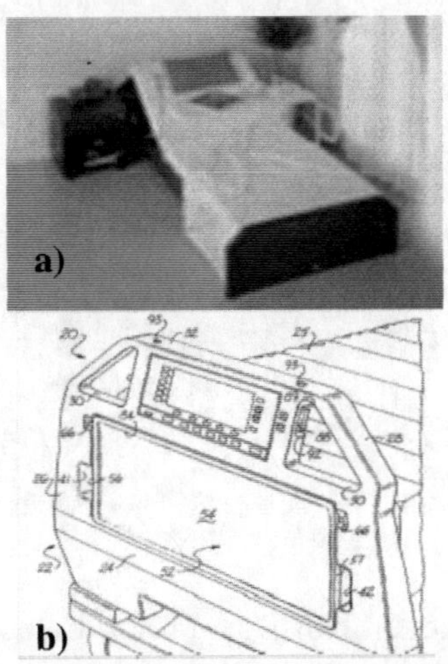

Figure 6. a) Electric bed for home use, 1983 [19]; **b)** Patent graph showing innovations in the front panel, 1993 [32];

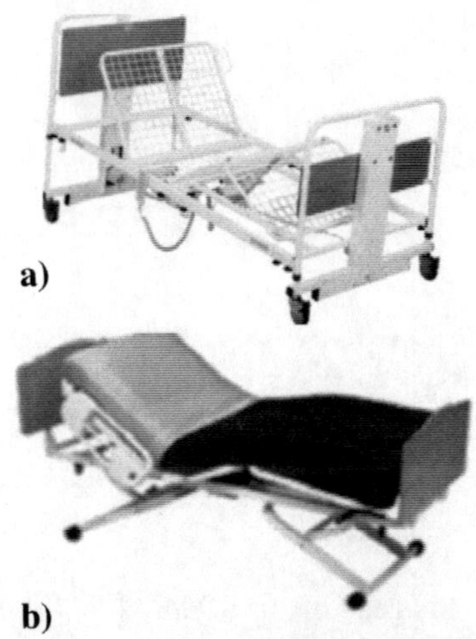

Figure 7. a) Sidhil Millenium Community Bed model, 1990 [22]. **b)** Joerns UltraCare model, 1999 [33].

Patents of the 80's show the interest in including means to weigh patients, avoiding their exit from the bed (see Figure 8), using incorporated or detachable control instruments. Other patents include alternative nurse-calling means for disabled patients [29] (Figure 5c)) and the monitoring of their position and bed exit, as well as special active and passive mattresses [34, 35].

During the 90's, more beds with varied and advanced mechanical functions appeared, such as the possibility of sitting in a chair position to exit the bed, visible in the Hill-Rom Total Care model of 1998 [11] (Figure 9). A domestic example is the mechatronic bed developed by the Industrial Design Research Center of the University of Buenos Aires *(Centro de Investigación en Diseño Industrial, Universidad de Buenos Aires)* [17].

At this time, mechatronic beds are a reality, and further technological and integrational advances allow consolidating the progress towards "intelligent" and integral equipment, with monitoring and advanced services, and integrated to accessories like active mattresses, under a unique control panel. The development of voice control as a potential advantage is mentioned in a 1991 patent, described as an environmental control unit for the hospital bed and its surroundings [36]. The effectiveness of voice recognition systems available in the 90's was limited if compared to present solutions, but its potential is still significant.

Concerning regulations on medical devices, the Council Directive 93/42/CEE on medical devices [37] appears in 1993, and the first particular standard for electrically operated hospital beds is published (60601-2-38:1996) [13], indicating the special care and relevance that this type of products, of increasing capacity and complexity, deserve in the context of medical electrical devices. Subjects covered by this standard included protection against electrical and mechanical hazards, electromagnetic compatibility, as well as indications and requisites for controls, stability, accuracy of operations and fault-condition tests. Towards the end of this timeframe, in the year 2000, the additional European standard EN 1979, was published. This standard regulated adjustable beds for disabled patients, under home-care environments [38]. Later on, the current ISO/IEC-approved particular standard, 60601-2-52, would be published in 2009 and enforced in 2013, integrating both prior standards, and dealing with basic safety and essential performance of medical beds, under a broad range of application environments (from hospital to homecare) and based on a new theoretical background, resulting in updated reference values and requisites [39].

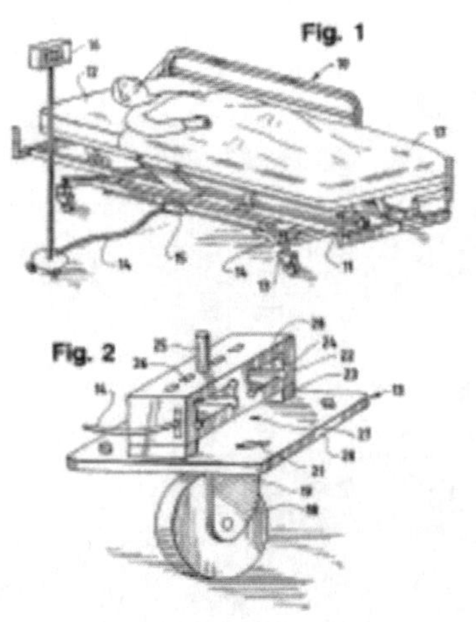

Figure 8. Images of a 1994 patent showing a weighing scale included in the wheels [40]

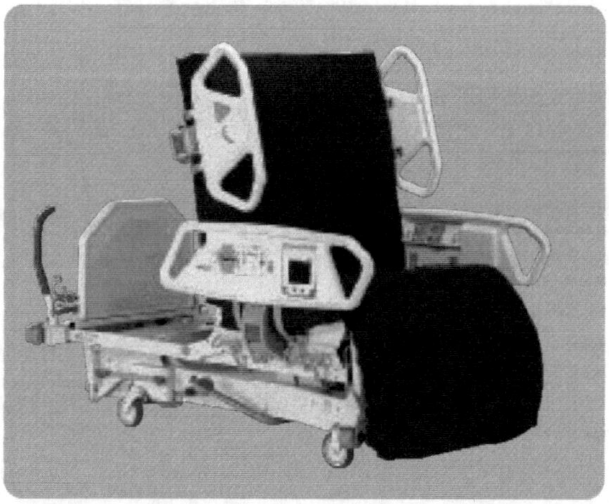

Figure 9. Hill-Rom Total Care model of 1998 [11].

3. Conclusions and future perspectives: Mechatronic beds of the 21st century

The analysis of this first historical progression of high complexity hospital beds allows outlining conclusions about elements that direct and determine growth and changes in these equipments:

- Many functions found in state-of-the-art automatic beds corresponds to counterparts found in their origins: from basic mechanical functions up to concepts like monitoring bed exit, communication with caregivers, and specific proposals for the care of critical patients or with variable disabilities. All of these are found throughout the mentioned timeframe, both as specific products as well as on registered inventions.
- The increasing "delegation" of tasks on this type of equipment is manifest, in line with the growth of the demand and the changes in the structure of the health system, and provides means for easing the work of the caregivers and, when possible, facilitating the autonomous care of the patient.
- There is a growing specificity in different models of electric medical beds, with morphologic, functional and aesthetic aspects orientated to specific environments and populations (intensive care in hospitals and domiciliary care, for the elderly, pediatric, bariatric patients, etc.)
- Finally, there is a remarkable trend towards relocating and re-discovering new implementations for the increasing range of technologies in dispersion. This trend will shape the advances for current market demands, such as: inclusion of tactile interfaces, functions of connectivity and complex real-time monitoring, as well as of new sensors and actuators, together with enhanced signal-processing capabilities. These elements allow the creation of new and improved versions of these devices, making available novel means for developers to propose new variants of added value for this family of products (focused in the experience of the user), while preserving the equipment's basic functions. This synthesis and technological growth has changed the visible face of the market of high complexity mechatronic beds in the last decades.

References

[1] *Who Invented The Hospital Bed?* 2014; Available in: http://www.askdeb.com/health/who-invented-the-hospital-bed/.
[2] National Association of Bedding Manufacturers, 1964 *Nation's Oldest Family-Held Bedding Firm: Adam Wuest, Inc.*
[3] *Is the Product a Medical Device?* ; Available in: http://www.fda.gov/medicaldevices/deviceregulationandguidance/overview/classifyyourdevice/ucm051512.htm.
[4] World Health Organization, 2010 *Medical Devices: Managing the Mismatch. An Outcome of the Priority Medical Devices Project.*
[5] USAID World Health Organization, *Joint position paper on the provision of mobility devices in less-resourced settings: a step towards implementation of the Convention on the Rights of Persons with Disabilities (CRPD) related to personal mobility*. 2011, World Health Organization.
[6] Lyder C H and Ayello E A 2008 Pressure Ulcers: A Patient Safety Issue, in Patient Safety and Quality: An Evidence-based Handbook for Nurses, AHRQ Publications.
[7] *Preventing Pressure Ulcers: An information booklet for patients*. Octubre 2004, Metropolitan Health and Aged Care Services Division: Melbourne, Australia.
[8] World Health Organization 2013 *International Perspectives on Spinal Cord Injury* Malta.
[9] Department of Economic and Social Affairs United Nations, *World Population Ageing*. 2013: Nueva York.
[10] ArjoHuntleigh International Ltd. *Enterprise 9000: Intensive Care Hospital Bed*. 2010; Available in: http://www.arjohuntleigh.com/products/medical-beds/hospital-beds/enterprise-range/enterprise-9000/downloads/.
[11] Hillenbrand Industries, 1998 *Hillenbrand Industries Annual Report 1998*.
[12] Hill-Rom Services Inc. 2007 *TotalCare Bariatric Plus Therapy System*.
[13] International Electrotechnical Commission. *IEC 60601-2-38:1996 Medical electrical equipment - Part 2: Particular requirements for the safety of electrically operated hospital beds* 1996; 1.0:[Available in: https://webstore.iec.ch/publication/16883.

[14] U.S. Food and Drug Administration. *A Guide to Bed Safety Bed Rails in Hospitals, Nursing Homes and Home Health Care: The Facts*. 2010 [cited 2015; Available in: http://www.fda.gov/MedicalDevices/ProductsandMedicalProcedures/GeneralHospitalDevicesandSupplies/HospitalBeds/ucm123676.htm.
[15] Mariño M 1995 *Tesis de Doctorado*, in *Facultad de Arquitectura, Diseño y Urbanismo*, Universidad de Buenos Aires.
[16] Mariño M H S 1997 *Ergonomic mechanism for use in hospitals*.
[17] Centro de Investigación en Diseño Industrial de Productos Complejos. *UBACyT AR050: Investigación y Desarrollo de una Interfase Mecatrónica para la Atención de Pacientes que Deben Permanecer Postrados por Largos Períodos de Tiempo.* 2015]; Available in: http://www.cidi.bime.com.ar/lineas2.php?id=4.
[18] *Push-Button Hospital Bed: Invention of Los Angeles doctor eliminates the bedpan*, in *LIFE Magazine*. 1945. p. 92-94.
[19] *Our History*. [cited 2014; Available in: http://www.paramount.co.jp/english/contents/106.
[20] Sweetland E. J. 1950 *Bed construction*.
[21] *Hill-Rom: Historia*. [cited 2014; Available in: http://www2.hill-rom.com/spain/History_long.htm.
[22] *The Sidhill Story*. [cited 2014; Available in: http://www.sidhil.com/about/the-sidhil-story/.
[23] Reiser S J 1981 *Telecommunication, automation and medical practice*, in *Medicine and the Reign of Technology*, Cambridge University Press.
[24] *Stryker History*. [cited 2014; Available in: http://www.stryker.com/en-us/corporate/aboutus/history/index.htm.
[25] Kalamazoo Public Library. *Dr. Homer Stryker: Inventor, orthopedic surgeon, businessman*. Available in: http://www.kpl.gov/local-history/biographies/homer-stryker.aspx.
[26] Black Warren R 1964 *Manual control for a motorized hospital bed*.
[27] Adams J S, Foster L D et al 1976 *Guard including electrical controls and slidable underneath the bed*.
[28] Drew W D, Foster L D et al 1980 *Side guard for bed including means for controlling remote electrical devices*.
[29] Natitus D P and Carlson C T 1981, *Portable patient call*.
[30] Brush B L and Capezuti E 2001 *Historical analysis of siderail use in American hospitals. J Nurs Scholarsh*. **33**(4): p. 381-5.
[31] Roberts W A, 1980 *Hospital bed monitor*.
[32] Ashcraft D N, et al. 1998 *Convertible footboard for a patient support*.
[33] Joerns Healthcare, *Joerns: Excellence and innovation for over 100 years*.
[34] Hasty C E, 1990 *Air-operated body support device*.
[35] Murphy M, 1992 *Decubitus ulcer mattress*.
[36] Douglas T L, 1994 *Voice-actuated, speaker-dependent control system for hospital bed*.
[37] Consejo de las Comunicaciones Europeas 1993 *Directiva del Consejo CEE 93/42 Relativa a los Productos Sanitarios*.
[38] HMS-VILGO Medical Device Manufacturers *New Standard for Medical Beds EN 60601-2-52*.
[39] International Electrotechnical Commission 2013 *IEC 60601-2-52:2009 Medical electrical equipment - Part 2-52: Particular requirements for the basic safety and essential performance of medical beds*.
[40] Geringer E R and Goldberg J 1998 *Caster mounted weighing system*.

Acknowledgments
This work was carried out under -and supported in part by- Project 20621201000011BA UBACyT 2013-2016, Facultad de Arquitectura, Diseño y Urbanismo, Universidad de Buenos Aires (FADU-UBA), and Facultad de Ciencias Fisicomatemáticas e Ingeniería, Pontificia Universidad Católica Argentina (UCA).

Processing of microCT implant-bone systems images using Fuzzy Mathematical Morphology

A Bouchet[1], L Colabella[2], S Omar[2], J Ballarre[2], J Pastore[1]
[1] Digital Image Processing Group. Electronics Dept. UNMDP. CONICET
[2] INTEMA. UNMDP. CONICET

E-mail: jpastore@fi.mdp.edu.ar

Abstract. The relationship between a metallic implant and the existing bone in a surgical permanent prosthesis is of great importance since the fixation and osseointegration of the system leads to the failure or success of the surgery. Micro Computed Tomography is a technique that helps to visualize the structure of the bone. In this study, the microCT is used to analyze implant-bone systems images. However, one of the problems presented in the reconstruction of these images is the effect of the iron based implants, with a halo or fluorescence scattering distorting the micro CT image and leading to bad 3D reconstructions. In this work we introduce an automatic method for eliminate the effect of AISI 316L iron materials in the implant-bone system based on the application of Compensatory Fuzzy Mathematical Morphology for future investigate about the structural and mechanical properties of bone and cancellous materials.

1. Introduction

The relationship between a metallic implant and the existing bone in a surgical permanent prosthesis is of great importance since the fixation and osseointegration of the system leads to the failure or success of the surgery. In the last decade, computational modeling has emerged as a powerful tool to investigate about the mechanical properties of bone and cancellous materials, as well to predict their mechanical strength with virtual or *in silico* experiments [1]. For this purpose, tridimensional models using the Finite Element Method (FEM) are built from micro Computed Tomography scans with a resolution of a few microns. These models and techniques could help to develop computational tools for predicting bone-implant adaptation, prosthesis stability and to design smart biomaterials for bone and tissue regeneration [2,3]. One of the problems presented in reconstruction of these images, is the effect of the iron based implants, with huge artifacts (interface scattering, beam hardening) distorting the micro CT image and leading to bad 3D reconstructions [4,5]. The solution generally to avoid or diminish these artifacts is use Synchrotron Radiation, several filters during the scans, or some standard commercial programs [6,7].
Therefore the goal of this work is to generate clear images from micro CT bone-implant system scans, applying image processing techniques, in particular Compensatory Fuzzy Mathematical Morphology (CFMM) [8-10], for then compile the processed images into a 3D reconstruction. This analysis could help to visualize the effect of implant inclusion into the marrow cavity, for example, and see the osseointegration in a tridimensional level. The images obtained with the conventional microCT

equipment usually have distorting artifacts related with the presence of the metallic implants, and are not easy to process to 3D images. Hence, conventional image processing methods for shape analysis cannot be applied in these cases [11-13]. In this paper we propose to apply CFMM techniques to solve the artifact effect generated by the presence of the implant, and to finally have clear tridimensional images of the bone-implant system, that could be use to different further analysis.

2. Materials and methods

2.1. In vivo samples

In vivo experiments were conducted with 6 Wistar – Hokkaido (WKAH/Hok) male adult rats (weight 350 ± 50 g), according to rules of ethical committee of the National University of Mar del Plata (Interdisciplinary Committee, April 2005/October 2010), taking care of surgical procedures, pain control, standards of living and appropriated death. Implants of stainless steel (AISI 316L) in wire shape were sterilized in autoclave for 20 minutes at 121 °C. Rats were anaesthetized with Ketamine and Xilacine (100mg/Kg, 10mg/Kg) according to their weight. The region of surgery surface was cleaned with antiseptic soap. The animals were placed in a supine position and the implantation site was exposed. A region of around 0.5 cm diameter was scraped in the femur plateau and a hole was drilled using a hand drill of 0.125 cm diameter at low speed. The wire implants were placed by press fit into the medullary canal of the femur. The animals were sacrificed after 8 weeks with an overdose of intraperitoneal sodic thiopental (120 mg/kg) with a previous anesthesia as described above and the femur with implant was retrieved. Conventional X-ray radiographs were taken before retrieving the samples for control of the correct position of the implant into the bone marrow cavity.
The retrieved samples were cleaned from surrounding soft tissues and fixed in neutral 10 wt% formaldehyde for at least 24 h. Then they were dehydrated in a series of alcohol-water mixture followed by a methacrylate solution and finally embedded in poly-methyl methacrylate (PMMA) solution and polymerized for 5 days at 32 ºC.

2.2. Micro CT images

In micro-computed tomography the microCT SkyScan 1072 system was used. The SkyScan 1072 represented high resolution desktop X-ray microtomography (Micro-CT) systems for *in vitro* scanning. Tomographic techniques were applied for the investigation of trabecular microarchitecture of the distal physis of the femur bone in all animals 8 weeks after surgery. Probes were located centrally on the tripod in the middle of the microtomography camera's area of sight and scanned using this parameters: camera pixel size 8.70 μm, Image Pixel Size 4.89 μm, Rotation step 0.150 degrees, exposure 950 ms. Raw scans were reconstructed with the nRecon program (Skyscan, Belgium).

2.3. Compensatory Fuzzy Mathematical Morphology for image processing

In this section it is described the Compensatory Fuzzy Mathematical Morphology (CFMM) that was used for the segmentation of the microstructure of the femoral trabecular bone. This morphology is a particular case of the Fuzzy Mathematical Morphology, based on the replacement of t-norms and s-norms by conjunctions and disjunctions of Compensatory Fuzzy Logic. This is possible by relaxing the constrains imposed by t-norms and s-norms [10]. By replacing t-norm and s-norm by conjunction and disjunction, respectively, was obtained the dilation and erosion operators for the CFMM, called compensatory dilation and compensatory erosion, respectively. For example, for two fuzzy sets μ and ν, the compensatory dilation of μ by ν is defined by:

$$\delta(\mu,\nu)(x)=\sup_{y\in U}\left[C\left(\mu(y),\nu(y-x)\right)\right] \quad (1)$$

where C is a compensatory conjunction.

For two fuzzy sets μ and ν, the compensatory erosion of μ by ν is defined by:

$$\varepsilon(\mu,\nu)(x)=\inf_{y\in U}\left[D\left(\mu(y),c\left(\nu(y-x)\right)\right)\right] \quad (2)$$

where D is a compensatory disjunction and $c(a)=1-a$, $\forall a\in[0,1]$.
Based on these two operators, more complex ones can be defined, as for example compensatory opening, closing and top-hat.

In this paper the top-hat by closing was used. This operator is defined by the following equation:

$$TH_{\nu}(\mu)=\mu-\varepsilon_{\nu}\left(\gamma_{\nu}(\mu)\right) \quad (3)$$

This transform stands out locally dark objects that have been eliminated in the closing filter in a gray scale image.

2.4. Segmentation Method

In this section, the algorithm proposed to segment automatically the implant-bone system is described. To show the effects of the algorithm, it was applied to one sequence of microCT. The proposed algorithm is based on two stages:

2.4.1 Background extraction in the bone-implant system

When segmenting microCT implant-bone systems images, a wide range of useless information arises, which has to be discarded as a prior step for classifying the different areas. To obtain effective results during the classification process, it is necessary to work with images containing the femoral bone and eliminate the effect of the iron based implants. The segmentation of the bone tissue is the first step in the trabecular bone classification. A correct segmentation will end, finally, in the success or failure of the upcoming segmentation processes, 3D reconstructions and structural analysis.

For this segmentation, an automatic algorithm was proposed. In a first step, Alternating Sequential Filters (ASFs) by reconstruction was applying using the sequence "CO" (closing-opening) with a cross structuring element and 7 iterations. In this way, a new image was obtained in which the bone zone is homogeneous and then a regions labeling criterion using geodesic distance was applied, which can evolve adapting to the topology of the objects in the image [14-15].

Figure 1 shows the result of the classification after the extraction of the femoral bone by the algorithm proposed.

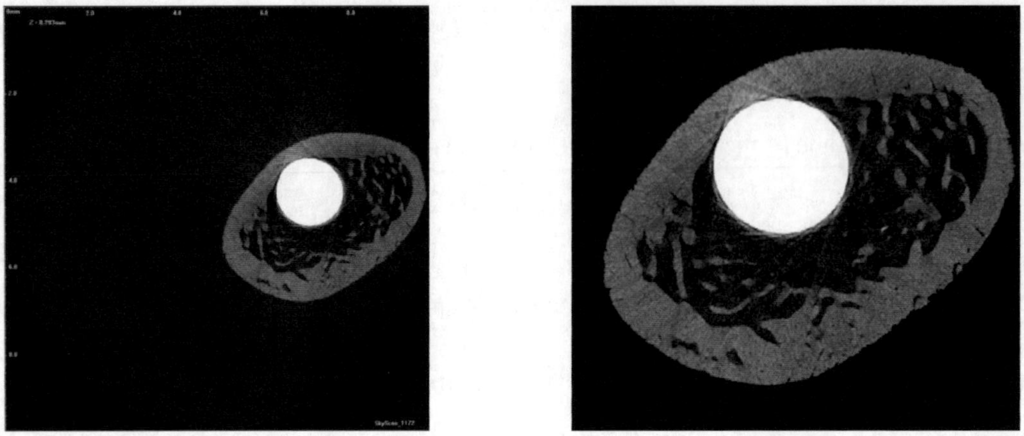

(a) (b)

Figure 1. Elimination of redundant information (a) Original image, (b) image obtained after applying the proposed algorithm.

2.4.2 Segmentation of bone tissue by Compensatory Fuzzy Mathematical Morphology

In this second step, the method for segmentation of the trabecular tissue is described. Each image stack is considered as a single image for processing.

Step 1: The first step consists in eliminate the implant from the images. A threshold is applied to eliminate the implant. In this case, 250 was used as an umbral because the implant zone is homogenous with 255 value.

Step 2: A compensatory Top–Hat was applied to the images. In this case, a conjunction and disjunction of compensatory logic based on the geometric average for defining the operators of the equations (1) and (2) was applied. A 3x3 Gaussian structuring element was used. This method allowed to emphasize the objects of interest homogenizing the image background, as can be seen in Figure 2.

Step 3: Otsu binarization was applied.

Step 4: In this last step, the morphological gradient, using a 3x3 cross structuring element, is applied to extract the boundaries.

In the next section, the results obtained through the proposed method are described.

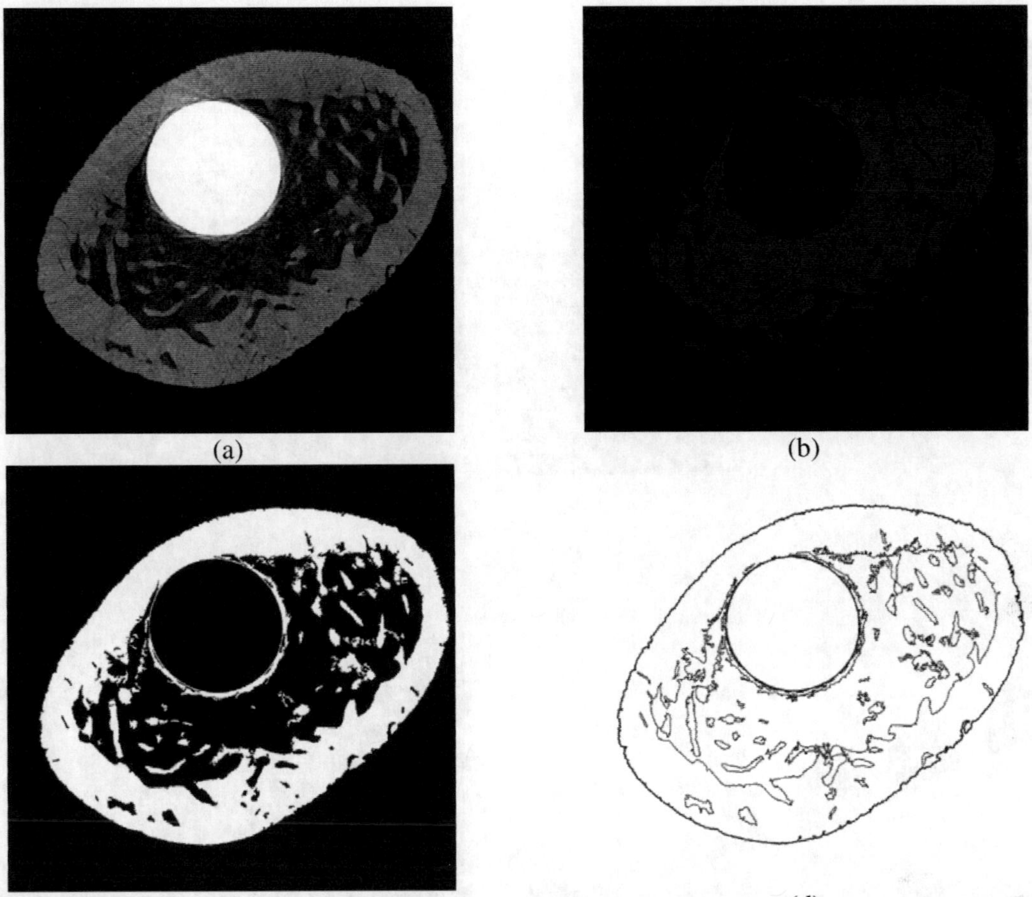

Figure 2. Segmentation of the trabecular tissue in microCT images. (a) Original image, (b) image obtained after applying compensatory Top–Hat, (c) image obtained after binarization, (d) boundary.

3. Results and discussion

The effect of the implant in the microCT images was studied for all six implant-bone samples. As the artifact is the same for all (is the same type of implant), only one sample is presented here.

Figure 3 show 3D reconstructions for the same stack of images, in the first case without processing and in the second case after applying the proposed method. The reconstruction was made by a tridimensional model using the Finite Element Method (FEM) [16]. The artifact effect related with the metal implant present in the images is showed in Figure 3b. Figure 3c shows a more precise 3D reconstruction obtained by applying techniques of image processing, in particular, CFMM operators. This kind of models have a particular interest, given that they help to predict the bone adaptation with the implant, the prosthesis stability and to design smart biomaterials for bone and tissues regeneration.

 doi:10.1088/1742-6596/705/1/012055

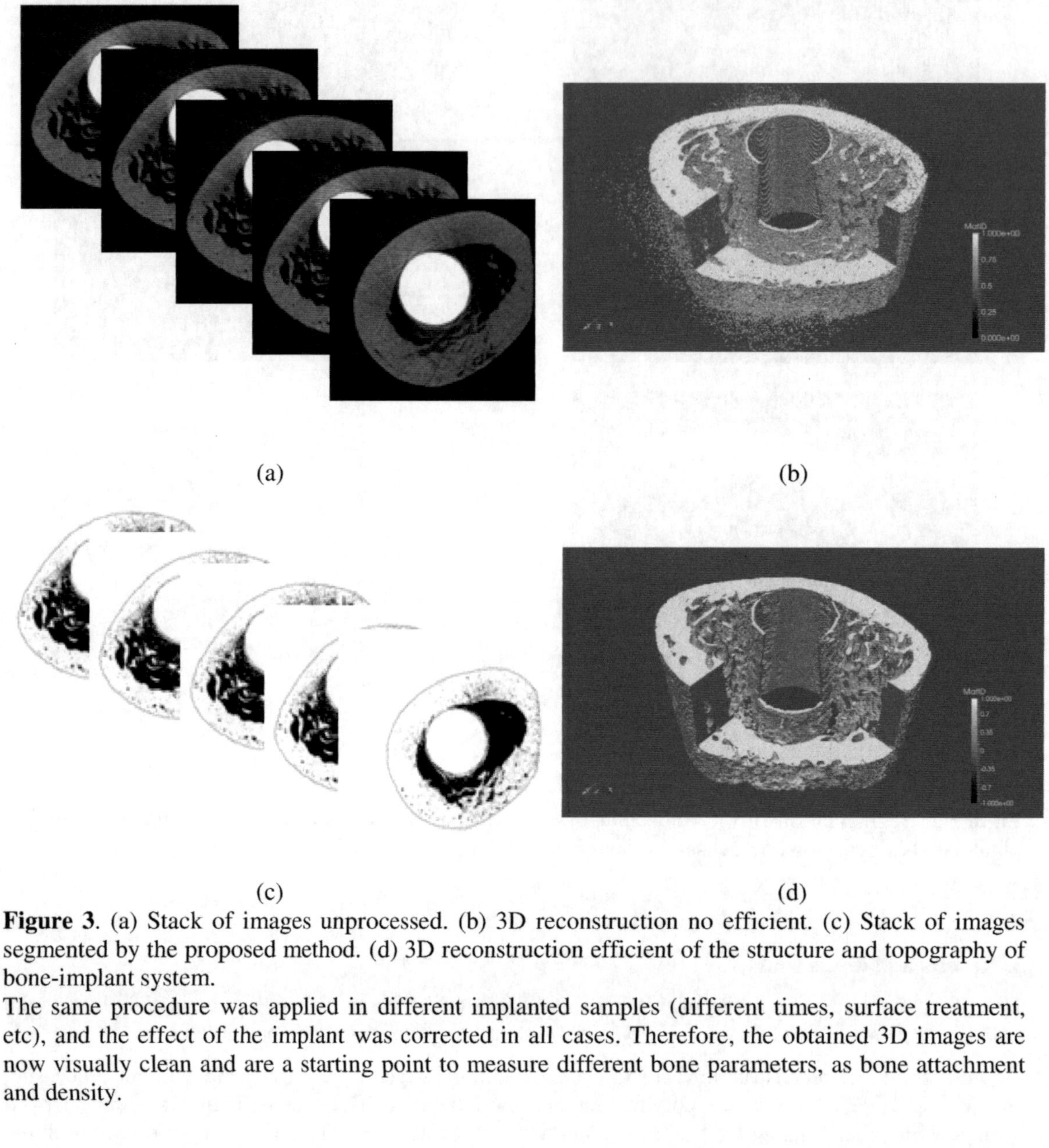

Figure 3. (a) Stack of images unprocessed. (b) 3D reconstruction no efficient. (c) Stack of images segmented by the proposed method. (d) 3D reconstruction efficient of the structure and topography of bone-implant system.

The same procedure was applied in different implanted samples (different times, surface treatment, etc), and the effect of the implant was corrected in all cases. Therefore, the obtained 3D images are now visually clean and are a starting point to measure different bone parameters, as bone attachment and density.

4. Conclusions

This work presents an automatic segmentation of the bone-implant system in conventional microCT images in a rat femur model. The Fuzzy Mathematical Morphology helps to extract the huge artifacts generated by iron based implants in microCT images. The method can be adapted to several implant cases and different types of metals since it is not a conventional filter, allowing the user to segment the trabecular bone at the periphery of metallic implants, obtaining clear images to process them and create 3D reconstructions.

As a future work, the proposed method will be validated with the goal to investigate about the structural and mechanical properties of trabecular bone and cancellous-like materials.

5. References

[1] Hart, R T. Bone Modelling and Remodelling: Theories and Computation. In *Bone Mechanics Handbook, Second Edition*. CRC Press, Boca Ratón, 2001.

[2] A. Gomez Sanchez, W. H. Schreiner, J. Ballarre, A. Cisilino, G. Duffó, S. Ceré. Surface modification of titanium by anodic oxidation in phosphoric acid at low potentials. Part 2. In vitro and in vivo study. Surface and Interface Analysis. ISSN 1096-991. (2013). DOI 10.1002/sia.5298.

[3] Miaraa, B, Rohanb, E, and Labatc, B. Piezomaterials for bone regeneration design – homogenization approach. *Journal of the Mechanics and Physics of Solids*, 53, 11 (2005), 2529-2556.

[4] R. Bernhardt, D. Scharnweber, B. Müller, P. Thurner, H. Schliephake, P. Wyss, F. Beckmann, J. Goebbels, H. Worch, J. Gasser, J. Kirkpatrick, B. Rahn, Comparison of microfocus- and synchrotron x-ray tomography for the analysis of osteointegration around TI6AL4V-implants, European Cells and Materials 7 (2004) 42-51.

[5] F Edward Boas & Dominik Fleischmann. CT artifacts: Causes and reduction techniques, Imaging Med. (2012) 4(2), 229-240.

[6] C.A. Neldam, E.M. Pinholt, Synchrotron μcT imaging of bone, titanium implants and bone substitutes - A systematic review of the literature, Journal of Cranio-Maxillofacial Surgery 42 (2014) 801-805.

[7] J.W. Song, J.Y. Cha, T.E. Bechtold, Y.C. Park, Influence of peri-implant artifacts on bone morphometric analysis with micro-computed tomography, The International journal of oral & maxillofacial implants 28 (2013) 519-525.

[8] Bloch I. (2005). Fuzzy spatial relationships for image processing and interpretation: a review, Image and Vision Computing 23: 89-110.

[9] Bloch I. and Maitre H. (1995). Fuzzy mathematical morphologies: A comparative study, Pattern Recognition 28: 1341-1387

[10] A. Bouchet, J.I. Pastore, R.E. Andrade, M. Brun, V. Ballarin, Compensatory Logic applied to Digital Image Processing, in: R.E. Andrade, J.M. Gómez, A.R. Valdéz (Eds.) Towards a trans-disciplinary technology of Business and Organizational Intelligence: Gathering Knowledge Discovery, Knowledge Management and Decision, Shaker Verlag, Aachen, University of Oldenburg, Alemania, 2011, pp. 226-239.

[11] J. Serra, Image analysis and mathematical morphology, Vol. 1, Academic Press, London, 1982.

[12] J. Serra, Image analysis and mathematical morphology, Vol. 2, Academic Press, London, 1988.

[13] R. González, R. Woods, Digital Image Processing, Addison Wesley1996.

[14] Pastore J, Moler E, & Ballarin V. "Procesamiento digital de imágenes a través de filtros morfológicos direccionales", Revista Argentina de Bioingeniería. Vol 14 N 1, pp 14-21, 2008.

[15] Pastore J, Moler E & Ballarin V. "Segmentation of brain magnetic resonance images through morphological operators and geodesic distance", Digital Signal Processing, Vol. 15, pp 153-60, 2005.

[16] Arbenz P, van Lenthe G H, Mennel U, Muller R, and Sala M. "Multi-level micro-finite element analysis for human bone structure", Workshop on State-of-the-Art in Scientific and Parallel Computing, Sweden, 2006

Acknowledgments
Authors wishing to acknowledge assistance of Dr Paul Zaslansky and Dr. Felix Repp (Charité-Berlin, Max Planck Institute of Colloids and Interfases, Golm, Germany) for the help with the microCT experimental data collection.

Videofluoroscopy Instrument to Identify the Tibiofemoral Contact Point Migration for Anterior Cruciate Ligament Reconstruction Follow-up: CINARTRO

F Simini [1], D Santos [2], L Francescoli [3]

[1]Núcleo de Ingeniería Biomédica, Facultades de Medicina e Ingeniería - Universidad de la República - Montevideo, Uruguay
[2]Departamento de Rehabilitación y Medicina Física (UIBLH) Hospital de Clínicas - Universidad de la República - Montevideo, Uruguay
[3]Clínica de Ortopedia y Traumatología, Facultad de Medicina - Universidad de la República - Montevideo, Uruguay

E-mail: simini@fing.edu.uy

Abstract. We measure the Tibiofemoral contact point migration to offer clinicians a tool to evaluate Anterior Cruciate Ligament reconstruction. The design of the tool includes a C arm with fluoroscopy, image acquisition and processing system, interactive software and report generation for the clinical record. The procedure samples 30 images from the videofluoroscopy describing 2 seconds movements of hanging-to-full-extension of the knee articulation. A geometrical routine implemented in the original equipment (CINARTRO) helps capture tibial plateau and femoral condile profile by interaction with the user. The tightness or looseness of the knee is expressed by the migration given in terms of movement of the femur along the tibial plateau, as a percentage. We automatically create clinical reports in standard Clinical Document Architecture or CDA format. A special phantom was developed to correct the "pin cushion effect" in Rx images. Five cases of broken ACL patients were measured giving meaningful results for clinical follow up. Tibiofemoral contact point migration was measured as 60% of the tibial plateau, with standard deviation of 6% for healthy knees, 4% when injured and 1% after reconstruction.

1. Introduction

The knee articulation (KA) is responsible for two contradictory functions: static stability during full extension and adaptation to ground irregularities during gait and running [1]. Clinical appraisal of movement is subjective, which has led to the need of objective measurements. To this end, several methods have been suggested, albeit all circumscribed within research settings and not currently in clinical use. Clinicians look at the looseness or tightness of the knee during flexion and extension but have no objective measure of either of them.

In the presence of Anterior Cruciate Ligament (ACL) injury there is a need for clinical assessment. According to Benjaminse et al. [2] the main function of the ACL is to prevent tibial translation forward, with respect to the femur. Therefore a loose KA is found for ACL injured patients. A reconstructed KA after ACL injury should recover at least part of the contention with no excess stiffness. Diagnostic tests to confirm ACL tears (and knee looseness) include the Lachman test, the pivot shift test and the use of the KT 1000/2000 arthrometer [3]. Magnetic resonance imaging is also used, because it provides the fine soft tissue detail necessary to confirm a diagnosis [4].

The normal knee bends with respect to a migrating fulcrum or centre of rotation. The ACL injured KA behaves differently. One method to objectively describe this state of tightness/looseness was described by Baltzopoulos [5] which consists of determining the contact point between the femur condile and the tibial plateau, which was called the "Tibiofemoral (TF) contact point". Baltzopoulos measures the migration of this TF contact point during flexo/extension on the tibial plateau [6]. This determination of knee artrokynematics could be used as an objective method to evaluate ACL reconstruction [7] and the first clinical applications have been described [8]. TF contact point migration data also helps to put in place rehabilitation strategies and to monitor them, as well as to give feedback to the surgeon. An alternative method to objectively record the tightness/looseness of the knee was published by Rouleaux, as the mechanical centre of rotation [9] derived from two adjacent videofluoroscopy images during flexo/extension. More related to mechanical engineering practice, this method of the geometrical centre of rotation has little clinical significance though, because the instantaneous fulcrum is located within the femoral condile. We set ourselves the goal of developing a methodology and a clinical instrument for ACL follow-up, based on our prior medical device development experience [10]. The present paper reports on the implementation of the Baltzopoulos method as a medical instrument prototype (which we called CINARTRO), while development of software according to the Rouleaux method are at present under way. Clinical application was performed on five patients before and after surgery. The goal of the present applied research is that of obtaining a dynamic and functional evidence of knee movement as opposed to conventional static imaging studies, such as MRI, CT and RX, to guide clinical management.

2. Rationale and specification of the instrument

2.1 Relevance of the instrument designed

The incidence of ACL lesions is estimated as 1 every 3000 population, which accounts for over a billion US dollars in surgical and rehabilitation costs only in the USA [11]. This overall cost is equivalent to US dollars 10.000 for every injured person including 6 months of rehabilitation [12]. Surgery and rehabilitation restore the stability and the complex movement of the knee. The main concern of the surgeon is to count on reliable and objective elements to evaluate the postoperative functional restoration. The physiotherapist, moreover, is interested in periodic knee functional measurements as his or her rehabilitation work progresses, and in particular the relative movement of femur and tibia, whether too tight or too loose.

The desired instrument to address these needs should add quantitative input to the traditional clinical approach of gait observation and tests such as the one described by Lachman using the arthrometre, (e.g.: KT-1000 y KT- 2000) [2]. For ACL reconstruction evaluation, usual clinical practice includes imaging studies such as X rays, CT scans and NMR, all of which are purely static evaluations. Although anatomical structures may be shown in great detail, no dynamic aspects are considered, which are part of the essence of knee functions. The instrument to be designed should record, evaluate and quantify kinematic characteristics of the KA.

2.2 Specification of the instrument to be designed

There are no instruments available today to spot quantitatively the TF contact point during knee joint movement, to be used in patient care. The instrument to be designed should compare the injured knee with the other knee and the injured knee prior and after surgery, as well as at set intervals during rehabilitation.

The instrument should determine TF contact point migration during flexo-extension and should create a document for the Electronic Clinical Record of the patient, as well as a hard copy. The details of the determination should be decided by the specialist during an interactive software application. The instrument should show both the injured and contra-lateral knee joints images and TF contact point determinations, allowing reference and comparison. Additionally, the instrument should include ways to follow-up a knee joint being reeducated over the months and years. The instrument was called CINARTRO to suggest kinematics ("CIN") and joint ("ARTROS"), to study KA movement.

2.3 Experimental set up and design of the instrument
In order to obtain a proof of concept and to validate the method, an experimental set up was build to analyze results on real data. To this end the following elements were connected:

- C-arm Xray equipment with video data acquisition
- Data acquisition hardware to digitize the videofluoroscopy images
- Software to determine the fiducial points in every image (tibial plateau and femur condile)
- Software to calculate TF contact point in every image
- Software to calculate the percentage of migration of TF contact point with respect to tibial plateau length

The method used consists of obtaining a series of 30 Xray images during extension and 30 images during flexion. The overall doses delivered by the C-arm was 250 microGy (J/Kg) absorbed every 2 seconds movement.

3. Project and building blocks

To tackle the clinical need for the instrument specified, we suggest a configuration such as the one represented in Figure 1. A C-arm apparatus gives standard sequences of images to be processed off line. The video film obtained is transformed into a series of 30 still images, evenly distributed from hanging leg position to full extension. Every image is shown to the operator in an off line environment with a Graphical User Interface (GUI) [13]. In this image the user is asked to identify the following five points: tibial plateau extremities and three points on the femoral condile profile. The central point is the furthest point on the femur "distalwise" and the other two are approximately one cm apart on either side, on the profile, as shown in Figure 4.

The five points for each image are processed by CINARTRO to find the shortest segment between the tibial plateau (straight line) and the curved femoral condile. The three femoral points are fitted to a circumference, and the software calculates by simple geometry the distance from the condile to the tibial plateau. Once the segment going from tibia to femur is known, its mid point is defined as the TF contact point. One TF contact point is calculated for each one of the 30 images. The projection of the TF contact point onto the tibial plateau is taken by the software as the position of TF contact point. We express this position as a percentage of the tibial plateau segment, from back to front (posterior-anterior wise). A projection at the very back of the tibial plateau is 0% and a projection at the foremost end is labelled 100% which is a way to unify measurements amongst anatomically different individuals.

CINARTRO builds a graph of the TF contact points in terms of percentage of tibial plateau with respect to the angle of extension (from 0 degrees to 90 degrees, the latter being full extension). The comparison of this graphical representation of (a) contralateral healthy, (b) injured and (c) reconstructed movements is the evidence that CINARTRO gives the clinician on the KA kinematics (Figure 2) of a single individual. Once the TF contact point migration is determined in the three situations -injured, healthy contralateral and reconstructed- data are stored and a clinical report is created, in which free text can be added by the operator.

We build a phantom consisting of a matrix of lead spheres (3 mm diameter, 1 cm apart) to be placed between the X ray emitter and the fluoroscope, thus allowing spatial calibration of the software. Figure 3 shows the phantom developed to calibrate videofluoroscopy derived images. In order to record exact dimensions, prior to the study of each patient, we obtain a phantom image and identify a few points at random to deduce a correction factor for all coordinates of the 30 images.

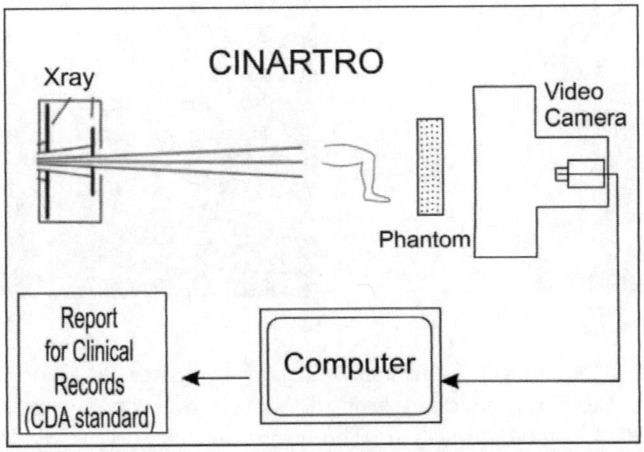

Figure 1. Block diagram of CINARTRO. Serial X ray images are obtained during extension. Fiducial points are marked for every image, allowing the Tibiofemoral (TF) contact point to be determined in each one of the 30 "hanging-to-extension" images. The clinical report shows the migration of the TF contact point projection along the tibial plateau.

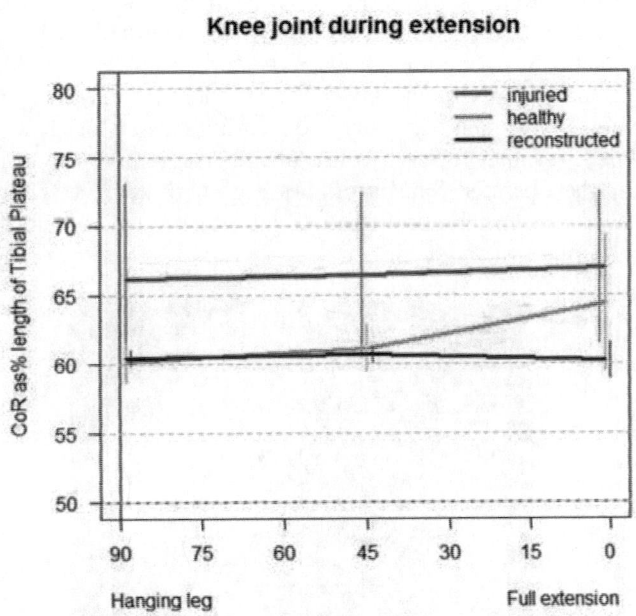

Figure 2. CINARTRO presentation of Tibiofemoral (TF) contact point migration as extension progresses, with respect to tibial plateau segment length. This case is the mean of several patients, but the same graphical format is created for each individual, with the same coordinates and values. Reprinted from Santos, et al. [14]

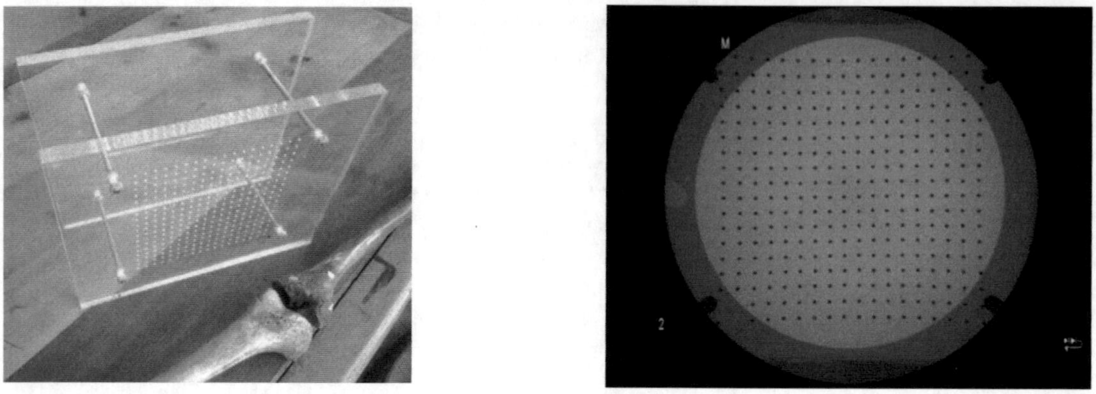

(a) **(b)**

Figure 3. (a) Phantom for CINARTRO made with hard acrylic material. Anatomical pieces (Femur and Tibia) are included in the foreground to show the scale. Note the 324 lead spheres of 1 mm, spaced 1 cm apart, embedded in the hard acrylic. The dimensions are 18 cm by 18 cm. **(b)** Rx image of the Phantom, used to calibrate and correct the C-arm distortion. Note the "pin cushion effect" as spheres -originally in a straight line- appear on a curved line showing the X-ray system distortion.

4. Clinical Results Using CINARTRO

CINARTRO was used to evaluate ACL reconstruction of five subjects, after securing Ethics Committee approval of the "Hospital de Clinicas". The patients were instructed to sit with their knee close to the C-arm. Patients performed extension/flexion movements in 2+2 seconds for a total of three repetitions, recorded by videofluoroscopy. This procedure was performed for the injured knee and the contralateral knee before surgery. Six months later, a third data set is obtained only for the reconstructed knee. Videofluoroscopy sequences were fragmented into 30 images, evenly distributed over the extension movement. No images are analyzed during flexion because we assume that the principal function of the ACL is tibial forward constraint [2]. Figure 4 shows one of the 30 images with the points marked by the operator. The information of both segments (straight line for tibial plateau and three points curve for the femoral condile profile) is used to determine the TF contact point according to the Baltzopoulos method [5].

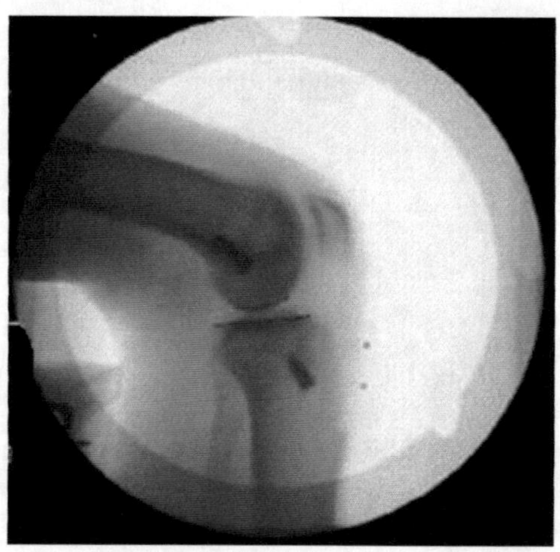

Figure 4. X ray of a knee joint taken by CINARTRO. Note the tibial plateau highlighted by 2 points selected by the user with the software and 3 points of the femur condile profile. The TF contact point is determined as the midpoint of the shortest segment between femur and tibial plateau.

Five male patients were studied (25 ± 3.6 years, range 18-35 years), all with an isolated tear of the ACL three months earlier (no ligament nor meniscal injury); all had a closed knee injury. As inclusion criterion, all patients had a score of the International Knee Documentation Committee (IKDC) of grade "A". Additionally all patients had a healthy contralateral knee.

Bone-patellar tendon-bone surgical technique was used for all patients. The first X-ray exploration was done the day before surgery and the second was performed six months after reconstruction only on the reconstructed knee [14].

In order to evaluate the surgical procedure and its effect on everyday life, the Lysholm-Tegner score [15] was enquired by questioning before and after reconstruction. The contralateral KA had a LT score of 100%, also assigned to the pre-lesion KA, both based on patient memory. Before and after reconstruction, the scores were 60% and 97%, respectively, suggesting that surgery was successful.

As a sample for the clinical use of CINARTRO, we estimated the mean TF contact point excursion of the five patient group. In healthy condition TF contact point was located 62% (SD 6%) with respect to the tibial plateau, as measured on the contralateral healthy knee (Table 1). After ACL lesion, this measurement is 61% (SD 4%) and finally, after ACL reconstruction surgery, it was 60% (SD 1%).

TABLE 1 – EVALUATION OF ANTERIOR CRUCIATE LIGAMENT (ACL) RECONSTRUCTION BY TIBIOFEMORAL CONTACT POINT MIGRATION PRIOR AND AFTER SURGERY.

	Healthy*	Injured	Reconstructed
Patient	MTFCP**	MTFCP	MTFCP
1	62%	57%	61%
2	56%	59%	60%
3	60%	61%	61%
4	61%	68%	61%
5	71%	59%	58%
mean	62% (SD 6%)	61% (SD 4%)	60% (SD 1%)

* Healthy contralateral knee. ** MTFCP: Migration of Tibiofemoral contact point (% with respect to tibial plateau from back to front in sagittal plane as shown in Figure. 3)

5. Discussion

Clinical practice of ACL reconstruction follow-up has demanded for many years a practical instrument to benefit from the results published by research in KA kinematics. On one side, current research has explained the dynamics of the consequences of ACL rupture and has described the partial reconstruction obtained with different surgical techniques. This explanation remained within the realm of academic work, while clinicians had no other way to record the evolution of a repaired ACL but static images and their observation of the patient with limited displacement tests such as the KT1000.

By developing CINARTRO we are suggesting an objective procedure to record the result of ACL reconstructive surgery immediately after inflammation disappears, and later at intervals during physiotherapy and rehabilitation. The first five patients studied here with our methodology follow a common pattern, that of being young amateur sportsmen with an ACL injury in an otherwise healthy organism. Subjective evaluation by the patients themselves gives a clear indication of the improvement of their quality of life (Table 1) after reconstructive surgery. Our study has added an objective measurement of knee kinematics by recording the TF contact point migration with respect to the tibial plateau. We are developing -for future versions of the CINARTRO software- the suggestion to the user of the fiducial points detected in the previous image as a first guess to be corrected by fine displacement of the mouse. This will greatly help the determination of the TF contact point on each image of the fluoroscopic series.

The X-ray apparatus used for the preliminary proof of concept of CINARTRO is cumbersome to use and expensive (C arm is originally intended for cardiac procedures). We are working towards the design of a dedicated compact X-ray device, derived from veterinary equipment [16] which will allow us to design a simple, portable CINARTRO.

Acknowledgment
The authors thank Alberto Leardini, of Istituto Rizzoli, Bologna and Thomas Andriacchi, Stanford University, for useful comments and advice on the present research. Guillermo Avendaño of the Universidad de Valparaiso, Chile, is also acknowledged for contributions referring to X-ray equipment being developed to substitute the C-arm in CINARTRO. Dr. Teresa Camarot is thanked here due to her continuous support within the Department of Rehabilitation aimed at improving patient care. The contributions of Braian Elliot and Mauricio Bouza are acknowledged in an early versión of CINARTRO software. Williams Olivera and Marcio Rodriguez have been programming parts of the current CINARTRO prototype, for which they are thanked by the authors.

References

[1] I. A. Kapandji, *The Physiology of the Joints: The knee*. Churchill Livingstone/Elsevier, 2010.
[2] A. Benjaminse, A. Gokeler, and C. P. van der Schans, "Clinical diagnosis of an anterior cruciate ligament rupture: a meta-analysis.," *J. Orthop. Sports Phys. Ther.*, vol. 36, no. 5, pp. 267–288, May 2006.
[3] P. B. Lewis, a D. Parameswaran, J.-P. H. Rue, and B. R. Bach, "Systematic review of single-bundle anterior cruciate ligament reconstruction outcomes: a baseline assessment for consideration of double-bundle techniques.," *Am. J. Sports Med.*, vol. 36, no. 10, pp. 2028–36, Oct. 2008.
[4] L. Herrington, C. Wrapson, M. Matthews, and H. Matthews, "Anterior Cruciate Ligament reconstruction, hamstring versus bone-patella tendon-bone grafts: A systematic literature review of outcome from surgery," *Knee*, vol. 12, no. 1, pp. 41–50, Jan. 2005.
[5] V. Baltzopoulos, "A videofluoroscopy method for optical distortion correction and measurement of knee-joint kinematics.," *Clin. Biomech. (Bristol, Avon)*, vol. 10, no. 2, pp. 85–92, Mar. 1995.
[6] V. Baltzopoulos, "Muscular and tibiofemoral joint forces during isokinetic concentric knee extension," *Clin. Biomech.*, vol. 10, no. 4, pp. 208–211, 1995.
[7] D. Santos, L. Francescoli, J. Loss, F. Arbío, and F. Simini, "A Tool to Assess Anterior Cruciate Ligament Recostruction by Quantitative Localization of the Knee Centre of Rotation," in *19th Congress of the European Society of Biomechanics (ESB2013)*, 2013.
[8] F. Simini and D. Santos, "Anterior Cruciate Ligament Reconstruction Follow-up Instrumentation based on Centre of Rotation Videofluoroscopy Determination: Development of an original equipment, CINARTRO, and first clinical use," in *Presented at I2MTC Congress, Montevideo - Uruguay, May*, 2014.
[9] R. van Dux, R. Huiskes, and G. Selvik, "Roentgen Stereophotogrammetric Methods for the Evaluation of the three Dimensional Kinematic Behaviour and Cruciate Ligament Length Patterns of the Human Knee Join," *J. Biomech.*, vol. 12, pp. 727–731, 1979.

[10] F. SIMINI, H. PIRIZ, and C. SCARONE, "Proyectos de Ingeniería Biomédica: tecnologías desarrolladas en la Universidad disponibles para el país," *Rev. Ing.*, vol. 49, pp. 16–21, 2004.

[11] L. Griffin, "Noncontact anterior cruciate ligament injuries: risk factors and prevention strategies," *J. Am. Acad. Orthop. Surg.*, vol. 8, no. 3, pp. 141–50, 2000.

[12] R. K. Flynn, C. L. Pedersen, T. B. Birmingharn, A. Kirkley, D. Jackowski, and P. J. Fowler, "The Familial Predisposition Toward Tearing the Anterior Cruciate Ligament: A Case Control Study.," *Am. J. Sports Med.*, vol. 33, no. 1, pp. 23–28, Jan. 2005.

[13] D. Santos, "Estudio de los Centros de Rotación Instantáneos de la Rodilla en Pacientes con Plastia del Ligamento Cruzado Anterior," Tesis de maestría en Ciencias Médicas; PROINBIO. Fac. de Medicina - UdelaR- Montevideo., 2014.

[14] D. Santos, F. Simini, L. Francescoli, F. Massa, A. Barquet, and T. Camarot, "Beyond traditional clinical evaluation of knee articulation movement to physiological assesment of dynamic ACL funtion during extension," in *XIII International Symposium on 3D Analysis of Human Movement. École Polytechnique Fédérale de Lausanne; Switzerland*, 2014.

[15] R. Marx, E. Jones, A. Allen, D. Altchek, S. O`Brien, and S. Rodeo, "Reliability, Validity, and Responsiveness of Four Knee Outcome Scales For athletic Patients," *J. BONE Jt. Surg.*, vol. 83, no. 10, pp. 1459–1469, 2001.

[16] G. Avendaño, "Expertos en Ingeniería Biomédica se reúnen en jornadas nacionales organizadas por la UV," 2015. [Online]. Available: http://www.uv.cl/pdn/?id=6104.

Journal of Physics: Conference Series **705** (2016) 012057 doi:10.1088/1742-6596/705/1/012057

Application of Oversampling to obtain the MTF of Digital Radiology Equipment.

M. Narváez, J. P. Graffigna, M. E. Gómez, R. Romo
Gabinete de Tecnología Médica, Facultad de Ingeniería.
Universidad Nacional de San Juan, San Juan, Argentina

E-mail: jgraffig@gateme.unsj.edu.ar
Abstract. Within the objectives of theproject Medical Image Processing for QualityAssessment ofX-Ray Imaging, the present research work is aimed at developinga phantomX-ray image and itsassociated processing algorithms in order to evaluatethe image quality rendered by digital X-ray equipment. These tools are used to measure various image parameters, among which spatial resolution shows afundamental property that can be characterized by the Modulation Transfer Function (MTF)of an imaging system [1]. After performing a thorough literature surveyon imaging quality control in digital X-film in Argentine and international publications, it was decided to adopt for this work the Norm IEC 62220-1:2003 that recommends using an image edge as a testingmethod. In order to obtain the characterizing MTF, a protocol was designedfor unifying the conditions under which the images are acquired for later evaluation. The protocol implied acquiring a radiography image by means of a specific referential technique, i.e. referred either to voltage, current, time, distance focus-plate (/film?) distance, or other referential parameter, and to interpret the image through a system of computed radiology or direct digital radiology. The contribution of the work stems from the fact that, even though the traditional way of evaluating an X-film image quality has relied mostly on subjective methods, this work presents an objective evaluative toolfor the images obtained with a givenequipment, followed by a contrastive analysis with the renderings from other X-filmimaging sets.Once the images were obtained, specific calculations were carried out. Though there exist some methods based on the subjective evaluation of the quality of image, this work offers an objective evaluation of the equipment under study. Finally, we present the results obtained on different equipment.

1. Introduction

The field of Digital X-Ray Imaging has gained wide acceptance on the grounds of stat-of-the-art advances, by incorporating new technologies that allow for readily obtaining improved imagesfor medical diagnosis, as contrasted with traditional methods.
Indeed, due to the development of digital image treatment with information extraction and enhanced visualization technologies, along with greater software power to discern the reaches of any given parameter of interest, it is now possible to easily improve the diagnosis renderings of the medical specialist.

In spite of this significant progress, it is still necessary to evaluate the quality of images. According to a definition, quality is understoodas the "Property or set of properties inherent toa given thing, which allowsfor judging its value". In the medical field, quality is associated to satisfying the expectations of the patient as regards diagnosis and treatment of his illness or condition, as performed by the physician, following the guidelines and rules established in national and international protocols, while trying to keep a reasonably lowest possible costfor the involved sanitary institution"[5].

In Medical Sciences, the subject of Medical Image Quality can be defined as the evaluation (objectively or subjectively) of the performance of a medical imaging system by means of measurement computations done on a given image. That is, it is technically possible to quantify and set values to image features with the aid of parameter data, e.g. spatial resolution, contrast, noise, and the like. Nevertheless, it is also possible to evaluate image quality by resorting to subjective estimations made on the overall performance of the X-ray

equipment. But, since this method relies on human capabilities for assessment, it may suffer from a certain variability degree on its conclusions. Therefore, it is sensible to seek an improved parameter definition that allows objectively defining the quality of X-ray imaging.
The concept of image quality is defined within the environment of medical diagnosis and consists of evaluating (objective or subjectively) the functioning of a generating system of medical images from measurements done on a resultant image. That is, technically it is possible to quantify or to value the characteristics of the image from parameters such as spatial resolution, contrast, and noise, between others. Nevertheless, it is also possible to evaluate the quality fromsubjective estimations of the global functioning of the equipment. This method of evaluation, dependent on human factor, presents certain degree of variability. Therefore it is important the parameters definition that characterize in an objective way the image quality.

Several authors[2] assert that image quality entails combining the effects of noise and spatial resolution. Therefore, if a straightforward, automatic analysis is made upon these parameters, the evaluation process will get free from both the object and the observer. This, in turn, will enable to do an unbiased quantitative and objective contrastive assessment among parametric characteristics of various detectors and technologies. Besides, when considering the current trend for incorporating digitally networked facilities into health centers, along with DICOM protocols capable of managing digitalized imaging, the conditions could then be optimally set for achieving faster and more reliable imaging analysis for improved diagnosis.In general, and in certain places, this renewal trend for digital X-ray equipment is being made progressively, but the need to re-design the image-quality control tests is an unquestionable fact.

Among other rmeans, one way to characterize objectively the performance of a digital imaging system is by using Modulation Transfer Function (MTF) as the evaluating parameter. In the literature, several measuring procedures can be found. However, in one such works, Neitze et al. have stated that the use of different evaluation routines may lead to sharp discrepancies in parameter determination, some of them reaching up to 30%, even when employing a single image for all measurements.

The early published works showed various image-quality evaluating methods that used, for example, bars patterns [11], and even a radio-transparent discs matrix [12]. In an effort to prevent the occurrence of calculation differences and to eliminate the variations amongadopted methods, the International Electrotechnical Commission IEC 62220-1 [4] published a standard in 2003 that set the procedure to determineimaging-quality parametersby usingthe MTF of the digital detectors of X-ray imaging equipment. Through this standardization, it is thus possible to unify the measuring procedures and, consequently, to do reliable contrastive studies in this field.

This work framed within the larger project Medical Imaging Processing for image quality evaluation of X-ray equipment. The general objective is to design and developa testing device having low cost and simple construction, for measuring standardized image quality parameters of digitalized X-ray equipment. The work continues into the development of an algorithm to calculate the MTF of a digital detector, following the above cited CEI standards, applied to images stored in DICOM-format files and using a protocol aimed at combining the conditions under which the images have been acquired for later evaluation.

1.1. MTF, Modulation Transfer Function.

MTF is a mathematical function that provides a measurement on how well an X-ray imaging system transfers the contrast details of an examined object onto the X-film image. Otherwise expressed, MTF is directly related to quantifying the capability of the X-ray equipment togenerate an image that accurately reflectsthe details of the explored object, with a scale ranging from 0 to 1 or 100 % as a maximum value [6]. MTF is considered the best indexfor measuring spatial resolution as well as a contrast index for different spatial frequencies. If the system were thought to be a filter (generally a lowpass filter) the MTF would representits frequency response.

In practice, MTF can be obtained by means of three methods: the slit method [7], the edge method [8] and the grid method (grids used at different spatial frequencies). The typical option has been to use the image of a slit to obtain a MTF perpendicularly to the object. In order to improve the brightness estimation for a given system, the concept of Line Spread Function (LSF)is used. The LSF is a contouror profile of the image obtained from a very thin split. The MTF is the Fourier transform of LSF (equation 1):

$$MTF(u) = |F[LSF(x)| = |\int_{-\infty}^{\infty}[LSF(X).e^{-2\pi j.x.u}]dx| \qquad (1)$$

being TF (u) = (output amplitude/input amplitude)$_u$, where u is the spatial frequency.

The implicit difficulty of this method is that using slits requires precise manufacturing and alignment, are highlyexposed to radiation to allow the transmission across the slit and a correction to take into account the small size of the slit.
Sameiet al. [9] have concluded that the edge method provides a better definition of MTF fall at low frequencies. For this purpose an image of anedge phantom, called theEdge Spread Function (ESF),is obtained (Fig.1).

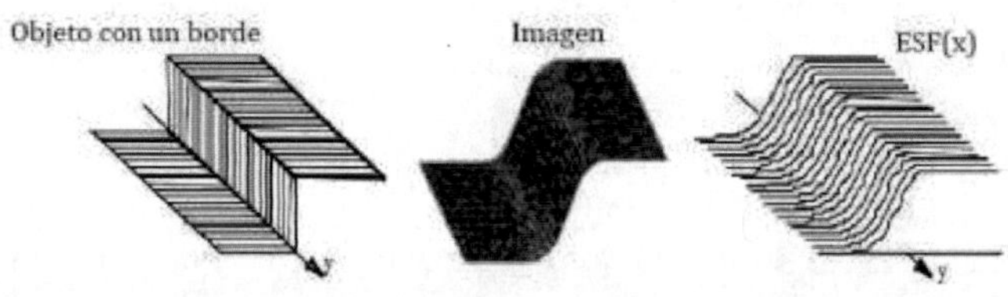

Figure1. The ESF is the image of an opaque object having an edge.

Mathematically, LSF is the first derivative of the ESF, as stated in the following equation:

$$LSF(x) = \frac{d[ESF(x)]}{dx} \qquad (2)$$

In practice, ESF can be easily obtained due to the simple form of the plate (←??) needed for the experience. Once the image is obtained, a perpendicular cut to the edge is made to obtain the ESF. Then, the LSF of the system is obtained by deriving the ESF. Now, Eq. 2 will be valid only when working on the linear range of the sensor.

Various analytical and experimental methods have been used to obtain the MTF from the LSF and the ESF. Fujita and cols [7] designed an oversampling method applied to LSF. Samei and cols. [8] also worked withoversampling but combined it with other image processing techniquesby using ESF.

Figure 2 shows a conceptual scheme of the MTF. Input signals to the system are represented by sinusoidal functionshaving different frequencies. Once they pass through the image system, the output signals showloweramplitudes thanthose of the input signals, but they keepthe frequency. It is also notedthat greaterinput frequencies cause loweroutput amplitudes, which means that the output amplitude is affected by the image system, and that this effect is dependent on the spatial frequency. In the figure it can also be noted that, at lower values of spatial frequencies (corresponding to large objects), the signalis transferred almost entirely, with no significant loss in contrast of the output image. As the spatial frequency increases (smaller objects), the amplitude of the output signal (and therefore, the contrast) decreases. This effect confirms the low-pass characteristics that X-ray imaging systems have.

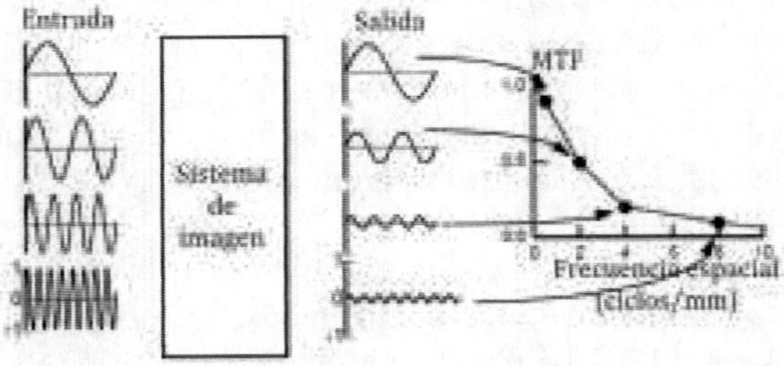

Figure 2. The MTF depictshowan X-ray system affects the detailsof an image.

2. Materials and Methods.

The method proposed for the obtaining of the MTF requires using a specific radiological technology, a normalized phantom and a set of algorithms forimage processing. In this section, the characteristics of each one of these elements are described.

A. Description of the Radiological Technology.

The geometry established by IEC 62220-1 [4] for acquiring the MTF is defined by choosing a radiation quality 5 (RQA 5) under 70KVp and 100mA, as well as a thin focus in a 18x24 chassis. Images were taken without any anti-diffusive grid, and 1.5 m source-detector distance, as shown in Fig. 3.

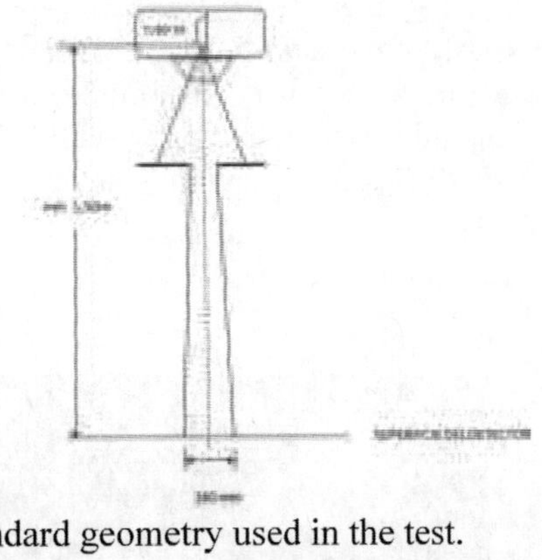

Figure 3. Standard geometry used in the test.

The Norm IEC 62220-1 [4] allows using other qualities, though it recommends using RQA5. The standard adopts the edge method, specifying the use of a radio-opaque object of tungsten of 1 mm of thickness with a straight edge and definite dimensions, asshown in Fig. 5.

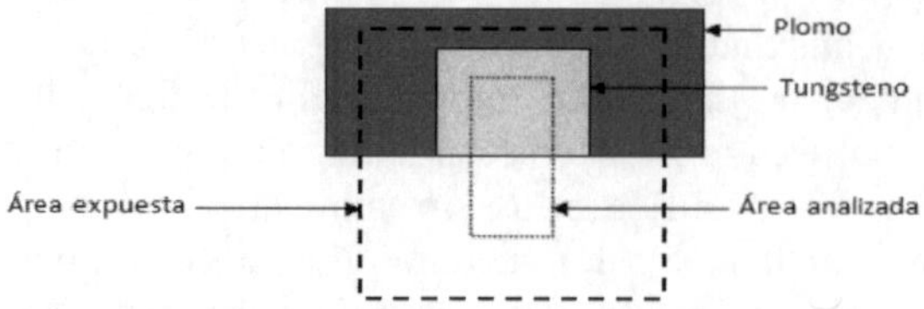

Figure 5.Phantom (specifications according to Norm IEC 62220-1:2003).

At present there are market-available edges, but, asNeitzelet al. [10] have demonstrated, a semitransparent edge may lead to an erroneous MTF estimation due to asmall amount of ed diffused radiation coming through the back of the edge and that reaches the detector. But this can be adapted for evidence tests, because the precision of the method is not altered, just its accuracy. In addition, a semitransparent edge is light and easier to align.

The proposed phantom in this work (Fig. 6a) is built with a1mm-thick copper plate (22,5% transmission, for an RQA5 bundle quality), sized75x100 mm, with a precision edge, contoured with a 3mm-thick lead fringe, with which an "infinitely long edge" is simulated.

(a) (b)

Figure 6. (a) Images of the constructed phantom. (b) Radiographic images of the constructed phantom.

The tungsten plate proposed in the standard is quite expensive; therefore, it has been replaced by a copper one, which is a material havingvery similar characteristics [13].

Since the edge must necessarily have a well-defined finish, a very precise cutting must be exercised. To this aim, the edge cut was performed with such care,and images were obtained through an Olympus microscope equipped with a 4x objective (Fig. 4a). Next, the cut was milled and polished (Fig. 7b), with images obtained with an Arcano 10x stereoscopic magnifying glass.

.

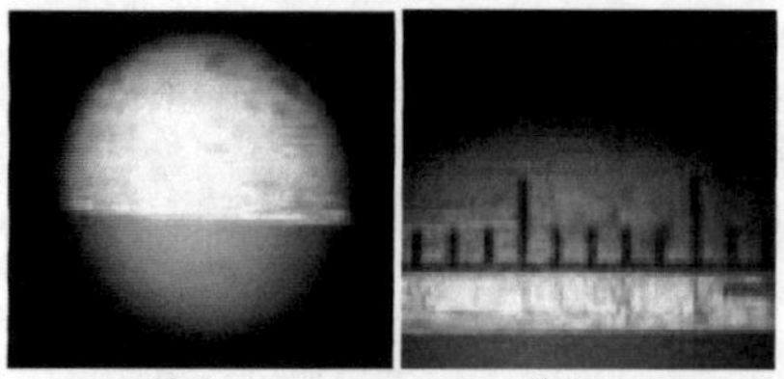

(a) (b)

Figure 7.Images of the constructed edge.

The method consists on acquiring phantom images taken with a light camera angle (1.5°-3°) respecting the detector matrix, so as to obtain an over sampling profile that prevents any

digitalization effect (Figure 6.b). Images on a direction parallel to the columns of the detector were firstly taken; then,on the direction of rows. The edge's center was positionedcoinciding with the center of the radiation bundle so as to prevent that anyoffsetting motion may alter the image.
The method consists of acquiring images of the phantom with a light camera angle (1,5° - 3°) with regard to the matrix of the detector, to obtain an oversampling profile that avoids the effects of the digitalization (Figure 6.b). First, images in a parallel direction to the columns of the detector were took, and then in the direction of the rows. The center of the edge was placed at the center of the bundle of radiation to prevent that the displacements of it could modify the image.

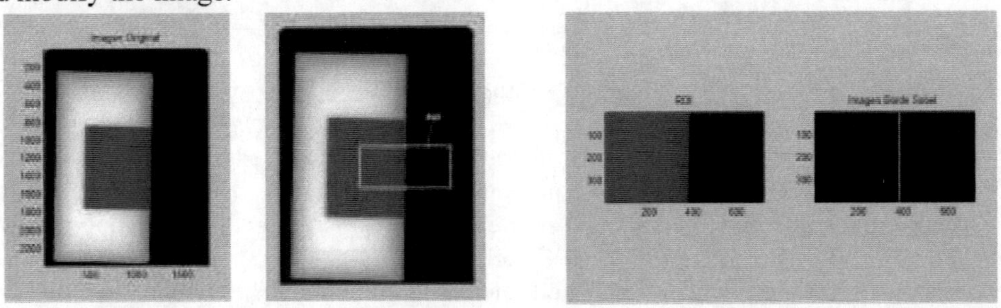

(a) (b) (c)

Figure 8. (a) Contours of the image. (b) Region of interest. (c) Determination of the edge.

The region of interest (ROI) used in finding the MTF is defined as a 50x100mm rectangle of the transition edge, asshown in Fig. 8b. To determine the edge, it was proceeded toextract the imagecontour by applying the Sobel's method for computing the image's gradient of the image, asshown in Fig. 8c. The small angulation needed to acquire the images allows for applyingoversamplingtechniques as we will *see hereinafter.However, is it a fundamental requirement to know accurately the value of the* angle ϴ. For this purpose the Hough line transformis used, which consists of obtaining an image of the parameters space of the lines contained in an image. In order to do this,all the lines that can pass through a specific pixel are cumulated in every parameter and for every pixel of edge. In the Figure 9a the polar coordinate of the Hough transform is shown.

The region of interest (ROI) used for the determination of the MTF is defined as a rectangle of 50 mm x 100 mm, of the edge of transition, as Figure 8b shows. For the determination of the edge it begun with the extraction of the image contour, for this there was used Sobel's method that calculates the gradient of the image, as it appears in Figure 8.c. The small angulation needed for the acquisition of the images allows applying oversampling techniques as we will see hereinafter. However, is it a fundamental requirement to know accurately the value of the angle ϴ. For this purpose the Hough line transform is used, which consists of obtaining an image of the parameters space of the lines contained in an image. In order to do this, all the lines that can pass through a specific pixel are cumulated in every parameter and for every pixel of edge. In the Figure 9a the polar coordinate of the Hough transform is shown.

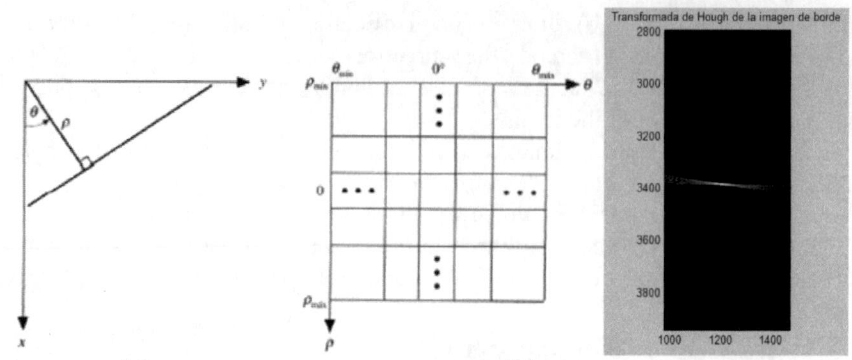

(a) (b)

Figure 9. HoughTransform.

In Figure 9b appears the Hough transform of the image of edge where it can be appreciated a maximum corresponding to the parameters of the line of edge of the phantom [14]. Then, every pixel was projected along a perpendicular line to the edge obtaining the Edge Spread Function (ESF) or Edge Response Function (ERF). A scheme of the obtaining of the ESF is observed in the Figure 10a. Each of the columns of the image has been located in consecutive way forming a unique vector.

Due to the fact that the edge is perfectly straight and the levels of gray are changing gradually, it is possible to construct a function of edge with a precision defined by the quantity of rows that are needed to advance a pixel in the columns. For the example N=4. N is the nearest point to 1/tgθ. The standard indicates that the range of values of the measured angle must be between 1,5° and 3°, which implies a value of N comprised between 40 and 20 respectively. It means that every N rows of the image is possible to obtain an ESFi(x). If the image has a high of m rows it can be divided equally m/N times the ESFi (x). It is necessary to bear in mind that before to the averaging it is necessary to do a displacement of each one of the functions (Figure 10b)

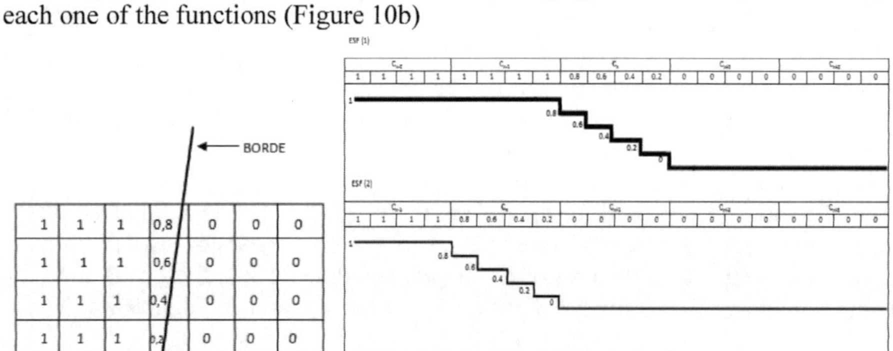

(a) (b)

Figure 10.Edge Scheme of the ESF

The Line Spread Function (LSF,) calculated from Equation 2is obtained in the digital domain by numerical differentiation of the ESF according to the expression:

$$LSF\ (x) = ESF\ (x-1) - ESF\ (x+1) \qquad (3)$$

The LSF (x) is a function with a high number of samples, superior to 18K samples. This causes noise in the MTF(u) where it is not possible to evaluate the quality of the image. For it, the method proposed by [17] is used which consists in decreasingthe quantity of samples according to the frequency resolution Δf needed and performing an apodization. It produces an MTF (u) sharper.

To define the quantity of samples n a frequency resolution is assumed, expressed in pl/mm (couples of lines for mm):

$$\Delta f = 0{,}1 \, {}^{pl}\!/_{mm}.$$

Considering thatΔf t is defined from the Fourier Transform as:

$$\Delta f = \frac{1}{\Delta x.n}$$

and the resolution of the ESF (x) is the quotient between the size of the pixel PS and N:

$$\Delta x = \frac{PS}{N}$$

thenthe quantity of samples n is defined for:

$$n = \frac{1}{\Delta x\,.\,\Delta f} = \frac{N}{0{,}1.PS}$$

The apodization consists to apply a window to the LSF (x) where$LSF_a(x)$is defined for:

$$LSF_a(x) = \begin{cases} 0 & x < -k \\ LSF(x) & -k < x < k \\ 0 & x > k \end{cases}$$

Finally MTF(u) is calculated using the equation (1) and it is normalizedapplying the following expression:

$$MTF_N(u) = \frac{MTF(u)}{\max\{MTF(u)\}}$$

B. Protocol Definition

The elaborated Protocol aimsto analyze the quality of images of the equipments of digital radiology, by means of the MTF [4]. It needs the employment of a phantom, the RX equipment and software for parameters evaluation.

The tests are realized with the following conditions of reference: 70kVp, 100mAs, thin area (0.6mm), without antidiffusive grid at a photo-detector distance of 1,50 m and the quality of used radiation was the RQA 5.

The methodology of work proposed for the calculation of the MTF comprises the following steps:

a) Annotate the code and the information of the movie in the "Data Schedule".

b) Place the chassis in carries chassis.

c) Places the phantom constructed on the chassis, inclined 1.5 - 3 degrees with respect to the axis of the detector to measure MTF.

d) Realize two exposures (from two different positions of the phantom), one with the edge of the phantom orientated in aperpendicular direction to the axis anode-cathode (see Figure 11a), and other orientated in a parallel direction (see Figure 11b). Do the exposures with the reference conditions mentioned previously.

e) Digitize the images.

f) Process each digital image to evaluate the quality of it.

g) Annotate the results in the "Quality index sheet".

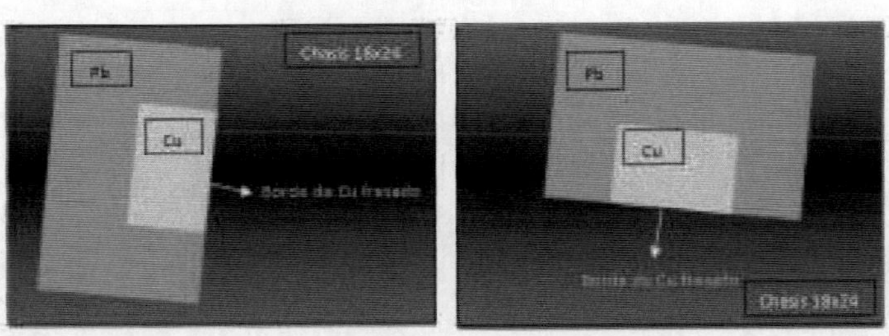

a) b)

Figure 11. Position of the phantom: a) perpendicular to the axis anode - cathode; b) parallel to the axis anode - cathode.

Results

The proposed methodology was evaluated on 4 computed radiology equipments.The KV, mA and time techniques described in the protocol were used. For each equipment 2 images of the phantom were obtained in parallel and perpendicular directions to the flow of electrons (cathode - anode).

In every image there was extracted the zone of analysis of 5cm x 10cm and the values of N, the average function of edge dispersion ESFprom (x), the function of dispersion of line apodizedLSFa(x) and the MTFN (x) were calculated. In Table 1 and in the Figure 12 the results are presented.

	MTF (X) (frecuencia espacial en pl/mm)			MTF (Y) (frecuencia espacial en pl/mm)		
	10%	50%	90%	10%	50%	90%
Equipo 1	2.8500	1.1500	0.3000	1.9000	0.9000	0.1000
Equipo 2	2.2000	0.8500	0.0500	2.3500	1	0.2000
Equipo 3	3.2500	1.6500	0.4500	2.8000	1.3000	0.3500
Equipo 4	2.9500	1.4500	0.4000	2.4000	1.1000	0.2000

Table 1.MTF's values.

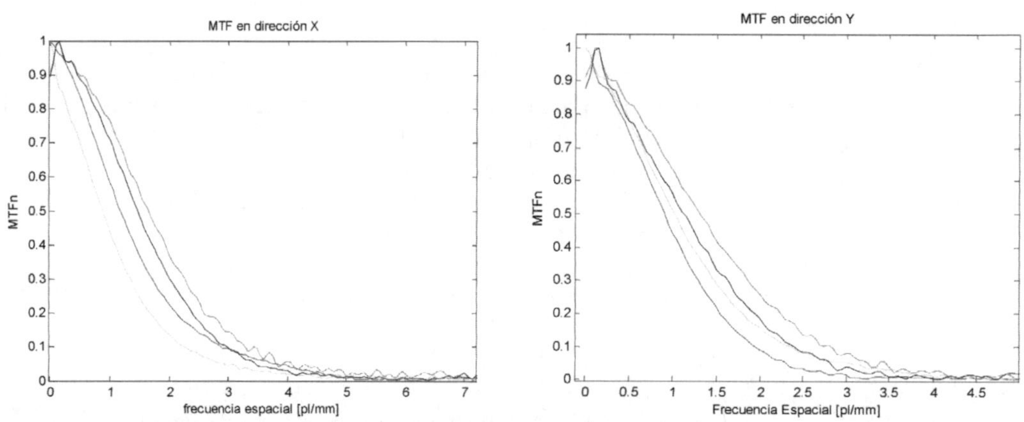

Figures 12: MTFn for the axes X and Y

From the graphs the frequency values of the MTF(X) are obtained at 10 %, 50 % and 90 %. Due to the fact that there are no values of acceptance in the standardfield, they have been compared with information published by [15 [16] [17], and it has be found an approximate error of 0.1 in the MTF at 50 %.

Conclusions

The previous sections have explained the theoretical aspects of the evaluation of the frequency response of a radiology system, which allows evaluating simultaneously aspects related to the contrast and to the resolution [18]. For thispurpose the standard IEC 62220-1:2003that describes the conditions for the acquisition of the image and then for the calculationtheModulation Transfer Function. Though the standard describes the general characteristics for the obtaining of the function, different methods exist for its practical obtaining. In this work techniques of oversampling, frequency resolution restriction and the apodization are combined to obtain a quality graph.

For the evaluation of the method images of 4 equipments of computed radiology of different centershave been registered, obtaining in a complete and a clear way the MTF. Later these values have been compared with values published of equipments of the same technology, obtaining similar values. The implemented method is part of the controls on radiology equipment that had been implemented in the Check and Calibration of Medical Equipment Laboratory of the GATEM-UNSJ. We expect in the future to integrate this measurement as a part of the protocols of valuation institutionally implemented.

References

[1] International Commission of Radiation Units and Measurements. "Medical imaging-the assessment of image quality", ICRU 54, 1995.

[2] Buhr E, Günther-Kohfahl S, Nietzel U. Accuracy of a simple method for deriving the presambled modulation transfer function of a digital radiographic system from an edge image. Med Phys 2003; 30: 2323-2331

[3] Neitzel U, Günther-Kohfahl S, Borasi G, Samei E. Determination of the detective quantum efficiency of a digital x-ray detector: Comparison of three evaluations using a common image data set. MedPhys 2004; 31:2205-2211.

[4] Comision Electrotécnica Internacional. "Equipos electromédicos. Características de los dispositivos digitales de imágenes de rayos X.Parte1" IEC 62220-1, 2003.

[5] Rivas Ballarín, María Ángeles. Garantía de calidad en radiodiagnóstico. Fundamentos de Física Médica. 2012. Volumen 2. Pág. 163. SEFM.

[6] Juan Pablo Graffigna - Raúl Romo. Fundamento de Imágenes Médicas. Ver 1.0-6/2003.

[7] Fujita H, Tsai D-Y, Itoh T, Doi K, Morishita J, Ueda K, Ohtsuka A. A simple method for determining the modulation transfer function in digital radiography. IEEE Trans Med Imag 1992; 11:34-39.

[8] Samei E, Flynn MJ, Reimann DA. A method for measuring the presampled MTF of digital radiographic systems using an edge test device.Med Phys 1998; 25: 102-113.

[9] Samei E, Ranger NT, Dobbins III JT, Chen Y. Intercomparasion of methods for image quality characterization. I. Modulation transfer function. Med Phys 2006; 33:1454-1465.

[10] Neitzel U, Buhr E, Hilgers G, Granfors PR. Determination of the modulation transfer function using the edge method. Influence of scattered radiation. Med Phys 2004; 31: 3485-3491

[11] Hillen W, Schiebel U, Zaengel T. Imaging performance of a digital storage phosphor systems. Med Phys 1987; 14:744-751.

[12] Båth M, Håkansson M, Månsson LG. Determination of the two-dimensional detective quantum efficiency of a computed radiography system. MedPhys 1003; 30:3172:3182.

[13] Buades MJ, González A, Tobarra, B. Implementacion de un programa informático para la determinación de la DQE de un sistema de radiología digital. Revista de Física Médica 2006; 7(2):57-67.

[14] Gonzales, RC; Woods, RE. "Tratamiento Digital de Imágenes". Editorial Addison-Wesley/Diaz de Santos. USA. 1996.

[15] Jong-Woong L, DaeCheolKweon, JiwonChoi, Eun-HoeGoo, Kyung-Rae D, Jae-Eun Jung, Yong-SeokChoi, Sun-KwangHwang, Hee-DooKang, Woon-KwanChung. Evaluation of the Response of the Modulation Transfer Function to a Computed Radiography Photostimulable Phosphor Imaging Plate for Various Numbers of X-ray Exposures.Journal of the Korean Physical Society, Vol. 58, No. 2, February 2011, pp. 326_333.

[16] NikunjDesaia, AbhinavSingha, Daniel J Valentinob. Practical Evaluation of Image Quality in Computed Radiographic (CR) Imaging Systems, Proc. SPIE 7622, Medical Imaging 2010: Physics of Medical Imaging, 76224Q (March 23,2010); doi:10.1117/12.844640

[17] Joseph D., La Veigne, "Comparison of Fourier transform methods for calculating MTF", In Proceeding of the SPIE 6941, pp. 69410Y-42-47, 2008.

[18] Morales, M.M.; Camargo, M.L.; Romo R.; Graffigna, J.P.; Echenique, A M. Estudio Multicéntrico para el Análisis de Calidad de Imagen de Equipos Mamográficos con Tecnología Convencional. ArtículoCompleto. Congreso.V Latin American Congress on Biomedical Engineering CLAIB 2011.CORAL.Consejo Regional de IngenieríaBiomédica para América Latina.

BCI-FES system for neuro-rehabilitation of stroke patients

Fabricio A Jure[1], Lucía C Carrere[1], Gerardo G Gentiletti[1] and Carolina B Tabernig[1]

[1]Laboratorio de Ingeniería en Rehabilitación e Investigaciones Neuromusculares y Sensoriales (LIRINS), Universidad Nacional de Entre Rios, Oro Verde, Entre Ríos, Argentina.

Email: ctabernig@bioingenieria.edu.ar

Abstract. Nowadays, strokes are a growing cause of mortality and many people remain with motor sequelae and troubles in the daily activities. To treat this sequelae, alternative rehabilitation techniques are needed. In this article a Brain Computer Interface (BCI) system to control a Functional Electrical Stimulation (FES) system is presented. It can be used as a novel tool in easy setup clinical routines, to improve the rehabilitation process by mean of detecting patient´s motor intention, performing it by FES and finally receiving appropriate feedback The BCI-FES system presented here, consists of three blocks: the first one decodes the patient´s intention and it is composed by the patient, the acquisition hardware and the processing software (Emotiv EPOC®). The second block, based on Arduino's technology, transforms the information into a valid command signal. The last one excites the patient´s neuromuscular system by means of a FES device. In order to evaluate the cerebral activity sensed by the device, topographic maps were obtained. The BCI-FES system was able to detect the patient´s motor intention and control the FES device. At the time of this publication, the system it's being employing in a rehabilitation program with patients post stroke.

Key Words: BCI; FES; Neuro-Rehabilitation; Stroke; Emotiv; Arduino.

1. Introduction

According to the 2010 Argentinean National Census of Population and Housing, 12% of the surveyed population has partial or permanent disabilities, of which 16% of those are motor type [1]. Paralysis associated with Strokes is among the leading causes of disability in adults. Moreover, strokes are the second cause of death known in the province of Entre Ríos (Argentina) [2].

There is scientific evidence of certain neurological rehabilitation strategies which help in the functional recovery of people who have had a stroke, but a significant number of them will keep sequelae which will require continuous assistance to succeed in the daily activities. Therefore, the implementation of new therapeutic tools which facilitate the reintegration, autonomy in daily life and the social inclusion of this important population is needed. These innovative therapeutic systems should be reliable and easy to setup, to facilitate its daily clinical use in health institutions.

Brain Computer Interfaces (BCI) can be used as a therapeutic tool to improve the recovery of the voluntary motor control affected by some illness or traumatism [3][4]. In order to perform this type of rehabilitation therapies, the BCI pose a training strategy in which patients evoke brain activity related to movements to facilitate cortical neuroplasticity [5]. Studies have shown that sensory signals of functional desired movement are a perfect feedback which has a positive impact in neuroplasticity [6].

As it is well known, Functional Electrical Stimulation (FES) is a rehabilitation technique that is used to restore function in people with disabilities by exciting the neuromuscular system and to give a proprioceptive and visual feedback of their paralyzed limb [7].

To develop a neuro-rehabilitation program based on a BCI-FES combination approach, we implemented an innovative therapeutic system. It records and processes the brain signals related to upper limb motor imagery/intention of patients. Then it generates a valid order, patient intention related, to command the FES device. In this paper, we present the implementation of this BCI-FES system and the topographic maps of the cortical regions which are activated during the imagination of specific motor tasks in a case of study.

2. Materials and Methods

2.1. System design

The proposed system is shown in figure 1. It consists of three main components: the patient with EEG acquisition device and its processing software, the BCI-FES interface and the FES system.

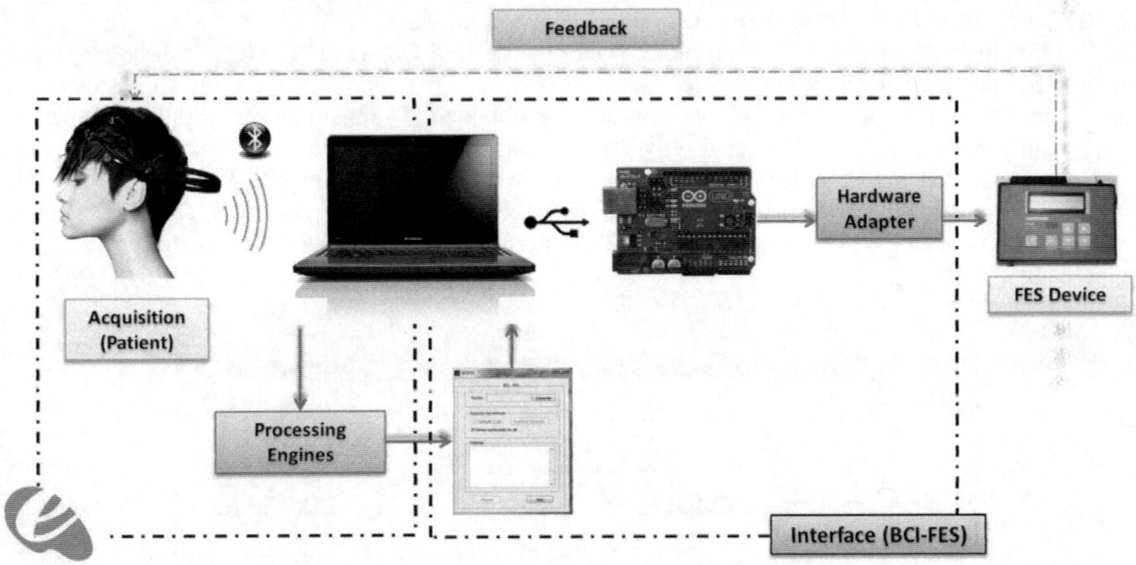

Figure 1. Block diagram of the BCI-FES proposed system. The first block which decodes the intention of the patient, is composed by the patient, the acquisition hardware and the processing software (Emotiv EPOC®). The second block, which transforms the information of the previous block into a valid command signal, is composed by the BCI-FES interface. The last block, which excites the muscular tissue, is composed by the FES device.

2.1.1. Acquisition and Processing.

To record EEG signal from the subject, an Emotiv EPOC EEG® headset (Emotiv Systems Inc., San Francisco, USA) was employed. This headset is a wireless, easy to setup and portable tool. It gives good flexibility and movement to the user compared with other systems.

This Emotiv EEG headset has 14 electrodes (saline sensors) locating at AF3, AF4, F3, F4, F7, F8, FC5, FC6, P7, P8, T7, T8, O1, O2 (figure 2) and two additional sensors that serve as CMS/DRL reference channels (one for the left and the other for the right hemisphere of the head). The Emotiv EEG's 14 data channels are spatially organized using the International 10–20 system. The sampling rate is 128 Hz, the bandwidth is 0.2–45 Hz, and the digital notch filters are at 50 Hz and 60 Hz [8].

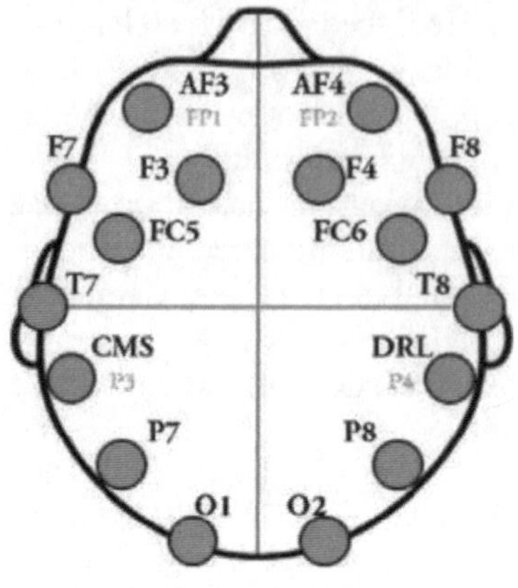

Figure 2. Sensor placement of the 14 data channels in the Emotiv EEG [8].

After setting the Emotiv system into the head of the patient, the system automatically recognized the impedance of the electrodes and indicates if the noise-signal level is acceptable. The signal is sent by Bluetooth to the PC where it is processed by the Emotiv software.

The Emotiv Software has a training block where the desired features in the EEG can be set. It has an user-friendly graphic interface where it can be seen a 'cube' and it is possible to set different movements depending on the number of features to be obtained. To obtain the features for this neuro-rehabilitation tool, it has been selected only 1 (one) feature (the extension of the wrist), which was associated with the movement of the 'cube' to bottom of the screen. Figure 3 shows the graphic interface of the system in which it can be found the electrodes condition (A), the different features to train (B) and the outcome of the system (C). The valid output of the system is a character which it is selected by the user.

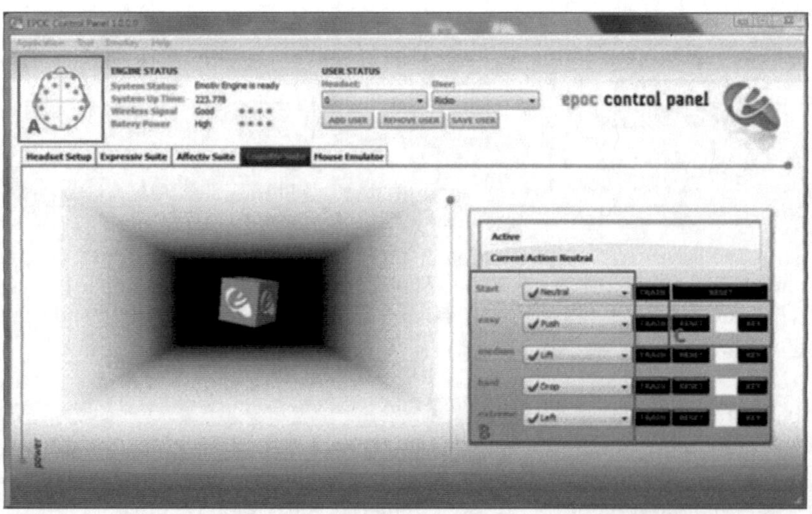

Figure 3. Software Emotiv® EPOC: Interaction screen (explanation in text).

The BCI's calibration is formed by 2 (two) sessions: the first session records the baseline level of the brain signals (in the software is indicated as 'Neutral'); the second session records the achievement of the movements of the cube linked with a real intention of the patient (in the software, it is indicated as 'Push, Lift, Drop, Left, etc.'). In this case the movement intention of the wrist extension is recorded.

This means that the calibration of the action that the software identifies as a movement of the cube is linked with the intention of the movement of the wrist. At the end of this calibration, a brain activity baseline is obtained. Then, each time that the software identifies this same pattern, the cube will move indicating that the intention is achieved.

2.1.2. BCI-FES Interface. The output of the first block is a character which is used to activate the FES device. For doing this, hardware based on Arduino's technology and software based on MatLab's environment have been designed.

The hardware is a portable embedded system with an Arduino UNO Rev3® (Arduino, Italy) which is a microcontroller board based on the ATmega328. It has 14 digital input/output pins (of which 6 can be used as PWM outputs), 6 analog inputs, a 16 MHz ceramic resonator, a USB connection, a power jack, an ICSP header, and a reset button. It simply connects to a computer with a USB cable. The Arduino UNO can be programmed with the open-source Arduino Software (IDE) and its environment is written in Java.

The BCI-FES software was written in C++ using MatLab® (MathWorks Inc., Massachusetts, USA) environment. This software converts the output of the Emotiv system into a valid command input for the ArduinoUNO board. It is made of three blocks: the first block evaluates the connection of the ArduinoUNO to the computer and identifies its connection port; the next block sets the electrical stimulus duration which helps the physiotherapist to control the stimulation; and the last block shows the status of the device. It also has a test button.

The output signal of this interface is a 5[V] pulse with a duration set by the user. This signal activates the FES device, which is a normal-closed system, so it is necessary that the ArduinoUNO generates a 'false value' as a valid command to activate the FES device. An adapter made of an opto-isolator (4N28, Motorola Inc., USA) has been designed which in 'activated condition' keeps the electrical loop of the FES device closed, and in 'non-activated condition' opens the electrical loop of the FES device allowing it to stimulate.

2.1.3. FES Device. The FES Stimulator® (Flexicar S.A., Argentina) is a two channels controlled stimulator designed to stimulate the neuromuscular system in order to generate a muscular contraction. This kind of stimulation is used in many injuries of the Central Nervous System where the muscles and the nerves are healthy. It has skin surface electrodes and the intensity and duration of the train pulse can be set easily by the user. The type of current is defined as biphasic symmetric rectangular. The rise and fall of the stimulation can be adjusted to prevent a sudden contraction that might induce a stretch reflex of the muscles.

Figure 4 shows the information pathways through the different blocks of the BCI-FES system.

2.2. Topographic Maps.

In order to evaluate the cerebral activity recorded by the Emotiv EEG headset®, cortical cerebral topographic maps were obtained from a subject who had no history of neurological diseases and was untrained in using BCI systems. Informed consent was obtained prior to experimentation.

EEG signal was recorded by g.MOBIlab$^+$ (g.tec Medical Engineering®, Austria) module acquisition system and BCI2000 platform. The electrodes were positioned using the EMOTIV® system electrode distribution (see figure 2 above).

The experiment was conducted using the following paradigm. Subject was seated in a comfortably chair, facing a screen and was instructed to relax, watch the screen and avoid blinking and swallowing. EEG recordings consisted in 20 minutes sessions approximately, divided in 4 runs with rest intervals between 1- 2 minutes long.

Each run included 3 different tasks which involved the imagined movement of right hand, left hand or both hands in response to a visual stimulus in the middle of the screen. Every task was repeated 10 times randomly during each run, separated by a random inter-trial interval of 5 to 6 seconds of duration. During the inter-trial intervals, the screen was blank, and subjects were asked to relax. The experiment included 5 sessions.

The EEG recording signals between 8 and 30 Hz were processed. Using the "Offline Analysis" tool available in BCI2000 platform®, spectrums analysis were computed for each task versus rest. Then, the topographic maps for the selected frequencies were performed. In these maps, each electrode position is properly marked and spatial distribution of values encoded in colors was plotted [9].

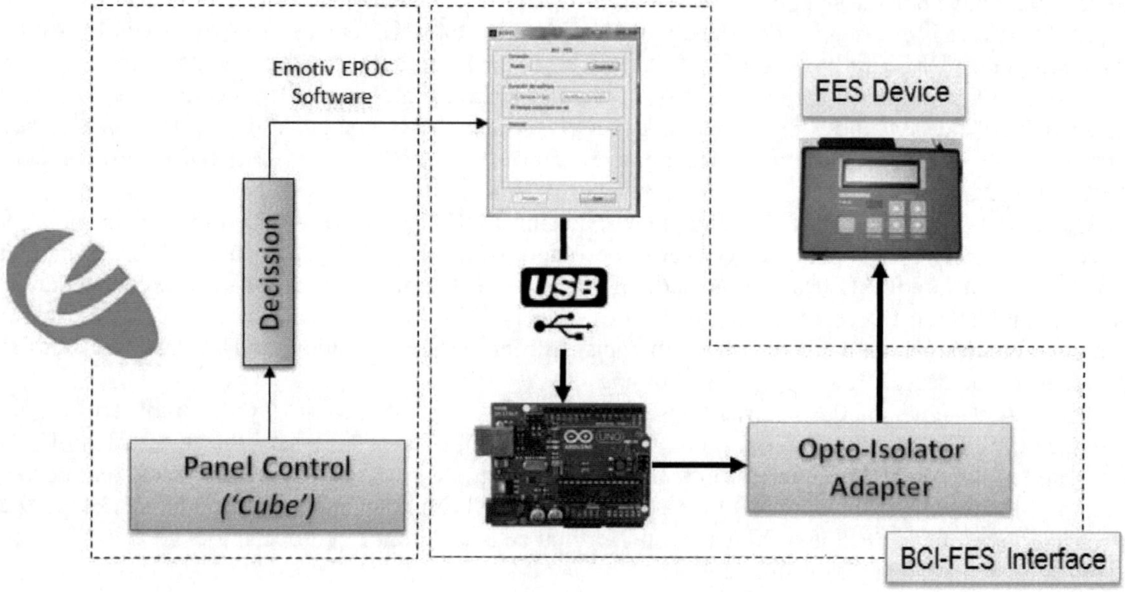

Figure 4. Information pathways through the BCI-FES system.

3. Results

3.1. BCI-FES System

In figure 5, photographic picture of the BCI-FES system is shown. It was able to sense the EEG, to detect the motor-related cerebral activity and to command the FES device. In figure 6, a photography of stroke patient using the system can be observed.

3.2. Topographic maps

Figure 7 presents the topographic maps of the analysed frequencies for the subject for each task. The results correspond to activities related to motor imagination of left/right and both hands. It can be observed that Event-Related Desynchronization (ERD) patterns associated to hand´s motor imagery are visualized for all tasks. Furthermore, it is observed that the localization of ERD patterns appears in different cortical regions, due to the special EMOTIV® system electrode distribution (see figure 2 above). Also it can be observed that the ERD frequencies for only one hand were located on *beta* rhythm range, while for both hands was located in *mu* rhythm one. Besides, during the imagination of right hand, cortical regions associated with other mental activities were also activated.

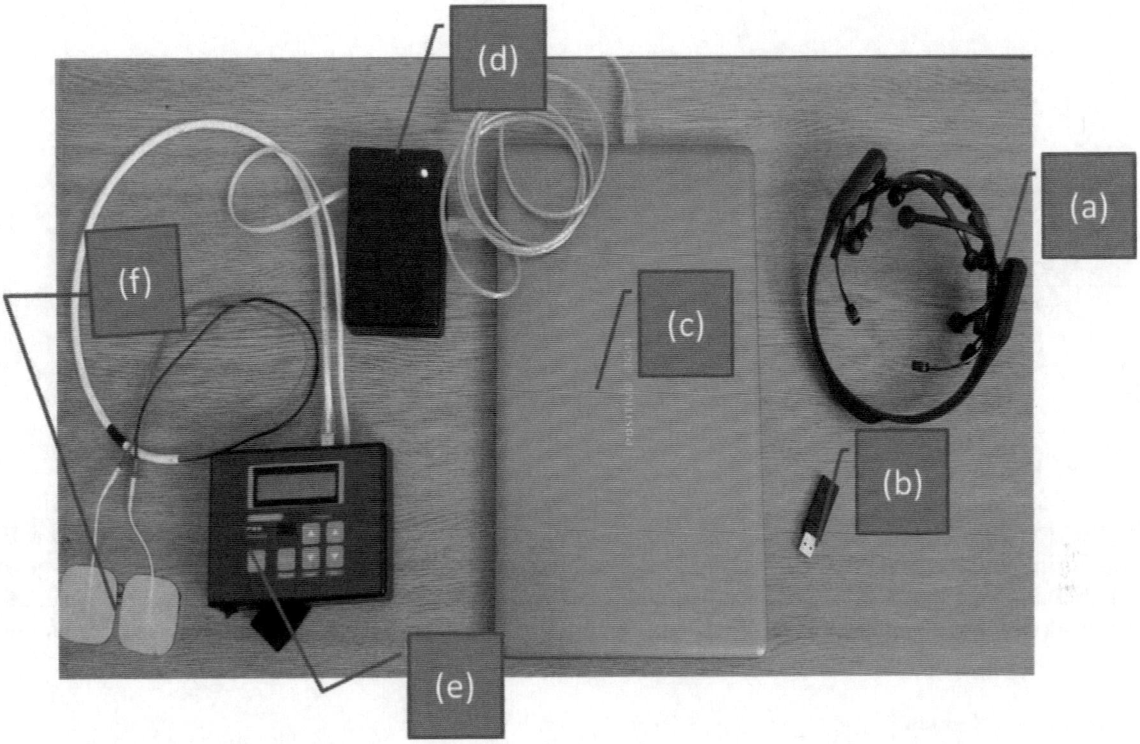

Figure 5. Photography of BCI-FES system: (a) Emotiv EEG headset®; (b) Bluetooth transmitter; (c): portable computer; (d): BCI-FES interface; (e): FES Stimulator®; (f): FES electrodes

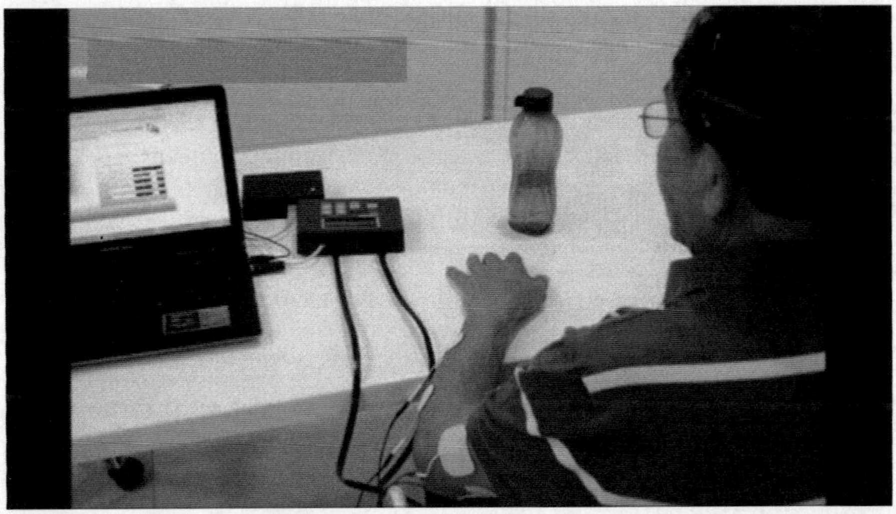

Figure 6. Photography of a stroke patient using the BCI-FES system during its treatment.

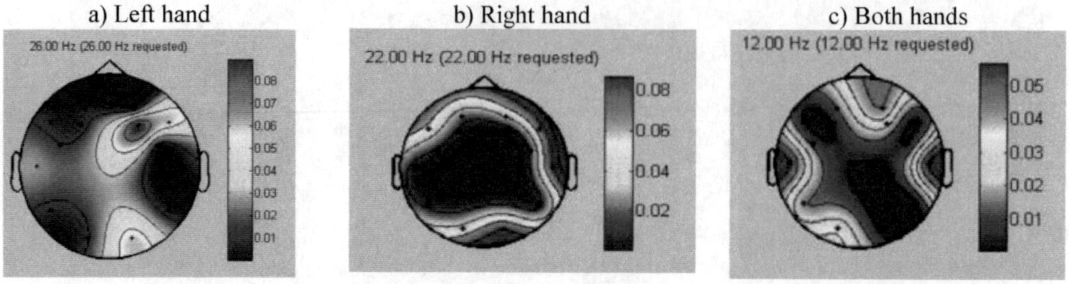

Figure 7. Topographic maps obtained by the processing of signals from Emotiv EEG headset® during related to the imagination of hands' motor activity.

4. Conclusion

The developed BCI-FES system reached the proposed aims. It is able to sense the brain activities related to motor tasks, to process it and to control a FES device. It is portable and easy to set up and use during daily clinical practice. Therefore at this moment, it is being used and evaluated in a new rehabilitation program, with post stroke patients. It is relevant to mention that EMOTIV® system's electrodes location is not the best suitable for recording EEG from motor cortex. This issue brings the need to design a headset with appropriate electrode location to detect more specifically motor ERD.

5. Acknowledgments

This work was supported by grant from "Fundación Rosarina de Neuro-Rehabilitación". Authors would like to specially thank Dr. Carlos BALLARIO and Lic. Camila LOPEZ for their labor implementing the rehabilitation program based on this BCI-FES.

6. Disclosures

No conflicts of interest, financial or otherwise, are declared by the authors.

References

[1] Censo Nacional de Población, Hogares y Viviendas 2010. Censo del Bicentenario. Resultados definitivos, Serie B Nº 2 Instituto Nacional de Estadística y Censos (INDEC) Buenos Aires, octubre de 2012. Disponible en http://www.censo2010.indec.gov.ar; abril 2014

[2] Mata Peña J y Gietz C, *Análisis de la situación de salud de la Provincia de Entre Ríos*, Informe presentado al Gobierno de la Provincia de Entre Ríos, marzo 2002.

[3] Daly J, Wolpaw JR, *Brain-Computer Interfaces in neurological rehabilitation*. Lancet Neurol. Vol VII pp 1032-1043 (2008).

[4] Gentiletti G, Tabernig C and Acevedo R.*Interfaz Cerebro - Computadora:Estado del arte y desarrollo en Argentina* Revista Argentina de Bioingeniería, Revista SABI, pp22-29: Vol 13, N° 1 Junio 2007.

[5] Wang W, et.al. *Neural Interface Technology for Rehabilitation: Exploiting and Promoting Neuroplasticity*. Phys. Med. Rehabil. Clin. N. Am. Vol XXI. pp 157-178 (2010)

[6] Jiang, Lijun, et al. *Motor imagery controlled wheelchair system*, Industrial Electronics and Applications (ICIEA), 2014 IEEE 9th Conference on. IEEE, 2014.

[7] Tabernig C. *Estimulación eléctrica funcional.* Study book of "Equipamiento para terapia y rehabilitación", Facultad de Ingeniería, Bioingeniería, UNER, abril 2011.

[8] McMahan, Timothy et al. *Modality specific assessment of video game player's experience using the Emotiv*. Entertainment Computing, vol. 7, p. 1-6, 2015.

[9] Schalk and J. Mellinger, (2010) *General-Purpose Software for Brain–Computer Interface Research*, Data Acquisition, Stimulus Presentation, and Brain Monitoring.Springer-Verlag London Limited, London ,pp. 9–35.

EEG Event-Related Desynchronization of patients with stroke during motor imagery of hand movement

Carolina B Tabernig[1], Lucía C Carrere[1], Camila A Lopez[2], Carlos Ballario[2]

[1] Laboratory of Rehabilitation Engineering and Neuromuscular and Sensorial Research (LIRINS), National University of Entre Ríos, Oro Verde, Entre Ríos, Argentina.

[2] Fundación Rosarina de Neuro-Rehabilitación, Rosario, Santa Fe, Argentina

E-mail: ctabernig@bioingenieria.edu.ar chballario@gmail.com

Abstract. Brain Computer Interfaces (BCI) can be used for therapeutic purposes to improve voluntary motor control that has been affected post stroke. For this purpose, desynchronization of sensorimotor rhythms of the electroencephalographic signal (EEG) can be used. But it is necessary to study what happens in the affected motor cortex of this people. In this article, we analyse EEG recordings of hemiplegic stroke patients to determine if it is possible to detect desynchronization in the affected motor cortex during the imagination of movements of the affected hand. Six patients were included in the study; four evidenced desynchronization in the affected hemisphere, one of them showed no results and the EEG recordings of the last patient presented high noise level. These results suggest that we could use the desynchronization of sensorimotor rhythms of the EEG signal as a BCI paradigm in a rehabilitation programme.

Keywords— **Stroke, Desynchronization, EEG, brain computer interface, BCI.**

1. Introduction

Paralysis associated with stroke is among the leading causes of disability in adults. About 25% of men and 20% of women will have a stroke if they live to age 85 or more, and between 25% and 40% of survivors develop sequelaes of the same hierarchy. While there is scientific evidence that some strategies of neuro-rehabilitation collaborate on functional recovery, a significant number of patients are left with sequela of this hierarchy that require ongoing assistance to conduct their daily lives. For this important patient population it is necessary to implement new therapeutic tools that will facilitate their reemployment, autonomy in daily life and social inclusion.

Brain Computer Interfaces (BCI) can be used for therapeutic purposes to improve voluntary motor control that has been affected by trauma or disease [1]. One of his paradigms raises the use of command signals caused by desynchronization of sensorimotor rhythms (SMR) of the electroencephalographic signal that occurs when the individual imagines and performs the movement of his/her hands [2].

SMR refer to oscillations recorded on brain activity in somatic sensorimotor areas, concentrated in the frequency bands of *mu* (8 a 12 Hz) and *beta* (12-30 Hz) [3]. Changes in the SMR produced by the motor imagination are exhibited as a power decrease in the EEG signal related to an internally or externally event and this decrease is known as event-related desynchronization. Similarly, a power increase in the signal is known as event-related synchronization (ERS). It is important to note that actual or imagined movement of a limb is reflected in an ERD pattern, which is characterized by its localized cortical (or scalp) topography and frequency specifity [4].

The cortical localization of the ERD patterns is a result of the somatotopic organization of sensory and motor cortices. In this arrangement, the hand representation is on the cortex, and is lateralized. This explains why left and right hand's ERD patterns can be spatially discriminated using the EEG. However, patients with stroke have damaged or displaced the motor cortical area which commands the movements of its affected limbs. This situation complicates the use of a BCI by ERD for neuro-rehabilitation therapy.

Different studies have investigated about ERD corresponding to motor imagery of upper limb in healthy people [5]. Nevertheless, there are few reports on its application to detect motor imagery of affected upper limbs of people with stroke. Scherer et al. [6] analyzed the EEG of hemiparetic stroke patients during left hand and right hand motor imagery in order to determine time-frequency maps of ERD and ERS. No common activation pattern was found over the damaged hemisphere. Then, it is necessary to explore what happens in each patient with stroke, individually.

Here, we analyze EEG recordings of hemiplegic stroke patients by ERD method to determine, for each, the most sensitive frequency components able to discriminate between a resting state and the imagination of affected hand movements.

2. Materials and methods

2.1. Patients

Six subjects with isquemic stroke participated of this study (1 female, 5 males, mean age 68.3 +- 8.7 years). Informed consent was obtained from all subjects prior to experimentation. All subjects suffered from unilateral lesion because of the cerebrovascular damage. In order to evaluate the patients' functional capabilities, Motor Activity Log (MAL) index was obtained [7]. Index MAL is a structured interview for stroke patients to assess the use of their paretic arm. Participants are asked standardized questions about the amount of use of their more-affected arm (AOU) and the quality of their movement (QOM) during the functional activities indicated. After administering the MAL, a mean MAL score is calculated for both scales by adding the rating scores for each scale and dividing by the number of items asked. In table 1, characteristics of subjects who participated in the study are shown. MAL equal zero indicates that a task it is actually not being performed because it is either very difficult for the participant, inconvenient, or requires increased time for completion.

Table 1. Characteristics of patients. (AOU: amount of use, QOM: quality of movement)

Patient	Affected arm	Stage	Gender	Lesion Description	MAL index	
					AOU	QOM
1	Left	chronic	M	infarction in the right sylvian territory; extensive cortico-subcortical; fronto-temporo-parietal lobe	1,03	1,48
2	Left	chronic	M	extensive cerebral infarction in the right sylvian territory; fronto-temporo-parietal lobe and lenticular nucleus	0	0
3	Left	acute	M	infarction in the right hemisphere	0,034	0,034
4	Left	chronic	M	infarction in the right sylvian territory; extensive cortico-subcortical; fronto-temporo-parietal lobe; with subsequent hemorrhagic transformation	0,68	0,62
5	Left	chronic	F	extensive cerebral infarction in the right fronto-temporo-parietal lobe; sylvian territory.	0,033	0,033
6	Right	chronic	M	infarction in the paraventricular nucleus and left basal ganglia.	0,13	0,13

2.2. EEG recordings

The EEG signal was recorded by means of g.MOBIlab+ module portable acquisition system and BCI2000 platform [8]. The data were recorded with 8 wet active electrodes placed on the scalp according to the international 10-20 system. The electrodes were positioned using the cap, g.GAMMAcap. Positions chosen for electrode placement were: C3, C4, T7, T8, Pz, F3, F4 and Cz (Figure 1). These positions were selected due to their cortical localization of interest for the study.

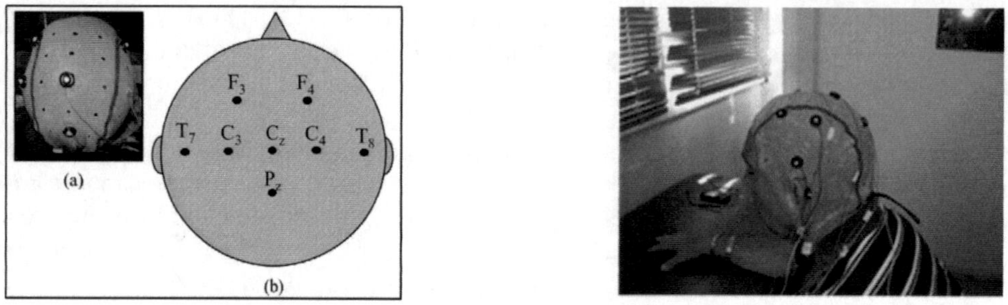

Figure. 1. Electrodes locations. (a) Electrodes positions with gGAMMAcap. (b) EEG derivations. (c) Photography of a patient during motor imagination trials

2.3. Experimental protocol

The experiment was conducted using the following paradigm. Subjects were seated in a comfortably chair and were instructed to relax. The EEG recordings consisted of sessions of 20 minutes approximately, divided in 4 runs with rest intervals between 1 or 2 minutes long. Each run included 3 different tasks which involved the actual or imagined movement of right hand, left hand or both hands in response to an auditive cue. Every task was repeated 10 times randomly during each run, separated by a random inter-trial interval of 5 to 6 seconds of duration. During the inter-trial intervals, subjects were asked to relax. The experiment included 3 sessions, which corresponded to the imagination of hands' motor to grip a glass.

2.4. Signal processing

The EEG recording signals between 8 and 30 Hz were processed. This frequency range was divided in two frequency bands of interest: *mu* rhythm and *beta* rhythm. Using the "Offline Analysis" tool available in BCI2000 platform, the spectrums of r^2were computed for each task (i.e, both hands imagery, left/right hand imagery) versus rest (i.e, the inter-trial interval) [8]. The spectrums were calculated for specific positions of electrodes, according to the somatotopic organization of the upper limbs in the motor cortex: C3 and C4. For each frequency band of interest, maximum difference in values of r^2 between electrodes Cz and C3-C4 was identified [9]. This maximum difference was associated with its corresponding frequency value, which was adopted as analysis frequency to represent the ERD patterns.

EEG signals from healthy hemisphere were processed to identify the frequency values with maximum difference in values of r^2 between electrodes Cz and C3-C4, for each band (*mu* and *beta*). These frequencies were employed to analyze the affected hemisphere.

The patterns were showed in topographic maps for the selected frequencies. In these maps, each electrode position is properly marked and spatial distribution of r^2 values encoded in colors is plotted.

3. Results

In figure 2 the selected topographic maps of 5 of 6 patients studied are shown. The processing signal of patient 6 did not show ERD due to excessive noise during the EEG recordings. Therefore, it is not shown in the results.

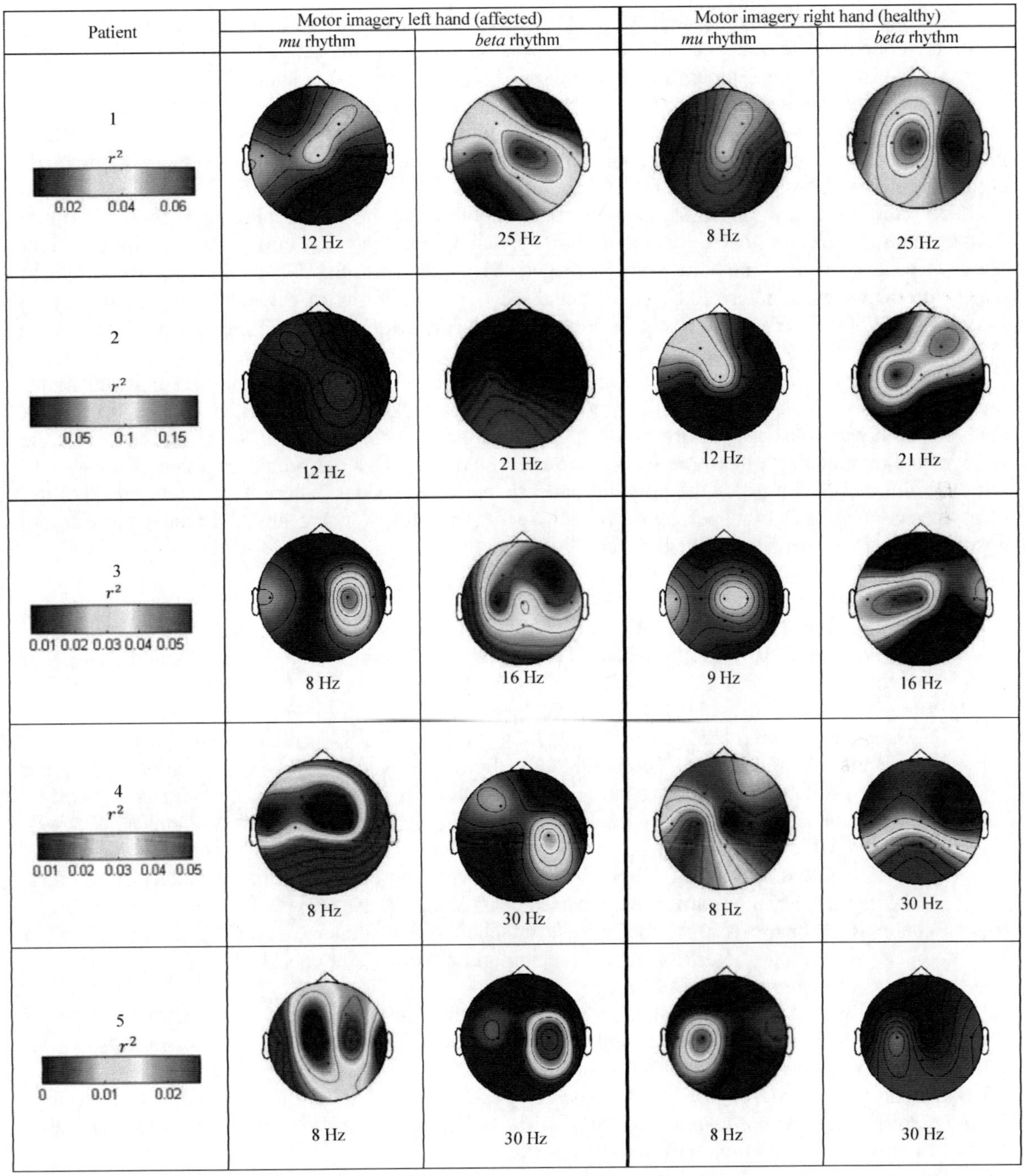

Figure. 2. Topographic maps during imagery motor tasks

4. Discussion and conclusions

In five of the six volunteers desynchronization related to motor imagery of the healthy hand was observed. Three of them showed desynchronization in *beta* rhythm, while the rest of volunteers exhibited ERD in *mu* rhythm. These results are coincident with those reported by McFarland et al. [10], where it was demonstrated that both movement and imagery were associated with desynchronization of *mu* and *beta* rhythms.

In one of the volunteers (Subject 2), no evidence of desynchronization related to motor imagery of affected hand was found. A null MAL index was obtained for this subject (table 1), furthermore it was reported by the therapist that this subject showed signs of depression and lack of motivation.

The results showed that the rhythm band for motor imagery ERD of the affected hand was opposite to the rhythm band obtained for the healthy limb. Subject 1 showed the same band of ERD in movement and motor imagery. As reported by Scherer et al. [6], there is no evidence of an ERD common pattern of stroke patients.

In reference to lesion localization described in table 1 and the topographic maps shown in figure 2, the findings suggest that even though the patients exhibited cerebral infarction in the right hemisphere it is feasible to obtain a motor imagery ERD pattern in the damaged cortex. Nevertheless these patterns were characterized by good spatial localization but low r^2 values for all the subjects, so more sessions of the experiment could be conducted in order to determine the possibility of improving them.

The results of this paper were encouraging since they confirmed possibility of using motor imagery BCI as a neuro-rehabilitation tool in stroke patients. Thus it elicited the importance of studying each particular case in order to design a personalized configuration of the device before the onset of the treatment. Furthermore it could be suggested to make a periodic calibration of the system with the aim of adjusting it to the possible patient's cortical changes. These experiences are the first steps in studying ERD of upper limb post stroke in Argentina. There are still many questions to answer which represent new challenges.

5. References

[1] Daly J, Wolpaw JR 2008 Brain-Computer Interfaces. *Neurol. Rehabil* **2** 1032-1043

[2] Wang W, et.al. 2010 Neural Interface Technology for Rehabilitation: Exploiting and Promoting Neuroplasticity. *Phys. Med. Rehabil. Clin. N. Am* **21** 157-178

[3] Pfurtscheller G and M. D. J. 2013 *BCIs That Use Sensorimotor Rhythms. Brain – Computer Interfaces : Principles and Practice* 227–241 (New York: Oxford University Press)

[4] Pfurtscheller G and L. da S. Fernando 2005 EEG Event-Related Desynchronization (ERD) and Event - Related Synchronization (ERS).*Electroencephalography: Basic Principles, Clinical Applications, and Related Fields* 1003–1016 (Philadelphia : Lippincott Williams & Wilkins)

[5] Pichiorri F, De Vico Fallani F, Cincotti F, Babiloni F, Molinari M, Kleih S C, Neuper C, Kübler A, and Mattia D 2011 Sensorimotor rhythm-based brain-computer interface training: the impact on motor cortical responsiveness *J. Neural Eng.* **8**(2) 025020

[6] Scherer R., Mohapp A., Grieshofer P., Pfurtscheller G., Neuper C. 2007 Sensorimotor EEG patterns during motor imagery in hemiparetic stroke patients *International Journal of Bioelectromagnetism* **9**(3) 155 – 162

[7] Doussoulin S. A, Saiz J.L, Blanton S. 2013 Psychometric properties of a Spanish version of Motor Activity Log-30 in patients with hemiparetic upper extremity due to stroke *Rev Chil Neuro-psiquiat* **51** (3) 201-210

[8] Schalk and J. Mellinger, 2010 *General-Purpose Software for Brain–Computer Interface Research, Data Acquisition, Stimulus Presentation, and Brain Monitoring* (London :Springer-Verlag London Limited) 9–35

[9] Carrere C, Tabernig C Detection of foot motor imagery using the coefficient of determination for neurorehabilitation based on BCI technology 2015 *IFMBE Proceedings–CLAIB 2014*, ISSN: 1680-0737, Ed. Springer

[10] McFarland D, Miner L,Vaughan T, and Wolpaw J. 2000 Mu and Beta Rhythm Topographies During Motor Imagery and Actual Movements *Brain Topography* **12**(3)

Conflict of Interest

The authors declare that they have no conflict of interest

Development of a Kinect-based exergaming system for motor rehabilitation in neurological disorders

A. Estepa[1], S. Sponton Piriz[2], E. Albornoz[1,3] and C. Martínez[1,4*]
[1]Research Institute for Signals, Systems and Computational Intelligence (sinc(i)), Facultad de Ingeniería y Ciencias Hídricas, Universidad Nacional Litoral
CC 217, Ciudad Universitaria Paraje El Pozo, S300, Santa Fe, Argentina.
[2] Centro de Rehabilitación e Investigación "Dr. Esteban L. Maradona", Santa Fe.
[3] CONICET, Argentina.
[4] Laboratorio de Cibernética, Facultad de Ingeniería, Universidad Nacional de Entre Ríos.
E-mail: *cmartinez@sinc.unl.edu.ar

Abstract. The development of videogames for physical therapy, known as *exergames*, has gained much interest in the last years. In this work, a sytem for rehabilitation and clinical evaluation of neurological patients is presented. The Microsoft Kinect device is used to track the full body of patients, and three games were developed to exercise and assess different aspects of balance and gait rehabilitation. The system provides visual feedback by means of an avatar that follows the movements of the patients, and sound and visual stimuli for giving orders during the experience. Also, the system includes a database and management tools for further analysis and monitoring of therapies. The results obtained show, on the one side, a great reception and interest of patients to use the system. On the other side, the specialists considered very useful the data collected and the quantitative analysis provided by the system, which was then adopted for the clinical routine.

1. Introduction

Since the advent of video game consoles, a large number of devices have been created to provide new forms of gameplay different to the traditional joystick. These devices (dance pads, bikes, motions sensors, etc.) seek to immerse the player in the virtual world through their body movements. These devices have gained great popularity in recent years in a new type of software so-called *exergames* [1]. This new kind of games encourages body movement and motor abilities by using the whole body as a control device. The exergames are part of the field of *serious videogames*, defined as applied computer applications designed for a primary purpose other than mere entertainment, such as training, education, health care and others [2, 3].

In the field of medical clinic, different studies indicate the positive effects of exergaming on health-related behavior changes [4]. Here, the combination of games and exercise equipment allows the creation of systems that collects data during the exercises. This turns out to be very interesting for the medical staff given that the movements and evolution of the patients can be evaluated in an objetive manner. The range of applications is very wide, from promoting physical activity in children, middle-aged and ederly people up to the rehabilitation of patients with brain damage and neurodegenerative diseases [5, 6].

Among the different input devices used to interface with the game systems, one that has received great attention is the Microsoft Kinect. This hardware features an infrared depth sensing camera that capture the full body movements of a person in a 3D space, an RGB camera that provides a color image of the scene and a microphone array for voice recording [7]. Basically, the applications carry out the markerless body tracking and this information directly controls a virtual character on screen, which represents the pose and movements of the user.

Using this device, different medical applications of exergames were proposed in the last years. Following spinal cord injuries, a number of systems were designed to train balance of the upper body [8, 9, 10]. In [11], the system is adapted to home-based rehabilitation, where it first records the movements in the hospital setup and then it evalutes the performance of the users in their homes. Two recent in-depth reviews of health care and rehabilitation applications were presented in [12, 13].

In this article, a Kinect-based sytem with three exergames is designed to evaluate different psychophysical rehabilitation exercises of neurological patients. The system provides visual feedback to the user by means of an avatar (virtualisation of the patient) in the screen that follows his/her movements. Also, it gives the orders using sound and visual stimuli, supplementing the benefits of the experience. Finally, a quantitative analysis of the movements is carried out by saving the joint positions of the skeleton in a database, which allows to follow the evolution of the therapy.

The article is organized as follows. Section 2 describes the clinical routines that were proposed by the therapeutic specialists, the interface and funcionalities of the developed system and the hardware setup in the medical centre. Section 3 presents the experimental conditions and results obtained by the patients. Finally, Section 4 resumes the conclusions of this work and gives some ideas about future work.

2. Methodology and developed tools

2.1. Medical institution and exercises

This work was carried out on the Rehabilitation and Research Centre "Dr. Esteban Laureano Maradona" located in Santa Fe city (province of Santa Fe, Argentina). One of the main tasks is the rehabilitation of sub-acute patients which are clinic and neurologically stable, who do not require hospitalization but need an intense rehabilitation plan. In this context, under the strict supervision of a physiotherapist of the institution, diverse routines for balance and gait rehabilitation were defined and developed [14, 15]. The studies were carried out at the facilities of the institution, which can be seen in Figure 1. All the patients signed informed consent for the studies.

2.1.1. Exercise #1

The therapeutic objective of this exercise is to train the patient to rest his/her weight on the hips, on either side. Historical records of this exercise can be used to measure the treatment evolution of hemiplegic patients.

For this exercise, the physiotherapist set the goal angles between the neck and the trunk. In order to perform the exercise in a ludic way, the proposed "game" presents to the patient a series of coming balls (placed at the desired angles) that he/she must avoid by leaning the trunk towards where a green arrow indicates. The patient should reach the goal angles to fulfil the exercise. The achieved angles are recorded at every time, and these records are later analysed by the specialists.

In Figure 2, the scene is presented and the balls arrangement at different angles (left and right sides) can be appreciated. The angles can be set every $12 - 15$ degrees, also the number of repetitions and the time among repetitions are configurable. In each repetition of the exercise,

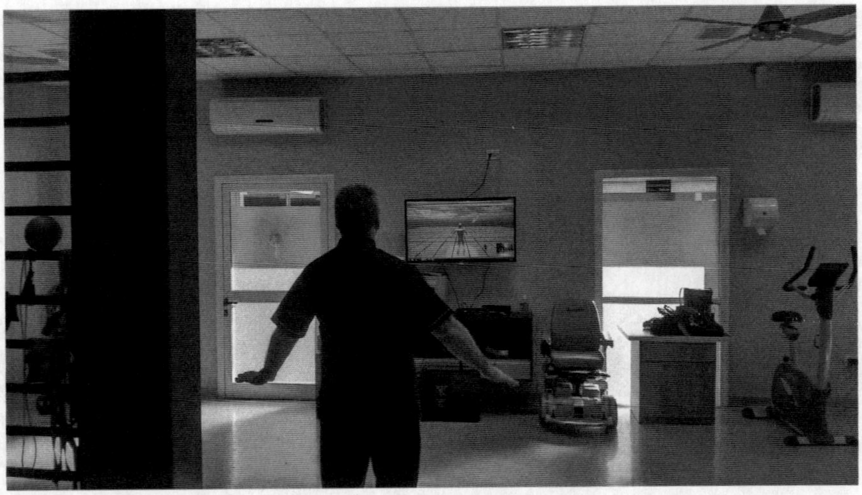

Figure 1: Facilities in the rehabilitation centre.

Figure 2: Scene configuration for exercise #1.

the performance is marked as successful, partially successful or not successful. The angle is calculated in the $x - y$ plane using the information of the spine and shoulder center points.

2.1.2. Exercise #2

In this exercise, the patient should train his/her stability while flexing the legs. The patient practices the gait positions in place, that is, displacements are not performed. This training is important in order to correct the foot drag, present in neurological patients. Furthermore, the exercise can be enhanced if the patient has upper-limbs motor skills. In this second scheme, the patient should raise the right hand together with the left leg, and vice versa. This provides a proprioceptive stimulus for scapulo-pelvic coordination.

The game presents an avatar in a scene, in which some balls are rolling towards the patient. The goal is to avoid the balls by raising one foot per time. To reach a successful performance, the foot should be raised up to the level previously fixed by the expert. All the session is registered for a more exhaustive evaluation, regardless the success in the game. Similar to Exercise #1, the number of repetitions and the time among repetitions are configurable (essential to prevent the patient to wear down quickly). If the motor skills will be evaluated, the scene can includes a ball that should be reached with the opposite hand to the raised foot, at the level of elbow,

Figure 3: Example for Exercise #2.

shoulder or head.

In addition, it is possible to configure an audible signal to indicate the right moment to raise the foot. This is useful to avoid overload in patients with stability problems and also to estimate the response time to the sound stimulus. Figure 3 shows a scene with the described elements.

2.1.3. Exercise #3

This exercise is oriented to train the pre-gait and it starts with the patient standing and both feet aligned. The patient should do one single step to the front and then, should step the other foot to line up the feet. This is repeated three times. Next, the patient should return to the initial position using the same sequence of steps. Depending on the patient, this can be performed in reverse or forward (after a 180° rotation). In both cases, the objectives are to generate a motor engram about the pre-gait process and to train the patient memory to remember step distances and sequence.

For this exercise, the scene has a series of colour platforms in which the patient should perform the pre-gait. The avatar is semi-transparent to improve the visualization and the starting foot (previouly fixed by the expert) is indicated on screen. Similar to previous exercises, the data are registered for later analyses. Figure 4 presents a scene of the Exercise # 3 in which the colour platforms can be seen.

2.2. Modules

The software of the complete system was conceived in three parts: the *administration* module to configure the exercises; the *registration* module to deal with medical record of patients and the *core* module that includes the processing of the acquired depth images during the rehabilitation routines. There is also a control block that monitors physical exercises fulfilment while collects relevant information.

2.2.1. Administration module The administration module allows the expert to select and fix the parameters of the exercises for each session, according with the type of patient, physical demand, etc. Here, the expert can configure some messages in order to graphically notify the performance to the patient while a physical exercise is running. Also, the information on successes and failures (shows when the routine ends) can be set.

Figure 4: Starting conditions for Exercise #3.

2.2.2. Registration module This module is very important to perform monitoring, because it allows to save the studies and look up the medical records of the patients. At present, every patient has a register that includes the name, age, gender and pathology (the information could be increased if needed). Furthermore, each session linked to a particular patient is recorded using the session date, the exercise and its settings, the tracking data, the obtained results and additional comments about the session.

2.2.3. Core module This module includes the software functions related to the processing of the data collected from Kinect device during the exercises. The particularities are detailed next.

In Figure 5a, a human body with the tracking points is presented in order to simplify the references, and to attain a better understanding of the developed exercises. Due to the capabilities of the Kinect device, the user should respect the usability ranges (Figure 5b). Each exercise in our system has a validation step in order to suggest the user the best location in the room.

The system was developed in `C++`.The database was implemented using SQLite[1] because it is the most widely deployed database engine and its source code is in the public domain. In order to develop the *virtual sceneries*, the Unity3D[2] development platform was used. This is a flexible and powerful tool for creating multiplatform 3D applications and its integration with Kinect is achieved by a wrapper developed by Rumen Filkov [16]. The Kinect device provides the orientation information for the skeleton as hierarchical rotation matrices based on the bone relationship. This information is used by the wrapper to perform the animation.

3. System tests and results

In this Section, the results of the system evaluation in real conditions are presented. Three patients with different pathologies were invited to perform the exercises under the strict monitoring of the physiotherapist. Brief descriptions of the patients are shown in Table 1. It is important to mention that patients B and C have important motor problems. The patients were requested to perform two initial attempts before registering the data, in order to learn the functioning and become familiar with the system.

[1] https://www.sqlite.org/.
[2] https://unity3d.com/.

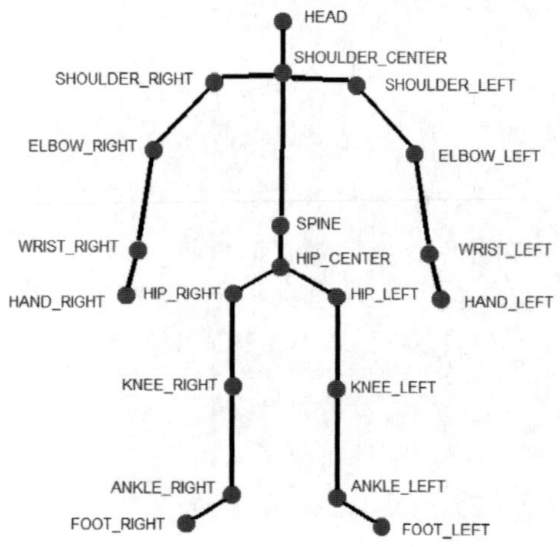

(a) Tracking point in a human body.

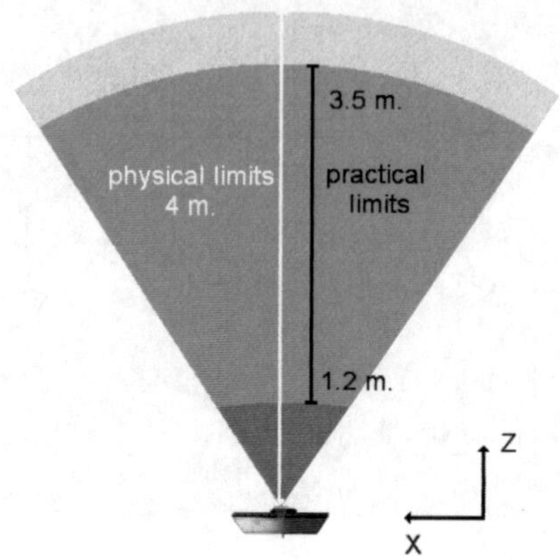

(b) Usability ranges for Kinect device.

Figure 5: (a) Tracked points and (b) general specifications for Kinect working conditions.

Table 1: Brief descriptions of patients.

Description	Gender	Age	Pathology
Patient A	Female	18	Brain tumor
Patient B	Female	38	Cerebrovascular accident
Patient C	Male	20	Dystrophy

3.1. Results of Exercise #1

For this exercise, the inclination angle was set in 12°, 24° and 36° for both sides and the number of repetitions was fixed to 10. Figure 6 shows the patients performances in time (sampling rate is 1/30 seconds). The red line represents the goal angles and the blue line is the reached angle. Positive angles represents inclinations towards left side and negative angles indicates inclinations towards right side.

As can be observed, patient “C” has many difficulties to accomplish the objective. The discrete scores computed for the exercise were: 8/10 for patient “A”, 6/10 for patient “B” and 4/10 for patient “C”.

3.2. Results of exercise #2

For this exercise, the parameters were fixed at: 20 cm for the level to which the foot should be raised up, 5 repetitions, the rigth foot as the starting foot (the superior balls take the opposite side of the lower balls).

Figure 7 presents the evolution of foot height in time (blue line), one foot per time alternately. The red line shows the threshold altitude. As can be seen, patient “A” has not problems to perform the exercise while the others patients cannot accomplish the objective in most cases. Patient “C” showed serious problems to raise the left leg up. The complete running of this performance can be seen at `https://youtu.be/-tt-3Cp1Nhw`.

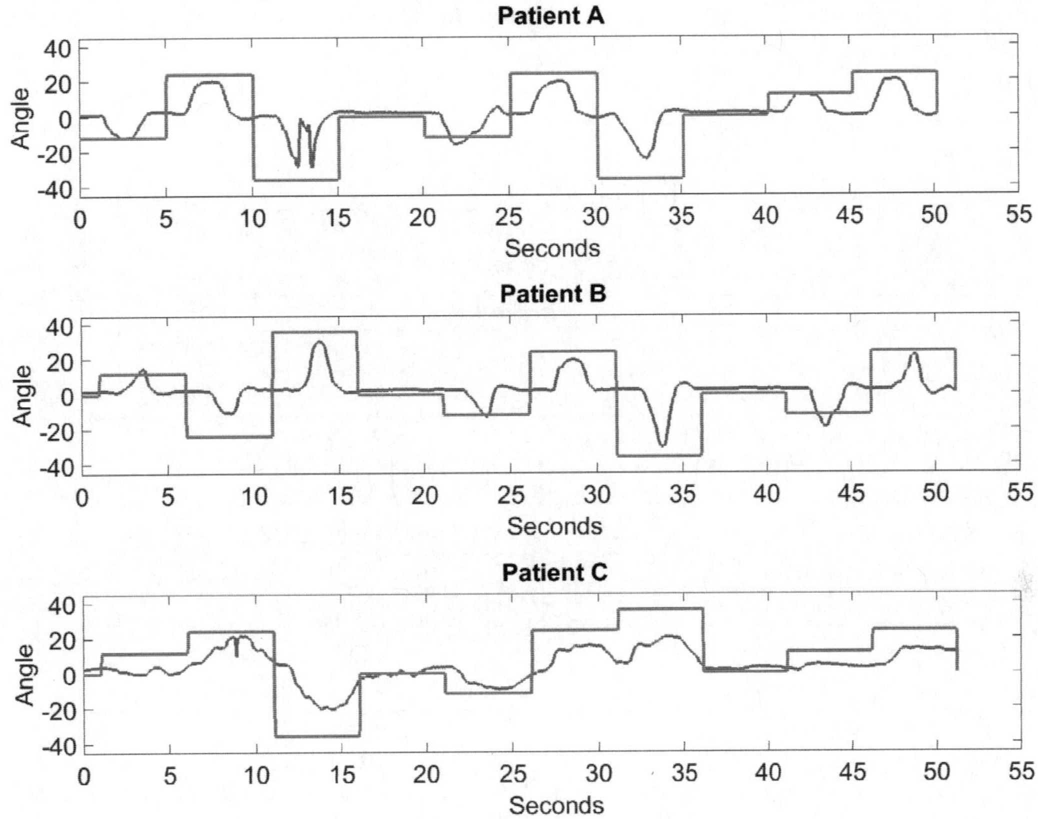

Figure 6: Results of Exercise #1.

3.3. Results of Exercise #3

Figure 8 exhibits the performance of patient "A" showing the traversed distance over time. The graph is presented as a user gait towards the Kinect device, within the usability ranges of Kinect. The green line shows the left foot trajectory while the red line represents the right foot. As can be seen, the patient started the exercise stand up at 3.7 metres from the camera. Using the collected information, the specialist can estimate the step length and speed, also they can check the sequence of the steps.

4. Conclusions and future work

In this paper, we have presented a system for rehabilitation of neurological patients using Kinect technology. A set of different routines were designed and developed according to requirements of expert physiotherapists.

Our system allows to collect data from movements during the exercises and save them in medical records. This information is very useful to evaluate the biomechanics of the patients and it allows subsequent and repetitive analyses, even in medical inter-consultation. Furthermore, the "games" were developed to be easily understandable and to provide information in order to assist the expert and user at every exercise. Preliminary results indicate that patients have a great predisposition and a notable interest to use the system. In addition, the specialists are favourably satisfied with the system performance and they consider very useful the monitoring data collected.

In future works, the amount and accuracy of the extracted data will be improved using arrays

Journal of Physics: Conference Series **705** (2016) 012060 doi:10.1088/1742-6596/705/1/012060

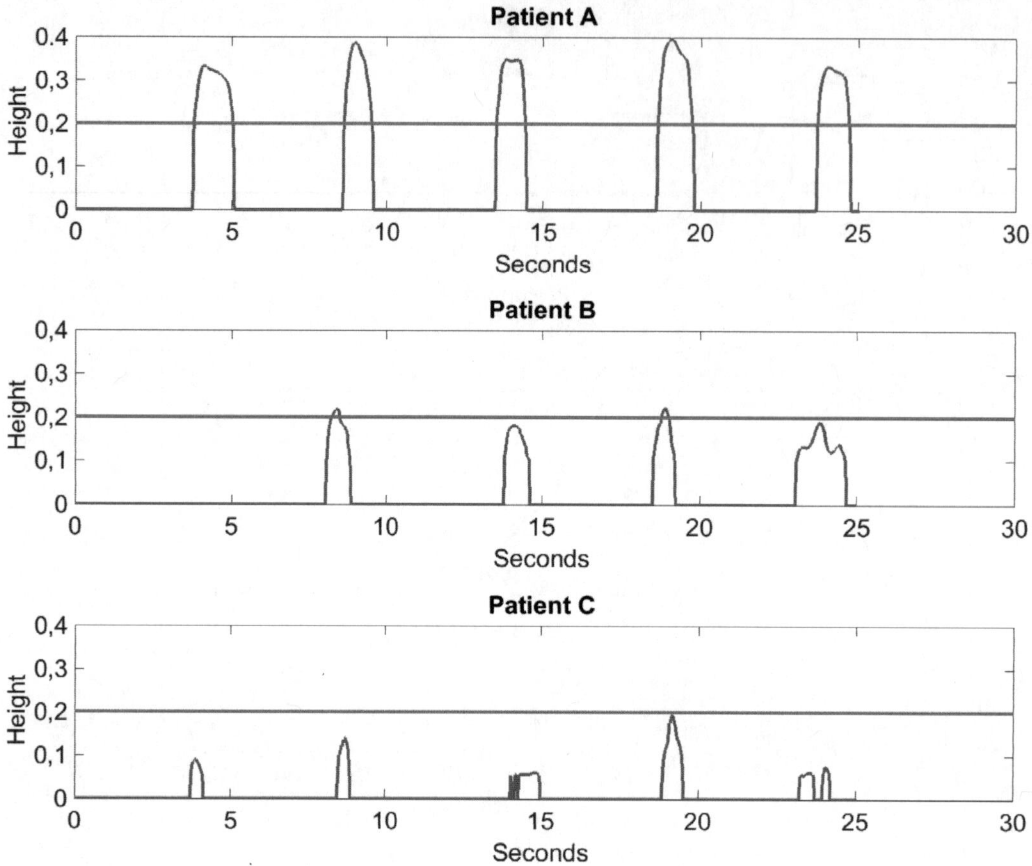

Figure 7: Results of Exercise #2 .

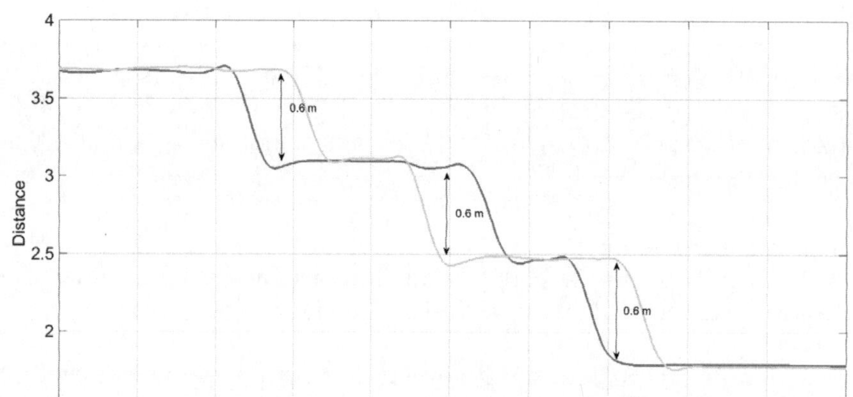

Figure 8: Results of Exercise #3 for patient "A".

of Kinect devices, which would make possible to extend the range of movement for Exercise #3. On the other hand, the actual system will be integrated with new technology as the Wii Balance Board. Furthermore, the management of medical records and its accessibility will be improved using intuitive graphical interfaces.

Acknowledgments

The authors would like to thank the *Universidad Nacional de Litoral* (with PACT 2011 #58 and CAI+D 2011 #58-511), as well as the *National Scientific and Technical Research Council* (CONICET), from Argentina, for their support. In addition, they would like to express a sincerely grateful to the director and staff of the Rehabilitation and Research Centre "Dr. Esteban Laureano Maradona" from Santa Fe (Argentina).

References

[1] Thompson D 2015 *Games for Health Journal* **4** 8–11
[2] Sinclair J, Hingston P and Masek M 2007 *Proceedings of the 5th International Conference on Computer graphics and Interactive Techniques in Australia and Southeast Asia* (ACM) pp 289–295
[3] Alvarez J and Djaouti D 2011 *Serious Games & Simulation for Risks Management* 11–15
[4] Göbel S, Hardy S, Wendel V, Mehm F and Steinmetz R 2010 *Proceedings ACM Multimedia* pp 1663–1666
[5] Garrido J E, Marset I, Penichet V M and Lozano M D 2013 *Proceedings of the 7th International Conference on Pervasive Computing Technologies for Healthcare* (ICST (Institute for Computer Sciences, Social-Informatics and Telecommunications Engineering)) pp 319–322
[6] Finco M and Maass R 2014 *Serious Games and Applications for Health (SeGAH), 2014 IEEE 3rd International Conference on* pp 1–6
[7] Microsoft Kinect for Windows `https://www.microsoft.com/en-us/kinectforwindows/` accessed: 2015-06-08
[8] Lange B, Chang C Y, Suma E, Newman B, Rizzo A S and Bolas M 2011 *Engineering in Medicine and Biology Society, EMBC, 2011 Annual International Conference of the IEEE* (IEEE) pp 1831–1834
[9] Lange B, Koenig S, McConnell E, Chang C, Juang R, Suma E, Bolas M and Rizzo A 2012 *Virtual Reality Short Papers and Posters (VRW), 2012 IEEE* (IEEE) pp 171–172
[10] Chang C Y, Lange B, Zhang M, Koenig S, Requejo P, Somboon N, Sawchuk A A and Rizzo A A 2012 *Pervasive Computing Technologies for Healthcare (PervasiveHealth), 2012 6th International Conference on* (IEEE) pp 159–162
[11] Su C J *et al.* 2013 *International Journal of Information and Education Technology* **3** 448–454
[12] Webster D and Celik O 2014 *J. Neuroeng. Rehabil* **11** 108
[13] Da Gama A, Fallavollita P, Teichrieb V and Navab N 2015 *Games for Health Journal* **4** 123–135
[14] DeLisa J A 1998 *Gait analysis in the science of rehabilitation* vol 2 (Diane Publishing)
[15] Bronstein A M, Brandt T, Woollacott M H and Nutt J G 2004 *Clinical disorders of balance, posture and gait* (Arnold)
[16] Kinect with MS-SDK `http://rfilkov.com/2013/12/16/kinect-with-ms-sdk/` accessed: 2015-04-24

Journal of Physics: Conference Series **705** (2016) 012061 doi:10.1088/1742-6596/705/1/012061

Analysis of the times involved in processing and communication in a lower limb simulation system controlled by SEMG

A. Profumieri[1], C. Bonell[1], P. Catalfamo[1,2], A. Cherniz[1]
[1]Laboratorio de Ingeniería en Rehabilitación e Investigaciones Neuromusculares y Sensoriales (LIRINS). Facultad de Ingeniería, Universidad Nacional de Entre Ríos,. http://www.bioingenieria.edu.ar/grupos/lirins/
[2]Centro de Investigaciones y Transferencia de Entre Ríos (CITER). Consejo Nacional de Investigaciones Cientificas y Técnicas (CONICET) y Universidad Nacional de Entre Ríos (UNER). Oro Verde, Entre Ríos, Argentina.

{acherniz,cbonell}@bioingenieria.edu.ar

Abstract. Virtual reality has been proposed for different applications, including the evaluation of new control strategies and training protocols for upper limb prostheses and for the study of new rehabilitation programs. In this study, a lower limb simulation environment commanded by surface electromyography signals is evaluated. The time delays generated by the acquisition and processing stages for the signals that would command the knee joint, were measured and different acquisition windows were analysed. The subjective perception of the quality of simulation was also evaluated when extra delays were added to the process. The results showed that the acquisition window is responsible for the longest delay. Also, the basic implemented processes allowed for the acquisition of three signal channels for commanding the simulation. Finally, the communication between different applications is arguably efficient, although it depends on the amount of data to be sent.

1. Introduction

Virtual reality has been defined as the man-machine interface technology that allows real time simulation of environments, activities or situations, where the subject interacts that allows for user interaction via multiple sensory channels [1]. It was proposed in conjunction with robotic systems, for the rehabilitation of patients with cerebrovascular accidents [2] and for training in patients wearing upper limbs prosthesis [3]. Experimental work showed that interfaces had positives effects in the rehabilitation process of lower limbs and demonstrated its efficacy for recovery of gait and balance [4]. In patients with lower limb amputation, the clinical viability of the prosthesis depends on a complex adaptation process, including the ability of the patient to accept the new corporal image [5]. In these patients, virtual environments could be used to evaluate the prosthesis control, for training and psychological adaptation. Also it might be use as an alternative or complementary tool for gait analysis [6, 7].

Nowadays, it is possible to access the hardware and software technology needed to implement different types of virtual environments, and also provide the user with a real time image projection. In our previous work, the implementation of a virtual environment that simulates specific movements of the lower limbs was achieved. Due to its versatility, this development allows the control of the simulation using different options and could be used in future research [8]. Surface electromyography (SEMG) could be used to control the simulation. However, the feasibility to command the simulation developed using this type of electrophysiological signals needs to be analysed. Also, the delay between the limb movement and the simulation movement must be regarded as imperceptible by users so that the response could be read as real time [9]. Hence, it was necessary to study the times involved in the signal acquisition and processing, in order to use the simulation in the investigation of new rehabilitation strategies.

In this work, the acquisition and processing of SEMG signals and the communication between the different applications of the proposed system were analysed, and the delays generated by each of stages were examined. To do that, two standard processes were used to control the simulation: detection of muscular activity and proportional control of movements. Also alternative solutions are proposed to optimize the simulation.

2. Materials and methods

In this section, the applications used to record the signal, build the simulation and perform the processing of the signal to control the model are described. Also, the tests performed to analyse both the execution times of the different processes and the subjective evaluation of the proposed system are indicated.

2.1 SEMG signal recording

Three channels of the SEMG signal from the rectus femoral muscle from 4 men volunteers, aged between 24 and 26 years, without lower limb pathologies, who gave their consent to participate were recorded. Volunteers were seated, with the thigh in horizontal position and the leg flexed 90°. They were asked to performed flexion-extension movements of knee. Disposable Ag/AgCl surface electrodes with conductive gel were used, which were placed on the muscle body midway between the superior iliac spine and the patella, following the SENIAM (Surface MyoGraphy for the Non-Invasive Assessment of Muscles) recommendations [10].

The registered signal was amplified and filtered analogically between 0.1 and 300 Hz using the Grass instrumentation amplifier, model 8-18-36, and was digitized with the A/D converter Data Translation ECON Series board, model DT9816. Matlab R2013b was used for the acquisition and processing of signals and the simulation command. This application is used for signal processing because of its versatility, since it allows to work efficiently with matrices, it has specific tools for acquisition and processing signals and it enables communication with others applications [11]. The signal was sampled at 1000 Hz, with an acquisition window of 210 ms. The signal was filtered using a digital fifth-order high-pass Butterworth filter with cut-off frequency of 10 Hz, to eliminate movement artefacts, and a fifth-order notch filter at 50 Hz, to remove power line noise [12].

2.2 Lower limb simulation

A basic model of the lower limb was built, with three segments (thigh, leg and foot) and three joints (hip, knee and ankle), using the 3D simulation software V-REP [13]. The simulation was made of elementary figures linked through joints that allow rotation in one direction. Joints were configured in order to be controlled from a remote application in Matlab. Figure 1 shows the aspect of this model.

Also, a simulation using the built in model "Bill" was used, as shown in figure 2, which was also controlled remotely with Matlab. This model allows moving the right leg joints only in one dimension and offers a better visual aspect.

Once the signals were acquired and processed with Matlab, the angles of the knee joint were obtained, and sent to V-REP to command the simulation. In order to send the data, a RemoteApi was used, using the function vrep.simxSetJointTargetPosition, in mode simx_opmode_oneshot. The amount of data sent to V-REP depends on the acquisition time and processing actions to obtain the simulation command.

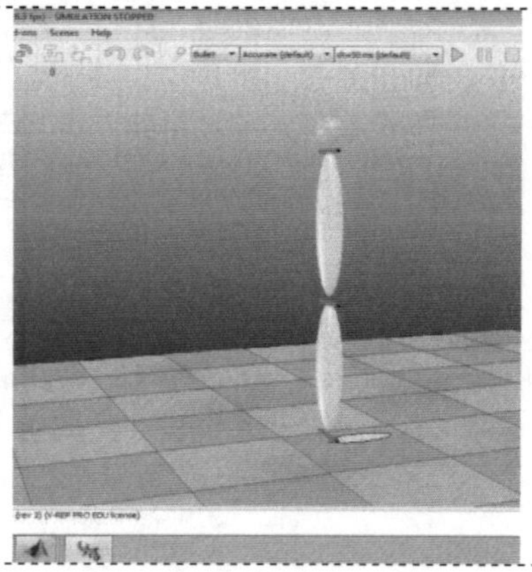

Figure 1. Basic model of the lower limb made by three segments (thigh, leg and foot) and three joints (hip, knee and ankle).

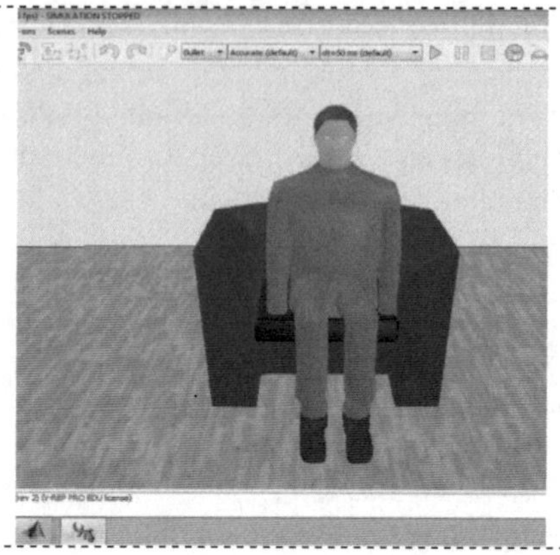

Figure 2. Model using the built in option Bill from the simulation program V-REP.

2.3. SEMG signal processing

Two common SEMG signal processing techniques were selected to control the simulation: a threshold-based control and a proportional control.

2.3.1. Initial processing. In order to perform either a threshold or proportional control, it is necessary to normalize and quantify the EMG signal, hence reference values are needed. The first reference value is the RMS of the EMG signal when the muscle is relaxed. To obtain this value, the SEMG signal was recorded at the beginning of the simulation with the muscle at rest (relaxed muscle) during one second, and the RMS value of this signal was computed according to equation (1), where $x[i]$ is the discrete SEMG signal, and termed as $\mathrm{RMS}_{\mathrm{rest}}$ for future reference.

$$\mathrm{RMS}_{\mathrm{rest}} = \sqrt{\frac{1}{N}\sum_{i=1}^{N} x_i^{\,2}}. \tag{1}$$

Then, the volunteers were asked to perform a contraction at maximal intensity during five seconds, only once, and the maximum voluntary contraction (MVC) was registered during one central second. Later, two processes were performed : on one hand, the signal was filtered, the envelope of the signal was calculated and ($x_f[i]$) and the maximum value of signal was obtained, called $\mathrm{EMG}_{\mathrm{MVC}}$. On the other hand, the RMS value named $\mathrm{RMS}_{\mathrm{MVC}}$ was computed using (1).

$$\mathrm{EMG}_{\mathrm{MCV}} = \max\left(x_f[i]\right), i = 0, \ldots, N \tag{2}$$

2.3.2. Normalization and envelope. The envelope of the signal was calculated in order to eliminate fast variations and the normalization (with the reference value of the MVC) was computed in order to make the comparison with the threshold value, and in the future to relate information from different muscle groups [12]. To obtain the signal envelope ($x_f[i]$), a fifth-order Butterworth low pass filter with cutoff frequency of 40 Hz was applied. The filter was implemented with the Matlab function filtfilt, after signal rectification. The normalization of the signal was obtained by dividing x_f *by* $\mathrm{EMG}_{\mathrm{MVC}}$. The x_n signal obtained is then used for detection of muscle activity.

2.3.3. Muscle activity detection. This process was used to determine the presence or absence of muscle activity and, a binary signal (one if muscle activity was detected or zero if not) was constructed. The values of the envelope of the normalized SEMG signal ($x_n[i]$) were compared against a threshold value. Muscle activity was considered present if the signal exceeded the threshold. The threshold value varied between zero and one, and was obtained heuristically in preliminary tests from which it was noted that very small or big values caused an improper behaviour of the simulation either by constant moving or not responding to movements, respectively. It was determined that 0.7 was an appropriate value for the threshold. The binary signal obtained from this processing was then used to command the extension or flexion of the knee lower limb in the V-REP model.

$$\theta[i] = \begin{cases} 90, & x_n[i] \geq 0.7 \\ 0, & x_n[i] < 0.7 \end{cases}, i = 1, \ldots n. \tag{3}$$

2.3.4. Muscle activity quantization. This process is performed to obtain the angle of the leg from the SEMG signal. The RMS value for the acquisition window of the signal $x[i]$ was obtained, as shown in (1), hereinafter $\mathrm{RMS}_{\mathrm{signal}}$, and then subtracting the normalized RMS value of muscular activity at rest and dividing all by the RMS value of maximum voluntary contraction [12].

$$v = \frac{\mathrm{RMS}_{signal} - \mathrm{RMS}_{rest}}{\mathrm{RMS}_{MCV}}. \tag{4}$$

For the quantization of data, an interval of 0° to 90° for the knee angle was considered, varying in steps of 10°. Thus, the information v obtained with the aforementioned process was assigned one of these ten values of angle θ. These angle values were then used as input to the simulator.

2.4. Command obtained from SEMG

Based on the processes indicated before, two methods were implemented to command the simulation objects. These methods were called simulation A and simulation B and their particular characteristics are described below. Both simulations share an initial stage where the A/D converter board and the communication between V-REP and Matlab are configured.

2.4.1. Simulation A. This simulation was controlled by a binary signal (ones and zeros), converted to angle values of 0° or 90° to indicate whether the simulated limb has a full extension or flexion of the knee. This simulation was started with the configuration stage, and then the MCV values of muscle were obtained. Later, the signal was acquired, digitally filtered and its envelope and normalization was obtained. Finally, the process described in Section 2.3.3 was applied, and the angles $\theta[i]$ that control the knee joint to V-REP (210 data per channel) were sent. This procedure was repeated from the step of acquisition while the simulation was running.

2.4.2. Simulation B. This procedure aimed at controlling the limb simulation so that it responded proportionally to the RMS value of the SEMG signal recorded. To do this, the simulation started with the configuration of the A/D converter board and applications and the initial stage where the $\mathrm{RMS}_{\mathrm{rest}}$ and $\mathrm{RMS}_{\mathrm{MVC}}$ were obtained (Section 2.3.1). Then, while simulation was running, the following actions were repeated: acquisition, filtering, normalization and quantization of SEMG signal (according to section 2.3.4), and sending of the value of the angle θ (1 data per channel) for controlling the movement of the knee joint.

2.5. Tests

Given the procedures described above, two types of tests, with different objectives were considered.

2.5.1. Test 1: Individual process analysis. The delays generated by each process were calculated in order to determine which of them was regarded as more time consuming. For time measurements a Matlab tool called tic-toc was used. Each process described in section 2.3 was executed 10 times and the average value of time accomplished was obtained. Then, some processes were selected to build the programs to get the commands to control the simulation (see section 2.4).

2.5.2. Test 2: Qualitative assessment of the simulation. In this stage, a poll was made to assess qualitatively the performance of the simulation, from the opinion of the volunteers about the time response.

Delays of 100, 150 and 200 ms were introduced to the system in order to evaluate which was the waiting limit regarded as acceptable in the simulation response.

The poll has four questions, where the last one asks about the simulation assessment when delays are added (a: 100 ms b: 150 ms and c: 200 ms):

- How do you evaluate the lower limb simulation? (Very Good-Good-Regular-Bad)
- How do you evaluate the time response of simulation A?(Acceptable-Not Acceptable)
- How do you evaluate the time response of simulation B? (Acceptable-Not Acceptable)
- Considering delays a, b and c in response simulation. How do you evaluate each delay? (Acceptable-No Acceptable)

3. Results

3.1. Analysis of Individual processes: results from test 1

Table 1 shows the time needed for configuration, storage, filtering (as described in section 2.1), normalization and envelope detection (as described in section 2.3.2) for each of the 10 trials made. All processes run using 3 acquisition channels.

Table 1. Time used for each of the processes: A/D converter configuration, write/read data, filtering, normalization and envelope detection (by Hilbert transformation and by a low pass filter technique).

	Time per trial (ms)										
Process	**1**	**2**	**3**	**4**	**5**	**6**	**7**	**8**	**9**	**10**	**Average**
Configuration	70	50	50	60	50	50	60	50	50	60	55
Read/write	90	60	32	33	42	32	43	32	35	32	43,1
Filtering	21	5	5	5	5	5	6	3	6	3	6,4
Hilbert	13	7	6	5	7	7	7	6	7	7	7,2
Low-pass	16	12	12	10	11	10	11	11	11	11	11,5

Table 2 shows the time demanded by the processes used for the detection of muscle activity (section 2.3.3) and quantization (section 2.3.4). In Matlab, the threshold method used for the detection of muscle activity, equation (3), can be carried out in two ways: using iterative structures (for or while) or with vector operations. The latest makes use of the advantages of working with vectors in Matlab, using relational operators to compare each signal element with the threshold. For the present paper, both implementations were used and their performances were compared. The latter method is shown in Table 2 with the name Op.relat.Vect.

Also, Table 2 shows the time needed to send the angle values obtained from the Matlab processing to the V-REP simulation. The first row shows the time required to send all the data for three channels and in the second row, the time demanded to send just 3 data to V-REP.

Table 2. Processing time demanded by muscle activity detection, quantization and data send from Matlab to VREP.

	Time per trial											
Process	**1**	**2**	**3**	**4**	**5**	**6**	**7**	**8**	**9**	**10**	**Average**	**Unit**
Iteration for	590	958	602	732	682	704	699	711	777	711	716,6	ms
Iteration while	768	851	827	800	867	875	732	869	933	775	829,7	ms
Op. relat. vect.	50	10	11	19	18	10	9	18	15	9	16,9	µs
Quantization	703	717	913	936	783	951	943	688	677	772	808,3	µs
Sending 630 data	108	74	75	79	81	78	84	75	68	77	79,9	ms
Sending 3 data	401	402	403	405	403	406	408	404	402	405	403,9	µs

3.2. Qualitative assessment of the simulation: results from test 2
The results from the enquiry made to volunteers showed that 75% considered the simulation as "good" meanwhile 25% consider it as "very good".

Seventy five percent of volunteers considered the response of simulation A as "acceptable" and 25% consider it as "not acceptable". However, for simulation B, 100% of volunteers consider the time response as "acceptable".

Figure 6 shows the opinions about time response with added delays of 100, 150 and 200 ms.

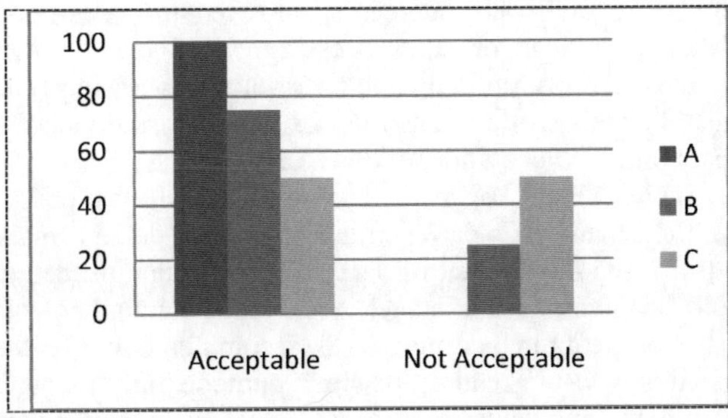

Figure 6. Opinion about the simulation performance, with up delays of 100 ms (dark blue, A), 150 ms (blue, B) y 200 ms (light blue, C).

4. Discussion
In this section tests results are discussed. All processes were performed for three channels. However, as the work objective was the study of processing time, not the information contained in the channels, only the flexion- extension of the knee was simulated.

4.1. Analysis of the Processes
The objective of this test was to analyze the time delays generated by the different stages corresponding to signal acquisition, data processing and communication between applications, in order to evaluate if it is possible to command a VREP simulation in real time using SEMG signals. Some processes cannot be avoided, as the configuration of acquisition board, filtering and data sending to V-REP. Others, as normalization, envelope detection and muscle activity detection, depend on the application and have different ways to be implemented. The proposed simulation is useful to study others algorithms for muscle activity detection, which are now under investigation [14]. Also, some processes can be eliminated, as the write/read of data in a file.

The analysis performed in this work aimed at evaluating the impact of different system stages on simulation. Different authors reported that the response time of the system should not be more than 300 ms approximately, in order to be considered imperceptible [9].

From Table 1 it can be seen that the acquisition board configuration and processes related to communication between the V-REP and Matlab applications, and request of handlers (variables that allow communication between applications) had adequate runtimes, especially considering that they should be run only once at the start of simulation. Furthermore, the writing and reading of data in a file consumes considerable time. This is a practice that takes place to keep recorded data permanently, but, if possible, it should be avoided. In this work, the data are temporarily stored in memory, directly as an array of Matlab. The time consumed, evaluated in preliminary tests, are in the order of µs.

Filtering was performed using a Butterworth filter due to its maximally flat magnitude response. Unfortunately, its phase response is not null, but it can be cancelled applying the filter in two opposite directions. Also, a notch filter in 50 Hz was used to eliminate line noise. The processing time demanded by the filtering was low and this stage can be eliminated if the acquisition system includes the filters.

The time demanded by the envelope detection stage using a low-pass filter is acceptable. In previous tests the envelope using the Hilbert transformation was obtained. This process was more efficient (demanded 7.2 ms), but the output signal characteristics had a worse quality than the output signal from the low pass filter.

Detection of muscle activity using a threshold and based on iterative cycles for or while consumed considerable times (well above the 300 ms proposed in [9]), limiting its application in real time systems. However, the method based on relational operations using vectors shows very good results, as shown in Table 2, so this method was chosen to perform this task. From this experience, it was concluded that all processes involving long cycles in Matlab slow down the execution of the program and makes it very difficult to work in real time. The use of arrangements, however, provides good results and is one of the main advantages that Matlab provides for data processing.

The RMS value calculation, normalization and quantization demanded very low processing time and also gave an acceptable simulation response. For this work, the angular excursion was divided in 10 levels. If the number of levels increases, the simulation behaviour becomes unstable.

The time needed to send the data to VREP (Table 2) is the longest. However, for a window of 210 ms it was possible to send the total data within the schedule time. When the amount of data is small, the time to send the data is in the order of µs. But when the amount of data grows, the time needed to send the data is longer, so special care should be given to this stage. A possible solution to this problem is to make a sub-sampling of data to be sent. In addition, it is recommend to use the simx_opmode_oneshot V-REP mode (as used in this work) instead of the simx_opmode_oneshot_wait mode, which is slower given that it sends the command and waits for a response, before placing a new communication.

It can be noted from the tables that the first trials seemed to require longer times than the following. A possible explanation to this is that the first time the program is run, the variables are created and memory spaces are reserved for them, which is an extra step that slows the process. In the second execution, variables are already created in memory and only access for reading or writing is performed.

It is worthwhile to note here that all processes could be implemented within the recommended ranges. Also, all processes were carried out on three channels which is important because often it is necessary to have information of more than one muscle to perform a more physiological simulation control. Despite of having three channels of data sent to V-REP, only the command of the knee joint was accomplished. For future work, models with control for other joints or knee control using different muscles information are proposed.

4.2. Qualitative assessment of the simulation: test 2 results

Volunteers in general evaluated the simulation as acceptable. Simulation A with an ON-OFF response had an acceptation of 75% while the simulation B had 100% of acceptance. The reason of this, could be that simulation B had a shorter delay (and allowed changing the leg position in function of SEMG signal amplitude registered (related to muscle force), which is a more natural representation of the movement.

The qualitative assessment of the simulation was made using the lower limb model built with elementary figures, so that volunteers focussed on simulation delays, rather than the aesthetic part of the simulation. It important to mention that the response time did not change when the Bill model was used.

The delays used for the assessment were greater than the recommended by bibliography, because some algorithms intended to be used for commanding the simulation may require higher processing times than the conventional processes. Hence, delays of 100, 150 and 200 ms were added. Figure 6 shows that the simulation acceptation decreases when delays increases. However these results are not conclusive because the sample of volunteers is very small.

Finally, the volunteers had some problems with the simulation control. Some of the possible reasons are: MVC measurement errors, the ability of the volunteers or the lack of practice. These problems are a system weakness that shows the strong necessity of user training in devices commanded by SEMG (simulation or prosthesis). This system offers a low cost possibility to train and in case of prosthesis adaptation, it could be used even without the physic device.

5. Conclusions

It was possible to control the knee flexion and extension in a simulation, created using VREP, and to command the simulation in real time by the SEMG signal obtained from the rectus femoral.

The evaluation of the processes implemented showed that the critical processes were window acquisition size and communication between Matlab and VREP (although this depend the amount of data to send).

Volunteer's tests indicated that simulation acceptance is good and when simulation delays are increased, acceptance decreased. This should be taken into account to propose new signal processes that required high computation cost or more channels to process.

References

[1] Adamovich SV, Fluet G G, Tunik E and Merians AS 2009 Sensorimotor training in virtual reality: a review*NeuroRehabilitation***25** 29

[2] Zhang X, Zhang J, Sun L and Li M2013 Rehabilitative motion planning for upper limb rehabilitation robot based on virtual reality *IEEE Int. Conf. on Robotics and Biomimetics* 746

[3] Dawson MR, Carey JP and Fahimi F 2011 Myoelectric training systems *Expert Rev. Med. Devices***8** 581

[4] Wagner J, Solis-Escalante T, Scherer R, Neuper C and Müller-Putz G 2014 Its how you get there: walking down a virtual alley activates premotor and parietal areas*Front. Hum. Neurosci.* **8** 93

[5] Mayer A, Kudar K, Bretz K, Tihanyi J and Mayer Á 2008 Body schema and body awareness of amputees *Prosthet. Orthot. Int.* **32** 363

[6] Ferrari A, Cutti AG, Garofalo P, Raggi M, Heijboer M, Cappello A and Davalli A2010 First in vivo assessment of Outwalk: a novel protocol for clinical gait analysis based on inertial and magnetic sensors *Med. Biol. Eng. Comput.***48** 1-15

[7] Lambrecht J, Pulliam C and KirschR 2011 Virtual reality environment for simulating tasks with a myoelectric prosthesis: an assessment and training tool *J. Prosthet. Orthot.* **23** 89

[8] Cardona N, Catalfamo P, Cherniz A and Bonell C2015 Virtual Platform Technologies for Evaluation of Rehabilitation Progress and for Prosthetic Control *VI Latin Am. Cong. on Biom. Eng.* 984

[9] Hermens H, Freriks B, Disselhorst-Klug C and Rau G 2000Development of recommendations for SEMG sensors and sensor placement procedures. *J. Electromyogr. Kines.***10**361

[10] MATLAB version 7.10.0 R2010a *The MathWorks, Inc.*

[11] DeLuca G 2003 Fundamental Concepts in EMG Signal Acquisition. *Copyright DelSys Inc.*

[12] V-REP Virtual Robot Experimentation Platform *Coppelia Robotics*
[13] Zecca M, Micera S, Carrozza M and Dario P 2002 Control of Multifunctional Prosthetic Hands by Processing the Electromyographic Signal*Critical ReviewsBiom. Eng.***30**459
[14] Bonell C, Tabernig C y Spaich E 2015 Evaluation of a double threshold algorithm to detect electromyographic activity in the healthy and paretic Tibialis Anterior muscle. *VI Latin Am. Cong. On Biomedical Engineering* (Springer International Publishing) p 940

Journal of Physics: Conference Series **705** (2016) 012062 doi:10.1088/1742-6596/705/1/012062

Somatic and movement inductions phantom limb in non-amputees

D M Casas[1], G G Gentiletti[2], A A Braidot[1]

[1]Laboratorio de Biomecánica, Facultad de Ingeniería, Universidad Nacional de Entre Ríos, Oro Verde, Argentina

[2]LIRINS, Facultad de Ingeniería, Universidad Nacional de Entre Ríos, Oro Verde, Argentina

Abstract. The illusion of the mirror box is a tool for phantom limb pain treatment; this article proposes the induction of phantom limb syndrome on non-amputees upper limb, with a neurological trick of the mirror box. With two study situations: a) Somatic Induction is a test of the literature reports qualitatively, and novel proposal b) Motor Induction, which is an objective report by recording surface EEG. There are 3 cases proposed for Motor illusion, for which grasped movement is used: 1) Control: movement is made, 2) illusion: the mirror box is used, and 3) Imagination: no movement is executed; the subject only imagines its execution. Three different tasks are registered for each one of them (left hand, right hand, and both of them). In 64% of the subjects for somatic experience, a clear response to the illusion was observed. In the experience of motor illusion, cortical activation is detected in both hemispheres of the primary motor cortex during the illusion, where the hidden hand remains motionless. These preliminary findings in phantom limb on non-amputees can be a tool for neuro-rehabilitation and neuro-prosthesis control training.

1. Introduction

The amputations of the upper limb in most of the cases occur from congenital malformation in the case of children and trauma in the case of adults, only in the United States there are about 40,000 people with upper limb amputation[1]. It is Known that about 90% of patients with some amputation reports having experiences of phantom limb such as pain, stinging, itching, tingling, voluntary and involuntary movements, these are typical of Phantom Limb (PL)[2][3].

There are two different concepts in these experiences: Somatic sensation and proprioception (corporeal awareness). Somatic sensations are those in which stimuli are applied to the body and produce feel pressure, heat, joint pain cold and others. Proprioception is related to the conscious awareness of the schema of body parts such as shape, size and position. There is another classification among the movement phantom limb sensations: voluntary, quasi involuntary (e.g. answering the telephone) and totally involuntary (e.g. new positions or spasms by tighten the fingers on the loss limb [4].

Treatments of Phantom Limb Pain (PLP) include the use of drug, which is directed to pain management with drugs such as anticonvulsants, muscle relaxants and antidepressants. However, since some years ago, one of the most common therapies for PLP has been the mirror box, in which body symmetry and proprioception are taken as an advantage, performing a neurological illusion where the subject sees the uninjured limb reflected as the limb loss, reducing pain and improving motor sensations, thus giving the patient a visual feedback of the movement[2].

Something similar occur in augmented reality system based on a brain computer interface (BCI) that provides feedback to the patient in the treatment of PLP syndrome [5]. These treatments are an illusion, which occur in afferent way in the Central Neural System CNS , while imaginary movement occur efferent way[6].

These neurologic illusions/tricks in recent research have been applied to induce phantom limb on non-amputated people. These techniques use the mirror box and tactile stimulation while looking at the mirrored limb, recording evoked potentials from illusion on somatosensory cortical activity of the reflected limb [7]. A second technique uses the rubber hand illusion, with tactile stimulation synchronously in the control limb and the rubber hand, evoking the withdrawal reflex while threatening the rubber hand, to observe the cortical activation with functional Magnetic Resonance Imaging (fMRI) and skin conductance response) [8][9][10][8][9][10]. Therefore, literature shows that the techniques employed only study somatosensory cortical activation and that the study of motor cortex activity by phantom limb are only studied with fMRI[11].

Moreover, the neuroprosthesis are still under development and current control methods for them require a great deal of mental concentration by the patient. Consequently, it requires the patient to do the control by imagination, of non-invasively with surface EEG or invasive, in which the array of microelectrodes are implanted on the cortex. For example, a quadriplegic patient with invasive BCI moves a robotic arm in a few weeks of training, with slow movements though [12]. Another invasive application, is method in which an array of microelectrode implantation in peripheral nerve provide a feedback of pressure and feel to the brain[13][14]. Relationship of motor imagery and movement of a hand orthosis, is performed in a study from BCI based on proprioceptive feedback control, with a 128 channels of sEEG [15].

There are systems to treat phantom limb pain with signals process EEG to move a virtual hand model[5]. Similarly, in another study of phantom limb pain with a the control of a virtual avatar based on EMG's signal from stump, thereafter this control is integrated into a prosthesis[16]. However, there were not found studies in literature linking the prosthetic control to neurological changes associated with the presence of PLP. Nevertheless, it is known that subjects preserve their intentions or natural movements even after amputees.

The present paper starts checking the techniques of Inducing-Somatic Phantom Limb (ISPL) in non-amputees. It is also proposed to study Induce-Movement Phantom Limb (IMPL) in non-amputees using the register of the activity of the primary motor (MI) cortex using electroencephalography (EEG). We propose the hypothesis that these results of the induction PLM in non-amputees could be used to generate knowledge bases that help neuroprostheses's control training. This training provides the use of a noninvasive method, with few electrodes from signals that occur naturally or intuitively in amputation patients or diseases peripheral nerve.

2. Methodology

2.1. Experimental setup

In this study, participants were seated in a comfortable chair and a table in front of the subject in which they rested both arms. In situations when mirror box is used (see section study situations), it was placed on the table in front of the participant in which the subject introduced one arm (hidden limb), while the outside limb is reflected in mirror box[2]. The box rested on the shoulder joint to ensure the reflected limb proprioception. This is because the location of the limb has a strong relationship with the visual information that the subject has of the body[10]. The figure 1 shows experimental setup.

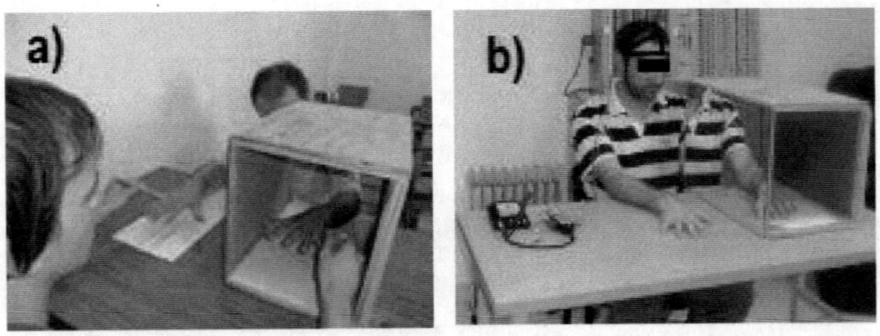

Figure 1 Experimental setup. A) Somatic test. B) Motor test

2.2. Study Conditions

The study has two different tests for induction of PL in non-amputees; the first is a checking of the techniques used for ISPL, while the second investigates IMPL. Each test is described below:

2.2.1. Test 1. Inducing-Somatic Phantom Limb (ISPL)

The subject was asked to stare the limb in the mirror, to rest his elbows on the table, to flex the wrist and keep his hands in the air. While the experimenter sitting opposite the subject, stimulates the fingers of the reflected and hidden limbs during 2 minutes, inside and outside the mirror box, applying synchronously brushstrokes with identical brushes. Then, the subject is asked to close his eyes for 10 sec and when opening the eyes receives an attack with a false knife in the limb outside of the mirror box. This is done to test is inducing withdrawal reflex in both limbs.

2.2.2. Test 2. Inducing-Movement Phantom Limb (IMPL)

We have three conditions in which the three auditory stimuli are applied, described in section EEG acquisition. These conditions arose in order to make a comparative analysis of each case, as described below:

- **Control** *Case 1*: Movement made without a mirror box. The subject placed both forearms on the table, he was asked to open and close their hands according to stimulus indicated. (See Table 1).
- **Illusion** *Case 2*: neurological trick (illusion) is performed by positioning the mirror box as described in the section above hiding his left hand. The subject was asked move his hands same as case 1, watching his hand in the mirror. (See Table 1).
- **Imaginary** *Case 3*: Imaginary move. In this case the subject does not execute any movement, only imagine move hand according to stimulus indicated. (See Table 1).

Table 1. Stimulus and cases

Stimulus / Case	1 Left	2 Right	3 Both
Control (Without mirror box)	Movement of the Left hand grip.	Movement of the Right hand grip.	Movement of the Both hands grip.
Illusion (With mirror box)	Movement of the Left hand grip hidden.	Movement of the Right hand grip. Looking hand in the mirror.	Movement of the Both hands grip. Looking Right hand in the mirror.
Imaginary (Without mirror box)	Imagine movement of the Left hand grip.	Imagine movement of the Right hand grip.	Imagine movement of both hands grip.

2.3. Subjects

The tests were applied to 19 subjects, healthy adult volunteers (7 females and 12 males) mean age 31.55, 2 left-handed, 16 right-handed and 1 ambidextrous. The group includes 16 subjects without prior knowledge (naive) of ISPL. In particular we select 6 subjects (2 female and 4 males), in which the withdrawal reflex was observed stronger in the ISPL experience. This group was applied the second method IMPL recorded with EEG. Within this group all subjects are right handed and the mean age is 23 years. All subjects given a written consent form for recording EEG signals and performing the experimental protocol.

2.4. EEG data acquisition

EEG was recorded from 10 Ag / AgCl, electrode positions F3, F4, T7, C3, Cz, C4, T8, Pz and two mastoid reference channels. The EEG-amplifier G.TecMobilab Plus has the following characteristics: Filters 0.5-100 Hz, sensitivity of 500μV, monopole, ADC with 16 Bits and 256 Hz sample rate and Bluetooth communication. Were programmed tasks in open source BCI2000 [17], in which auditory stimuli, indicating the subject the task, each stimulus for 1.7 s + 0.2 s, with 10 random repetitions of each stimulus were recorded and four runs each session. The auditory stimuli are "right", "left" and "both", the actions shown in Table 1. The rest condition occurs between stimulus and stimulus, the subject should not make none task, remaining in a relaxed state.

2.5. Analysis data

A questionnaire for experience ISLE analysis was performed, although it was a subjective measure, it provides an initial assessment of the existence of PL in non-amputees. After the experience, the subject answers a questionnaire consisting of 6 closed questions (yes or no) additionally the subject freely expresses his/her experience. The questionnaire is adapted from the somatic experience of White *et al*[18].

Pr 1. Did you feel comfortable or weird during the experiment? (Comfortable: yes, weird: no)
Pr 2. Did you feel any unnatural sensation in your hands?
Pr 3. Did you feel at any time having another hand?
Pr 4 Was your hidden hand longer than usual?
Pr 5. Did you feel that the hidden hand's movements were entirely normal?
Pr 6. Have you ever seen something similar to this experiment before?
Pr 7. Is there anything you want to add?

Questions 1 and 2 enable the assessment of the feeling that the subject experiences in front of the mirror box; 3 to 5 evaluate the suggestion that the subject may have as regards the hidden hand, question 6 assesses whether the prior knowledge of the experiment affects the withdrawal reflex response.

Moreover, BCI2000 Offline Analysis graphical tool was used in the analysis of Inducing-Movement Phantom Limb, which is based on characteristics. It operates in two steps, the first, turns time signal to features, such as: EEG amplitude, time after some stimulus and channel. After transformation the raw data EEG to characteristics, the data is selected and organized into two groups, rest or stimulus.

Therefore, there is a number of samples for each feature from two different conditions, the coefficient of determination (r-squared - r^2) is calculated between the two conditions on the signal. r^2 values provide a measure of magnitude in which certain characteristic of EEG (frequency and amplitude of a particular location) is influenced by the stimulus or task performed (condition). The tool provides 3 graphics, the first relates r^2 values and frequencies; the second is a histogram (not shown), which relates r^2 values with frequencies and the channels being used, finally the last is a topographic map that displays the spatial distribution of r^2 values for a selected frequency.

The graph of the frequency spectrum and r^2 were used to manually verify frequency at which occurs event of desynchronization of each individual with the maximum value of r^2, because each subject has its own frequency range in which occur non-synchronization and synchronization in their EEG signals and located between the μ and β rhythms (8 to 30 Hz) when doing the proposed tasks. The frequency determined was used as input to generate the respective topographic map and a visual inspection of motor activity of the cerebral cortex for each subject and each condition.

3. Results

3.1. Test 1 ISPL

The expected response to the stimulus is the withdrawal reflex, evoked by attacking with a false knife. Figure 2 shows the percentage of three types of response: non response to the stimulus, undefined and a clear response in withdrawal reflex. A considerable percentage of subjects show a clear withdrawal reflex response to the attack with the false knife.

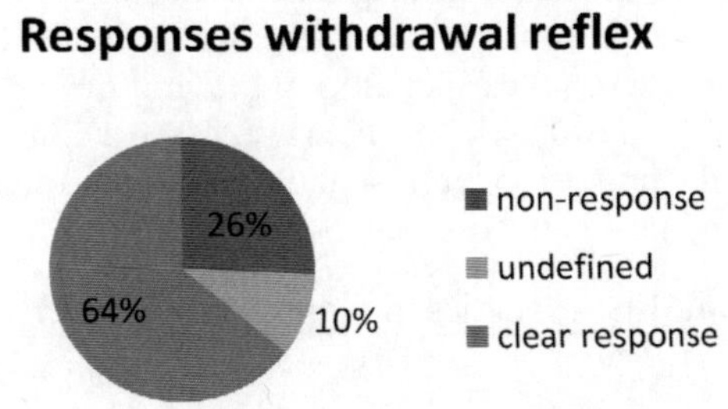

Figure 2. Results of responses withdrawal reflex.

The frequency of responses was obtained in the questionnaire to all subjects in the ISPL test are shown in Figure 3.

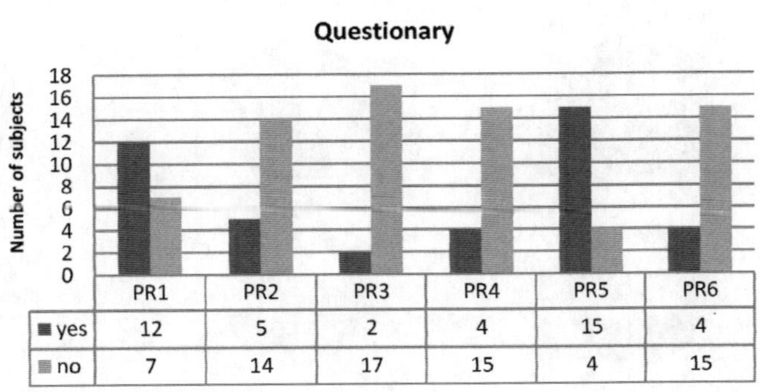

Figure 3. Graphic representation of the responses to the questionnaire applied.

3.2. Test 2 IMPL

Figure 4 shows the frequency found for subjects 1 and 2 in the control case with the maximum value of r2. While the frequency between individuals is different, for each subject it persists for all case studies.

Journal of Physics: Conference Series **705** (2016) 012062 doi:10.1088/1742-6596/705/1/012062

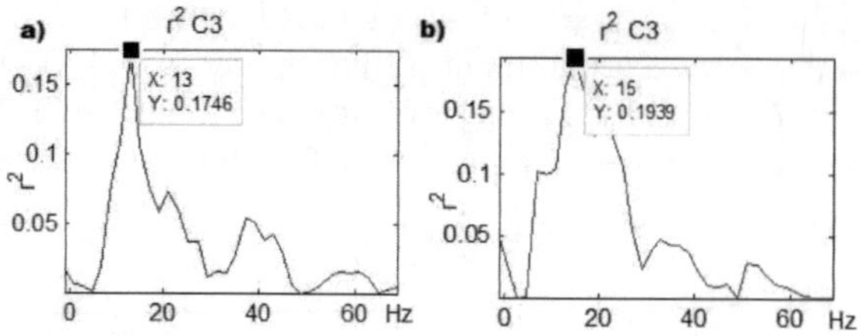

Figure 4. Spectrum of r^2 a) Subject 1, b) Subject 2

Topographic distributions of the activity of the cerebral cortex are shown in Table 2. For each case study test 2 IMPL motor activation, the corresponding motor activation area is clearly observed. Activation is an important distinction in the contra lateral area at which the movement occurs. However, in the case of illusion and right hand movement in the mirror it generates a signal in the right hemisphere (which corresponds to the left hand still hidden), similar to the signal magnitude of the left hemisphere (corresponding to moving the right hand). Also, the signal in the right hemisphere (still hidden left hand) for the same case of illusion also has a comparable scale with the signal of the right hemisphere (corresponding to the left hand) when moving both hands or when moving only the hand left, in the case of control.

Table1 Case studies in a single subject for test 2 IMPL

Case \ Stimulus	Left	Right	Both
Control	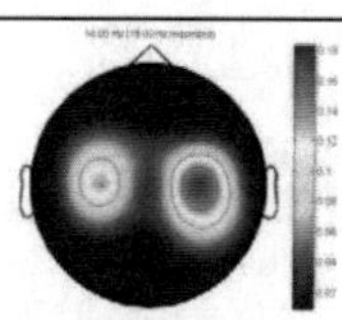	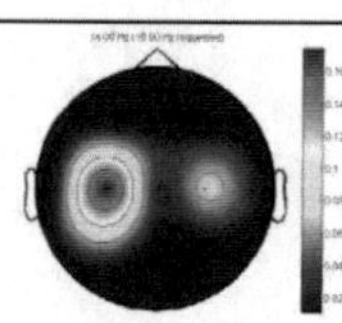	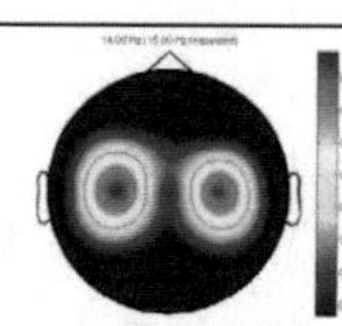
Illusion	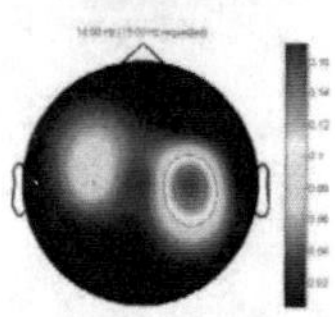	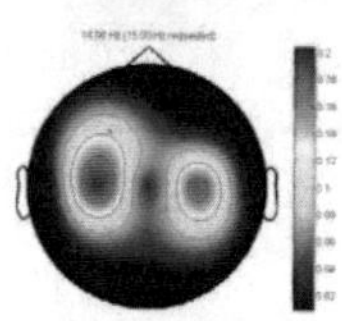	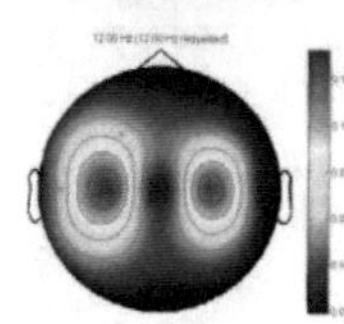
Imaginary	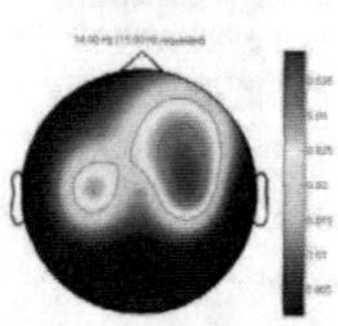	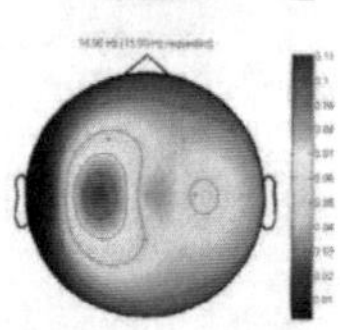	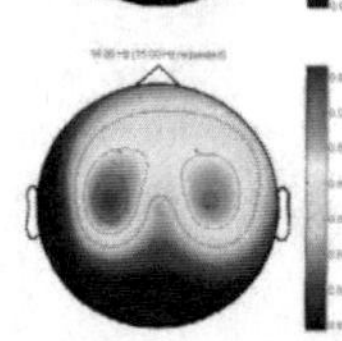

Table 3 Illusion case in six different subjects

Case \ Stimulus	Left	Right	Both
Subject 1	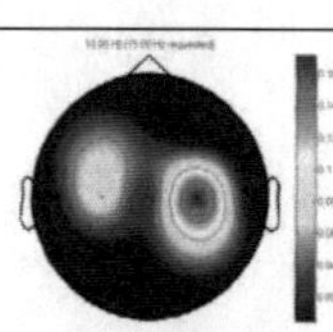	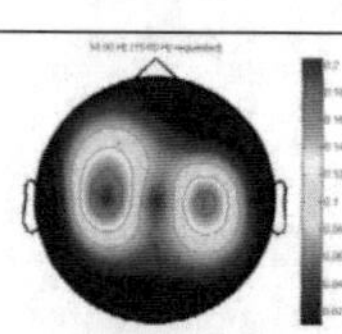	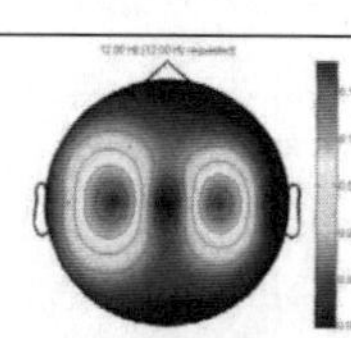
Subject 2	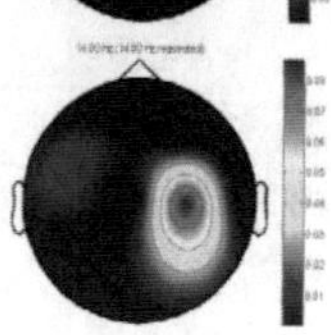	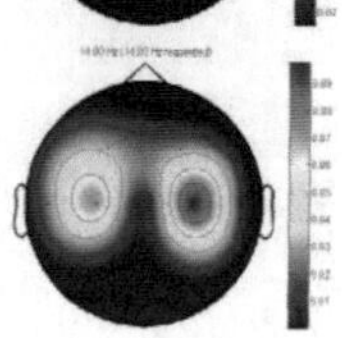	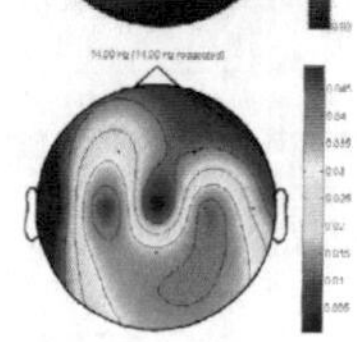

Subject 3

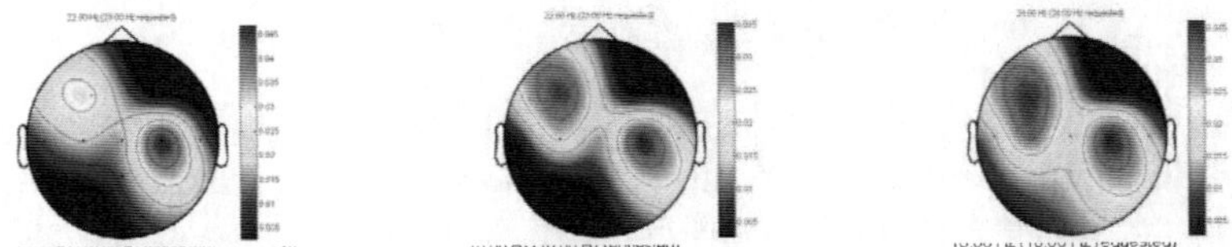

Subject 4

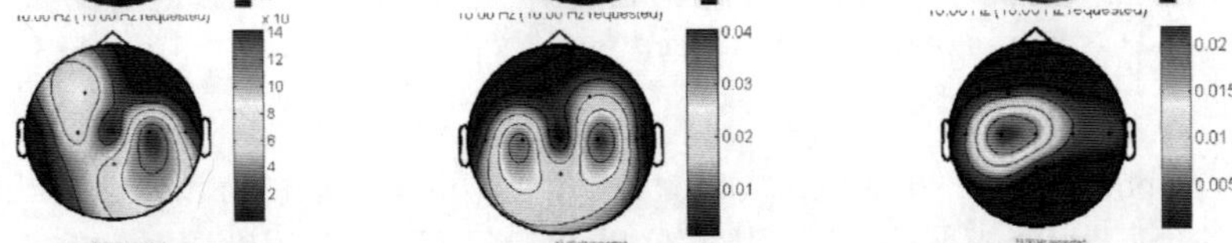

Subject 5

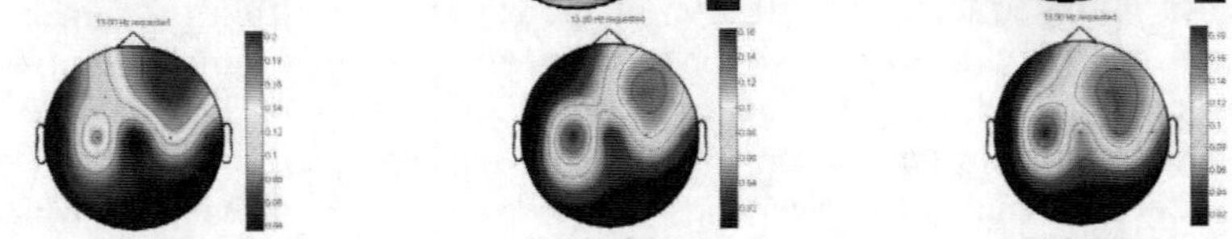

Subject 6

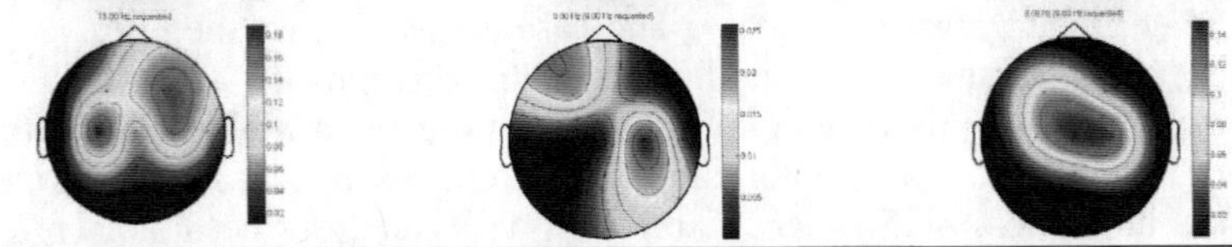

Table 3 shows the 6 subjects who participated in the trial in case IMPL illusion, under 3 stimuli applied in each case study. It is observed that the **illusion** (without movement of the left hand) in subjects 1, 3, 5 and 6 generate a signal to the right hemisphere that is just weaker than the corresponding right hand moving (left hemisphere) signal . In subjects 2 and 4 **illusion** is even more intense for the right hemisphere than the corresponding right hand moving (left hemisphere) signal. In subjects 4 and 6, illusion shows a representation with different characteristics for the movement of both hands.

4. Discussion

4.1. Test 1 ISPL

The withdrawal reflex response to the stimulus in each person is different, in some subjects may be stronger than others. Non-response suggests that like the amputees have the PL 90% of subjects, in non-amputees could also assume that in a small percentage no PL induction is possible. This is consistent with a slightly different experiment by Ramachandran *et. al*[2], who found that the phantom nose is not induced in all subjects in the experiment on non-amputees. In their experiment, they found a 40% non-illusion, which is close to 36% of this work, considering 26% of non-response and 10% of indefinite answer.

In the analysis of the somatic test other researchers made questionnaires staggered for a comprehensive statistical analysis [18][10]. However, somatic experience that is recorded in this work, aims to check qualitatively whether it is possible or not to induce PL in non-amputees, as a pre-illusion assay to IMPL.

The analysis of the responses to question 1 (PR1), shows that the majority of subjects concerned have felt comfortable during the experiment. As regards question 2, it suggests that there is not abnormal sensation during the experiment (PR2). This may be interpreted that the environment is natural for the subject

In reference to the suggestion that it can get to have the subject and felt a third hand (PR3), a high percentage of the subjects say they did not feel to have a third hand. In accordance to the literature to [18][10],when asked if a hand was longer than the other (PR4), negative response was obtained. Additionally the hidden hand movements were completely normal (pr5). The present result suggests that there was no suggestion felt by the subject when neurological trick was done with the mirror box.

The negative answer to Question 6 (pr6 shown that having seen the experiment previously does not influence the results, because in the 3 subjects who knew the experiment the withdrawal reflex is produced all the same). This is particularly important because it shows that neurological trick can be induced even in people experienced with the mirror box.

4.2. Test 2 IMPL

The response to the stimulus 1, of the left hand (hidden hand in the case of illusion) in the three case studies it is linked to the registration of the cortical activation of primary motor area of the right hemisphere (contra lateral activation), typical for this type of movement, as seen in a conventional EEG recording [19], since the left hand is hidden does not occurs the neurological trick. In the case of imaginary the register is also common for these tasks [20].

Moreover, stimulus 2 corresponds to the grip of the right hand, in which the activity of the left hemisphere motor area is observed. The case control shows the corresponding activation only on the right hand; the same occurs in the case of imagination.

The most interesting case in this study occurs under stimulus 2 grip of the right hand in case 2, where the activation of the motor cortex in both hemispheres appears during registration of illusion, with the motionless hidden hand. Activation is located on the motor cortex in the same area in a common register where both hands are moved. This indicates that the position of the phantom hand on the motor cortex is established on the spot that corresponds to the motionless hand. This is similar to the activation cortical seen with fMRI on amputees with phantom limb, in which the cortical activity of the phantom hand is recorded where the missing limb was, replacing in the brain's perception of the lost limb by the perception of phantom limb [11].

Something similar also occurs in cases of patients with supernumerary phantom limb after a brain stroke, in which cortical activity is recorded using fMRI when the patient is asked to "move" his or her supernumerary phantom limb [21]. Likewise happens when induced inducing-Somatic Phantom Limb by touch in non amputated and registering EEG [7] or fMRI images that are obtained of amputees and non-amputees under ISPL [9]. This work is being extended this finding of results to movement, it is estimated that in the future control signals may be generated intuitively in amputees. These signals could command neuroprosthesis drivers with a relatively shorter learning.

The activity in the grasping movement of both hands in all cases are similar, but are somewhat different in the value of their r^2 being the most intense for case 1, where there is no neurological trick or imagination and smaller values for the case 3 registration imagination. This is consistent with the free and voluntary manifestation of the participants who said that the case of imagery requires more effort. Nevertheless, imagery can be recorded without prior training [19], for best results training is required, that helps to make more easily and without mental exhaustion this task [22]. Illusion can become a mechanism that facilitates learning to achieve in less time the control capabilities of neuroprosthesis.

The EEG is performed with few electrodes accordingly it is premature to say that it is not established a parallel neuroplasticity and hence it is not possible to observe a cortical activation if it occurs in a different area the brain dedicates to the attention of movement hands. In other studies where the rubber hand as a third arm is used to generate the ISPL using the tactile illusion [10][18], could also be neuroplasticity. Comparatively in a future work with more electrodes it could be established whether neuroplasticity is being generated in this condition parallel or some other type of neuroplasticity.

In the next step of this research it is planned statistical signal processing, which allows extract those characteristics of the pattern found in the illusion to generate base of intentions phantom as control strategies and training in the use of prosthetics with natural intentions of the patient.

Journal of Physics: Conference Series **705** (2016) 012062 doi:10.1088/1742-6596/705/1/012062

5. Conclusion

The registers of case 2, where the setup illusion (neurological trick) was implemented, show that it is possible to induce phantom limb movement in non-amputees. Literature clearly shows that the phantom limb can be induced sensitively; however it was not entirely clear whether it could be done by movement. In this paper we verify that the induction-movement can also be registered with a non-invasive, inexpensive and easily accessible compared to fMRI study. In this context, it should be clarified that the experience of case 2 (illusion) can be understood as a sensorimotor trickery, since it explores the driving nature, hand movement reflected and vision that deceives proprioception inducing member phantom (reflected member).

For future work it is important to expand the number of subjects registered for better study window. It also remains to be established whether the proposed method can be applied to the rehabilitation of patients with brain diseases related to correct motion or the proposed solutions for quadriplegics noninvasively [12].

It also remains to consider whether this induction in people with brain injuries (eg. Stroke), they can reach neuroplasticity in less time with the proposed method. Thus, the illusion of the mirror box is not only an aid in the treatment of phantom limb pain but can become a tool for neuro-rehabilitation and training of neuro-prostheses controlled with patient's natural signals amputation, and that could be applied with few electrodes, following these preliminary findings [8][23], instead of having a maladaptive neuroplasticity that generates the phantom limb pain.

References

[1] K. Ziegler-Graham, E. J. MacKenzie, P. L. Ephraim, T. G. Travison, and R. Brookmeyer, "Estimating the prevalence of limb loss in the United States: 2005 to 2050.," *Arch. Phys. Med. Rehabil.*, vol. 89, no. 3, pp. 422–9, Mar. 2008.

[2] T. D. O. Hebb, V. S. Ramachandran, and W. Hirstein, "The perception of phantom limbs," *Brain*, vol. 121, pp. 1603–1630, 1998.

[3] M. G. Treviño, S. Salazar Marioni, C. E. Escamilla Ocañas, S. Daniel, H. R. Martínez, and G. Rivera, "Síndrome del miembro fantasma, dolor real," *Rev. Médica Md*, vol. 4, no. 04, pp. 32–36, 2012.

[4] J. P. Hunter, "The effect of tactile and visual sensory inputs on phantom limb awareness," *Brain*, vol. 126, no. 3, pp. 579–589, Mar. 2003.

[5] J. E. Arango, J. C. Mazo, and A. PeñaPalacio, "Sistema para Rehabilitación del Síndrome del Miembro Fantasma utilizando Interfaz Cerebro-Computador y Realidad Aumentada," *Iber. J. Inf. Syst. Technol.*, no. 11, pp. 93–106, Jun. 2013.

[6] R. Kitada, E. Naito, and M. Matsumura, "Perceptual changes in illusory wrist flexion angles resulting from motor imagery of the same wrist movements.," *Neuroscience*, vol. 109, no. 4, pp. 701–7, Jan. 2002.

[7] L. L. Egsgaard, L. Petrini, G. Christoffersen, and L. Arendt-Nielsen, "Cortical responses to the mirror box illusion: a high-resolution EEG study.," *Exp. brain Res.*, vol. 215, no. 3–4, pp. 345–57, Dec. 2011.

[8] M. Botvinick and J. Cohen, "Rubber hands 'feel' touch that eyes see.," *Nature*, vol. 391, no. 6669, p. 756, Feb. 1998.

[9] A. Guterstam, G. Gentile, and H. H. Ehrsson, "The Invisible Hand Illusion: Multisensory Integration Leads to the Embodiment of a Discrete Volume of Empty Space," *J. Cogn. Neurosci.*, pp. 1–22, 2010.

[10] A. Guterstam, V. I. Petkova, and H. H. Ehrsson, "The illusion of owning a third arm.," *PLoS One*, vol. 6, no. 2, p. e17208, Jan. 2011.

[11] M. Lotze, H. Flor, W. Grodd, W. Larbig, and N. Birbaumer, "Phantom movements and pain. An fMRI study in upper limb amputees.," *Brain*, vol. 124, no. Pt 11, pp. 2268–77, Nov.

2001.

[12] J. L. Collinger, B. Wodlinger, J. E. Downey, W. Wang, E. C. Tyler-Kabara, D. J. Weber, A. J. C. McMorland, M. Velliste, M. L. Boninger, and A. B. Schwartz, "High-performance neuroprosthetic control by an individual with tetraplegia.," *Lancet*, vol. 381, no. 9866, pp. 557–64, Feb. 2013.

[13] N. Kameswaran, D. K. Cullen, B. J. Pfister, N. J. Ranalli, J. H. Huang, E. L. Zager, and D. H. Smith, "A novel neuroprosthetic interface with the peripheral nervous system using artificially engineered axonal tracts.," *Neurol. Res.*, vol. 30, no. 10, pp. 1063–7, Dec. 2008.

[14] X. Navarro, T. B. Krueger, N. Lago, S. Micera, T. Stieglitz, and P. Dario, "A critical review of interfaces with the peripheral nervous system for the control of neuroprostheses and hybrid bionic systems.," *J. Peripher. Nerv. Syst.*, vol. 10, no. 3, pp. 229–58, Sep. 2005.

[15] A. Ramos-Murguialday, M. Schürholz, V. Caggiano, M. Wildgruber, A. Caria, E. M. Hammer, S. Halder, and N. Birbaumer, "Proprioceptive feedback and brain computer interface (BCI) based neuroprostheses.," *PLoS One*, vol. 7, no. 10, p. e47048, Jan. 2012.

[16] B. N. Perry, A. L. Alphonso, J. W. Tsao, P. F. Pasquina, R. S. Armiger, and C. W. Moran, "A Virtual Integrated Environment for phantom limb pain treatment and Modular Prosthetic Limb training," *2013 Int. Conf. Virtual Rehabil. IEEE*, pp. 153–157, Aug. 2013.

[17] G. Schalk, D. J. Mcfarland, T. Hinterberger, N. Birbaumer, J. R. Wolpaw, and A. B. I. B. C. I. Technology, "BCI2000 : A General-Purpose Brain-Computer Interface (BCI) System," *IEEE Trans. Biomed. Eng.*, vol. 51, no. 6, pp. 1034–1043, 2004.

[18] R. C. White, A. M. Aimola Davies, and M. Davies, "Two hands are better than one: a new assessment method and a new interpretation of the non-visual illusion of self-touch.," *Conscious. Cogn.*, vol. 20, no. 3, pp. 956–64, Sep. 2011.

[19] B. Blankertz, G. Dornhege, M. Krauledat, K. Müller, V. Kunzmann, and F. Losch, "The Berlin Brain – Computer Interface : EEG-Based," *Brain*, vol. 14, no. 2, pp. 147–152, 2006.

[20] T. Wang, J. Deng, and B. He, "Classifying EEG-based motor imagery tasks by means of time-frequency synthesized spatial patterns.," *Clin. Neurophysiol.*, vol. 115, no. 12, pp. 2744–53, Dec. 2004.

[21] D. J. McGonigle, R. Hänninen, S. Salenius, R. Hari, R. S. J. Frackowiak, and C. D. Frith, "Whose arm is it anyway? An fMRI case study of supernumerary phantom limb.," *Brain*, vol. 125, no. Pt 6, pp. 1265–74, Jun. 2002.

[22] T. Wang, J. Deng, and B. He, "Classifying EEG-based motor imagery tasks by means of time-frequency synthesized spatial patterns.," *Clin. Neurophysiol.*, vol. 115, no. 12, pp. 2744–2753, 2004.

[23] G. R. Müller-Putz, R. Scherer, G. Pfurtscheller, and R. Rupp, "EEG-based neuroprosthesis control: a step towards clinical practice.," *Neurosci. Lett.*, vol. 382, no. 1–2, pp. 169–74, 2005.

Triaxial MMG investigation during FES-induced cycle movements

M C Gelain[1], G N Nogueira-Neto[2], P Nohama[3] and P M Gewehr[4]

[1]Master's student of Electrical Engineering and Industrial Informatics at UTFPR (Rehabilitation engineering laboratory phone: +55 (41) 3310-4489), Brazil.
[2]PhD in Biomedical Engineering, researcher and professor of Computer Engineering at Pontifícia Universidade Católica do Paraná (PUCPR), Brazil (e-mail: nogueira.g@pucpr.br).
[3]PhD in Biomedical Engineering, researcher and full professor at the PPGTS *stricto sensu* program at Pontifícia Universidade Católica do Paraná (PUCPR), Brazil (e-mail: percy.nohama@gmail.com).
[4]PhD in Medical Physics Bioengineering and professor of Biomedical engineering at UTFPR, Brazil (e-mail: gewehr@utfpr.edu.br)

E-mail: gelainfisio@gmail.com

Abstract. Spinal cord injury is damage to the spinal cord that results in loss of functions, causes muscle, joint and bone deficits, moreover it increases the muscle tone level. FES-cycling is an alternative rehabilitation process to conserve the musculoskeletal system affected by the injury in the functional state. The aim of this study is to get the legs bilaterally synchronized by FES and compare MMG signal responses. The preliminary protocol consists of five consecutive contractions. The first and the last contractions were discarded and the three intermediate contractions were analyzed. A ratio between MMG values of muscle responses during propulsion and return of pedalling phases was calculated for each cycle and each leg. The applied FES profile consisted of pulse and burst on time of 250 us and 5 ms and frequencies of 1000 Hz and 20 Hz, respectively. Results showed that greater MMG RMS values tend to occur during the propulsion phase than in the return phase. MMG Z axis of RLL and MMG X axis of LLL were the axes that presented the greatest ratios. The MMG RMS signal showed that greater values tend to occur during the propulsion phase than in the return phase of the same thigh.

1. Introduction

Spinal cord injury (SCI) is damage to the spinal cord that results in loss of functions, causes muscle, joint and bone deficits and, moreover, it increases the muscle level [1]. Such tissues could be preserved through the long-term rehabilitation utilizing functional tasks such as treadmill training, leg cycling and functional electrical stimulation (FES) that also optimizes the plasticity mechanisms [2]. The association of leg cycling with FES is an alternative strategy to train the impaired muscles, allowing improvements in muscle mass and strength, blood flow, bone density, decreasing the abdominal adipose tissues by cardiopulmonary fitness and influencing the psychological health [3] and social integration [4, 5].

However, it is possible to raise at least three complications in this rehabilitation strategy: (i) finding appropriate parameters to provide the synchronization timing that generates a coordinated pedalling movement [6]; (ii) producing pedalling movements elicited by FES while minimizing the effects of muscle fatigue that acts as a muscle performance suppressor mechanism [2, 7, 8]; (iii) determining the required stimulation amplitude to contract the muscles accordingly without causing tissue damage and pain [9, 10]. Regarding the third complication, it is particularly important to be aware of this problem when healthy persons participate in the research.

The SCI classification (paraplegic or quadriplegic and complete or incomplete) is important to the rehabilitation process, as well as the injury extension like Peripheral Nervous System (PNS) integrity, the specification of lower motor neuron, allowing the lower limb movements by neuromuscular stimulation [9].

The employed system [11] was already used in other studies, analyzing primarily isometric contractions without bilateral stimulation.

Mechanomyography (MMG) is a non-invasive technique used for registering the muscle vibrations sensed on the muscle surface. Inside the muscle, the myofibrils contract generating intra-tissular pressure waves. Conversely to an electromyography (EMG) device, an MMG acquisition system can be easily attached to an electrical stimulator because the high voltage electrical stimuli do not interfere with the vibrations recorded by the accelerometer embedded in the sensor, a problem that hinders continuously monitoring signals during FES application. However, MMG is employed with isometric contractions, instead of dynamic contractions. Regarding leg bilateral stimulation to continuously propel a cycle ergometer's crown, MMG can be used to investigate the muscle contractive activity during both FES-elicited contraction and return phases. Therefore, the aim of this study is to produce pedalling movements using synchronized FES bilaterally applied to the lower limbs and compare the MMG signal amplitude response.

2.Methods

The preliminary protocol was performed at PUCPR, Brazil, in the Rehabilitation Engineering Laboratory, using Movement's® LX130 cycle ergometer. Three healthy male participants were seated on the ergometer and the required adjustments were made according to their leg size. Because recumbent cycle ergometers did not provide the best ergonomics to develop strength in the propulsion phase [12], the seat height was increased i.e. the hip and knee angles were set around 100 degrees of flexion, increasing the length of rectus femoris.

Three electrodes were used, one placed over the belly of rectus femoris (between the superior border patella and anterior superior iliac spine) [13], other placed over femoris nerve (femoral triangle) [14] and the last one over the belly of gluteus maximus (between the great trochanter and sacrum) [13].

The protocol consisted of two parts. In the first one (warm up), individuals cycled during 2 min. In the second part, they ceased pedalling and FES was applied over rectus femoris and gluteus maximus muscles of both legs, to allow five consecutive contractions. Data obtained in the first and the last contractions were discarded. The amplitudes of MMG signals of both legs were compared.

The FES profile applied consisted in pulse and burst on time of 250 us and 5 ms, respectively, pulse and burst frequencies of 1000 and 20 Hz, respectively, square wave, four channels. Multichannel FES reduces the number of electrodes to be employed for eliciting electrical stimulation once it requires one single common return electrode near the motor point for each leg [9].

The amplitude modulated FES profile consisted of 10 s divided in three phases: propulsion, consisting of ramp up phase (4.5 s); plateau (0.5 s); and return phase representing the last 5 s, as illustrated in Fig.1. The FES amplitude values were set according to two factors: (i) sensitivity response of participants i.e. how affordable they perceived the sensation and (ii) the required energy to let the leg artificially propel the cycling.

Journal of Physics: Conference Series **705** (2016) 012063 doi:10.1088/1742-6596/705/1/012063

Figure 1 – FES amplitude modulated profile.

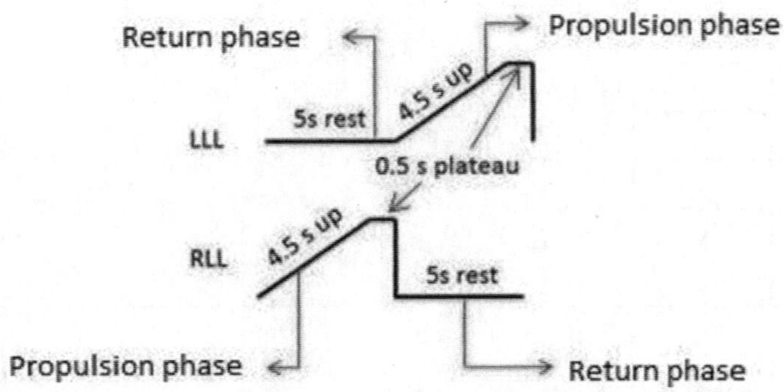

LLL = Left Lower Limb; RLL = Right Lower Limb

The MMG sensor uses a triaxial accelerometer (Freescale MMA7260Q MEMS [13x18 mm, 0.94 g] with sensibility equal to 800 mV/G in 1.5 G [G is gravity acceleration]). The hardware amplification was 2.2x and a Butterworth 3rd order bandpass filter was built. The software has been developed in the NI LabVIEW® platform to acquire and process all signals. The acquisition system consisted in an acquisition board (NI-USB 6221), using a frequency of 1 kHz to signal acquisition [11].

Data were analyzed for each contraction during four continuous seconds that corresponded to the increasing stimulation ramp and plateau phases. The three intermediate contractions were taken from the five contractions and a simple ratio between MMG RMS (propulsion phase / return phase) was calculated for each cycle and each leg.

3.Results

The second, third and fourth propulsion phase contractions of persons 1, 2 and 3 are indicated in Tables I and II by the MMG RMS ratios calculated for each cycle and each leg. One can visualize in Table II the mean of the three subjects.

Table I. Second, third and fourth contraction using FES of Person 1 and 2

			2nd	3rd	4th	Mean
Person 1	RLL	X	1.56	1.48	0.31	1.11
		Y	1.64	1.30	3.45	2.13
		Z	2.55	2.90	2.52	2.66
	LLL	X	1.79	2.14	1.01	1.65
		Y	1.48	1.41	0.64	1.18
		Z	1.48	1.41	0.64	1.18
Person 2	RLL	X	1.00	1.63	1.88	1.50
		Y	0.93	1.64	2.36	1.64
		Z	1.64	2.21	4.24	2.70
	LLL	X	1.95	1.17	1.19	1.44
		Y	1.14	1.23	0.60	0.99
		Z	1.14	1.23	0.60	0.99

RLL = right lower limb; LLL = left lower limb; X, Y and Z represents the three sensor axes.

Table II. Second, third and fourth contraction using FES of Person 2 and Mean

			2nd	3rd	4th	Mean
Person 3		X	0.78	0.84	0.89	0.84
		Y	0.71	0.64	0.81	0.72
	RLL	Z	1.26	0.84	0.99	1.03
		X	2.47	1.67	1.64	1.93
		Y	1.59	2.18	1.89	1.89
	LLL	Z	1.59	2.18	1.89	1.89
Mean		X	1.10	1.29	0.64	1.01
		Y	1.06	1.18	1.94	1.39
	RLL	Z	1.79	1.80	2.25	1.95
		X	2.03	1.60	1.28	1.64
		Y	1.35	1.52	0.91	1.26
	LLL	Z	1.35	1.52	0.91	1.26

RLL = right lower limb; LLL = left lower limb; X, Y and Z represents the three sensor axes.

The results in Tables I and II lead to the assumption that greater MMG RMS values tend to occur in the contraction phase in spite of the return phase of the same thigh. This can be observed by values greater than one. MMG Z axis to right lower limb (RLL) and MMG X axis to left lower limb (LLL) were the axes that presented the greatest ratios.

4.Discussion

FES-cycling might be an option to socially include SCI people, providing recreational activities as proposed by Hunt et al. who developed a tricycle [4], or activities of daily living (ADL) using a wheelchair with pedals like the one built by Watanabe et al. [5]. Before FES-cycling become a marketable product it is necessary to focus on research outcomes.

An important factor is the electric stimulation perception by subjects. Besides not being presented in the table, individuals informed a perception of an intensity difference between legs during FES application. The sensation was stronger on the right thigh than on the left thigh, however, the stimulus intensity needed to be adjusted until the sensation on both legs seemed equal. The same fact was observed by [5] and the delay of sensation would cause a delay on muscle answer, being necessary to adjust the pedalling time and velocity. Gfohler and Lugner [6] emphasized that the sensibility between legs occurs due to physiological differences and each individual has his/her own parameters profile for FES-cycling.

From a sensor's triaxial perspective, it is possible to observe that there were differences between the responses of MMG signals between legs. In Table I and II, the mean column and mean rows show that the Z axis presented greater average RMS values for the pedalling propulsion phase than for the returning phase to the RLL. To the LLL, conversely, the X axis presented similar response. FES intensity adjustments could be the reason for the existence of such particular differences since they were performed according to individual perception [10, 15]. However, another factor can be strength asymmetry [16] in both thighs and gluteus, as well as position, size and shape of the electrodes [17].

The adjusted seat position of the participants shortens the rectus femoris and since it is a biarticular muscle this fact compromises the propulsion phase, moreover it caused more discomfort during FES application. The difference between angles and muscles involved in pedalling with recumbent and vertical cycle ergometers are explained in [18]. However, Lopes et al. [19] affirm that differences in muscles activation do not exist using the two types of cycle ergometer.

The most correct posture for pedalling is obtained with a vertical cycle ergometer [20]. During a kinematic test the cyclist posture and joint position were verified, between 5-10 initial minutes, the fatigue was observed, because there were some joint alterations changing the initial posture. The adopted new musculoskeletal body position was trunk lean towards the front, increasing the hip flexion to facilitate the gluteus maximus contraction, consequently altering the knee joint and decreasing the knee flexion angle on the pushing phase [12], finally promoting an easier propulsion.

One difficulty that happens when working with FES application to healthy people is whether the contraction is happening solely due to the electric stimuli or also with the help of the participant. So there are two physiological reflexes to FES: H-reflex and flexion afferent reflex. During a muscle contraction elicited by FES, the H-reflex mechanism occurs due to external stimuli, sending an afferent signal to the spinal cord which in turn returns the signal by efferent pathway. However, two responses to neutralization of the H-reflex signals could be received when there are high levels of stimulus intensity: (i) major firing quantity of motor fibers inhibiting the motoneurones and (ii) the efferent response occurring at the same time as the afferent stimuli, creating a collision between them [21]. Another possibility is the flexion afferent reflex [22], because the organism could interpret the stimulus intensity as harmful and block the voluntary movement. The inverse can also happen, when a healthy individual aids the movement voluntarily blocking the reflex. Variations in the efferent stimulus conduction velocity can vary between 20-80 ms [22] and to the reflex between 20-30 ms [21], further the FES OFF time (pedalling return phase) is shorter than the time it takes for the CNS stimulus to adapt, therefore the CNS can interrupt or sum with H-reflex signal.

5.Conclusion

The MMG RMS signal from FES stimulation showed that greater values tend to occur during the propulsion phase than in the return phase of the same thigh. The signals recorded by different MMG axes can respond with different trends. Future research will occur with SCI individuals using personalized amplitude patterns in order to avoid the muscle unbalance as a result of disproportionate stimulation intensity.

References

[1] Mazzoleni S, Stampacchia G, Gerini A, Tombini T, Carrozza M C 2013 FES-cycling training in spinal cord injured patients. *35th Annual International Conference of the IEEE Engineering in Medicine and Biology Society (EMBC)* 5339-5341

[2] Sadowsky *C L*, Hammond E R, Strohl A B, Commean P K, Eby S A, Damiano D L, Wingert J R, Bae K T, McDonald J W 2013 Lower extremity functional electrical stimulation cycling promotes physical and functional recovery in chronic spinal cord injury. *The journal of spinal cord medicine.* **36** 623-631

[3] Fornusek C, Davis G M, Baek I 2012 Stimulation of shank muscles during functional electrical stimulation cycling increases ankle excursion in individuals with spinal cord injury *Archives of physical medicine and rehabilitation,* **93** 1930-1936

[4] Hunt K, Ferrario C, Grant S, Stone B, McLean A N, Fraser M H, Allan D B 2006 Comparison of stimulation patterns for FES-cycling using measures of oxygen cost and stimulation cost. *Medical engineering & physics,* **28** 710-718

[5] Watanabe T, Murakami T, Handa Y 2013 Preliminary tests of a prototype FES control system for cycling wheelchair rehabilitation *IEEE International Conference on Rehabilitation Robotics (ICORR)* 1-6

[6] Gfohler M, Lugner, P 2004 Dynamic simulation of FES-cycling: influence of individual parameters *IEEE Transactions on Neural Systems and Rehabilitation Engineering* **12** 398-405

[7] Hunt K J, Fang J, Saengsuwan J, Grob M, Laubacher M 2012 On the efficiency of FES cycling: A framework and systematic review *Technology and Health Care* **20** 395-422

[8] Griffin L, Decker M J, Hwang J Y, Wang B, Kitchen K, Ding Z, Ivy J L 2009 Functional electrical stimulation cycling improves body composition, metabolic and neural factors in persons with spinal cord injury *Journal of electromyography and Kinesiology* **19** 614-622

[9] Peckham P H, Knutson J S 2005 Functional electrical stimulation for neuromuscular applications* *Annu. Rev. Biomed. Eng.* **7** 327-360

[10] Ambrosini E, Ferrante S, Ferrigno G, Molteni F, Pedrocchi A 2012 Cycling induced by electrical stimulation improves muscle activation and symmetry during pedaling in hemiparetic patients *Neural Systems and Rehabilitation Engineering, IEEE Transactions on,* **20** 320-330

[11] Silva L R 2002 Sistema de controle de eletroestimulador em malha fechada utilizando lógica fuzzy MSc dissertation, CPGEI, UTFPR, Curitiba

[12] So R C H, Ng J K F, Ng G Y F 2005 Muscle recruitment pattern in cycling: a review *Physical Therapy in Sport* **6** 89-96

[13] A. O. Perotto 2011 *Anatomical Guide for the Eletromyographer: The Limbs and Trunk*. Springfield, Illinois, USA: Charles C. Thomas

[14] E. Krueger-Beck 2010 Influência dos parâmetros estimulatórios na eficiência da contração muscular evocada pela estimulação elétrica funcional em pessoas hígidas ou com lesão medular MSc dissertation, CPGEI, UTFPR, Curitiba

[15] Gelain M C, Santos E L, Krueger-Beck E, Nogueira-Neto G N, Nohama P 2014 Avaliação térmica e vibracional da fadiga muscular durante o uso do cicloergômetro: estudo piloto presented at the *XXIV Congresso Brasileiro de Engenharia Biomédica*

[16] Smak W, Neptune R R, Hull M L 1999 The influence of pedaling rate on bilateral asymmetry in cycling *Journal of biomechanics* **32** 899-906

[17] Hermens H J, Freriks B, Disserlhorst-Klung C, Rau G 2000 Development of recommendations for SEMG sensors and sensor placement procedures *Journal of electromyography and Kinesiology* **10** 361-374

[18] Fujiwara T, Liu M, Tanuma A, Hase K, Tsuji T 2005 Pedaling exercise for neuromuscular re-education: a review *Critical Reviews™ in Physical and Rehabilitation Medicine* **17**

[19] Lopes A D, Alouche S R, Hakansson N, Cohen M 2014 Electromyography during pedaling on upright and recumbent ergometer *International journal of sports physical therapy* **9** 76

[20] Dingwell J B, Joubert J E, Diefenthaeler F, Trinity J D. 2008 Changes in muscle activity and kinematics of highly trained cyclists during fatigue *Biomedical Engineering, IEEE Transactions* **55** 2666-2674

[21] Rothwell J C 1987 Functional Consequences of Activity in Spinal Reflex Pathways *Control of Human Voluntary Movement* ed: Springer 126-157

[22] Koppen B M, Staton B A 2008 *Berne and Levy Physiology*: Elsevier Brasil

The valuable use of Microsoft Kinect™ sensor 3D kinematic in the rehabilitation process in basketball.

Ariel Braidot,[1]* Guillermo Favaretto[1], Melisa Frisoli[1], Diego Gemignani[2], Gustavo Gumpel[3], Roberto Massuh[1], Josefina Rayan[1], Matías Turin[1].
[1]Laboratorio de Biomecánica, Facultad de Ingeniería, Universidad Nacional de Entre Ríos, Oro Verde, Entre Ríos, Argentina
[2] Instituto de Ciencias del Deporte, Villaguay 20, Paraná, Entre Ríos, Argentina
[3] Instituto del Chañar Monte Caseros 363, Paraná, Entre Ríos, Argentina
[1]* E-mail: abraidot@bioingenieria.edu.ar

Abstract. Subjects who practice sports either as professionals or amateurs, have a high incidence of knee injuries. There are a few publications that show studies from a kinematic point of view of lateral-structure-knee injuries, including meniscal (meniscal tears or chondral injury), without anterior cruciate ligament rupture. The use of standard motion capture systems for measuring outdoors sport is hard to implement due to many operative reasons. Recently released, the Microsoft Kinect™ is a sensor that was developed to track movements for gaming purposes and has seen an increased use in clinical applications. The fact that this device is a simple and portable tool allows the acquisition of data of sport common movements in the field. The development and testing of a set of protocols for 3D kinematic measurement using the Microsoft Kinect™ system is presented in this paper. The 3D kinematic evaluation algorithms were developed from information available and with the use of Microsoft's Software Development Kit 1.8 (SDK). Along with this, an algorithm for calculating the lower limb joints angles was implemented. Thirty healthy adult volunteers were measured, using five different recording protocols for sport characteristic gestures which involve high knee injury risk in athletes.

1. Introduction

Knee injuries are very frequent in people who practice sports, either professionaly or in an amateur way. 1 in 3,000 people suffer from a knee injury, and of those injuries, 70% is due to the practice of sports. There are a significant number of individuals with this injury [1] particularly in basketball.

Marqueta and Tarrero made an research in the NBA (National Basketball Association) and the ACB (Spanish Basketball Professional League), where they quantified basketball injuries in the various anatomical structures of the body. In both leagues, the incidence of knee injury is in second place, 9.4% and 14.2% respectively in NBA and ACB[2].

The incidence of injury in the external compartment of the knee (osteochondral and / or meniscal injuries) in basketball players is more common than those found in the internal compartment. In recent work, Gumpel raises the possibility that this damage is due to the transmission of loads to the lateral compartment due to the position valgus knee in certain gestures own of game [3].

[1]* To whom any correspondence should be addressed.

It is believed that this chondral injury is caused by a direct trauma to the knee flexed at 40 degrees, predominantly on the medial femoral condyle (including tennis, football and rugby). Injury occurs as chondral disruption by the direct impact with tissue repair and secondary reactive hypertrophic synovium that is located overlying the original injury. In basketball, unlike most sports, the chondral injury occurs most frequently in the external compartment of the knee. It is stated that this event is due to the pronounced position of knee valgus that athletes develop in this sport during every game and every practice session [3] [4]. The logical consequence of this type of damage is a loss of mobility due to the pain and therefore the inability to carry out the activity completely. In the professional basketball field, this is a problem that not only affects the health of the player, but it also presents both an economic and sporting problem to the institution where the athlete belongs.

Even though the injury as a whole as ACL, MCL and medial meniscus is widely known and is assumed to have high incidence, it has been found that the number of lesions in external compartment knee greatly exceeds the internal compartment knee injuries in basketball and other sports [5].

3D kinematic measurements of lower limb and ground reaction force in professional basketball players showed that greater knee valgus moment occurs during a change of direction when both hip flexion and internal knee rotation are incremented [6]. Colby et al. characterized the activation of the hamstrings and quadriceps, also determining the angle of knee flexion by analyzing four gestures: change of direction, jump with and without change of direction, run and stop[1, 7]. Other authors performed a 3D kinematic analysis of the differences in valgus knee between men and women when landing from a platform jump immediately followed by a vertical jump [8, 9].

Studies of Crossley et al. [10] indicate that hip muscle function is compromised in people with anterior knee pain (AKP). They found strong evidence of deficits in hip muscle strength (abduction, external rotation and extension) in patients with AKP compared to healthy people. A measurement of hip muscle function is given by the patterns of activation using electromyography (EMG). A single-leg squat is often an exercise used for clinical evaluation for determining whether the hip muscle function is compromised or not. Nowadays, there is visual analysis of movement patterns used to evaluate lower limb neuromuscular control. Previous studies conducted at gait laboratories using sophisticated equipment have shown movements of the joint in the hip and knee. Consequently, there is a need to develop a more simpler and clinically applicable method for evaluating the performance of one-leg squats and other movements of interest in assessing muscle function. In addition, analysis of the performance of a leg squat is also used to assess rehabilitation after knee injury. A continued development of a method in this field could bring an innovative aspect, which is the quantification of this movement during rehabilitation.

The Microsoft Kinect® sensor has been used to record movement in recent years. Specifically, Stone et al. conducted an analysis of the gesture of the jump from a platform on the ground and a jump back up. The results indicate that the skeletal model Kinect® Microsoft has acceptable accuracy for use as a tool to identify the risk of ACL injury [11].

From this analysis, it is possible to establish that knee injuries may be caused by the transmission of loads to the lateral compartment due to the position of valgus knee in certain own gestures of basketball game. Even though there have been some studies done, the causes have not yet been determined as well as the ability to generate optimal strategies for the prevention and rehabilitation of this traumatologic pathology. It is therefore presented in this work the proposal for the development of a system of registration and testing protocol for measuring kinematic 3D using the Microsoft Kinect®, a simple and portable tool for field use. This tool was used to record and analyze some common movements made during basketball practice, which served to develop a system that uses the data output of Microsoft Kinect® sensor system. The algorithms needed for the 3D kinematic assessment were calculated, while also obtaining the relationship between global and local coordinate systems using additionally the Software Development Kit (SDK) additionally which is available for free.

2. Material and Methods

2.1. Implementation of the registration system
In this presentation, the implementation of a system of capturing motion without markers (MMCS-Markerless Motion Capture System) is based on the Microsoft Kinect® sensor for Xbox 360. This is a portable and easy to assemble system, capable of acquiring and storing 3D kinematic data records in different scenarios for subsequent statistical analysis. This developed system is used to evaluate several representative sporting gestures and protocols.

2.1.1. Hardware: In this development we worked with Microsoft Kinect® sensor, device developed by PrimeSense® [9] and distributed by Microsoft [12] for the Xbox 360 console, and since June 2011, available for PC. It has a VGA color camera 640 × 480 (CMOS) with Bayer color filter, an infrared camera 640 × 480 (CMOS), an infrared laser projector, microphone Multi Array, a motor to adjust the tilt angle of the device and a accelerometer for sensing the spatial orientation (figure 1) [13].

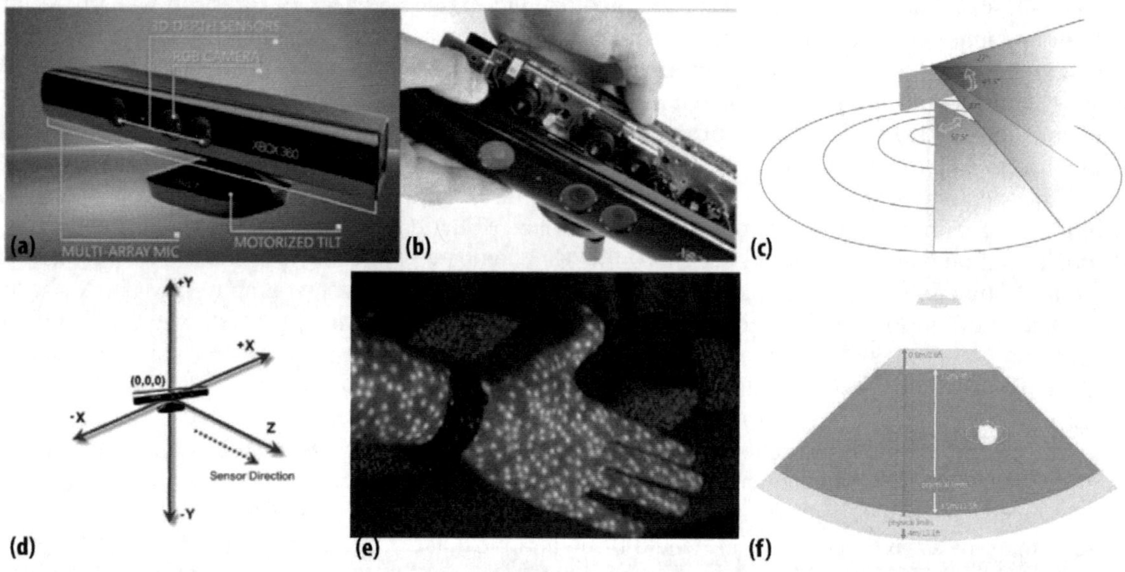

Figure 1. Features of Microsoft Kinect® sensor. (a) - (b) image sensor. (c) - (f) outline of vision and depth ranges. (d) the MMCS global coordinate system. (e) image of points infrared laser pattern projected and captured by the sensor [13].

The sensor can capture RGB images and depth at different resolutions at a maximum rate of 30 frames per second (CPS) depending on the desired resolution. The vision cameras range, both RGB and depth, has a horizontal angle of 57.5 ° and a vertical angle of 43.5 ° with the possibility of displacement of 27 ° down to 27 ° upward through motor possessing tilt (figure 1a.) [13].

This device combines infrared structured light laser and computer vision algorithms to obtain a depth chart from a single image. The entire process consists on projecting infrared laser pseudorandom points pattern at the scene (Figure 1e.). The infrared sensor then captures an image obtained thereof and then, through a deformation process, it estimates how far away objects are located in the scene. All this is done at the hardware level and a depth chart is obtained from grayscale of 16 bits, with a maximum resolution of 640 x 480, where each pixel encodes the distance between the object and the sensor in millimeters. From the 16 bits of each pixel, 13 of these contain distance information, giving a resolution of 1 mm in depth for optimal sensor range (800-4000 [mm]), this is shown in Figure 1f [14].

2.1.2. Software: The development of the 3D kinematic system uses Microsoft Kinect® SDK (SDK - Software Development Kit) with its drivers, tools, native APIs (Application Programming Interface) and code available APIs for the development of Windows applications based on Kinect. This SDK has extensive technical documentation and open code samples optimized for the exploration and experimentation of the main features offered by the sensor. One of its highlights is the pipeline offered by the Natural User Interface-API (NUI - API) for recognition and automatic tracking of people in real time (Skeleton Tracking), which gives applications the ability to record and interact with those present in the field of view of the sensor.

This process is accomplished through classification algorithms (randomized decision forest) trained to quickly and accurately predict the 3D positions of the joints of the body from a depth image without using time information (Figure 2a.). Once the "runtime core" has determined body parts and the more likely skeleton is chosen, then the space data for each local coordinate system of the skeleton are generated (Figure 2b.). By not using time information for each picture, it can achieve detection of up to 6 people in the scene (Figure 2c.), with the possibility of active capture of two people selected [13].

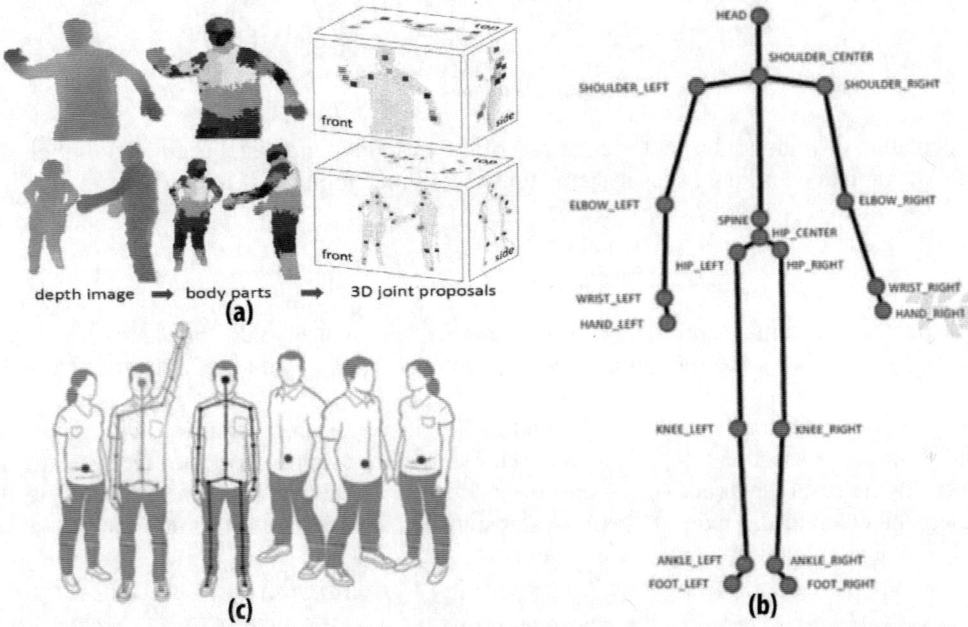

Figure 2.a. Input and Output of algorithm Randomized Decision Forest for recognizing people [14]. **b.** 3D kinematic model used by the Microsoft Kinect SDK. **c.** Skeleton Tracking.

Thus through this API the spatial position of the markers 20 (joints) can be obtained, as well as the spatial orientation, relative to the sensor, of the segments of the skeleton (Bones) to a maximum frequency of 30 Hz.

The C# language oriented objects and integrated development environment (IDE) Microsoft Visual Studio 2012 Express are used for the development of the acquisition software, processing and recording. This set up provides great advantages in the integration of libraries and components provided by the SDK when developing applications based on the Microsoft Windows Kinect® sensor as well as extensive documentation and code samples.

A program using MatLab® was used to process the recorded data of the position and orientation of segments. This can also do the graphical presentation and evaluate parameters of interest of the different registers that allow evaluating the proposed tool. An algorithm for calculating the angles of the joints of the lower limb was implemented. The angles of the joints of the lower limbs in the three anatomical planes for each recorded frame were obtained from the guidelines regarding the origin of coordinates provided by the SDK (Figure 1d.). Local segment coordinate systems are defined to evaluate the angles in each of the joints, as shown in Figure 3a [15].

Journal of Physics: Conference Series **705** (2016) 012064 doi:10.1088/1742-6596/705/1/012064

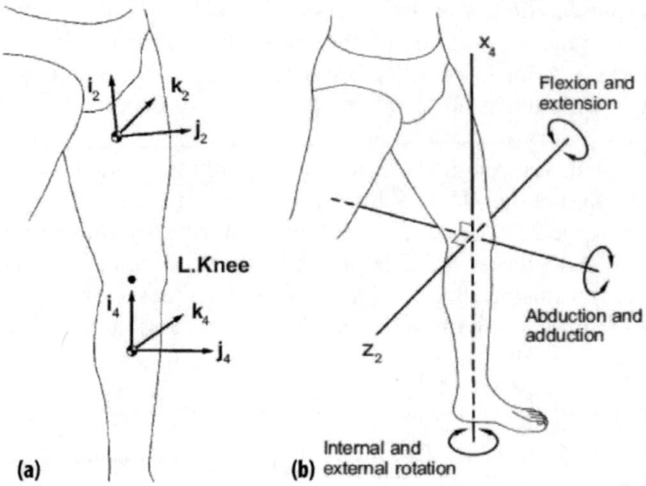

Figure 3. Calculation of joint angles. a. Orientations of the proximal and distal segments to the left knee joint. b. Anatomical coordinate axes systems for the left knee joint [15].

Joint axis is defined to evaluate each articulation,

$$I_{Joint} = \frac{k_{proximal} * i_{distal}}{|k_{proximal} * i_{distal}|} \quad (1)$$

linking two segments: proximal and distal. For example, for knee joint, the versor ($\boldsymbol{i}_{proximal}$) corresponds to the versor ($\boldsymbol{i}_{thigh}$) and the versor ($\boldsymbol{k}_{distal}$) to the versor ($\boldsymbol{k}_{leg}$) and similarly for each joint [15].

Likewise, the axes of rotation of joints are defined as: flexion-extension occurring with respect to the medial-lateral proximal segment ($\boldsymbol{k}_{proximal}$), internal-external rotation axis occurs relative to the longitudinal axis of the distal segment ($\boldsymbol{i}_{distal}$) and the abduction-adduction is defined relative to the floating axis perpendicular to the axis of flexion-extension and internal-external rotation ($\boldsymbol{I}_{Joint}$), see figure 3-b [15]. The anatomical joint angles are obtained.

$$\alpha_{Joint} = sin^{-1}[\boldsymbol{I}_{Joint} \cdot \boldsymbol{i}_{Proximal}] = flexion\ angle\ (+)/extension\ (-) \quad (2)$$

$$\beta_{Joint} = sin^{-1}[\boldsymbol{k}_{Proximal} \cdot \boldsymbol{i}_{distal}] = abduction\ angle(+)/adduction(-) \quad (3)$$

$$\gamma_{Joint} = sin^{-1}[\boldsymbol{I}_{Joint} \cdot \boldsymbol{k}_{distal}] = internal\ rotation\ angle(+)/external(-) \quad (4)$$

2.2. Participants

The study was conducted through the registration of 30 participants (11 women and 19 men) who gave their informed consent; the average age is 25.53 ± 4.13 years, 27 right- dominant, 3 left-dominant, without knee injuries and 22 of the participants practiced amateur sports of some form. The average height is 1.73 ± 0.10 [m], and the weight is 73.62 ± 14.76 [kg].

2.3. Experimental set

Participants performed five different tasks, the first consisted on static position and the remaining four were pre-set movements. In the following sections the movements registered with software developed in this work and using the Microsoft Kinect that was described below. Three repetitions of each gesture were registered.

2.3.1. Standup posture: In the first record the subject is positioned in the capture area of the Microsoft Kinect. The acquisition volume was marked with white tape on the floor. Each participant was instructed to remain in a natural upright position for a few seconds while the recording was made.

2.3.2. Free throw with jump: 2.3.3. The participant was positioned in the acquisition volume of the Microsoft Kinect and makes a free throw of basket. participant was instructed to perform the exercise of jump and shot to hoops without race. The collection of data began with the participant standing and ended when returning to the start position.

2.3.3. Free throw with jump and run: In each record the participant must begin the movement outside the acquisition volume of Microsoft Kinect. The participant performed a race in the longitudinal direction of the axis of acquisition by 3-5 steps as fast as possible. When reaching the acquisition volume, the participant must stop abruptly and make the jump and shoot hoops, then cushion his fall and return to a standing position.

2.3.4. Jump from a 30 cm high platform: In this study, a platform of 30 cm high was used. The participant was positioned standing up on the platform in the center of the acquisition volume marked on the ground. The participant must do a running jump "Squat Jump". The record was made starting from the standing position on the platform by placing the legs at an equal distance shoulder-width, and then the participant had to make a leap moving his arms to get impulse. At the moment of falling over the front of the platform, he had to cushion his fall and return to the upright position. The recording of data was taken during the complete gesture.

2.3.5. Single-Leg Squat: The participant was at the acquisition volume and performed a squat supported on a lower limb meanwhile other limb remained straight forward, with the arms folded across the chest and hands resting on opposite shoulders. In the exercise, the participant performed squats lowering down as much as possible, doing five repetitions while keeping a balance and performing each squat every two seconds in a controlled manner. The movement on both lower limbs was captured with the developed system.

3. Results

This initial work presents the results and initial analysis of the data recorded by several tasks used in rehabilitation and sports are presented, which allow us to appreciate the value of the registration tool Microsoft Kinect® in its applicability to this type of field studies and in rehabilitation.

A static postural analysis of patients is important, considering the fact that it provides many clinical data that are useful for designing an effective treatment. This study evaluated the tendency to varus or valgus knee in a static standing posture and the differences between men and women. The varus-valgus angle average was obtained in a standing posture of the left and right knees of all participants. An analysis of these results shows that the use of this measurement system reveals that participants in a static position have a tendency to varus angle at knee. The average valgus-varus in men and women are presented in Table 1.

Table 1 presents the average population showing a slight varus angle at knee for both genders, showing a higher incidence in men. Dispersion occurs in small angular values, which is due to the variability between participants, thus it presents values close to the standard deviation values. It is believed that this men-women difference is because women have a physiologically wider pelvis, and therefore a greater Q angle compared to men. In other studies, it has been found that this difference is 2-4 degrees [16, 17] therefore expected to possess less varus angle. These results are consistent with those presented in Table 1.

Table 1. Mean values and standard deviation of varus-valgus of left and right knee in static posture for men and women.

	Left Knee		Right Knee	
	Mean	Standard Deviation	Mean	Standard Deviation
varus (-) / valgus (+) in women	-1,60	± 2,16	-0,53	± 1,18
varus (-) / valgus (+) in men	-2,98	± 2,46	-3,41	± 2,61

Figure 4 shows, as an example, knee and hip angles of one of the records of a participant who did a free throw with jump that is usually performed during a basketball game. The peak knee flexion can be seen during the impulse and the landing after the free throw and also the asymmetry of the movement that is characteristic of each player. The participant can do advances with one leg or with the other, put one leg in front of the other or put both legs together to jump or tilt either side. To perform a first preliminary assessment, the peaks varus-valgus angle were calculated during impulse and landing phases. In order to evaluate these phases (impulse and landing), the knee rotation was used. This is because when the knee is extended there is no rotation due to a blockage at this joint (see Figure 45). Since a more detailed analysis of the gesture is not the aim of this study, results and preliminary assessments are presented only for illustrative purposes.

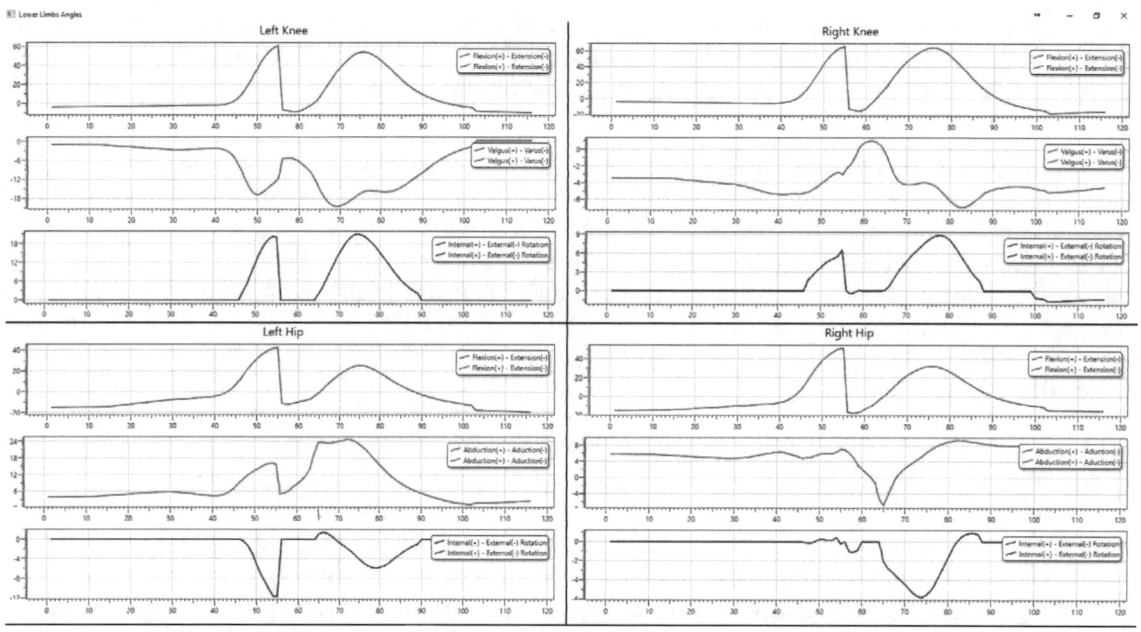

Figure 4. Knee and hip angle of one of the records of a free throw with jump in basketball.

Table 2 shows the average and standard deviation of varus-valgus knees peak values during the impulse and landing phases. It can also be observed the time difference between the occurrence of the flexion peak and the maximum angle of knee valgus-varus.

To make a quantitative analysis of valgus-varus motion in all records, we divided the qualified participants into the ones that did an internal rotation of the knee during a basketball shot and the others who did an external one, knowing that the two movements are linked and that the participants inherently make one of the two rotations when shooting. Peak angles of each record are shown in Figure 5: impulse and landing phases, and the rotation of the knee (internal or external). The distinction between internal or external rotation is made to distinguish whether the knee had-made a varus or valgus correlated movement. The external compartment knee injury is associated with valgus motion, then higher valgus angle amplitude, higher the risk of injury. It is expected that a further continuation of this study can bring a more complete analysis of this gesture.

Table 2. Varus-valgus knee average and standard deviation of angle peak values in a single throw.

	Varus-valgus peak angle of impulse phase with knee internal rotation (°)	Varus-valgus peak angle of impulse phase with knee external rotation (°)	Varus-valgus peak angle of landing phase with knee internal rotation (°)	Varus-valgus peak angle of landing phase with knee external rotation (°)
Mean	-6,99	3,82	-8,90	2,76
Standard Deviation	4,17	3,06	6,61	6,23

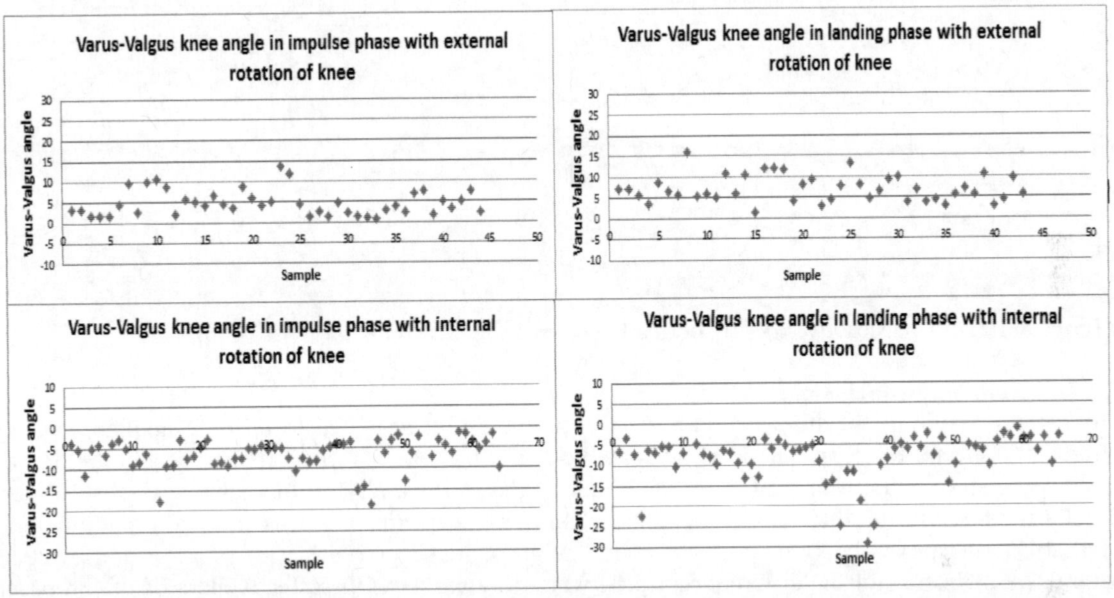

Figure 5. Peak angles of varus-valgus knees in impulse and landing phase in free throw with jump in basketball, separated by the type of knee rotation (internal or external).

Free throw with jump and run from a participant is shown in Figure 6 as an example. The flexion-extension angles, varus-valgus and internal-external rotation of the knee and hip of each lower limb were plotted. The flexion peaks corresponding to the impulse and landing phases can be seen. In addition, all participants that performed a jump from a platform and five repetitions of the single leg squat were evaluated, in both instances three sets of each record. For length purposes, these data are not presented in this study, but the quality of measuring 3D kinematics is the same as shown in Figure 6.

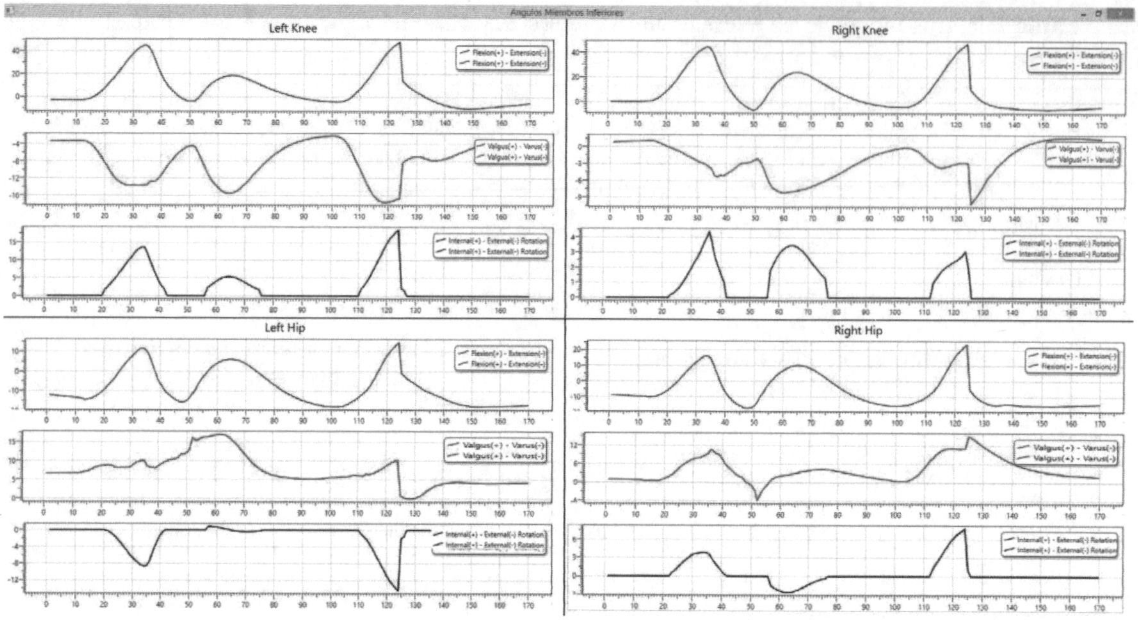

Figure 6. Results of records showing both a basketball shot with run and with jump.

4. Discussion and Conclusions

The system developed in this work is a simple, portable tool designed to on-field appliance, using Microsoft Kinect® to record and analyze some common movements in sports and rehabilitation. 3D kinematics algorithms from data available using the Software Development Kit (SDK) were developed. Records of the position and orientation of segments were used in a program using MatLab®. This program allows the processing of data, make graphical presentation and evaluate parameters of interest of the different registers, which in turn show the effectiveness of the proposed tool. It also provides visual feedback when using its software during rehabilitation, which serves to assess the treatment progress and can be used as feedback to the patient.

In order to show the usefulness of the proposed system, measurements of 30 participants of 5 different protocol of records used in sports training and rehabilitation were made. Relevant results showing the capture quality and a preliminary analysis of the obtained results in a standing posture and free throw with jump tasks were done. Due to the limited length of this study, only a record of free throw with jump and run is presented, with the omission of registers of jump from a platform of 30 cm high and registers of single leg squat. However, it is emphasized that the quality of the measurements made with Microsoft Kinect® for the five types of records used as a first test of the device allow proper quantification of movement. Also, a future study using this proposed tool could show a comparison of simultaneous captures of Microsoft Kinect® with commercial equipment available in our laboratory. In this context, a comparison can be made between the two systems, and then the Microsoft Kinect® can be definitely validated for on-field. In summary, the use of the Microsoft Kinect® for field use is of better value in instances when the use of a commercial laboratory is not effective due to several operational reasons.

5. References

[1] Colby, S., Francisco, A., Yu, B., Kirkendall, D., Finch, M., & Garrett, W.. Electromyographic and kinematic analysis of cutting maneuvers implications for anterior cruciate ligament injury. 2000 *The American journal of sports medicine*, **28**(2), 234-240.

[2] Manonelles Marqueta, P., Tárrega Tarrero, L. Epidemiología de las lesiones en el baloncesto. 1998 Archivos de Medicina del Deporte. **15** (68), 479-483

[3] Gumpel, G., & Rosello, G. Frecuencia de lesiones en el compartimiento lateral de la rodilla en jugadores profesionales de básquet. 2014 Rev. Asoc. Argent. Traumatol. Deporte, **21**(1), 31-33.
[4] Berenstein, M., Raijman, M.. Lesión condral de la rodilla. Síndrome seudomeniscal. 1994 Artroscopía, **1**(1), 32-37.
[5] Shelbourne, K. D., & Nitz, P. A.. The O'Donoghue triad revisited Combined knee injuries involving anterior cruciate and medial collateral ligament tears. 1991 *The American journal of sports medicine*, **19**(5), 474-477.
[6] McLean, S. G., Huang, X., & van den Bogert, A. J.. Association between lower extremity posture at contact and peak knee valgus moment during sidestepping: implications for ACL injury. 2005 *Clinical Biomechanics*, **20**(**8**), 863-870.
[7] Wilderman, D. R., Ross, S. E., & Padua, D. A.. Thigh muscle activity, knee motion, and impact force during side-step pivoting in agility-trained female basketball players. 2009 *Journal of athletic training*, 44(1), 14.
[8] Ford, K. R., Myer, G. D., & Hewett, T. E.. Valgus knee motion during landing in high school female and male basketball players. 2003 *Medicine and Science in Sports and Exercise*, **35** (10), 1745-1750.
[9] Hewett, T. E., Myer, G. D., Ford, K. R., Heidt, R. S., Colosimo, A. J., McLean, S. G., & Succop, P.. Biomechanical measures of neuromuscular control and valgus loading of the knee predict anterior cruciate ligament injury risk in female athletes a prospective study. 2005 *The American journal of sports medicine*, **33** (4), 492-501.
[10] Crossley KM, Zhang WJ, Schache AG, Bryant A, Cowan SM. Performance on the single-leg squat task indicates hip abductor muscle function. 2011 *Am J Sports Med.*;**39** (4), 866–873.
[11] Stone, E. E., Butler, M., McRuer, A., Gray, A., Marks, J., & Skubic, M. (2013, July). Evaluation of the Microsoft Kinect for screening ACL injury. In Engineering in Medicine and Biology Society (EMBC), 2013 *35th Annual International Conference of the IEEE* (pp. 4152-4155). IEEE.
[12] Kato, S., Urabe, Y., & Kawamura, K.. Alignment control exercise changes lower extremity movement during stop movements in female basketball players. 2008 *The Knee*, **15** (4), 299-304.
[13] Shotton, J., Sharp, T., Kipman, A., Fitzgibbon, A., Finocchio, M., Blake, A., ... & Moore, R. Real-time human pose recognition in parts from single depth images. 2013. *Communications of the ACM*, **56**(1), 116-124.
[14] Jana, A. Kinect for Windows SDK Programming Guide. 2012 Packt Publishing Ltd.
[15] Vaughan, C. L., Davis, B. L., & O'connor, J. C. Dynamics of human gait 1992. Champaign, Illinois: Human Kinetics Publishers.
[16] Grelsamer, R. P., Dubey, A., & Weinstein, C. H.. Men and women have similar Q angles A clinical and trigonometric evaluation. 2005 *Journal of Bone & Joint Surgery*, British Volume, **87**(11), 1498-1501.
[17] Byl, T., Cole, J. A., & Livingston, L. A. What determines the magnitude of the Q angle? A preliminary study of selected skeletal and muscular measures. 2000. *Journal of Sport Rehabilitation*, 9(1), 26-34

High efficiency and simple technique for controlling mechanisms by EMG signals

N. Dugarte[1,2], A. Álvarez[2], J. Balacco[2], G. Mercado[2], A. Gonzalez[2], E. Dugarte[1,2], F. Javier[1], G. Ceballos[1], Olivares A[1]

E-mail: ndj0227@gmail.com, irb@frm.utn.edu.ar

Summary - This article reports the development of a simple and efficient system that allows control of mechanisms through electromyography (EMG) signals. The novelty about this instrument is focused on individual control of each motion vector mechanism through independent electronic circuits. Each of electronic circuit does positions a motor according to intensity of EMG signal captured. This action defines movement in one mechanical axis considered from an initial point, based on increased muscle tension. The final displacement of mechanism depends on individual's ability to handle the levels of muscle tension at different body parts. This is the design of a robotic arm where each degree of freedom is handled with a specific microcontroller that responds to signals taken from a defined muscle. The biophysical interaction between the person and the final positioning of the robotic arm is used as feedback. Preliminary tests showed that the control operates with minimal positioning error margins. The constant use of system with the same operator showed that the person adapts and progressively improves at control technique.

1. Introduction

From the beginnings of history, human beings have used the knowledge acquired to develop some kind of technology to improve their health and living conditions. The implementation of mechanisms has improved and extended motor skills of people [1]. The engineering problem is to design techniques for controlling these machines. In some cases is not possible to implement manual control and in other cases the implemented designs are so complex that generate a relatively slow response [2].

This project makes a contribution on subject of robotic equipment control using advantage muscle activity [2] [3]. The knowledge development on design of cybernetic systems efficient [3] contributes to improving the applicable technology in many fields. Among other applications, the medical problems solution in people with disabilities [4]. The benefits of implementing these mechanisms in therapeutic recovery techniques could be of great importance.

With the available technical capacity at regional level, a biological-mechanical interface has been designed for positioning a robotic arm (RA) with movement in two axes [5]. The RA final positioning depends on EMG signals [3] [6] captured of person who acts as system operator.

The main goal of this project is to control each RA axis independently. In that sense the movement on each axis is executed from the amplitude level of signal captured at particular muscle section. Thus, a

[1] Grupo de Ingeniería Biomédica (GIBULA) de la Universidad de Los Andes (ULA), Mérida, Venezuela.
[2] Instituto Regional de Bioingeniería, Universidad Tecnológica Nacional – Facultad Regional Mendoza.

specific muscle tension is reflected as RA positioning in the respective axis. The idea is to train the operator to achieve the desired positioning from control exercised over the corresponding muscles.

2. Methodology

The RA consists of a mechanical structure with two freedom degrees. The mechanism allows to arm termination move at any point on a plane defined by Cartesian axes (X, Y) [7], within a plane space of (20,20) centimeters (cm). The travel in each axis is performed independently and depends on the action of independent motors. One motor for each axis.

The control system works with two identical and independent electronic circuits. Each electronic circuit drives a motor and is responsible for the movement in one axis. Given that the circuits for controlling movement on the two axes are same; this paper presents the explanation of components that handle the movement in one axis.

The circuit response depends of electromyography signal picked up from muscle specific on system operator. The Fig. 1 represents the sequence of different steps making up the control of a freedom degree.

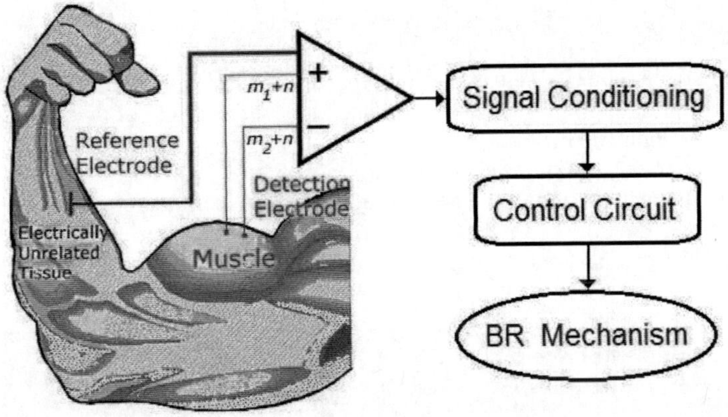

Fig. 1, RA control for a freedom degree.

2.1. Working principle

As seen in Fig 1, the system consists of four parts: the biopotential amplifier, the signal conditioning stage, the control circuit and the RA mechanism. The different elements are arranged for acquiring and conditioning of EMG signal. Then the system checks and generates the corresponding mechanical movement.

2.2. Biopotential Amplifier

This step consists of a differential amplifier [8] [9] with a high common-mode rejection ratio (CMRR) [10], designed for capturing EMG signal [3]. This circuit can receive signals of amplitude between 0.01 and 100 millivolts (mV) in a bandwidth from 10 to 2000 Hz. The electrical circuit of Fig. 2 is formed by three operational amplifiers: A1, A2 and A3. A1 and A2 carry out the amplification of the muscle signal which are simultaneously taken, with surface electrodes placed on the skin.

This circuit improves of rejection by the device of unwanted input signals common to both input leads. Given that m1+n and m2+n are input signals, the difference between m1 and m2 signals make up the EMG and the n is the common mode voltage. A3 amplifies the potential difference between the two measurements to obtain only the desired signal.

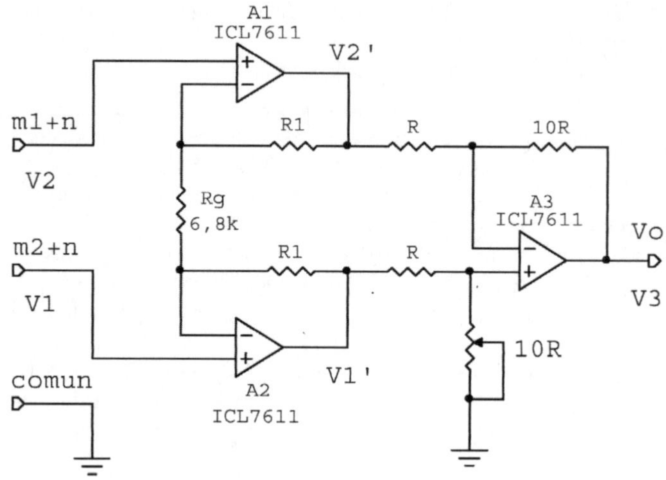

Fig. 2, Biopotential amplifier circuit

In performance tests the signals received from the pectoral muscles were used for driving instrument operator. The selection of these muscles was motivated to test the ability of disabled persons who have only motor activity in theirs torso.

Fig. 3 shows 28 seconds of EMG signal acquisition picked up with the Fig. 2 circuit. The signal was taken from the right pectoral muscle. The figure presents 6 contraction and relaxation efforts of different intensity and duration. These muscles are used to verify the usefulness of instrument on people with disabilities and have motor activity only in his torso.

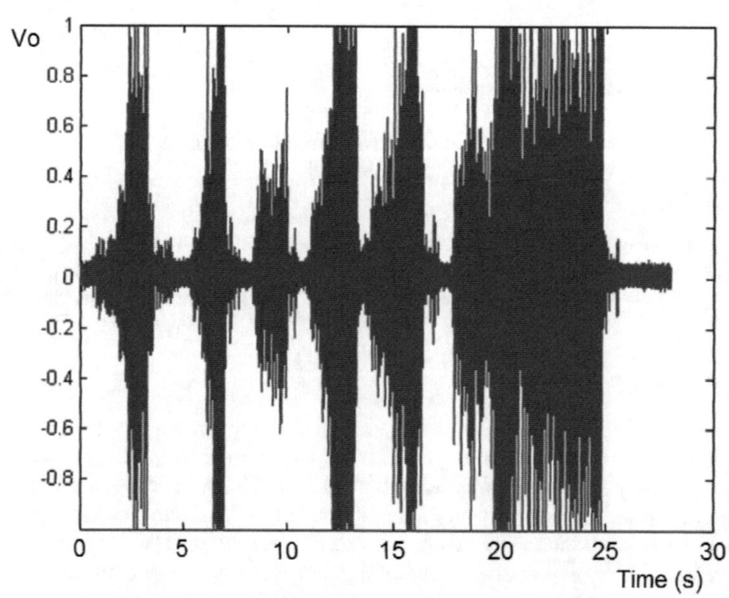

Fig. 3, EMG signal captured.

The digitization signal displayed in Fig. 3 was done with a data acquisition board at 20,000 samples per second with 16 resolution bits. The signal digitization is carried out in order to certify the system working through its computer analysis [11].

2.3. Signal conditioning

The signal conditioning is performed in order to capture only the generated amplitude-variation due to muscular tension. First the signal is rectified to obtain the positive components. Then low-pass filtered retrieves only the envelope [11]. The Fig. 4 diagram shows the electrical circuit designed for this purpose.

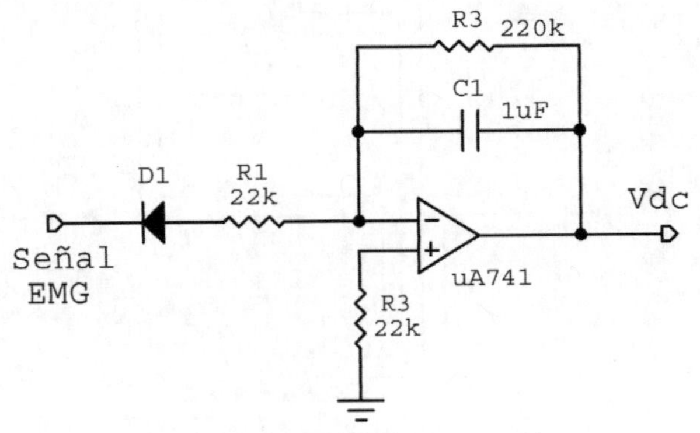

Fig. 4, Signal conditioning circuit.

The captured signal allows certify the performance of each stage. The signal identified as Xn in Fig. 5, shows the result of passing the Fig. 3 signal by rectifier diode, D1 identified in the circuit Fig. 4. The signal Yn in Fig. 5, presents the output defined as Vdc in Fig. 4 circuit.

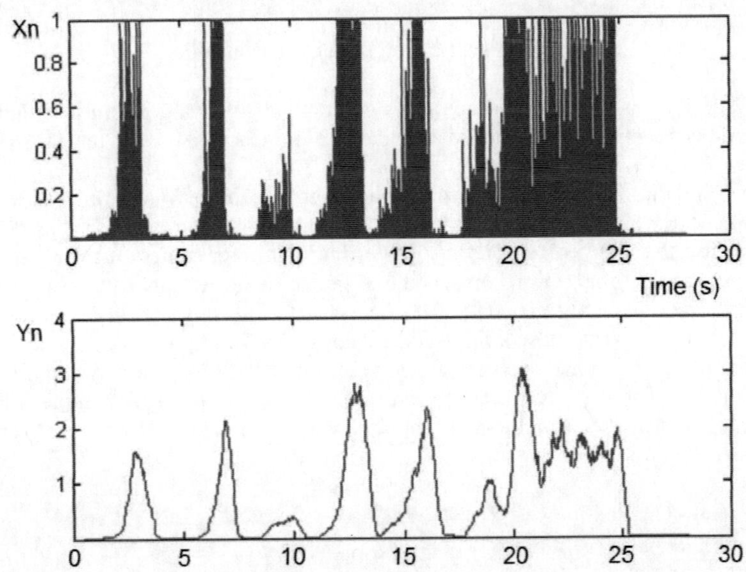

Fig. 5, Rectified and filtered EMG signal.

2.4. Control stage

The circuit showed in Fig. 6, consists of an electronic control of closed loop. It works by continuously correcting of difference generated between the input signal and the provided signal by potentiometer identified as RV1. The RV1 potentiometer is coupled to corresponding axis of mechanism, thus its resistance value is adjusted with changes in position of RA. This loop is to be able to define the position of the RA location continuously in the corresponding axis.

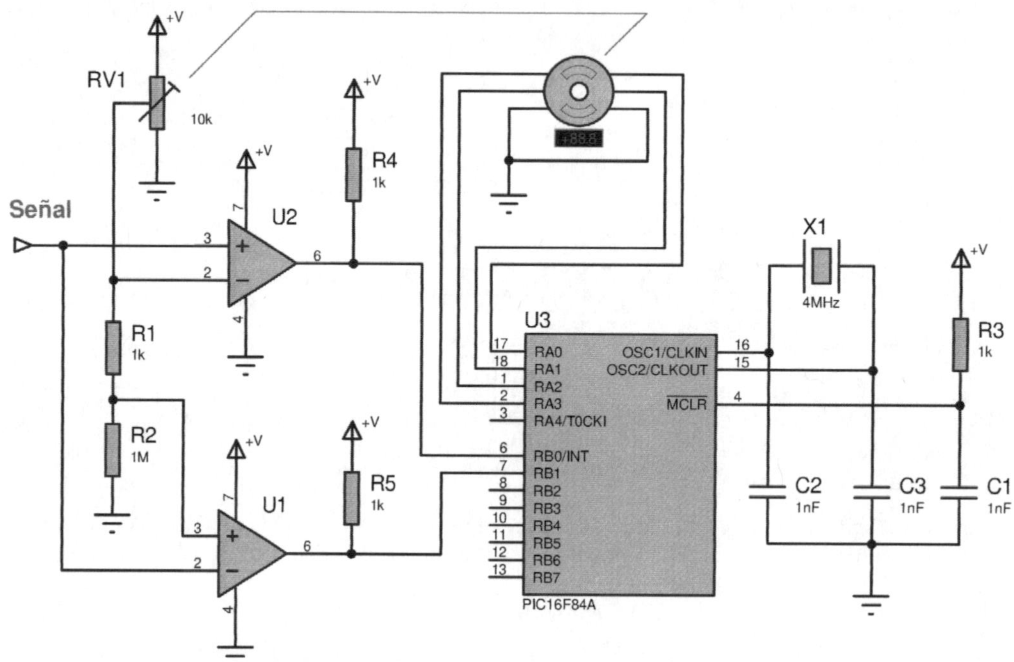

Fig. 6, Control circuit with microcontroller.

The U1 and U2 devices at Fig. 6 are comparators circuits [11]. The algorithm installed in PIC16F84 microcontroller [12] identifies the state of U1 and U2 comparators, with the aim of locating the position of AR.

The Fig. 7 shows a simplified diagram of microcontroller algorithm. When this software is running, checks the RB0 and RB1 inputs corresponding to U1 and U2 comparators. This is done in order to increase or decrease the pole corresponding to stepper motor for generating forward or backward movement on mechanism. This action corrects the difference between amplitude value of input signal and RV1 voltage. The microcontroller output is connected to four power transistors for motor current drives. To simplify the diagram, these transistors it is not shown in Fig. 6.

The control circuit, which in turn it is working in connection with the other system elements, allows that the mechanical activity of RA can be constantly adjusted for EMG signal. This condition increases the motor steps when input signal amplitude is greater than reference and return the motor rotation when signal amplitude decreases.

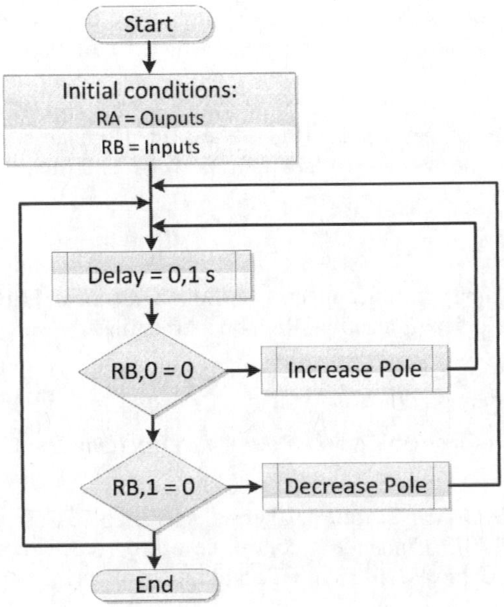

Fig. 7, Block diagram of control algorithm.

3. Results

When the subject under test performs a muscle voluntary contraction, it is noted a movement of RA in corresponding axis. The positioning depends directly of muscle tension intensity from the subject. When the operator uses the two control circuits connected to two individual body muscles it is observed that is capable of positioning the RA at any point on the plane (X,Y) the mechanism.

4. Discussion of results

Preliminary results showed it is feasible make up a control for handling mechanical devices from myographic signals.
The most interesting of test emphasizes that the simplification of system has achieved high efficiency levels at real-time responses with minimal risk of electronics failures.
Is a design with few components enabling high miniaturization, thereby is saving space and reducing weight.
In initial stage is noted that the system operator presented difficult for coordinate muscle activity at RA control. But, constant practice allowed efficiency increase at conducting control.

5. Conclusion

From the depths of history, the Man has dreamed about of controlling machines which only use human will and desires to move parts and mechanisms.
Preliminary tests showed that the implemented method can be used in the management of motorized machines using EMG signals.
It also was possible to obtain a response to voluntary muscular effort without having first running physical movement. Therefore, it was showed that it is really possible handle any mechanical device even if the operator is not to be able to do joint movement.

Day after day, the number of disabled is rising through over the world, for this reason is very important to develop new technologies of prosthesis which fit of the best way to biomechanical functions of human being.
Additionally, case studies are presented where analysis factors are critical, such as myographic applications in relation to athletic performance [13]. It should also take into account the technological challenge for develop control system to facilitate the use and management of very complex instruments and mechanisms.

6. References

[1] Schvab L 2011 Maquinas y Herramientas – Guía Didáctica, Education Ministry, Institute National of Technological Education, Republic Argentina.

[2] Loaiza J and Arzola N 2011 Evolución y tendencias en el desarrollo de prótesis de mano, Dyna, year 78, Number 169, pp: 191-200, Medellin.

[3] Barea R 2011 Tema 5 - Electromiografía, Electronic Departament – University Alcalá, Republic Argentina.

[4] Tabernig C, Escobar S, Bonell C, Reta J, Cherniz A, Spaich E, Cerrato M, Massafra A, Zunino F and Caamaño M 2012 Comando y control de neuroprótesis motoras por electromiografía de superficie, Rev. Ciencia, Docencia y Tecnología, Vol. 2 (2).

[5] Lizcano R, Puentes J and Valenzuela C 2005 Control para un robot articulado con tres grados de libertad que simule el movimiento de pata, Thesis, Engineering Faculty, Pontificia University Javeriana, Bogota - Colombia.

[6] Dorador J 2004 Robótica y prótesis inteligentes, University Digital Journal, Vol. 6 (1)

[7] Angel J 2008 El plano, MathCon, Available in: www.math.com.mx, last accessed: August 2015.

[8] Drake J., "El amplificador de instrumentación", Publication of the Department of Electronic Instrumentation and Communications, Cantabria University, Spain, 2005.

[9] Kugelstadt T., "Getting the most out of your instrumentation amplifier design", Analog Applications Journal, Vol. 4, pp: 25-30, 2005.

[10] Eamon Nash, "A Practical Review of Common Mode and Instrumentation Amplifiers", edit Analog Devices, Copyright by Helmers Publishing Inc., USA, 1998.

[11] Boylestand R., Nashelsky L., "Electrónica: Teoría de circuitos", sixth edition, Pearson Publishing, Mexico, 1997.

[12] Angulo J., Angulo I., "Microcontroladores "PIC" Diseño práctico de aplicaciones, primera parte", Edit McGraw-Hill, third edition, Madrid, 2003.

[13] Massó N, Rey F, Romero D, Gual G, Costa Ll y Ana Germán 2010 Aplicaciones de la electromiografía de superficie en el deporte, Apunts Med Esport Journal; 45(165):127-136.

Institute of Physics
Dirac House, Temple Back
Bristol BS1 6BE UK

ISSN: 1742-6588
ISBN 978-1-5108-2416-4